Instrumental Analysis

CHARLES K. MANN
THOMAS J. VICKERS
WILSON M. GULICK

Florida State University

Harper & Row, Publishers
New York Evanston
San Francisco London

Sponsoring Editor: John A. Woods
Project Editor: Lois Wernick
Designer: Ben Kann
Production Supervisor: Bernice Krawczyk

Instrumental Analysis

Library of Congress Cataloging in Publication Data
Mann, Charles Kenneth.
Instrumental analysis.
Includes bibliographical references.
1. Instrumental analysis. I. Vickers, Thomas J., 1939- joint author. II. Gulick, Wilson M., 1939- joint author. III. Title.
QD79.I5M36 543'.08 73-16540
ISBN 0-06-042535-0

Contents

Preface

Instructors who offer courses in applied technology must continually seek the proper balance between utilitarian and fundamental subject matter. Heavy emphasis on immediate usefulness may amount to indoctrination about current methods. Such courses generally benefit students in only a very narrow specialty, and the useful lifespan of the information may be quite short. On the other hand, too heavy emphasis on fundamentals may obscure the basic purpose of the course. Many topics have great fundamental significance without being very helpful to a person wishing to use a technique in practice. In selecting the topics discussed in this book, we have attempted to achieve this balance, with the hope that the material will serve both as an aid in learning effective use of existing instruments and as a basis for incorporating deeper insights as well as information about future developments.

In ordering the topics, a section consisting of nine chapters that discuss electronics has been placed at the beginning because it is material that is fundamental to all types of instrumentation. This is followed by a group of nine chapters dealing with optical spectroscopy. Chapters on microwave spectroscopy, magnetic resonance, mass spectroscopy, chromatography, and electrochemistry then follow in order.

The proliferation of instruments often leads students to perceive them as a collection of unrelated pieces of equipment. Actually, a relatively small number of basic concepts is used repeatedly in various combinations, and an understanding of these concepts greatly facilitates the intelligent use of equipment. In writing the first nine chapters of this book, an attempt has been made to select those concepts that appear repeatedly and to discuss them on a level appropriate to chemistry juniors who do not have strong mathematical backgrounds.

This material is presented at the beginning because it is fundamental to all types of instrumentation, and it is intended to be presented as a unit.

However, since presentation of concepts in the abstract is generally ineffective, it is intended that reference be made to this material as specific types of instruments are discussed. As an aid, specific references to the electronics section have been included where appropriate in the later chapters.

After the introduction to electricity, the Thévenin model, which can be used to understand the transfer of information encoded as voltage or current levels, is presented. This is not only helpful in discussing subsequent electronics topics, but it is also fundamental to successful performance in many types of equipment when instruments are interconnected. Behavior of reactive networks is presented separately from the introductory material because of the fundamental importance of reactive phenomena in determining the operating limitations of electronic equipment in general.

The process of amplification is discussed with the intent of providing enough familiarity with the operation and with the components used to enable the student to appreciate its function in instruments. This section is not intended to teach amplifier design, but instead to bring the student to a level at which he can easily read the more detailed treatments that are readily available. Following the discussion of amplifiers and of amplifier performance, the reader's attention is directed to the performance of systems, with a brief introduction to some concepts from information theory which leads to a discussion of noise and of techniques that may be used to enhance the signal-to-noise ratio.

The discussion of fundamental measuring operations, which is frequently placed at the beginning in a text of this sort, has been instead delayed until Chapter 7. A reader who has some knowledge of the problems involved in accurate transfer of voltage and current can readily appreciate the limitations that are imposed on measurement systems. Therefore, the discussion of models from Chapter 2 is briefly focused in Section 7.1 to provide the basis for discussion of measuring devices. The increasing use of linear amplifiers in voltmeters and the wide applicability of oscilloscopes makes it appropriate for this section to follow, rather than precede, that dealing with amplifiers and feedback. Transducers are discussed at this point because, like measuring devices, an appreciation of factors important to information transfer is required. Centralizing the discussion of transducers focuses attention on similarities that exist in diverse areas and provides a useful tie-in between the electronics and instrument applications sections of the book.

Analog and digital instrument design is reserved for the end of the electronics section because whole systems are involved. The discussion of operational amplifiers is incorporated in Chapter 8, "Analog Instrument Design," rather than in chapters dealing with amplifiers because operational amplifiers are so intimately associated with analog computers and analog instruments generally. Their use in scientific instruments merits a discussion that would have significantly increased the length of the earlier chapters. In addition, it may be useful to reinforce the earlier discussion by continuing the exposition after other topics have been presented.

While the nature of the subject makes it feasible to discuss analog design in some detail, this is certainly not possible for digital design. The purpose of Chapter 9, "Digital Instruments," is to present a discussion that will be meaningful in light of the experience expected of junior-level science majors. It is intended to serve as a starting point, either for more detailed discussion of specific topics of interest to the instructor, or for additional reading on the part of the student. It is hoped that as the reader moves on to more specialized text material or to experiences with digital equipment, this chapter will provide him with a frame of reference that will be of value in correlating new information as he receives it.

The section dealing with spectroscopy opens with Chapter 10, which provides a framework for spectroscopic methods by describing the electromagnetic spectrum and the general properties of electromagnetic radiation and its interaction with matter. Succeeding chapters (11–20), arranged in order of decreasing energy of the spectroscopic transition, are devoted to instrumentation and techniques for specific regions of the electromagnetic spectrum.

Since techniques of optical spectroscopy are perhaps most familiar to the student, these techniques are considered first. Chapter 11 describes instrumentation and measurement principles common to all types of optical spectroscopy. This is a long chapter, but the length is justified because in subsequent chapters duplication of material is avoided and the presentation of material is simplified. Moreover, there are important principles that apply equally well to several spectroscopic methods and hence do not fit logically into any single chapter. For example, characterization of spectrometer performance, as described in Section 11.5, is essential if an analyst is to be assured of making valid measurements, but this is a topic that traditionally receives scant attention when the discussion of spectroscopic instrumentation is dispersed among several chapters.

Chapters 12–14 deal with analytical methods based on atomic spectroscopy. The basic theory is presented in Chapter 12. Chapter 13 describes the three closely related techniques of flame atomic emission, absorption, and fluorescence spectroscopy. Chapter 14 describes arc/spark emission spectroscopy. Chapters 15–18 describe methods based on molecular spectroscopy. Chapter 15 discusses briefly the fundamental principles of electronic, vibrational, and rotational spectroscopy. Chapter 16 presents ultraviolet-visible absorption spectroscopy. Because of the wide utility of UV-visible techniques, this is a rather lengthy chapter. Chapter 17 describes fluorescence and phosphorescence spectroscopy. At present, these techniques are of rather limited use, and the treatment is correspondingly brief. The description of optical spectroscopic techniques concludes in Chapter 18 with infrared and Raman spectroscopy. Currently available laser–Raman instrumentation assures that this technique will become more fully exploited than in the immediate past; hence, it has been given more emphasis than in previous texts.

This arrangement allows grouping of techniques that provide similar information and that have similar instrumental requirements. For example, atomic emission and atomic absorption, which provide complementary data and utilize practically identical instrumentation, are considered in the same chapter. Arc/spark emission spectroscopy has the same theoretical basis as flame emission, but utilizes very different instrumentation. Thus, it is useful to consider the theoretical basis for the two techniques together in Chapter 12, but to describe the analytical methods in separate, but sequential, chapters. Likewise, UV-visible absorption and fluorescence spectroscopy, and infrared and Raman spectroscopy should be considered together.

A chapter on microwave spectroscopy is included in recognition of the potential of this method as an analytical tool, and the rapid growth in utilization of this technique that is likely as commercial instruments become more available. The amount of space in Chapter 19 devoted to klystron oscillators and wave guides, which may seem excessive, is justified because these devices are also used in electron spin resonance, and their description can thus be omitted from the chapter on magnetic resonance (Chapter 20).

The organization of Chapter 20 is different from that of any other chapter. Similarities between electron spin resonance (esr) and nuclear magnetic resonance (nmr) are stressed, since the apparent differences are ones of engineering, not of principle. The chapter begins with an introduction to the absorption of radio frequency energy by spin systems, stressing the point that these are magnetic dipole interactions. Chemical applications of proton nmr are then given rather detailed consideration. This is the single most important area of current nmr use; it is presented early in the chapter for the benefit of the instructor who chooses to omit further sections due to shortage of time. Other important nmr nuclei are then discussed, with the most attention being given to ^{13}C.

Those who have the time and interest may proceed to the sections that discuss relaxation and line shapes, including exchange phenomena and double resonance. These topics are developed within the framework of the Bloch equations which, while not derived, are presented in a way that is not ad hoc; the treatment does not omit any crucial steps except the tedious transformation to the rotating coordinate system. The transformed equations are simply stated. With this much theory available, discussion of instrumentation is facilitated. The function of a cross-coil probe can be explained quantitatively using Bloch formulation, an approach that is impossible unless the theory precedes the instrumentation section. Finally, magnetic field modulation is discussed, and the similarities between nmr and esr are again stressed; the apparent difference is shown to be a result only of the difference in the widths of the spectral lines that are observed.

In the discussion of mass spectrometry, we have included the types of spectrometers that are in frequent current use or serve to demonstrate a particular principle. Some designs are omitted, and those that are not in widespread use are mentioned only briefly. While not comprehensive, sufficient attention to the interpretation of medium-resolution spectra is given that the chemist may be able to interpret a spectrum. The section on high-resolution spectrometry stresses the role of computers in data reduction and acquisition and includes a discussion of inorganic applications of spark–source spectrometers.

Chromatography is introduced in Chapter 22 by a discussion of the basic process involved, and the factors that affect it, without regard to type of chromatography. A brief discussion of the plate model is given, and methods of measuring HETP are explained.

With this introduction, gas chromatography is discussed in some detail. Emphasis is given to factors that determine chromatographic efficiency. Thermal conductivity and ionization detectors are described. Liquid chromatography is discussed, with emphasis given to high-pressure operation of columns and to the thin-layer form of open-bed chromatography. Factors that determine the selection of stationary and moving phases are discussed.

The chapter dealing with electrochemistry (Chapter 23) is organized according to the amount of current used in a particular experiment, ranging from zero (ideal potentiometry) to large (macroscopic coulometry). While fundamentals are briefly reviewed, it is assumed that the student has had some introduction to electrochemistry. The mass-transport problem is handled in some detail. A knowledge of calculus is assumed, and Fick's second law is developed rather than postulated. The solution to this equation is included; however, one may omit this section and take up with the solution at the beginning of the following section if desired.

After discussion of mass transport and utilization of that concept in the description of several techniques, the chapter concludes with a brief discussion of electrode kinetics. It is hoped that at this point the student will be able to grasp what "slow" and "fast" processes mean in electrochemical terms and be in a position to understand the various meanings of "reversibility." The section on kinetics attempts to summarize and bring together some of this material.

The 23 chapters of this book provide more material than can conveniently be treated in a single term, but the material is organized, as should be apparent from the preceding description, so that major sections are sufficiently independent that they may be presented separately. Thus, we have attempted to offer the user considerable flexibility in the selection of topics and in the order of their presentation.

C. K. M.
T. J. V.
W. M. G.

chapter 1 / Introduction to electricity and electrical signals

chapter 1

To accomplish the aims of this text, it is necessary that the reader have some background knowledge of electricity and electrical circuits and be able to understand the terminology. This first chapter therefore presents a brief discussion of simple DC and AC circuits. It is not intended as a basis for detailed study of the subject but rather as a minimal introduction of terms and concepts. Comprehension of the chapters that follow requires an understanding of only very simple circuit calculations; accordingly, the emphasis is primarily qualitative.

1.1 / CURRENT, VOLTAGE, AND OHM'S LAW

1.1.1 / Current

Electrical current is produced by the movement of charge in a conductor. Chemists generally think of it in terms of electron movement, although positive charge carriers may be responsible for current flow in gases and liquids. When one deals with electrical circuits and electronics, current is often considered, by convention, to be the flow of positive charge, and this conventional meaning, implied in the symbolic notation used for semiconductor components, will be used exclusively in this text.

The practical unit of electrical charge is the coulomb (C), related to chemical units by the Faraday constant, 96,494 C/equivalent, which corresponds to Avogadro's number of electrons. Current is the rate of charge movement, measured in coulombs per second or amperes.

1.1.2 / Voltage

Electrical charges exert mutual attractive or repulsive forces as described by Coulomb's law:

$$F = \left(\frac{1}{4\pi\varepsilon}\right)\left(\frac{q_1 q_2}{r^2}\right) \tag{1.1}$$

In the rationalized meter–kilogram–second system, F is force in newtons, r is the charge separation in meters, and q_1 and q_2 are the respective electrical charges. The proportionality constant ε has the dimensions coulomb^2 per newton-meter^2. A charge sets up an electrical field, which exerts a force on a second charge. Therefore, the movement of a charge in an electrical field involves the performance of work. The work done in moving a unit charge from one point to another in an electrical field is the potential difference between the points. This is expressed in newton-meters (joules, J) per coulomb. The practical unit is the volt (V), defined as 1 J/C. The terms *voltage*, *potential difference*, and *electromotive force* (emf) have similar meanings; some consider them to be synonymous, although others make a distinction. In discussing electricity and electronics here, voltage and potential difference will be considered to be synonymous; in discussing electrochemistry, the term used will be electromotive force.

1.1.3 / Ohm's law and resistance

The imposition of a potential difference between two points in a conductor results in a flow of current proportional to that potential difference. This relationship, Ohm's law, is given as

$$E = IR \tag{1.2}$$

where E is the potential difference in volts (V), I is the current in amperes (A), and R is the resistance in ohms (Ω). For a given type of material, the resistance is directly proportional to the conductor length l and inversely proportional to the cross-sectional area a,

$$R = \frac{\rho l}{a} \tag{1.3}$$

where proportionality constant ρ is the resistivity, which has the dimensions ohm-meters. Resistivity is a convenient parameter for comparing the resistive properties of different materials. The room-temperature resistivity values of some commonly used conductors are shown in Table 1.1.

An understanding of the behavior of electrical circuits may be aided by an

TABLE 1.1 / *Resistivities of conductors*

Material	*Resistivity* ($\mu\Omega$-*cm*)
silver	1.5
copper	1.7
aluminum	2.6
tungsten	5.6
brass	6
nichrome	100
carbon	350

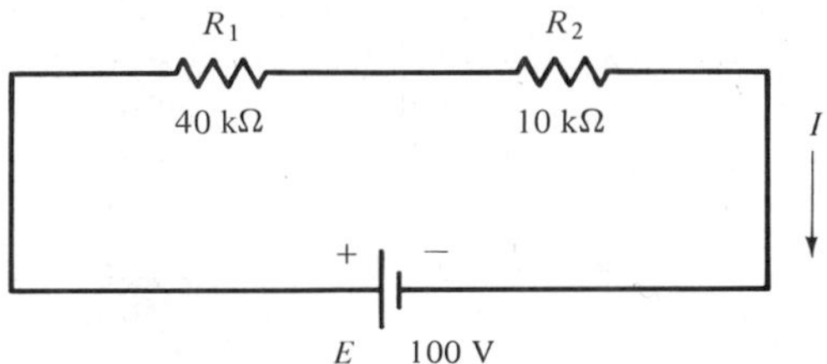

Figure 1.1

analogy to hydraulic systems. Voltage may be considered analogous to the fluid pressure in a pipe, the quantity of charge to the fluid volume, current to the flow rate, and resistance to the friction.

1.1.4 / Conductance

In certain applications it is convenient to focus attention on the conduction of current rather than on resistance. Conductance is defined as the reciprocal of resistance, with dimensions of reciprocal ohms (Ω^{-1}); the symbol G is frequently used. In terms of conductance Ohm's law may be stated as

$$I = EG \tag{1.4}$$

1.1.5 / Power

A potential difference must be imposed in order to cause a flow of current in a conductor. One implication of this fact is that the work in the process is directly related to the resistance of the conductor. Examination of the units shows that the product of current and voltage is the rate at which work is done, called power:

$$[\text{current (coulombs per second)}][\text{voltage (joules per coulomb)}] \\ = \text{power (joules per second)} \tag{1.5}$$

The unit joules per second is the watt (W). If voltage equal to IR, from Ohm's law, is substituted in Eq. (1.5), then

$$\text{power} = I^2R \tag{1.6}$$

Similarly, current can be eliminated between this and Ohm's law:

$$\text{power} = \frac{E^2}{R} \tag{1.7}$$

1.2 / ELECTRICAL CIRCUITS

1.2.1 / Series circuits

When circuit components are connected so that the same current is drawn by more than one component, they are said to be in series. Figure 1.1 shows two resistances in series with a constant voltage source. The conventional symbols for the components are used; resistance is measured in kilohms (kΩ)

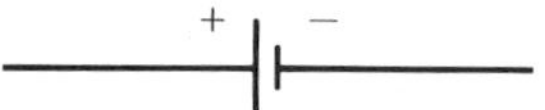

Figure 1.2

and voltage in volts (V). The flow of conventional current is shown to be clockwise, from positive to negative. When the conventional voltage-source symbol, Fig. 1.2, is used, it is understood that the positive terminal is identified by the longer bar and that positive charge moves out of that terminal to the circuit and into the other terminal from the circuit. Polarities are usually not stated explicitly in circuit diagrams. In series connections, the resistances are additive; hence, the current is obtained by the use of Ohm's law,

$$I = \frac{E}{R_1 + R_2} = \frac{100 \text{ V}}{50 \text{ k}\Omega} = 2 \text{ mA} \tag{1.8}$$

where I is measured in milliamperes (mA). Similarly, voltages in a series connection are additive, but with polarities taken into account. In Fig. 1.3, two voltage sources are shown in series opposition (called *bucking*). The net voltage is 10 V, responsible for clockwise current flow:

$$I = \frac{20 - 10 \text{ V}}{100 \text{ k}\Omega} = 0.10 \text{ mA} \tag{1.9}$$

Resistances in series can be used as a voltage divider, as shown in Fig. 1.4. This circuit may be analyzed by a calculation of the total current:

$$I = \frac{E}{R_1 + R_2} = \frac{60 \text{ V}}{150 \text{ k}\Omega} = 0.40 \text{ mA} \tag{1.10}$$

Then the fraction of total voltage dropped across resistance R_2 can be obtained:

$$e_{\text{out}} = IR_2 = (0.40 \text{ mA})(50 \text{ k}\Omega) = 20 \text{ V} \tag{1.11}$$

The fact that the resistances of a voltage divider are in series and therefore carry the same current makes it possible to simplify the calculation of the output voltage. For example, from Fig. 1.5, the calculation is

$$e_{\text{out}} = e_{\text{in}}\left(\frac{R_2}{R_1 + R_2 + R_3}\right) = \frac{(20 \text{ V})(0.2 \text{ k}\Omega)}{3.2 \text{ k}\Omega} = 1.2 \text{ V} \tag{1.12}$$

Figure 1.3

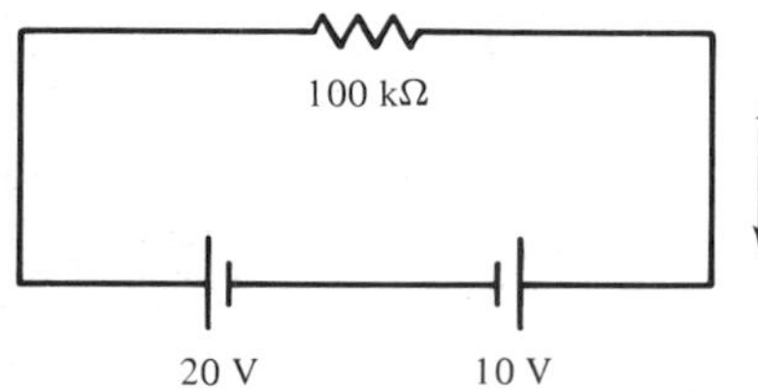

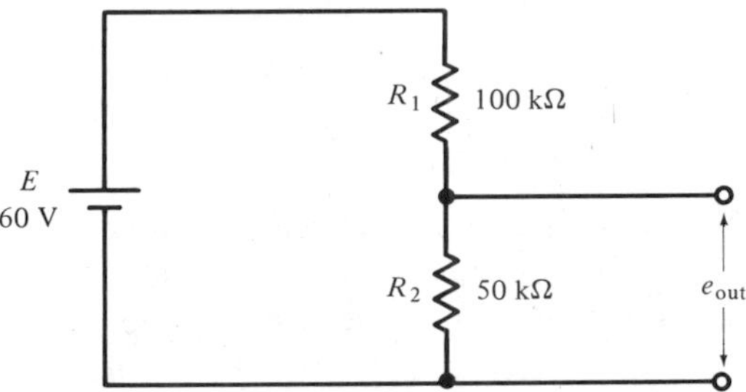

Figure 1.4

Equation (1.12) emphasizes that a simple relationship can be established between e_{in} and e_{out}. Because of this, voltage dividers are widely used as attenuators; that is, they are used to produce a signal with an accurately known fractional value of another signal. This is the basic component in sensitivity adjustments of instruments such as gas chromatographs, recorders, and oscilloscopes.

This circuit is so widely used that it is worthwhile to learn to recognize by inspection the solution in the form of Eq. (1.12) rather than to solve the problem in two steps, as in Eqs. (1.10) and (1.11). It should be noted, however, that Eq. (1.12) is valid only when no current is drawn in the output terminals; if current is drawn, the observed value of e_{out} is smaller than that predicted by the equation. This situation, which is of great practical importance, is discussed in detail in Section 2.3.3.

1.2.2 / Parallel circuits

Components connected so that a single voltage is dropped across all of them while the total current is divided between them are said to be connected in *parallel*. This is illustrated for an array of resistors in Fig. 1.6. The fact that

Figure 1.5

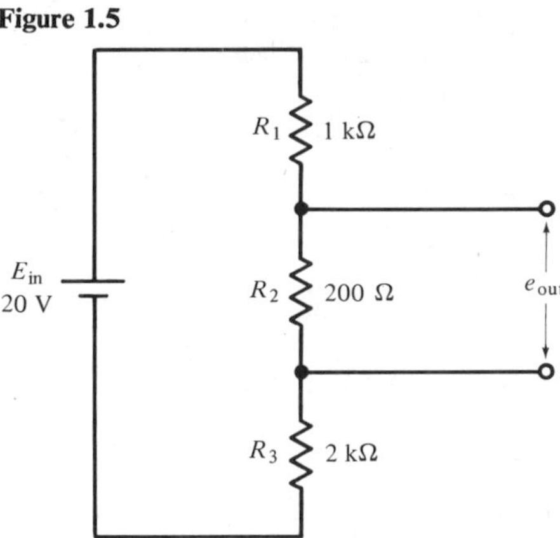

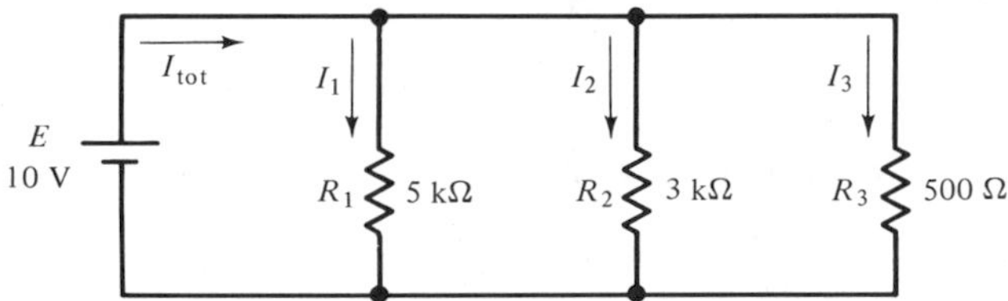

Figure 1.6

the imposed voltage E is common to all while the total current is divided makes it convenient to use Ohm's law in the form shown in Eq. (1.4). Since the total current is the sum of the currents shown in each parallel arm, it may be expressed as

$$\begin{aligned} I_{tot} &= I_1 + I_2 + I_3 = EG_1 + EG_2 + EG_3 \\ &= (10\ \text{V})(2.0 \times 10^{-4}\ \Omega^{-1}) + (10\ \text{V})(3.3 \times 10^{-4}\ \Omega^{-1}) \\ &\quad + (10\ \text{V})(20 \times 10^{-4}\ \Omega^{-1}) \\ &= 25.3\ \text{mA} \end{aligned} \tag{1.13}$$

This may be written

$$\begin{aligned} I_{tot} &= E(G_1 + G_2 + G_3) = EG_{tot} = (10\ \text{V})(25.3 \times 10^{-4}\ \Omega^{-1}) \\ &= 25.3\ \text{mA} \end{aligned} \tag{1.14}$$

Consider the relationship between resistance and conductance:

$$\frac{1}{R_{tot}} = \frac{1}{R_1} + \frac{1}{R_2} + \frac{1}{R_3} \tag{1.15}$$

If a large resistance is connected in series with much smaller resistances, the total resistance is dominated by the largest, and most of the total voltage is dropped across this component. When a small resistance is connected in parallel with larger ones, the smallest contributes most heavily to the total conductance and draws the largest current.

When only two parallel arms are involved, as in Fig. 1.7, it is convenient to calculate individual currents as a function of total current without taking reciprocals:

$$I_2 = I_{tot}\left(\frac{R_1}{R_1 + R_2}\right) = (10\ \text{mA})\left(\frac{20\ \text{k}\Omega}{21\ \text{k}\Omega}\right) = 9.52\ \text{mA} \tag{1.16}$$

Voltage sources can be connected in parallel so as to increase the current available in a load, but only if they are exactly identical in magnitude. If they are not, then large currents may flow from one source to the other and thereby damage them.

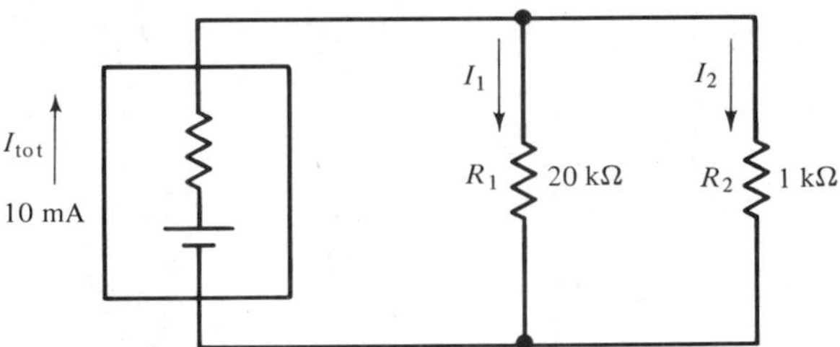

Figure 1.7

The examples given above illustrate the duality that exists between series and parallel circuits. The series circuit is used as a voltage divider, described by an equation of the form of Eq. (1.12). The parallel circuit may be used as a practical current divider, described by an equation of the form of Eq. (1.16). These two equations have the same form, with current and voltage terms interchanged.

This duality is encountered repeatedly. For example, it may be seen in the voltage and current source models discussed in Section 2.3.1 and in the small-signal equivalent circuits discussed in Section 4.10.

1.2.3 / Series-parallel circuits

Networks that consist of resistances in series and in parallel may be treated by condensing them to simpler equivalents, as indicated in Fig. 1.8, where R_6 is the parallel equivalent of R_2 and R_3, and R_7 is the parallel equivalent of R_4 and R_5. The total current may be obtained from the series equivalent circuit in Fig. 1.8(b):

$$I_{tot} = \frac{E}{R_1 + R_6 + R_7} = \frac{25 \text{ V}}{21.1 \text{ k}\Omega} = 1.2 \text{ mA} \tag{1.17}$$

Figure 1.8

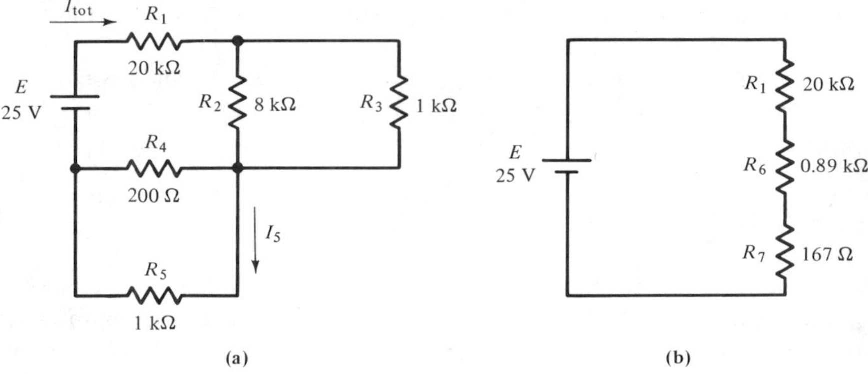

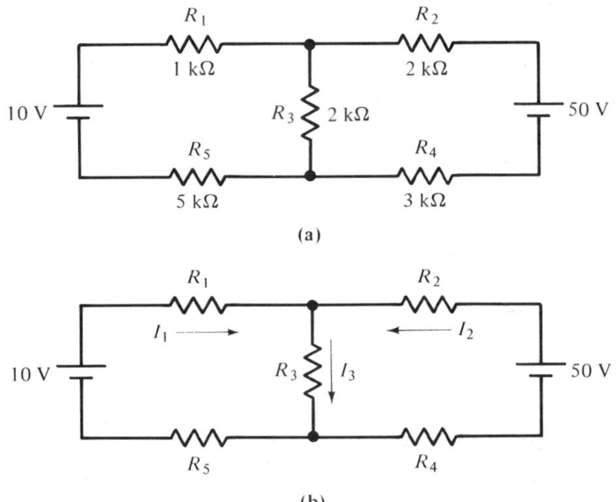

Figure 1.9

If the current in one parallel branch were desired, say, I_5, it could be obtained from the total current I_{tot} by noting that I_{tot} divides between R_4 and R_5:

$$I_5 = I_{tot}\left(\frac{R_4}{R_4 + R_5}\right) = (1.2\ \text{mA})\left(\frac{0.2\ \text{k}\Omega}{1.2\ \text{k}\Omega}\right) = 0.2\ \text{mA} \tag{1.18}$$

1.2.4 / Kirchhoff's laws

After all possible simplifications of the type described above are made, more than one loop may remain. This type of problem can be handled by an application of Kirchhoff's laws:

1. In a network, the algebraic sum of voltage drops around a loop is zero.
2. The algebraic sum of currents directed toward a junction point in a network is zero.

To solve a problem, the current directions are assumed in each branch of the network. Then the current and voltage laws are applied to formulate enough equations to solve for the unknown terms. If there are j junctions (nodes) and n currents in a network, it will be possible to write $j - 1$ current equations and $n - (j - 1)$ voltage equations that are independent. If incorrect assumptions of the current directions are made, the values will show a negative sign in the solution.

In the network in Fig. 1.9, the current directions will be assumed as indicated. There are two nodes and three currents; therefore, it is possible to write one independent current equation and two independent voltage equations. Equation (1.19) is obtained by applying the current law to the upper node:

$$I_1 + I_2 - I_3 = 0 \tag{1.19}$$

In applying the voltage law a voltage source is given a positive sign if, in going around the loop, one passes from negative to positive through the source. The voltage drop in a resistance is given a negative sign if one passes in the direction of current flow, that is, if the current flow is from positive to negative. By taking the left loop of Fig. 1.9(b), Eq. (1.20) is obtained:

$$10\ \text{V} - I_1R_1 - I_3R_3 - I_1R_5 = 0 \tag{1.20}$$

By assuming current in milliamperes and noticing that the same current is drawn by both the 1-kΩ and the 5-kΩ resistances, Eq. (1.21) is obtained:

$$10\ \text{V} - 6I_1 - 2I_3 = 0 \tag{1.21}$$

Similarly, Eqs. (1.22) and (1.23) may be written for the right loop of Fig. 1.9(b):

$$50 - I_2R_2 - I_3R_3 - I_2R_4 = 0 \tag{1.22}$$

$$50 - 5I_2 - 2I_3 = 0 \tag{1.23}$$

Substitution of Eq. (1.19) into Eqs. (1.21) and (1.23) produces

$$10 - 8I_1 - 2I_2 = 0 \tag{1.24}$$

$$50 - 2I_1 - 7I_2 = 0 \tag{1.25}$$

The solution of these equations yields $I_1 = -0.55$ mA, $I_2 = 7.30$ mA, and $I_3 = 6.75$ mA, and so an incorrect assumption for the direction of I_1 is indicated.

1.2.5 / Superposition

The labor involved in solving network problems often can be minimized. When there is more than one current or voltage source, the sources may be considered one at a time, and the desired currents or voltages solved for. The actual total current or voltage is obtained as the algebraic sum of the individual values by superposition of the individual solutions. This method will be applied to the network of Fig. 1.9. To deal with the 10-V source, a hypothetical conductor is substituted for the 50-V source; that is, it is "shorted out." This yields the network shown in Fig. 1.10(a). This can be simplified by combining the pairs of resistances in series, R_1 with R_5 and R_2 with R_4, to give the network in Fig. 1.10(b). This can be further simplified by combining the resulting pair in parallel in Fig. 1.10(b) to give the 1.43 kΩ shown in Fig. 1.10(c). Finally, this can be simplified to the single resistance shown in Fig. 1.10(d). The total current can be obtained by applying Ohm's law:

$$I_1' = \frac{E}{R} = \frac{10\ \text{V}}{7.43\ \text{k}\Omega} = 1.34\ \text{mA} \tag{1.26}$$

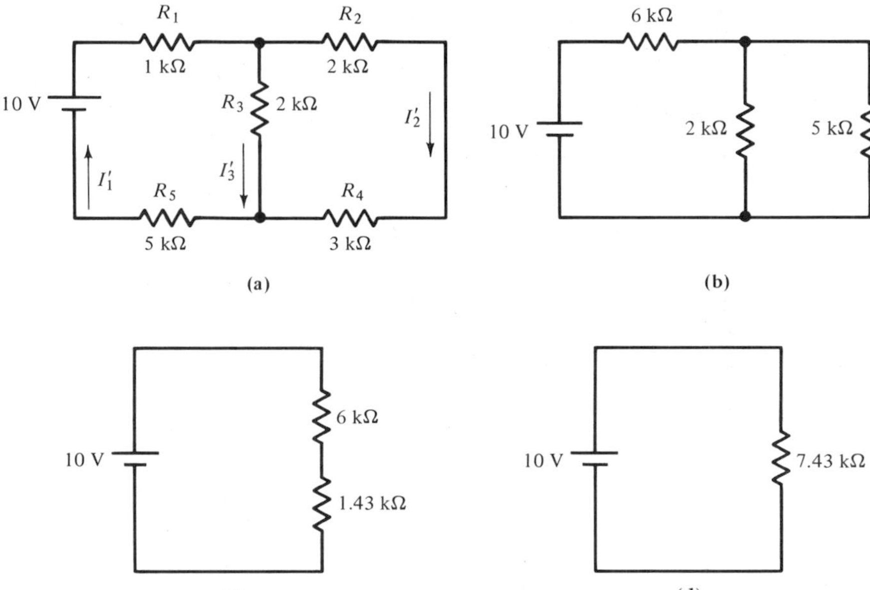

Figure 1.10

The currents in the parallel branches of Fig. 1.10(b) can then be determined:

$$I_3' = \tfrac{5}{7} I_1' = 0.96 \text{ mA} \qquad (1.27)$$

$$I_2' = \tfrac{2}{7} I_1' = 0.38 \text{ mA} \qquad (1.28)$$

The second voltage source can be treated similarly, as indicated in Fig. 1.11, giving the values $I_1'' = 1.93$ mA, $I_2'' = 7.7$ mA, and $I_3'' = 5.78$ mA, and the current directions shown in the figure. Total currents can then be obtained by superposition of the values in Figs. 1.10(a) and 1.11(a), as shown in Fig. 1.12. The result is the same as that obtained by applications of Kirchhoff's laws; however, a direct solution of simultaneous equations has been avoided.

1.3 / ELECTRICAL SIGNALS

1.3.1 / DC and AC signals

The terms DC and AC, direct current and alternating current, are employed in describing electrical signals. Taken within the context of electrical power transmission, a DC signal is one of constant polarity, whereas an AC signal is one of periodically changing polarity. In the field of electronic instrument design and performance these terms are retained, but the meanings are somewhat different from those of ordinary usage. It is more meaningful for the purposes of this text to understand that a DC signal is one whose polarity and magnitude are constant for the duration of the observation, and an AC signal is one whose magnitude changes with time but averages to zero over a

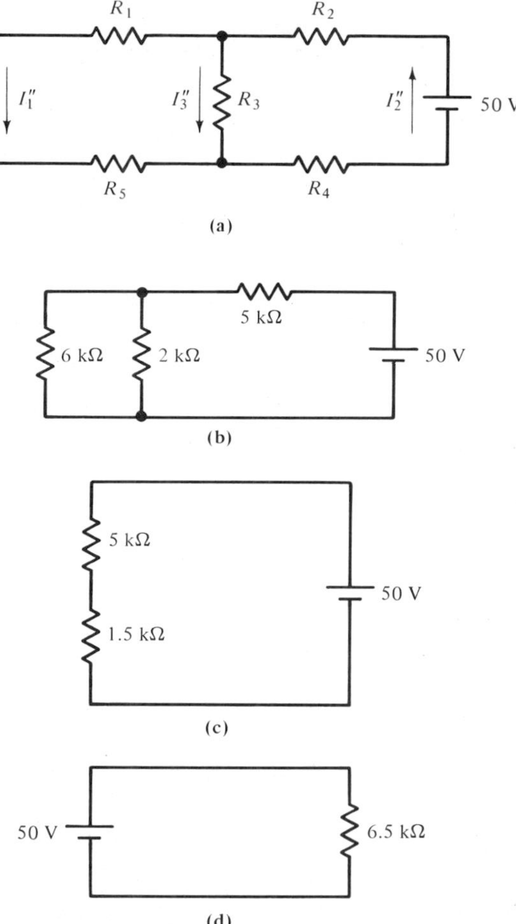

Figure 1.11

period of time. By using these restricted definitions of the terms, the signal illustrated as the instantaneous value of the magnitude as a function of time, curve 1, in Fig. 1.13, may be seen to consist of a constant DC component, curve 2, and a variable AC component, curve 3. Notice that the DC component is the average value of the entire signal over the period of the observation and that the average value of the AC component is zero. A signal such as the conventional domestic "110 VAC," which has an average value of zero, has no DC component.

The concept of separating a signal into DC and AC components is not limited to signals having sinusoidal AC components nor to those having periodic AC components. A nonsinusoidal signal with AC and DC components is illustrated in curve 1 of Fig. 1.14; again, curve 2 represents the DC component and curve 3 represents the AC component.

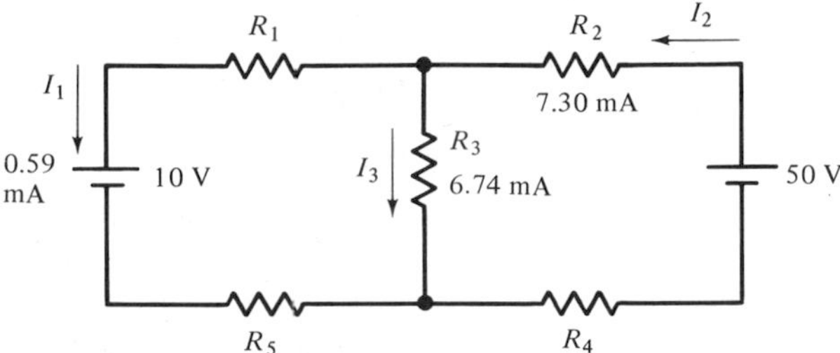

Figure 1.12

These definitions may appear arbitrary, but they are of considerable practical importance. It is easy to separate the AC and DC components of signals by means of reactive circuit components such as capacitors, inductors, and transformers (these are discussed in the following sections of this chapter and in Chapter 3). Furthermore, situations often occur in which separation is desirable. Examples are encountered in signal waveform shaping (Section 3.4) and in establishing steady state operating conditions for amplifiers (Section 4.4.4).

It is implied in the illustrations in Figs. 1.13 and 1.14 that the signals persist for periods of time that are long relative to the period of observation. Although this is a frequently encountered situation, it need not be true. If a signal, such as curve 1 in Fig. 1.15, that consists of a sinusoidal AC component added to a DC component but that persists only from time T_a to T_b, were observed under conditions such that the AC and DC components were separated, the AC component, curve 2, would show severe distortion in order to exhibit a zero average value.

1.3.2 / Sinusoidal signals

In considerations of signal-handling capabilities, it is common practice to restrict attention to sinusoidal signals. This practice is attractive because

Figure 1.13

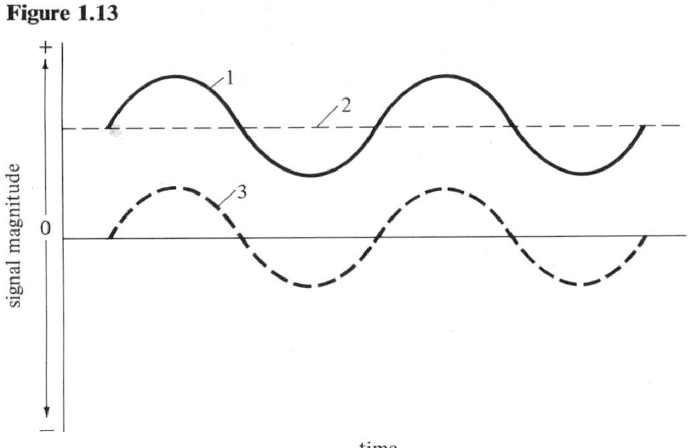

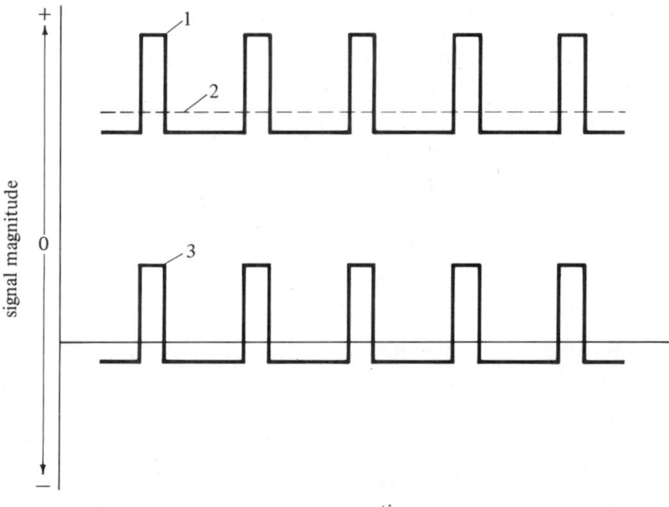

Figure 1.14

sine waves are easily described and because mathematical operations often are more easily performed with them than with other signal forms. The description of a DC signal is complete when magnitude and polarity relative to a standard have been given. A complete description of a sinusoidal signal can be given in terms of three quantities: the magnitude, the frequency, and the phase, relative to a standard.

In a graphical presentation of instantaneous signal value vs. time, all three types of information can be presented simultaneously. The time elapsing between adjacent comparable conditions of a periodic function, such as that illustrated in Fig. 1.16, is termed the *period*. For example, the period of signal I or II in the figure is the time lapse from 1 to 8, the adjacent positive crossings. This is also the time between adjacent positive or negative peaks or between any two adjacent comparable points. The reciprocal of the period is the frequency, f, which is ordinarily expressed in units of hertz (Hz), where 1 Hz = 1 cps, or in angular frequency (ω), expressed in radians per second,

$$\omega = 2\pi f \tag{1.29}$$

and 1 cycle is 2π radians. Period, then, may be expressed in either of two ways:

$$\text{period} = \frac{1}{f}\ \text{sec} \tag{1.30}$$

$$= \frac{2\pi}{\omega}\ \text{sec} \tag{1.31}$$

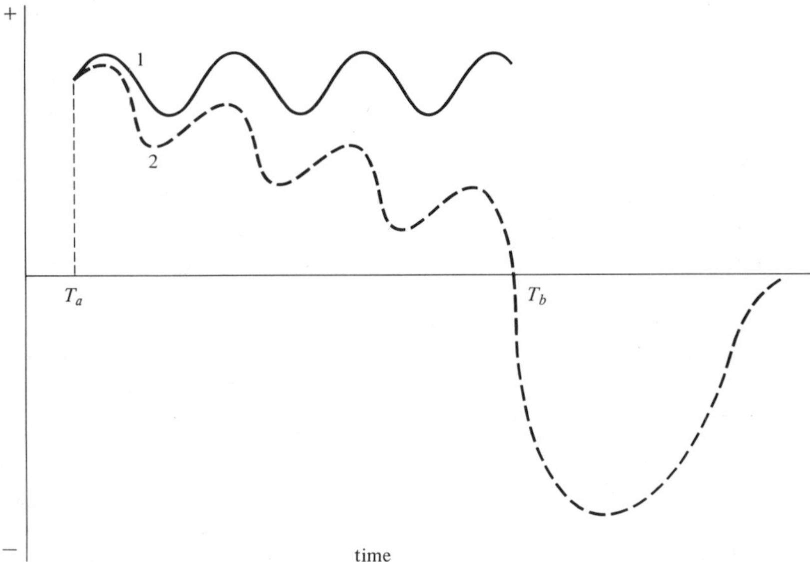

Figure 1.15

For the purposes of graphical presentation of periodic waveforms, it is often convenient to use the product of angular frequency and time, rather than time, as a coordinate.

The magnitude, or signal intensity, which is completely defined by the graphical presentation, may be expressed more concisely by various non-graphical presentations. Having identified a signal as a sine wave of specified frequency, its magnitude may then be defined by stating the maximum instantaneous value, for example, 5 V for signal I in Fig. 1.16(a). The peak-to-peak value, 10 V in the illustration, is also used. Another useful

Figure 1.16

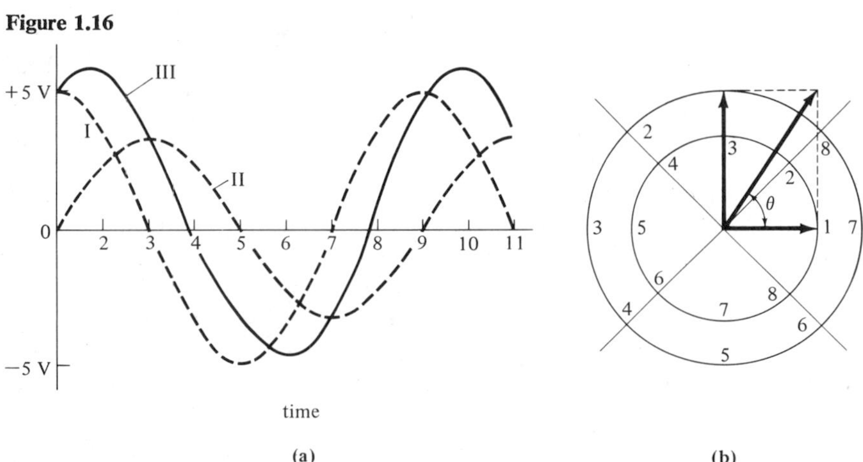

notation for designating signal amplitude is the root-mean-square (rms) value. This is defined as

$$\bar{E}^2 = \frac{1}{T}\int_0^T e^2\,dt \tag{1.32}$$

where $\bar{E}$ is the rms voltage, e is the instantaneous voltage, and T is the period of time over which the integral is averaged. The period of time is chosen so as to be large enough for a representative average to be obtained. If the signal magnitude is defined as

$$e = E_{\max} \sin \omega t \tag{1.33}$$

where $E_{\max}$ is the peak voltage and ω is the frequency in radians per second, substituting Eq. (1.33) into Eq. (1.32) and solving for $\bar{E}$ gives

$$\bar{E} = 0.707E_{\max} \tag{1.34}$$

The rms voltage is the effective voltage. For example, if a DC signal is imposed across a resistance, the power dissipated is given by Eq. (1.7). One volt rms has the same effect as 1 V DC. The rms notation is used to describe commercial and domestic AC electricity. The use of rms notation is not limited to sinusoidal signals; however, the use of Eq. (1.34) is limited to sine waves because of its origin.

For such purposes as ordinary domestic electrical installations the specification of the waveform, frequency, and amplitude, as in Eq. (1.33), is adequate. However, it is incomplete in that it does not permit the determination of the signal value at a given instant. It is established that the signal magnitude varies sinusoidally with time at a rate specified by ω within limits defined by $E_{\max}$, but it cannot be determined when, for example, the signal passes through a maximum. The description could be completed by specifying the time at which the signal has a characteristic value. Generally, however, it is sufficient to describe its relationship with respect to a defined reference signal such as that from the AC power distribution line or from one generated by the motion of a mechanical switch within an instrument. This is ordinarily done by specifying the phase relationship between the signals.

Such a designation is meaningful only when the signals being compared are of the same frequency. This may be done graphically as in Fig. 1.16. A more concise description of frequency relationships is provided by phasor notation. A phasor is a conventionalized vector, the length of which represents the amplitude of the signal, peak value, peak-to-peak, or rms. The phasor may be visualized as rotating counterclockwise about its origin at the frequency of the waveform. The projection of the phasor on the vertical axis, plotted against time, will generate the waveform.

The phasors of the waves in Fig. 1.16(a), if stopped at a particular instant, might be represented as shown in Fig. 1.16(b). Since by convention the phasor rotates counterclockwise, the fact that wave II lags behind wave I is denoted by their relative positions. To generate the relationship shown in Fig. 1.16(a), the phasors would remain just 90° apart as they rotate; this is also indicated in Fig. 1.16(b). If the phasor convention is used, then Fig. 1.16(b) gives the necessary quantitative information about amplitudes and the phase relationship. Phase relationships are ordinarily expressed by the angle between the phasors. By convention, phasors are indicated by boldface notation; thus, $\mathbf{E}_{\mathrm{III}}$ implies sine wave III with amplitude E_{III}, frequency ω, at phase angle θ relative to a stated reference:

$$\mathbf{E}_{\mathrm{III}} = E_{\mathrm{III}} \sin(\omega t + \theta) \tag{1.35}$$

If two voltage sources are in series, their equivalent can be obtained by addition, polarities being observed. When the sources furnish AC rather than DC signals, their instantaneous values are additive, as are DC voltages. This is shown in graphical form in Fig. 1.16(a), where curve III is the sum of curves I and II. Since signals I and III may be represented by phasors $\mathbf{E}_{\mathrm{I}}$ and $\mathbf{E}_{\mathrm{III}}$, vector addition, shown in Fig. 1.16(b), produces the phasor $\mathbf{E}_{\mathrm{III}}$. The signal amplitude is equal to the phasor length, and the phase of signal III, relative to signal II, is given by angle θ.

1.3.3 / Nonsinusoidal signals

The ease with which a sinusoidal signal may be described and treated mathematically makes it convenient in discussions of the performance of instruments. When the nature of common chemical signals is considered, however, it may appear that the extensive use of sine waves represents an entirely unrealistic simplification. In fact this is not true, because any signal may be considered the sum of one or more sinusoidal signals if the frequency and amplitude are chosen correctly. Specifically, it was shown by Fourier that a periodic waveform may be represented by a series of the following type:

$$\begin{aligned} f(\omega t) = {} & a_0 + a_1 \cos(\omega t) + a_2 \cos(2\omega t) + a_3 \cos(3\omega t) + \cdots \\ & + a_n \cos(n\omega t) + b_1 \sin(\omega t) + b_2 \sin(2\omega t) \\ & + b_3 \sin(3\omega t) + \cdots + b_n \sin(n\omega t) \end{aligned} \tag{1.36}$$

In this series $f(\omega t)$ represents the periodic waveform and ω represents the angular frequency of the fundamental component of the waveform. The multiples 2ω, 3ω, and so on, are harmonics of the fundamental. The constant a_0 describes the DC component of the signal, if any, and a_1, a_2, and so on, are coefficients that determine the extent to which each harmonic is included.

Any periodic function expressed in mathematical terms may, in principle,

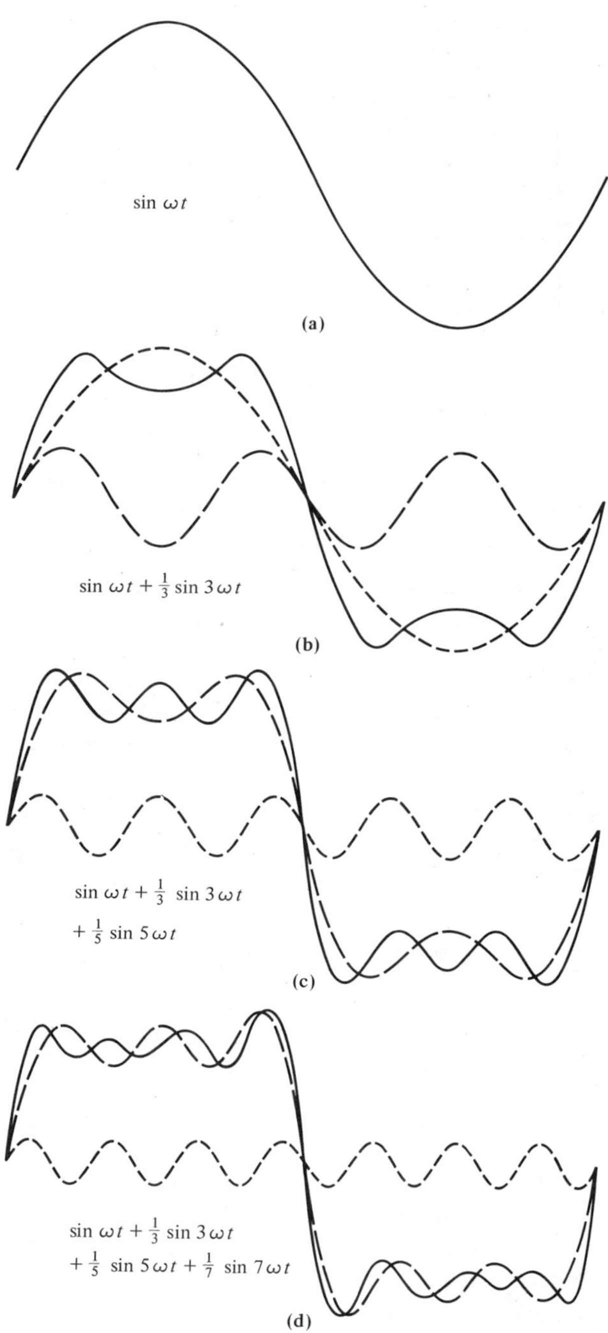

Figure 1.17 / *Approximation of a square wave by summation of a few sinusoidal components.*

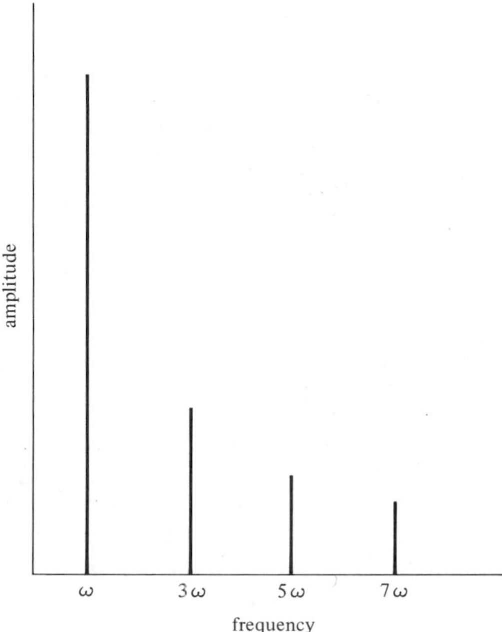

Figure 1.18 / *Spectrum of the wave illustrated in Fig. 1.17(d).*

be subjected to Fourier analysis for an evaluation of the terms in an expression comparable to Eq. (1.36). For example, a square wave may be represented by Eq. (1.37), which contains only odd harmonics (sine waves):

$$f(\omega t) = \frac{4}{\pi}\left[\sin(\omega t) + \frac{1}{3}\sin(3\omega t) + \frac{1}{5}\sin(5\omega t) + \frac{1}{7}\sin(7\omega t) + \cdots + \frac{1}{n}\sin(n\omega t)\right] \tag{1.37}$$

This is illustrated graphically in Fig. 1.17. In Fig. 1.17(a) the fundamental wave is shown; in Figs. 1.17(b)–1.17(d) the third, fifth, and seventh harmonics, suitably attenuated, are added. The resultant wave, in Fig. 1.17(d), is a recognizable approximation of a square wave. It is evident that the addition of higher harmonics causes sharpening of the edges and flattening of the tops of the square wave. An ideal square wave can be approximated to any desired degree of accuracy by the inclusion of a suitable number of high-frequency components. A somewhat more concise graphical presentation may be made by plotting the coefficients of the terms of the appropriate form of Eq. (1.36) against frequency; this produces a spectrum, indicated in Fig. 1.18, for a square wave.

By extension of this method, nonperiodic time functions may be represented by a Fourier *integral*, in which the sinusoidal components have all possible frequencies from zero to infinity. It is therefore possible to represent

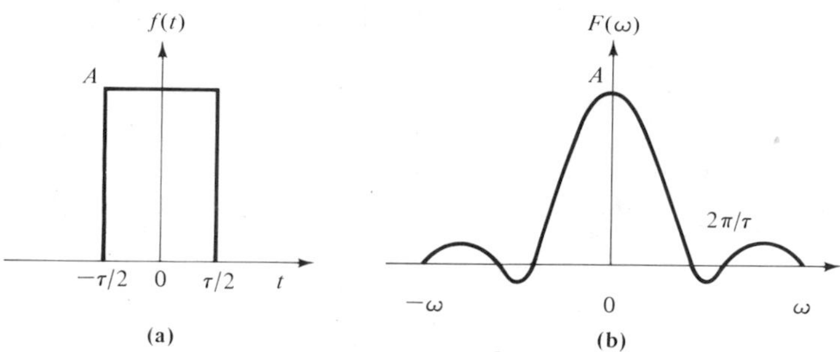

Figure 1.19 / (a) *Time–amplitude, $f(t)$.*
(b) *Frequency–amplitude, $F(\omega)$, for a square pulse.*

any real-time response function as a continuous frequency spectrum. For compactness it is usual to express the sines and cosines in complex exponential form. The lower limit of integration usually is extended from zero to negative infinity, which permits the use of only one exponential. The integrals given in the following equations are known as Fourier *transforms*, since they permit transformation of a time-response function to a frequency spectrum and the reverse:

$$f(t) = \int_{-\infty}^{\infty} F(\omega)e^{j\omega t}\,d\omega \tag{1.38}$$

$$F(\omega) = \int_{-\infty}^{\infty} f(t)e^{-j\omega t}\,dt \tag{1.39}$$

For example, consider the case in which $f(t)$ is a square pulse of amplitude A and duration τ. Equation (1.39) may be used to show that

$$F(\omega) = \frac{A\tau \sin \omega\tau/2}{\omega\tau/2} \tag{1.40}$$

The time–amplitude plot and the frequency–amplitude plot for this example are compared in Fig. 1.19. The occurrence of negative frequencies is an artifact of mathematical treatment and causes no practical difficulty.

The use of spectrum analysis on nonsinusoidal signals is of great assistance in assessing the ability of an instrument component to handle that signal without distortion. For example, suppose that the fundamental frequency in Fig. 1.17(a) is 100 kHz. An amplifier capable of passing 700-kHz signals undistorted would be able to reproduce the waveform shown in Fig. 1.17(d). An amplifier capable of passing 500 kHz but that attenuates 700 kHz could produce the wave in Fig. 1.17(c) without distortion, but an input comparable to that in Fig. 1.17(d) would be reproduced approximately in the form of Fig. 1.17(c) as well. Factors that bear on the ability of instruments to transmit high- and low-frequency signals are discussed in Chapters 3 and 4.

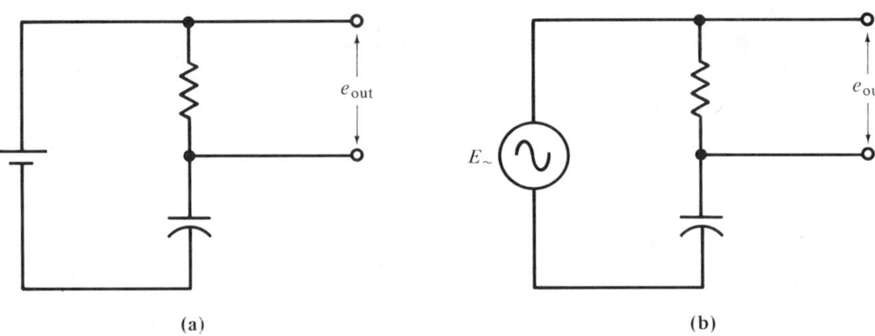

Figure 1.20

Another area in which spectrum analysis is important is Fourier spectroscopy. In this technique, a conventional frequency–amplitude plot is produced by performing a Fourier transform on the experimental data. This subject is discussed in Section 6.4.4.

1.4 / REACTIVE COMPONENTS

1.4.1 / Capacitance

When signals that vary with time are used, two electrical quantities must be taken into consideration in addition to those discussed above. These are capacitance and inductance. A capacitor is a device that contains two conductors, insulated from each other but physically located so that there is interaction between their electrical fields. When a charge is placed on one conductor, an equal and opposite charge will be induced on the other. Specifically, when a voltage is imposed across a capacitor, as in Fig. 1.20, the relationship between voltage and charge is given by

$$q = CE \tag{1.41}$$

where the proportionality constant C is the capacitance, which has the unit farad (F) when volts and coulombs are used. In an actual device, the conductors take the form of plates. Other factors remaining constant, capacitance is increased by increasing the area of the plates, by decreasing the separation between the plates, and by increasing the dielectric constant of the material between the plates.

A suitable dielectric material must be a nonconductor and must exhibit a large dielectric constant. The dielectric constant, which is the proportionality constant ε of Eq. (1.1), is a measure of the ability of a material to minimize the electrostatic interaction of charges. Thus, an increase in the dielectric constant of the material separating two charges would reduce the force acting on them and, hence, the potential difference between them as, indicated in Eq. (1.1). Accordingly, it is an implication of Eq. (1.41) that an increase in dielectric constant causes an increase in capacitance C and a decrease in potential difference E for a given charge q.

Capacitors are fabricated with air, mica, various ceramics, paper, and various polymers as dielectrics. Electrolytic capacitors are made with a metal foil and a conductive solution as plates, with a nonconductive oxide on a foil to serve as dielectric; large capacitance results because of the very small effective separation between the plates of the capacitor.

As a unit of capacitance, the farad is inconveniently large; more appropriate for use with the capacitors ordinarily encountered are the microfarad (μF or 10^{-6} F), the nanofarad (nF or 10^{-9} F), and the picofarad (pF or 10^{-12} F).

1.4.2 / Effect of capacitance on signals

When a DC signal is imposed on a capacitor, the plates charge until the relationship in Eq. (1.41) is satisfied. Then, if there is no significant leakage, nothing further happens. If a charged capacitor is disconnected from the voltage source without being discharged, it will remain charged, exhibiting a static voltage across its terminals until the charge has leaked off, a process that may be very slow. This provides a mechanism for the storage of information coded as charge; it also may cause a safety hazard if large voltages and capacitances are involved. To a DC signal, therefore, a capacitance is simply an open circuit once it has been charged to the imposed voltage.

With AC signals, however, the situation is different. During one half-cycle, the capacitor is charged with an appropriate polarity. During the next half-cycle, it is discharged and then charged again, but the polarity is now reversed. The result is a continuous flow of current into and out of the device. Although there is no actual conduction of current across the dielectric, there is, from the point of view of the external circuit, a transmission of the AC signal "through" the capacitor. Thus, in the case illustrated in Fig. 1.20(a), e_{out} would be zero after the capacitor has been charged completely, because no current would flow and all of the voltage would be dropped across the capacitor. With the AC source in Fig. 1.20(b), however, current would flow in each half-cycle, with the result that there would be a variable voltage drop observed as e_{out}.

A useful relationship can be obtained from Eq. (1.41) by noting that current is charge flow per unit time:

$$i = \frac{dq}{dt} \tag{1.42}$$

Then

$$i = \frac{d(CE)}{dt} = \frac{C\,dE}{dt} \tag{1.43}$$

Taking

$$E = E_{\max} \sin \omega t \tag{1.44}$$

and substituting in Eq. (1.43), one obtains

$$i = CE_{\max} \frac{d(\sin \omega t)}{dt} = \omega CE_{\max} \cos \omega t$$

$$= \omega CE_{\max} \sin\left(\omega t + \frac{\pi}{2}\right) \tag{1.45}$$

This states quantitatively, for sinusoidal signals, the important relationship between current and voltage in a capacitive circuit. Current leads voltage, and in a purely capacitive circuit (no resistance or inductance) it leads by 90° when the waveform is sinusoidal. The qualitative relationship can readily be determined by inspection. When a voltage is suddenly applied to a circuit that contains a capacitance, as in Fig. 1.20(a), current must flow before the capacitance can be charged to the applied voltage.

It is implicit in Eq. (1.45) that when an AC signal is imposed on a circuit such as that in Fig. 1.20(b), current will flow, but the capacitor will affect the fraction of imposed voltage that drops across the resistance. The circuit will act as an AC voltage divider; this follows because current is out of phase with the imposed voltage. The voltage dropped across the resistor will be equal to the product of current and resistance. It will therefore be in phase with the current and out of phase with $E_{\sim}$. Since the imposed voltage is equal to the sum of voltages dropped in all series components, and since it is not equal to that dropped in the resistance, there must be a voltage, not in phase with either the total voltage $E_{\sim}$ or e_{out}, dropped across the capacitor.

The quantity that describes the effect of a capacitor on the flow of current is the capacitive reactance, X_c, which is related to capacitance in the following way:

$$X_c = \frac{1}{\omega C} = \frac{1}{2\pi f C} \tag{1.46}$$

Here X_c is the reactance in ohms, C is the capacitance in farads, ω is the angular frequency in radians per second, and f is the frequency in hertz. Reactance is inversely proportional to capacitance. Although it has units of ohms, it differs from resistance in being frequency dependent. If, in the circuit of Fig. 1.20(b), $E_{\sim}$ is a 10-V rms signal at 1000 Hz, R is 1.0 kΩ, and C is 0.010 μF, then

$$X_c = \frac{1}{2\pi(10^3\ \text{Hz})(10^{-8}\ \text{F})} = 16\ \text{k}\Omega \tag{1.47}$$

Many of the signals of interest in chemical instruments have AC components. Furthermore, it is fundamentally impossible to avoid at least small (picofarad) capacitances in all circuits, and often capacitors are

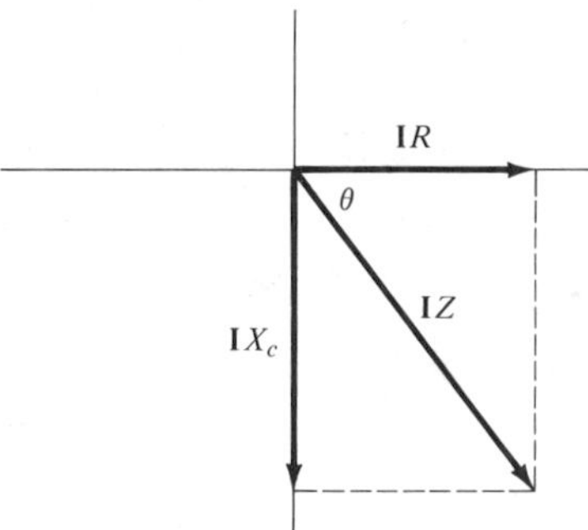

Figure 1.21

purposely included. Accordingly, current–voltage magnitude and phase relationships in capacitive circuits are basic to an understanding of the behavior of many types of instruments. They are especially important where amplifiers (Chapter 4) and feedback circuits (Chapter 5) are involved.

If one wishes to solve for the currents and voltages involved in the circuit of Fig. 1.20(b), then the use of Ohm's law in the form shown in Eq. (1.2) will be convenient, because this is a series hookup. The total voltage will be the sum of the voltages dropped across the resistor and the capacitor. If instantaneous values are taken, they can be added together, as indicated in Fig. 1.16(a). However, if one prefers not to reproduce graphically the entire waveforms, then AC currents and voltages cannot be treated as scalar quantities. One must specify both the phase relationship and the magnitudes. This is ordinarily done by using phasor notation, described above, in which a phase relationship is encoded as an angle between two phasors that represent signals having the same frequency. By convention, signals in resistive circuits are oriented along the horizontal axis and those in reactive circuits are oriented along the vertical axis. In the terminology of vector analysis, these are called the real and imaginary axes, respectively, and these terms are carried over for use in manipulating phasors.

Thus, the voltage dropped across the resistance of Fig. 1.20(b) would be represented by $\mathbf{I}R$, with the phasor symbol $\mathbf{I}$ used to designate a sinusoidal current. Since this is a resistive signal, it is oriented along the positive real axis, as in Fig. 1.21. We have already seen in Eq. (1.45) that voltage lags current in a capacitance; therefore, the $\mathbf{I}X_c$ phasor that represents the voltage drop across the capacitance must be along the negative imaginary axis. Vector addition gives a third phasor, $\mathbf{I}Z$, the total voltage applied to the circuit,

$$E_{\sim} = \mathbf{I}Z \tag{1.48}$$

where Z is impedance in ohms. This solution provides the magnitude from the phasor lengths and the phase relationships from the angles in the diagram. Specifically, it points out that when a sinusoidal signal is imposed on a

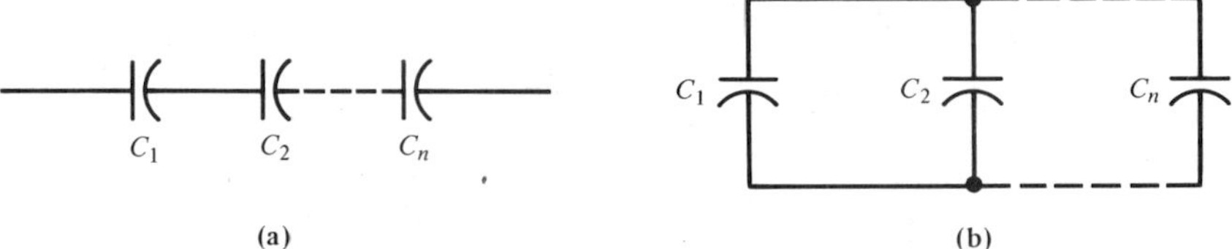

Figure 1.22

series R–C circuit, the voltage drop in the resistance leads the imposed voltage, and the drop in the capacitance lags the imposed voltage. This becomes important if one considers the type and origin of distortion caused by the process of amplification.

From the point of view of actual calculation of voltages or currents, this is incomplete. Since the components are in series,

$$E_{\sim} \equiv \mathbf{I}Z = \mathbf{I}R + \mathbf{I}X_c \tag{1.49}$$

Consider the orientation of the phasors in Fig. 1.21; $\mathbf{I}Z$ can be calculated by considering it to be the hypotenuse of a right triangle having $\mathbf{I}R$ and $\mathbf{I}X_c$ for sides, as

$$(IZ)^2 = (IR)^2 + (IX_c)^2 \tag{1.50}$$

or, with I removed, as

$$\begin{aligned} Z &= \sqrt{R^2 + X_c^2} = \sqrt{(1 \times 10^3)^2 + (1.6 \times 10^3)^2} \\ &= 1.89 \times 10^3\ \Omega = 1.89\ \text{k}\Omega \end{aligned} \tag{1.51}$$

The values for R and X_c having been obtained, the phase angle θ can be calculated:

$$\begin{aligned} \theta &= \tan^{-1}\frac{IX_c}{IR} = \tan^{-1}\frac{X_c}{R} \\ &= \tan^{-1}\frac{1.6 \times 10^3}{1.0 \times 10^3} = 58.0^\circ \end{aligned} \tag{1.52}$$

1.4.3 / Capacitances in series and in parallel

It was mentioned above that resistances in series are directly additive and that the reciprocals of resistances in parallel are directly additive. This is true regardless of whether the imposed signal is DC or AC, because resistance is not frequency dependent.

When capacitances are arranged in series, as in Fig. 1.22(a), they behave as though a single capacitance were present. The relationship between the individual capacitances in series and their equivalent takes the same form as that for resistances in parallel:

$$\frac{1}{C_{\text{tot}}} = \frac{1}{C_1} + \frac{1}{C_2} + \cdots + \frac{1}{C_n} \tag{1.53}$$

When capacitances are connected in parallel, as in Fig. 1.22(b), their equivalent is simply the sum of the individual values:

$$C_{\text{tot}} = C_1 + C_2 + \cdots + C_n \tag{1.54}$$

When series or parallel capacitances appear, they may be lumped together before reactance is calculated.

1.4.4 / Inductance

If an AC current flows through a conductor, the resulting magnetic field will show a time-dependent variation that corresponds to that of the current. If this fluctuating field cuts across a conductor, a proportional voltage will be induced. It can be shown that when this process involves a single conductor —that is, when the voltage is induced in the conductor through which the initial current flows—the induced voltage is in opposition to the externally imposed voltage responsible for the current flow. This is obvious without specific proof because, if the converse were true, this could constitute the active component of a perpetual-motion machine. The effect of an inductance is to impede the flow of AC current in the component, because the self-induced voltage is effectively subtracted from the externally applied voltage.

Inductance ordinarily is introduced into a circuit by means of a coil of wire. This ensures maximum interaction between the conductor and the magnetic field induced around it and, hence, maximum inductance for a given length of wire.

1.4.5 / Effect of inductance on signals

The induced voltage may be related to the current by

$$e = \frac{L\,di}{dt} \tag{1.55}$$

The proportionality constant L is the inductance, the value of which may depend on the geometry of the conductor, the nature of the material surrounding it, or the magnitude of the current. Inductance is measured in units of henrys when current and voltage are expressed in amperes and volts. Inductance is generally a constant only over limited ranges of operation.

Some important properties of inductance may be illustrated by starting with Eq. (1.55) and assuming a sinusoidal AC signal as in Eq. (1.44). Solving Eq. (1.55) for current gives

$$i = \frac{1}{L}\int e\,dt \tag{1.56}$$

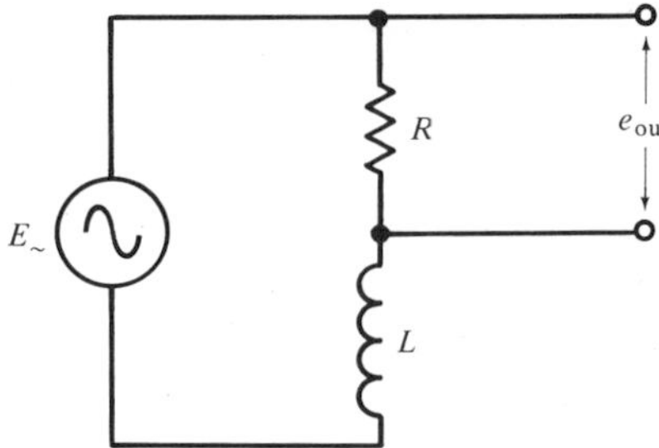

Figure 1.23

Substituting the value of e from Eq. (1.44) in Eq. (1.56) gives

$$i = \frac{1}{L}\int E_{max}\sin(\omega t)\,dt = -\frac{E_{max}}{\omega L}\cos(\omega t) \tag{1.57}$$

$$= \frac{E_{max}}{\omega L}\sin\left(\omega t - \frac{\pi}{2}\right) \tag{1.58}$$

From this equation it may be seen that when a sinusoidal voltage is imposed on an inductance, the resulting current shows a sinusoidal waveform that is 90° out of phase with the voltage and that lags the voltage.

To compare inductance and capacitance: To a DC signal an inductance is a dead short (shows zero resistance), whereas a capacitance is an open circuit after the initial charging; the current leads the voltage by 90° in a capacitance but lags by 90° in an inductance.

The inductance of a circuit component can be handled like the capacitance. The effect may be described in terms of inductive reactance X_L:

$$X_L = \omega L = 2\pi f L \tag{1.59}$$

Inductive reactance is proportional to inductance and to frequency.

For the inductor, Ohm's law may be expressed as

$$\mathbf{E}_L = \mathbf{I}X_L \tag{1.60}$$

with the dimensions ohms, volts, and amperes, as before. Thus, as frequency increases in a circuit such as that in Fig. 1.23, the current decreases, whereas just the reverse is true of the circuit in Fig. 1.20(b). This has important implications in determinations of the maximum frequency at which electronic equipment can operate, because it is impossible to avoid small, stray inductive and capacitive couplings within an instrument, which are caused by magnetic and electrostatic interactions and by insulation leakage. This is illustrated in Fig. 1.24, in which R_s represents insulation leakage, which should be of the order of megohms. The components L_s and C_s represent stray inductance and capacitance caused by coupling of adjacent components. These stray signal paths do not, in practice, cause much trouble with DC or low-frequency AC signals, because R_s and X_{L_s} can be made very large so that they

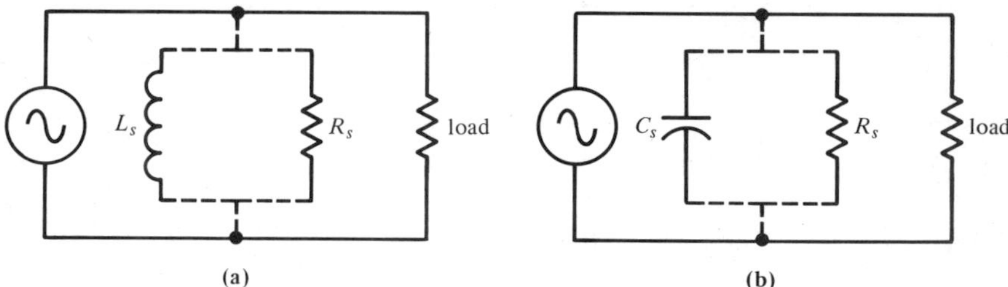

Figure 1.24

conduct only small stray currents. Similarly, X_{C_s} can be made very large at low frequencies because C_s can be quite small. At high frequencies, however, stray capacitance furnishes a significant leakage pathway. This shorting of the signal around a component often constitutes the limiting factor that determines the upper frequency limit at which an instrument can operate.

To analyze a circuit such as the L–R series combination in Fig. 1.23, a graphical solution may be used. This is shown in Fig. 1.25. Here the voltage drop in the resistance $\mathbf{I}R$ is oriented with the positive real axis. The drop in the inductance, $\mathbf{I}X_L$, will therefore go along the positive imaginary axis. This is true because the location of the $\mathbf{I}R$ phasor has defined the orientation for both current and voltage in the resistance, but only for current in the inductance. Since the current lags the voltage, the $\mathbf{I}X_L$ phasor will be placed as shown. Again, impedance can be obtained by vector addition, as indicated graphically, or by

$$Z = \sqrt{R^2 + X_L{}^2} \tag{1.61}$$

1.4.6 / Inductances in series and in parallel

Inductances in series are additive just as are resistances. The equivalent of inductances in series, Fig. 1.26(a), is a single inductance with a value equal to the sum of the individual values:

$$L_{\text{tot}} = L_1 + L_2 + \cdots + L_n \tag{1.62}$$

Figure 1.25

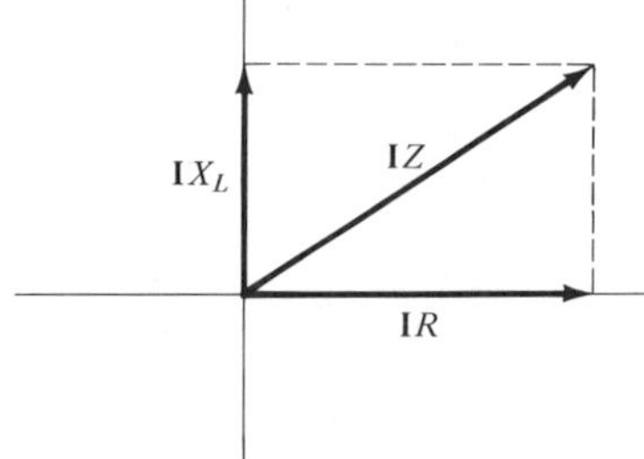

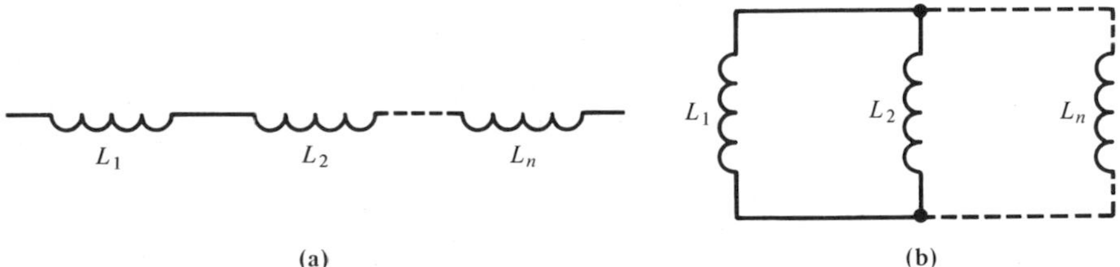

Figure 1.26

For a parallel hookup, as in Fig. 1.26(b), the equivalent value can be obtained in the same way that parallel resistances are:

$$\frac{1}{L_{\text{tot}}} = \frac{1}{L_1} + \frac{1}{L_2} + \cdots + \frac{1}{L_n} \tag{1.63}$$

Like the equivalent of capacitances, the equivalent of inductances can be obtained before the reactance is calculated.

1.4.7 / Series *L–C* and *L–R–C* circuits

When a sinusoidal signal is imposed on a capacitor, the current and voltage waveforms are out of phase. The result is that power is alternately transferred between the capacitor and the circuit to which it is connected, energy being stored for periods of time as electrostatic charge on the capacitor. This can be seen in Fig. 1.27. In Fig. 1.27(a), the voltage, current, and power waveforms, curves 1, 2, and 3, respectively, are shown for the case of a sinusoidal signal applied to a resistance. Current and voltage are in phase. The power curve represents a plot of the products of instantaneous values of current and voltage. It takes the form of a sine wave of twice the frequency of the imposed voltage or current and with only positive values. Power is dissipated in the resistance, but none is returned to the circuit. In Fig. 1.27(b), a similar presentation is made for the case of a sine-wave voltage imposed across a capacitance. Current (curve 2) leads voltage (curve 1), and power (curve 3) takes, as before, the form of a sine wave of twice the imposed frequency, but in this case it is symmetrical with respect to zero. Power is absorbed during a quarter-cycle of the imposed voltage and is transferred back to the circuit during the next quarter-cycle, with the result that there is no dissipation of power in the capacitance.

A similar result would be obtained for an inductance, with the power curve 180° out of phase from that in Fig. 1.27(b). In this case, energy is stored half the time in the magnetic field of the inductance. It follows, therefore, that if an inductance and a capacitance are connected together, as indicated in Fig. 1.28, and a signal is imposed, energy can be transferred back

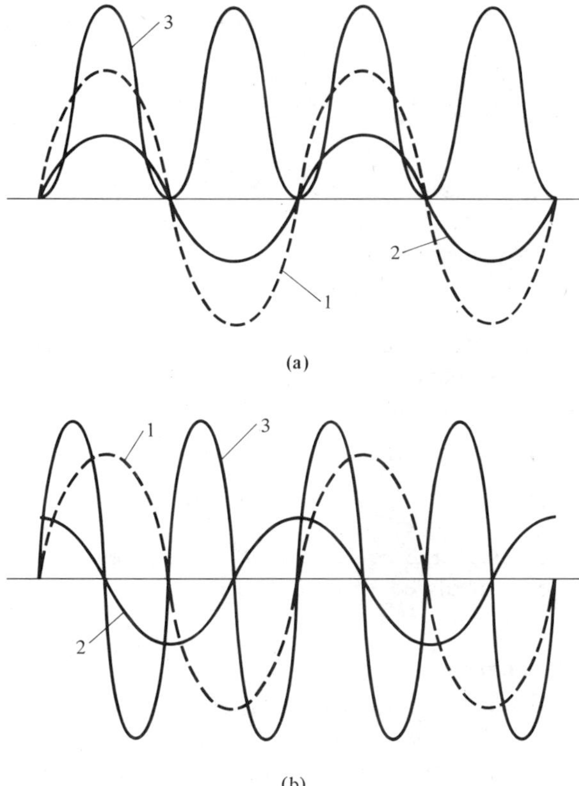

Figure 1.27

and forth between the two reactive components indefinitely with no loss. The oscillation would occur at a characteristic frequency ω_0, for which

$$X_C = X_L \tag{1.64}$$

$$\frac{1}{\omega_0 C} = \omega_0 L \tag{1.65}$$

$$\omega_0 = \frac{1}{\sqrt{LC}} \tag{1.66}$$

The analysis of L–C circuits can be carried out by the procedure used above. If the circuit of Fig. 1.29(a) is considered, the phasor diagram in Fig. 1.29(b) could be obtained. This can also be expressed as

$$Z = X_L + X_C \tag{1.67}$$

and at frequency ω_0 the impedance would be zero.

Actually, of course, this represents an idealization, because it is not possible to fabricate a real device that exhibits either capacitance or

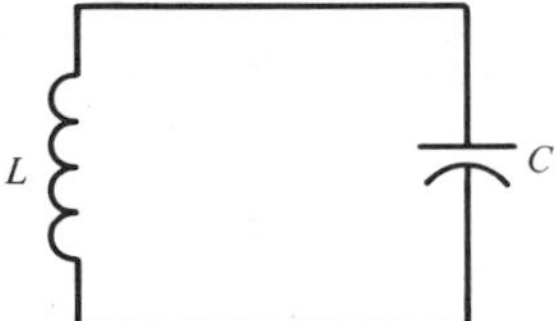

Figure 1.28

inductance without also having resistance; real L–C circuits do transfer energy alternately from one component to the other, but with a continual loss because of the resistance. This situation may be treated as in Fig. 1.30, in which resistance is placed in series with the reactive components. This may represent a discrete resistor placed deliberately in the circuit, or it may symbolize the resistance of the material that makes up the inductor, the capacitor, and the wires connecting them. The phasor diagram in Fig. 1.30(b) is constructed in the same way as those described above. There are two phasors along the imaginary axis, the difference of which must be added to the $\mathbf{I}R$ phasor to give $\mathbf{I}Z$. Impedance may be calculated as

$$Z = \sqrt{R^2 + (X_L - X_C)^2} \tag{1.68}$$

For the circuit illustrated in Fig. 1.30(a), the frequency is

$$\omega_0 = \frac{1}{\sqrt{LC}} = \frac{1}{\sqrt{(0.5 \text{ H})(300 \text{ pF})}} = 8.16 \times 10^4 \text{ rad/sec} \tag{1.69}$$

where the units are henrys and picofarads, and at this frequency

$$X_C = \frac{1}{\omega_0 C} = X_L = \omega_0 L = 40.8 \text{ k}\Omega \tag{1.70}$$

The network impedance at ω_0 obtained from Eq. (1.68) would be 50 kΩ, equal to the resistance, as indicated in Fig. 1.30(b). The phase angle between current and voltage would be zero at ω_0.

The series L–C circuit has considerable practical utility because of its ability to oscillate and to show a minimum impedance at the resonance

Figure 1.29

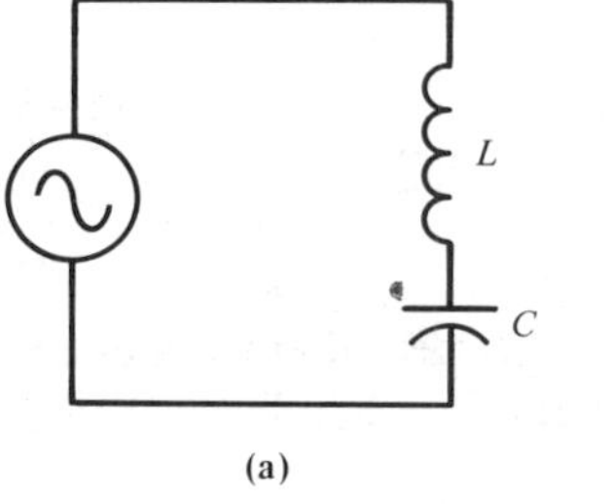

(a)

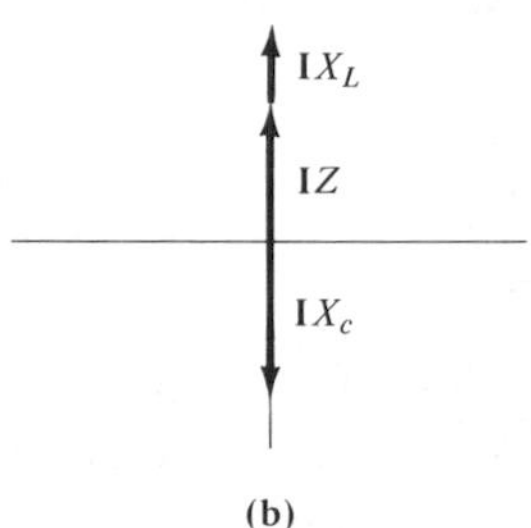

(b)

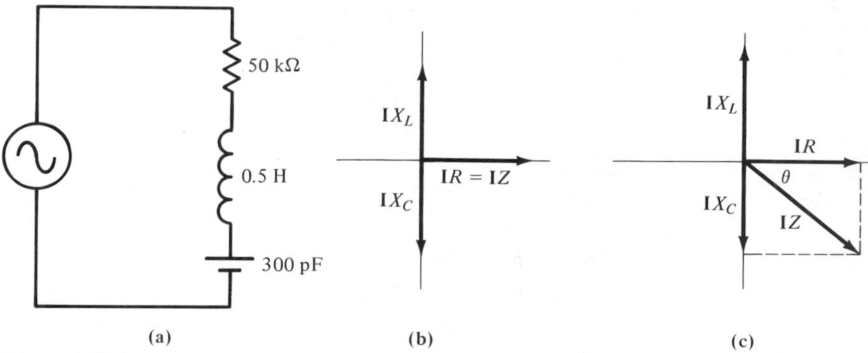

Figure 1.30

frequency. When combined with an amplifier, it can be used to discriminate against signals with frequency components near ω_0. The effectiveness of this circuit as an oscillator is diminished by power loss in any resistance. A parameter has been defined to serve as a figure of merit for this aspect of the behavior of a resonating circuit:

$$Q = \frac{X_0}{R} \tag{1.71}$$

Here X_0 is the reactance at ω_0. It is evident that a large Q value is desired to minimize power loss. In addition, if impedance is considered a function of frequency, a large Q implies a small minimum value of impedance and, hence, an increased effectiveness in frequency discrimination.

At other frequencies, the network impedance would be larger than that at ω_0. For example, at 10 kHz

$$X_L = 2\pi fL = 2\pi(10^4 \text{ Hz})(0.5 \text{ H}) = 31.4 \text{ k}\Omega \tag{1.72}$$

$$X_C = \frac{1}{2\pi fC} = \frac{1}{2\pi(10^4 \text{ Hz})(3 \times 10^{-10} \text{ F})} = 53.3 \text{ k}\Omega \tag{1.73}$$

$$Z = \sqrt{R^2 + (X_L - X_C)^2} = 54.7 \text{ k}\Omega \tag{1.74}$$

The phasor diagram pertaining to this solution is shown in Fig. 1.30(c). At this frequency, X_C is larger than X_L, and voltage accordingly lags current:

$$\theta = \tan^{-1}\frac{X_L - X_C}{R} = 23.67^\circ \tag{1.75}$$

1.4.8 / Parallel circuits

When reactances and impedances are connected in series, their values are directly additive, phase shift being taken into account, as has been demon-

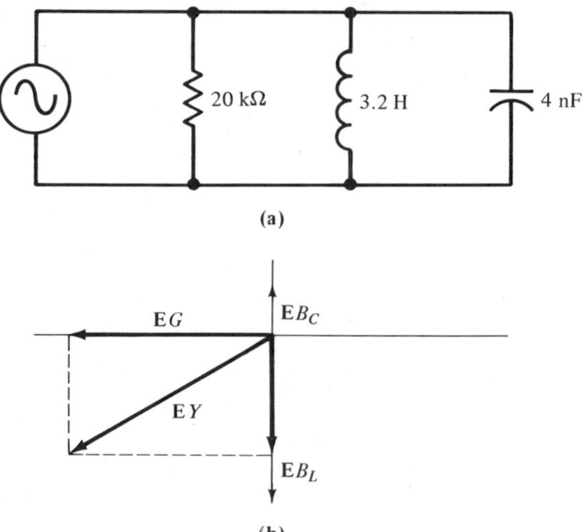

Figure 1.31

strated above. Impedances and reactances in parallel show behavior analogous to that of resistances; the reciprocal of the total impedance is equal to the sum of the reciprocals of the individual impedances or reactances:

$$\frac{1}{Z_{\text{tot}}} = \frac{1}{Z_1} + \frac{1}{Z_2} + \cdots + \frac{1}{Z_n} \tag{1.76}$$

With resistances in parallel hookup, it is frequently convenient to use conductances, since they are directly additive. Similarly, reciprocal quantities are defined for dealing with parallel hookups of reactive components with AC signals. The reciprocal of impedance is admittance:

$$\frac{1}{Z} = Y \tag{1.77}$$

The reciprocal of reactance is susceptance, which may be either capacitive or inductive:

$$\frac{1}{X_C} = B_C \quad \text{and} \quad \frac{1}{X_L} = B_L \tag{1.78}$$

Both admittance and susceptance have dimensions of reciprocal ohms, as does conductance.

Circuit analyses can be performed either graphically or analytically, as in the case of series circuits. For example, the total impedance and the phase angle between current and voltage waveforms are calculated for the circuit in Fig. 1.31(a), assuming a 1000 Hz signal, as follows:

$$G = \frac{1}{R} = \frac{1}{20\ \text{k}\Omega} = 5.0 \times 10^{-5}\ \Omega^{-1} \tag{1.79}$$

$$B_L = \frac{1}{2\pi f L} = \frac{1}{2\pi(10^3\ \text{Hz})(3.2\ \text{H})} = 5.0 \times 10^{-5}\ \Omega^{-1} \tag{1.80}$$

$$B_C = 2\pi f C = 2\pi(10^3\ \text{Hz})(4 \times 10^{-9}\ \text{F}) = 2.5 \times 10^{-5}\ \Omega^{-1} \tag{1.81}$$

$$Y = \sqrt{G^2 + (B_C - B_L)^2} = 5.6 \times 10^{-5}\ \Omega^{-1} \tag{1.82}$$

$$Z = \frac{1}{Y} = 18\ \text{k}\Omega \tag{1.83}$$

The phase relationship between current and voltage waveforms can be determined as before:

$$\theta = \tan^{-1}\frac{B_C - B_L}{G} = 26.6^\circ \tag{1.84}$$

A graphical solution of this problem, using a phasor diagram, is given in Fig. 1.31(b). To show conductance and susceptance as reciprocals of resistance and reactance, the reciprocal of the magnitude is taken, and the direction of the vector is changed by 180°. Therefore, conductance is oriented along the negative real axis, capacitive susceptance along the positive imaginary axis, and inductive susceptance along the negative imaginary axis.

The parallel *L–C* circuit, like the series *L–C* circuit, will undergo resonance at a characteristic frequency, which is defined by Eq. (1.66). The parallel circuit differs from the series circuit in that at ω_0 admittance is at a minimum (ideally zero, if there were no resistance) and, accordingly, impedance is at a maximum. This is the converse of the situation that obtains with series circuits. The parallel *L–C* circuit is called the "tank" circuit because of its ability to store energy by allowing the energy to be transferred back and forth between the electrostatic field of the capacitance and the magnetic field of the inductance. As before, the effectiveness of the tank circuit is limited by resistive power loss, which is described in terms of the *Q* parameter of Eq. (1.71). The tank circuit is widely used to control the frequency of sine-wave signal generators and as the frequency-discriminating component in tuning circuits.

1.4.9 / The transformer

Flow of current in a conductor results in the establishment of a proportionate magnetic field surrounding the conductor. Conversely, a moving magnetic field in the space surrounding a conductor, caused either by growth or decay of the field or by movement of the conductor, induces a voltage in it, which will cause current flow if a complete circuit exists. The reactive effect of a coil is caused by induction of a voltage by the magnetic field set up by the externally imposed AC signal. This is *self-induction*, since the imposed and induced electrical signals occur in the same conductor. If a second conductor is placed in close physical proximity to the first, a signal will similarly be induced in it by the process of *mutual inductance*; a device of this type is a *transformer*.

A transformer, therefore, provides for the transmission of an AC signal from one circuit to another to which it is linked only by its induced magnetic field. In practice, the two conductors are in the form of coils, usually termed the *primary* and the *secondary*, and are wound on a common iron core to ensure the most effective magnetic interaction. In an ideal transformer, the entire magnetic field established by the primary interacts with the secondary. Magnetic flux density, proportional to primary current, is inversely proportional to primary inductance, hence inversely proportional to the number of turns of wire, n_1, in the primary. Voltage is induced in the secondary in proportion to the interaction between the field and the coil; hence, secondary voltage is ideally proportional to the number of turns of wire, n_2, in the secondary. It therefore follows that the relationship between primary and secondary voltages is

$$\frac{E_1}{E_2} = \frac{n_1}{n_2} \tag{1.85}$$

If there is no power loss,

$$E_1 I_1 = E_2 I_2 \tag{1.86}$$

and therefore substitution of Eq. (1.85) into Eq. (1.86) gives

$$\frac{I_2}{I_1} = \frac{n_1}{n_2} \tag{1.87}$$

By the use of a larger number of turns in the secondary than in the primary, $n_2 > n_1$, the voltage can be stepped up and the current stepped down; conversely, if $n_2 < n_1$, voltage is stepped down and the current is stepped up. Note that, according to Eq. (1.86), if the secondary is open circuited, the power dissipated in it should ideally be zero and, accordingly, the primary should draw no current.

Transformer technology has been developed to such a point that the ideal relationships are sufficiently accurate for many purposes. To a good approximation the device is self-regulating; power is delivered to the primary only as it is removed by loading the secondary.

The effect of loading a transformer can therefore be calculated simply. In Fig. 1.32, the current and voltage in the secondary circuit may be expressed by

$$\frac{E_2}{I_2} = R_2 \tag{1.88}$$

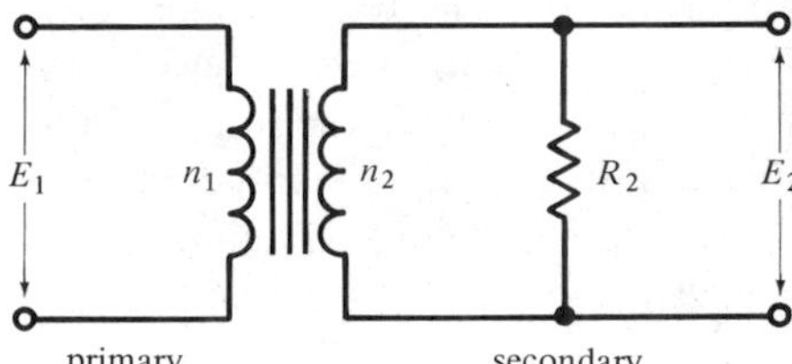

Figure 1.32

Substitution of Eqs. (1.85) and (1.86) into Eq. (1.88) gives

$$\frac{E_1}{I_1}\left(\frac{n_2}{n_1}\right)^2 = R_2 \tag{1.89}$$

or

$$R_{1(\text{eq})} = R_2\left(\frac{n_1}{n_2}\right)^2 \tag{1.90}$$

In an ideal transformer, the effect in the primary of loading the secondary, $R_{1(\text{eq})}$, is proportional to the *square* of the turns ratio. This relationship has practical importance in many applications in which AC signals must be moved from one component to another. One example is in audio systems. Effective power transfer from the amplifier to the loudspeaker can be achieved only when the amplifier and speaker impedances are matched. This can be achieved readily by the use of a coupling transformer with turns ratio properly selected, according to Eq. (1.90). The ability to transform impedances can also be used in some cases to improve the signal-to-noise ratio of noisy signals.

PROBLEMS

Calculate the resistance between points A and B for each of the following:

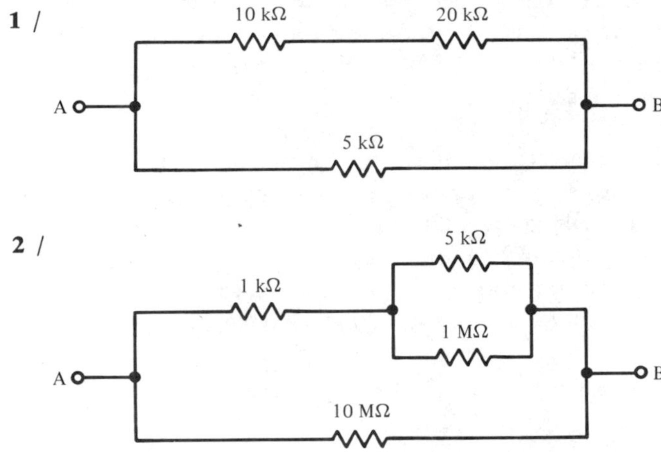

3 /

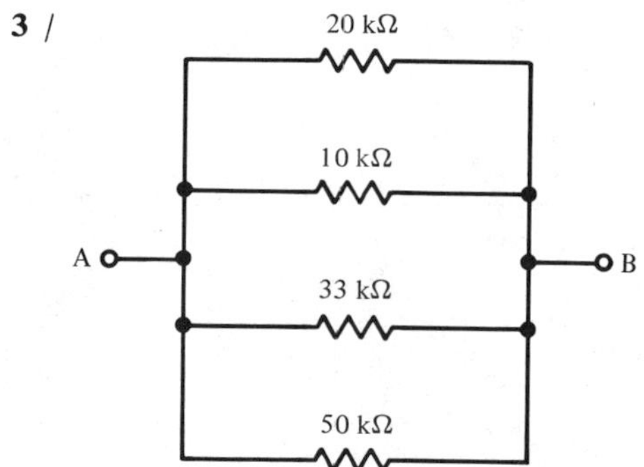

4 /

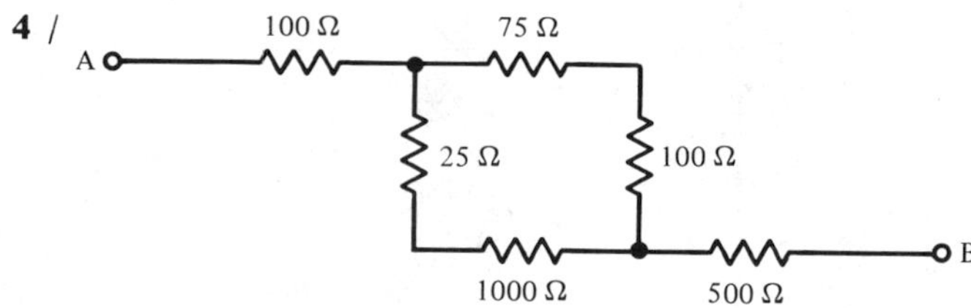

Calculate the voltage between points A and B for each of the following:

5 /

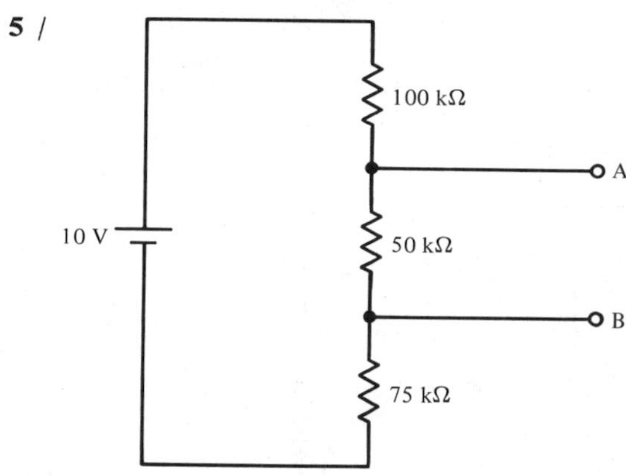

6 /

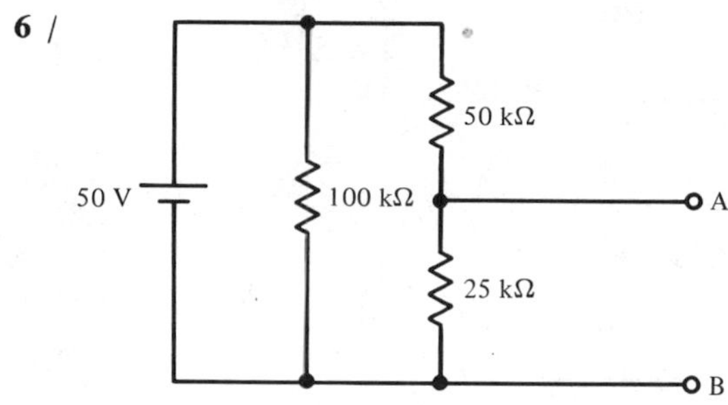

7 /

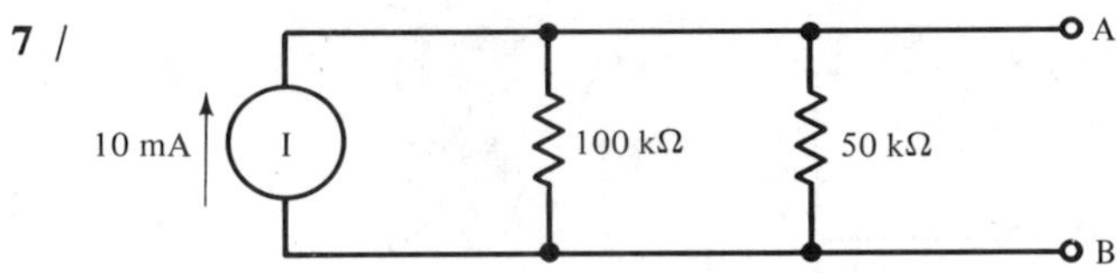

8 /

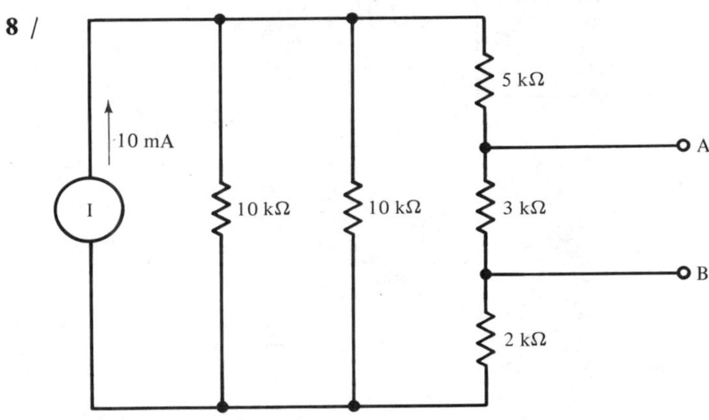

9 /

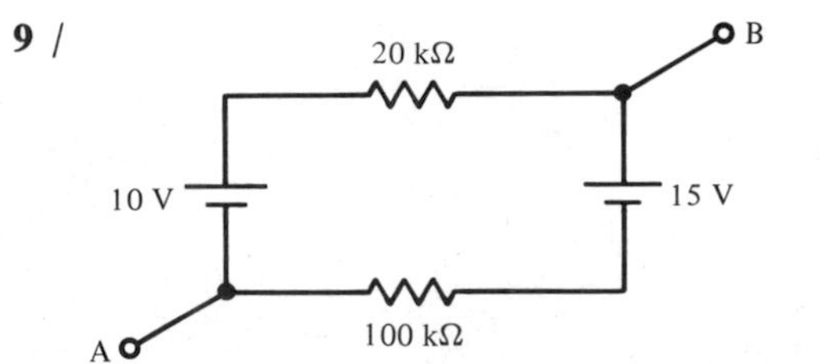

Calculate the current designated I for each of the following:

10 /

11 /

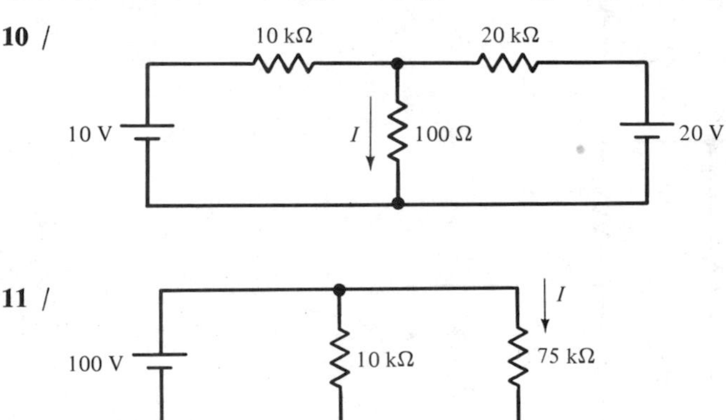

12 /

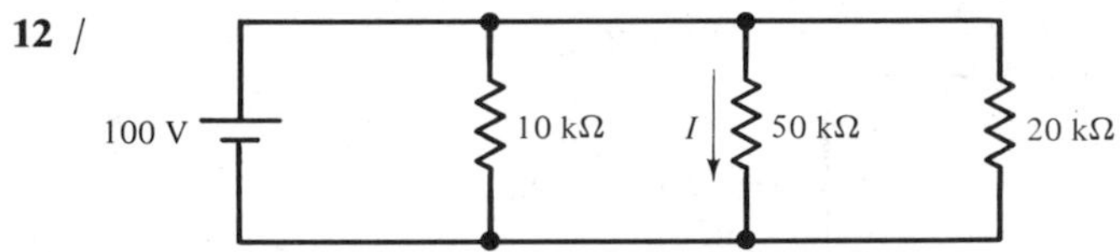

13 /

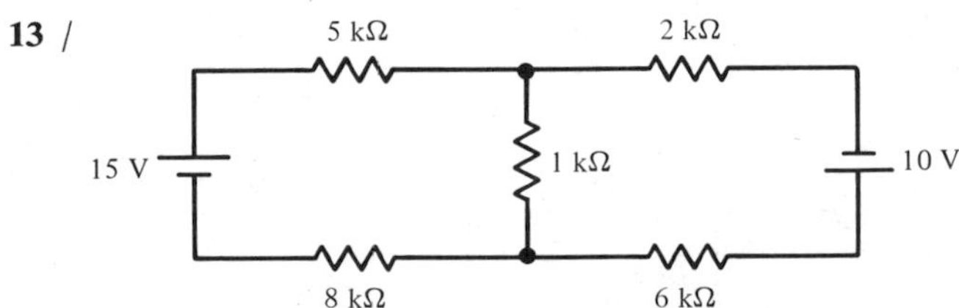

14 /

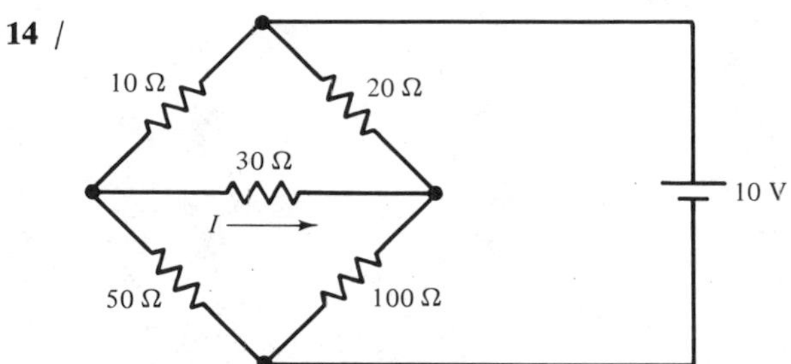

Calculate the impedance and the phase angle between current and voltage for each of the following, assuming a 10-kHz sine-wave signal:

15 /

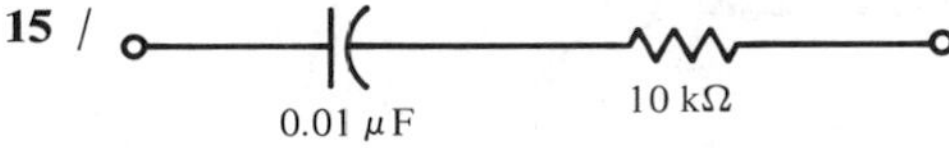

16 /

10 kΩ
25 pF

17 /

10 kΩ
1 H

18 /

10 kΩ
1 mH

19 /

100 kΩ
1 nF
0.1 H

20 /

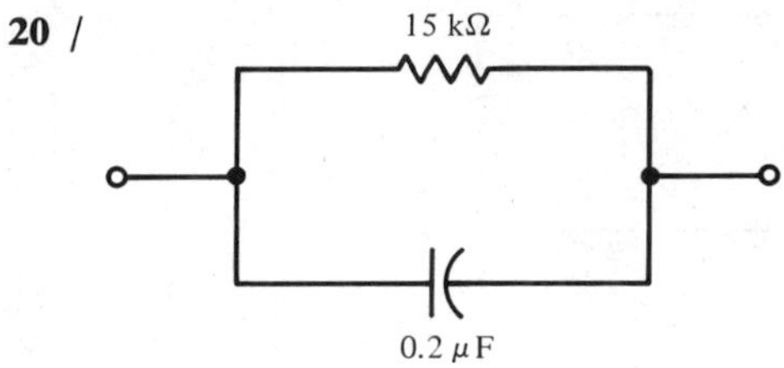

21 /

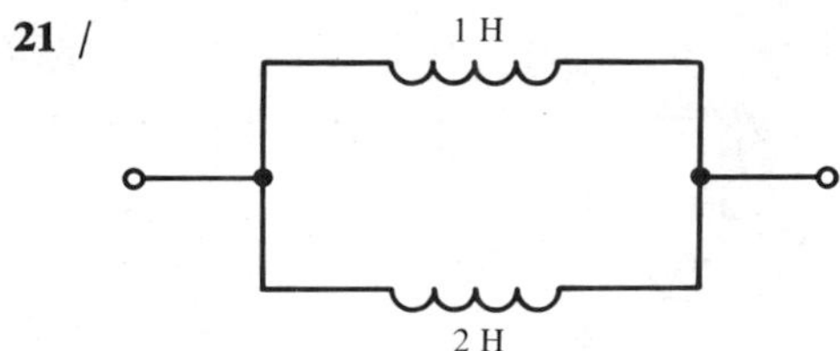

22 /

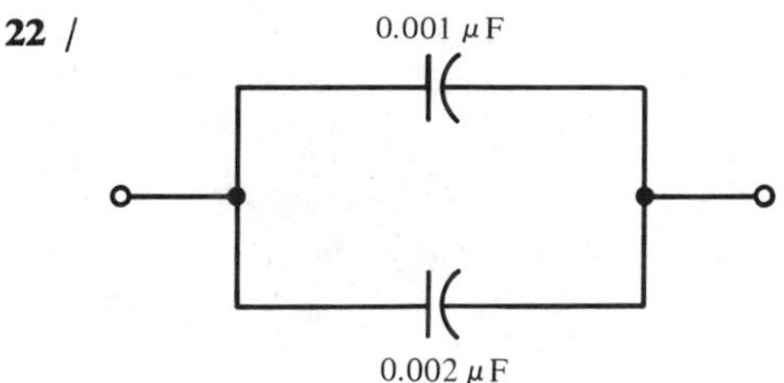

23 /

24 /

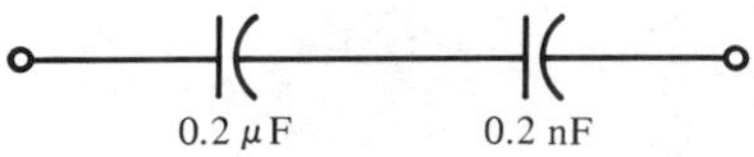

Calculate the resonance frequencies for each of the following:

25 /

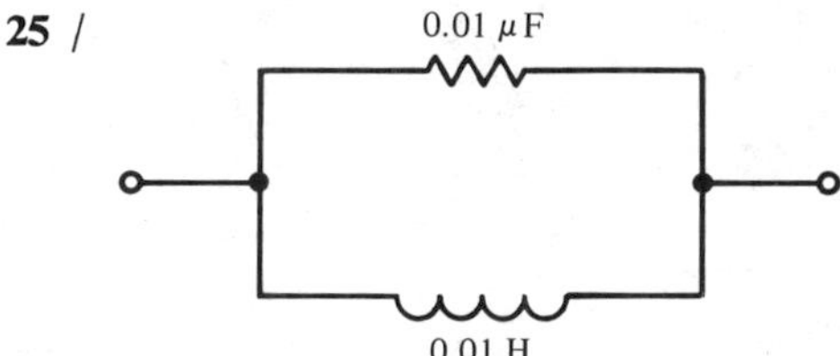

26 /

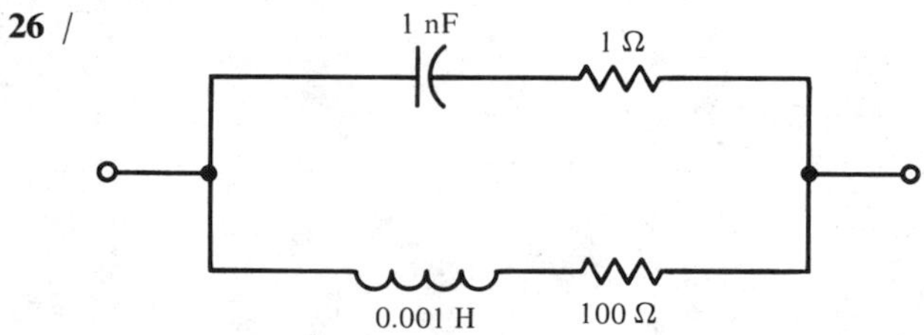

27 /

28 /

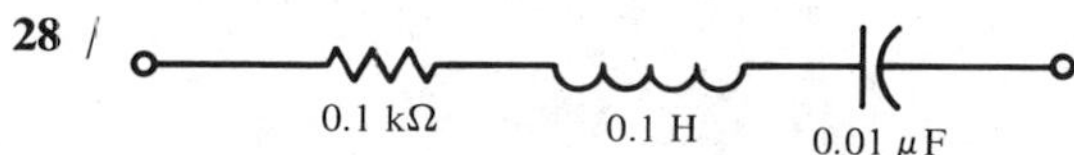

Calculate the impedance for each of the following at DC, at 1 kHz and then at 100 kHz:

29 /

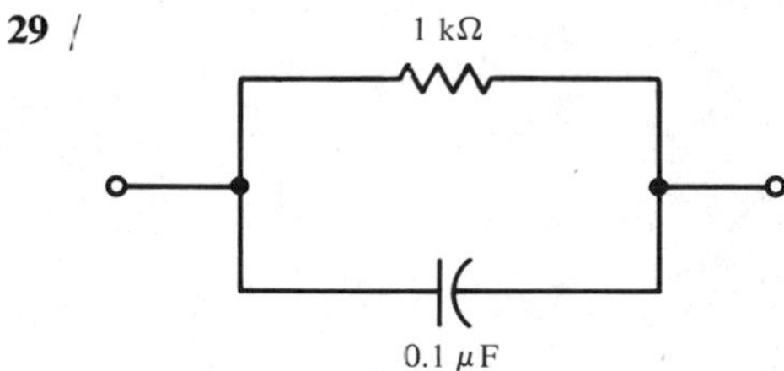

30 /

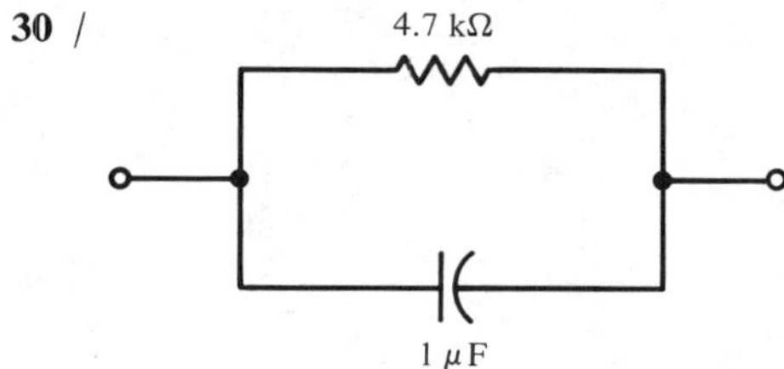

chapter 2 / Models of electronic systems

chapter 2

In dealing with electronic systems, as in other areas, it is customary to use models to describe systems and to facilitate the transfer of information about them. Quite generally, newcomers to the field experience difficulties in understanding electronic systems until the ability to comprehend the models has been mastered. In this chapter we shall undertake to discuss the use of the following: schematic diagrams, wiring diagrams, and equivalent circuit diagrams. We shall consider them in a general way before attempting to describe any of the areas in which they may be of use. This approach is taken because the models are very generally applicable to the subjects that are to be discussed. In effect, we are asking the reader to consider the subject initially in the abstract, with the realization that it may not be immediately apparent to him why he should be interested. The concepts introduced here are used repeatedly in the chapters that follow.

2.1 / THE NATURE OF ELECTRONIC MODELS

The three types of models to be discussed—wiring diagrams, schematic diagrams, and equivalent circuit diagrams—are similar in appearance in that they are graphical and are composed primarily of stylized, nonmathematical symbols. They serve quite different purposes, however, and a failure to make the distinction between them can cause confusion.

2.1.1 / Wiring diagrams

Wiring diagrams are literal representations of the device in question. Each component is represented by a symbol, and the diagram indicates exactly how connections are made between the components. The function of a wiring diagram is to assist the user in identifying the parts of his equipment.

2.1.2 / Schematic diagrams

Schematic diagrams are similar to wiring diagrams in that each component is represented by a symbol and that interconnections between components are shown. However, the purpose of a schematic is to provide a concise explanation of the *function* of the parts of an instrument. The schematic is arranged so as to facilitate comprehension of function without regard to the actual physical construction of the instrument.

Efficient use of instrument systems, not only by the designer and repair technician but also by the operator, requires the ability to read schematic diagrams. Without this ability, the user will experience great difficulty in gaining a solid understanding of the capabilities and limitations of his instrument. He will also be severely handicapped in attempting to communicate with technical people regarding repair, modification, and acquisition of equipment. He must, of course, be able to follow a schematic if he intends to do any work on his equipment himself.

The knack of reading a schematic is somewhat similar to, although very much simpler than, translating from a foreign language. One needs to know not only how to provide English equivalents of individual words, but also how to grasp the meanings of phrases and sentences. Similarly, although it is necessary to be able to identify the symbols and to be aware of the conventions followed in drawing schematics, this alone will not suffice. One must also be able to recognize the functions of groups of components in order to know what they are supposed to do. Perhaps the most important single point is that the user must be able to trace the path of a signal through an instrument and to recognize the direction of transmission of information in a circuit.

Considering the extremely large number of different types of components presently produced by the electronics industry, the task of acquiring a vocabulary of basic units may appear to be formidable. In fact, however, certain concepts have extremely wide applicability, so that scientific instrumentation consists to a great degree of varying combinations of a relatively small number of concepts. Furthermore, the concepts do not change rapidly with the ongoing progress of technology but, rather, evolve as new types of devices become available, the same basic idea appearing at one time with vacuum-tube components, at another with bipolar transistors, and at still another with integrated circuits. It is therefore feasible for the instrument user to familiarize himself, at least qualitatively, with the more widely used concepts and to learn to recognize the more frequently used representations of them. In this and the following chapters, an attempt is made to provide useful sketches of these basic concepts and to give some indication of their role in contemporary instrumentation.

2.1.3 / Equivalent circuits

Equivalent circuits are models that serve to explain the behavior of a device without necessarily attempting to explain how that behavior is carried out. Equivalent circuits may be entirely empirical. Very often they constitute such strikingly oversimplified descriptions of the real device that it is difficult for the uninitiated reader to appreciate their usefulness. They are, nevertheless, extraordinarily useful in that they provide a simple and precise means of quantitatively describing specifically defined aspects of behavior.

Familiarity with one type, the Thévenin equivalent circuit, is of particular importance to the instrument user because he can specify in terms of its components the factors that control the accuracy with which information may be moved from one instrument to another. This matter of the transfer of information, which may be a voltage measurement made with a meter or an oscilloscope, a current measurement in polarography, or a recording of pH by means of a strip-chart recorder, is one of prime importance to the laboratory scientist, because it determines the accuracy with which he receives data from his instruments. This problem is considered from different perspectives in Sections 7.1 and 8.2.

From a different point of view, it is quite possible for a researcher who is not an engineer to become sufficiently knowledgeable to be able to modify equipment so that it will better serve specialized functions. This, of course, requires serious background study. Some feeling for the uses of equivalent circuits will greatly facilitate this study, and a discussion that, hopefully, will serve this purpose is included in Section 4.10.

2.2 / ELECTRONIC SYMBOLS AND CONVENTIONS

A number of the more common symbols used in schematics are identified in Fig. 2.1. Electronic devices consist of components connected by conductors: wires, cables, and printed-circuit conductor strips. On a schematic, these will be shown as lines joining symbols. It is assumed in a schematic that conductors exhibit zero resistance; therefore, all points on a single conductor, say points A–F in Fig. 2.2, are at exactly the same potential. This is obviously not literally true, but conductor resistance is usually small enough that it does not influence circuit performance. Conductors that cross, with electrical contact between them, are indicated by a dot at the connection, as shown at point A in Fig. 2.2. Crossed conductors that are not in electrical contact, because of insulation or physical separation, do not have the dot. In certain parts of a circuit, especially at the input of an amplifier, difficulty may be encountered because of unwanted electrical or magnetic coupling between conductors, which may result in a stray signal appearing where it ought not to be. This can often be avoided by surrounding the sensitive part of the circuit with a grounded metal shield, which may be a wire braid placed outside the insulation of a wire or a metal box surrounding the circuit. It is shown on a schematic by enclosure in dotted lines, as indicated at the input in Fig. 2.2.

In laying out schematics, it is common practice to arrange for the signal to move from left to right. Figure 2.2 is a typical representation of a two-stage amplifier. The signal source e_s is connected to the first stage, transistor Q_1, and associated components; the first-stage output is connected to the second stage, transistor Q_2, and its associated components. Use of the

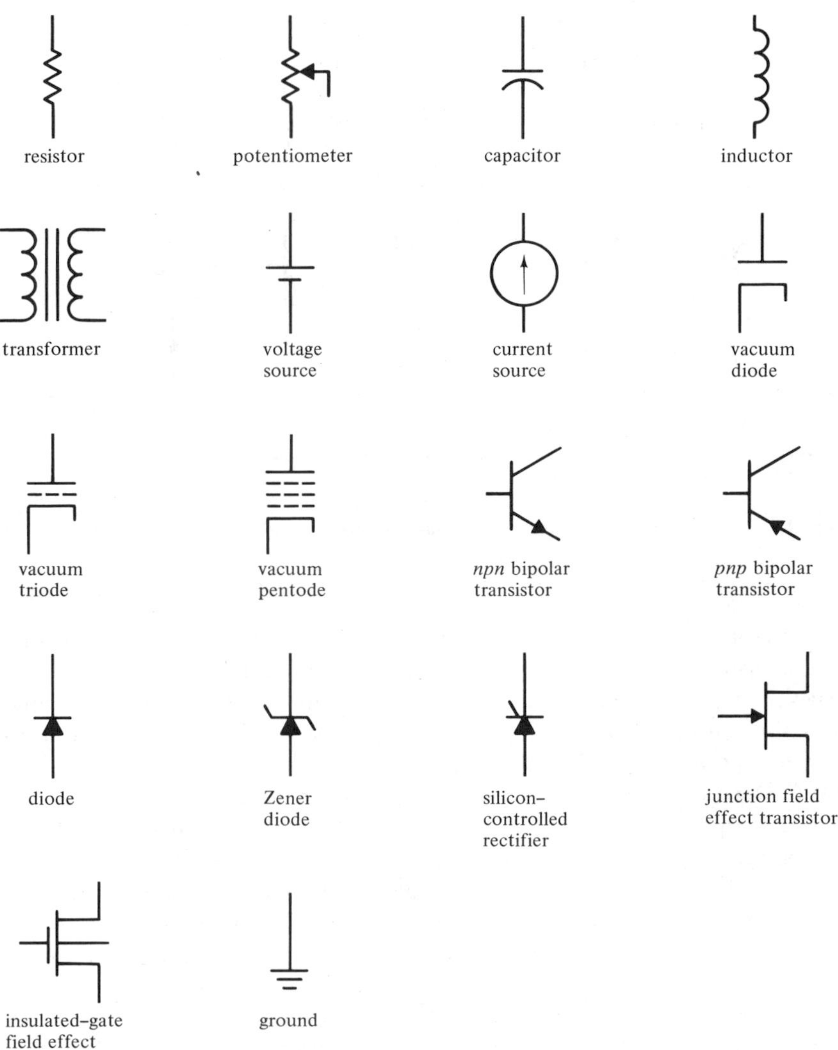

Figure 2.1 / *Symbols for frequently used components.*

ground symbol indicates that the conductor in question is connected to a reference voltage level. For instruments that are operated from AC mains, this ordinarily will be the earth ground provided with the electrical connections, possibly the third wire in the line cord, to which the instrument chassis is usually connected. Battery-operated instruments may be entirely isolated, in which case the ground symbol refers to the common conductor of the circuit, which is connected to the chassis. If such an instrument is connected to another, the instrument grounds, often termed *chassis grounds*, are ordinarily connected.

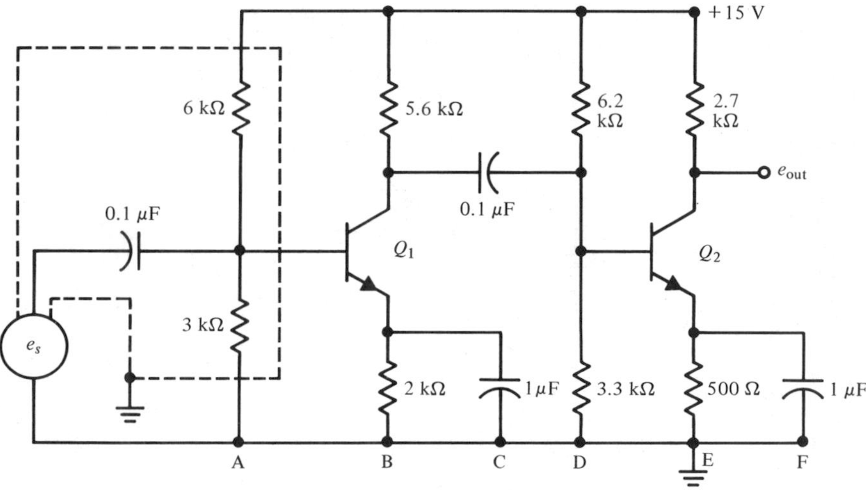

Figure 2.2 / *Schematic diagram for a two-stage amplifier.*

Schematic diagrams are usually labeled with component values, as in Fig. 2.2, and usually these numbers are printed on the components themselves. However, certain types of components, especially resistors and capacitors, have values shown in color codes; these color codes are based on the standard color designations given in Table 2.1.

Of the various codes, that for resistors is so widely used that one must be familiar with it to be able to examine equipment. For axial lead resistors, the type for which coding is most common, the resistance and the tolerance are encoded in four colored bands, illustrated in Fig. 2.3. Resistance is shown as a two-digit number multiplied by 10 raised to an appropriate power. The two digits and the power of 10 are encoded. Some examples are given in Table 2.2.

To accommodate values less than 1 Ω and values between 1 and 10 Ω, silver is used for the 10^{-2} multiplier and gold is used for the 10^{-1} multiplier. Resistors are commonly made to tolerances of either 5 percent or 10 percent. These are indicated by gold and silver, respectively, for the fourth band.

TABLE 2.1 / *Standard color-code designations*

Number	*Color*	*Number*	*Color*
0	black	5	green
1	brown	6	blue
2	red	7	violet
3	orange	8	gray
4	yellow	9	white

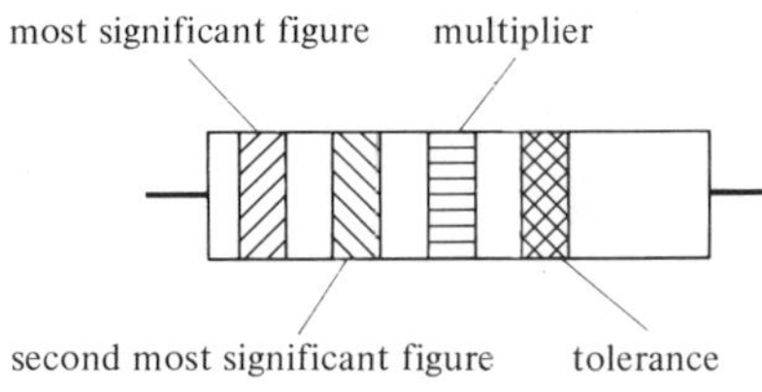

Figure 2.3 / *Resistor color code.*

2.3 / SIGNAL SOURCE MODELS

It is noted in Section 2.1.3 that equivalent circuits provide an effective basis for examining the problems encountered in moving signals between instruments. The present discussion pertains to the transmission of an analog voltage signal, a signal that contains information encoded as a voltage proportional to the magnitude of the quantity being measured. This is a commonly encountered situation in a chemistry laboratory, and it presents special problems, because the consequence of an incorrect decision is very often not the failure to transmit a signal but, instead, the transmission of a distorted signal. Furthermore, it may not be at all obvious that the distortion is occurring.

2.3.1 / Ideal current and voltage sources

For the purpose of constructing models, it is convenient to refer to two additional circuit components, the ideal current source and the ideal voltage source. An ideal current source is a device that is capable of delivering its characteristic current under all conditions. By this definition the load (R_{load}) in Fig. 2.4(a), into which the signal is delivered, can have any value from zero to infinity. This implies that the power source responsible for current flow is capable of furnishing an infinite voltage to push a finite current through an infinite resistance—obviously only hypothetically possible. The ideal voltage source has an analogous definition. It is a device

TABLE 2.2 / *Application of the resistor color-code scheme*

Resistance (Ω)	*First band*	*Second band*	*Third band*
0.15	brown	green	silver
2.2	red	red	gold
33	orange	orange	black
680	blue	gray	brown
8,200	gray	red	red
10,000	brown	black	orange
270,000	red	violet	yellow
1,500,000	brown	green	green
22,000,000	red	red	blue

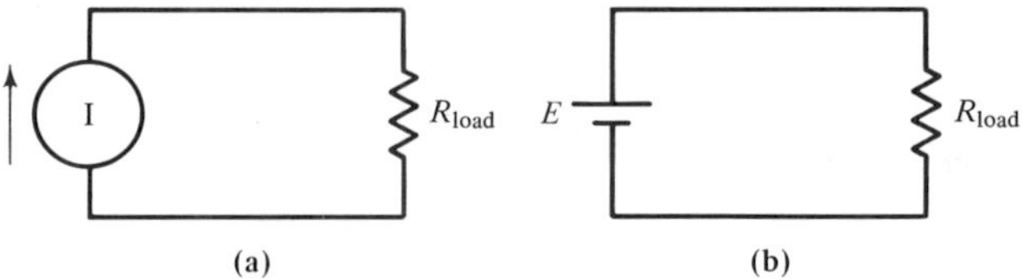

Figure 2.4 / (a) *Ideal current.*
(b) *Ideal voltage sources.*

capable of imposing its characteristic voltage under all conditions. In Fig. 2.4(b), R_{load} can have any value from zero to infinity, which implies that the voltage source is capable of delivering an infinite current into a dead short.

Thus, both the current and the voltage sources are idealized concepts rather than physically real devices. However, they are extremely useful in constructing models because of their predictability. They may be combined with other concepts, such as resistance or capacitance, so as to form models that simulate the behavior of real devices. The simplicity of the theoretical components makes it possible to calculate the exact behavior of the models. In defining these components, it is not necessary to place any restrictions on their time-dependent behavior. The characteristic voltage or current may be DC or an AC signal with any desired waveform.

2.3.2 / Thévenin's theorem

A very widely used model can be constructed by applying Thévenin's theorem, which states that any pair of terminals of any device may be replaced by a model that consists of an ideal voltage source in series with an impedance. Thus, a device of any degree of complexity, from a pH meter to a digital computer, containing any number of terminals, may be replaced by this model, which consists of just two components. As a practical matter, of course, the theorem is ordinarily not applied randomly to pairs of terminals, but only to those that serve as signal sources or as signal receivers.

The equivalent circuit voltage is equal to the voltage difference between the chosen terminals when no current is drawn. The resistance is that which would be observed between the terminals with all internal power sources inactivated. For example, consider the circuit in Fig. 2.5(a), a voltage divider with the output taken across R_2. The Thévenin equivalent for this or, for that matter, any other device has the form shown in Fig. 2.5(b). Applying the two parts of the theorem, the equivalent voltage e_{ter} would have a value equal to the drop across R_2 with no current drawn to the external load at terminals X and Y:

$$e_{ter} = E_{int}\left(\frac{R_2}{R_1 + R_2}\right) = (25\text{ V})\left(\frac{20\text{ k}\Omega}{50\text{ k}\Omega}\right) = 10\text{ V} \tag{2.1}$$

The equivalent resistance would be the resistance that would be observed if the internal source E_{int} were inactivated, as indicated in Fig. 2.5(c). Note

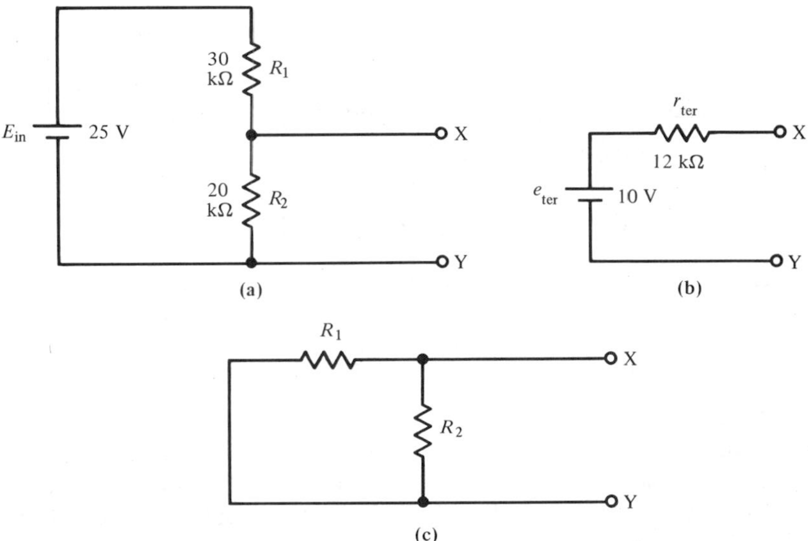

Figure 2.5 / *Thévenin equivalent circuit for a voltage divider.*

that the internal source is replaced by a short, not by an open circuit. Under these conditions the resistance between terminals X and Y is the parallel equivalent of R_1 and R_2:

$$r_{ter} = \frac{R_1 R_2}{R_1 + R_2} = \frac{(30\text{ k}\Omega)(20\text{ k}\Omega)}{50\text{ k}\Omega} = 12\text{ k}\Omega \tag{2.2}$$

One might expect that a calculation of the values of equivalent circuit components often would not be practical, because instruments that produce signals of interest are too complex for this type of calculation. However, the illustration does show quite directly what is meant by the components of the equivalent circuit. As far as terminals X and Y are concerned, the circuit in Fig. 2.5(b) is identical with that in Fig. 2.5(a). Furthermore, even in this case, it is simpler.

2.3.3 / Output impedance

The Thévenin equivalent circuit points to the distinction between a real and an ideal voltage source. The real source has a finite internal resistance associated with it, whereas the ideal source exhibits a zero internal resistance. It should be noted that internal resistance—or, more often, output resistance when signal sources are discussed—is a most significant quantity in describing behavior. A statement of output impedance, together with a knowledge of the voltage level of the signal, defines the quantity of power that a signal generator can deliver and allows a quantitative estimation of the error that will occur when the signal is delivered into another device.

Suppose that E in Fig. 2.6(a) is a DC voltage and that E_{XY} is to be

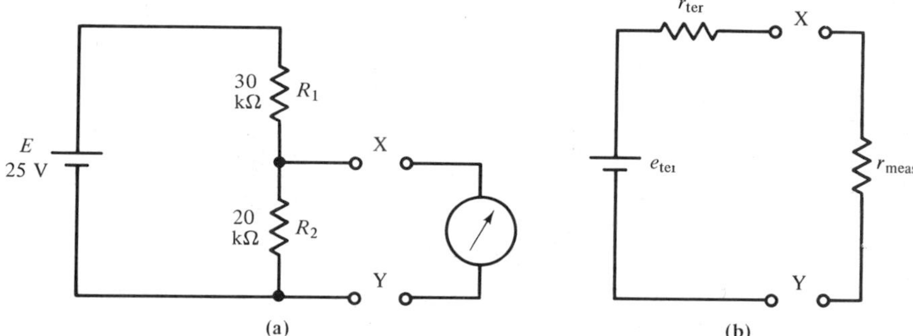

Figure 2.6

measured by using first a moving-coil voltmeter with a 10-kΩ input resistance, then a vacuum tube voltmeter with a 1-MΩ input resistance, and finally, an electrometer that shows a 100-MΩ input resistance. The measurement is indicated in Fig. 2.6(b), in which r_{meas} indicates the input resistance of the measuring device, and r_{ter} indicates the signal-generator output resistance of the Thévenin equivalent circuit. If we assume that the meters are accurate, each will register the voltage actually dropped across its terminals:

$$\text{measured voltage} = e_{ter} \frac{r_{meas}}{r_{ter} + r_{meas}} \tag{2.3}$$

This will amount to 4.55 V with the moving-coil meter, 9.90 V with the vacuum tube voltmeter, and 9.999 V with the electrometer.

This illustration is intended to emphasize several important points. The output impedance of a device is its Thévenin equivalent circuit impedance. It determines the extent of voltage attenuation that will occur when an observation of the signals is made. The observation must be determined from an independent measurement or calculation, because a single observation of the signal is distorted in the process of making the observation. A further implication is that when an accurate measurement of voltage is desired, one must minimize the source impedance and maximize the receiver impedance. That is, one approaches as nearly as possible the use of an ideal voltage source (zero output impedance) and an ideal voltage-measuring device (infinite input impedance). Note that these desiderata exist only when the desired information is contained in the magnitude of the voltage. If a voltage measurement is being made for another purpose—for example, to observe high-frequency waveform shapes—the impedance relationships described above would give very bad results.

Measurements on which construction of an equivalent circuit will be based must be made while the instrument is in operation. One cannot obtain a useful value for the equivalent circuit resistance simply by cutting off the power and measuring the resistance between the terminals, because active components—transistors and vacuum tubes—show very different behavior

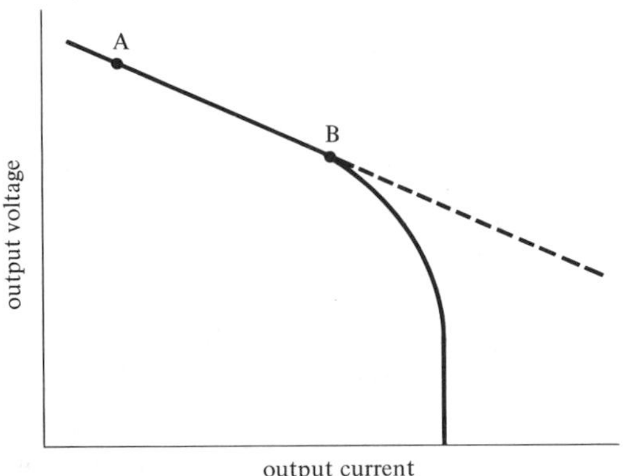

Figure 2.7 / *Experimental measurement of Thévenin equivalent circuit components.*

when they are in operation and when they are disconnected, and it is the active behavior that is of interest. Electronic devices are not usually designed to undergo drastic changes in operating conditions. For example, a power supply might show a relationship between output voltage and current such as that in Fig. 2.7. The output resistance would be the slope of the curve, and it is evidently not constant. In the usual case, the supply will be intended to operate only with a limited current flow, perhaps in the region between no current and point B; drawing larger current may, in fact, cause damage. The equivalent circuit could be determined by measuring current and output voltage at two points such as A and B. The slope of the line connecting A and B would give the output impedance; an extrapolation to zero current would give the equivalent circuit voltage.

2.3.4 / Norton's theorem

The Thévenin equivalent circuit can simulate, within its limitations, any signal source. It is, however, not the only type that can be used, and the choice of equivalent circuits is a matter of convenience. The Thévenin circuit is a useful model for devices that behave like voltage sources. This includes signal generators that do not have extraordinarily high output impedances and that produce signals with the desired information contained in the voltage. If the output impedance is large, or if, in the context of the problem, current rather than voltage is of interest, the Thévenin circuit is not as helpful. In such a case, it may be desirable to invoke Norton's theorem.

Norton's theorem states that a circuit consisting of an ideal voltage source in series with a resistance can be replaced by an ideal current source in parallel with the same resistance without causing any other change. The relationship between the magnitudes of the current and voltage sources is Ohm's law. Thus, the circuits in Fig. 2.8 would have identical effects on an external load if $E = IR$.

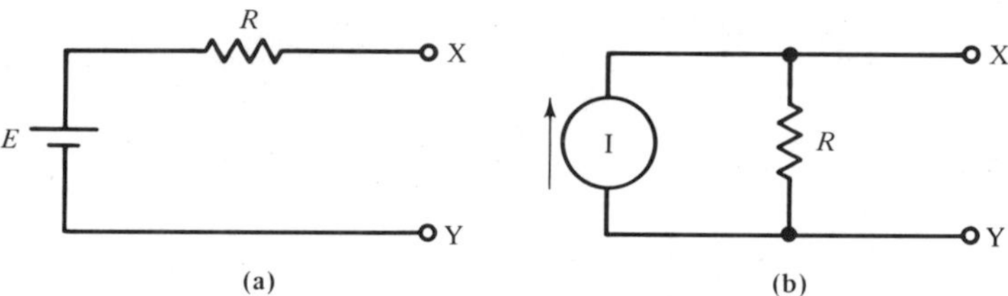

Figure 2.8

In the Thévenin circuit, the source voltage is divided between the internal and the external resistances; the smaller the internal resistance, the more nearly ideal the behavior of the device as a voltage source. In the Norton circuit, the source current is divided between the internal and the external resistances; the larger the internal resistance, the more nearly ideal the behavior of the device as a current source. An ideal current source would exhibit infinite internal resistance, thereby delivering all its characteristic current to any external load, regardless of its size. One important implication of this is that current measurements will be more accurate as the source impedance is made large and as the receiver impedance is made small. This is discussed in more detail in Section 7.1.

PROBLEMS

Show the Thévenin equivalent circuits, including values for components, for each of the following:

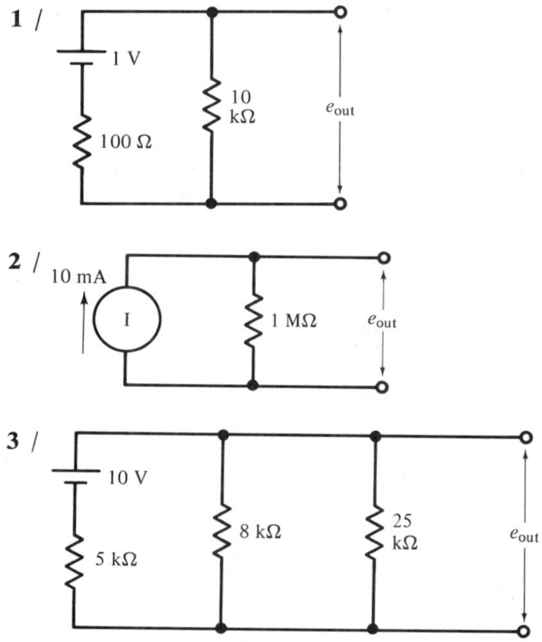

4 /

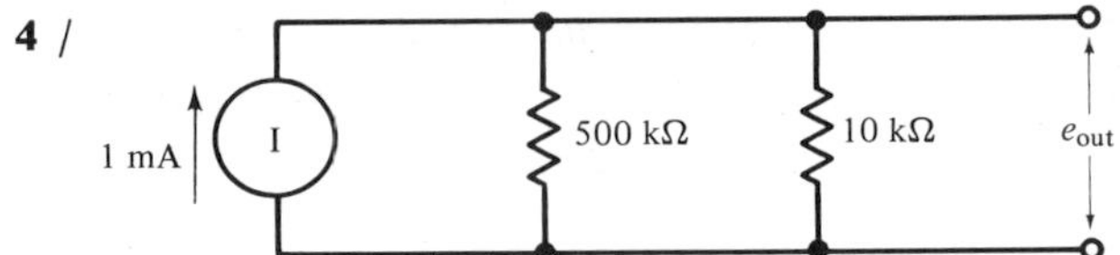

Calculate the input and output impedances for each of the following:

5 /

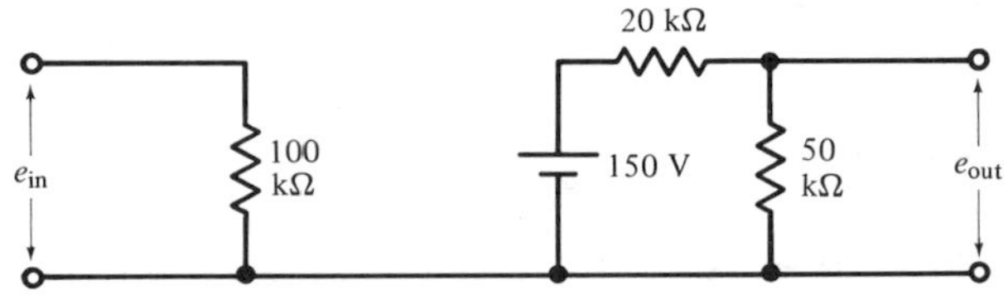

6 /

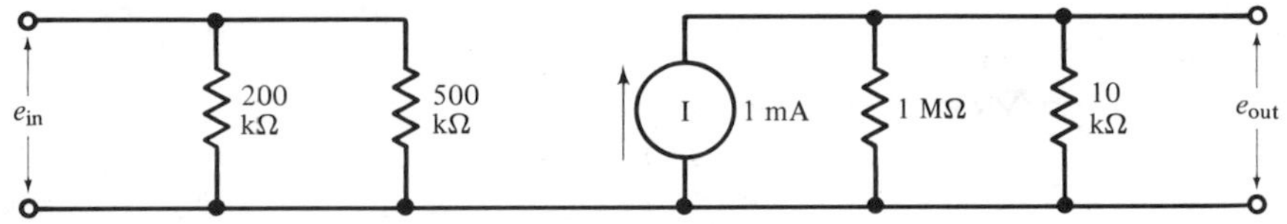

7 /

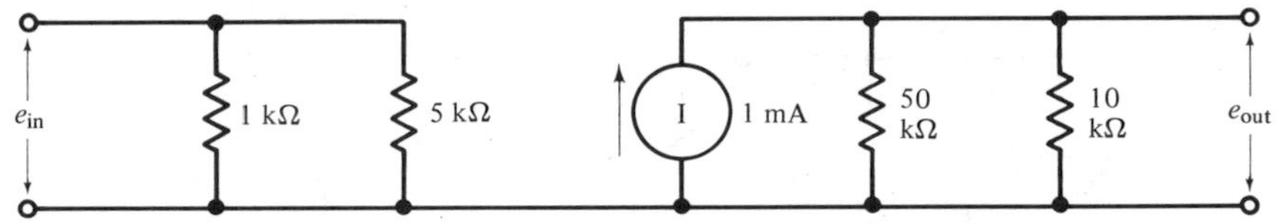

8 /

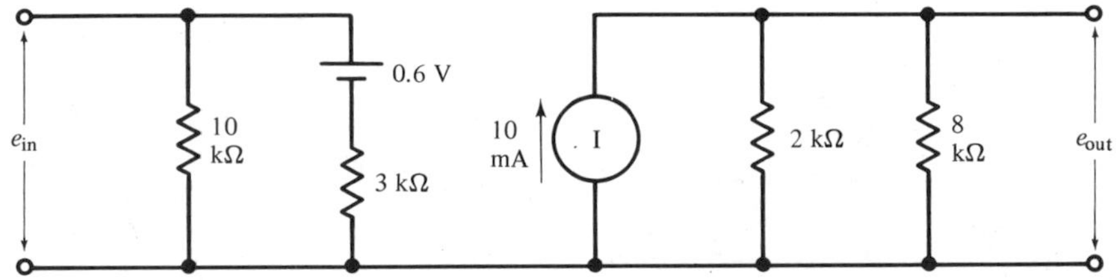

9 /

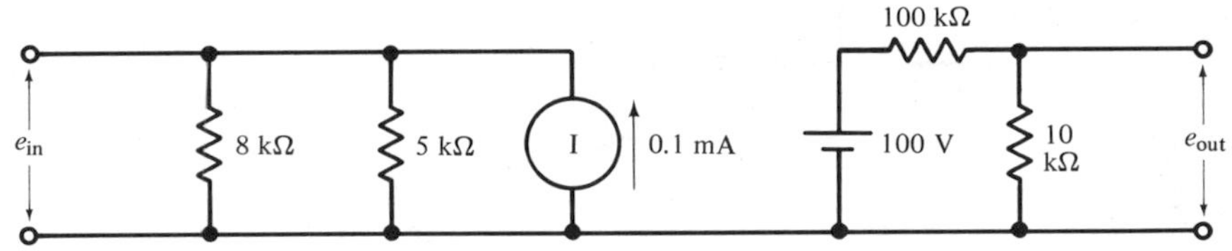

Calculate the percentage error in each of the following if r_{meas} represents the input resistance of either a voltage- or a current-measuring device, as appropriate to the source shown.

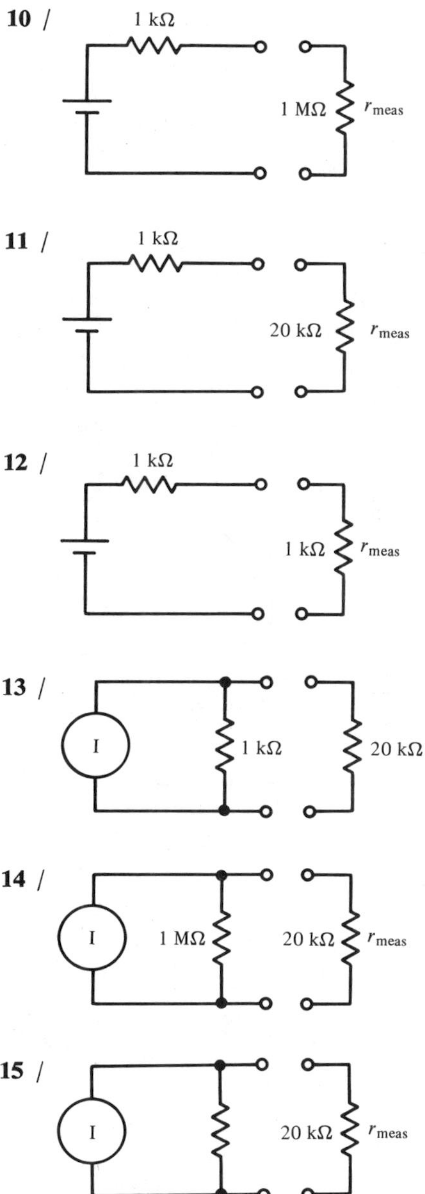

chapter 3 / Frequency response of inactive networks

chapter 3

The imposition of a signal on a network that consists entirely of resistive components can be adequately described by Ohm's law. Purely resistive circuits do not show frequency-dependent response; a sudden change in the applied voltage is accompanied by a simultaneous proportional change in current. Accordingly, current and voltage waveforms will be exactly in phase.

However, if a circuit contains reactive components, this will not be true. In Section 1.4, we noted that current and voltage are not in phase and that, furthermore, both the phase angle and the magnitude of reactance are functions of frequency. Capacitors show reduced reactance at high frequencies [Eq. (1.46)], and inductors show increased reactance at high frequencies [Eq. (1.59)].

These aspects of behavior of reactive circuits have extremely important implications for instrument performance. The fact that all circuit components inevitably exhibit at least small capacitances constitutes one of the fundamental factors limiting the operating speed of instruments. The fact that reactive components exhibit frequency-dependent behavior makes it possible to design circuit components that are selective in their response over a range of frequencies. Signals can be sorted out and unwanted ones rejected, discriminating in favor of those that are desired. Another implication of the phenomenon that causes current–voltage phase shift is that information encoded as signal voltage or current can be stored for periods of time that may vary from microseconds to hours, depending on the nature of the problem.

Reactive circuit behavior is involved in virtually every aspect of scientific instrumentation, and it is therefore desirable to be familiar with at least the qualitative aspects of such behavior before considering specific examples.

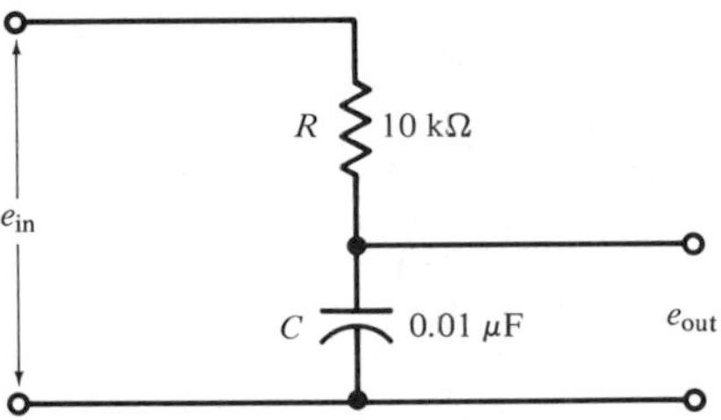

Figure 3.1 / *Low-pass RC filter.*

3.1 / CAPACITIVE FILTERS

3.1.1 / Low-pass *RC* filters

The frequency dependence of the behavior of reactive circuits, such as the series *RC* combination in Fig. 3.1, may be determined by network analyses, as described in Chapter 1. The circuit in Fig. 3.1 is drawn to emphasize its function as a simple voltage divider. The relationship between input and output voltage is

$$e_{\text{out}} = \left(\frac{X_c}{Z_{\text{tot}}}\right) e_{\text{in}} \tag{3.1}$$

where

$$X_c = \frac{1}{\omega C} \tag{3.2}$$

and

$$Z_{\text{tot}} = \sqrt{R^2 + X_c^{\,2}} \tag{3.3}$$

Capacitive reactance for the circuit in Fig. 3.1, defined by Eq. (3.2), varies with frequency, as indicated in Fig. 3.2(a). The gain of the circuit, defined for this purpose,

$$\text{gain} = \frac{e_{\text{out}}}{e_{\text{in}}} = \frac{X_c}{\sqrt{R^2 + X_c^{\,2}}} \tag{3.4}$$

is less than one since no active components (amplifiers) are involved. The variation of gain with change in frequency, defined by Eq. (3.4), is illustrated for the same circuit in Fig. 3.2(b). Notice that in the region below 1 kHz, the reactance of a 0.01-μF capacitor is so large that essentially all of the input voltage is dropped across the capacitance and e_{out} is very nearly the same as e_{in}. By contrast, above 100 kHz, the reactance is so small that most of the input drops across the resistance; hence, in this frequency range, the output

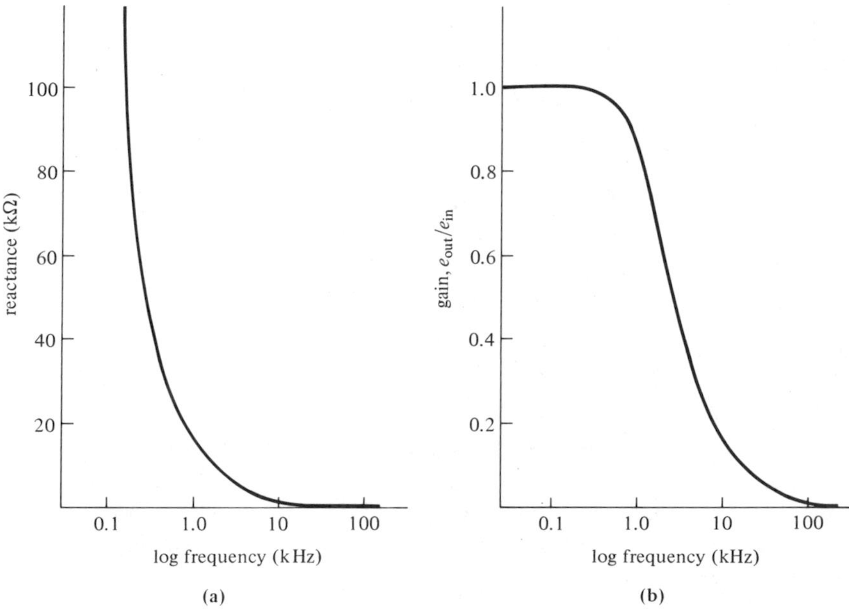

Figure 3.2 / *Low-pass RC filter performance vs. log frequency.*

is nearly zero. Therefore, the type of network shown in Fig. 3.1 serves very efficiently to transmit low-frequency signals but effectively blocks high-frequency signals. Because of this behavior, this network is termed a *low-pass filter*. The frequency range of efficient signal transmission, 0 to 1 kHz in this illustration, is termed the *bandpass* of the filter.

3.1.2 / Response to a step function

The effect of the filter is described above in terms of amplitude and frequency, where sine waves or signals composed of more than one sinusoidal component are assumed. It may also be convenient to consider the time required for a given change in signal magnitude, rather than expressing this concept in terms of frequency or reciprocal time. Suppose that a step function is imposed on the network in Fig. 3.1. By definition, a step function has a zero value before a certain time and a constant nonzero value after that time. An instantaneous change in signal magnitude takes place, with the finite signal persisting indefinitely. As will become evident, a step function is an idealization that can only be approximated in practice. In Fig. 3.3, the result of application of a step-voltage function is shown. At any time, the total voltage e_{in} is the sum of the voltage dropped across the resistance and that dropped across the capacitance:

$$e_{in} = e_R + e_{out} \tag{3.5}$$

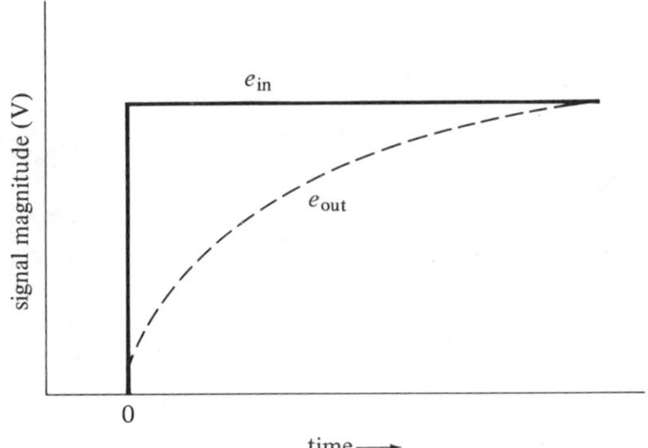

Figure 3.3 / *Response of a low-pass filter to a step function.*

The voltage drop across the capacitance may be obtained by solving Eq. (1.43) for voltage:

$$e = \frac{1}{C}\int i\,dt = e_{out} \tag{3.6}$$

Substituting Eq. (3.6) into Eq. (3.5) and expressing the drop across the resistance as the product of instantaneous current and resistance:

$$e_{in} = iR + \frac{1}{C}\int i\,dt \tag{3.7}$$

The solution for this equation is

$$i = \frac{e_{in}}{R}\exp\left(\frac{-t}{RC}\right) \tag{3.8}$$

The value of e_{out} will be equal to the input less that dropped across the resistance:

$$e_{out} = e_{in} - iR \tag{3.9}$$

$$= e_{in} - \left[\frac{e_{in}}{R}\exp\left(\frac{-t}{RC}\right)\right] \tag{3.10}$$

$$= e_{in}\left[1 - \exp\left(\frac{-t}{RC}\right)\right] \tag{3.11}$$

It should be noted that the product RC has dimensions of seconds, making the exponential term in Eq. (3.8) dimensionless. This term, a characteristic of the network, is called the *time constant*. It is useful because it facilitates a simple, semiquantitative estimation of the behavior of RC filters. For imposition of a step function, Eq. (3.11) implies that the output voltage will have 63.2 percent of its ultimate value after RC sec. Similarly, the current

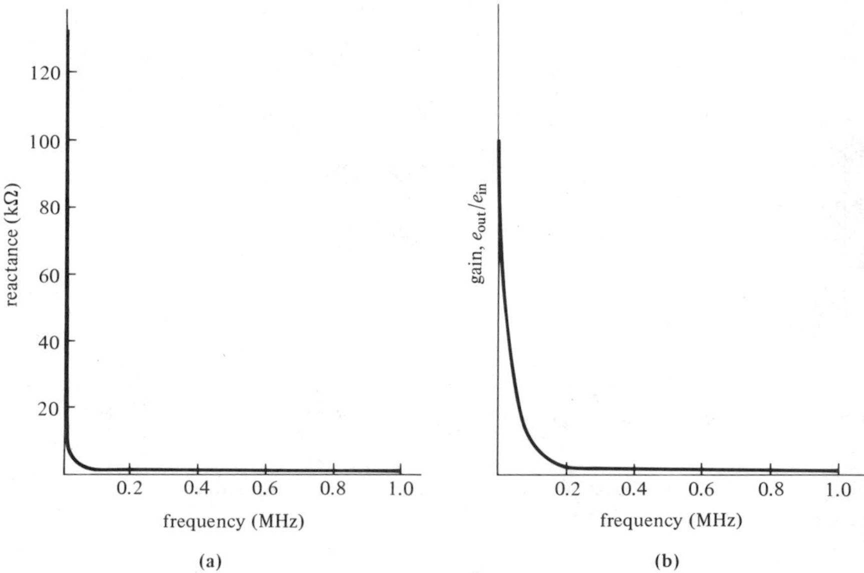

Figure 3.4 / *Low-pass RC filter performance vs. frequency.*

drawn by the network will have dropped to 36.8 percent of its initial value after *RC* sec. The 90 percent and 10 percent levels, respectively, occur after 2.3*RC* sec.

From Eq. (3.6), it is evident that the voltage across the capacitance of a low-pass filter is proportional to the integral of input current with respect to time. This is the basis for integration in analog computers discussed in Section 8.3.4.

A common use is the generation of a signal that increases linearly with respect to time, called a *ramp function.* This may be done by using a constant current to charge a capacitor. As an approximation to this, a constant input voltage can be imposed in a circuit analogous to Fig. 3.1. As indicated by Eq. (3.11) and Fig. 3.3, an exponential function is produced. This closely approximates a linear function during the first portion of the charging curve of the capacitance. The ramp function is widely used in timing operations and as an actuating signal in the automatic operation of instruments.

3.1.3 / Bode diagrams

In its various forms, the filter is often the limiting factor on the performance of instruments and systems; accordingly, the quantitative description of filter behavior is a matter of considerable importance. A concise notation is available, which is generally accepted and which should be familiar to instrument users. The information in Fig. 3.2 was presented in the form of reactance or gain, defined by Eq. (3.4), plotted against the logarithm of frequency. The same data are plotted against frequency in Fig. 3.4. This mode of presentation emphasizes that, over almost all of the frequency range considered, the capacitor exhibits a negligibly small reactance; hence, over this range, the filter blocks the input signal. However, it conveys almost no information about the region in which the reactance is appreciable. In particular, it entirely hides the fact, which is obvious in Fig. 3.2(a), that there is a limited band in which the network gain is essentially unity.

A more convenient presentation for many purposes is obtained by plotting a logarithmic function of gain against the logarithm of frequency. The customary logarithmic unit of gain is the decibel (dB), defined in terms of power ratios:

$$\text{gain} = \log \frac{\text{output power}}{\text{input power}} \tag{3.12}$$

$$= \log \frac{e_{\text{out}}^2/R_{\text{out}}}{e_{\text{in}}^2/R_{\text{in}}} \tag{3.13}$$

If $R_{\text{out}} = R_{\text{in}}$, then Eq. (3.13) can be simplified:

$$\text{gain} = 20 \log \frac{e_{\text{out}}}{e_{\text{in}}} \text{ dB} \tag{3.14}$$

Although this simplifying assumption will frequently not be literally valid, it is accepted by convention and Eq. (3.14) may be taken as a practical definition of gain, expressed in decibels.

When manufacturers' equipment specifications or descriptions of amplifiers are to be read, a facile relation of gain expressed in decibels to the corresponding ratio may be useful. From Eq. (3.14), a factor of 2 corresponds to 6 dB, an order of magnitude to 20 dB, 100 to 40 dB, 1000 to 60 dB, and so on. When the signal is attentuated, a negative sign may be used to indicate this.

The data used for Figs. 3.2 and 3.4 are replotted in Fig. 3.5 with the log-log presentation. In Fig. 3.5(a), log reactance is plotted against log frequency. Figure 3.5(b) shows gain in decibels against log frequency. This presentation is called a *Bode diagram*, after its originator; it is especially useful because it makes possible a very concise description of the behavior of a filter. The curve, which has qualitatively the same shape for all low-pass filters, consists of a segment with zero slope joined to a segment with a negative slope or rolloff of 20 dB/decade. Furthermore, if the diagram is approximated by asymptotes in the curved region, their intersection at frequency ω_0 corresponds to the reciprocal of the filter time constant. That is, a linear approximation of the Bode diagram for a low-pass filter can be obtained by taking a line at 0 dB up to a breakpoint at a frequency corresponding to $1/RC$ Hz. From that point the curve drops off linearly at a rate of 20 dB/decade.

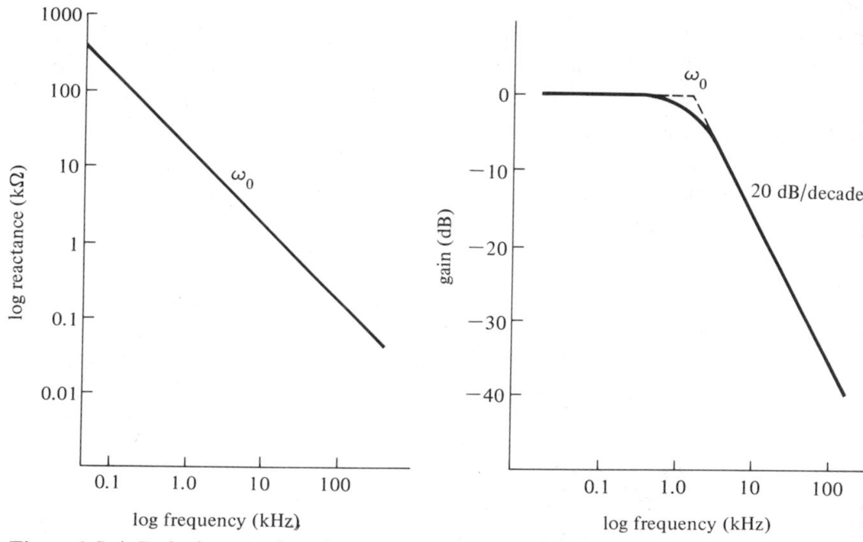

Figure 3.5 / *Bode diagram for a low-pass RC filter.*

From Fig. 3.5(a) it can be seen that at ω_0, the reactance amounts to 10 kΩ, equal to the resistance. For this reason, ω_0 is often referred to as the *half-power frequency*. Half of the signal is dropped across the resistance and half across the capacitance. It is an implication of Eq. (3.14) that, at the half-power frequency where voltage gain from Eq. (3.14) is 0.71, the gain expressed in decibels is −6.0. The linear approximation to the Bode diagram, which can be constructed with a knowledge of one piece of data, the time constant, will show a maximum error of 6 dB at the breakpoint. For many purposes, this approximation is entirely adequate; accordingly, an understanding of the implications of the filter time constant permits an extremely concise statement of the performance characteristics of a filter.

3.1.4 / Phase relationships

In addition to the effect of filters on signal amplitude, it is necessary to consider also their effect on the phase relationship between input and output voltages. Equation (1.45) shows the relationship between current and voltage in a capacitance. Current leads a sine wave voltage by 90°, as indicated by Fig. 1.27(b). In dealing with a low-pass filter, it is not directly a question of the phase angle between current and voltage, but one of that between input and output voltages. For the circuit illustrated in Fig. 3.1, the input voltage is the total voltage drop across the series combination of resistance and capacitance. The output voltage is that dropped across the capacitance. This can be described graphically by a diagram such as Fig. 3.6(a). This diagram is redrawn in Fig. 3.6(b) with the input voltage, $\mathbf{I}Z$, oriented along the positive real axis. For a low-pass filter, the output voltage

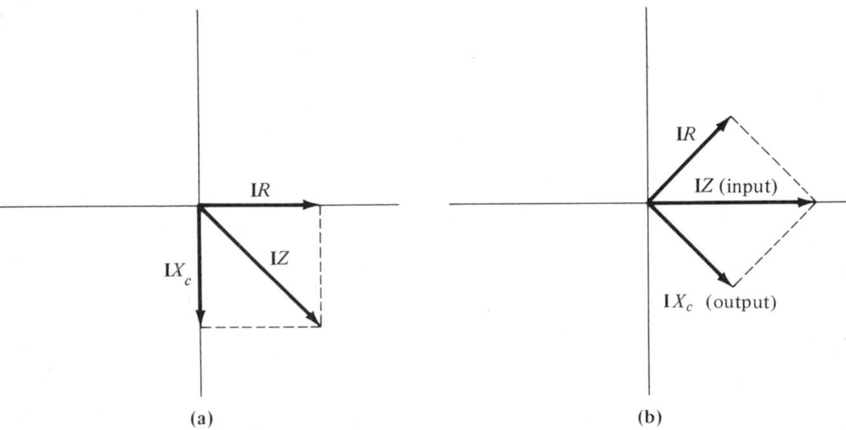

Figure 3.6 / *Input–output phase relationships for low-pass filters.*

corresponds to $\mathbf{I}X_c$, as indicated in Fig. 3.6(b), which evidently lags the input. Since X_c is frequency dependent and R is not, the phase angle between input and output voltage must also be frequency dependent, as shown in Eq. (3.15):

$$\theta = \tan^{-1} \frac{\mathbf{I}R}{\mathbf{I}X_c} \tag{3.15}$$

For the frequency ω_0, at which $X_c = R$, the angle would be 45°. The relationship is shown graphically as curve 1 of Fig. 3.7 for the network shown in Fig. 3.1.

It is noteworthy that relatively small changes in gain—for example, 0 to −6 dB in Fig. 3.5(b)—are accompanied by large phase shifts, which amount to 45° over this same frequency range. Furthermore, when the normalized log-log presentation of Fig. 3.7 is used, curve 1 represents *all* low-pass filters. It, like the linear Bode diagram approximation, can be drawn if the filter time constant is known.

3.1.5 / High-pass *RC* filter

If the network in Fig. 3.1 is simply inverted, taking the output across the resistance, as indicated in Fig. 3.8, rather than across the capacitance, it is converted from a low-pass filter to a high-pass filter. In this case, the relationship between input and output voltages is

$$e_{\text{out}} = \left(\frac{R}{Z_{\text{tot}}}\right) e_{\text{in}} \tag{3.16}$$

It can be seen that the gain would approach unity at high frequencies, in this case above 100 kHz, in which region the resistance would be much larger than the reactance. At low frequencies, the gain would be small, approaching zero in this case below 1 kHz, because the reactance would be much larger than the resistance. The performance of this filter can be described

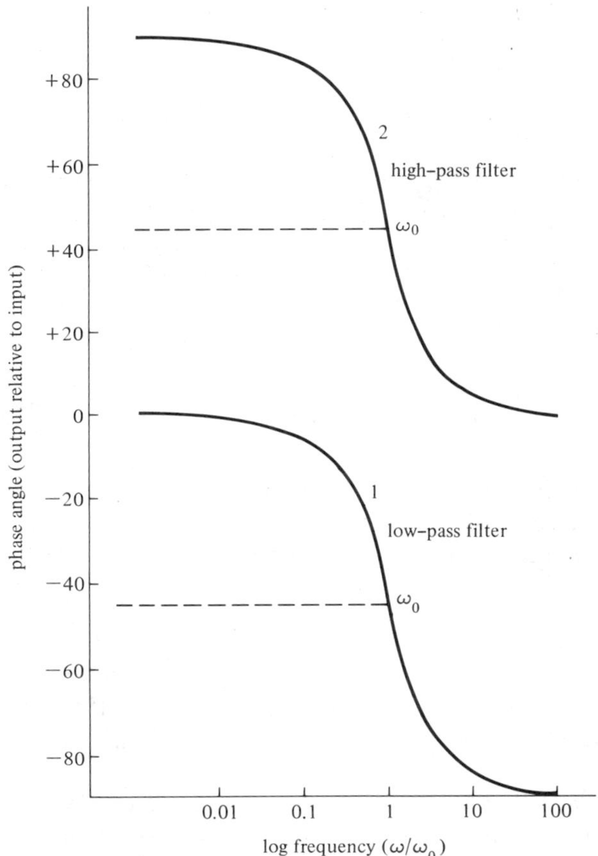

Figure 3.7 / *Input–output phase relationships for filters.*

graphically, as was done for the low-pass filter. A Bode diagram is presented in Fig. 3.9. The procedure for obtaining a linear approximation is the same as before. The breakpoint occurs at $1/RC$ rad/sec and the rolloff is at 20 dB/decade, but with positive slope. The actual curve will again be 6 dB down, at the half-power frequency ω_0.

When a step function is imposed on the high-pass filter in Fig. 3.8, the result, will of course, be the same as for the low-pass filter in Fig. 3.1, since the same components are used. The only difference is the point at which the observation of output is made. Therefore, both Eqs. (3.7) and (3.8) are useful. However, since output is taken across the resistance,

$$e_{out} = iR \tag{3.17}$$

Substituting Eq. (3.8) into Eq. (3.17) gives

$$e_{out} = e_{in} \exp\left(\frac{-t}{RC}\right) \tag{3.18}$$

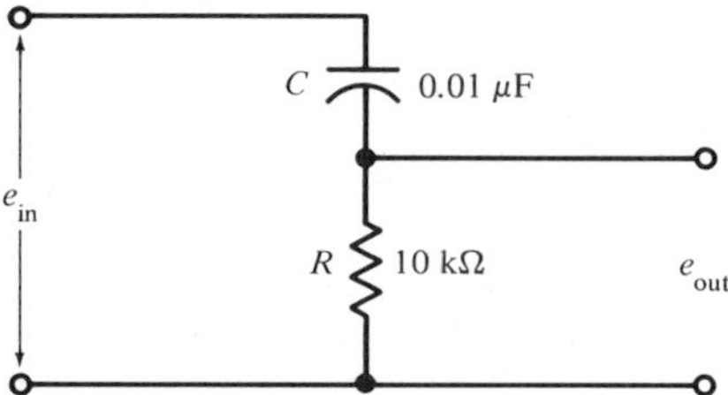

Figure 3.8 / *High-pass RC filter.*

This will vary with time, as indicated in Fig. 3.10. At the time of imposition of the step, input and output voltages will be equal. After *RC* sec, the output will have dropped to 36.8 percent of its initial value and to 10 percent after 2.3 *RC* sec.

The output from a high-pass filter is taken across the resistance. Accordingly, the output leads the input, as indicated in Fig. 3.11. In this case, the phase angle is

$$\theta = \tan^{-1} \frac{\mathbf{I}X_c}{\mathbf{I}R} \tag{3.19}$$

This is shown as curve 2 of Fig. 3.7 for the network of Fig. 3.8. For both types of filter, the phase shift approaches zero as the gain approaches unity; phase shift approaches 90° as the gain approaches zero.

Figure 3.9 / *Bode diagram for a high-pass filter.*

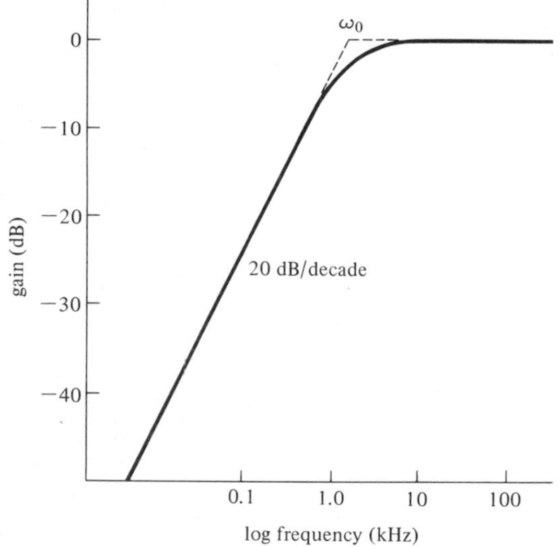

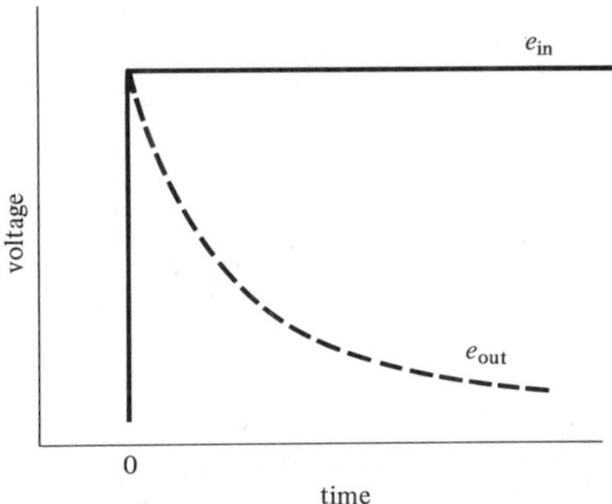

Figure 3.10 / *Response of a high-pass filter to a step function.*

3.2 / INDUCTIVE FILTERS

3.2.1 / Low-pass *LR* filters

It is generally true that the functions of *RC* filters can be duplicated by *LR* filters. The relationship between capacitance and inductance in filter networks can be seen by examining the equations relating current to voltage for each type of reactive component. For capacitance, this is

$$e = \frac{1}{C} \int i \, dt \tag{3.6}$$

while for inductance it is

$$i = \frac{1}{L} \int e \, dt \tag{3.20}$$

These two equations have the same form, except that the roles of current and voltage are inverted. Thus, a low-pass *RC* network, Fig. 3.1, operates by accepting a flow of current into the capacitance, with the resulting voltage across the capacitance being taken as output. The reactive component is

Figure 3.11 / *Input–output phase relationships for high-pass filters.*

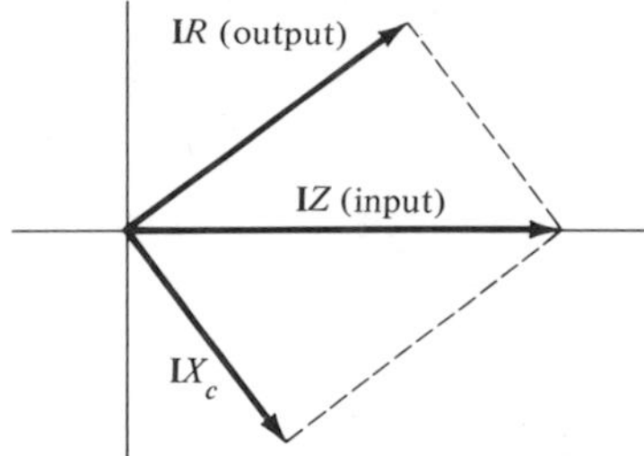

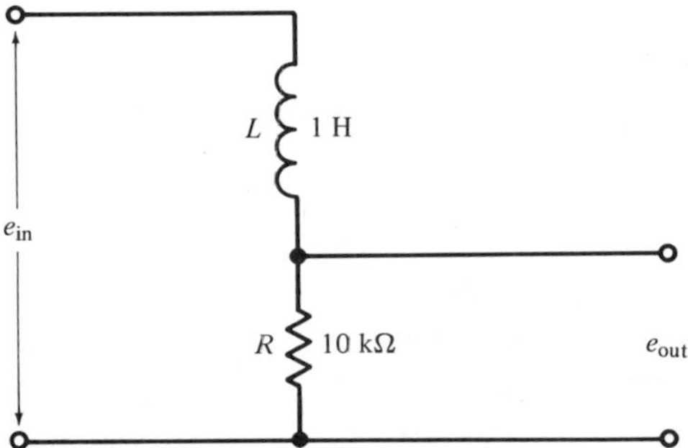

Figure 3.12 / *Low-pass LR filter.*

connected in parallel with the load, and the result is to discriminate against high-frequency input signal components. By contrast, in the low-pass *LR* filter, Fig. 3.12, the reactive component is in series with the load because an inductance presents a large reactance to high-frequency signals:

$$X_L = \omega L \tag{3.21}$$

The variation of reactance with angular frequency is shown graphically for a 1-H inductance in Fig. 3.13(a). The log-log presentation is shown in Fig. 3.13(b).

Figure 3.13 / *Inductive reactance vs. frequency.*

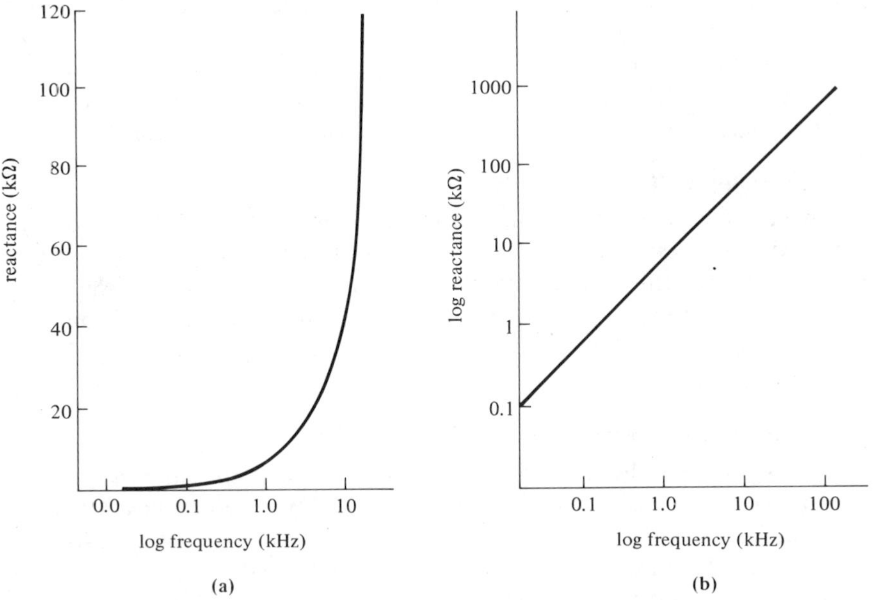

3.2.2 / Response to a step function

The equation for voltage across an inductance, corresponding to Eq. (3.6) for capacitance, may be obtained by solving Eq. (3.20) for voltage:

$$e = L\frac{di}{dt} \tag{3.22}$$

The relationship between input and output voltage for Fig. 3.12 may be obtained by setting input voltage equal to the sum of the voltages dropped across the inductance, e_L, and the resistance, e_{out}:

$$e_{\text{in}} = e_L + e_{\text{out}} \tag{3.23}$$

Substituting Eq. (3.22) into Eq. (3.23) and expressing the drop across the resistance as the product of instantaneous current and voltage,

$$e_{\text{in}} = L\left(\frac{di}{dt}\right) + iR \tag{3.24}$$

for which the solution is

$$i = \left(\frac{e_{\text{in}}}{R}\right)\left[1 - \exp\left(\frac{-tR}{L}\right)\right] \tag{3.25}$$

$$e_{\text{out}} = iR = e_{\text{in}}\left[1 - \exp\left(\frac{-tR}{L}\right)\right] \tag{3.26}$$

Equation (3.26) has exactly the same form as Eq. (3.11), which describes the low-pass *RC* filter, except that the term *L*/*R* replaces *RC*. For inductive filters, the *L*/*R* time constant has the same implications as the *RC* time constant in capacitive filters. Thus, the circuits in Figs. 3.1 and 3.12 have the same time constant and accordingly show identical behavior. In particular, Figs. 3.2(b), 3.4(b), and 3.5(b) apply to both circuits. The Bode diagram may be used to describe *LR* filters just as it is used for *RC* filters.

3.2.3 / High-pass *LR* filters

A low-pass *LR* filter can be converted to a high-pass filter by taking output across the inductance rather than across resistance. This is illustrated in Fig. 3.14. Since the time constant for the circuit in Fig. 3.14 is the same, 10^{-4} sec, as for the circuit in Fig. 3.8, the relationship between input and output signals is the same. Accordingly, the Bode diagram in Fig. 3.9 may be used equally well for the *LR* filter.

3.2.4 / Phase relationships

LR filters, like *RC* filters, cause a phase shift between input and output that is frequency dependent. As indicated by Eq. (1.58), voltage leads current in an inductive circuit. A graphical solution for a series *LR* circuit is shown in Fig. 3.15(a). In Figs. 3.15(b) and 3.15(c), the voltage drop across the entire circuit, **I***Z*, which is the input, is shown oriented along the positive real axis.

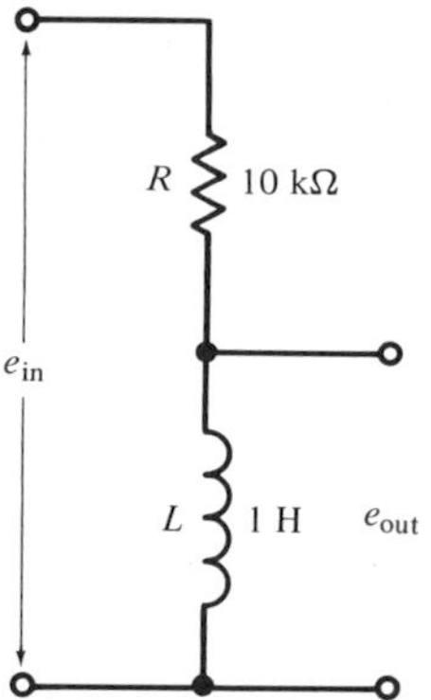

Figure 3.14 / *High-pass LR filter.*

In a low-pass filter, output is taken across the resistance, making **I**R the output voltage as indicated in Fig. 3.15(b). Output lags input, as it does for a low-pass RC filter. Similarly, since the output is taken across the inductance of a high-pass filter, output leads input, as shown in Fig. 3.15(c). The phase-shift behavior of LR filters is therefore identical with that of RC filters. Figure 3.7 may be applied equally well to the circuits in Figs. 3.12 and 3.13 and to those in Figs. 3.1 and 3.8, since the time constants are the same.

3.3 / FILTERS AND INSTRUMENT BEHAVIOR

3.3.1 / Low-pass behavior

It was noted above that filter behavior is important because it represents a limiting factor in determining the behavior of instruments. Specifically, all instruments behave as though the output contained one or more low-pass filters. These are not necessarily filters deliberately introduced by the designer, although this is often done in order to tailor the high-frequency response; it may be caused by capacitance, which is unavoidably associated with all circuit components, usually referred to as stray capacitance. Whatever the source of the capacitance, discrete components or stray capacitance.

Figure 3.15 / *Input–output phase relationships for LR filters.*

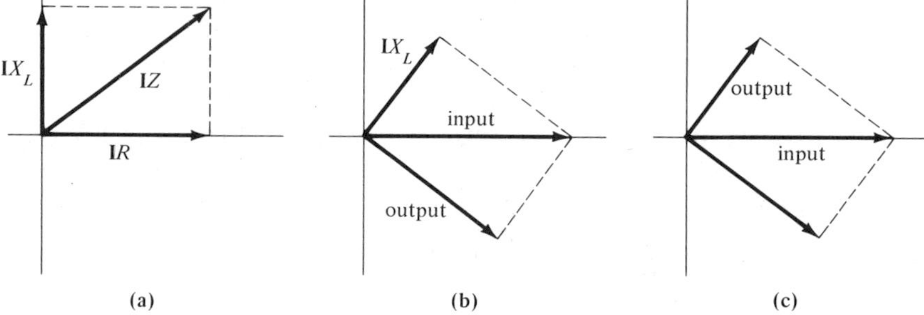

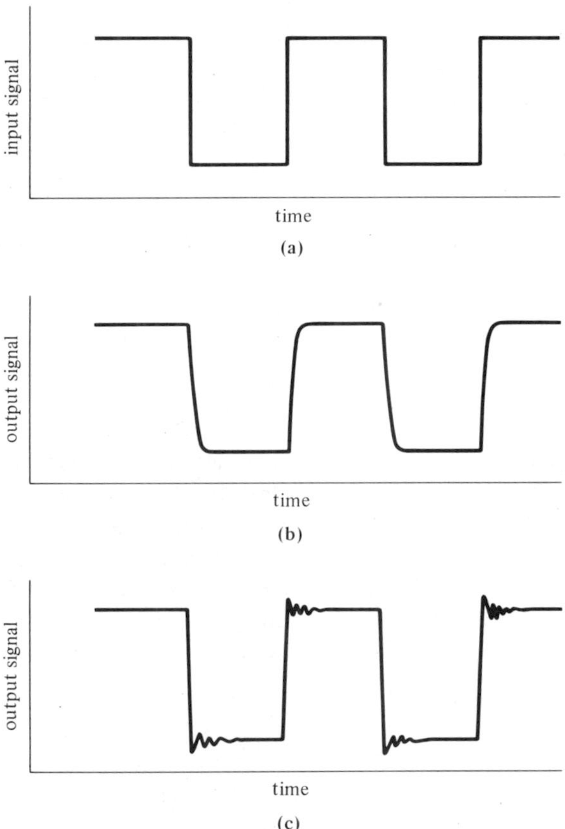

Figure 3.16 / *Signal distortions caused by limited high-frequency response.*

the upper frequency limit beyond which the signal is attenuated is determined by the time constant, and it is customary to use Bode diagrams to describe amplifier performance. These diagrams have a shape in the high-frequency region that is similar to that of Fig. 3.5(b), but with positive rather than negative gains, since the output of an amplifier is larger than the input.

An additional important implication of the high-frequency behavior of instruments is that it is fundamentally impossible to observe or transmit a signal that contains instantaneous changes in magnitude. All instruments exhibit low-pass filter behavior with finite time constants, although upper half-power frequencies of instruments vary widely, of course, depending on the intended use. In addition, the "state of the art" capability is continually changing with evolution of technology. Regardless of the actual design, a finite time constant implies a finite time for a change in signal magnitude. Real step functions and square waves always exhibit sloping leading and trailing edges. In Fig. 3.16(a) an ideal square wave is illustrated. The types of distortion that must be expected because of limited high-frequency bandpass are illustrated in Figs. 3.16(b) and 3.16(c). The leading and trailing edges slope because of discrimination against the very high-frequency signal components of the input signal, which are needed to cause the most rapid change. A truly discontinuous change implies infinite frequency components. In addition, there will be more or less distortion of the corners of the signal, as indicated in Fig. 3.16(c). This occurs when an instrument exhibits high-frequency rolloff at a rate substantially greater than 20 dB/decade. In this case, a significant time is required for the output to settle down after the sudden change; this behavior is termed *ringing*.

3.3.2 / High-pass behavior

High-pass *RC* filters are very commonly used to transmit signals between instruments or between various sections within an instrument. This is done primarily to avoid DC interactions between signal sources and receivers. A network like that of Fig. 3.8 can be connected between two points of differing DC voltage with no effect other than initial charging of the capacitor. An AC signal, however, will be transmitted, providing that the range of frequencies is correctly matched with the bandpass of the filter. Having eliminated the original DC component by means of the filter, it is possible to establish a new one arbitrarily. In Fig. 3.8, this would be zero (ground), but it could in principle be any other voltage level. It is apparent, however, that the low-frequency response of an instrument that contains *RC* coupling networks will be limited. This behavior is determined by the time constants of the coupling networks and is usually described by means of a Bode diagram similar to Fig. 3.9, but with the gain greater than unity.

The DC component of a signal can also be eliminated by using transformer coupling. This approach provides for complete electrical isolation of one segment of a circuit from another with only magnetic coupling between them. It may be helpful in eliminating noise from signals, and it constitutes an important safety feature when instrument circuits are thus isolated from power lines.

3.4 / WAVEFORM SHAPING WITH FILTERS

The ability of filter networks to discriminate against certain frequency components of a signal makes them useful for waveform shaping. For example, the low-pass filter is widely used in power supplies to remove high-frequency components from rectified AC signals in order to produce a DC voltage. Rectification is discussed in Section 4.3.3, and a rectifier circuit is shown in Fig. 4.5(b). A simplified form of the rectifier circuit is shown in Fig. 3.17(a). Here the components e_s and R_s represent the Thévenin equivalent of a full-wave rectifier.

The output of e_s is represented by the waveform of curve 1 in Fig. 3.18. It is obtained by performing a switching operation, described in connection

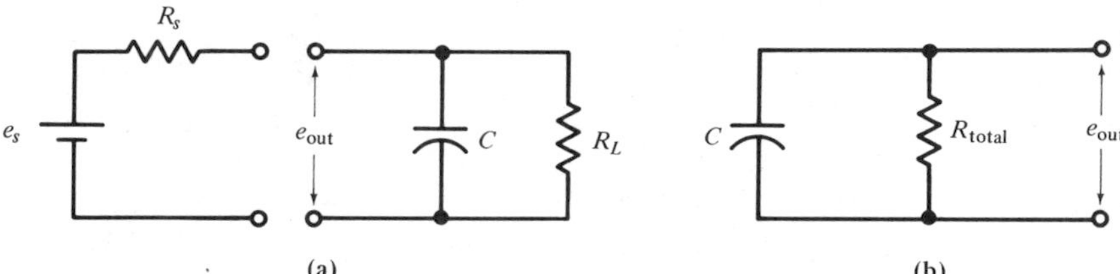

Figure 3.17 / *Power supply low-pass RC filter.*

with Fig. 4.4, on a sine wave of frequency ω. This introduces higher-frequency components, which cause the signal amplitude to change rapidly at intervals corresponding to a frequency of 2ω. R_s represents the resistance of the rectifier. Toward signals that cause a flow of current out of the rectifier, from the voltage source toward the filter in this illustration, R_s is a small value (rectifier forward biased). Toward signals that cause a flow of current back into the rectifier, R_s exhibits a large value (rectifier reverse biased).

Filtering, that is, the removal or attentuation of the high-frequency components of the signal shown in Fig. 3.18, curve 1, may produce a signal with a waveform similar to Fig. 3.18, curve 2. The filtering action of the *RC* network occurs because, when charged, the capacitor must discharge through whatever resistance is connected across its terminals. When the input voltage of the filter is less than the output voltage—for example, time A to time B in Fig. 3.18—the capacitor discharges through the parallel equivalent of R_L and R_S, shown in Fig. 3.17(b) as R_{tot}. The rate of discharge, hence the extent of deviation of e_{out} from a DC voltage, is determined by the time constant, $R_{tot}C$. Since the resistance looking back into a rectifier is very large, the effective time constant is approximately R_LC. Accordingly, when R_L is small, meaning that considerable current is being supplied, the ripple on the output voltage is appreciable. If R_L is made larger, corresponding to a reduced load, the ripple is reduced.

Figure 3.18 / *Waveforms produced by a full-wave rectifier and low-pass filter.*

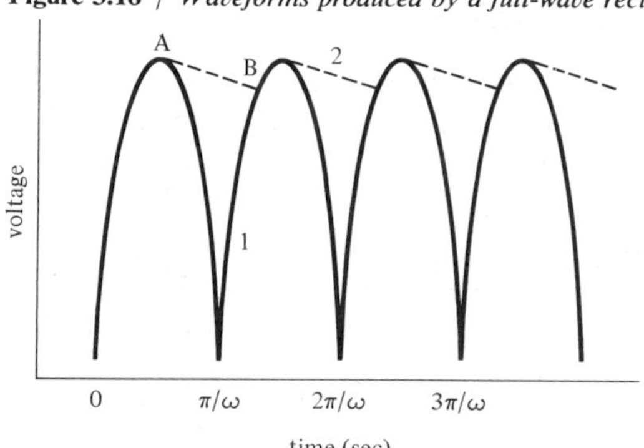

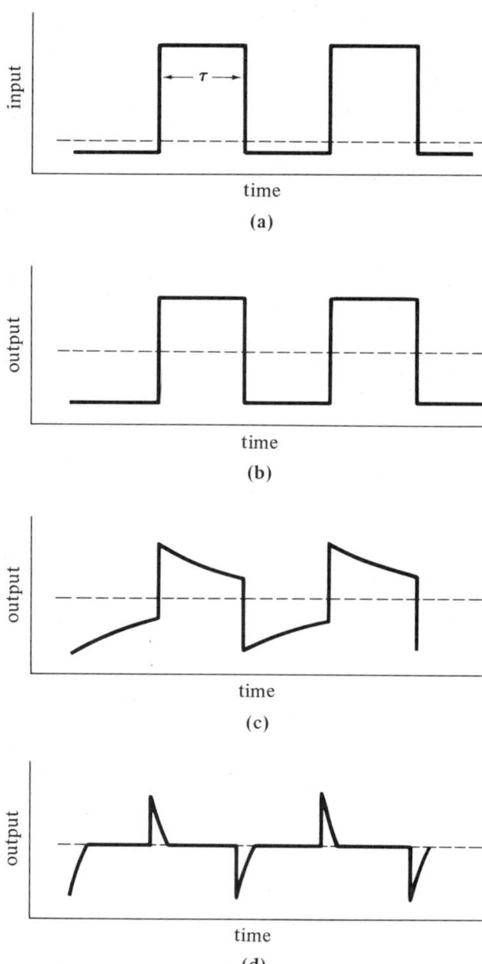

Figure 3.19 / (a) *Response of a high-pass filter to a square wave.*
(b) $RC \gg \tau$.
(c) $RC \simeq \tau$.
(d) $RC \ll \tau$.

The response of high- and low-pass networks to the step function has been described. This can readily be extended to include the very important case of response to a square wave. The input waveform is illustrated in Fig. 3.19(a). The shape of the output that would be obtained from various high-pass filters is illustrated in Figs. 3.19(b)–3.19(d). It is evident from Eq. (3.18) that the output waveform will be determined by the size of the filter time constant relative to the half-period, τ, of the input signal. With change of input voltage level, the output voltage will show essentially the same change, since the rising and falling portions of the square wave consist of high-frequency components for which filter gain is unity. The flat segments consist of low-frequency components, against which the filter discriminates to varying extents determined by the time constant. Thus, if $RC \gg \tau$, the capacitor will not have time to discharge appreciably before the next level change. The result, shown in Fig. 3.19(b), is that although the average value of the signal has been made zero, as compared with the input in Fig. 3.19(a), the shape of the signal is distorted very little. If $RC \simeq \tau$, then the capacitor does discharge appreciably during this time, leading to the considerable distortion shown in Fig. 3.19(c). With $RC \ll \tau$, the capacitor discharges completely during a half-period, causing the output signal to appear as a series of spikes, as in Fig. 3.19(d).

The effect of low-pass filters on square waves has been discussed above in

Figure 3.20 / (a) *Response of a low-pass filter to a square wave.*
(b) $RC \ll \tau$.
(c) $RC \simeq \tau$.
(d) $RC \gg \tau$.

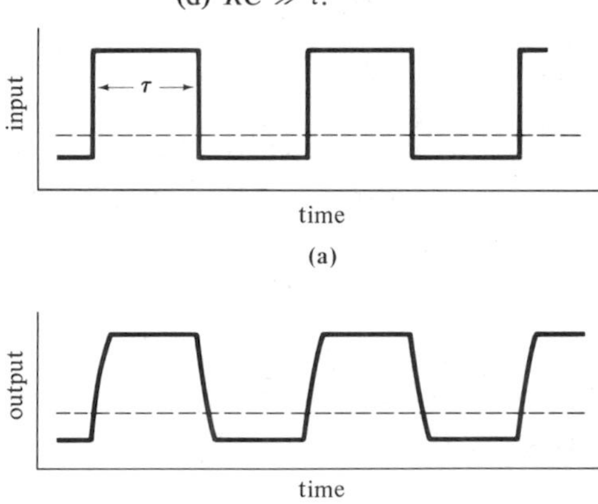

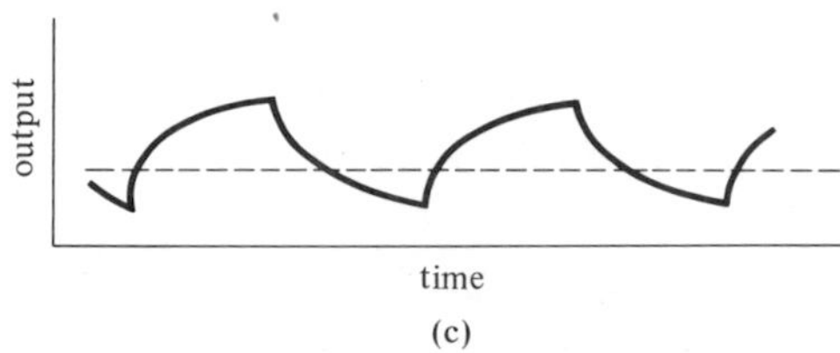

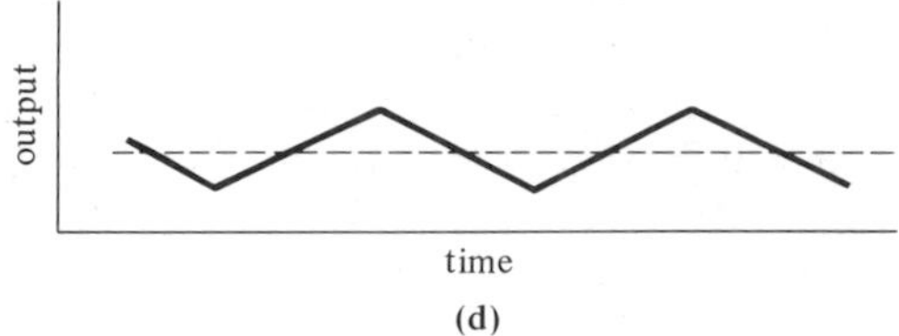

connection with instrument behavior. While it is ordinarily advantageous to minimize the low-pass time constant associated with an instrument in order to extend the frequency range of operation, a signal can be integrated if desired by making the low-pass time constant large as compared with the signal half-period. The relationship between input and output for a low-pass filter as a function of time constant is illustrated in Fig. 3.20.

The high-pass *RC* filter is very widely used because of its ability to transmit signals between circuits that are at different DC voltage levels. In this use, it is desirable for the coupling network to cause the smallest possible distortion of the signals. However, *RC* networks are also used, as illustrated in Fig. 3.19(d), to differentiate signals with discontinuities in order to produce voltage spikes that can readily serve as time markers. In this case, signal distortion is intentional. In examining a schematic, the intended purpose of an *RC* coupling can usually be deduced by considering the size of the time constant within the context of the speed of operation of the instrument, bearing in mind that signal frequencies higher than $1/RC$ will not be appreciably distorted.

The waveforms in Fig. 3.19 and 3.20 are presented with the assumption that only a portion of a continuing process is being shown. That is, the input signal has been imposed at least for a time that corresponds to several cycles before time zero in the figures and that the input has not been interrupted during the time represented by the figures. If this is not true, then distorted waveforms may be transmitted, because the average signal value must be zero for any given period of time. This is illustrated in Fig. 1.15.

PROBLEMS

For each of the following, calculate the gain, expressed (a) as the ratio e_{out}/e_{in}, and (b) in decibels, if a 10-kHz signal is imposed.

1 /

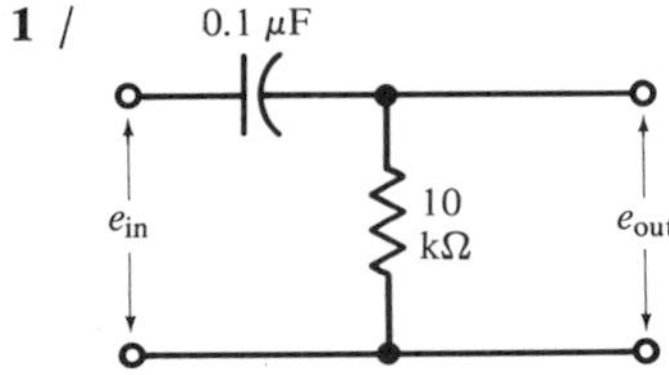

2 /

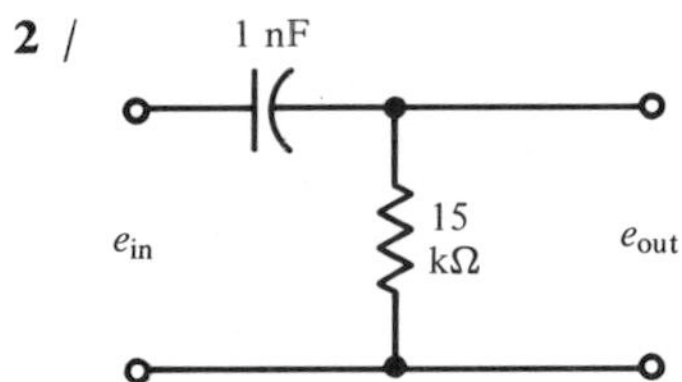

3 /

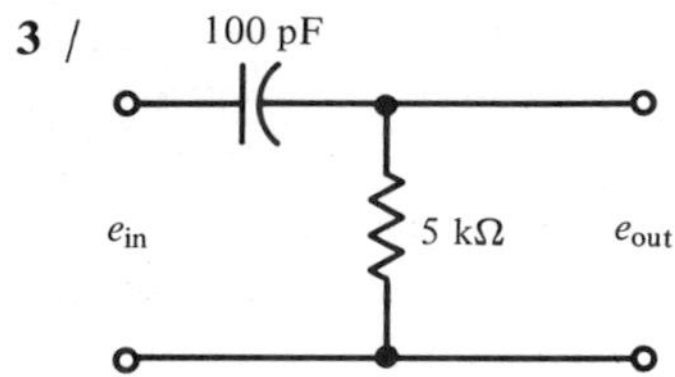
100 pF
e_{in}
5 kΩ
e_{out}

4 /

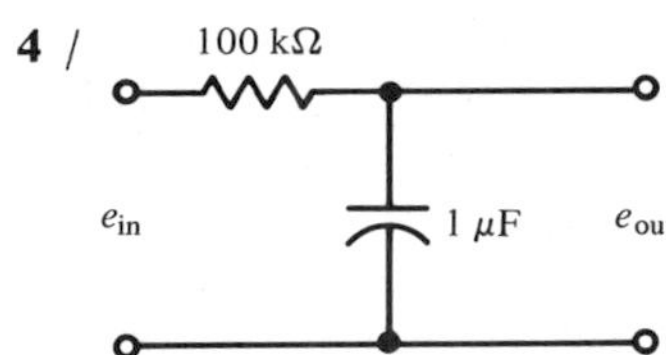
100 kΩ
e_{in}
1 μF
e_{out}

5 /

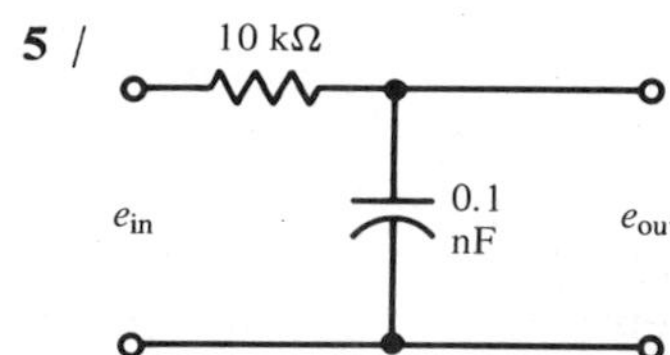
10 kΩ
e_{in}
0.1
nF
e_{out}

6 /

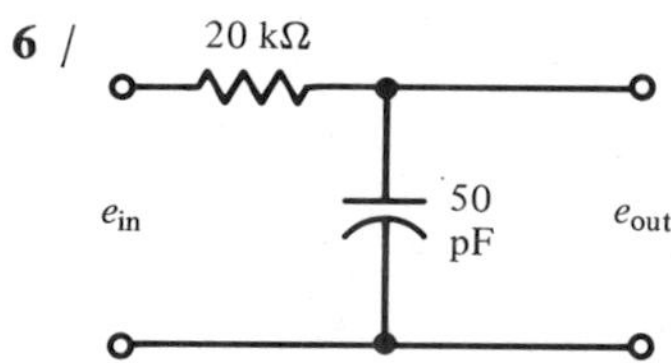
20 kΩ
e_{in}
50
pF
e_{out}

7 /

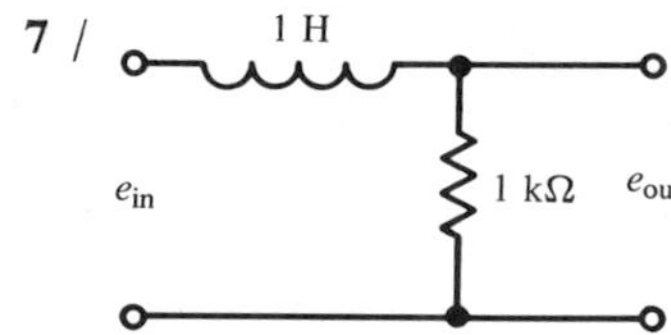
1 H
e_{in}
1 kΩ
e_{out}

8 /

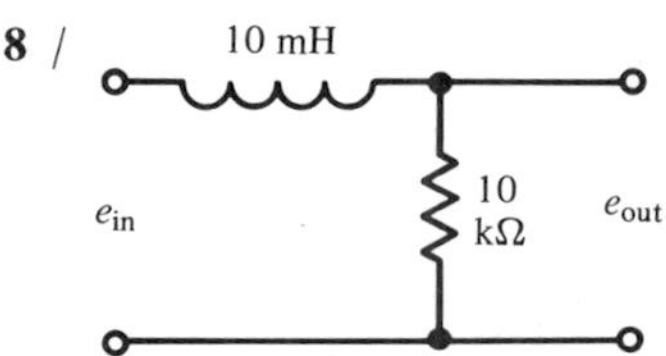
10 mH
e_{in}
10
kΩ
e_{out}

9 /

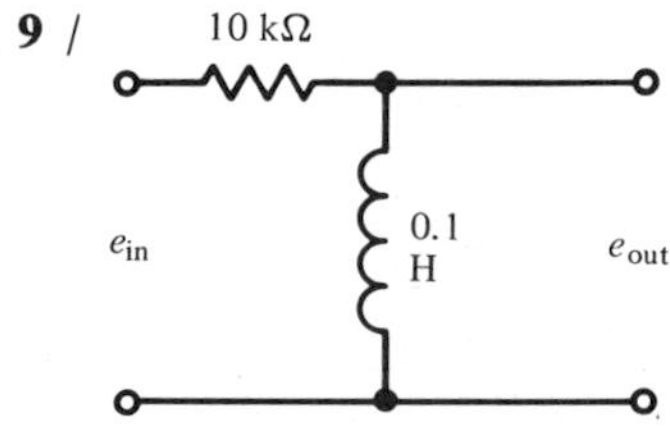

10 /

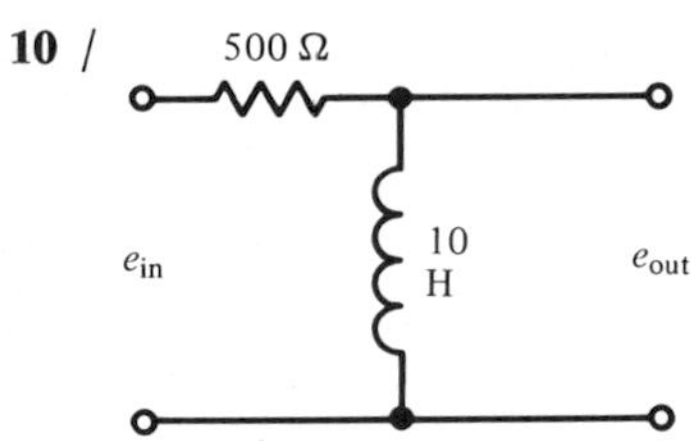

11–20 / For each of the circuits given in Problems 1–10, sketch the Bode diagrams, showing the coordinates.

21–30 / For each of the circuits given in Problems 1–10, calculate the phase angle between e_{in} and e_{out}. Show specifically which signal is leading.

31 / A voltage pulse of 10-μsec duration is imposed on the filters shown in (a) Problems 1, 2, 3; (b) Problems 4 and 5; (c) Problems 7 and 8; (d) Problems 9 and 10. Sketch the approximate shape of the output signal waveform.

chapter 4 / Amplification and amplifiers

chapter 4

Of the various functions of electronic components in scientific instruments, that of amplification is certainly the most important. Amplifiers are not only necessary to the operation of very many types of instruments, but they often constitute the limiting factor that determines the quality of performance. This is particularly true with respect to speed of operation, accuracy, and ease of interconnection of instruments. Therefore, while the task of amplifier design is ordinarily reserved to the specialist, it is important for the instrument user to be aware of the principles and limitations of operation of amplifiers in order to make realistic appraisals of the potentialities and limitations of equipment. In this chapter, the operation of three widely used amplifying devices, the field effect transistor (FET), the bipolar transistor, and the vacuum tube is first discussed in some detail. With this background, the design and performance of AC and DC amplifiers is discussed.

4.1 / THE PROCESS OF AMPLIFICATION

An amplifying device may be considered to be a variable resistor, the resistance of which is controlled by an electrical signal. In use, a voltage is imposed across the device and the signal to be amplified serves as the control, as indicated in Fig. 4.1. Fluctuations in the resistance result in corresponding fluctuations in current flow, providing that the driving voltage remains constant. This modulated current, which constitutes the output from the device, is therefore controlled by the intensity of the input signal. The relationship is, however, not a linear one with transistors, FETs, or vacuum tubes unless the variations in output current are carefully limited.

The device in Fig. 4.1(a) provides an output current; however, as shown, there is no provision for observation of the current fluctuations. The most common way of doing this is by placing a resistor, R_L, in series with the

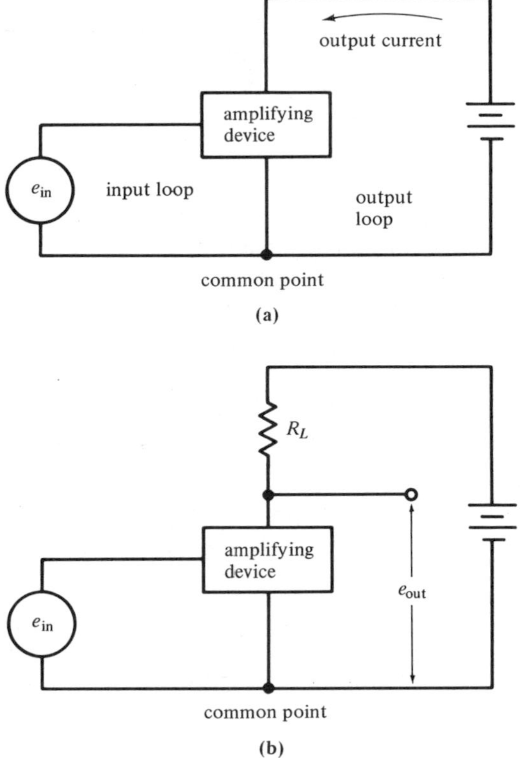

Figure 4.1 / *Generalized representation of the use of amplifying devices.*

amplifying device in the output loop, as indicated in Fig. 4.1(b). Output current will cause a proportional voltage drop across this resistor, which can be conveniently observed. The circuit in Fig. 4.1(b) may be viewed as a voltage divider, consisting of the tube or transistor and the load resistor that has the power supply voltage dropped across it. Output voltage is taken as indicated across the amplifying device. Since a change in the control signal causes an increase in the resistance of the device, an increasing fraction of the power supply voltage will be dropped across it and hence appear as output voltage. If the variable resistance becomes large compared with R_L, the output voltage will approach the value of the power supply as a limit. If the variable resistance becomes very small, essentially the entire power supply voltage is dropped across R_L; output voltage approaches zero, and

$$i_{\text{out(max)}} = \frac{R_L}{V_{\text{power}}} \tag{4.1}$$

An amplifier accepts an input signal, which may be either a current or a voltage, and produces an output signal, which may be observed either as a current or a voltage. The output may be larger or smaller than the input, although the usual purpose of an amplifier is to increase the signal amplitude. The output signal is not an exact scaled replica of the input, but will instead be somewhat distorted.

4.2 / NOTATION USED WITH AMPLIFIERS

To describe the behavior of an amplifier, several different classes of signals must be considered. These include the absolute magnitude of the instantaneous value of a signal; the average value, which corresponds to the operating or *Q*-point value for the variable; and the small-signal value. The meanings and uses of these terms are discussed in the body of the text below. A complete list of the symbols that are used is presented in Section 4.13.

The notation used with vacuum tubes has been standardized; that for transistors and FETs has not. In this text, the voltages applied to transistors and FETs are designated by the letter V; currents are designated I. Uppercase letters are used to designate average or *Q*-point values; lowercase letters are reserved for instantaneous values and for small-signal values. Current identities, emitter, base, collector, drain, source, and gate are indicated by subscript. An uppercase subscript is used to designate large-signal or absolute-magnitude values of current. A lowercase subscript indicates a small-signal value. Thus:

I_E = *Q*-point or average value of emitter current
i_D = instantaneous value of drain current
i_b = small-signal value of base current

This scheme is carried over to voltages; however, to avoid ambiguity, it is necessary to indicate the voltage level to which the measurement is referenced. The following examples illustrate the scheme:

V_{CE} = *Q*-point or average voltage of collector relative to that of the emitter
V_C = *Q*-point or average collector voltage relative to ground
v_{BE} = instantaneous value of base voltage relative to that of the emitter
v_B = instantaneous value of base voltage relative to ground
v_{ce} = small-signal value of collector voltage relative to that of the emitter
v_c = small-signal value of collector voltage relative to ground
V_{cc} = collector power supply voltage

In drawing equivalent circuits, physically discrete components are designated by uppercase italics, for example, R and C for resistors and capacitors. Equivalent circuit symbols that correspond to characteristics of a device, rather than to an actual device, are designated by lowercase italics. For example, r_{in} and r_p represent input resistance and vacuum tube plate resistance, respectively.

4.3 / SOLID STATE ELECTRONIC COMPONENTS

Modern amplifiers are now normally designed to use amplifying devices fabricated of semiconductor materials. These are designated as *solid state devices*, in order to make a distinction from vacuum tubes.

4.3.1 / Conduction in doped semiconductor materials

The addition of a very small amount of either a group III or of a group V element to highly purified semiconductor material, such as germanium or silicon, will substantially increase its conductivity. Addition of a group V element to the host lattice introduces outer-shell electrons in excess of those required for completion of octets. These can be caused to move in the crystal very readily by imposition of a voltage across it. This type of material, called an *n-type semiconductor*, conducts current by movement of electrons as *majority carriers*. Introduction of controlled concentrations of impurity is called *doping*. If a group III dopant is used, the lattice will have imperfections due to unfilled octets. Imposition of a voltage across this material, termed a *p-type semiconductor*, causes a flow of current by movement of electrons to the vacancies. On an atomic scale, this process can be visualized as amounting to a movement of the vacancies, or *holes*. Thus, current flow in *p*-type semiconductor material is attributed to a movement of holes as majority carriers. In *p*-type material, a certain fraction of electrons, described by the Boltzmann distribution, will be sufficiently energetic to move on imposition of a voltage. These *minority carriers* will exist at a concentration level that is exponentially related to temperature and will move in the opposite direction to that of the majority carriers. Similarly, holes will function as minority carriers in *n*-type material. Minority-carrier excitation is an important factor in causing semiconductor components to show temperature-sensitive behavior.

4.3.2 / The *np* junction

Movement of charge carriers can be described by the diffusion laws that are applied to movement of ions in electrolytic solutions. When pieces of *n*- and *p*-type material are placed in physical contact, there is a tendency for electrons to move from the *n*-type material across the junction into the *p*-type material, where their concentration is small and where they combine with holes. Similarly, holes will be transferred in the opposite direction. This causes a net accumulation of negative charge in the *p*-type material and of positive charge in the *n*-type material. A potential difference corresponding to this separation of charge builds up and serves to establish a stable equilibrium condition. At equilibrium, there is a *depletion zone* on either side of the *np* junction, in which there is a deficiency of majority carriers relative to their concentration outside of the zone. Material within this zone exhibits a much larger resistance than that without. If an external voltage

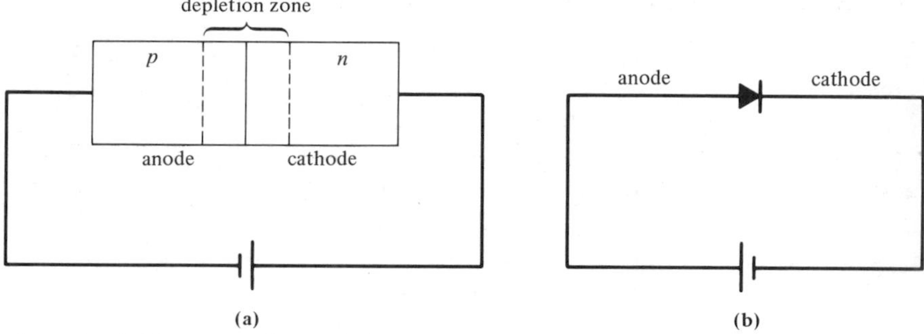

Figure 4.2 / *The semiconductor diode.*

is applied as indicated in Fig. 4.2(a), this causes an increase in the extent of separation of charge that already exists. As a result, the material exhibits a large resistance toward current flow and the potential across the junction cancels the externally applied voltage. The junction is said to be *reverse biased.* If the polarity of imposed voltage is reversed, as indicated in Fig. 4.2(b), the external power supply will, in delivering current, tend to minimize the majority carrier deficiency at the *np* junction. The junction then exhibits a relatively small resistance toward current flow and is said to be *forward biased.* The *np* junction therefore shows a large resistance to current flow in one direction and a small resistance to current flow in the opposite direction and can be used as a rectifying diode. The symbol for a semiconductor diode is given in Fig. 4.2(b); the *p*-type material is the anode and the *n*-type material is the cathode. The arrow in the symbol points in the direction of the flow of conventional current when the diode is forward biased.

4.3.3 / Diodes as rectifiers

One of the important uses for diodes is service as the active elements of electronic switches. A common example is the switching operation used to convert AC signals to DC. The arrangement that performs *half-wave rectification* is illustrated in Fig. 4.3(a), using a single-pole, single-throw switch.

It is assumed that the action of the switch is synchronized with the input waveform $e_{\sim}$, so that the switch is closed when the polarity is that shown in Fig. 4.3(a), corresponding to time intervals A–B in Fig. 4.3(b), and is open during the alternate half-cycles that correspond to time intervals B–A. This will permit transmission of power to the load only half of the time; but it does ensure that current flows through the load in only one direction. The solid waveform in Fig. 4.3(b) would be observed across the load.

A smoother delivery of power can be attained by *full-wave rectification,* which requires the equivalent of two single-pole, double-throw switches with the action of both synchronized with input waveform. This is illustrated in

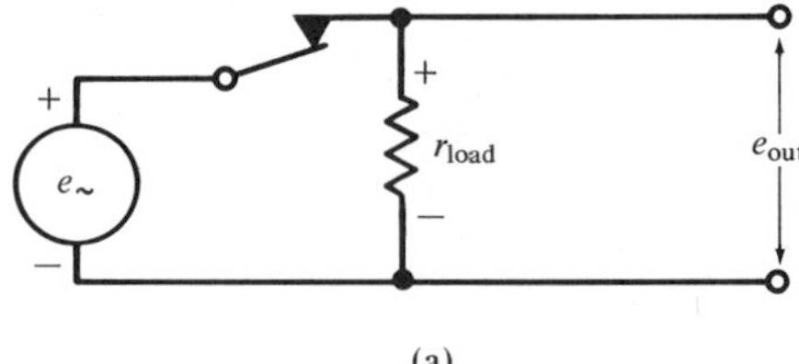

(a)

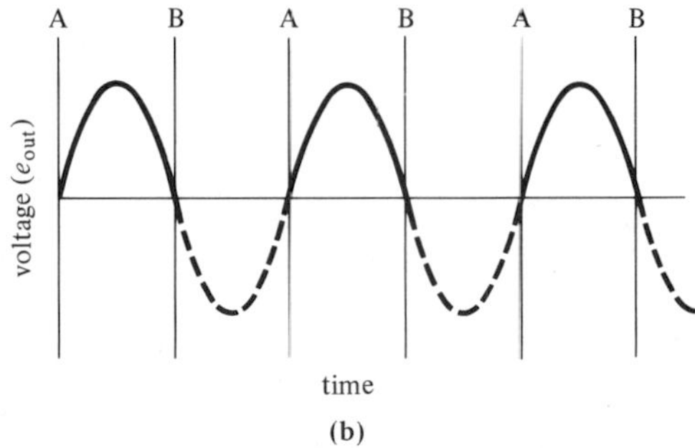

(b)

Figure 4.3 / *Half-wave rectification.*

Fig. 4.4(a). During time intervals A–B, when the polarity is that indicated in the diagram, the two switches are in the upper position. The resulting current flow makes the top of the load resistance positive relative to the bottom. At the start of the alternate half-cycles, B–A, the switches move to the lower positions to connect what is now the positive side of the signal to the top of the load resistance. Fluctuating current flows in one direction only through the load. The solid waveform in Fig. 4.4(b) will be observed across the load.

Rectification will occur only if the switching action is synchronized with the waveform. To convert AC power to DC, this can be achieved very simply with diodes by arranging for the input signal alternately to forward bias and to reverse bias them. In Fig. 4.5(a), half-wave rectification is accomplished by using one switching diode. Power is transferred to the load during the half-cycles in which the diode is forward biased.

Full-wave rectification can be obtained by using four diodes in the bridge configuration illustrated in Fig. 4.5(b). With signal polarity as marked, diodes D_1 and D_4 are forward biased, while diodes D_2 and D_3 are reverse biased. Current flow is indicated by the arrows. In alternate half-cycles, D_2 and D_3 are forward biased. Current flow is from the source through D_2 to the load and through D_3 to return to the power source. To produce smooth DC from the rectified AC, the high-frequency components must be filtered out. This was discussed in Section 3.4.

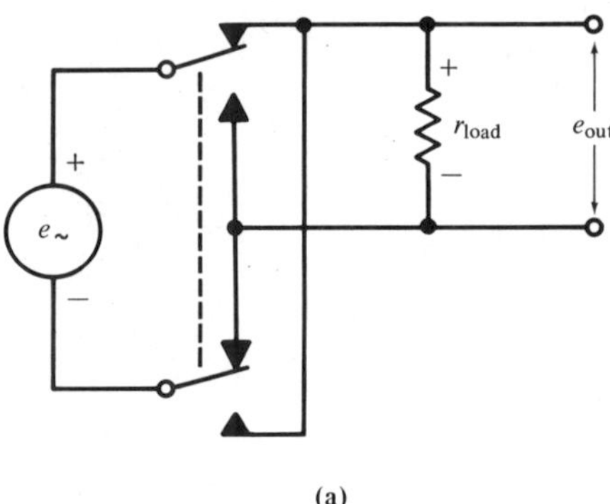

(a)

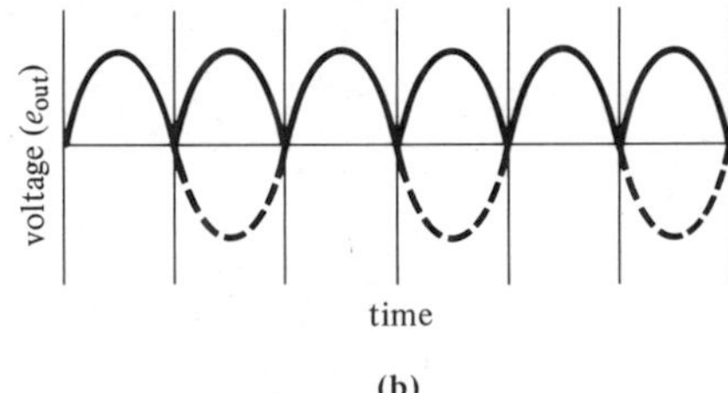

(b)

Figure 4.4 / *Full-wave rectification.*

4.4 / THE JUNCTION FIELD EFFECT TRANSISTOR

A simplified schematic representation of a junction field effect transistor (JFET) is given in Fig. 4.6. It consists of a bar of doped semiconductor, *n*-type in the illustration, which has two regions of *p*-type material embedded in opposite sides. The *p*-type regions, which are connected electrically, constitute the *gate*. One end of the bar is the *drain*, the other is the *source*.

Figure 4.5 / *Rectifiers.*
(a) *Diode half-wave rectifier.*
(b) *Diode bridge full-wave rectifier.*

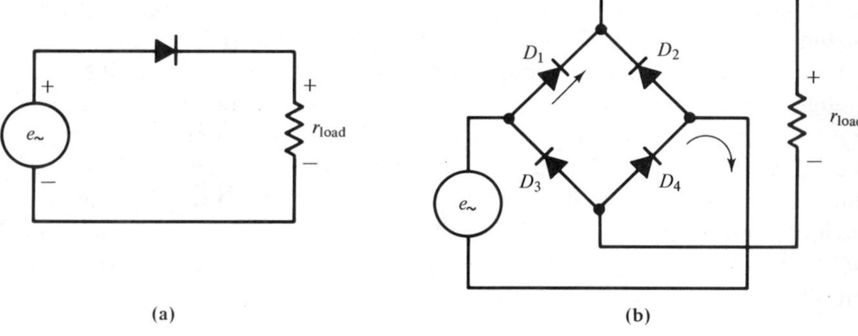

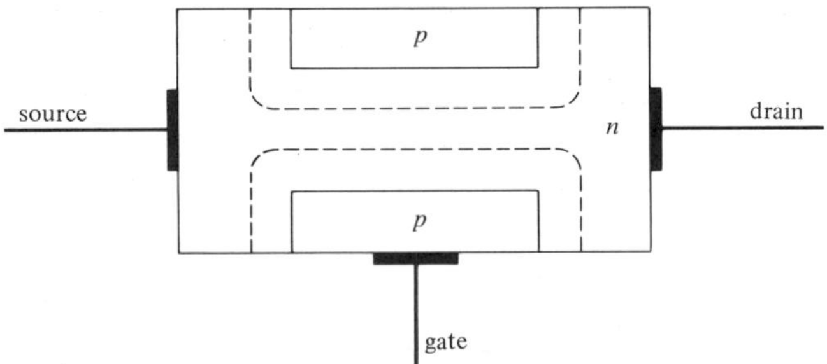

Figure 4.6 / *n-Channel field effect transistor.*

The region between the two gates is the *channel*. In operation, a voltage is imposed across the drain–source terminals that is of the correct polarity to cause movement of channel majority carriers from source to drain. For an *n*-channel device, the drain would be positive relative to the source; conventional current would flow from drain to source. A *p*-channel FET would be operated with reverse polarity, causing conventional current flow from source to drain. These two types are said to be *complementary*. The discussion here uses *n*-channel devices in illustrations; however, the same explanation holds for *p*-channel FETs if voltages and directions of current flow are reversed.

Application of a positive gate voltage relative to the source in Fig. 4.6 would forward bias the gate–channel junction, which is not a useful condition. With the junction reverse biased, a depletion zone would occur on either side of the junction. In JFETs, the channel is more lightly doped than the gate, causing a much more extensive development of the depletion zone in the channel than in the gate. Application of an increasingly negative gate voltage causes extension of the depletion zone into the channel. This restricts the effective cross-sectional area available for conduction of current, since the resistance of the depletion zone is very large. There exists for each device a characteristic gate–source voltage at which the channel area is zero. This is the *pinch-off voltage*, V_P, which if applied between the gate and the source will produce a very high resistance between the drain and the source.

In discussing the operation of the JFET, an additional important factor must be considered. The flow of current in the channel causes a voltage drop along the channel because of its resistance. For an *n*-channel FET, the drain is positive relative to the source; accordingly, even if v_{GS} were zero, application of v_{DS} has the effect of reverse biasing the gate–channel junction. The channel at the drain end is made positive relative to the gate. Therefore, the channel can be pinched off at the drain end by imposition of v_{DS} equal to V_P even when v_{GS} is zero. The relationship is

$$v_{GS} - v_{DS} = V_P \tag{4.2}$$

V_P is negative for an *n*-channel FET and positive for a *p*-channel FET. The

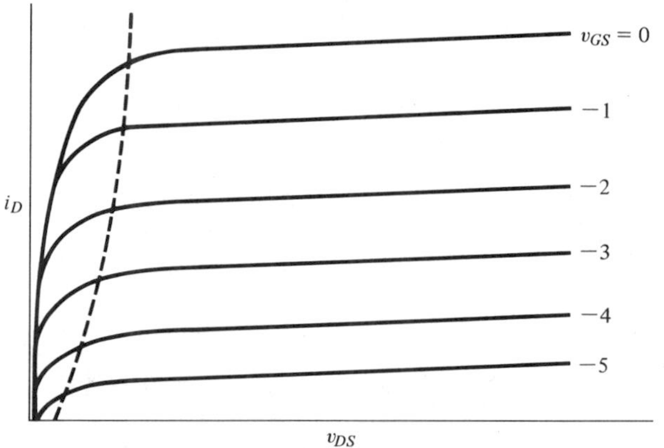

Figure 4.7 / *Current–voltage characteristic for the n-channel JFET.*

value of v_{DS} just necessary to pinch off the channel at the drain end is therefore a function of both V_P and of v_{GS}; it is termed the *threshold voltage*. If, for a given imposed v_{GS}, v_{DS} is increased from zero to the pinch-off threshold, i_D will increase, as indicated on the rising portion of Fig. 4.7. Beyond the point of pinch-off, an increase in v_{DS} causes the depletion zone to extend toward the drain, with the effect that, ideally, any increase in v_{DS} above the pinch-off threshold is exactly matched by an increase in drain–source resistance because of the extension of the depletion zone. This would

Figure 4.8 / *Schematic representation of an n-channel JFET for which $v_{DS} > v_{GS} - V_p$.*

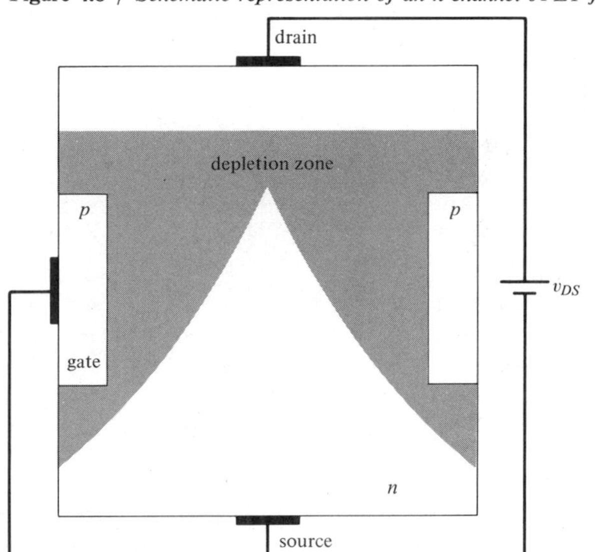

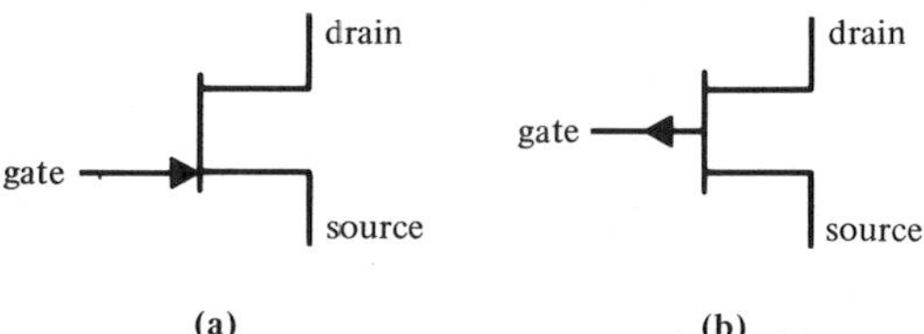

Figure 4.9 / *Junction field effect transistor symbols.*
(a) *n-Channel JFET.*
(b) *p-Channel JFET.*

cause the FET to pass a constant drain current above the threshold voltage. On Fig. 4.7, the dotted line indicates threshold voltages for the values of v_{GS} depicted. Figure 4.8 illustrates the condition that exists when v_{DS} exceeds the threshold value. When a JFET is operating in the constant-current region, the drain current is ideally a function of gate–source voltage only. This is the normal condition for use as an amplifier.

It should be noted that this description is somewhat oversimplified. Because of second-order effects, actual JFETs do not behave as true constant-current devices above the pinch-off threshold. At v_{DS} values above the pinch-off threshold, curves such as those in Fig. 4.7 show more or less slope, depending on the type of device chosen.

The symbol used for an *n*-channel JFET is shown in Fig. 4.9(a), that for a *p*-channel device in Fig. 4.9(b). Distinction between types is made on the basis of the direction of the arrow representing the gate. Very often the source is indicated by offsetting the gate as shown.

A single-stage amplifier using a JFET is illustrated in Fig. 4.10. The input signal is applied between the gate and the source, and the output is taken between the drain and the source. This is described as a *common-source* or a *grounded-source* amplifier, because the source is connected to the common point in the circuit. Suppose that the input is at some negative value and, because of the choice of power supply voltage, V_{dd}, and load resistance, it

Figure 4.10 / *Common-source amplifier.*

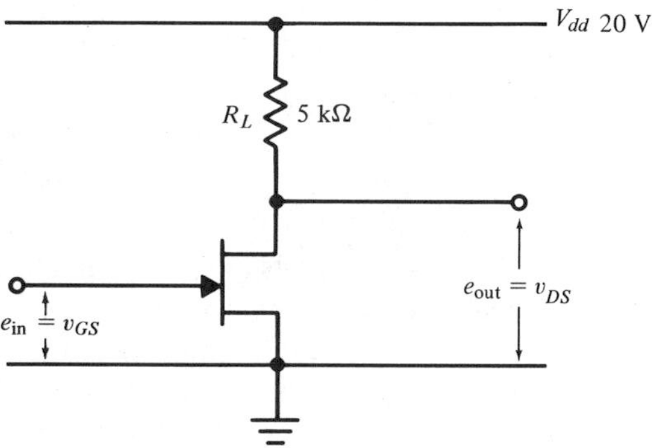

TABLE 4.1

Gate voltage		Drain current		Drain voltage	
v_{GS_1} (V)	v_{GS_2} (V)	i_{D_1} (mA)	i_{D_2} (mA)	v_{DS_1} (V)	v_{DS_2} (V)
−2.0	−1.9	3.0	4.5	7.0	5.5

causes a certain flow of drain current. Typical values might be those shown as v_{GS_1}, i_{D_1}, and so on in Table 4.1.

If the input voltage were changed to some more positive value, v_{GS_2}, this would cause an increase in i_D and a consequent decrease in v_{DS}. The device therefore acts as an inverter. Amplification is achieved because the change in v_{DS} can be made larger than that of v_{GS}. For example, a change in v_{GS} of +0.1 V might cause the changes indicated in Table 4.1. In this illustration, the voltage gain amounts to a factor of 15.

4.4.1 / The load line

To facilitate the design of an amplifier, a system for describing the performance of the FET is needed. This can take the form of a set of equations relating the relevant variables, or it can be in graphical form. For illustration, suppose that it is desired to design a common-source amplifier from a graphical description such as Fig. 4.7. It describes the behavior of a typical JFET, showing the variation in drain current with change in gate–source voltage and in drain–source voltage. The information typically available to anyone wishing to design an amplifier, in addition to the FET characteristic curve, is the power supply voltage, together with some knowledge of the nature of the signal that is to be amplified. Values of v_{DS} or i_D are, however, not available; accordingly, it is not possible to make immediate use of Fig. 4.7. This difficulty can be overcome by viewing the FET and the load resistor as a voltage divider (as in Fig. 4.11) made up of one fixed resistor and one variable resistor with the output taken as indicated. Drain–source voltage can be expressed as a simple function of power supply voltage:

$$v_{DS} = V_{dd}\left(\frac{\text{FET resistance}}{R_L + \text{FET resistance}}\right) \tag{4.3}$$

Drain current is also determined by FET resistance:

$$i_D = \frac{V_{dd}}{R_L + \text{FET resistance}} \tag{4.4}$$

For this purpose, let us assume that the effective FET resistance can change from zero to infinity with sufficient change in gate–source voltage. When the FET resistance is zero, drain–source voltage is zero and drain current is 4.0 mA. When the FET resistance is infinitely large, drain–source voltage is equal to V_{dd} and the drain current is zero. If values of drain current and

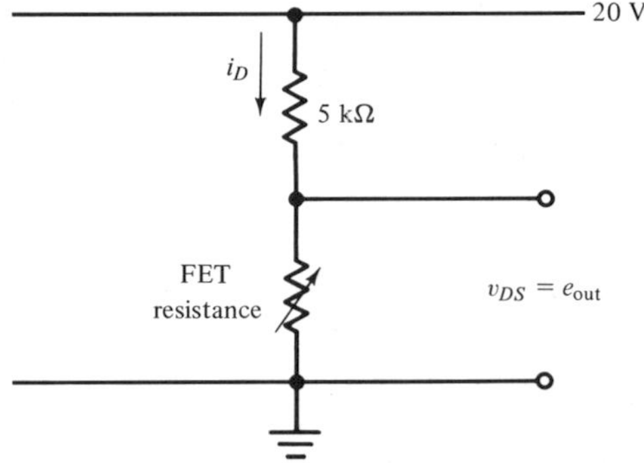

Figure 4.11 / *Voltage-divider model for an inverter amplifier.*

voltage are plotted on a FET characteristic curve as indicated in Fig. 4.12, the two points defined so far locate the extremities of a line (ABC) which is the locus of all possible values of drain current and of drain–source voltage for this particular load resistance and power supply voltage. This is known as the *load line* for the FET.

In actual operation, the portion of the line between points A and B in Fig. 4.12 cannot be used, because as gate voltage is increased from -5 V at cutoff (point C) to -2 V at point B, any further increase would cause the

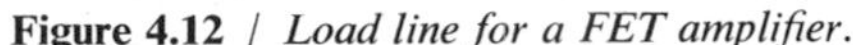

Figure 4.12 / *Load line for a FET amplifier.*

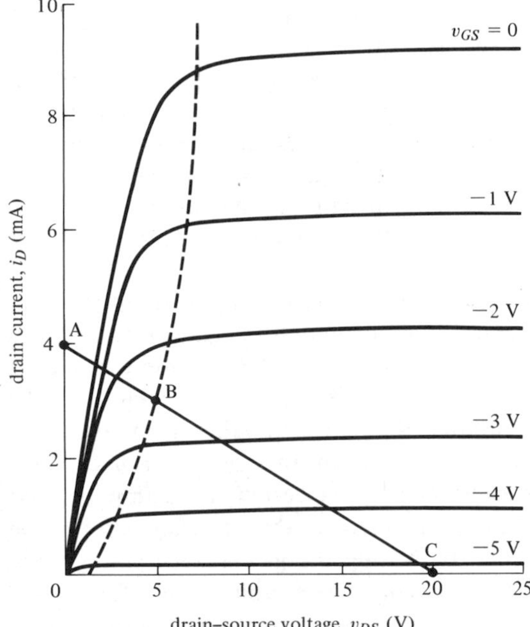

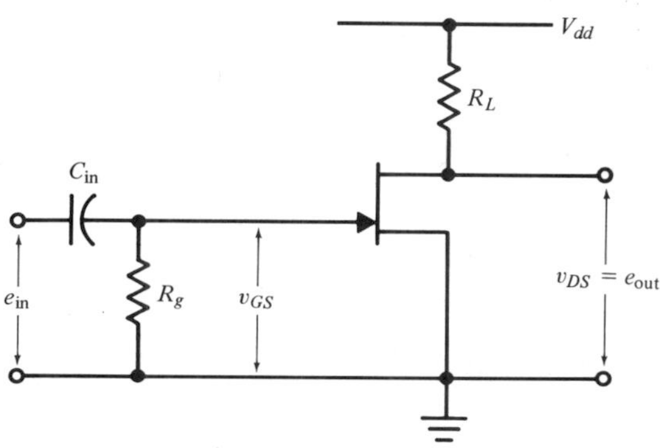

Figure 4.13 / *Common-source FET amplifier with AC input.*

drain–source voltage to drop below the pinch-off threshold and lead to gross nonlinearity of operation; so the operating range for the FET connected as shown is from B to C on the load line.

4.4.2 / The operating point

The possible range of input voltages is from -2 to -5 V. If a DC signal is to be handled, it must fall within these limits. If, however, an AC signal that can be transmitted by an *RC* high-pass filter is being amplified, then the DC level of the input signal is immaterial, since the filter will remove it. Only the magnitude of the excursions of the signal is of concern. This type of input arrangement is shown in Fig. 4.13. The resistor R_g is required to provide a definite DC level for the gate voltage for two reasons: If it is omitted, the gate will be electrically isolated by the input capacitor and the high resistance of the reverse-biased gate–channel junction. It would therefore be very subject to the influence of stray signals picked up by electrostatic interaction with other conductors. These stray signals would be amplified and would appear in the output signal. In addition, a small residual current flows in the gate circuit, usually of the order of 10^{-9} A, because of movement of minority carriers. If R_g is omitted, the filter capacitor will be gradually charged.

As shown in Fig. 4.13, in the absence of an externally applied input signal, the gate will be at ground voltage if it is assumed that the gate current causes a negligible drop across R_g. The FET will therefore be outside the desired operating range in Fig. 4.12. Accordingly, it is necessary to include in the circuit components that will ensure that the quiescent values (in the absence of an input signal) correspond to a point in the desired range. This equilibrium condition, which is determined by the design of the amplifier and which obtains in the absence of an input, is termed the *operating point* or the *Q point* for the FET. It is defined by stating the gate–source voltage, the drain–source voltage, and the drain current. The establishment of a defined operating point is termed *biasing*. In principle, it involves the addition of a

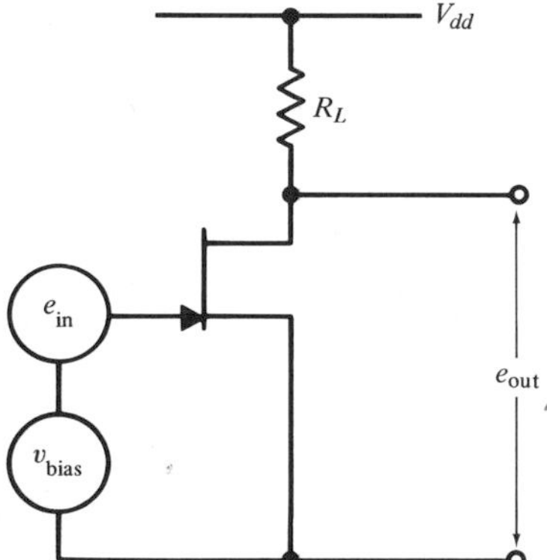

Figure 4.14 / *Relationship of biasing signal to input signal.*

DC signal to the input signal as in Fig. 4.14. This could be accomplished by placing a battery in the circuit, but in practice this is usually not done because cheaper and more satisfactory techniques are available.

A very commonly used method involves placing a resistor between the source and the common point of the circuit. It has the effect of raising the source voltage above that of the common point by the amount of the drop in R_S. If the gate voltage remains unchanged, it is accordingly biased negatively with respect to the source, since

$$v_{GS} = -i_D R_S \tag{4.5}$$

This arrangement is shown in Fig. 4.15. Qualitatively, the effect of increasing the value of R_S is to move the operating point to the right along the load line, that is, toward cutoff. This method cannot be used to achieve actual cutoff, however, since if i_D were to become zero, the bias voltage $i_D R_S$ would also be zero.

Figure 4.15 / *Establishment of bias by the use of a source resistor.*

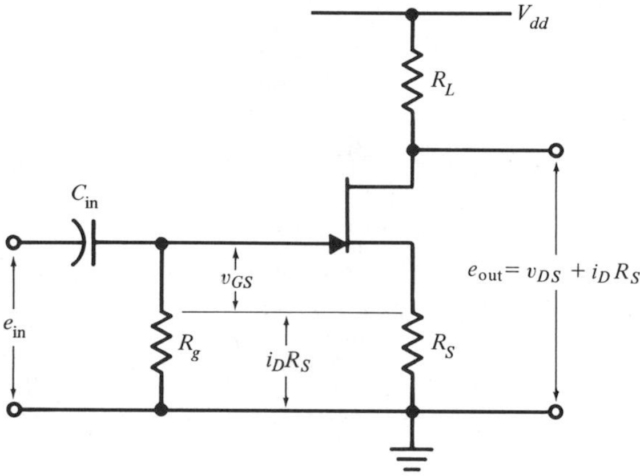

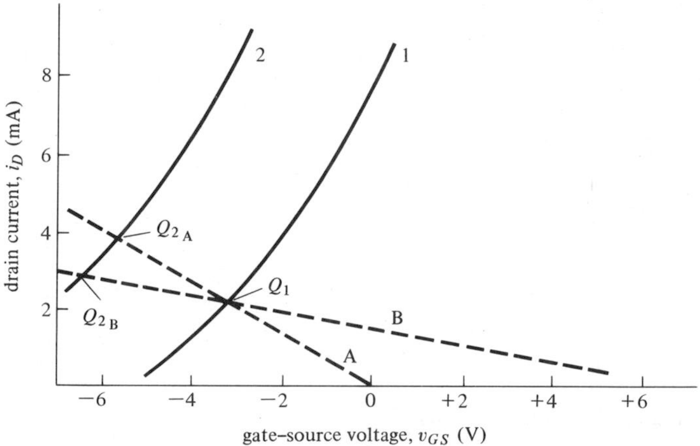

Figure 4.16 / *Common-source transfer characteristic.*

The establishment of a specific operating point can be carried out with the aid of a common-source transfer characteristic such as is shown in Fig. 4.16. If source resistor biasing is used, Eq. (4.5) can be plotted as a line on the transfer characteristic with a slope equal to the reciprocal of R_S; the line corresponding to $R_S = 1.5\ \text{k}\Omega$ is plotted in Fig. 4.16 as dashed line A. The intersection of this with the FET transfer curve 1 occurs at $v_{GS} = -3.2$ V, which corresponds to the Q point noted in Fig. 4.17. If a curve such as Fig. 4.16 is unavailable, essentially the same result can be obtained by noting that if an operating point near the center of the permissible operating range is desired, it is evident from Fig. 4.17 that at a v_{GS} value of about -3 V, i_D should be about 2.2 mA. This would be achieved by use of 1.3 kΩ for R_S. In either case, the load resistor value would have to be adjusted to account for the source resistor, since the slope of the load line is the reciprocal of the sum of R_L and R_S.

4.4.3 / Semiconductor component variability

The procedure outlined above is an oversimplification when applied to FETs, because the curves furnished by the manufacturer correspond to a *typical* unit, not necessarily to a particular FET of a given type. At the present stage of development of semiconductor technology, there are very appreciable variations from one unit to another in a nominally identical batch. Possible variability is represented in Fig. 4.16 by the location of transfer curve 2 relative to curve 1. With the same load and source resistor, FET 2 conducts heavily and would probably be at an operating point below the pinch-off threshold.

This difficulty can be circumvented by making measurements on individual FETs and designing the biasing networks accordingly; however, this procedure is generally unsatisfactory, because it makes mass production

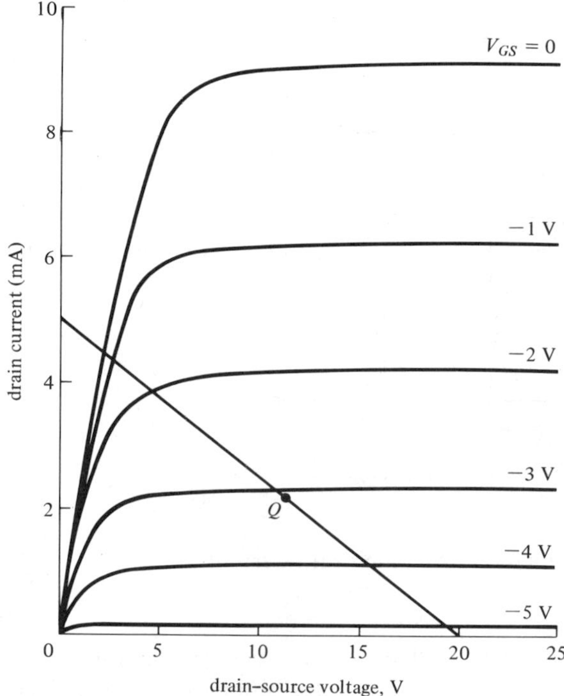

Figure 4.17 / *Selection of the operating point with the aid of a current–voltage characteristic.*

impossible and because it would necessitate a complete redesign if it were ever necessary to replace the FET during repair work. Accordingly, it is normally necessary to elaborate upon the design procedure for either FETs or transistors by incorporating large amounts of negative feedback in order to produce circuits that can compensate for the variability of the amplifying device. More detailed discussion of negative feedback is reserved to Chapter 5. Let it suffice here to point out that the use of R_S represents one way of introducing negative feedback. Increasing the size of R_S would reduce the slope of the bias line plotted in Fig. 4.16 and therefore reduce the effect on the quiescent value of i_D of the change in FET behavior. It would, however, tend to cut off the FET by moving the Q point to the right in Fig. 4.17. This could be avoided by introducing additional external biasing as indicated in Fig. 4.18.

In Fig. 4.18, the gate voltage, relative to ground, is

$$v_G = V_{dd}\left(\frac{R_2}{R_1 + R_2}\right) \tag{4.6}$$

and the source voltage is

$$v_S = i_D R_S \tag{4.7}$$

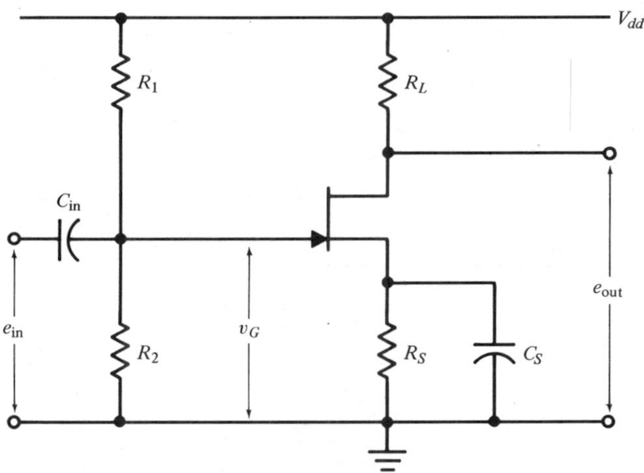

Figure 4.18 / *A practical FET common-source amplifier.*

The gate–source voltage, which controls the drain current, is

$$v_{GS} = v_G - v_S < 0 \tag{4.8}$$

The gate–source voltage is negative; that is, the junction is automatically reverse biased for any value of v_G or of R_S. This becomes apparent if one considers what happens when the biasing bleeder, R_1–R_2, is connected. If $v_{GS} \geq 0$, this will cause an increase in whatever drain current is flowing. Increasing the drain current increases the source voltage ($i_D R_S$) but does not change v_G; accordingly, v_{GS} will be negative at equilibrium.

The equilibrium value can be obtained from a curve such as Fig. 4.16 if the value of v_G, which is positive, is used as the origin of the plot as indicated for line B. The use of a positive value for v_G, rather than zero as in Fig. 4.16, makes it possible to reduce the slope of the bias curve ($1/R_S$) substantially and still maintain a reasonable drain current.

4.4.4 / The function of the bypass capacitor

Establishment of bias by use of a source resistor, together with a gate bleeder, is attractive because of its simplicity and effectiveness. However, one reason for its effectiveness is that negative feedback is introduced. This serves to compensate for FET variations, which is good; but it causes a reduction in the gain of the amplifier, which is probably bad.

This difficulty can be avoided if the signal being amplified is an AC signal with no extremely low-frequency components. If this is true, it is possible to place a capacitor C_S in parallel with R_S, as indicated in Fig. 4.18. The capacitance is chosen to offer a very small reactance at the frequencies of interest. When this is done, the source is connected to the common point through an impedance that behaves somewhat like a low-pass filter. At DC

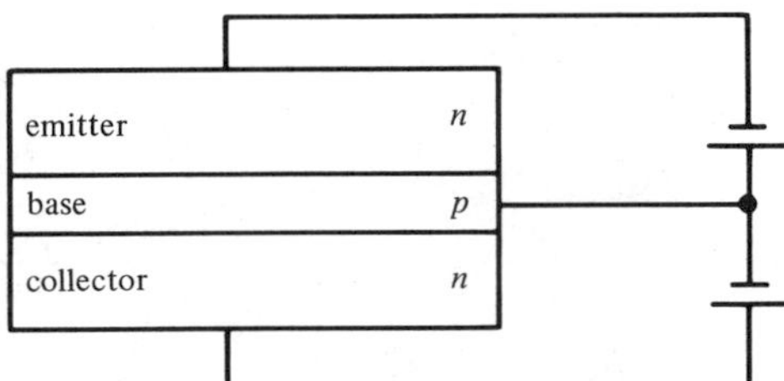

Figure 4.19 / *The npn bipolar junction transistor.*

and low frequencies, it has a value of R_S Ω. At signal frequency equal to the breakpoint frequency of the *RC* network, that is, $[1/(2\pi R_S C_S)]$ Hz, the impedance is $(R_S/2)$ Ω; at higher frequencies it drops off as indicated in a Bode plot. A bypassed source resistance, accordingly, functions as a simple resistance at DC and low frequencies, providing bias voltage but greatly reducing the gain of the FET. At a frequency equal to 100 times the breakpoint frequency, this impedance amounts to approximately 1 percent of its DC value and therefore has much less effect on the gain. A quantitative treatment is given in Chapter 5.

4.5 / THE BIPOLAR TRANSISTOR

When three sections of doped semiconductor material are arranged as indicated in Fig. 4.19, the resulting device may be used as a bipolar junction transistor. Throughout this text, the unqualified term *transistor* is reserved for the bipolar junction transistor. The three sections, the *emitter*, the *base*, and the *collector*, must either be of *n*-, *p*-, and *n*-type material, respectively, or of *p*-, *n*-, and *p*-type material, respectively. The arrangement shown in Fig. 4.19 is an *npn* transistor; the other possibility is a *pnp* device. In normal operation, the emitter–base junction is forward biased and the base–collector junction is reverse biased, as shown. As a result, emitter majority carriers (electrons for an *npn* transistor) cross to the base, where they may be expected to encounter holes. In a transistor, however, the availability of base majority carriers is restricted by controlling the thickness and composition of the base material. Consequently, most of the emitter majority carriers that cross the emitter–base junction diffuse entirely through the base without encountering holes.

Back biasing the base–collector junction prevents the movement of majority carriers across the junction. It does not, however, prevent the movement of electrons from the base to the collector; on the contrary, it facilitates this movement. As a result, emitter majority carriers are emitted into the base and collected by the collector. Moreover, the fraction of emitter majority carriers that reaches the collector is determined primarily by the composition and geometry of the device and is therefore approximately a constant.

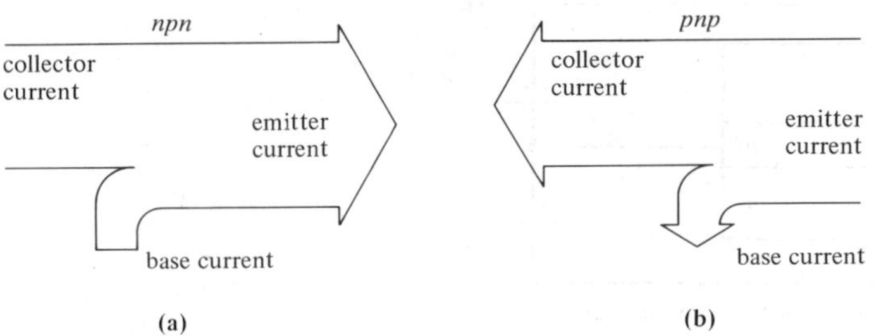

Figure 4.20 / *Current flow in bipolar transistors.*

To aid comprehension of the operation of a transistor, it may be helpful to note that the major current through the device is between the emitter and the collector; a small but constant fraction of this is drawn by the base. This is shown diagrammatically in Fig. 4.20(a) for an *npn* transistor and in Fig. 4.20(b) for a *pnp* transistor. The arrows indicate the direction of flow of conventional current.

The amplifying action of a transistor is based on the fact that the base current is a small but constant fraction of the emitter current. In normal operation, if the small base current is controlled, this effects control of the much larger emitter current. The mechanism of control of the emitter current by the base current stems from the fact that some of the emitter majority carriers do encounter and combine with base majority carriers, causing, for an *npn* transistor, a net transfer of negative charge from the emitter to the base. If this excess charge is not removed from the base, it cancels the externally applied voltage responsible for the flow of emitter current. To the extent that it is removed, further emitter–base charge transfer can occur. Since the fraction of emitter carriers trapped in the base is fairly constant, this constitutes an approximately linear control mechanism.

In addition to the effects of externally imposed base, collector, and emitter currents, the effect of the movement of thermally excited minority carriers must be considered. In the usual case, this type of current is small and is simply added to a much larger external current without causing difficulty. When movement of minority carriers from the base of a transistor is considered, however, the situation is different. Consider an *npn* transistor. Electrons are minority carriers in the base, and they will readily move into the collector under the influence of the external power supply voltage; this current is given the symbol I_{CO}. It is also true that electrons that have diffused from the emitter are attracted to the collector. The difference is that transfer of an electron from the emitter through the base to the collector causes no net change in the charge density in the base; on the other hand, thermal excitation of an electron in the base and transfer to the collector removes one negative charge from the base. Thus, I_{CO} has exactly the same effect as flow of base current from an external source since, for an *npn* transistor, conventional (positive) current is directed inward to the base of

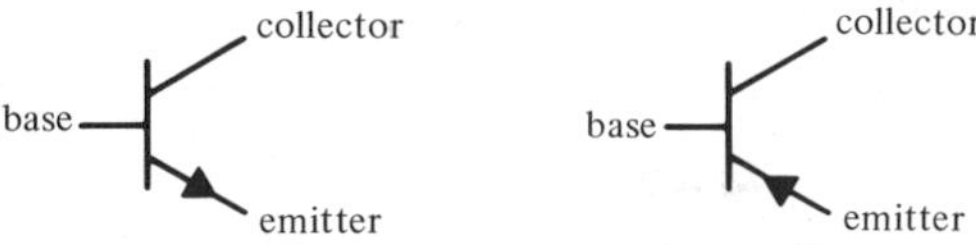

Figure 4.21 / *Bipolar transistor symbols.*
(a) *npn Transistor.*
(b) *pnp Transistor.*

the device. The current I_{CO} is amplified just as is i_B; an increase of temperature causes an increase in collector current. Moreover, I_{CO} bears an exponential relationship to temperature, so that sufficient temperature increase will cause an uncontrolled increase in collector current. Silicon devices exhibit lower I_{CO} values than germanium devices; as a result, they show significantly less temperature dependence and are therefore preferred for many applications.

The symbols used for *pnp* and *npn* transistors are shown in Fig. 4.21. Note that the arrows in the symbols show the direction of the flow of conventional current, not the direction of motion of majority carriers.

A rigorous quantitative treatment of transistor behavior is quite complex; however, some useful approximations can be made on the basis of the following simple statements:

$$i_E = i_C + i_B \tag{4.9}$$

$$\frac{i_C}{i_E} \simeq 0.95\text{–}0.99 \tag{4.10}$$

$$v_{BE} \simeq \text{constant} \tag{4.11}$$

As mentioned above, the fraction of total emitter majority carriers trapped in the base is primarily a property of the transistor; hence, i_C is a substantially linear function of i_E and is only slightly smaller than i_E. In the operation of a transistor, the base–emitter junction is forward biased. One implication of this is that the base–emitter voltage is subject to little variation and is actually determined mainly by the identity of the semiconductor material. It is approximately 0.2 V for a germanium transistor and approximately 0.6 V for a silicon transistor.

4.5.1 / The common-emitter configuration

If the bipolar transistor is considered to be a box provided with three terminals that can be identified, there are six different ways in which it can be connected in an arrangement similar to that shown in Fig. 4.1(b). However, three of the possibilities are entirely useless, and one offers only very limited utility. Of the remaining two, one will be discussed here in some detail because of its very great importance; this is the *common-emitter* configuration. The other is the *common-collector*—or *emitter-follower*—configuration, and it is discussed in Section 4.8.

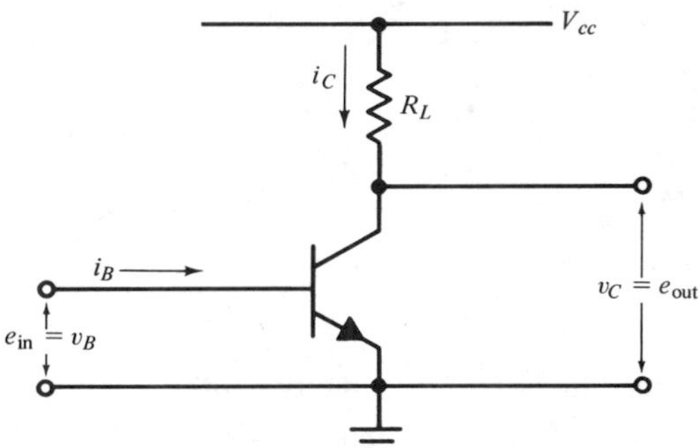

Figure 4.22 / *Common-emitter transistor amplifier.*

In the common-emitter configuration, the input signal is imposed between the base and the emitter, and the output is taken between the collector and the emitter. This is shown in Fig. 4.22. The input signal is shown as a voltage source because this is commonly the case; however, the transistor responds to the base current that flows as a result of imposition of the input signal. Suppose that the device is in operation as in Fig. 4.22. Steady state values of i_B, i_C, and i_E have been established according to the imposed voltages, e_{in} and V_{cc}, and the relationships given in Eqs. (4.9)–(4.11). If v_B is increased slightly, then i_B is increased. This will cause an increase in i_C and in i_E. Increasing i_E will necessarily increase the voltage dropped across R_L. Since one end of R_L is arbitrarily held at the voltage of the power supply, V_{cc}, an increase in the voltage dropped across it must cause the voltage at the other end, which is connected to the collector, to assume a more negative value than it had before the change in v_B occurred. A positive change in v_B, or an increase in i_B, will cause a decrease in v_{CE}, and therefore a negative change in v_C. A negative-going input voltage will, of course, cause just the opposite changes in the other variables.

The behavior of a common-emitter bipolar transistor is in certain respects similar to that of an FET inverter. It can furnish voltage gain larger than unity, and it causes inversion of the input voltage signal. The behavior of bipolar transistors differs, as mentioned above, in that they are current amplifiers. In addition, the ON–OFF behavior of transistors is quite different from that of FETs. When the gate voltage of a FET is driven sufficiently negative with respect to the source, the flow of drain current can be entirely cut off; similarly, the emitter current of transistors can be cut off by reducing the base current to zero. However, there is a considerable difference between FETs and common-emitter transistors when their ON characteristics are considered. A JFET is not normally operated with a forward-biased junction, because it then draws gate current and interacts with the input signal source in an entirely different way from its normal behavior. By contrast, a transistor always draws base current when it is

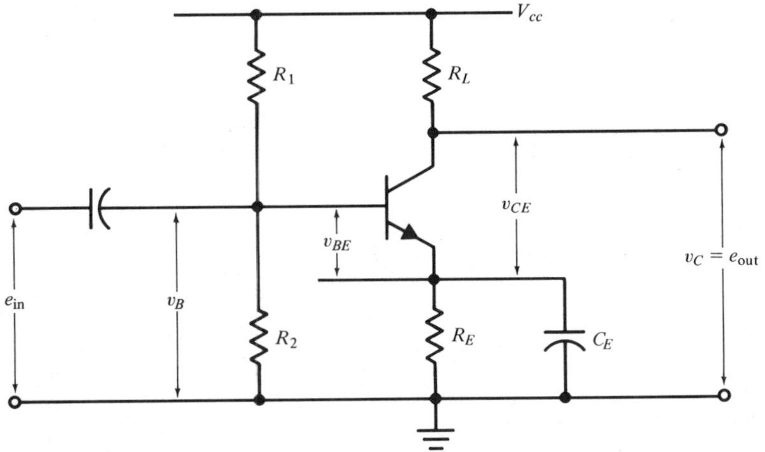

Figure 4.23 / *A practical common-emitter amplifier.*

operating, and input circuitry must therefore be designed to accommodate this requirement. Accordingly, no radical change in behavior occurs when a transistor is turned completely on. The device goes into *saturation*, in which condition the collector–emitter voltage is only about 0.02 V and the collector current is ordinarily limited by the load resistor. Transistors can accordingly be used very efficiently as switches, because the effective collector–emitter resistance can be varied by control of the base current from megohms to a few ohms.

4.5.2 / The operating point

For satisfactory operation it is necessary with transistors, as with FETs, to furnish bias in order to establish a suitable operating point. The factors discussed in relation to biasing the FET are important in connection with transistors. The biasing network must compensate for device variability and for temperature sensitivity. Accordingly, negative feedback must be used. A commonly used circuit is illustrated in Fig. 4.23. This looks identical to the circuit in Fig. 4.18, except for changes of notation and of the transistor symbol. In many respects it is very similar. Negative feedback is introduced by placing R_E between the emitter and the common point. The bleeder (R_1–R_2) serves to furnish base–emitter voltage, and the bypass capacitor, C_E, has the same function as does C_S in Fig. 4.18. This arrangement, like that in Fig. 4.18, is useful only for signal frequencies well above the breakpoint frequency of $R_E C_E$.

The biasing arrangement in Fig. 4.23 differs greatly from that in Fig. 4.18, however, in that the bias signal is a *current* rather than a voltage. The bleeder in Fig. 4.23 actually functions as a current source; that in Fig. 4.18 serves as a simple voltage divider to maintain gate voltage. The Thévenin equivalent of a bleeder is shown connected to the base of a transistor in

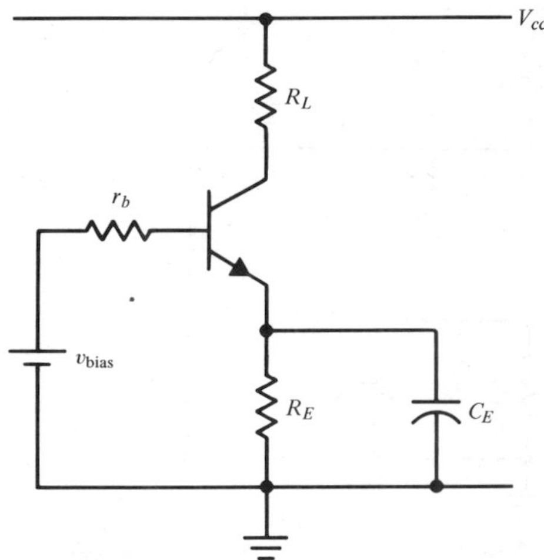

Figure 4.24 / *Operation of the bias network.*

Fig. 4.24. The voltage source, v_{bias}, is defined by Eq. (4.12), and the equivalent resistance is defined by Eq. (4.13):

$$v_{bias} = V_{cc}\left(\frac{R_2}{R_1 + R_2}\right) \tag{4.12}$$

$$r_b = \left(\frac{R_1 R_2}{R_1 + R_2}\right) \tag{4.13}$$

If the effective input resistance at the base were constant, the bias current furnished by the bleeder would evidently be constant and it would be possible to establish an operating point. Actually, this would be unsatisfactory, because the required bias current is appreciably affected by changes in transistor characteristics, such as those caused by temperature fluctuations. Accordingly, provision must be made for the bias current furnished by the bleeder to vary automatically to compensate for the effects of temperature fluctuations, as well as for other changes, such as transistor replacement. This automatic compensation is accomplished through the use of negative feedback, introduced in this circuit by the action of the emitter resistor. This is discussed further in Section 5.2.1.

4.6 / VACUUM TUBES

An important group of electronic components, consisting of the various types of vacuum tubes, is based on the transmission of a current as an electron beam across an evacuated space between electrodes. The electron beam is produced by imposition of a voltage across a pair of electrodes. One of these, the *cathode*, is fabricated of a material that readily emits electrons

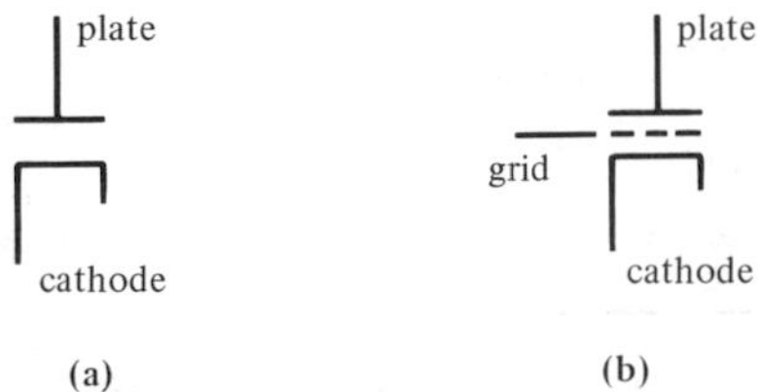

Figure 4.25 / *Vacuum tube symbols.*
(a) *The vacuum diode.*
(b) *The vacuum triode.*

when it is heated. When the other electrode, the anode or *plate*, is held at a positive voltage relative to the cathode, electrons emitted from the heated cathode are attracted to the plate. If, however, the polarity of the imposed voltage is reversed, there is no flow of current because the plate does not readily emit electrons.

Accordingly, this type of device, called a *vacuum diode*, is capable of conducting current in only one direction, a flow of electrons from cathode to plate. The symbol for a vacuum diode is given in Fig. 4.25(a).

4.6.1 / The vacuum triode

To use a vacuum tube as an amplifying device, a wire screen or *grid* is introduced between the cathode and the plate, producing a *triode*. A voltage applied to the grid, negative with respect to the cathode, causes attenuation or even complete interruption of the electron beam. This grid voltage constitutes the input signal to the device. The beam, modulated by the input voltage, constitutes the output signal from the device. Should the grid assume a positive voltage relative to the cathode, then electrons would be attracted to it rather than to the plate. Although this is allowed to happen in certain types of amplifiers, we shall restrict this discussion to what is known as *class A* operation, in which the grid is prevented from going positive with respect to the cathode. The symbol used in schematic diagrams to represent a vacuum triode is shown in Fig. 4.25(b).

A single stage of amplification is shown in Fig. 4.26. Here the input signal is applied between the grid and the cathode, and the output is taken between the plate and the cathode. This configuration may be described as a common-cathode arrangement, because the cathode is connected to the common point of the circuit. For a qualitative examination of the performance of the system, suppose that the input signal is at -1.0 V relative to ground. This means that the grid is 1 V more negative than the cathode, since it is grounded. This will cause the tube to pass a plate current of about 1.4 mA, considering the characteristics of this tube (Fig. 4.27), the load resistor, and the power supply voltage shown; the plate voltage, e_b, will be 160 V. The voltage drop in the load resistor will be

$$i_b R_L = (1.4 \text{ mA})(100 \text{ k}\Omega) = 140 \text{ V} \qquad (4.14)$$

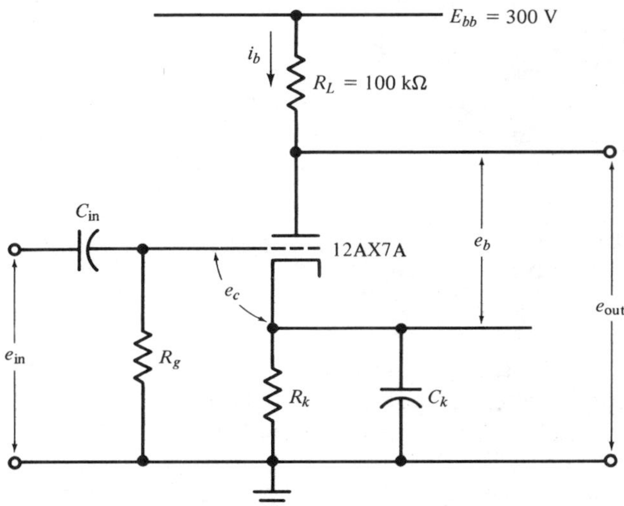

Figure 4.26 / *A practical common-cathode amplifier.*

If the input signal changes to -2 V, the more negative grid voltage will cause plate current to decrease to about 0.8 mA. The drop in the load resistor will then be 80 V, and the plate voltage will be increased to 220 V.

As the grid voltage moves to more negative values, plate current decreases and the plate voltage becomes more positive. This is responsible for the fact that the grounded cathode amplifier acts as an inverter. Accordingly, the arrangement in Fig. 4.26 is referred to as a triode inverter.

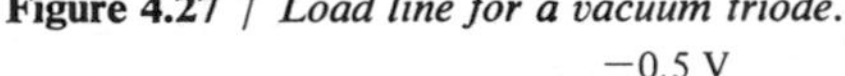

Figure 4.27 / *Load line for a vacuum triode.*

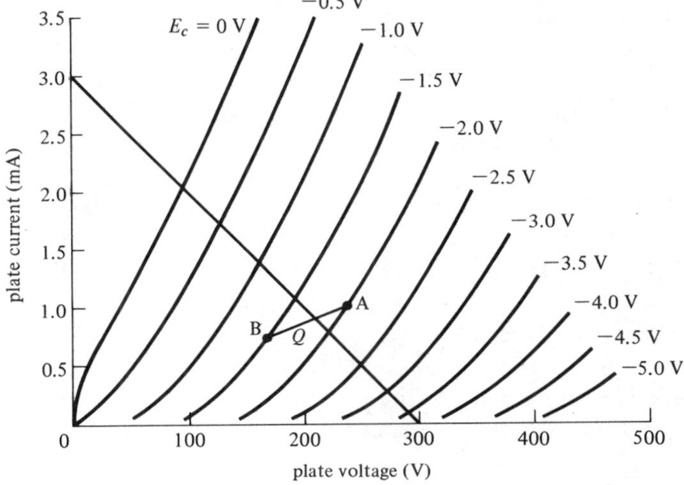

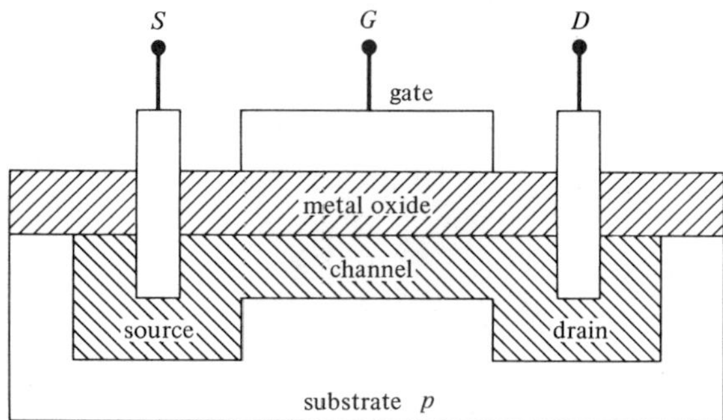

Figure 4.28 / *n-Channel MOSFET.*

4.6.2 / The operating point

The operating point for a vacuum tube can be established by the use of the graphical treatment described for the *n*-channel FET. In the case of vacuum tubes, however, individual units show sufficient similarity to make it feasible to rely on the manufacturer's curves in design. Accordingly, the biasing arrangement arrived at by drawing a load line and calculating the effect of the cathode resistor, R_k, does give satisfactory results. As is true of the transistor and the FET, the bypass capacitor serves to avoid negative feedback at signal frequencies.

4.7 / THE MOSFET

The principle involved in the operation of the JFET, control of the concentration of majority carriers by imposition of a voltage in a direction perpendicular to current flow in the device, is applied in a different way in the *metal oxide semiconductor FET*, or *MOSFET*. In this type of device, the gate is one plate of a capacitor; the body constitutes the other plate, and an insulating layer of metal oxide serves as the dielectric. An *n*-channel MOSFET is shown schematically in Fig. 4.28.

If a negative voltage, relative to the source, is imposed on the gate, the negative charge on the gate induces a corresponding positive charge in the channel, with consequent decrease in concentration of channel majority carriers and increase in resistance. A positive gate voltage will have the opposite effect. This device is, accordingly, capable of serving as an amplifier, just as is the JFET. One major difference is that the signal applied to the gate is insulated from the channel, so that the input impedance of a MOSFET is very large; it is therefore possible to accept both positive and negative inputs, since the action of the gate does not involve an *np* junction.

A variety of symbols has been used for the MOSFET, and industry-wide standardization has not occurred. Symbols that are frequently used are illustrated in Fig. 4.29.

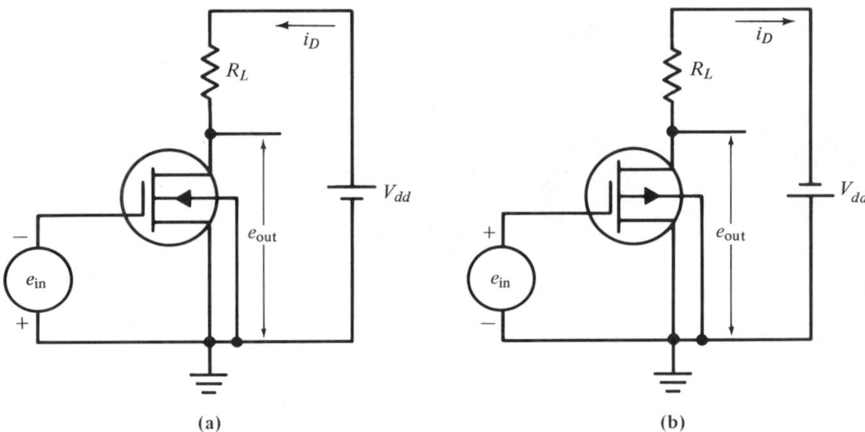

Figure 4.29 / *MOSFET symbols.*
(a) *n-Channel MOSFET inverter.*
(b) *p-Channel MOSFET inverter.*

4.8 / VACUUM TUBES VS. TRANSISTORS

In recent years, transistors have to a considerable extent supplanted vacuum tubes in newly designed instruments. It appears probable that, in the future, solid state devices will replace tubes entirely in most types of laboratory equipment. The reasons for this change and some of the implications of it are apparent when the characteristics of the various devices are compared. Of the several attributes that may be compared, perhaps the most important are space requirements, reliability, and input impedance.

The reduced space demands of solid state electronics compared with corresponding vacuum tube circuits not only makes possible substantial reduction in the size of instruments, but it also makes certain types of applications possible that would otherwise not be feasible. The reduced size of solid state circuitry stems partly from the fact that transistors can be fabricated on a very much smaller scale than can tubes and partly because the heat dissipated by transistors is very much less than that dissipated by tubes.

Vacuum tubes must be fabricated by mounting metal electrodes within a glass or metal vacuum chamber. The smallest of those commonly used has a volume of about 0.15 in.^{3}. A comparable value for discrete transistors, those produced as individual components, is 0.006 in.^{3}. Integrated circuits, in which transistors, resistors, and capacitors are fabricated and interconnected in large numbers on an extremely small scale, are available with thousands of transistors per cubic inch. Thus, possible reduction in space and cost is truly dramatic.

A vacuum tube must dissipate the energy produced in heating the cathode as well as that evolved from conduction of plate current. For a typical low-power, high-gain voltage amplifier, this would involve more than a watt per tube. By contrast, a typical value for an equivalent operation with discrete transistors is 0.01 W. Modern integrated circuits are operated with only a fraction of a milliwatt per unit. It is therefore possible not only to fabricate these circuits on a very small scale, but to operate them without elaborate cooling arrangements, which would increase greatly the size of equipment.

It is an implication of the comparison above that smaller power supplies suffice to operate transistorized equipment. This makes possible a significant reduction in the weight of instruments. It also makes it feasible to operate many instruments from batteries rather than from the AC mains. In addition to the obvious advantage where field use is required, battery-operated equipment is a great convenience even in the laboratory. Small instruments can be moved about without concern for line cords and, more important, interactions that may occur between instruments operated from a common power source can be easily avoided.

Solid state devices enjoy a substantial advantage over vacuum tubes in terms of reliability and lifetime. This has made it feasible to construct and operate very complex equipment such as computers. Where many thousands of components are involved, the demands of space and cooling capacity make vacuum tube devices extremely expensive; however, at least in principle it would be possible to supply larger buildings and larger air conditioning plants to accommodate more complex equipment, if these were the only factors. However, given a certain probability of failure of components, the probability of failure of the whole unit increases geometrically with the number of components. The failure rate of tubes is such as to preclude their use in extremely complex equipment. However, in the context of experience with ordinary laboratory equipment, one may expect that tubes will operate with only occasional replacement and, if conservative design has been followed, may last for many years.

The differences in input impedance between vacuum tubes and bipolar transistors is a consequence of their fundamental design. A vacuum tube responds to input voltage imposed at a grid in a vacuum chamber, while a transistor responds to input current; the tube naturally exhibits a much greater input impedance. If an amplifier is to accept a signal from a voltage source without causing distortion, its input impedance must be large as compared with the output impedance of the signal source. With correct design, for instance, by the use of emitter followers, it is possible to achieve quite low input current demands with bipolar transistors; however, so long as the choice must be between these and vacuum tubes, a significant number of applications exist for which transistors are unsuited. This has led to the design of hybrid equipment, containing both tubes and transistors.

With the introduction of reasonably priced FETs, vacuum tubes are gradually being eliminated as new laboratory equipment is designed. The FET offers most of the advantages of bipolar transistors along with an input impedance as high or higher than that of vacuum tubes. FETs are available

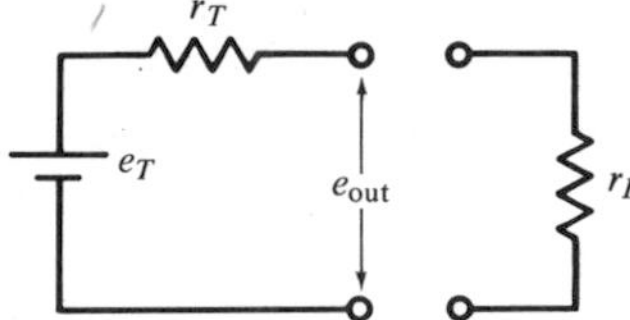

Figure 4.30 / *Thévenin equivalent circuit of an amplifier.*

both as discrete components and as units in integrated circuits. A typical use of individual components is at the input of pH meters. The high input impedance of the MOSFET makes it well suited to accept the voltage signal from a glass electrode. The MOSFET also can be used very efficiently as a switching device. This has led to its incorporation into logic circuitry, for which it is well suited because of convenient fabrication techniques.

4.9 / FOLLOWER AMPLIFIERS

All of the amplifiers discussed so far have in common the fact that the signal is inverted and that, with appropriate design, voltage gains larger than unity can be obtained. A class of amplifiers, termed *followers*, exists that differs in that the signal is not inverted and the voltage gain is slightly less than unity, ordinarily between 0.95 and 0.99. These amplifiers are useful because they offer a larger input impedance and a much smaller output impedance than inverters constructed with the same amplifying devices. Followers are, accordingly, used as buffer amplifiers, to couple a high-impedance source, which is incapable of furnishing much power, with a low-impedance load, which demands considerable power. This situation is illustrated in Fig. 4.30. Here the signal source is represented by its Thévenin equivalent, e_T–r_T, and the load is represented by r_L. The information to be transferred by the signal is encoded in e_T. The output voltage, to be imposed on an external load, is e_{out}. It differs from the basic signal by the voltage drop in r_T. If r_L is not large as compared with r_T, then e_{out} will be significantly distorted in comparison with e_T. A follower may be placed between the output and the load to avoid causing this distortion.

The term *follower* is used because the output follows the input; that is, the voltage gain is positive and almost unity. The systematic names for these are common collector for transistors, common drain for FETs, and common plate for tubes. More often, the names used are emitter follower, source follower, and cathode follower, respectively. They are illustrated in Fig. 4.31. Further discussion of the emitter follower is given in Chapter 5.

4.10 / SMALL-SIGNAL EQUIVALENT CIRCUITS

In dealing with amplifiers, it is desirable to be able to use an analytical approach to predict the performance of the device without having recourse either to experimentation or to a graphical treatment. To have practical

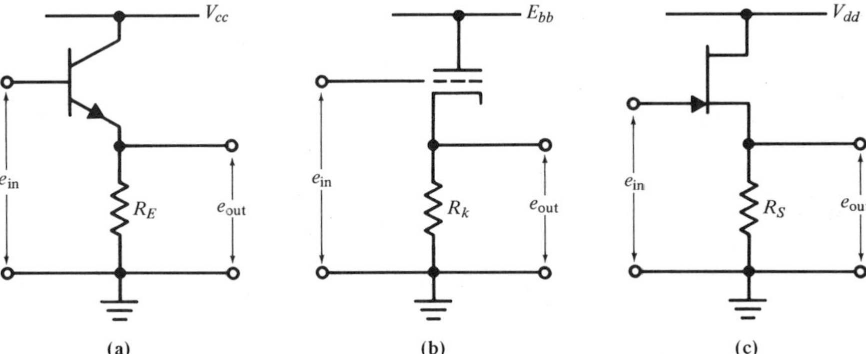

Figure 4.31 / *Follower-type amplifiers.*
(a) *Emitter follower.*
(b) *Cathode follower.*
(c) *Source follower.*

value, such a treatment must be simple and reasonably accurate. It has been pointed out that all types of amplifying devices—FETs, tubes, and transistors—show nonlinear outputs; therefore, sets of equations that accurately describe the input–output relationship necessarily are too complex for this purpose. However, it is true that any curve can be approximated, over a limited range, by a straight line. This is the basis for the small-signal model, which is based on a set of linear equations intended to apply over only a limited range. In this context, the term *small-signal* has the meaning assigned to *AC* in Chapter 1: that component of the whole signal which changes intensity during the period of observation. In addition, the amplitude of the change must be restricted to ensure the validity of the linear approximations.

4.10.1 / Small-signal model for the FET

The defining equations for the small-signal model can be obtained by an entirely general approach. Consider the amplifier represented in Fig. 4.32. Regardless of the type of device or its principle of operation, the signals of interest to the user are those pertaining to the input and to the output, as indicated in Eq. (4.15). This statement is not based on any assumed knowledge about the device; it simply states that I_{out} has *some* defined relationship to the other quantities:

$$I_{out} = f(V_{in}, V_{out}) \tag{4.15}$$

Figure 4.32 / *Generalized block diagram of an amplifier.*

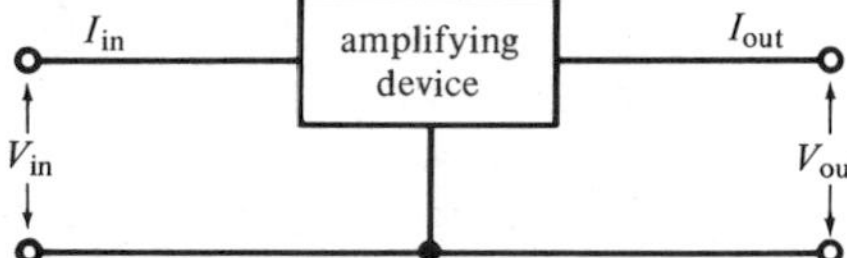

Other, similarly general statements can be made concerning any combination of the four variables given. This one is cited only because it leads to a useful expression.

Equation (4.15) may be restated in the notation appropriate to the FET:

$$i_D = f(v_{GS}, v_{DS}) \tag{4.16}$$

Taking the total differential of i_D:

$$di_D = \left(\frac{\partial i_D}{\partial v_{GS}}\right)_{v_{DS}} dv_{GS} + \left(\frac{\partial i_D}{\partial v_{DS}}\right)_{v_{GS}} dv_{DS} \tag{4.17}$$

This states that the change in drain current, di_D, is equal to the effect on drain current of an adjustment in gate–source voltage $(\partial i_D/\partial v_{GS})$ at constant drain–source voltage, times whatever change in gate–source voltage may occur, plus the effect on drain current of a change in drain–source voltage $(\partial i_D/\partial v_{DS})$ at constant gate–source voltage, times any change in drain–source voltage.

If it were possible to evaluate the coefficients $(\partial i_D/\partial v_{GS})$ and $(\partial i_D/\partial v_{DS})$, then this equation would give a complete description of the behavior of the FET. The nonlinear behavior of the device makes it impractical to do this for the unrestricted case, because the coefficients are not constants and the resulting equation would be cumbersome. Therefore, at this point the assumption regarding linear performance over a restricted range is made by taking the two coefficients as constants. Operationally, di_D does not have its usual mathematical meaning, but is understood to apply over the restricted range in which the coefficients are assumed to be constant. The limits of this range must be determined experimentally. Accordingly, it is desirable to restate Eq. (4.17):

$$\Delta i_D = \left(\frac{\partial i_D}{\partial v_{GS}}\right)_{v_{DS}} \Delta v_{GS} + \left(\frac{\partial i_D}{\partial v_{DS}}\right)_{v_{GS}} \Delta v_{DS} \tag{4.18}$$

In Eq. (4.18), Δi_D is the AC component of drain current and is designated i_d. Similarly, AC values of gate–source voltage and drain–source voltage are designated v_{gs} and v_{ds}, respectively. The coefficients are parameters that describe the behavior of a specific FET. They are obtained by measurement and are furnished to the user by the manufacturer of the device. For convenience, they are given specific names and symbols.

The *transconductance*, g_m, is defined by

$$g_m = \left(\frac{\partial i_D}{\partial v_{GS}}\right)_{v_{DS}} \tag{4.19}$$

It may be seen from the defining equation to have dimensions of conductance, but to involve a transfer of information from the input to the output of the device: hence, *trans*conductance. A restricted version, the transconductance

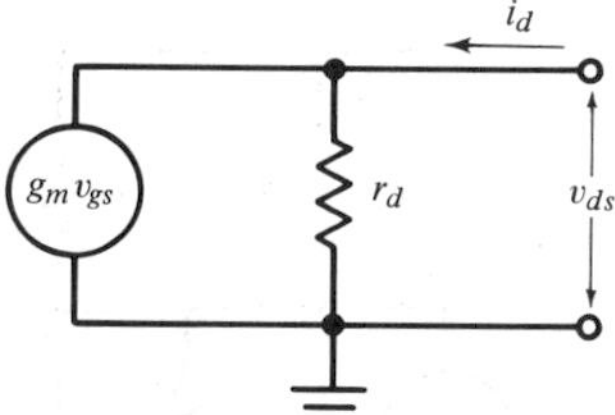

Figure 4.33 / *FET current source small-signal model.*

when $v_{GS} = 0$ and v_{DS} is at least equal to the pinch-off voltage, is identified as the *common-source output admittance* and is given the symbol y_{fs}. This is very often the form in which manufacturers' specifications are given.

The *drain resistance*, r_d, is defined by

$$\frac{1}{r_d} = \left(\frac{\partial i_D}{\partial v_{DS}}\right)_{v_{GS}} \tag{4.20}$$

and has dimensions of ohms. With these definitions, Eq. (4.18) can be restated:

$$i_d = g_m v_{gs} + \frac{v_{ds}}{r_d} \tag{4.21}$$

Equation (4.21) is the fundamental relationship that can be used as the basis for an equivalent circuit of the FET. In formulating a circuit to represent this equation, two currents must be shown in parallel, since they are added together in Eq. (4.21). It is appropriate to represent the first term on the right side of Eq. (4.21) as a current source that is controlled by the input voltage. The second term may be represented by a resistance in parallel with the current source, with i_d the sum of these two currents. The model is shown in Fig. 4.33, in which r_d and $g_m v_{gs}$ can be understood to represent the FET.

4.10.2 / Small-signal equivalent circuit conventions

In conventional form, a small-signal equivalent circuit differs considerably from the schematic diagram of an amplifier. This occurs as indicated in Section 2.1.3, partly because the model contains hypothetical circuit components that represent the attributes of the device, rather than the device itself, and partly because the model simulates only a restricted range of operation. Components that are part of the real device but that do not *directly* affect behavior in the restricted range are omitted. This difference in appearance between the schematic and the equivalent circuit may be responsible for the occurrence of a mental block against the use of the model. It may appear to be too unrealistic.

To consider one aspect of this question, let us formulate a small-signal model for the inverter shown in Fig. 4.34(a). The FET will be represented by

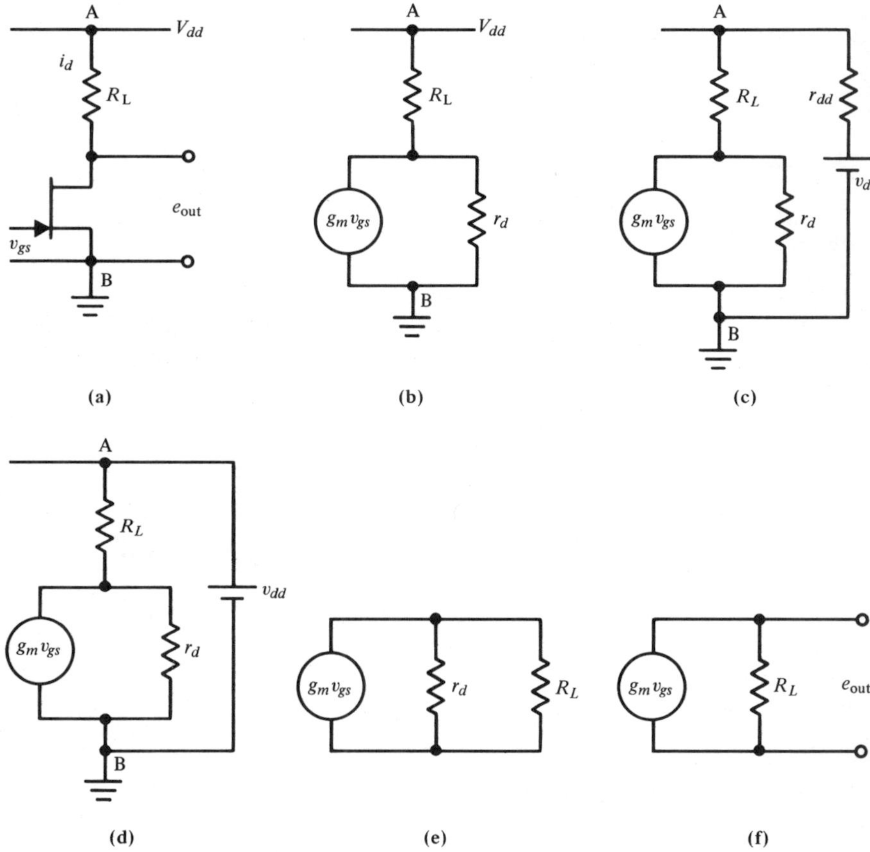

Figure 4.34 / *Formulation of a small-signal equivalent circuit for a FET amplifier.*

the hypothetical current source and resistor, allowing conversion of Fig. 4.34(a) to Fig. 4.34(b). The power supply will, in most cases, not be a part of the small-signal model, and it is important to understand why this is true. The power supply is evidently necessary for the operation of the amplifier. It is one of the variables that determines the location of the operating point; however, once the operating point is established, the power supply has no direct effect on the AC performance.

In Fig. 4.34(c), the supply is represented by its Thévenin equivalent, an ideal voltage source v_{dd}, and the internal resistance, r_{dd}. In the usual case, the power supply resistance is much smaller than other resistances in the circuit and, since it is in series with them, it has negligible effect. It is therefore realistic to eliminate it, as in Fig. 4.34(d). In Fig. 4.34(d), points A and B are held v_{dd} V apart by the power supply, but they are at almost the *same* AC voltage levels because we have assumed that the internal power supply resistance is negligible. If the voltage at one terminal of an ideal DC voltage source is made to fluctuate, then that of the other must undergo identical fluctuations. In this example, point B is grounded and thus prevented from showing a voltage change; accordingly, point A cannot either, and is said to be at *AC ground*. Therefore, we can further simplify the model by eliminating v_{dd} and substituting a short circuit. This is done in Fig. 4.34(e) and has the effect of placing r_d and R_L in parallel.

The behavior of a FET is such that r_d is almost always very large compared with R_L, and accordingly, makes little contribution to the performance of the model. Thus, a final simplification, shown in Fig. 4.34(f), is possible.

This extremely simple circuit may, if the restrictions stated in its formulation are observed, be used to predict two very important aspects of the behavior of the FET amplifier: the voltage gain and the output impedance. The AC component of drain current is given by

$$i_d \simeq -g_m v_{gs} \tag{4.22}$$

where the negative sign indicates signal inversion. Output voltage is given by

$$e_{\text{out}} = i_d R_L = -v_{gs} g_m R_L \tag{4.23}$$

Since v_{gs} is the input, voltage gain is given by

$$\text{voltage gain} = g_m R_L \tag{4.24}$$

The output impedance can be estimated by inspection of Fig. 4.34(f). Looking back into the output terminals, one sees R_L in parallel with the ideal current source, $g_m v_{gs}$. Since an ideal current source exhibits infinite internal resistance, R_L is the effective total value:

$$Z_{\text{out}} = R_L \tag{4.25}$$

It should be noted that in formulating this model, no account has been taken of capacitances that exist within the FET and between its terminals. These are small (in the picofarad range) and accordingly of no concern at low and medium frequencies. This model would, however, show serious inaccuracies if applied to signals with components in the megahertz region. It should be noted that Fig. 4.35 simulates an activated FET, one that has been connected to a power supply and properly biased. It does not describe the isolated FET itself.

Other components besides power supplies may be omitted from small-signal models, because they affect only large-signal behavior. For example, in Fig. 4.35(a), C_{in} serves to block the DC level of the signal source from influencing the gate. C_{in} would normally be large enough to make its reactance negligible at the frequencies of interest. If so, it can be omitted from the small-signal model, as in Fig. 4.35(b).

Resistors R_1 and R_2 of Fig. 4.35(a) bias the gate. The lower end of R_2 is grounded, and the upper end of R_1 will be at AC ground if it is held at a fixed voltage by the power supply, which can happen only if the power

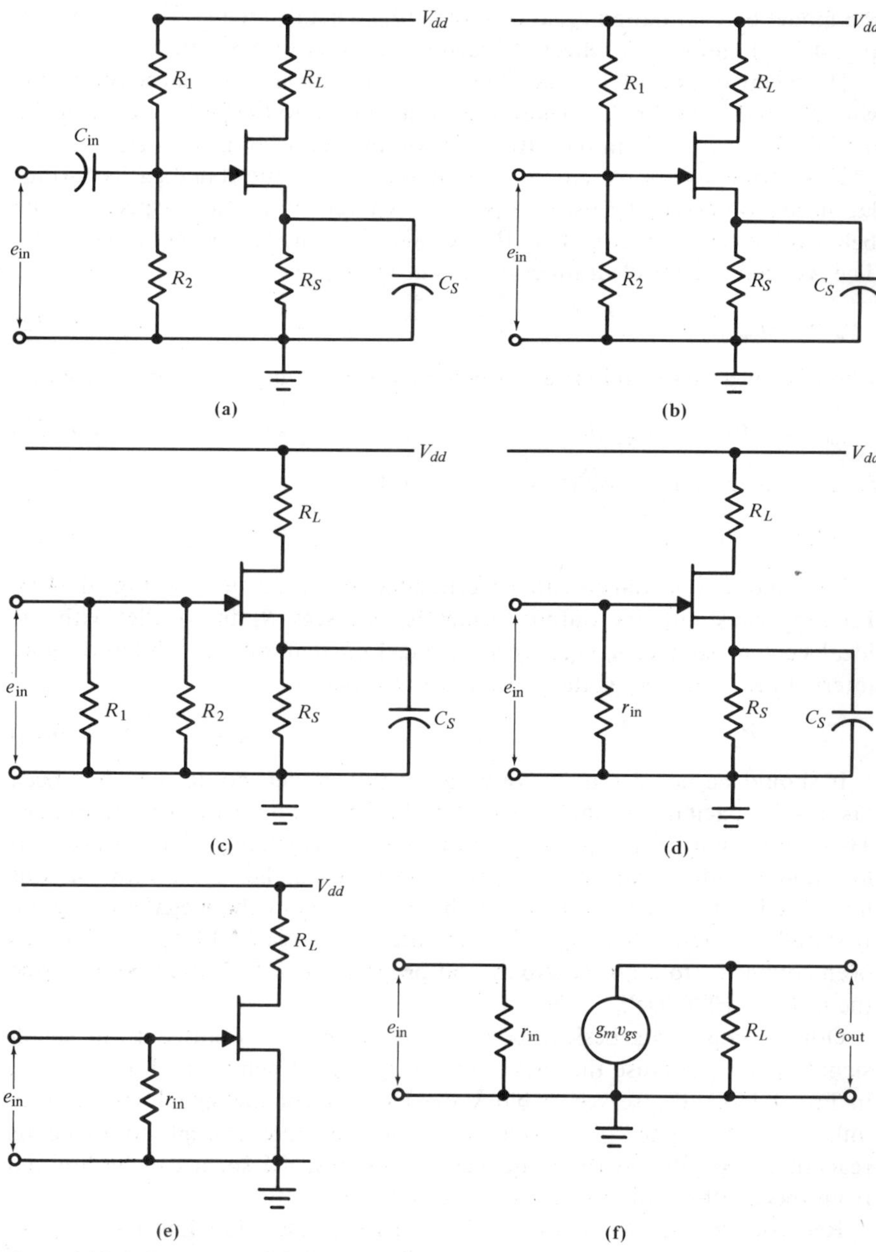

Figure 4.35 / *Conversion of the FET inverter schematic to the small-signal equivalent.*

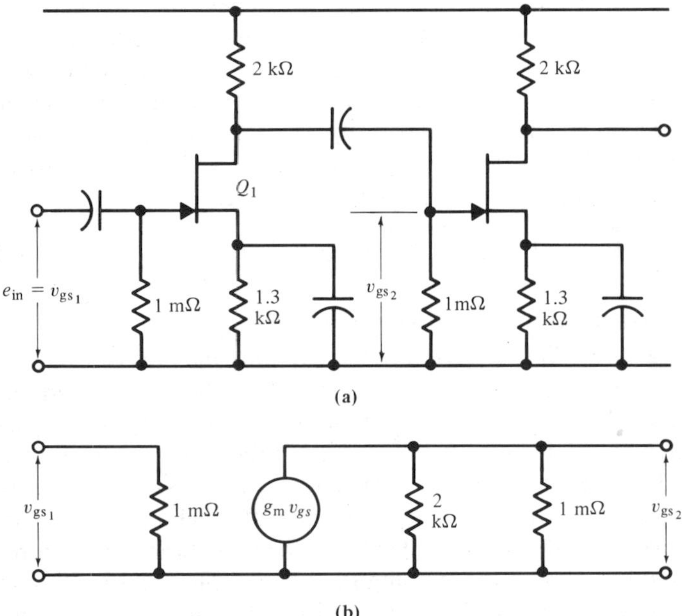

(b)

Figure 4.36 / *Input and output resistances for a FET inverter.*

supply internal resistance is small relative to R_1. Assuming this to be the case, these resistors are effectively connected in parallel to ground, as shown in Fig. 4.35(c). For simplicity, they can be lumped as r_{in} in Fig. 4.35(d). Note that in Fig. 4.35(d), r_{in} is placed in parallel with the gate resistance of the FET. Actually, however, gate resistance values are usually large compared with bias resistances. If so, the bias resistance will determine the input resistance, and this is indicated in Fig. 4.35(f).

It was shown in Section 4.4.4 that a bypass capacitor [C_S in Fig. 4.35(a)] can eliminate the effect of a biasing resistor, R_S, on amplifier gain if the reactance of the capacitor is small at signal frequencies. A correct choice of C_S causes the source to be held at AC ground. The fluctuating component of drain current, i_d, results in only a negligible AC voltage component at the source, even though there is an appreciable DC voltage, $I_D R_S$.

The completed medium-frequency small-signal equivalent circuit for Fig. 4.35(a) is shown in Fig. 4.35(f). It can be seen that the relationships for voltage gain and output impedance that are given in Eqs. (4.24) and (4.25) can be applied to this circuit.

To illustrate the utility of this information, suppose that it is desired to know the small-signal gain of the first stage of the amplifier illustrated in Fig. 4.36(a). Specifically, for a change at e_{in} of 1 V, what value of v_{gs} would be applied to the gate of Q_2? Let us assume that coupling and bypass capacitors have negligible reactance at signal frequency. If so, then Fig. 4.36(a) can be simplified to Fig. 4.36(b) in the same way that Fig. 4.35(a) was converted to Fig. 4.35(f). Looking out from the drain of Q_1, the signal sees the 2-kΩ load resistor of the first stage in parallel with the 1-MΩ gate resistor of the second stage, which amounts very nearly to 2 kΩ. Therefore, the gain may be calculated from Eq. (4.26), if the value of transconductance is known, either from the manufacturers' data or from measurement:

$$\text{voltage gain} = g_m R_L = (4 \times 10^{-3}\ \Omega^{-1})(2 \times 10^{2}\ \Omega) = 8 \qquad (4.26)$$

4.10.3 / Small-signal model for the vacuum tube

The current source model for the FET was derived after an entirely arbitrary choice from Fig. 4.32 of variables for Eq. (4.15). This choice was made in order to arrive at a voltage-controlled current source model that corresponds to the operation of a FET. A suitable vacuum tube model will also be voltage controlled; however, for a vacuum triode, a current source model is not convenient, because the effective resistance of vacuum triodes in operation is generally not large compared with the load resistance. Therefore, the very useful simplification made in going from Fig. 4.34(e) to Fig. 4.34(f) would not be valid. For this reason, instead of using Eq. (4.15) as a starting point, a different set of variables is appropriate:

$$e_b = f(e_c, i_b) \qquad (4.27)$$

Taking the differential of Eq. (4.27) and substituting incrementals for differentials:

$$\Delta e_b = \left(\frac{\partial e_b}{\partial e_c}\right)_{i_b} \Delta e_c + \left(\frac{\partial e_b}{\partial i_b}\right)_{e_c} \Delta i_b \qquad (4.28)$$

This equation provides the definition for the amplification factor, μ, and for the plate resistance, r_p:

$$\mu = -\left(\frac{\partial e_b}{\partial e_c}\right)_{i_b} \qquad (4.29)$$

$$r_p = \left(\frac{\partial e_b}{\partial i_b}\right)_{e_c} \qquad (4.30)$$

When Eqs. (4.29) and (4.30) are substituted into Eq. (4.28) and the incremental variables are replaced by small-signal terms, the relationship in Eq. (4.31) is obtained:

$$e_p = -\mu e_g + i_p r_p \qquad (4.31)$$

Equation (4.31) may serve as the basis for the triode inverter small-signal equivalent circuit. The schematic diagram and the small-signal model are

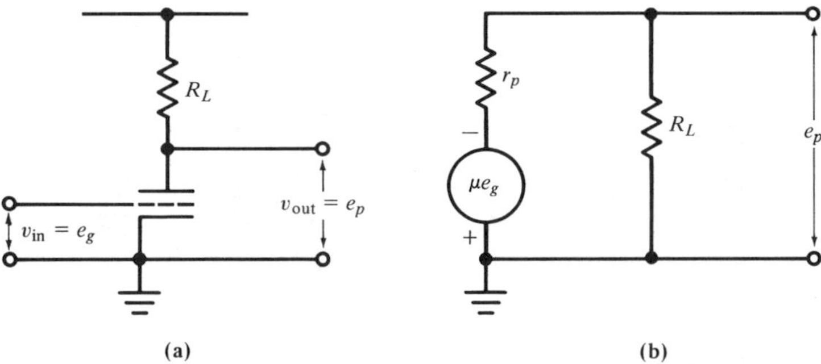

Figure 4.37 / *Vacuum triode inverter and corresponding small-signal model.*

shown in Fig. 4.37. From this, expressions for voltage gain and output impedance may be obtained:

$$\text{voltage gain} = \frac{V_{\text{out}}}{V_{\text{in}}} = \frac{\mu R_L}{R_L + r_p} \tag{4.32}$$

$$\text{output impedance} = \frac{r_p R_L}{R_L + r_p} \tag{4.33}$$

The small-signal models for the FET and vacuum tube are examples of the dual relationship between current and voltage dividers mentioned in Section 1.2.2.

4.10.4 / Small-signal model for the bipolar transistor

The transistor is a current-controlled device which in operation offers an output resistance that is generally large when compared with the load resistance. Accordingly, the small-signal model of choice is a current-controlled current source.

The bipolar transistor differs from the FET and the vacuum tube in that it shows bidirectional behavior to an appreciable extent. With either a FET or a tube, a change in the input signal causes a predictable change in the output signal; however, perturbations in the output circuitry have little effect on the input signal and can ordinarily be neglected. The bipolar transistor provides much less effective insulation of the input circuit from perturbations in the output, and the equivalent circuit must take this into account.

Accordingly, a small-signal transistor model is derived from two initial relationships:

$$I_{\text{out}} = f(I_{\text{in}}, V_{\text{out}}) \tag{4.34}$$

$$V_{\text{in}} = f(I_{\text{in}}, V_{\text{out}}) \tag{4.35}$$

From these relationships, two equations may be obtained, using the same general procedure that was described for the FET model. From Eq. (4.34)

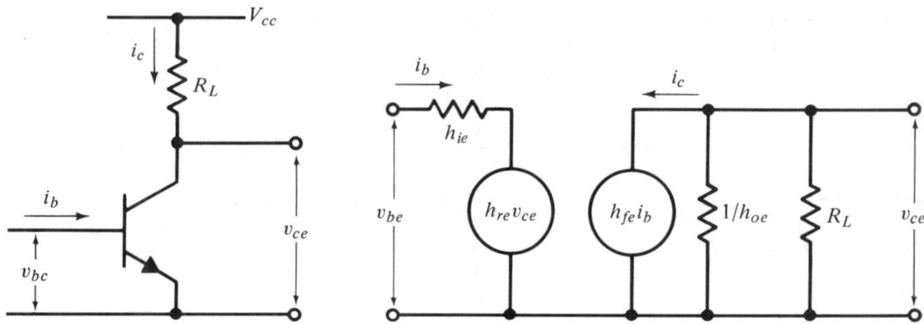

Figure 4.38 / *Common-emitter amplifier.*
(a) *Schematic diagram.*
(b) *Small-signal equivalent circuit.*

one gets Eq. (4.36), which is expressed in the notation appropriate to the common-emitter configuration of Fig. 4.38(a):

$$i_c = \left(\frac{\partial i_C}{\partial i_B}\right)_{v_{CE}} i_b + \left(\frac{\partial i_C}{\partial v_{CE}}\right)_{i_B} v_{ce} \tag{4.36}$$

From Eq. (4.35), Eq. (4.37) may be derived:

$$v_{be} = \left(\frac{\partial v_{BE}}{\partial i_B}\right)_{v_{CE}} i_b + \left(\frac{\partial v_{BE}}{\partial v_{CE}}\right)_{i_B} v_{ce} \tag{4.37}$$

As before, the coefficients serve as the basis for defining parameters that are assumed to be constant under the small-signal approximation.

$$\left(\frac{\partial i_C}{\partial i_B}\right)_{v_{CE}} = h_{fe} \quad \text{forward current transfer ratio} \tag{4.38}$$

$$\left(\frac{\partial i_C}{\partial v_{CE}}\right)_{i_B} = h_{oe} \quad \text{output conductance} \tag{4.39}$$

$$\left(\frac{\partial v_{BE}}{\partial i_B}\right)_{v_{CE}} = h_{ie} \quad \text{input resistance} \tag{4.40}$$

$$\left(\frac{\partial v_{BE}}{\partial v_{CE}}\right)_{i_B} = h_{re} \quad \text{reverse voltage transfer ratio} \tag{4.41}$$

Substituting Eqs. (4.38)–(4.41) into Eq. (4.36) and Eq. (4.37) gives

$$v_{be} = h_{ie}i_b + h_{re}v_{ce} \tag{4.42}$$

$$i_c = h_{fe}i_b + h_{oe}v_{ce} \tag{4.43}$$

Equation (4.42) equates v_{be}, the input voltage in Fig. 4.38(a), with terms that contain h_{ie} and h_{re}. From the defining relationship for h_{ie}, Eq. (4.40), one can see that it has dimensions of resistance and, furthermore, that it is defined as the ratio of input voltage to input current; it therefore amounts to input resistance. In Eq. (4.41), h_{re} can be seen to be dimensionless, because it is the ratio of input to output voltage. Therefore, in Eq. (4.42), input voltage is set equal to input current times input resistance plus output voltage times the reverse voltage transfer. It is reasonable to draw the input portion of the small-signal model as shown in Fig. 4.38(b), a hypothetical resistance in series with an ideal voltage source. An input signal at the base of a transistor encounters behavior similar to that which would be caused by an internal resistance (h_{ie}) and an internal battery ($h_{re}v_{ce}$), with the battery voltage determined by the *output* voltage. This is in accord with the observed fact that changes at the output reflect back on the input.

In Eq. (4.43), the output current, i_c, is equated with terms that contain h_{fe} and h_{oe}. Equation (4.38) defines h_{fe} as the ratio between output and input currents. The fact that a transistor is a current amplifier makes this the most significant parameter. We shall see below that it primarily determines both current and voltage gain. Equation (4.39) defines h_{oe} as the ratio of output current to output voltage, which makes it the output conductance. The first term on the right side of Eq. (4.43), $h_{fe}i_b$, must be symbolized by an ideal current source controlled by the magnitude of the input current. It was pointed out in Section 2.3.4 that useful models are obtained if internal resistance is placed in series with a voltage source or in parallel with a current source. This derivation of parameters has followed from the experimental observation that the behavior of an activated transistor resembles that of an ideal current source controlled by input current. Therefore, the reciprocal of h_{oe} is shown in parallel with $h_{fe}i_b$ in the output side of Fig. 4.38(b).

The transistor model shown in Fig. 4.38(b), known as the *hybrid model*, is subject to the same types of restrictions that pertain to the FET and vacuum tube models described above. It describes, as an approximation, the behavior of a transistor that is connected with a power supply and a suitable bias arrangement. It permits a rather simple estimation of the *incremental* output change resulting from an *incremental* input change. It does not describe the behavior of an isolated transistor, and it does not imply *anything* about the mechanism of transistor physics. There are no actual current or voltage sources located within the transistor; this is an entirely empirical, although very useful, model. Other transistor models are used. This one was chosen for illustration because it is most often used by semiconductor manufacturers in specifying transistors for sale.

In using the hybrid model, some further simplifications over Fig. 4.38(b) may often be made. The voltage simulated by the source $h_{re}v_{ce}$ is generally quite constant, having values of about 0.6 V for silicon and 0.2 V for germanium devices. These are, in fact, sufficiently reliable to serve as practical diagnostic criteria for identification of transistors and diodes in operating circuits. Therefore, we may usefully visualize the input to a

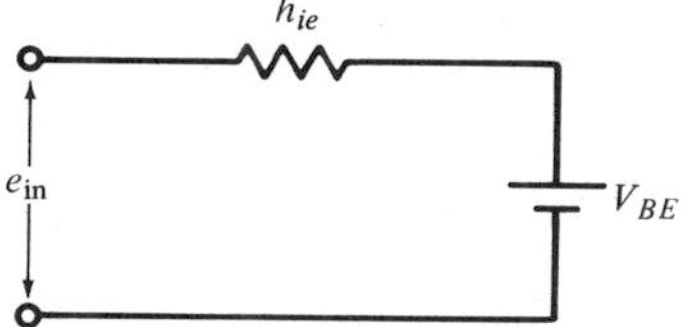

Figure 4.39 / *Representation of the input to a common-emitter transistor.*

common-emitter transistor as it is shown in Fig. 4.39. The input resistance, h_{ie}, is a function of the class of transistor being used and may vary from a few hundred to a few thousand ohms. The input voltage may, as a first approximation, be taken as a constant determined by the type of semiconductor material being used, plus the voltage drop across h_{ie}. The total (large-signal) input voltage will not, in fact, be very different from the fixed value noted above. Therefore, if the internal voltage source is considered to be constant, then it may be omitted from the small-signal model, as shown in Fig. 4.40. The h_{oe} parameter generally shows a reciprocal value in the range of tens to hundreds of kilohms, which is large compared to the usual values of R_L. Accordingly, the output portion of the model can be simplified as shown in Fig. 4.40.

4.10.5 / Estimation of common-emitter small-signal performance

To provide some insight into the use of the transistor small-signal model, consider the problem of calculating the voltage and current gain of the single-stage amplifier in Fig. 4.41(a). For this purpose, current gain will be understood to be i_{out}/i_{in} and voltage gain will be e_{out}/e_{in}, all AC components of the signals identified on Fig. 4.41(a). We shall assume that the bypass and coupling capacitors exhibit negligible reactance, which permits us to eliminate C_{in}, C_{out}, and the $R_E C_E$ network from consideration in this problem. The simplified circuit is shown in Fig. 4.41(b).

This circuit may be treated in the same manner as that of Fig. 4.35(a); the power supply may be considered to hold its positive terminal at AC ground. This permits the further simplification of Fig. 4.41(b) to Fig. 4.41(c), after which the transistor may be replaced by its small-signal model, as indicated in Fig. 4.41(d). In this, the two bias resistors are replaced by their parallel

Figure 4.40 / *Simplified hybrid model for a common-emitter transistor.*

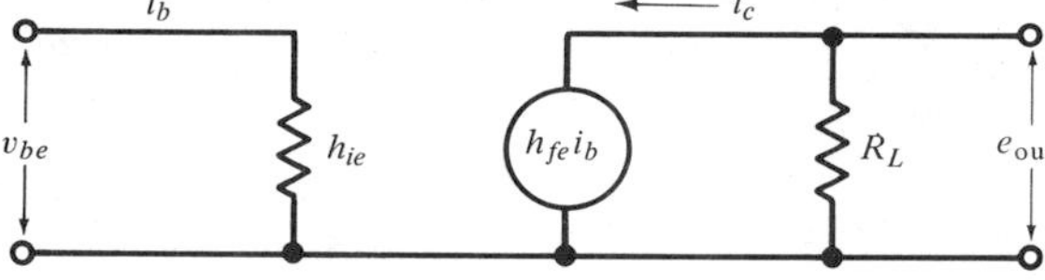

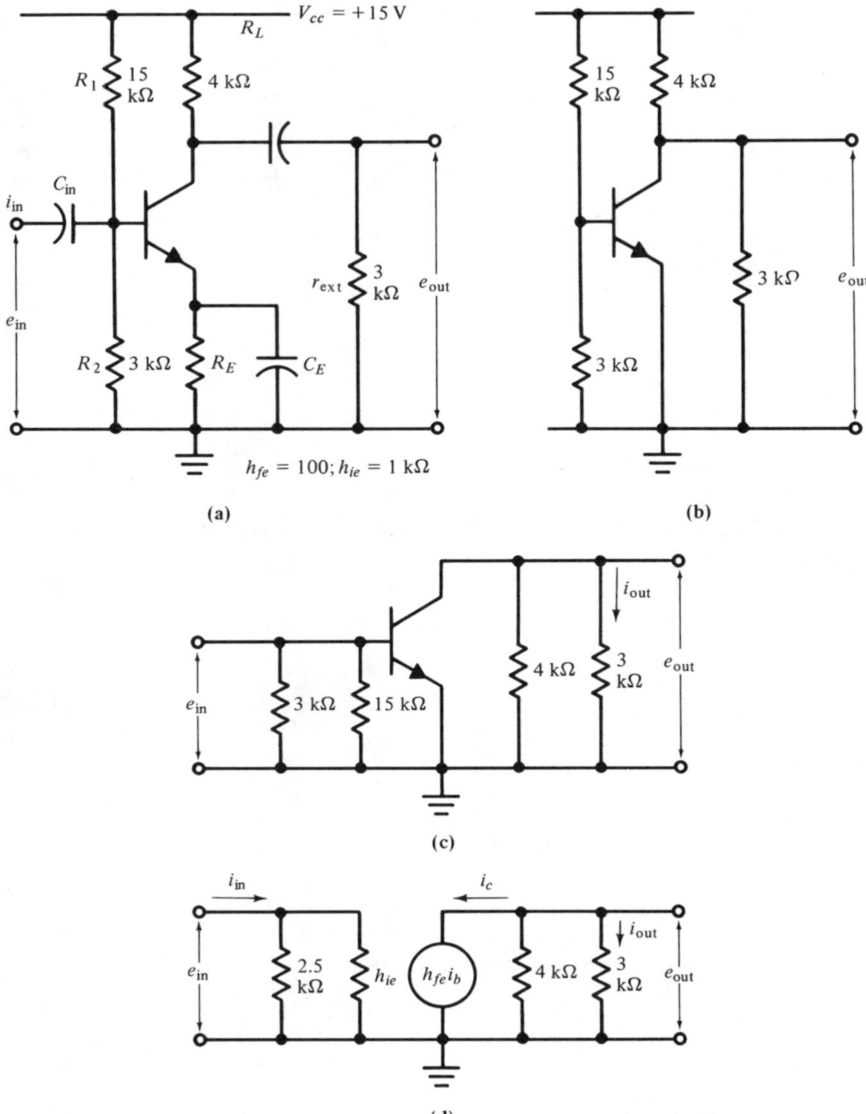

Figure 4.41 / *Single-stage common-emitter amplifier, schematic and equivalent circuits.*

equivalent, 2.5 kΩ. To calculate current gain, i_{out} must be related to i_{in}. At the input, i_{in} divides between the biasing resistors and the transistor base, with only base current being amplified:

$$i_b = i_{in}\left(\frac{2.5\ \text{k}\Omega}{3.5\ \text{k}\Omega}\right) = 0.71 i_{in} \tag{4.44}$$

Small-signal collector and base currents are related as indicated by Eq. (4.38). Combining it with Eq. (4.44) relates i_c to i_{in}:

$$i_c = i_b h_{fe} = 100 i_b = 71 i_{\text{in}} \tag{4.45}$$

Only that fraction of i_c drawn by the external load is useful. This can be related to i_{in} by combining the current-divider relationship for the two resistors in the output with Eq. (4.45):

$$i_{\text{out}} = \left(\frac{4000}{7000}\right) i_c = 41 i_{\text{in}} \tag{4.46}$$

and

$$\text{current gain} = \frac{i_{\text{out}}}{i_{\text{in}}} = 41 \tag{4.47}$$

It can be seen that the single most important factor in determining current gain is h_{fe}, which may vary in the range from 10 to 1000.

Having calculated current gain, the voltage gain is readily obtained. The input voltage is given by

$$e_{\text{in}} = i_{\text{in}} r_{\text{in}} = i_{\text{in}} \left[\frac{(2.5\ \text{k}\Omega)(1.0\ \text{k}\Omega)}{3.5\ \text{k}\Omega}\right] = 710(i_{\text{in}})\ \text{V} \tag{4.48}$$

and the output voltage by

$$e_{\text{out}} = i_{\text{out}}(3\ \text{k}\Omega) = (41)(i_{\text{in}})(3 \times 10^3\ \text{V}) = 1.23 \times 10^5 (i_{\text{in}})\ \text{V} \tag{4.49}$$

Therefore,

$$\text{voltage gain} = \frac{e_{\text{out}}}{e_{\text{in}}} = \frac{1.23 \times 10^5 (i_{\text{in}})}{7.10 \times 10^2 (i_{\text{in}})} = 173 \tag{4.50}$$

As is true of current gain, h_{fe} is the single most important factor determining the size of the voltage gain.

The small-signal model also makes it possible to estimate the input and output resistances in order to evaluate the interactions that may occur between a stage of amplification and a signal source or a receiver. Suppose that the common emitter in Fig. 4.42(a) is considered. If ideal behavior by the bypass and coupling capacitors is assumed, the equivalent circuit in Fig. 4.42(b) may be drawn. In it, the bleeder resistors are shown in parallel with the transistor input resistance, h_{ie}, because small-signal behavior is involved. Therefore, the input resistance will be the parallel equivalent of the two bleeder resistors and h_{ie}, shown as r_{in} in Fig. 4.42(c). This permits an estimation of the effect of this amplifier on a signal source, as discussed in Section 7.1.1. The output resistance of this amplifier is simply the load resistance, as was noted in discussing Fig. 4.40. If the input resistance of a signal receiver is known, it will be possible to estimate the effect of signal transfer to it.

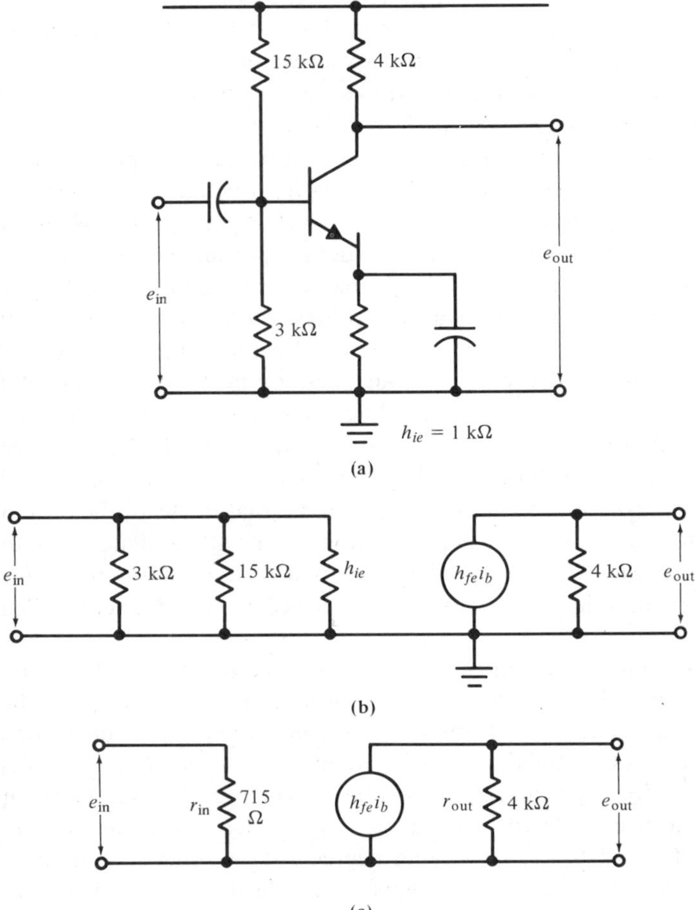

Figure 4.42 / *Single-stage common-emitter amplifier.*

4.11 / AMPLIFIERS

The discussion so far has dealt with the operation of individual amplifying devices, that is, of single *stages* of amplification; to provide adequate voltage and power gain, however, amplifiers commonly consist of more than one stage. As an approximation, the voltage gain of an amplifier that consists of n similar stages is equal to the gain of a single stage raised to the power n. Often three stages are used because this constitutes an optimum design with available amplifiers, using a larger number of stages being somewhat analogous to using too large an engine for the size of the body of an automobile. An odd number of stages is generally used because the output is nominally 180° out of phase with the input, assuming that each stage inverts the signal. This avoids the difficulty that would be caused by inadvertant feedback of the output signal to the input. Since the output may be orders of magnitude larger than the input, this may occur to a significant extent because of conduction by stray capacitive or inductive paths. If output is in phase with the input, this will cause the amplifier to go out of control. This point is discussed in detail in Chapter 5.

It is evident that voltage and power gain constitute vital performance characteristics of an amplifier. In addition, however, one must be concerned with reliability, linearity, stability, offset, noise level, and drift. By *reliability* is meant the ability of an amplifier to provide a reproducible output for a given input over an extended period of time without adjustment by the operator. *Linearity* refers to the relationship between input and output. It is generally desirable that a plot of output signal against input signal level show a linear relationship that passes through zero on both axes. If this plot does not go through the origin, the amplifier is said to exhibit *offset*. To provide accurate performance, an amplifier must be reliable, linear, and free from offset. An amplifier will necessarily provide an output that is controlled by the input. If it responds to internally generated signals instead, the output will either go to a positive or a negative limit or it will oscillate without control by the input. The term *stability* is used, not as a synonym for reliability, but to describe the ability of the device to avoid uncontrolled oscillations.

Noise is a spurious AC signal that is superimposed on the amplified signal; the noise signal ordinarily has components at all frequencies within the bandpass of the amplifier. Very low-frequency noise may be termed *drift*.

The performance of amplifying devices, when used in the simple circuits described above, does not make it possible to achieve the standards of accuracy that are considered acceptable for modern instrumentation. Transistors, FETs, and tubes, in varying degrees, show inherent nonlinear behavior. They, and the other components used with them, are sensitive to environmental changes, especially of temperature, and are affected by aging. To achieve acceptable levels of performance, it is necessary either to make use of the technique of negative feedback in amplifier design or to arrange that the performance of the instrument as a whole is not critically dependent on individual amplifier behavior. Further discussion of these important points is presented in Chapter 5. A discussion of the subject of noise in signals is presented in Chapter 6. The problems of offset and drift are discussed in Section 4.12.

4.12 / AC AND DC AMPLIFIERS

4.12.1 / AC amplifiers

An AC amplifier is one that is capable of responding only to the AC components of an input signal. For example, such an amplifier would produce the same output in response to either signal 1 or signal 3 of Fig. 1.13 or 1.14. By contrast, a DC amplifier produces an output that is a

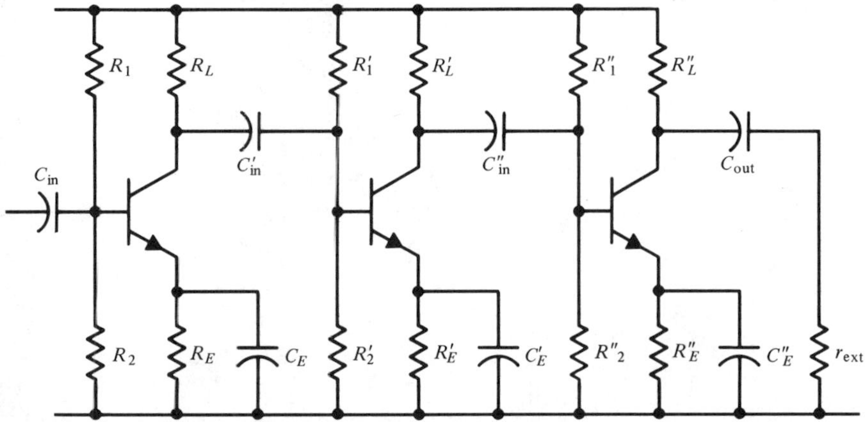

Figure 4.43 / *Three-stage RC-coupled common-emitter amplifier.*

function of the DC level of the input as well. A DC amplifier generally responds both to the DC level of the input and to at least a limited range of AC frequency components. The major distinction in terms of design between DC and AC amplifiers is in the biasing and interstage coupling that is used. DC amplifiers utilize resistive interstage coupling, which does not show frequency-dependent behavior. AC amplifiers are built with capacitor or transformer coupling which, as noted in Sections 3.3.2 and 1.4.9, respond only to AC signal components. The schematic diagram for a typical capacitor-coupled AC amplifier is presented in Fig. 4.43.

The relationship between gain and signal frequency for this type of amplifier is presented, somewhat idealized, as curve 1 in Fig. 4.44. Gain is highest in the midfrequency range (ω_1 to ω_2) and rolls off at both high and low frequencies. The midfrequency gain is determined primarily by the characteristics of the amplifying devices used. The gain in the low-frequency rolloff region (below ω_1) is primarily a function of the time constants for the

Figure 4.44 / *Bode diagrams for amplifiers. Curve 1: AC coupled; curve 2: DC coupled.*

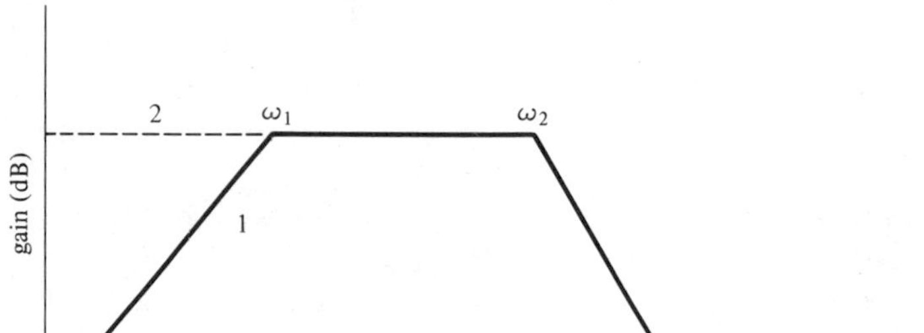

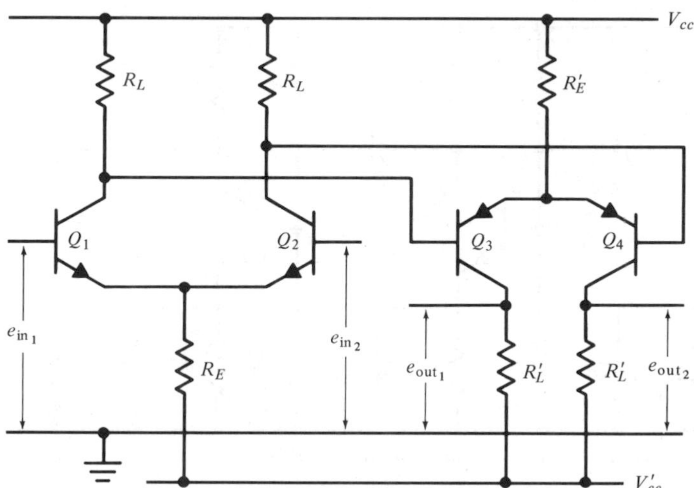

Figure 4.45 / *Two-stage emitter-coupled difference amplifier.*

high-pass interstage coupling capacitors and associated resistances—for example, C_{in}, R_1, R_2, and the transistor input for the first stage in Fig. 4.43. The behavior of high-pass filters is discussed in Section 3.3.2.

In the frequency range above ω_2, gain rolls off as would be expected for a low-pass filter, the behavior of which is discussed in Section 3.3.1. This filter may actually not be one deliberately included by the designer, but may be the result of unavoidable distributed capacitances associated with wires and other components. In many cases, however, low-pass filters are included deliberately in order to restrict the range of response.

4.12.2 / DC amplifiers

The signals produced in many chemical applications are in too low a frequency range to be handled by simple AC-coupled amplifiers; hence, DC-coupled amplifiers are widely used. The fact that they can respond to low-frequency signals makes them very susceptible to inaccuracies caused by drift. The difference amplifier, which produces an output as a difference between the responses of two amplifying devices operating in parallel, is relatively unaffected by perturbations that cause drift in DC-coupled circuits. Accordingly, it is very often found in instrument amplifiers for applications such as pH meters, recorders, and oscilloscopes and is chosen as an example of a DC amplifier for discussion here.

A two-stage emitter-coupled difference amplifier is illustrated in Fig. 4.45. Each stage consists of two transistors, Q_1–Q_2 and Q_3–Q_4; all components are directly connected to permit response to DC signals. A qualitative understanding of the operation of a difference amplifier can be obtained by first examining the behavior of an ideal amplifier and then considering some of the most important deviations from ideality exhibited by real amplifiers.

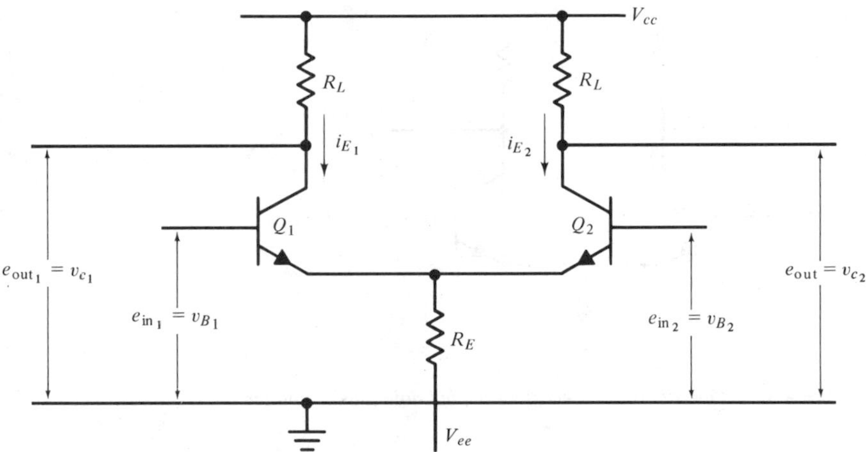

Figure 4.46 / *Single-stage difference amplifier.*

The ideal difference amplifier (Fig. 4.46) is constructed with two transistors that are identical in every respect. They are connected in parallel between two DC power supply levels. In operation, the current drawn by R_E is the sum of the emitter currents for both transistors. To ensure the desired operation of a difference amplifier, it is necessary that this sum remain constant. If i_{E_1} increases, then i_{E_2} must decrease by the same amount. This causes the output to be proportional to the difference between the two inputs.

To understand how coupling the emitters makes this happen, suppose that the amplifier is in operation and e_{in_1} increases, thereby increasing v_{B_1} and v_{BE_1}, which causes an increase in i_{B_1}, and hence in i_{E_1}. An increase in i_{E_1} results in increased voltage drop across R_E, which raises the voltage level of the emitters. If e_{in_2} is constant, increasing v_{E_2} amounts to *decreasing* v_{BE_2}, which partly cuts off Q_2 (reduces i_{B_2} and i_{E_2}). The result is that i_{E_2} decreases by exactly the same amount as i_{E_1} increases (in an ideal amplifier). Qualitatively, we can see (Fig. 4.47) that application of a positive-going signal to the base of Q_1 causes a negative change of v_{C_1} and a positive change of v_{C_2}. The system is symmetrical, so an analogous statement can be made about the other input. Therefore, the output is proportional to the *difference* between the input voltage levels.

Output may be taken between either collector and ground, in which case the output is said to be *single ended.* However, output may be taken between the two collectors, in which case it is said to be *floating*. Similarly, the input may be a signal referenced to ground and imposed on one amplifier input while the other is grounded. This is *single-ended input.* It is also possible to impose the two ends of a signal that is not referenced to ground on the two inputs of the difference amplifier. This situation, which has very important practical applications, is termed a *floating input.*

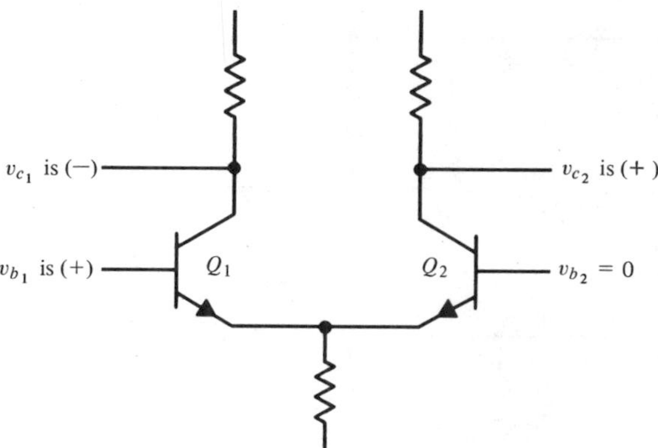

Figure 4.47 / *Operation of an emitter-coupled difference amplifier.*

To illustrate, suppose that it is desired to amplify the voltage V_{AB} of Fig. 4.48. A conventional amplifier connected at either point A or point B would see either V_{AC} or V_{BC}. A difference amplifier could be connected to points A and B and would amplify V_{AB}, a signal floating on V_{BC}.

The voltage V_{AB} is termed the *difference mode* input. The voltage V_{BC}, which would be applied to both inputs simultaneously, is the *common mode* input. The response by the amplifier to V_{AB} is its *difference gain*; the response to V_{BC} is its *common mode gain.* An ideal difference amplifier shows a finite difference gain, but it shows a zero common mode gain. Real difference amplifiers show common mode gains that are finite but much smaller than the difference gain. The ability to discriminate between common mode and difference signals is a sufficiently important characteristic to have led to the establishment of a convention for expressing it. This is the *common mode rejection ratio*, defined as the ratio of difference gain to common mode gain.

Figure 4.48 / *A floating signal: V_{AB}.*

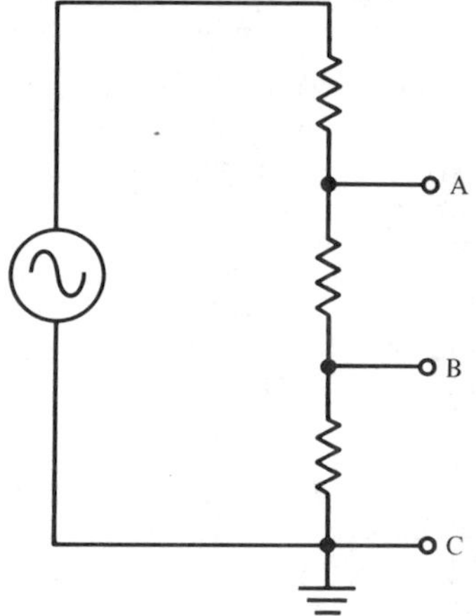

Amplifiers exhibiting common mode rejection ratios in excess of 10^4 are available.

4.12.3 / Comparison of AC and DC amplifiers

It is much easier to achieve a high level of performance with respect to drift and offset with AC amplifiers than it is with DC amplifiers. This is true because the reactive coupling components used in AC amplifiers do not transmit slow voltage changes that constitute drift from one stage to the next. By contrast, drift in one stage of a DC amplifier is amplified by each successive stage, just as is the desired signal. Similarly, the interstage coupling in AC amplifiers is incapable of transmitting any average DC level other than zero, so offset does not occur in the output.

Where circumstances permit, therefore, AC amplification is preferred to DC. However, there are many applications involving constant or very slowly changing signals that cannot be handled by AC amplifiers. It is accordingly necessary to provide high-quality DC amplification. Two different approaches are used. One involves taking sufficient care with all factors to provide satisfactory performance from DC amplifiers. The other involves the conversion of the DC signal to an AC signal, amplification of the AC signal, and then reconversion to DC. This is referred to as *chopper stabilization*. Both techniques are widely used; both can give quite good results. A discussion of chopper stabilization is presented in Section 8.2.5.

4.13 / SYMBOLS USED TO DESCRIBE AMPLIFIERS

4.13.1 / Vacuum tube notation

I_b, E_b = Q-point or average plate current and voltage
E_c = Q-point or average grid voltage
i_b, e_b = instantaneous plate current and voltage
e_c = instantaneous grid voltage
i_p, e_p = small-signal plate current and voltage
e_g = small-signal grid voltage
E_{bb} = plate power supply voltage
μ = amplification factor
g_m = transconductance

4.13.2 / FET notation

I_D = Q-point or average drain–source current
V_{DS}, V_{GS} = Q-point or average drain–source and gate–source voltage
V_D, V_G, V_S = Q-point or average drain, gate, or source voltage relative to ground
i_D = instantaneous drain–source current
v_{DS}, v_{GS} = instantaneous drain–source and gate–source voltage
v_D, v_G, v_S = instantaneous drain, gate, or source voltage relative to ground
i_d = small-signal drain–source current
v_{ds}, v_{gs} = small-signal drain–source and gate–source voltage
v_d, v_g, v_s = small-signal drain, gate, or source voltage relative to ground
g_m = transconductance
y_{fs} = small-signal common-source forward transfer admittance
V_P = pinch-off voltage

4.13.3 / Bipolar transistor notation

I_E, I_B, I_C = Q-point or average emitter, base, or collector current
V_{CE}, V_{BE} = Q-point or average collector–emitter or base–emitter voltage
V_E, V_B, V_C = Q-point or average emitter, base, or collector voltage relative to ground
i_E, i_B, i_C = instantaneous emitter, base, or collector current
v_{CE}, v_{BE} = instantaneous collector–emitter or base–emitter voltage
v_E, v_B, v_C = instantaneous emitter, base, or collector voltage relative to ground
i_e, i_b, i_c = small-signal emitter, base, or collector currents
v_{ce}, v_{be} = small-signal collector–emitter or base–emitter voltage
v_e, v_b, v_c = small-signal emitter, base, or collector voltage relative to ground
h_{fe} = β or small-signal common-emitter forward current transfer ratio
h_{ie} = small-signal common-emitter input resistance
h_{oe} = small-signal common-emitter output conductance
h_{re} = small-signal common-emitter reverse voltage transfer ratio

4.13.4 / Component notation

V_{dd} = FET drain power supply voltage
V_{cc} = transistor collector power supply voltage
E_{bb} = vacuum tube plate supply voltage

PROBLEMS

For each of the following, calculate the current drawn by the diode and the voltage at point A. Assume ideal voltage sources and ideal diodes (zero resistance if forward biased and infinite resistance if reverse biased).

1 /

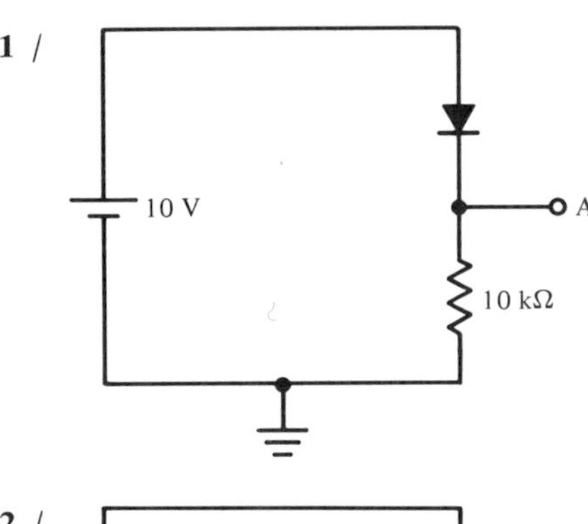

2 /

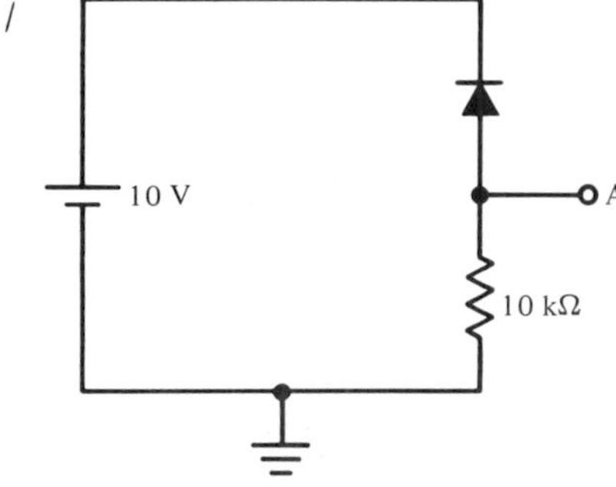

3 /

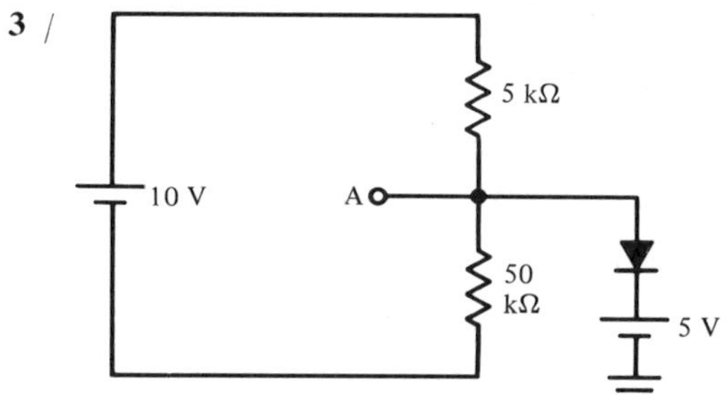
5 kΩ
10 V
A
50
kΩ
5 V

4 /

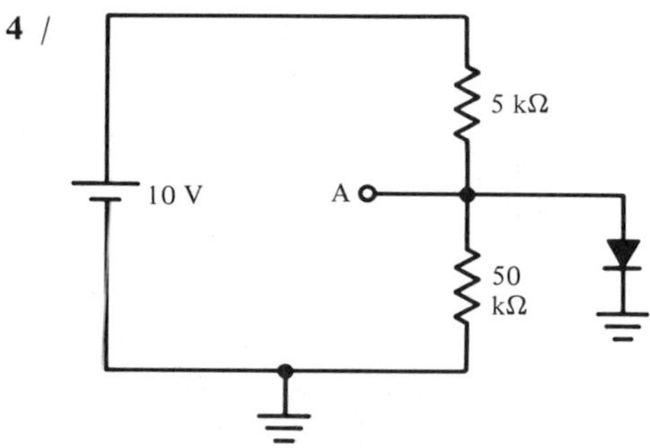
5 kΩ
10 V
A
50
kΩ

5 /

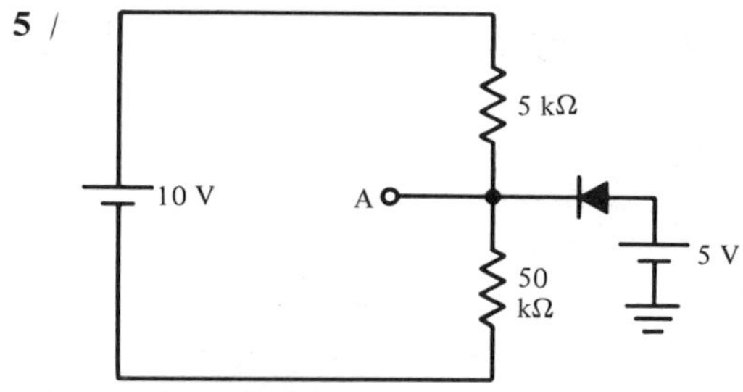
5 kΩ
10 V
A
50
kΩ
5 V

6 /

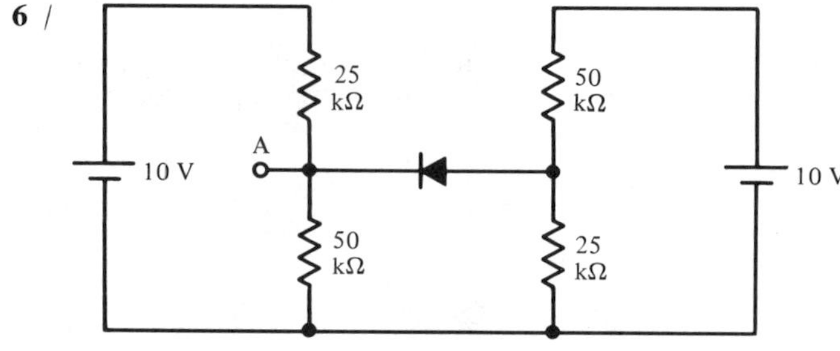
25
kΩ
50
kΩ
A
10 V
10 V
50
kΩ
25
kΩ

7 /

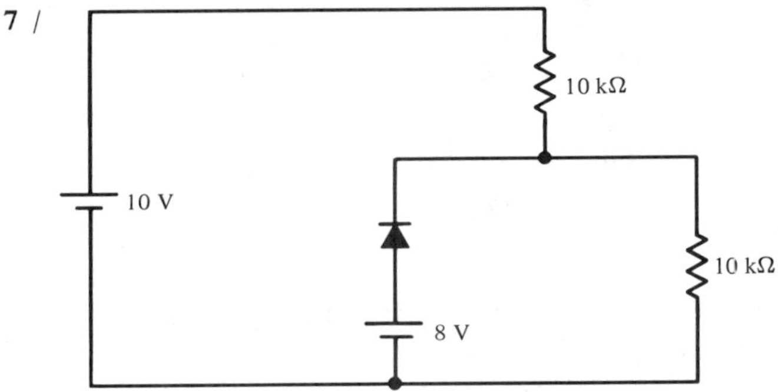

Figure 4.49

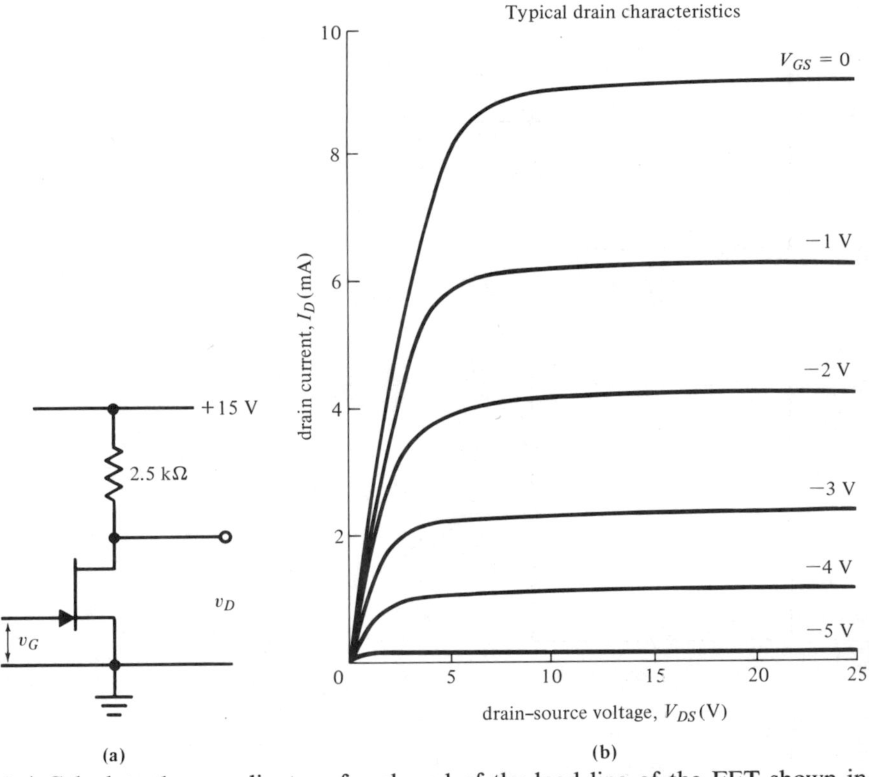

8 / Calculate the coordinates of each end of the load line of the FET shown in Fig. 4.49(a). If the characteristic in Fig. 4.49(b) is valid, what would be the values of i_D and v_D for the following values of v_G?

(*a*) −1.0 V
(*b*) −2.0 V
(*c*) −3.0 V

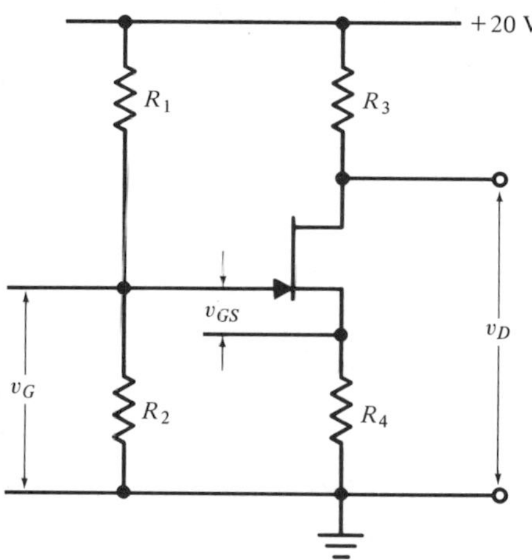

Figure 4.50

In Fig. 4.50, the resistor values are as follows:

Problem	R_1 ($k\Omega$)	R_2 ($k\Omega$)	R_3 ($k\Omega$)	R_4 ($k\Omega$)
9	9	1	1.5	1.0
10	9	1	1.0	1.5
11	9	1	0.5	2.0

9–11 / (*a*) Calculate the value of v_G.
(*b*) Assuming each of the following values for i_D, calculate the corresponding values for v_{GS}: 1.0 mA; 2.0 mA; 3.0 mA; 4.0 mA; 5.0 mA.
(*c*) After locating the load line corresponding to the values of R_3 and R_4 specified for each problem on Fig. 4.49(*b*), estimate the approximate equilibrium values of V_{GS}, V_D, and I_D.

12 / If the transistor in Fig. 4.51 is made of silicon, estimate the values of v_E, v_{CE}, and i_E.

Figure 4.51

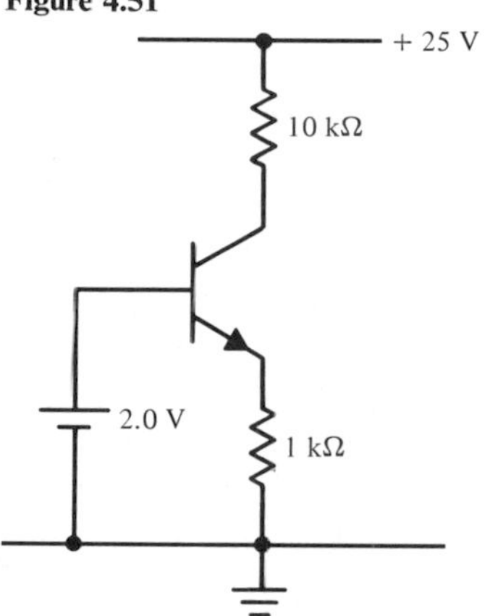

chapter 5 / Feedback and amplifier performance

chapter 5

5.1 / POSITIVE AND NEGATIVE FEEDBACK

It is in certain cases desirable to mix an amplifier output signal with the input signal, that is, to introduce *feedback*. If the feedback arrangement causes an increase in the magnitude of the net input signal, it constitutes positive feedback. Positive feedback causes an increase in the effective amplifier gain, which is often desirable, but it also causes instability. A system that is subject to positive feedback may go to a limit or may oscillate uncontrollably. A familiar example is the response of public address systems when the microphone is positioned to pick up the signal from the loudspeaker. This signal is amplified and broadcast, picked up, amplified and broadcast, and so on. The usual result is a noise of rapidly increasing intensity. Because of this type of difficulty, positive feedback is used only with care to increase the gain of an amplifier.

By contrast, negative feedback, which involves mixing a portion of the output in such a way as to cancel part of the input, causes a decrease in the useful gain. In addition to that uninteresting effect, introducing negative feedback also makes it possible to design amplifiers that are stable, linear, free from offset, and that show reduced sensitivity toward changing environmental conditions and toward variation in component characteristics. The use of negative feedback is essential to the design of mass-produced amplifiers. Reduction in gain is the price that must be paid to obtain the other desirable characteristics.

5.1.1 / Feedback and gain

A generalized statement that identifies the important factors involved in using feedback can be obtained from Fig. 5.1. In Fig. 5.1(a), an input signal is amplified by a system showing a gain G:

$$G = \frac{e_{\text{out}}}{e_{\text{in}}} \tag{5.1}$$

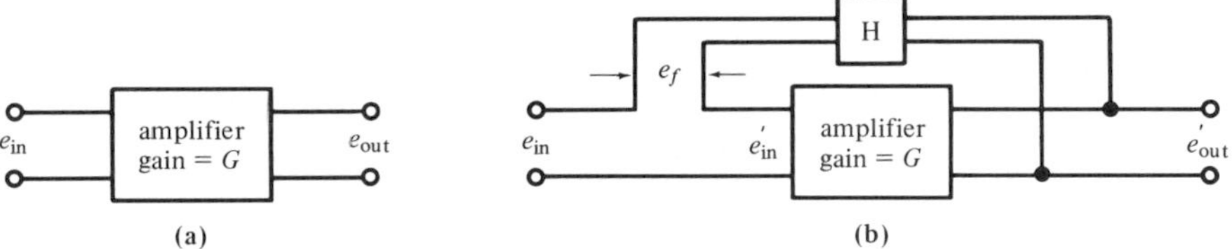

Figure 5.1 / *Generalized feedback arrangement.*

This parameter, characteristic of the amplifier, is termed the *open-loop gain.* In Fig. 5.1(b), a fraction, H, of the output signal from the same amplifier is fed back and mixed with the input signal. This has the effect of changing the output, which is therefore designated e'_{out}. The input signal will then be

$$e'_{in} = e_{in} + He'_{out} \tag{5.2}$$

and

$$G = \frac{e'_{out}}{e'_{in}} \tag{5.3}$$

Since the application of feedback changes the output signal but not the input, this amounts to a change in the effective gain of the system. Equation (5.4) defines the *closed-loop gain*, G_f, which is the effective gain with feedback:

$$G_f = \frac{e'_{out}}{e_{in}} \tag{5.4}$$

Substituting the value for e_{in} from Eq. (5.2) into Eq. (5.4) gives Eq. (5.5), which can be further simplified by substitution of the value for e'_{out} from Eq. (5.3):

$$G_f = \frac{e'_{out}}{e'_{in} + He'_{out}} \tag{5.5}$$

$$= \frac{Ge'_{in}}{e'_{in} + GHe'_{in}} = \frac{G}{1 + GH} \tag{5.6}$$

If $GH \gg 1$,

$$G_f \simeq \frac{1}{H} \tag{5.7}$$

From Eq. (5.7), it is seen that closed-loop gain becomes entirely a function of the feedback ratio H, providing that $GH \gg 1$. One implication of this is that closed-loop gain must be very much less than open-loop gain. This reduction in gain on introducing negative feedback, which may amount to 40–80 dB, is termed the *loop gain.*

One very important benefit derived from the use of negative feedback stems from the fact that the feedback ratio can be established by the use of

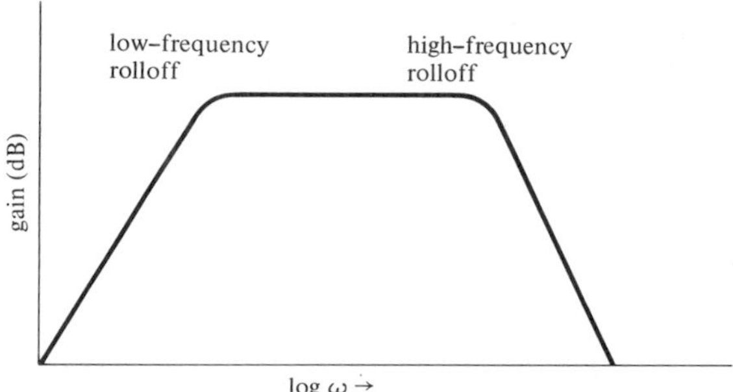

Figure 5.2 / *Bode plot for an AC-coupled amplifier.*

resistors. These are components that show linear behavior, which can be controlled within very close tolerances and which are subject to only slight variations in value as a result of aging or temperature change. Therefore, if GH is large, the critical components of the amplifier need not be the ill-behaved transistor or tube, but can be a set of accurate and reliable resistors. The present development of scientific instrumentation has been possible because of the availability of accurate, reliable, and inexpensive amplifiers, which in turn are dependent on the use of negative feedback in amplifier design.

5.1.2 / Feedback and phase shift

Negative feedback circuits are designed to provide a 180° total phase shift between the input signal and the signal that is fed back. If each stage shows low-pass and high-pass behavior, as discussed in Section 4.12.1 and shown in Fig. 5.2, the nominal phase shift of 180° per stage can be changed by a maximum of 90° per stage. In the low-frequency range, the angle can vary from 180° to 270° per stage. In the high-frequency rolloff range, it can vary from 180° down to 90° per stage.

If the circuit is designed to afford a total shift, θ, between e_{in} and e_f of Fig. 5.1(b), of 180° in the midfrequency range, the actual phase shift could be as large as 270° for one stage, 360° for two stages, and 450° for three stages. It is evidently true that a change in θ from 180° to 360° amounts to a conversion of negative feedback to positive feedback. This could occur in the low-frequency rolloff range. A similar effect is observed in the high-frequency rolloff range, with maximum change of θ from 180° to 90° for one stage, to 0° for two stages, and to 270° for three stages.

It is therefore true that filter behavior of amplifiers, in the high-frequency range for DC amplifiers and in both high- and low-frequency ranges for AC amplifiers, will cause a shift in phase, $\Delta\theta$, of $\pm 180°$. In Eq. (5.6), a change of θ by $\pm$ 180° would be noted by a change in the sign in the denominator. Accordingly, if the product $GH > 1$ at the frequency at which $\Delta\theta = \pm 180°$,

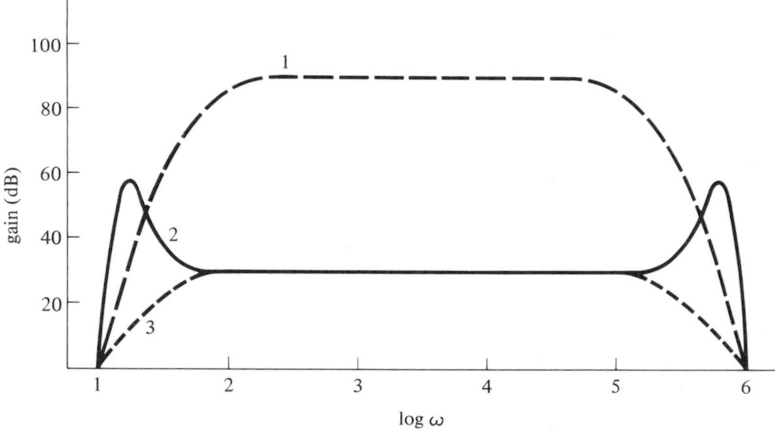

Figure 5.3 / *Response curve for a three-stage amplifier. Curve 1: without feedback; curve 2: with feedback; curve 3: with feedback, but with response at extreme frequencies altered by filters.*

the amplifier will exhibit self-sustained oscillation at that frequency and will not respond to an input signal.

It should be noted from Figs. 3.5, 3.7, and 3.10 that the frequency range in which a filter causes phase shift is identical with that in which gain is attenuated (the low- and high-frequency rolloff regions of Fig. 5.2). Therefore, as θ is changed, the effective value of G is reduced. For stable operation, GH must be reduced to a value less than unity at the frequencies for which $\Delta\theta = \pm 180°$.

It follows from this that one would anticipate difficulty from instability in amplifiers that exhibit large open-loop gains (G is large), in those with which large amounts of negative feedback are used (H is large), and in those with at least three stages of amplification ($\Delta\theta_{max} > 180°$). Demands on amplifier performance make it necessary to use heavily fed-back, high-gain, multistage amplifiers, so that precautions against instability are in fact quite important. The most common precaution against oscillation involves altering the output by placing filters in the circuit to make the rolloff curve similar to that of a single filter. This ensures that the maximum $\Delta\theta$ is 90°, which precludes oscillation.

This is illustrated in the Bode plot in Fig. 5.3. With no feedback, the rolloff is shown in curve 1 of Fig. 5.3 as 60 dB/decade because there are three stages. Use of some feedback reduces the midfrequency gain and somewhat extends the bandpass because there is less effective negative feedback at extreme frequencies. This extension of the bandpass may be desirable. With heavy feedback, as in curve 2 of Fig. 5.3, the midband gain is reduced further and that at high and low frequencies is increased, with the effect that it may be large enough to cause oscillation. This is avoided by including filters that

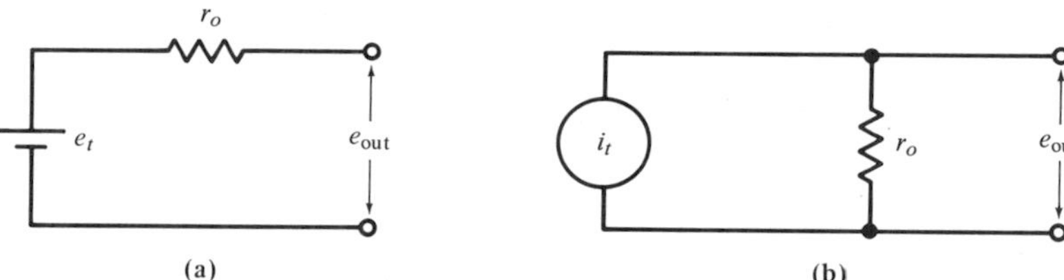

Figure 5.4 / *Amplifier output equivalent circuits.*
(a) *Voltage source equivalent:* $e_t = f(e_{in})$.
(b) *Current source equivalent:* $i_t = e_t/r_o$.

restrict the rolloff at high and low frequencies to 20 dB/decade, as in curve 3 of Fig. 5.3. This removes the "ears" from the response curve and greatly limits the bandpass. However, it permits the use of large amounts of negative feedback to produce linear gain without instability.

5.2 / TYPES OF FEEDBACK

Negative feedback can be introduced in a number of ways; the statements in Section 5.1 represent some generalizations that are applicable regardless of the technique used. There are, however, important characteristics that are peculiar to the type of feedback used, which makes it desirable to consider them in more detail. They are classified according to the type of signal that is fed back.

5.2.1 / Current feedback

One approach is to feed back a voltage that is proportional to the output current. When this is done, the effect of the feedback will be to minimize fluctuations in the output current. If the output current increases for any reason, then the magnitude of the fed-back signal increases, canceling out a larger fraction of the input signal. Conversely, any decrease in output current results in a corresponding decrease in the size of the fed-back signal and reduction in the extent to which the input signal is canceled. Current feedback tends to cause an amplifier to act as a constant-current source.

Another way of describing this behavior is to say that current feedback increases the output impedance of the amplifier. If the amplifier is considered as a box that provides a signal, it can be represented by a Thévenin equivalent circuit with either a voltage source [Fig. 5.4(a)] or the equivalent current source model [Fig. 5.4(b)]. The effect of current feedback in stabilizing output current can be simulated in the current source model by an increase in r_o relative to the external load, because as r_o increases, the overall behavior is more nearly that of the ideal current source. Stabilization of output current is ordinarily quantified in terms of the output impedance. This convention is used because it facilitates an estimation of the effect on the transmission of a signal from one instrument to another.

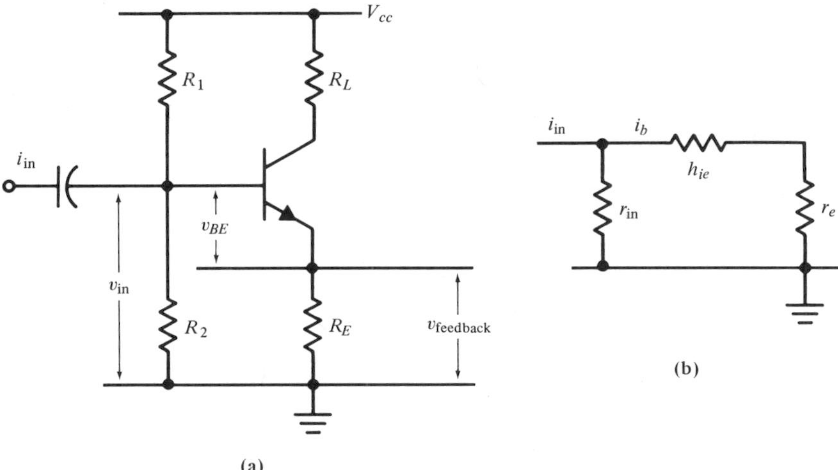

Figure 5.5 / *Common-emitter amplifier.*
(a) *Schematic diagram.*
(b) *Small-signal equivalent circuit of input.*

Another effect of output current stabilization is that it causes an increase in input impedance. This is not noticeable with FETs and vacuum tubes at low and medium frequencies because of their inherent high input resistances. With transistors, however, it is very important; moreover, current feedback is very often used to stabilize the operation of transistor amplifiers. Accordingly, it will be discussed in some detail for a common-emitter amplifier.

Current feedback is introduced by placing a resistance, R_E, between the emitter and the common point of the circuit, as indicated in Fig. 5.5(a). That this is current feedback can be seen from the fact that the voltage that drives current into the base, v_{BE}, is the difference between the input signal and the voltage dropped across R_E. The voltage across R_E, the feedback signal, is $i_e R_E$, which is a function of output current.

The small-signal equivalent of the input to Fig. 5.5(a) is given in Fig. 5.5(b). This is comparable to the conversion made between Figs. 4.37(a) and 4.38, except that a resistance r_e is introduced between h_{ie} and the common point. Note that this resistance is different from R_E. It is a hypothetical resistance that represents the effect of R_E on the input signal.

One cannot simply add the values of the components of Fig. 5.5(a) to evaluate input resistance because in operation they behave as though they are of different size than is indicated by their static values. It is, however, possible to relate input voltage and current to input impedance:

$$Z_{in} = \frac{e_{in}}{i_{in}} = \frac{i_b h_{ie} + i_e R_E}{i_b} \tag{5.8}$$

If i_e is expressed in terms of i_b, currents can be eliminated from Eq. (5.8):

$$i_e = i_c + i_b \tag{5.9}$$

$$i_c = h_{fe} i_b \tag{5.10}$$

Combining Eqs. (5.9) and (5.10) gives

$$i_e = i_b(h_{fe} + 1) \tag{5.11}$$

Therefore,

$$Z_{\text{in}} = h_{ie} + (h_{fe} + 1)R_E \tag{5.12}$$

or, in Fig. 5.5(b),

$$r_e = (h_{fe} + 1)R_E \tag{5.13}$$

R_E affects the input circuit as though it had a value that may be hundreds of times as large as the static value because it draws the larger emitter current, which is controlled by the smaller base current.

The necessity of minimizing the effects of variations in transistor characteristics caused by temperature fluctuations was mentioned in Section 4.5.2. Unless precautions are taken, an increase in temperature causes an exponential increase in I_{CO}, the base minority carrier current. This current is amplified just as is externally imposed base current and can therefore be a serious source of drift in the operating point of a transistor. If current feedback is used, an increase in emitter current will cause an increase in $i_E R_E$, hence a decrease in v_{BE} and a corresponding decrease in i_B. Accordingly, current feedback permits automatic stabilization of the performance of a transistor amplifier toward the effect of temperature change and component aging. It also minimizes the effects of the variations in characteristics that occur in nominally similar transistors, making possible the mass production of accurate amplifiers and simplifying repairs when component replacement is required.

It has been pointed out that negative feedback reduces gain. This can be evaluated quantitatively for this circuit by using the approach described in Section 4.10.5 with h_{ie} replaced by $[h_{ie} + (h_{fe} + 1)R_E]$. Since $(h_{fe} + 1)R_E$ may be much larger than h_{ie}, it can have an important effect on voltage and current gain. As was pointed out in Section 4.4.4, the effects of the feedback resistor can be eliminated at certain signal frequencies by a correct choice of bypass capacitor.

5.2.2 / The emitter follower

The emitter follower, discussed briefly in Section 4.9 and illustrated in Fig. 5.6, has the characteristics of not inverting the signal, of providing a gain slightly less than unity, and of offering a very high impedance to an input signal and a low impedance at the output. These properties make it especially useful as a buffer amplifier, which can couple a high-impedance signal source to a low input impedance receiver.

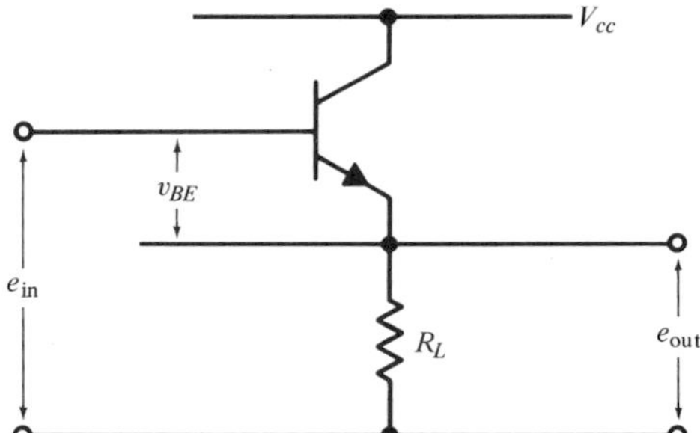

Figure 5.6 / *Emitter-follower (common-collector) amplifier.*

The high input impedance is in part caused by placing the load resistor between the emitter and ground. As was true for the circuit in Fig. 5.5(a), the effective value of the load at the input will be $(1 + h_{fe})R_L$, which is much larger than the static value. Absence of the Miller effect, discussed in Section 5.2.5, is also an important factor affecting the input impedance.

This device exhibits low output impedance because drawing current from the emitter tends automatically to stabilize the output voltage against perturbations in the load current. A qualitative understanding of this action can be obtained by considering the action of the components of the device. It is evident from inspection that v_{BE}, the voltage responsible for flow of base current, is

$$v_{BE} = e_{in} - v_E \tag{5.14}$$

Any change that causes an increase in v_{BE} will cause an increase in i_B and hence in i_E. Suppose that the amplifier is operating in some equilibrium condition and that a load, r_{ext} is connected. If no other change occurred, the effect of r_{ext} would be to reduce the effective emitter resistance to that of the parallel combination of R_E and r_{ext}. If i_E remained unchanged, this would cause a reduction of the value of v_E. We can see from Eq. (5.14), however, that a decrease in v_E causes an increase in v_{BE}, hence an increase in i_B and in i_E that will tend to offset the effect of reduction of emitter resistance. With this configuration, drawing output current to the external load has the effect of turning the transistor on to make more output current available and to stabilize the output voltage.

5.2.3 / Voltage feedback

When a signal that is proportional to output voltage rather than to output current is fed back, very significant differences in amplifier performance are noted. The amplifier will attempt to maintain a constant output voltage rather than a constant output current. Just as current stabilization implies a high output impedance, voltage stabilization implies a low output impedance. An example of voltage feedback is presented in Fig. 5.7. It is

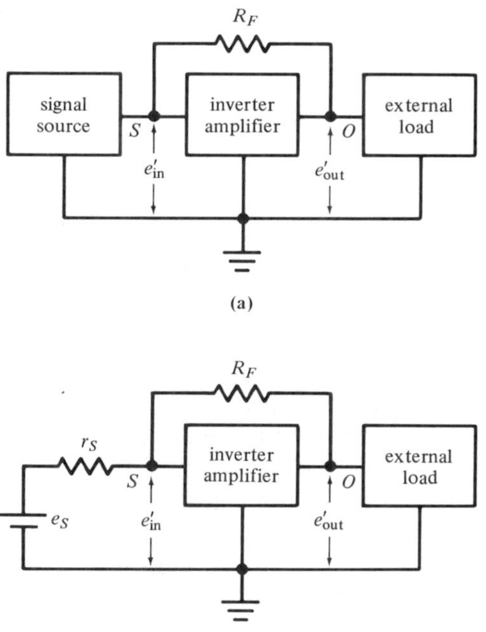

Figure 5.7 / *Voltage feedback.*

discussed in some detail because of its wide application in chemical instrumentation. In Fig. 5.7(a), a block presentation shows a signal source working into an inverter, which is bypassed by a feedback resistor R_F and which drives an external load.

The amplifier responds to an input signal at S:

$$e_{\text{out}} = -Ge_{\text{in}} \tag{5.15}$$

However, since the input and output terminals are connected by R_F, a given applied signal $+e_{\text{in}}$ does not drive point S to $+e_{\text{in}}$ V from ground, but instead (assuming that gain G is large) a small change at S is reflected in a large negative excursion at O. The effect of the feedback resistor is to cause the equilibrium input voltage e'_{in} to be smaller than it would be in the absence of the feedback resistor, which is the expected effect of negative feedback.

The mechanism of feedback in this circuit can be understood readily if the signal source is replaced by its Thévenin model as in Fig. 5.7(b). The input signal with feedback is smaller than it would be without feedback, because the current drawn by the feedback resistor causes a voltage drop in the source resistance, r_S. This explanation is based on the assumption that the amplifier input resistance is much larger than R_F, a restriction that is normally valid.

If the amplifier is in operation and some perturbation causes a positive excursion of the output voltage from its equilibrium value, the drop across

R_F is reduced, hence the input current is reduced. This leads to an increase in e'_{in}, which tends to cancel the change in e_{out}. A negative excursion from the equilibrium value of e'_{out} would, of course, have the opposite effect. The arrangement, therefore, does exhibit the stabilizing effect that is characteristic of negative feedback. It is interesting to note that the feedback mechanism is a function of the source impedance and that if r_S is made very small relative to the effective input impedance, the feedback mechanism does not work.

Accordingly, the effective input impedance is a characteristic of considerable importance. The input impedance here is the total impedance seen by the source, looking in at point S. It consists of the amplifier in parallel with R_F; as mentioned above, the amplifier impedance must be much larger than R_F, so the behavior at the input is controlled entirely by R_F. The input impedance (neglecting reactive behavior, which should be significant only at high frequencies) can be expressed by Eq. (5.16).

$$r_{\text{in}} = \frac{e'_{\text{in}}}{i_{\text{in}}} = \frac{e'_{\text{in}}}{i_F} \tag{5.16}$$

$$i_{\text{in}} = \frac{e'_{\text{in}} - e'_{\text{out}}}{R_F} \tag{5.17}$$

$$e'_{\text{out}} = -Ge'_{\text{in}} \tag{5.18}$$

Combining Eqs. (5.17) and (5.18):

$$i_{\text{in}} = \frac{e'_{\text{in}} + Ge'_{\text{in}}}{R_F} = e'_{\text{in}}\left(\frac{1 + G}{R_F}\right) \tag{5.19}$$

In the usual case, G is much greater than one, so

$$i_{\text{in}} = e'_{\text{in}}\left(\frac{G}{R_F}\right) \tag{5.20}$$

$$r_{\text{in}} = \frac{e'_{\text{in}}}{i_{\text{in}}} = \frac{R_F}{G} \tag{5.21}$$

5.2.4 / Operational feedback

It has been shown that amplifiers with voltage feedback exhibit small input impedances and require signal sources with appreciable output impedances to operate. Current sources, as was pointed out in Section 2.3.4, require measuring devices with input impedances much smaller than the source output impedance in order to make an accurate measurement. It is evident, then, that amplifiers with voltage feedback tend to be suitable current-measuring devices. The great importance of this application has led to the design of special-purpose amplifiers that permit emphasis of these attributes of voltage feedback and to the recognition of a type of feedback termed *operational feedback*. The amplifiers, which are almost useless for other purposes, are called *operational amplifiers*. A detailed discussion of this type of feedback and its uses is given in Chapter 8.

5.2.5 / The Miller effect

The behavior of the feedback resistor in Fig. 5.7 is an example of a component of an active circuit exhibiting an effective value that is different from its static value. The resistor behaves as though its conductance were G times larger than the static value because of the action of the amplifier in causing the voltage of the output end of the feedback resistor to undergo a change that is $-G$ times the change at the input end. A similar type of behavior is observed if a capacitance is placed between the input and output of an amplifier. The effective value of reactance is reduced by a factor of G. Considering that reactance is inversely proportional to capacitance, Eq. (1.46), this means that a capacitance in this position exhibits a dynamic value that is G times greater than its static value. This phenomenon is known as the *Miller effect*.

The Miller effect is used deliberately in various contexts to provide the performance of a large capacitance without having to furnish a large capacitor. More important, however, it constitutes a major factor that limits the high-frequency performance of many amplifiers. Amplifying devices, transistors or tubes, exhibit small, unwanted capacitances between their terminals. When the device is used as an inverter, the stray capacitance between the input and output terminals is subject to the Miller effect. Since reactance decreases with increasing frequency, ultimately constituting a short circuit around the amplifier, the Miller multiplying effect may be devastating at high frequencies. This is the reason for the widespread use of pentodes, rather than triodes, at the input of vacuum tube amplifiers. The former show very small grid–plate capacitance. It is also one reason for the use of follower amplifiers; the input and output of a follower are in phase and accordingly are not subject to the Miller effect.

PROBLEMS

1 / Figure 5.8(b) represents the circuit of Fig. 5.8(a) with the bleeder replaced by its Thévenin equivalent.

(*a*) For $v_b = 1.50$ V, $r_b = 15$ kΩ, and $h_{fe} = 100$, calculate the corresponding value of V_{CE}. Assume that V_{BE} is 0.60 V throughout.

(*b*) If the transistor in part (*a*) is replaced by one for which $h_{fe} = 50$, what would be the value of V_{CE}, other component values remaining unchanged?

(*c*) If the bleeder were changed to effect a 10 percent increase in v_b, other values remaining as in part (*a*), calculate the value of V_{CE}.

2 / Figure 5.9(b) represents the circuit of Fig. 5.9(a) with the bleeder replaced by its Thévenin equivalent.

(*a*) Calculate the value of v_b that would be required to give $V_{CE} = 5.0$ V if $r_b = 15$ kΩ, $V_{BE} = 0.60$ V, and $h_{fe} = 100$.

(*b*) If the transistor in part (*a*) is replaced by one for which $h_{fe} = 50$, what would be the value of V_{CE}, other component values remaining unchanged? Compare this with the answer to Problem 1(*b*).

(*c*) If the bleeder were changed to effect a 10 percent increase in v_b, other values remaining as in part (*a*), calculate the value of V_{CE}. Compare this with the answer for Problem 1(*c*).

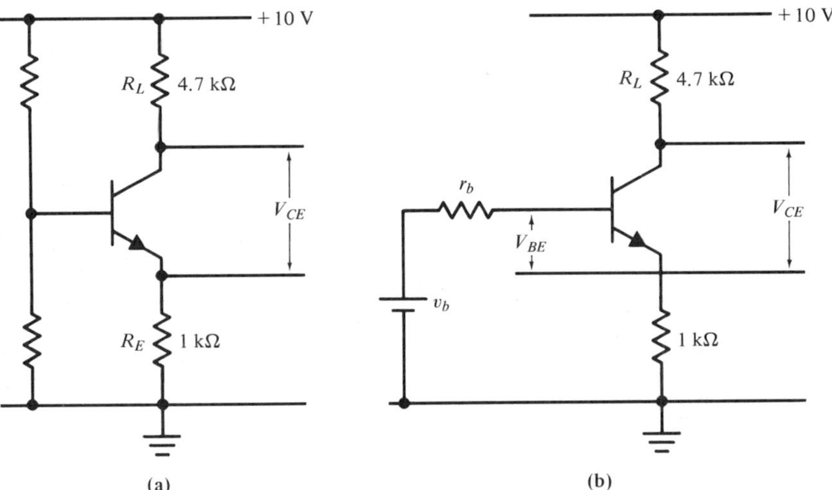

Figure 5.8

3 / Calculate the input resistance for each of the circuits shown in Fig. 5.10. Assume a transistor with $h_{ie} = 500\,\Omega$, $h_{fe} = 200$, and that bypass and coupling capacitors exhibit negligible reactances at frequencies of interest.

4 / A generator that has an output resistance of 1000 Ω is used to drive the amplifiers shown in Fig. 5.10. Assuming that bypass and coupling capacitors show negligibly small reactances, that stray capacitance is not important, and that the amplifiers are not overdriven, calculate the percentage amplitude distortion at the input that would be caused by each.

5 / An amplifier is constructed from three identical DC-coupled stages, each with a high-frequency rolloff of 20 dB/decade, starting at a breakpoint at 10 kHz and each exhibiting maximum gain of 20 dB.

Figure 5.9

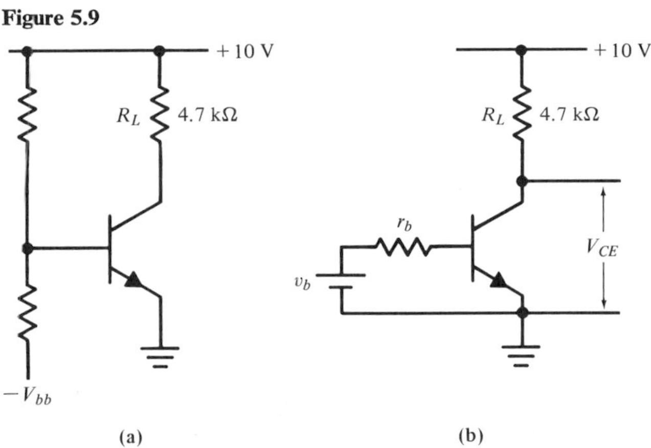

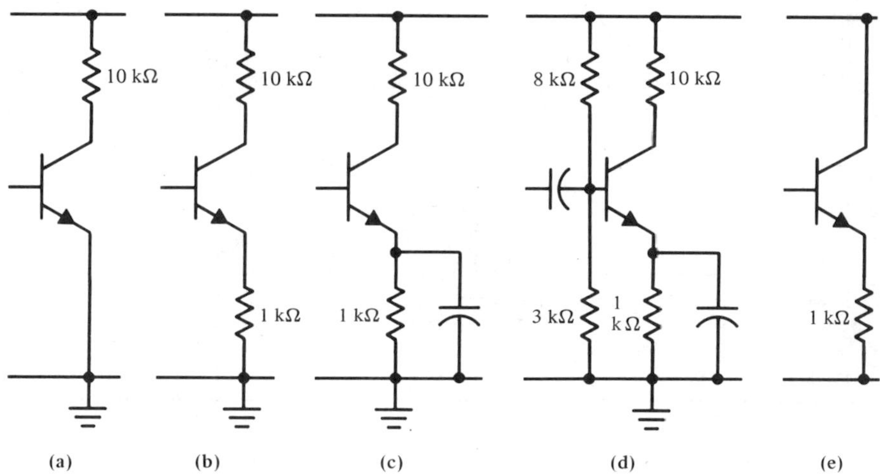

Figure 5.10

(*a*) At what frequency would the amplifier be liable to go into oscillation if connected with negative feedback?

(*b*) What would be the maximum feedback ratio, *H*, that could be used without the possibility of oscillation?

chapter 6 / Signal processing

chapter 6

6.1 / SIGNALS AND TRANSMISSION SYSTEMS

A scientific instrument may be considered to be a system made up of various components which have the following functions: detecting, transducing, signal processing, amplification, and readout. The transducer responds to the phenomenon of interest and produces an electrical signal that carries the desired information encoded as a voltage, current, frequency, or phase angle. This signal is amplified, processed as required in the context of the application, and used to drive the readout, which may be a meter, recorder, or other instrument.

The instrument accepts information from a detector and presents it in a convenient form to the user. From this perspective, an instrument is a communication system, and many techniques first used in communications equipment are widely used in instrument design. In addition, study of communication systems has led to the development of a body of knowledge dealing with information content of signals and the problems of information retrieval from noisy signals. While detailed discussion of these topics is beyond the scope of this book, consideration of some of the results is useful in focusing attention on critical factors and in providing an understanding of the reasons that certain components are very widely used.

6.1.1 / Information content of a signal

We intuitively recognize that the information contents of various kinds of signals differ. It is obvious, for example, that the information content of the signal used to transmit a television picture for 1 min must be greater than that of a signal that transmits only a 1-min teletype message. Similarly, the information content of the signal produced by a radiation counter measuring a "hot" sample or of an electrocardiograph following a heartbeat will be greater than that produced by a pH meter following the course of a manual titration. In selecting experimental equipment, it is therefore necessary to ensure that the system capacity is sufficiently great to handle the information content of the signal without unacceptable losses.

To do this, one must be able to determine both the information content of a signal and the capacity of a transmission system. Actual quantification of these is beyond the scope of this text; however, we can examine the question in a general way and note some of the most important factors.

Fundamentally, the information content of a signal is determined by its predictability. If a signal is received that could be predicted in advance, no new information would have been obtained. The more unpredictable the signal, the greater will be the information content. Specifically, the information content of an electrical signal can be related to the number of discernable intensity levels exhibited by the signal and to the rate at which the transition from one level to another is accomplished. For example, suppose that information is encoded as a voltage magnitude and that 5 V is the maximum signal intensity. If disturbances in the system make it impossible to distinguish with certainty between voltage levels that differ by less than 5 V, then information can be encoded only in terms of presence or absence of the signal.

This type of system could be used to transmit graphical information. This might be accomplished by subdividing the image to be transmitted into a large number of small areas, regularly arranged. Each of these would be scanned, assigning either black or white (signal on or off) to each one. Use of the transmitted signal to lay down an array of dots corresponding to the areas for which black was assigned could produce a facsimile resembling a newspaper halftone.

The information content of this signal could be increased by reducing the size of the unit areas of the original image being scanned. This would increase the number of samples taken and lead to improved resolution of detail in the facsimile. Assuming that the time required for the sampling process remained unchanged, the increased information content would be obtained by increasing the time of transmission. The rate of information transmission would not be affected, and the demands made on a transmission system would be unchanged.

Another way of increasing the information content would be to improve the system to permit unambiguous distinction to be made between smaller signal intensity increments. For each sample scanned, it would be possible to assign various shades of gray, as well as black and white. This improvement in resolution of the process would also make possible the transmission of a more accurate facsimile; if the sampling rate remained unchanged, the rate of information transmission would be increased, and the demands on the transmission system would be greater.

In general, the rate of information transmission may be described by

$$\text{rate} = \frac{1}{t}(\log_2 n) \tag{6.1}$$

where n represents the number of discernable signal levels available and t is the time required by the system to change from one level to another. The unit of information content is the *bit*, and the rate is usually expressed in bits per second. Base 2 logarithms are arbitrarily selected for convenience because of the widespread use of binary information coding, in which only two signal levels are used. Accordingly, the rate of transmission is simply the reciprocal of t for the system.

6.1.2 / Signal coding

In this context, the demands made on a transmission system, or on a readout device such as a recorder, may be stated in terms of the rate of information transmission. Irrespective of the nature of the signal, it is possible, in principle at least, to express it quantitatively in bits per unit time. In practice, the procedure used is determined by the scheme employed to encode the information.

Encoding of information for transmission by electrical signals can be accomplished in a number of ways. The simplest in concept is to provide a voltage or current, the instantaneous value of which is proportional to the instantaneous value of the measured quantity. This is termed an *analog signal*; it is an electrical analog of the measured quantity. Analog encoding is the usual choice for simple instruments and systems. To specify the information content of an analog signal, three quantities must be known. One is the range of signal levels that corresponds to the full range of response of the transducer. Another is the uncertainty inherent in the instantaneous measurement, which is usually expressed as a fraction of the full-scale signal. The third is the speed with which the signal may change in intensity. This last is often expressed in terms of the range of frequencies of the components that comprise the signal and is termed the *bandwidth* of the signal. Although one can obtain analog information rates in bits per unit time, it is more common for instrument application to specify the three quantities explicitly.

Besides analog coding, there is a variety of ways in which information transmitted as electrical signals can be encoded as pulses. That most commonly used involves distinguishing between two defined voltage levels. As will be pointed out in Section 9.1, these codes usually involve binary notation and are termed *digital signals*.

6.1.3 / Transmission system capacity

Having determined the information rate characteristic of a signal, one can specify the transmission capacity needed to handle it. The relationship given in Eq. (6.2) permits calculation of theoretical system capacity in bits per second:

$$C = B \log_2 \left(1 + \frac{S}{N}\right) \tag{6.2}$$

Equation (6.2) is of interest here primarily because it identifies the factors needed to characterize a communication system: The bandpass, B (Hz), and the ratio of signal power to noise power, S/N.

Generally, in selecting and using equipment, one is concerned with its effect on the signal of interest rather than with the theoretical capacity. To transmit a signal waveform without distortion, the bandpass of the system must encompass all of the frequency components of the signal that contribute significantly to signal power. This range can be determined quantitatively by obtaining a Fourier analysis of the signal. The bandpass of electronic equipment is operationally defined as the range between the upper and lower half-power frequencies.

Often the bandwidth of recording equipment is expressed in terms of the *rise time*, which is defined as the time required for a signal to change from one-tenth to nine-tenths of the final value. Equation (3.11) permits the calculation of the upper breakpoint frequency for a specified rise time, t_r. The result is shown in Eq. (6.3):

$$t_r = 2.3RC \tag{6.3}$$

As an example, let us consider the bandpass required for transmission of a pulse. An approximate indication of the results of inadequate bandpass can be obtained from the curves given in Figs. 3.19 and 3.20. These show the effects of high-pass and low-pass behavior on a square wave and relate the half-period (pulse width) to the RC time constant. The upper breakpoint frequency can be related to the pulse rise time by Eq. (6.3). For a pulse to be transmitted without gross distortion, the lower breakpoint time constant must be larger than the pulse width, for example, Fig. 3.19(b). The extent of distortion caused by system high-pass filter behavior, termed the *tilt*, can be calculated starting with Eq. (3.18). The result is given in Eq. (6.4), where τ is the pulse duration in seconds:

$$\%\ \text{tilt} = 100\left(\frac{\tau}{2RC}\right) \tag{6.4}$$

Thus, if it were desired to transmit 1-μsec pulses with a rise time of 0.1 μsec and 10 percent maximum tilt, it can be determined by use of Eqs. (6.3) and (6.4) that a bandpass of 3.3 MHz, from 32 kHz to 3.6 MHz, would be required.

6.2 / NOISE

Sufficiently careful observation of electrical signals invariably reveals the presence of components that are unrelated to the phenomenon being measured. Spurious signal components, which may occur at any frequency within the capabilities of the system, are termed *noise*. It is a major goal of instrument design to minimize noise in order to facilitate accurate and sensitive measurement. Some knowledge of the origins and characteristics of various types of noise is useful in understanding the measures that can be taken to minimize noise or, as it is often put, to optimize the signal-to-noise power ratio (S/N).

For purposes of discussion, noise may be considered in two categories: that which is caused by environmental disturbance and that which is continuous and apparently spontaneous. Noise of environmental origin—electrostatic or electromagnetic induction, microphonic pickup of mechanical vibration, influence of voltage surges in power systems, and so on—may in principle be eliminated by proper design. In practice, however, this is not completely achieved. The spontaneous varieties, thermal noise and shot noise, arise because of the discontinuous nature of matter and energy and are therefore not subject to elimination even in principle. Their interference constitutes a fundamental limitation on the possible sensitivity of electrical measurements.

6.2.1 / Thermal noise

Thermal noise is a voltage having its origin in thermally induced motion of charge carriers. It is known as *Johnson noise*, after the original investigator. The problem was studied by H. Nyquist, who derived the relationship given in Eq. (6.5):

$$\bar{e}_R^{\,2} = 4kTR\,\Delta f \qquad (6.5)$$

Here $\bar{e}_R^{\,2}$ is the mean square voltage, k is the Boltzmann constant, T is the absolute temperature, R is the resistance of the noise source, and Δf is the bandwidth over which the measurement is made.

Examination of Eq. (6.5) shows that thermal noise is a function of temperature, resistance, and effective bandwidth. These are the factors that must receive attention if noise is to be minimized. In practice, the complications attending the use of low-temperature thermostats generally cause major emphasis to be placed on control of resistance and bandpass.

Noise has, by definition, a random phase relationship with other components of a signal. Noise voltage or current cannot be treated mathematically, as would be the case with coherent signals. Mean-square values of current or voltage (rms), which are a measure of the effect of the signal and do not carry an implication of polarity, may be used for calculations involving noise signals. For example, if the rms voltage $\bar{e}_{AB}$ in Fig. 6.1 were measured using a meter with a 5-kHz bandpass, the signal would be the sum of those generated by the two resistors. At 25°C, these would be 13 μV for R_1 and 9 μV for R_2, calculated using Eq. (6.5). The total voltage would be

$$\bar{e}_{AB} = (\bar{e}_1^{\,2} + \bar{e}_2^{\,2})^{1/2} = 16\ \mu\text{V} \qquad (6.6)$$

As indicated in the illustration, thermal noise levels are likely to be of the order of magnitude of microvolts. Accordingly, difficulty from this source would be expected when a large resistance is placed at the input of an

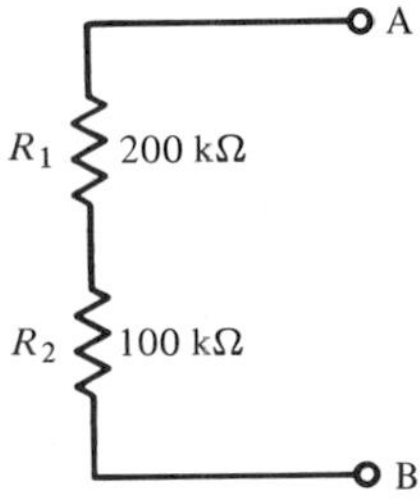

Figure 6.1

amplifier, where the noise will be amplified along with the signal. Resistance noise in other parts of a circuit is generally too small to cause trouble.

Thermal noise voltage is proportional to the square root of the bandpass, but it is not a function of frequency. This means that the noise level is constant over all observable frequencies, and is termed *white noise*. Moreover, the noise content of a signal can be reduced merely by restricting the bandpass of the device used to measure it. For example, suppose that with a meter that responds to signals from DC to 1 MHz, a signal that occurs in the region below 10 kHz is observed with S/N of 10. Reduction of the meter response range from 1 MHz to 10 kHz would amount to a 100-fold bandpass reduction and would provide a 10-fold improvement in S/N. This is, of course, useful only if the information occurs within the reduced frequency range. This point is discussed in greater detail in Section 6.4.1.

6.2.2 / Shot noise

Shot noise arises when charge carriers pass a barrier, as electrons emitted from vacuum tube cathodes or crossing an *np* junction. Charge carriers cross these barriers one at a time, causing small random fluctuations in current. Quantitative relationships vary depending on the type of device used, but noise levels from this source are likely to be of the order of magnitude of microvolts. Furthermore, shot noise is white and is therefore proportional to the square root of bandwidth. Accordingly, signal enhancement by bandpass reduction is an effective measure for controlling shot noise as well as thermal noise.

6.2.3 / Environmental effects

Circuits are commonly subject to various electrical, magnetic, thermal, and mechanical perturbations that cause signal noise. The following categories will be considered: drift and flicker noise, surges, and environmental noise.

Drift and flicker noise are disturbances of environmental origin, the intensities of which are approximately inversely proportional to frequency. They can be dealt with most effectively by avoiding the use of extremely low frequencies. This is discussed in Sections 4.12.3 and 6.3.1. Surges can be dealt with either by protecting the equipment from them or by choosing a design that is relatively immune to them. The power and ventilation systems furnished in modern laboratory construction serve as a buffer against many electrical and thermal events that might otherwise cause trouble. Aspects of design such as regulated power supplies and the use of negative feedback serve further to minimize the effects of surges. The instrument user can take precautions against overloading the power lines that serve his equipment. It is quite normal to make use of the kinds of measures alluded to here to minimize the effects of disturbing events in the low- and medium-frequency regions and to have a sufficiently low-frequency response readout to reject the high-frequency components.

Signal components of the type designated here as environmental noise are also commonly caused by electrostatic and electromagnetic pickup, by grounding problems, and by microphonic pickup of mechanical vibrations.

With well-designed equipment, mechanical vibrations are usually troublesome only in critical applications or in unusual locations. Magnetic pickup will be particularly noticeable in the vicinity of transformers or other equipment that produce substantial field intensities. It can be minimized by metallic shielding; for example, this is routinely done with oscilloscope picture tubes. In the usual case, the cumulative effect of noise from all sources is of the order of magnitude of 1 mV. Very often, difficulties stemming from improper use of instruments are caused by electrostatic pickup and from electrical interactions between instruments. These may lead to introduction of various noise components, but the power line frequency will generally dominate. Precautions may involve proper positioning of components, shielding sensitive circuits, and establishing reliable ground connections. Instrument users must be particularly concerned with shielding and grounding problems.

A conductor can be prevented from picking up stray signals by surrounding it by a metallic shield; grounding the shield short circuits stray signals and prevents the occurrence of appreciable voltages in the vicinity of the conductor. It is especially important that signals taken to low-level voltage-measuring devices—such as laboratory recorders, pH meters, and oscilloscopes—be shielded, and that nearby metallic objects such as furniture, mounting racks, and so on, be grounded. These precautions are routinely taken by equipment manufacturers and usually become of concern to the user only when modifications are made.

Additional routine precautions include establishment of a single, high-quality ground lead for an instrument or group of instruments working as a system. This is never the low side of the AC main. It is brought into a convenient connector or *bus* in the instrument. All components, mounting chasses, and so on are connected at one place to the ground bus. Multiple connections of one object to the bus provide paths of differing resistances, or *ground loops*, for the return of stray currents, which in turn establish undesirable voltage gradients around the equipment. Each instrument or subunit should have a single identified ground connection, which should be attached to the ground bus. This is usually through the power cord or a marked connector at the input. Each isolated circuit, for example, one powered from the secondary of a transformer, must be attached to the bus at one point only. This is routinely done by equipment manufacturers, but must be borne in mind when modifications are made.

6.3 / MODULATION

Modulation is a process by which the information content of a signal is moved from one frequency range to another as a result of nonlinear interaction between two time-varying signals. It is widely used in instrument design to achieve a variety of purposes; however, two applications especially are encountered: the presentation of information in a convenient frequency range for readout, and the separation of desired signals from noise.

For example, nuclear magnetic resonance spectrometers of necessity operate in the radio frequency range. It is convenient to bring the information out to an electromechanical recorder which ordinarily operates at less than 1 Hz. To allow this, the signal undergoes modulation in stages to get down to the desired frequency range. The process of modulation is basically the same, regardless of whether the information is being shifted to higher or to lower frequencies; however, the term *modulation* is often reserved for shifts to higher frequencies. Shifts to lower frequencies are termed *demodulation* or *detection*.

Certain frequency ranges may be expected to be especially troublesome. Below 1 Hz, drift and flicker noise cause difficulties because AC coupling cannot be used. Very often, therefore, low-frequency signals are modulated in order to move them to higher frequencies and so permit the use of AC amplifiers. This approach is generally combined with the technique of phase-sensitive detection, which is discussed in Section 6.4.3, to achieve very efficient discrimination between desired signals and noise.

6.3.1 / Modulation, a nonlinear process

In the process of modulation, the information-bearing signal is used to modify a second signal, which is termed the *carrier*. The modification may involve encoding of information in terms of variations of carrier amplitude, *amplitude modulation* (AM). Another commonly used approach involves encoding in terms of change in carrier frequency, which is *frequency modulation* (FM).

Whatever scheme is chosen, it is essential that the actual modulation step be nonlinear in order that new frequency components are formed that contain the information. The output from a linear process involving two signals will be their sum or difference, but it will not contain any new frequency components. If, for example, information in the DC to 1-kHz region is added to a 1-MHz signal for the purpose of making it possible to minimize the effects of 60-Hz noise by filtering out all components in the composite signal below 100 Hz, the experiment will fail. The process of addition merely produces a signal having components corresponding to each of the inputs.

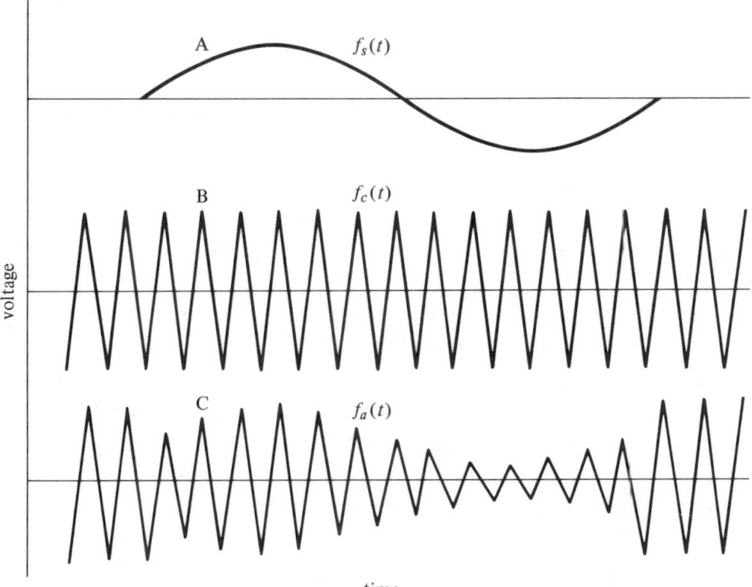

Figure 6.2 / *AM waveforms.*

Filtering out the DC–100 Hz segment will destroy all information in that region. The same result would be obtained for any other linear procedure of mixing. By contrast, a nonlinear mixing process produces signal components at frequencies other than those of the input signals. If the new frequency components carry the desired information, the spectral region of the original signal can be discarded without loss.

6.3.2 / Amplitude modulation

As mentioned above, there are a number of ways of carrying out continuous carrier modulation. This discussion will be limited to one illustration of AM, which is widely used and which possesses the general attributes of any modulation process.

Suppose that the signal containing information, $f_s(t)$, is represented for illustration by the single sine wave of Eq. (6.7), which is also illustrated as curve A of Fig. 6.2.

$$f_s(t) = E_s \sin(\omega_s t) \tag{6.7}$$

If it is used to modulate the carrier, $f_c(t)$, curve B of Fig. 6.2,

$$f_c(t) = E_c \sin(\omega_c t) \tag{6.8}$$

the signal, $f_a(t)$, Eq. (6.9) is produced. From curve C of Fig. 6.2,

$$f_a(t) = [E_d + E_s \sin(\omega_s t)] \sin(\omega_c t) \tag{6.9}$$

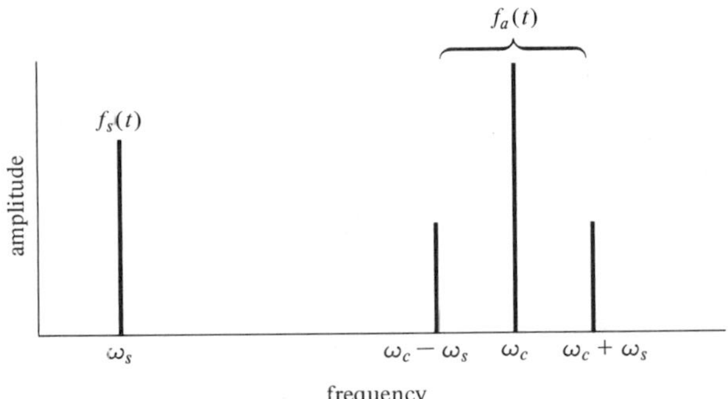

Figure 6.3 / *AM with a single modulating frequency.*

it can be seen that the modulating signal, $f_s(t)$, appears as the envelope of the modulated signal, $f_a(t)$.

Equation (6.9) can be simplified by defining m, the *degree of modulation*, as indicated in Eq. (6.10):

$$m = \frac{E_s}{E_c} \tag{6.10}$$

Combining Eq. (6.9) with Eq. (6.10) gives Eq. (6.11),

$$f_a(t) = E_c[1 + m \sin(\omega_s t)] \sin(\omega_c t) \tag{6.11}$$

which can be expanded as shown in Eq. (6.12):

$$f_a(t) = E_c\left[\sin(\omega_c t) + \left(\frac{m}{2}\right)\cos(\omega_c - \omega_s)t - \left(\frac{m}{2}\right)\cos(\omega_c + \omega_s)t\right] \tag{6.12}$$

Equation (6.12) makes it clear that the modulated signal, curve C of Fig. 6.2, actually consists of three sinusoidal signals, one at the carrier frequency and two symmetrically displaced about it, as indicated in Fig. 6.3. If $\omega_c \gg \omega_s$, as is implied here, then the information initially appearing in the low-frequency range has been moved to the range of the carrier frequency.

In this example, it was not indicated how modulation was achieved. There are actually many ways to do this, the necessary condition being that it be nonlinear. Consider the arrangement shown in Fig. 6.4. The modulating signal and the carrier are added together, and the sum of their voltages is applied to the diode. The diode is a nonlinear device; the current flowing can be represented by Eq. (6.13), in which e_i is the sum of $f_s(t)$ and $f_c(t)$:

$$i = a_1 e_i + a_2 e_i^2 + a_3 e_i^3 + \cdots + a_n e_n^n \tag{6.13}$$

As implied by Eq. (6.13), the output voltage appearing across the series resistance would be a complex waveform, owing to the various terms of the

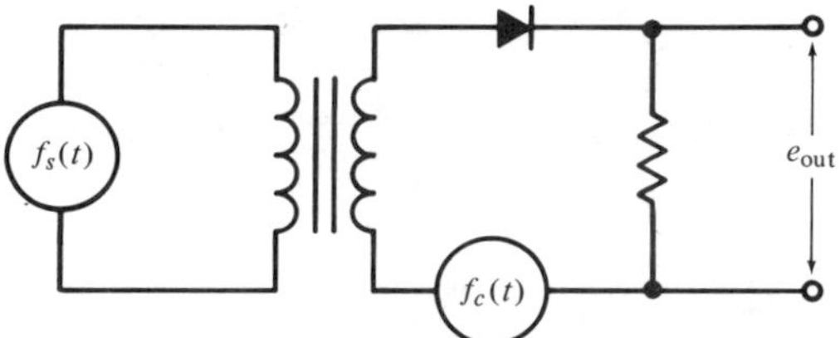

Figure 6.4 / *Use of a diode as a modulator.*

series. In fact, the process can be carried out so that only the first two or three terms make significant contributions:

$$
\begin{aligned}
i = a_1 e_i + a_2 e_i^2 &+ a_1[E_s \sin(\omega_s t) + \underline{E_c \sin(\omega_c r)} \\
&+ a_2 E_s^2 \sin^2(\omega_s t) + \underline{2E_s E_c \sin(\omega_c t)\sin(\omega_s t)} \\
&+ E_c^2 \sin^2(\omega_c t)]
\end{aligned}
\tag{6.14}
$$

If the output signal is filtered so as to retain only the components represented by the underscored terms of Eq. (6.14), it can be seen to have the same form as Eq. (6.9) and therefore to constitute amplitude modulation.

If, instead of a signal frequency, the modulating signal consisted of a band of frequencies, as is ordinarily the case, then the result might be that shown in Fig. 6.5. This emphasizes the main point that the information contained in the modulating signal is shifted into the *sidebands*, one above and one below the carrier frequency. If either one of the sidebands is retained, all other signal components can be discarded without loss of information. This is a general attribute of modulation processes.

If the carrier frequency is much greater than the highest component of the modulating signal, then the sidebands will appear relatively close to the carrier, as illustrated. If $\omega_c \simeq \omega_s$, then the sum and difference will be widely separated, which is the situation when the information in high-frequency signals is detected for use at lower frequencies.

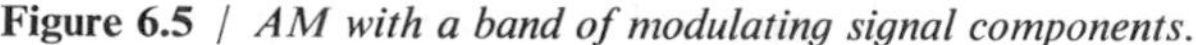

Figure 6.5 / *AM with a band of modulating signal components.*

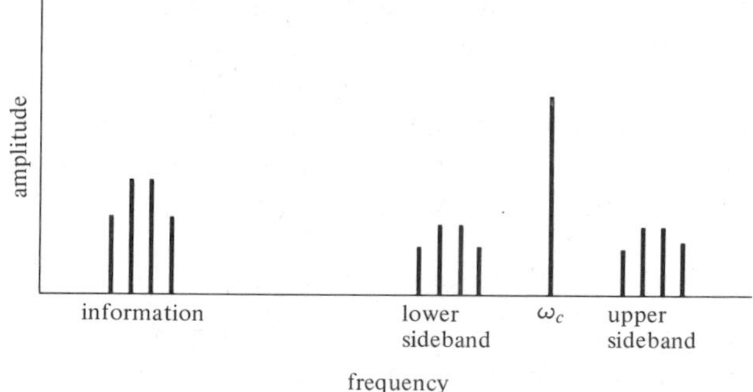

6.4 / SIGNAL ENHANCEMENT

It is universally true that electrical signals are accompanied by some noise. Accordingly, the extraction of information from noisy signals becomes a problem whenever low signal intensities must be handled, and the design of instruments quite often reflects this. An instrument user needs some acquaintance with the more widely used techniques in order to be able to understand the function of these components of his equipment. Accordingly, a brief discussion of S/N enhancement is given below.

The various approaches fundamentally amount to extending the period of observation during which the signal is effectively averaged. Random noise components are thereby minimized, since they have zero average values.

6.4.1 / Bandwidth reduction

It was pointed out in Section 6.2 that thermal and shot noise show uniform intensities over the accessible range of the spectrum; accordingly, noise content can be reduced by restricting the bandpass of the system. Noise reduction should be proportional to the square root of the bandwidth. Thus, reduction of the bandpass of an instrument by a factor of four, as from DC to 400 kHz down to DC to 100 kHz, would be expected to increase the S/N ratio by a factor of two. Actual results of bandwidth reduction may be either better or worse than this, because white noise may not be the dominant factor. If bandwidth reduction leads to exclusion of an especially noisy region, such as that at 60 Hz, the result may be very good. However, if the reduced bandwidth lies at low frequencies, the result may be unsatisfactory because of $1/f$ behavior of drift and flicker noise.

6.4.2 / Signal averaging

Bandwidth reduction by means of filters is ineffective if S/N in the restricted bandpass is around unity, because random noise fluctuations will determine the phase of the signal observed at any particular instant. This is not necessarily true, however, if the signal is averaged over a period that is long compared with the periods of the interfering noise components. If this can be done, the average of a noise signal will tend toward zero, while that of nonrandom information will be its actual value, even though signal and noise may have the same frequencies.

If the experiment of interest can be made to yield an unchanging signal for a defined period of time, it can simply be integrated to average out noise (Section 8.3.4). Such a case might be encountered in observing the output from an absorption spectrometer. The instrument might be set at a particular wavelength and the output monitored for a period of time.

The use of photographic recording in emission spectroscopy is another example of signal averaging. If the instantaneous light intensity corresponding to a line were observed, it would show very substantial and rapid fluctuations. Photographic recording produces a response that is approximately the same as would be obtained from steady illumination by a light

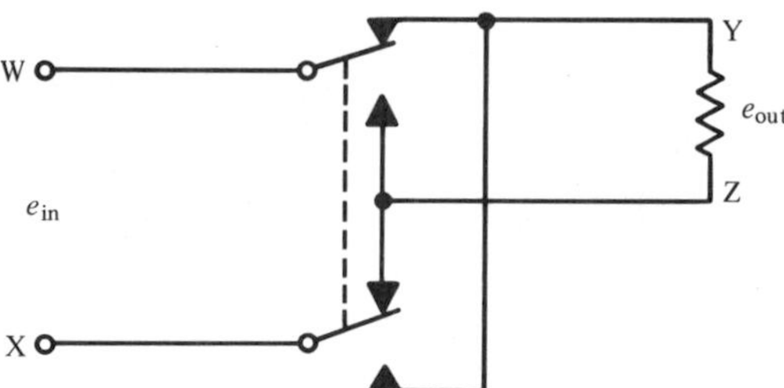

Figure 6.6 / *Switching arrangement for phase-sensitive detection.*

source with intensity equal to the average of the fluctuating source. Moreover, the film can simultaneously record all lines over an appreciable spectral region.

Measurement of low-intensity nuclear magnetic resonance signals produced by naturally occurring ^{13}C is an example of an application of signal averaging involving more complex instrumentation. In the usual case, the operator repeatedly scans the region of interest, and the spectrometer output is sampled at preselected intervals. The signal samples are stored and averaged with corresponding samples obtained on successive scans. The success of this approach is largely dependent on the ability of the operator to reproduce very accurately the conditions of observation on successive scans.

6.4.3 / Phase-sensitive detection

Phase-sensitive detection (also called synchronous rectification) is a method of demodulation by which only signals that are coherent with (have a fixed phase relationship with) a reference signal are extracted from a modulated signal. It can provide effective S/N enhancement under conditions for which bandwidth reduction is ineffective—as, for example, if the reduced bandpass does not reproducibly include the signal of interest or if the resulting S/N is still around unity after reduction. With the signal frequency accurately controlled, it is possible to use sharply tuned amplifiers to achieve a high degree of bandwidth reduction. The information, even though buried in noise, can be correlated with the reference and extracted by phase-sensitive detection. This technique makes it possible to eliminate even those noise components occurring at signal frequency, since noise is not coherent with other signals.

The operation of phase-sensitive detection can be visualized by considering the circuit shown in Fig. 6.6. In it, two ganged single-pole, double-throw switches are shown connected so that the input signal can be connected to the load in two ways, W–Y and X–Z or W–Z and X–Y. Figure 6.7(a) is intended to represent only the information in the input waveform. If the switches are driven by the reference signal and its phase is adjusted relative to the information so that they move to the *up* position at times A and to the *down* position at times B, the output will be that shown in Fig. 6.7(b). The

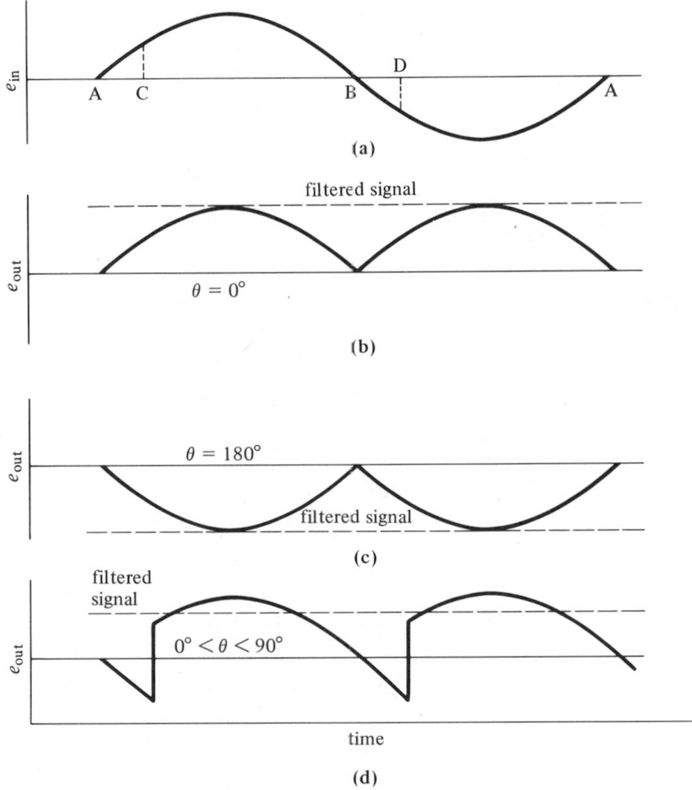

Figure 6.7 / *Phase-sensitive detection: effect of variation of phase angle θ between reference and information signals.*

use of a low-pass filter would produce a positive DC signal proportional to the amplitude of the input, which could be handled by a high-gain amplifier without its being saturated by noise. If the information and reference waveforms were 180° out of phase, the switches would move to the *up* position at times B and to the *down* position at times A, giving the output shown in Fig. 6.7(c). Filtering this signal would give a negative DC signal proportional to the amplitude of the input signal. If the reference-information phase relationship is something other than 0° or 180°—if, for example, switching occurred at times C and D—the result would be that shown in Fig. 6.7(d). Filtering would give an output equal to that from Fig. 6.7(b) multiplied by the cosine of the phase angle.

This operation transfers information from its original frequency to DC, which is demodulation or detection. Only signals that are coherent with the reference produce nonzero outputs; signals of frequencies other than that of the reference are rejected, since they have no fixed phase relationship with it and their average rectified value produced by filtering is zero.

In a wide variety of applications, phase-sensitive detection can be combined with signal modulation to provide a technique for extracting information from noisy signals. It is particularly useful in eliminating drift from low-frequency signals. To produce a smooth output after rectification, an effective low-pass filter must be used. The filter breakpoint frequency would ordinarily be at least two orders of magnitude smaller than the modulation frequency and would determine the upper frequency limit for the system. For example, suppose that information is encoded as the amplitude of a 1-kHz sine wave. The filter bandwidth would exclude changes in that amplitude that occur more rapidly than 10 Hz.

When this technique is used, it is essential that the information-bearing signal be modulated before the noise is encountered, because if the noise is modulated, it will not be rejected when the signal is detected.

Modulation is very often performed by interrupting or *chopping* the signal. A common example is the mechanical periodic blocking of a light beam in a spectrometer by a rotating disc with a hole in it. The beam is abruptly switched on and off, an operation that is nonlinear and therefore meets the criterion of a modulating process.

For example, consider modulation by chopping in atomic absorption spectroscopy. The experiment involves measurement of the attenuation of the intensity of a monochromatic light beam when it traverses a flame into which a sample is aspirated. The extent of reduction in transmitted light intensity is an exponential function of the concentration of absorbing sample atoms in the flame. Major noise sources are the light source and the flame. Chopping the beam before it enters the flame produces a fluctuating signal from the detector having an amplitude proportional to the transmitted light intensity. The frequency and phase are determined by the chopper. Detection might be achieved by driving the rectifying switches by the same motor as is used for chopping, thus ensuring synchronization. In this arrangement, noise arising in the flame, the detector, or the amplifier serving the rectifier will be rejected because it is not coherent with the reference signal. Noise originating in the light source is modulated and therefore is not rejected.

6.4.4 / Instrumental applications of the Fourier transform

Conventional spectroscopic technique involves the use of a monochromator to isolate restricted spectral bands for the purpose of making absorption or emission measurements. By this approach, the data needed for an amplitude–frequency spectrum of radiation absorbed or emitted can be obtained. The procedure affords a measurement of only a fraction of the total radiation at a time; this constitutes a severe limitation because, other factors being equal, improvement of resolution by reducing the monochromator bandpass causes reduction in measured signal intensity. This degrades the S/N ratio and accordingly limits the sensitivity of the measurement by restricting the application of signal enhancement techniques. Furthermore, the sequential nature of the process is time consuming.

It is true that if, in absorption spectroscopy, a beam consisting of an array of frequencies broad enough to cover the desired spectrum were used instead of monochromated light, absorption would occur and all of the information obtained in a conventional scanning experiment would be present. This is an attractive possibility, since it would afford a significant increase in signal intensity and a decrease in measurement time. However, this type of experiment is normally not useful, because the information would be in the form of a time–amplitude, rather than a frequency–amplitude, presentation. For spectroscopic purposes, only the latter can be used directly. If the time–amplitude plot were made, the spectroscopist would be in a position somewhat analogous to that of an observer who attempts to deduce by inspection the spectrum in Fig. 1.18 from the waveform in Fig. 1.17(d). The desired conversion is that of a Fourier transform, which can be performed practically only with the aid of a computer.

There are additional difficulties when this type of experiment is attempted in the optical spectroscopic region. Detectors presently available do not produce electrical signals at optical frequencies. The process of detecting with an optical transducer is in fact detection, in the sense that this term is used in discussing modulation. The monochromated beam is analogous to curve B in Fig. 6.2, the beam transmitted through a sample to the detector is analogous to curve C, and the detected signal is analogous to curve A. The "white light" experiment would correspond to an attempt to recover information from the envelopes of many modulated carriers, such as curve C, simultaneously using just one detector. The Fourier transform procedure permits a mathematical resolution of the array of modulated carriers, but only if a method is found of recording, effectively, the instantaneous values of light intensity vs. time at optical frequency. This difficulty is circumvented in optical spectroscopy by using an interferometer to modulate the beam to shift the information to a sufficiently low-frequency range to permit direct conversion to proportional electrical signals.

6.4.5 / Conclusions

Of the various signal enhancement techniques that have been discussed, bandwidth reduction is the simplest in that it involves no change in those signal components that are of interest; also, no change in the time scale of the observation is implied. The S/N ratio is improved by eliminating unused segments of the spectrum, providing that $S/N > 1$ in the limited bandwidth of interest.

In considering the implications of bandwidth reduction, it is evident that if the observation is limited to a low-frequency region, such as from DC to 1 Hz, only slowly changing signals can be observed. It should be noted, however, that this is true irrespective of the frequency range passed. From Eq. (6.2) we see that the capacity of a system is a function of bandpass, not of the absolute magnitude of frequency. Accordingly, a 1-Hz bandpass provides the same capacity at 1 MHz as at DC. A 1-MHz signal with a 1-Hz width could show changes in the amplitude of its envelope only at the rate of 1 Hz.

Signal averaging is a form of bandwidth reduction that permits observation of signals even when the *S/N* ratio in the frequency range of interest is less than unity. However, to use this technique, the observation time must be long compared with the period of the lowest-frequency signal component of interest.

This limitation can be avoided by the use of signal modulation and phase-sensitive detection. The detected signal is averaged, but since the information has been shifted to higher frequencies by modulation, the period of the observation need not exceed the period of the lowest-frequency component of the signal.

Both the requirement of extended observation period and of prior modulation can be circumvented by the use of the Fourier transform technique. However, one must have a signal that is "white" with respect to the frequencies of interest, and the equipment required to perform the transform must be available.

PROBLEMS

1 / A 10-mV strip-chart recorder is sufficiently accurate to distinguish unambiguously between signal levels that differ by 0.1 mV. Its response time is 1 sec full scale, with 0.1 sec additional to settle at an equilibrium position. Estimate the rate at which it can accept information.

2 / (*a*) Calculate the signal bandwidth that can be accepted by a recorder that can show full-scale response in 1 sec.

(*b*) If this recorder exhibits 0.1-mV uncertainty in its measurement for a 10-mV span, calculate its theoretical information capacity. Note that *S/N* refers to the ratio of signal power, which is proportional to the square of voltage.

3 / The output from a spectrophotometer is found to show a *S/N* ratio of two when observed by means of an oscilloscope that has a 500-kHz bandpass. Estimate the *S/N* ratio to be expected theoretically if a recorder having a 0.5-Hz bandpass is used to observe the signal.

4 / The signal from an electrocardiograph typically has components in the range 0.05–100 Hz. Suppose that these signals are to be sent by telephone to a central observation station and that the telephone lines offer a 10-kHz bandpass. If it were decided to modulate the signals in order to serve more than one patient with a single telephone line, how many could in principle be accommodated?

5 / Suppose that it takes 30 min to scan a desired region to produce a nuclear magnetic resonance spectrum. The signals are recorded and successive scans are averaged to improve *S/N*. How long would the process take if the *S/N* of the average were to be an order of magnitude better than that of a single pass?

chapter 7 / Fundamental measuring operations

chapter 7

The operation of an instrument involves transfer of information across two interfaces: that between the phenomenon of interest and the instrument and that between the instrument and the user. The details of these interfaces are determined by the nature of the instrument, whether a spectrophotometer or a polarograph or a mass spectrograph, and by the type of design that is chosen, analog or digital, automatic or manual operation.

The input interface generally includes a component that produces an electrical signal carrying information about the phenomenon under observation This type of component is called a *transducer*; a photoemissive cell is an example. When illuminated by light of suitable wavelength, a photocell passes a current proportional to light intensity, providing a signal that can be amplified and measured. The output interface, which serves to transmit information from the instrument to a person or to another instrument, must include some form of *readout*. This may be a meter, recorder, or a data printer to communicate with humans, or a suitable signal generator to pass information to other instruments.

While the details of input and output interfaces vary greatly, depending on the nature of the instrument, their basic function is that of measurement of an electrical signal. It is therefore useful to consider the problems associated with measurement of electrical signals before attempting to deal with specific examples in chemical instrumentation.

7.1 / PRINCIPLES OF ELECTRICAL MEASUREMENT

Information measured by instrument transducers is often encoded in terms of current intensity, less often in terms of voltage level or of frequency or signal phase relative to a standard. Most commonly, the critical factors determining the accuracy of measurement are correct impedance relationships and proper frequency response of the coupled components.

7.1.1 / Impedance relationships

It has been pointed out in Section 2.3.3 that to make an accurate measurement of voltage, the input resistance of the measuring device must be large compared with the output resistance of the signal source. With this relationship of input and output resistances, the error that would occur if an appreciable fraction of the signal were dropped across the source resistance is avoided. An important exception to this generalization regarding voltage transfer occurs if the information being transmitted is encoded in the form of a very high-frequency voltage wave shape, such as a train of pulses with pulse shape the critical factor. Here the conditions for most efficient power transfer, that is, source impedance matching receiver impedance, would be necessary.

In general, however, if a voltage is to be measured, it is necessary to arrange that the receiver input resistance be much larger than the source output resistance. Accordingly, the use of voltage sources with large output impedances places a restriction on the input impedance of the components coupled to them.

Conversely, accurate measurement of current source signals requires that the source output impedance be large compared with the receiver input impedance. In the ideal case, a receiver that exhibits a zero input resistance would be incapable of causing any perturbation of the source current, regardless of how little power is available to effect current flow. It follows, therefore, that high-impedance current sources are more easily dealt with than low-impedance current sources.

7.1.2 / Frequency response

It is generally necessary that the frequency response of the system be compatible with the operating range of the signal being measured. Three rather different situations are frequently encountered. One is the case in which the signal can be expected to contain only relatively low-frequency components, say of the order of 1 Hz as an upper limit. Another is that in which frequency components are in the range of capability of electronic components, that is, up to the gigahertz range. The third is the case in which frequencies in the optical range must be handled.

When only low-frequency signal components will be encountered, the designer may select components that incorporate mechanical working parts, such as moving-coil meters and electromechanical servo machines, if this is desirable. For signals in the intermediate range, he is constrained in most instances to use electronic devices. If mechanical components are to be used, the information must be shifted to low frequencies, probably by demodulation. For signals in the optical frequency range, no alternative presently exists other than a shift of information to lower frequencies by suitable detection devices. This is the function of optical transducers such as photocells and thermocouples.

7.1.3 / Voltage measurements

Voltage measurements can be made by direct application of the source to an amplifier that responds to voltage signals, ordinarily one having a FET input in modern equipment. More often, however, they are made by connecting the signal source to a very sensitive current-measuring device that exhibits a large input impedance. Amplifiers utilizing transistor inputs, which are very common "voltage" amplifiers, in fact respond to currents. The measurement is of the current that flows as a result of the application of the voltage. Moving-coil meters, discussed in Section 7.2.1, also fall in this category. However, they are generally not connected directly to transducers because of inadequate input impedance and frequently inadequate sensitivity.

7.1.4 / Current measurements

Owing to the nature of available transducers, the bulk of instrumental measurements are basically current measurements. Probably the most widely used instrumental current-measuring arrangement is that in which a resistor of known value is placed in series with the current source and the voltage dropped across it is observed. It is necessary, of course, that the source resistance be much larger than the dropping resistor; however, there are many useful transducers (see Section 7.3.1) that exhibit sufficiently high source resistances to permit the use of megohm-dropping resistors. This makes it possible to make current measurements in the submicroampere range with quite ordinary voltage-measuring devices. It is interesting to note that the current output of a transducer is often measured by dropping it across a resistor to produce a voltage that is measured by observing the current that it will drive into a transistorized amplifier.

Current measurements, of course, are also made by means of moving-coil meters. In many cases, however, the current output available from transducers is too small to make direct readout practical. Accordingly, moving-coil meters are often used as amplifier readouts.

When impedance-matching problems preclude the use of dropping resistors, low input impedance current-to-voltage conversion can be accomplished by using fedback amplifiers. A discussion of the application of operational feedback to current amplification is given in Section 8.2.2.

7.2 / DEVICES FOR CURRENT AND VOLTAGE MEASUREMENT

7.2.1 / Moving-coil meters

The most widely used electrical measuring device is the moving-coil meter, which can be used to make either current or voltage measurements. In operation, the meter coil is placed in series with the current source load, Fig. 7.1(a), or in parallel with the voltage source load, Fig. 7.1(b). In either case, current drawn by the meter passes through a coil of wire, setting up a magnetic field around the coil, which is mounted within the field of a perma-

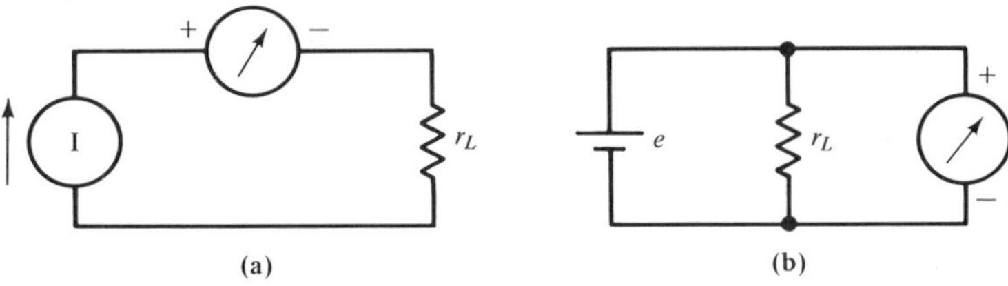

Figure 7.1 / (a) *Current measurement.*
(b) *Voltage measurement.*

nent magnet. With proper orientation of the parts, interaction between the fields causes a force to be exerted between the coil and the magnet that is proportional to the current. The coil is mounted in a bearing that permits it to rotate, and it is spring-loaded and fitted with a pointer or a mirror that indicates the extent of deflection.

The sensitivity of this type of device, determined by the number of turns in the coil, the strength of the permanent magnet, and the nature of the mounting arrangement, can vary from amperes to fractions of microamperes for full-scale deflection. To achieve the degree of ruggedness that permits design of a conveniently portable instrument, sensitivities are ordinarily in the milliampere range. Adjustable sensitivity for current measurement is arranged by connecting variable shunting resistors in parallel with the meter movement. Given the basic meter sensitivity in amperes per scale division and the coil resistance, the effective sensitivity with any specific shunt resistance can be readily calculated. The accuracy of a measurement may be expected to be in the range of 1–5 percent of the full-scale value.

For any measuring device, the speed of response is an important characteristic; those utilizing moving mechanical parts are necessarily quite limited, considering the frequencies that may be encountered in electrical measurements. Response time of moving-coil meters is limited by bearing friction and inertia of the moving parts. Where it is important, weight is generally minimized by using a light beam reflected from a small coil-mounted mirror, rather than a pointer, to indicate coil deflection. Instruments are available that follow signals at frequencies up to several hundred hertz; however, most moving-coil meters require at least a substantial fraction of a second for full-scale response.

Magnetic damping is another factor that may significantly influence response time. In the absence of other considerations, a spring-loaded coil mounted in a good bearing would be expected to oscillate for a protracted time if displaced from its equilibrium position. However, if the coil is within a magnetic field and if it is part of a complete circuit, any movement causes induction of a current in the coil in a direction such that the magnetic field induced by that current interacts with the external magnetic field to hinder the coil movement. Accordingly, the lower the resistance to which the coil

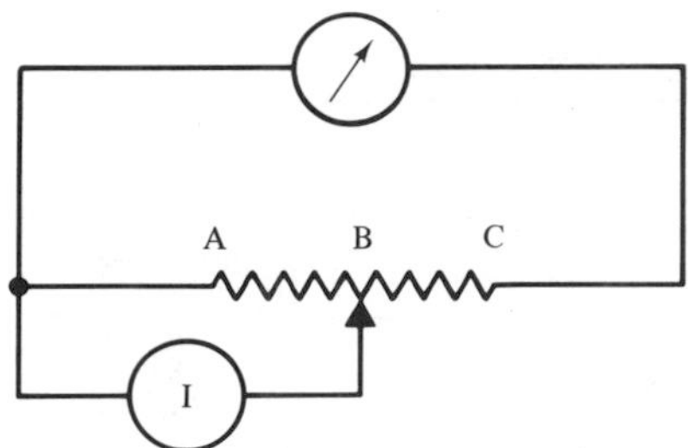

Figure 7.2 / *Use of an Ayrton shunt to avoid variation in meter response time when sensitivity is changed.*

is connected, the more effective will be the damping action. This is why the terminals of a very sensitive meter are shorted before it is moved. The damping action with a low-resistance shorting strap permits only very slow coil movement in response to vibration, even though the spring mounting may be very delicate.

On deflection, a meter shunted by a large resistance will respond rapidly and will tend to oscillate; with a small shunt resistance, it may respond very slowly. A definite shunt resistance therefore exists for each meter which permits maximum speed of deflection to a new equilibrium position without any overshoot. This is termed the *critical damping resistance* and is specified by the manufacturer for meters that are sufficiently sensitive to exhibit this type of behavior. When such a meter is used, for example, to measure polarographic currents, variation of shunt resistance to change sensitivity may be expected to affect response time. This may be avoided by the use of an Ayrton shunt, which is shown in Fig. 7.2. Here R_{AC} may be taken equal to the critical damping resistance. If it is much smaller than the signal source resistance, then the meter is always critically damped. Effective meter sensitivity can be adjusted by varying R_{AB} and R_{BC}.

Another important characteristic of meters is the input resistance. For moving-coil meters, this is simply the coil resistance, possibly modified by shunt or series resistances included to adjust sensitivity. Coil resistance values are typically in the range from a few hundred to a few thousand ohms. For very sensitive meters, values larger than 10 kΩ are encountered.

The discussion above pertains to the use of moving-coil meters for current measurements. Instruments intended for this use are termed *ammeters* or *galvanometers.* Ammeters are provided with necessary shunts and are calibrated directly in current units. Galvanometers ordinarily are calibrated in arbitrary scale divisions, actual sensitivity being determined by the type of shunt used with the meter.

If, as indicated in Fig. 7.1(b), the meter is placed in parallel, rather than in series, with the load, it can serve as a voltmeter. In this case, the meter does not measure the current drawn by the load directly, but instead measures the current that the voltage dropped across the load is capable of driving through the meter. If the meter sensitivity and resistance are known, then the voltage can be calculated. For example, a meter having a 1-kΩ coil and a 1-mA full-scale sensitivity would show full-scale deflection when 1 V was imposed. It could therefore be calibrated as a 0–1-V meter. The sensitivity could be made to be 0–10 V by placing a 9-kΩ resistor in series with the coil or 0–100 V by using a 99-kΩ resistor. Irrespective of the range chosen, this meter requires 1 mA for full-scale deflection. For this reason, voltmeter resistance would be specified as 1000 Ω/V. Meter input resistances usually fall in the range of 1–100 kΩ/V.

This type of meter can also be used as an ohmmeter for resistance measurements. To do this, an independent power supply is applied across the unknown resistance, which is in series with a known resistance. The meter is used to measure the voltage drops across the two resistances, the results of which could be used to calculate the unknown resistance. In practice, switching arrangements and dial calibration are provided so that actual calculation is unnecessary.

Moving-coil meters are often used to measure AC signals by arranging to rectify the input before the measurement is made. Such meters may measure the peak value or some reproducible fraction of the peak value and be calibrated in terms of rms values on the assumption that sinusoidal signals are measured. For nonsinusoidal signals, other types of measuring devices, such as oscilloscopes, must be used.

From the foregoing discussion it is apparent that DC current and voltage, AC voltage, and resistance measurements can be made with the same basic meter movement. It is therefore feasible to combine these functions in a single instrument by providing suitable dial calibrations and switching arrangements. The resulting instrument is termed a volt-ohm-meter (VOM) and is a very commonly used laboratory test instrument.

7.2.2 / Electronic meters

The moving-coil meter is simple and versatile. When it is inadequate for a particular measurement, the problem is likely to involve frequency response, input impedance, or accuracy. Performance is often improved by placing a linear amplifier between the meter and the signal source. This type of arrangement may be built into instruments to provide a dial readout. It may stand alone as an electronic voltmeter or an electrometer, a high input impedance voltmeter.

These instruments show the characteristics of the amplifier at the input and those of the meter at the output. It is therefore possible to provide the large input impedance characteristic of FETs or vacuum tubes, which is useful for voltage measurements, or by proper use of negative feedback (Section 5.2.4) to make the resistance very small to facilitate accurate current measurements.

While incorporation of an amplifier with a moving-coil meter cannot improve the accuracy inherent in the meter movement, greater accuracy and ease of reading can be achieved by using a digital rather than an analog

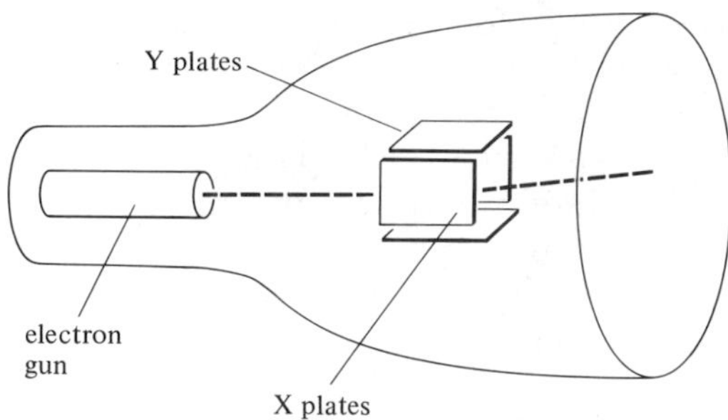

Figure 7.3 / *Simplified representation of an oscilloscope cathode-ray tube.*

meter. The operation of a digital meter involves periodic sampling of the input signal. Each sample is converted to a digital coded form which is used to control a readout display.

Thus, instead of determining the position of a needle with respect to a calibrated dial, the measurement is read directly in digital form. Analog meters are inherently limited in accuracy, as was mentioned above. By contrast, a digital meter could in principle be constructed with any number of digits. In fact, meters with two- to six-digit resolution and accuracy are commonly used.

7.2.3 / Oscilloscopes

The problem of increasing the applicable frequency range is ordinarily resolved in one of two ways: by elimination of mechanical moving parts from the instrument, or by using detection to shift the information of interest to a frequency range that can be handled by electromechanical devices. Most commonly, mechanical parts are eliminated from the readout by using a cathode-ray tube for visual presentation of information. This tube, of which a TV picture tube is one example, consists of an electron gun mounted to direct its beam at a phosphor-coated screen inside the face of the tube. A simplified diagram is given in Fig. 7.3. The electron beam produced by the gun is deflected by either electrostatic or magnetic fields which are generated by an amplifier driven by the signal to be examined. The small inertia of an electron beam permits very fast response.

The oscilloscope, which is a widely used voltage-measuring device for high-frequency signals, utilizes electrostatic deflection through two pairs of plates as indicated in Fig. 7.3. In the usual case, a linear "sawtooth" signal is imposed on the X plates, which causes the beam to move at a constant rate horizontally across the tube face and then to fly back abruptly. The signal to be observed is imposed on the Y plates.

The oscilloscope can be used to observe signals in the range of hundreds of megahertz. It does not, however, offer any improvement in accuracy by comparison with a moving-coil meter.

While cathode-ray presentation and also more elegant schemes involving high-speed computers may be used to examine high-frequency signals, more often some form of detection is used to shift the information to the frequency range appropriate to a moving-coil meter. This approach is used in all forms of optical spectroscopy.

7.2.4 / The potentiometer

The readout devices discussed above are examples of direct measurement. The output is directly dependent on the accuracy, linearity, and reproducibility of the amplifier or meter used. The development of negative feedback techniques has made it feasible to produce amplifiers with performance adequate for direct measurement for almost any application; however, if the experiment involves an indirect comparison rather than a direct measurement, much simpler components can be used to make possible significant reductions in construction and maintenance cost.

Fundamentally, an indirect measurement involves the comparison of the unknown to be measured with a standard. If a selection of standards is available so that one can be chosen to match the unknown exactly, then the measurement is complete. The advantage of this procedure, as compared with a direct measurement, lies in the fact that the measuring device need only be capable of detecting differences between the standard and the unknown. It need not be capable of linear response, nor of any great accuracy. In using a comparison measurement, the requirements for linearity and accuracy can be shifted from the amplifier, which is inherently nonlinear and inaccurate, to a standard that can much more easily be made highly accurate and reproducible.

The concept of comparison measurements is extremely versatile; it can be applied to measurements of many types of unknowns, with electronic, mechanical, or electromechanical equipment. In the present discussion, we shall confine our attention to measurements of electrical signals.

The potentiometer is a widely used example of this type of instrument. Two voltage sources are connected in series opposition, as indicated in Fig. 7.4. Arrangements are made to measure either their voltage difference (error voltage) or the current that flows as a result of it. When the standard is adjusted to be equal to the unknown, the potentiometer is balanced and the error voltage and the current are zero. The accuracy and reliability of the measurement is primarily a function of the performance of the standard, that is, of the accuracy of e_{std} and of the linearity of the slidewire used to pick off a known fraction of e_{std} to compare with e_{unk}. The reliability of these components can be very precisely controlled. When the instrument is balanced, the current drawn from the unknown is zero, which means that the input impedance is infinite, the ideal value for a voltage-measuring device. It should be noted that input resistance is not infinity when the potentiometer is not balanced; accordingly, the output impedance of the signal source must be sufficiently low that its characteristic voltage may be imposed on the

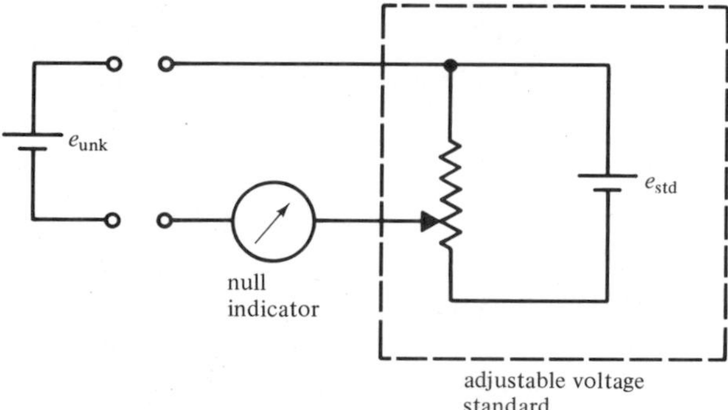

Figure 7.4 / *Manual potentiometer.*

unbalanced potentiometer input resistance. Otherwise, the potentiometer cannot be balanced. Ordinarily, imbalance is detected by means of a sensitive galvanometer, as indicated in Fig. 7.4; however, when high signal source impedances must be handled, as in pH measurements, a high input impedance amplifier is used to detect the error voltage.

7.2.5 / The servo mechanism

Scientific instrumentation is automated at present to a considerable extent. For example, automatic continuous readout to chart recorders is very generally used. Increasingly, automatic operation, by means of control loops in which some form of computer monitors the instrument and makes continuous adjustment of variables, is coming into use. These developments are based on advances in many areas, but one type of device, the servo-driven potentiometer, is truly ubiquitous. Its function is to produce mechanical motion that is proportional to a voltage. It can therefore communicate with humans by drawing lines on charts. It can control instruments or other machinery by adjusting variable resistors, by blocking light paths, or by controlling the flow of fluids.

The design of the servo-driven potentiometer is a combination of two basic concepts, comparison measurement and phase-sensitive detection, which has the fortunate effect of making minimal demands on the components. As a result, it is possible to construct instruments that provide highly accurate operation at a cost that is modest compared to that of other items of equipment. Moreover, the overall performance of the instrument is less sensitive to deterioration in the condition of many of its components than is generally the case, and therefore provides enhanced reliability.

A simplified version of a servo-operated potentiometer is shown in Fig. 7.5(a). The unknown voltage is compared with e_s, that fraction of the internal standard voltage, e_{std}, that is picked off by the slidewire wiper. If e_{unk} is not equal to e_s, the difference, or error signal, is received by the servo amplifier, which functions as a null indicator. A nonzero error signal is amplified and

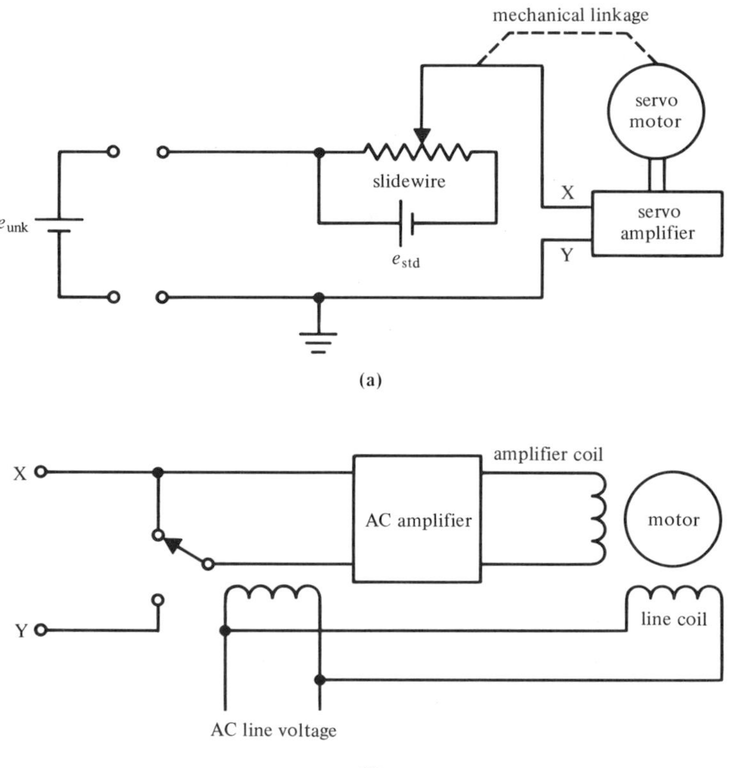

Figure 7.5 / *The servo-driven potentiometers.*
(a) *Block diagram.*
(b) *Some details of the servo amplifier and motor.*

applied to the servo motor, which is mechanically linked to the wiper of the potentiometer. Connections are made so that the motor moves the wiper to the right when X is positive relative to Y and to the left when Y is positive relative to X.

For satisfactory operation, the null detector must be capable of detecting very small error signals that exist when the instrument is not balanced and of distinguishing between positive and negative signals. It must be free of drift and offset. The fact that a comparison measurement is used, however, means that the control apparatus does not have to show linear or accurate gain.

This required performance is achieved in an elegantly simple manner by using the error signal to modulate a carrier. The modulated signal is amplified by a simple, high-gain, AC-coupled amplifier and subjected to phase-sensitive detection. The use of phase-sensitive detection makes the system highly insensitive to drift, offset, and other noise that may be introduced while the information is encoded in the modulated signal. Only

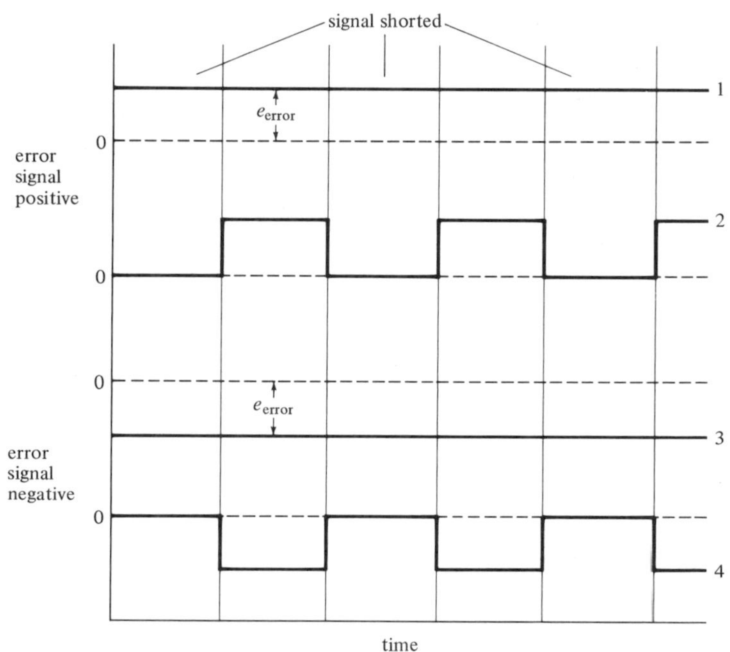

Figure 7.6 / *Modulation by chopping. Effect of polarity change on chopped signals.*

noise present before the modulation step will be detected, and the only active component within the instrument that can contribute is the adjustable standard, which can be made almost noise free. Noise introduced after detection, however, may cause trouble; it can be introduced by the mechanical linkage between the motor and the slidewire.

Modulation is accomplished by chopping, that is, by periodically interrupting the servo amplifier input and shorting X and Y of Fig. 7.5(a) together. This is ordinarily done by a mechanical switch driven by the AC line voltage. This operation constitutes modulation, since the on–off switching action is nonlinear. It produces a signal that consists of a carrier of line frequency with its amplitude modulated by the error signal. The modulated signal has a precisely defined phase relationship with the AC line voltage, which is a necessary requirement for phase-sensitive detection.

The servo chopper, amplifier, and motor are shown in somewhat greater detail in Fig. 7.5(b). The magnetically driven chopper produces the signals shown in the curves of Fig. 7.6. Curve 2 illustrates the effect of chopping the positive signal of curve 1, and curve 4 illustrates the effect of chopping the negative signal of curve 3. The chopped signals differ in two ways: Values of curve 2 are either zero or positive, while those of curve 4 are either zero or negative, and there is a 180° phase angle between them. An AC-coupled servo amplifier is used to take advantage of its simplicity of design. This makes it impossible to distinguish between positive and negative error

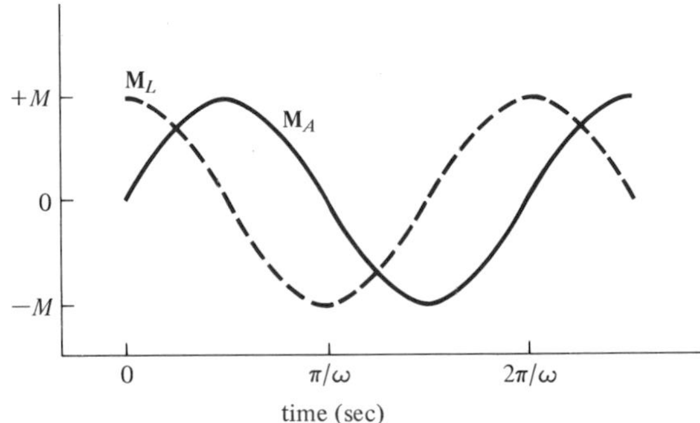

Figure 7.7 / *Servo motor field coil energizing signals.*

signals at the amplifier output on the basis of the polarity of the signal. Regardless of the input, the output will show an average value of zero.

Nevertheless, the information regarding the sign of the error signal is not lost; it is encoded as the phase of the chopped signal, relative to that of the AC line waveform as a reference. If a positive error signal decreases to zero and then becomes negative, the amplifier output decreases to zero and then increases again, but with a 180° phase shift.

Phase-sensitive detection is accomplished by a two-phase induction motor. This consists of an armature of ferromagnetic metal mounted to rotate within the magnetic fields of two coils. One of these is energized by the servo amplifier output, which is the modulated signal; the other is energized by the AC line signal, which serves as the switching signal for the detector. The coils are oriented with their magnetic fields at right angles, as indicated in Fig. 7.5(b). From this figure, the AC amplifier output and the line voltage would apparently be either in phase or 180° out of phase. Actually, a phase-shifting network is included to ensure that the two signals are always 90° out of phase. If a positive error signal causes the amplifier output to lead the AC line signal, a negative error signal will cause it to lag.

The motor serves as a phase-sensitive detector in that its direction of rotation is controlled by the phase of the amplified chopped error signal and its speed is very approximately controlled by the amplitude of the error signal. The motor is unresponsive to signal components that are not coherent with the error signal, which is the expected behavior of a phase-sensitive detector.

The motor armature is acted upon by the magnetic field that is the resultant of the fields produced by the two coils. If the resultant field rotates, the armature will follow it. Let us assume that the field strengths of the amplifier and line coils are given by Eqs. (7.1) and (7.2) and shown graphically in Fig. 7.7.

$$\mathbf{M}_A = M \sin \omega t \tag{7.1}$$

$$\mathbf{M}_L = M \sin\left(\omega t - \frac{\pi}{2}\right) \tag{7.2}$$

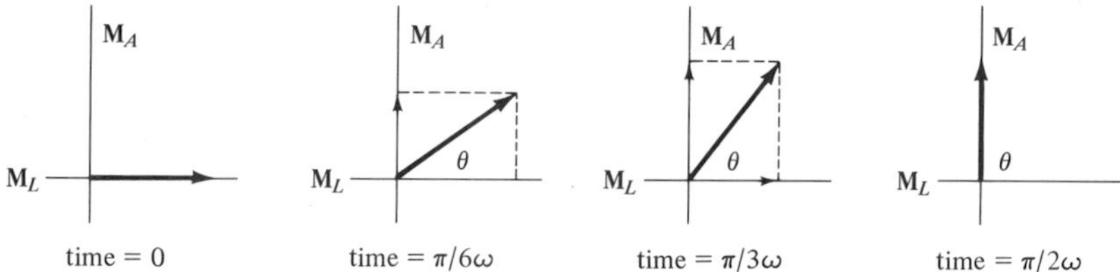

Figure 7.8 / *Coil and resultant magnetic fields for two-phase induction motor operation.*

Here the signal energizing the line coil lags that in the amplifier coil.

Values for $\mathbf{M}_L$, $\mathbf{M}_A$, their resultant, and the physical angle between the resultant and the axis of the line field coil, θ, are given in Table 7.1 for various times. In Fig. 7.8, the axis of the line coil is oriented in the horizontal and that of the amplifier coil in the vertical. Vectors representing instantaneous values of the two coil fields are plotted accordingly.

At time zero, the amplifier coil field is zero and the line coil field is at a positive maximum. As indicated in Table 7.1 and in Fig. 7.8(a), the resultant lies along the axis of the line coil. After $\pi/6\omega$ sec (30 electrical degrees) the fields are as indicated in Fig. 7.8(b) and the resultant has shifted to make $\theta = 30°$. With passage of time, the resultant continues to move. As indicated in Fig. 7.8(c) and 7.8(d), the physical angle between the resultant field and the reference coil axis follows the electrical angle ωt, rotating counterclockwise, and the armature follows the resultant field. The resultant and the armature rotate clockwise if, instead of lagging as in Eq. (7.1), $\mathbf{M}_A$ leads $\mathbf{M}_L$.

7.2.6 / The impedance bridge

A variety of phenomena can be observed by measurement of impedance or resistance. These are discussed in Section 7.3.2. Measurements are usually made by comparing the unknown impedance with standards arranged in a bridge array. As an illustration, the measurement of resistance by a DC Wheatstone bridge will be described here. The Wheatstone bridge consists of four resistances arranged as indicated in Fig. 7.9. Measurement is per-

Table 7.1

Time (sec)	*Amplifier coil field strength,* $\mathbf{M}_A$	*Line coil field strength,* $\mathbf{M}_L$	*Resultant field strength*	*Resultant–line coil angle,* θ *(deg)*
0	0	M	M	0
$\pi/6\omega$	$0.500M$	$0.866M$	M	30
$\pi/3\omega$	$0.866M$	$0.500M$	M	60
$\pi/2\omega$	M	0	M	90

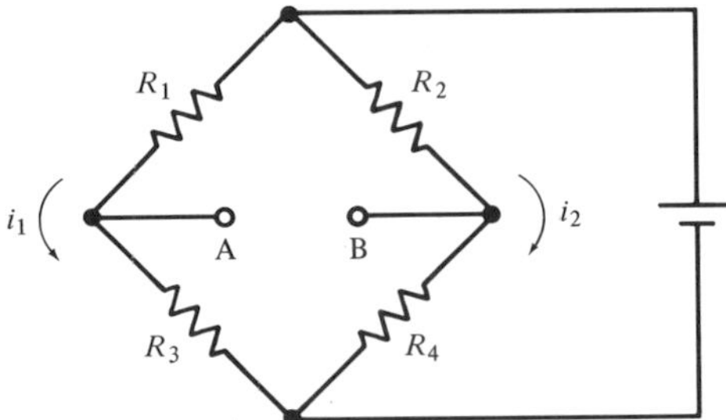

Figure 7.9 / *DC Wheatstone bridge.*

formed by observing the voltage difference e_{AB}. The bridge is said to be balanced when $e_{AB} = 0$, therefore

$$i_1R_1 = i_2R_2 \tag{7.3}$$

$$i_1R_3 = i_2R_4 \tag{7.4}$$

Accordingly,

$$\frac{R_1}{R_3} = \frac{R_2}{R_4} \tag{7.5}$$

Very often, two arms of the bridge, for example, R_3 and R_4, are made accurately equal and constant; therefore, when the bridge is balanced, R_1 and R_2 are equal. Measurement amounts to a comparison of R_1 with R_2 and is detected by the use of a null detector across the arms of the bridge. These general conditions are identical to those encountered in using a potentiometer; as might be expected a servo mechanism also can be readily adapted to this measurement.

Very often, an actual value of resistance is not required. For example, in gas chromatography, only an indication of the resistance change is needed. This is actually obtained by bringing the bridge initially into balance and observing Δe_{AB} by means of a chart recorder to provide an indication of change in thermal conductivity as a function of time.

Bridge measurements are applicable to the measurement of impedances containing capacitive or inductive components. To deal with reactances, it is necessary to energize the bridge with an AC signal, rather than with the DC source shown in Fig. 7.9. Frequently, 1 kHz is used, which is high enough to afford considerable separation from 60-Hz leakage from the power lines but not so high as to lead to severe difficulties caused by stray capacitances. The null detector must be capable of responding to an AC voltage imbalance. Earphones, AC voltmeters, and oscilloscopes are used. The standard impedance used to match the unknown must contain suitably chosen reactances. For example, if the unknown contains a capacitance in parallel with a resistance, the impedance put in place of R_2 would be a parallel R–C

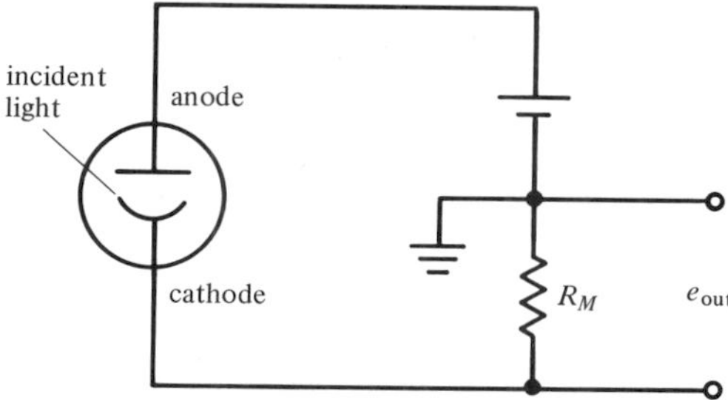

Figure 7.10 / *Vacuum photoelectric cell.*

combination. Both resistance and capacitance are adjusted to balance the bridge.

It should be noted that ohmmeters, described in Section 7.2.1, cannot be used to measure impedances that have reactive components. They can be used for the resistive component of an inductor but not of a capacitor. In particular, they are not useful for measurements on electrochemical systems, such as solution, electrode, or battery resistances, without special precautions. For these applications, bridge measurements are ordinarily used.

7.3 / TRANSDUCERS AND THEIR USES

There are a great many different transducers used in scientific instruments. For this discussion of electrical measurements, it is convenient to consider them in terms of their behavior as signal sources, rather than of the phenomenon being measured. The most commonly encountered transducers may be grouped in three categories: current sources, variable impedances, and voltage sources.

7.3.1 / Current source transducers

Many of the widely used current source transducers involve flow of current between electrodes, usually in an evacuated or gas-filled space. These are operated so that the interelectrode space exhibits a very high resistance. Thus, the component provides a high output resistance which, for a current source, facilitates coupling to an amplifier. Common examples are vacuum photoemissive cells used for measuring light intensity, gas-filled devices used to measure ionizing radiation, and those used to monitor the composition of flowing gas streams.

A / *Photoemissive tubes*

Absorption of photons transfers the characteristic photon energy to the absorber. Under certain conditions, this may cause the escape of an electron from the absorbing surface. If this process occurs in a device such as that indicated in Fig. 7.10, the emitted electrons may move away from the cathode

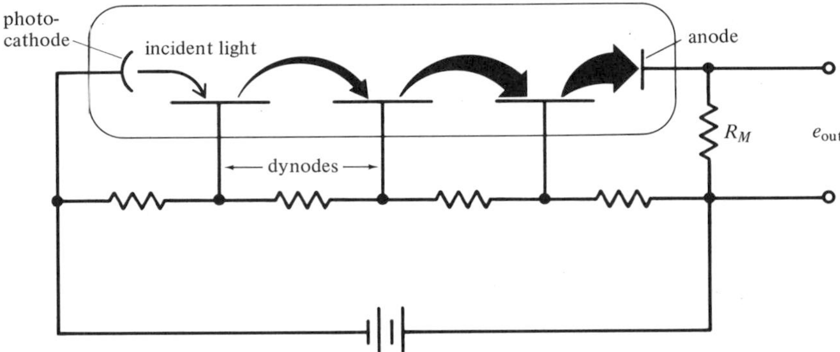

Figure 7.11 / *Photomultiplier tube.*

to an anode under the influence of an applied voltage gradient and thus produce a current controlled by the intensity of the incident light.

In practice, the cathodes of photoemissive tubes are made of materials that require minimal energies for electron emission, such as silver oxide coated with cesium metal. These cathodes respond to light in the visible and ultraviolet regions. They are placed within evacuated chambers to avoid collision between molecules and photoelectrons which would alter performance or even cause uncontrolled discharge. Simple vacuum photoelectric cells offer a sensitivity of approximately 100 μA/lumen and produce currents of the order of magnitude of a few microamperes.

The sensitivity of a photoemissive tube can be increased by allowing the photoelectrons to impinge upon an electrode after having been accelerated sufficiently to cause secondary ionization. Such a device, a *photomultiplier tube*, is shown schematically in Fig. 7.11. With proper operating conditions, the impact of an electron on a dynode causes emission of several electrons. This process is carried out repeatedly as indicated. Commercially available photomultiplier tubes ordinarily have 10 or more stages and provide sensitivities in the range of amperes per lumen. They are used to measure very low light intensities, for which anode currents range in the order of microamperes to milliamperes. The power supply needed to operate a photomultiplier tube must furnish voltage of the order of 1 kV in order to energize the sequence of dynode stages. By contrast, a photoelectric cell requires only around 100 V.

With either photoelectric cells or photomultiplier tubes, the currents to be measured are of the order of milliamperes. The light intensities represented are, of course, very different; nevertheless, coupling can conveniently be made by placing a dropping resistor, R_M in Fig. 7.10 and Fig. 7.11, in series with the anode and applying the resulting voltage drop to a suitable amplifier. The measuring resistor is typically of the order of a megohm; accordingly, the signal level may be in the order of a volt or more. The amplifier must show a reasonably high input resistance, but usually need not have a very large voltage gain. The ease with which photoemissive cells can be coupled with other components has led to their very wide application in many types of instruments.

B / *Ionizing radiation detectors*

Various types of radiation, alpha and beta particles, gamma and X rays, can be detected by observing the effects of ionization occurring when the radiation is absorbed. These detectors may take the form of gas-filled tubes in which electrodes are mounted. A voltage is imposed between the electrodes in order to collect the ions produced by the radiation being measured.

If a voltage is imposed that is sufficient merely to collect the ions formed, the device functions as an *ionization chamber*. The current is proportional to the number of ionizing events. Imposition of an increased voltage gradient, ordinarily by using a wire for the anode, makes it possible to cause the formation of secondary ions by collision of the primary ions with the fill gas. In this type of operation, the current level is proportional to the number of ionizing events; in addition, the current pulses produced by individual absorption events may be proportional to the number of ions produced in each event, hence to the energy of the photon or particle absorbed. A detector operating in this mode is termed a *proportional counter*.

With a sufficient increase in the voltage gradient, the secondary ionization can become so extensive that all ionizing events that occur produce an avalanche of electrons. Such a detector, a *Geiger counter*, produces a large uniform pulse for all ionizing events, but is incapable of distinguishing between various types of ionizing radiation.

Detection analogous to that of a gas-filled proportional counter can be achieved by use of a reverse-biased *np* junction as the absorber. Absorption of ionizing radiation—for example, gamma or X radiation—in the depletion zone of the junction causes formation of charge carriers that can cross the junction. The number of charge carriers freed by a single ionizing event is a function of photon energy. These detectors are used for gamma and X-ray spectrometry because they provide better resolution of photon energies than is possible with a gas-filled proportional counter.

C / *Flame ionization detectors*

Flame ionization detectors are widely used with gas–liquid chromatographs. The chromatograph effluent stream is introduced into a hydrogen–air flame maintained between two electrodes. When a carbon compound enters the flame, the combustion process produces a concentration of ions that is proportional over wide limits to the concentration of the carbon compound. The detector operates by measuring the current produced by collecting the ions at the electrodes.

D / *Photoconductive transducers*

The relatively high resistance characteristic of pure semiconductors is reduced in fabrication of solid state electronic components, such as transistors, by doping with controlled concentrations of impurities to form *n*- or *p*-type material. Reduced resistance is associated with a reduction in the energy required to excite a charge carrier from an equilibrium position in the crystal lattice. A reduction in resistance can also be achieved by irradiating the semiconductor with light of sufficient energy to excite charge carriers from the valence band to the conduction band. The resistance of the material is then a function of light intensity.

This phenomenon is the basis for photoconductive radiation detectors, such as lead sulfide detectors, which are used in the near infrared region of the spectrum (1–3 μm). These devices show a current output that is a function of the incident radiant power and exhibit an output impedance that is high enough to allow convenient signal amplification.

7.3.2 / Variable-resistance transducers

Several types of measurements can be made based on variation of resistance or impedance. These include temperature, thermal conductivity, infrared radiation intensity, and electrolytic conductivity.

A / *Resistance thermometers*

Very accurate temperature measurements can be made by using the temperature dependence of resistance of various materials. In particular, resistance thermometers are made of platinum, nickel, and certain refractory semiconductors (thermistors). The resistance of metals increases with increasing temperature, that of thermistors decreases. By means of a platinum resistance thermometer, temperature variations as small as 0.0001° can be measured over the range of $-190°$ to $+660°$C. Thermistors exhibit larger temperature coefficients than metal thermometers, but are less stable and accurate. With either type, measurements are ordinarily made by means of a bridge, as discussed in Section 7.2.6.

B / *Bolometers*

The *bolometer* is a device somewhat similar to the resistance thermometer; it is used to measure small heating effects, especially for measurement of infrared radiation. It consists of an absorbing material, often a thin strip of platinized platinum, which is made one element of a Wheatstone bridge. Radiation falling on this element raises its temperature, hence its resistance. This application, in contrast to the operation of a resistance thermometer, does not involve a determination of a value of temperature, but only observation of the relative change in resistance, which can be related to infrared radiant power in a comparison measurement, such as is used in a spectrophotometer.

C / *Thermal conductivity*

Measurement of variations in thermal conductivity of gases can be made by observing resistance changes of metallic filaments. A filament is heated by passing a current through it. Under proper conditions, equilibrium can be established between heat input and heat loss via thermal conductivity by gas surrounding the filament; filament resistance is constant as long as thermal equilibrium is maintained. Any change in thermal conductivity will cause the filament temperature, hence its resistance, to change. This device is used in detectors for gas–liquid chromatographs. Response actually is an indication of a change in thermal conductivity, rather than a determination of an absolute value. If either hydrogen or helium, which have very large thermal conductivities, is used as carrier gas, then all other compounds that are chromatographed may be expected to show smaller conductivities. Accordingly, the detector responds to all compounds, with the sensitivity being affected by the difference in the thermal conductivity of the carrier and the eluate compound. As is generally true of resistance measurements, a bridge arrangement is used.

D / *Electrolytic conductivity*

Measurement of electrolytic conductivity is an example of a resistance measurement that is ordinarily made with an AC-energized bridge rather than a DC bridge. An electrolytic conductivity cell consists of two inert electrodes, frequently platinum, immersed in the test liquid. If a voltage is applied, a current can be observed; however, current flow is necessarily associated with electrochemical reactions at each electrode which, if allowed to occur for appreciable periods, will change the conductivity of the system. Use of an AC voltage, usually of 1000 Hz, circumvents the problem. The electrochemical reactions may be reversible, in which case no net reaction occurs. Introduction of an AC signal across the electrode–solution interface may to a considerable extent occur via the capacitance associated with the interface; this minimizes the extent of electrochemical reaction at the interface for a given current flow. If conductivity measurements are made in the radio frequency range, electrode reactions are almost completely eliminated and it is not necessary to bring the electrodes in contact with the solution. Capacitive pathways are effective even if the electrodes are mounted outside the glass container holding the solution.

7.3.3 / Voltage source transducers

Of the tranducers that are widely used, relatively few involve direct generation of a voltage. Two types, thermocouples and potentiometric electrodes, are quite important.

A / *Thermocouples*

When two dissimilar metals are brought together, as indicated in Fig. 7.12(a), a difference in temperature between the junction points, A and B,

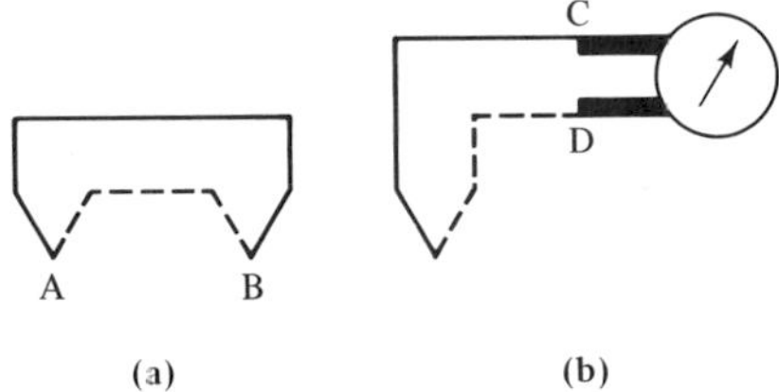

Figure 7.12 / (a) *A simple thermocouple.*
(b) *A thermocouple with readout.*

gives rise to a contact potential that will drive current around the loop. Within restricted temperature limits, thermal voltages are a linear function of temperature. Even over wide ranges, the voltage is accurately reproducible and represents an excellent basis for precise temperature measurements. Calibration values are determined by the choice of metals, but typically amount to several millivolts for a 100° temperature difference.

Although a contact potential can be expected with any pair of metals, certain pairs are widely used because of inertness to chemical attack and high sensitivity of output. A couple made of platinum paired with platinum–rhodium (90–10 percent) is especially useful in the range 0°–1450°C. The sensitivity is inadequate at temperatures below 0°C. Copper–constantan (Cu and Ni, with small amounts of Fe, Mn, and C) shows better low-temperature sensitivity and is useful from −200° to +350°C. Iron–constantan can be used from −200° to +800°C and is very often encountered in instrumental applications where extremely high accuracy is not needed.

The response of a thermocouple is a function of the difference between junction temperatures. To make a measurement of an unknown temperature, one must know not only the calibration for the pair in question, but also the temperature of the junction that serves as reference. For careful measurements, the reference junction is placed in a thermostat, often at the ice point. For many applications, such as oven and furnace control, a few degrees error is tolerable, and ambient temperature is used as the reference.

The low internal resistance of a typical thermocouple makes it feasible to use either a voltage or a current readout. Sufficient power is available to drive moving-coil meters without amplification. In the usual case, the thermocouple loop is opened and the readout leads are connected, as shown in Fig. 7.12(b), thereby introducing additional bimetallic junctions. For example, the meter leads in Fig. 7.12(b) may be of a material different from either one of the thermocouple pair. However, so long as the junctions at C and D are at the same temperature, no new potentials of significant magnitude are introduced.

To achieve highest accuracy, a high-impedance voltage readout is used, with either a potentiometer or a sensitive voltmeter. The readout can be calibrated to indicate temperature directly. If a current readout is used, then the resistance of the entire loop, including the meter, must be set before the readout can be calibrated. For highest sensitivity, the meter resistance should be very small. However, if this is true, the resulting current in the thermocouple wire may cause heating, hence an error. For this reason, the meter resistance is made larger than that of the wire, even though this causes a loss of sensitivity. Commercially constructed readouts, such as oven pyrometers, are calibrated for the type and resistance of thermocouple wire.

B / *Potentiometric electrode systems*

If two dissimilar wires are immersed in an electrolytic solution, a potential difference can usually be observed between them; this potential may change in a systematic way with variation in solution composition. However, an electrical measurement that involves any direct current flow in a solution will of necessity be accompanied by electrochemical reactions at the electrode–solution interfaces. Accordingly, no rigorously defined meaning can be assigned to solution potential measurements, such as those suggested above with two wires, unless both electrodes show reversible electrochemical behavior toward components of the solution. Taking the reaction involving reversible interconversion of the two common oxidation states of silver,

$$Ag^+ \rightleftharpoons Ag^0 + e^- \tag{7.6}$$

a silver wire immersed in a solution containing silver ions would be expected to show the half-reaction potential defined by the Nernst relationship:

$$E_{Ag^0} = E^0_{Ag^+ - Ag^0} - \frac{RT}{F} \ln \frac{(Ag^0)}{(Ag^+)} \tag{7.7}$$

where E^0 is the standard electrode potential for the half-reaction in question, R is the gas constant, T is the absolute temperature, and F is the Faraday constant. Enclosure of symbols in parentheses indicates the activity of the species designated. Thus, the potential measured at a silver wire would have a defined relationship to the concentration (more precisely, activity) of silver in solution because it constitutes one-half of a reversible redox couple.

Potentiometric measurements must, of course, involve two electrodes in order to make electrical contact with the solution. One electrode, the *indicator electrode*, must show reversible behavior toward the ion that is to be measured. Silver would be the choice if Ag^+ were to be measured. The other electrode, the reference, is designed to maintain a constant and reproducible potential between the electrical circuit and the solution. This is accomplished by incorporating large, stable concentrations of both halves of a suitable redox couple in the design of the electrode. When an observation is made, as indicated in Fig. 7.13, the meter will measure the difference in voltage between the two electrodes, which is the sum of all of the voltages around the circuit. It is therefore not possible to measure a single half-reaction potential as was implied in Eq. (7.7); however, this is not a severe

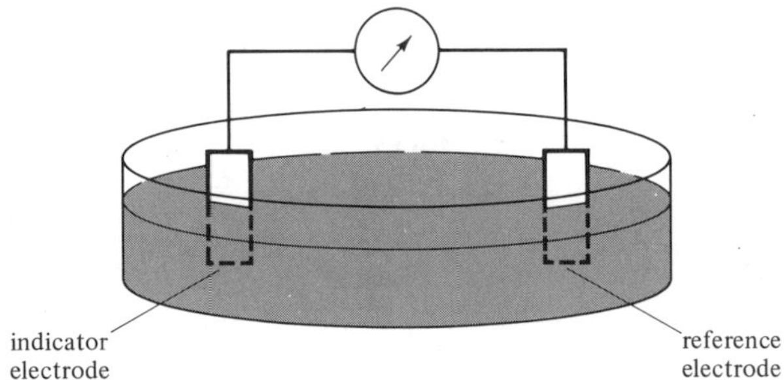

Figure 7.13 / *Schematic representation of a potentiometric measurement.*

practical limitation, since an arbitrarily defined reproducible value can be assigned to the reference and any observed difference can be attributed to the indicator electrode.

Potentiometric electrodes are available that respond selectively to a large variety of ions. They afford an extraordinarily useful chemical analysis technique, which is rapid and nondestructive and usually requires no preparation of the sample. One of the most widely used is the pH electrode, based on the reversible hydrogen couple,

$$2H^+ + 2e^- \rightleftharpoons H_2 \tag{7.8}$$

from which

$$E = E^0 - \frac{RT}{2F} \ln \frac{(H_2)}{(H^+)^2} \tag{7.9}$$

The value of E^0 for this couple is arbitrarily taken to be zero, and the activity of hydrogen gas at atmospheric pressure is defined as unity. Therefore, an electrode that shows Nernstian behavior for hydrogen would provide the response indicated in Eq. (7.10):

$$E = -\frac{RT}{F} \ln (H^+) \tag{7.10}$$

which can be recast in terms of pH:

$$\text{pH} = \frac{F}{RT} E \tag{7.11}$$

From Eq. (7.11), it is evident that a meter could be calibrated to indicate pH directly if provision were made for an adjustment of the sensitivity to account for the expected range of fluctuations in temperature. This measurement can be made on a practical basis because platinum shows Nernstian behavior toward the hydrogen couple. This measurement is rarely encountered, however, because of the inconvenience of providing hydrogen gas compared with the ease of use of glass electrodes.

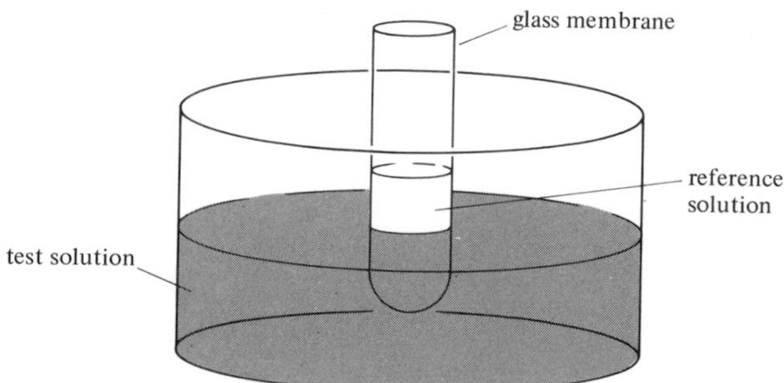

Figure 7.14 / *Glass electrode for pH measurement.*

When certain types of glasses are formed into membranes and arranged so that they separate solutions having different pH values, it is found that a potential is set up across the membrane that has a Nernstian relationship to the difference in pH. Such an arrangement is shown schematically in Fig. 7.14. A measurement of the potential difference across the membrane would produce a voltage that is related to the pH difference by Eq. (7.11). This could be accomplished by attaching inert metallic contacts to each face of the membrane. In fact, the practical difficulties of fabrication lead to the use of two reference electrodes to bring the signal out to a meter. The user of a pH meter is ordinarily aware of only one reference electrode because the second is located within the sealed glass membrane electrode.

Most potentiometric electrode systems exhibit fairly low internal resistances and therefore do not present problems in making voltage measurements. The voltage range for this type of experiment is typically 0–2 V, DC. Measurement to the nearest millivolt is ordinarily sufficient; however, certain applications require 0.1-mV accuracy. In either case, no rigorous demand is made on the accuracy, precision, or reproducibility of available DC voltmeters.

The glass electrode, however, has an extremely large internal resistance, since the glass membrane is a circuit component, and it does require quite a special readout instrument for satisfactory performance. A pH meter differs from conventional voltmeters in that its scale is suitably calibrated and provision is made for either manual or automatic temperature compensation. The major factor, however, is that a pH meter must be capable of accurate measurement of the signal produced by a very high-impedance voltage source. The required input resistance can be achieved in a variety of ways. One is to use specially designed vacuum tubes, electrometer tubes, which draw very small grid currents.

Another method is to place a capacitor in series with the input with provision for periodic motion of the plates of the capacitor relative to each other. This causes a periodic change in the capacitance of the component. If the voltage applied to both plates is the same, a change in capacitance has no effect; if it is different, it can be seen from Eqs. (1.41) and (1.42) that the capacitor will periodically charge and discharge. Although there is no net current flow, there will be an AC signal produced with an amplitude controlled by the DC voltage. This principle, which is used in vibrating-reed electrometers, affords very high input resistance. The DC input effectively modulates a carrier at the frequency of vibration. An AC amplifier followed by a phase-sensitive detector can be used for very low-level voltage measurements.

For most applications, pH meters can be constructed with multistage DC-coupled difference amplifiers with insulated gate FETs at the input. These are operated with a sufficient degree of feedback to afford adequately linear and reproducible response. The readout is most often through a moving-coil meter with a suitably calibrated dial; digital readouts are used in certain applications.

chapter 8 / Analog instrument design

chapter 8

8.1 / ANALOG VS. DIGITAL

The techniques for handling signals in instruments may be categorized as either analog or digital. This distinction has such far-reaching implications that it is used as a basic descriptive criterion for instruments or systems.

An analog signal magnitude has a simple relationship to the phenomenon being described. The signal is thus an analog of the phenomenon. Restricting this discussion to electrical signals, the signal magnitude, either current or voltage, is usually proportional to the measured quantity. The digital signal represents a code, usually in binary form, which can be transmitted electronically based on two signal levels.

Accordingly, the critical performance factor in an analog instrument is ability to transmit and reproduce accurate current or voltage levels. The critical factor in a digital instrument is ability to transmit or reproduce unambiguously a signal based on the two defined binary signal forms. This fundamental difference in performance requirement is reflected broadly in instrument design and in utilization.

Adequate performance of an analog instrument is directly determined by the accuracy of signal handling. Linearity, stability, reproducibility, and long-term reliability of signal-handling components is required. By contrast, a digital instrument need only be sufficiently accurate to avoid obscuring the distinction between the two defined signal levels that represent the binary code.

It follows that analog design imposes much more rigorous demands on individual component performance than does digital design. Meeting the higher standards makes analog components larger and more expensive than digital components. Furthermore, the accuracy that is possible with analog design, although entirely adequate for many applications, is much less than can be offered by digital equipment.

On the other hand, many transducers, including some that are very widely used in scientific instruments, produce signals that are inherently analog and directly acceptable by analog instruments. For many of the smaller tasks, the interface between transducer and digital instrument may be as expensive as the entire comparable analog system.

Analog presentation of data for human use is often much more satisfactory than a digital presentation. For example, it is easier to evaluate the similarities between two spectra if they are presented in graphical form rather than as two tables of numbers identifying the intensities and frequencies of peaks. This factor is so pressing that it has become general practice to provide, with more elaborate digital equipment, subsystems that reconvert the data to analog form for the benefit of the user.

As a broad generalization, analog equipment is less expensive than digital equipment. Analog design will generally be used when there is a need for special-purpose equipment to perform relatively simple tasks. Digital design is generally reserved for applications that involve large amounts of data, very high-speed operation, very high accuracy, or very complex operations.

8.2 / THE OPERATIONAL AMPLIFIER

Satisfactory performance of an analog instrument is critically dependent on the response of each of its components, because information is encoded as a signal magnitude. Therefore, the system noise level directly affects precision, and offset or nonlinearity causes proportional degradation of accuracy. To achieve acceptable performance at a reasonable cost, analog instruments generally utilize negative feedback systems, as described in Chapter 5. To a considerable extent, they are built either with servo mechanisms or with operational amplifiers. A discussion of servo mechanisms was presented in Section 7.2.5. Operational feedback was mentioned as a type of voltage feedback in Section 5.2.4. A discussion of its application follows.

It was pointed out in Section 5.2 that in order to use voltage feedback, the amplifier feedback loop must draw current from a source that exhibits an appreciable output impedance. This is the basic condition for a successful current measurement; and voltage feedback is widely used for this purpose, especially with operational amplifiers designed specifically for this function.

An operational amplifier is a high-gain, wide-band, DC-coupled difference amplifier. The conventional presentation of an operational amplifier is shown in Fig. 8.1. The amplifier itself is usually represented by the triangle with inputs and output as indicated in Fig. 8.1(a). To achieve stable operation, this type of amplifier must be placed in a feedback loop. An arrangement that is used for many purposes is shown in Fig. 8.1(b). The loop must connect the output to the inverting input as shown in order to set up negative feedback.

Qualitatively, the function of the amplifier is to impose whatever voltage is required at the output to hold, by means of the feedback loop, the inverting input at a voltage close to that of the noninverting input. In practice, the

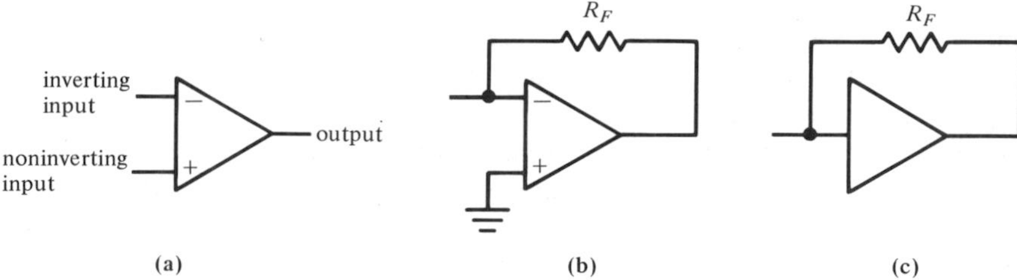

Figure 8.1 / *Operational amplifier symbols.*

inverting input voltage is usually only a few millivolts different from that of the noninverting input, which is often grounded. Accordingly, the condition of the inverting input in a properly functioning operational amplifier is described as being at *virtual ground.* That circuit point is not actually grounded, but the amplifier holds it near ground, or near the reference voltage level imposed on the noninverting input. For certain purposes, it behaves as though it were grounded. When the inverting input is held at virtual ground, this is often referred to as the *summing point* of the circuit.

Although the noninverting input is often grounded as shown in Fig. 8.1(b), it can serve as an active input in certain types of applications, which are discussed below. In the remainder of this discussion, if the positive input is grounded, it will be omitted from the schematic diagram, as is done in Fig. 8.1(c). It will be shown specifically when it is an active input.

For the present purposes, we need not be greatly concerned about the details of the amplifier represented by the triangular symbol. The two inputs are the result of the use of a difference amplifier, Fig. 4.46, in at least the first stage of the device. These may involve transistors, FETs, or tubes in older equipment. The output is usually some form of follower that provides at least modest (e.g., 2 mA) current capability. Operational amplifiers are usually purchased as sealed units; the user treats them as "black boxes" that show the attributes listed above.

8.2.1 / Activating an operational amplifier

We are dealing with a DC-coupled device that exhibits a very large open-loop gain. Suppose, for example, we consider the amplifier in Fig. 8.2(a). With gain,[1] $A = 80$ dB, and a maximum output swing of ± 10 V, the maximum tolerable input voltage from Eq. (8.1) would be only 1 mV.

$$e_{\text{out}} = -Ae_{\text{in}} \tag{8.1}$$

[1] In Chapter 5, the symbol G was used for amplifier gain, following the usual practice in the literature dealing with gain–frequency relationships. However, the symbol A is generally used when operational amplifiers are discussed, and this convention is followed here.

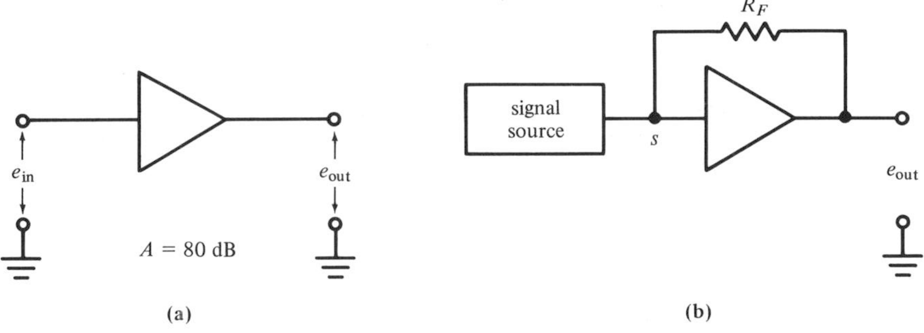

Figure 8.2 / *Activation of the operational amplifier.*
(a) *Open-loop operation.*
(b) *Closed-loop operation.*

Since this is the magnitude of uncertainty due to noise that is ordinarily experienced, it is evident that open-loop operation is of little value.

Suppose, however, that a signal is applied as shown in Fig. 8.2(b), so that a voltage e_s is imposed on the inverting input. The output voltage will then be

$$e_{out} = -Ae_s \tag{8.2}$$

However, with the loop completed by R_F, e_{out} is fed back to reduce the effective input voltage. R_F will accordingly have a voltage e_F dropped across it,

$$e_F = e_s - e_{out} \tag{8.3}$$

and will draw current i_F:

$$i_F = \frac{e_F}{R_F} \tag{8.4}$$

Substituting Eqs. (8.2) and (8.3) into Eq. (8.4) and rearranging gives

$$e_s = \frac{i_F R_F}{1 + A} \tag{8.5}$$

8.2.2 / Current measurement

If the amplifier open-loop input impedance is large, the current delivered by the signal source will be equal to the current drawn by the feedback loop:

$$i_{in} = i_F \tag{8.6}$$

Substituting the value of i_{in} from Eq. (8.6) and the value of e_s from Eq. (8.5) into Eq. (8.2) gives an expression for output voltage in terms of amplifier gain, feedback resistance, and input current:

$$e_{out} = \frac{A}{1 + A}(i_{in} R_F) \tag{8.7}$$

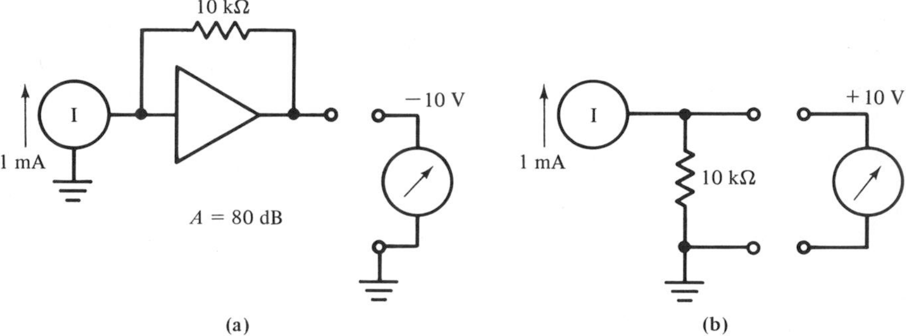

Figure 8.3 / *Current measurement with* 10-V/mA *sensitivity.*

We can also calculate the effective input resistance, r_{in}, from Eq. (8.5):

$$r_{in} = \frac{e_{in}}{i_{in}} = \frac{R_F}{1 + A} \tag{8.8}$$

The application of operational feedback to current amplification is described quantitatively by Eqs. (8.7) and (8.8). If the open-loop gain is large, so that

$$A \gg 1 \tag{8.9}$$

then these relationships can be simplified:

$$e_{out} = -i_{in}R_F \tag{8.10}$$

$$r_{in} = \frac{R_F}{A} \tag{8.11}$$

It is therefore possible to construct a current amplifier that shows an effective gain that is determined by the size, accuracy, and long-term reliability of the feedback resistor and that can present a very low-input resistance to a signal source.

Suppose that the amplifier in Fig. 8.3(a) is to be used to provide an output response that corresponds to 10 V for 1-mA input current. Assuming that Eq. (8.9) is valid at the signal frequency, a 10-kΩ feedback resistor would be required. Similar results can, of course, be obtained with the measuring resistor alone, with no amplifier. It is therefore worth indicating just why the use of the amplifier is advantageous. Either with or without the amplifier, a 1-mA current should cause a 10-V drop across the resistor. As indicated in the diagram, this will be observed as negative output in one case and as positive output in the other. In fact, the direct measurement of Fig. 8.3(b) will be successful only if the signal source is capable of imposing a 10-V output. With A = 80 dB, the input impedance in Fig. 8.3(a) will be 1 Ω, compared with 10 kΩ for the circuit of Fig. 8.3(b). The permissible lower

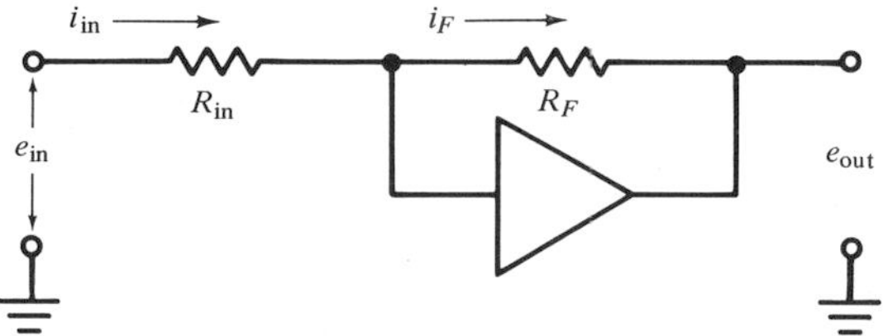

Figure 8.4 / *Voltage measurement by an operational amplifier.*

limit of source impedance is therefore reduced by four orders of magnitude by the amplifier without loss of accuracy in the measurement. In addition, the amplifier will be capable of serving as a buffer between the signal source and the meter used to show the output. Without the amplifier, the meter input impedance must be large compared with the measuring resistor, often a difficult requirement to meet.

8.2.3 / Voltage measurement

Operational feedback produces the low-input resistance that is appropriate to a current-measuring device; however, this type of device can be used for voltage measurements in just the same way that a moving-coil meter is used—by placing a known resistance in series with the input and measuring the current driven through it by the unknown voltage. This arrangement is shown in Fig. 8.4, where R_{in} is the measuring resistor. Assuming that the open-loop input resistance is large (the input resistance of the triangle itself):

$$i_{in} = i_F = \frac{e_{in} - e_s}{R_{in}} = \frac{e_s - e_{out}}{R_F} \tag{8.12}$$

$$e_{out} = -Ae_s \tag{8.13}$$

Substituting e_s from Eq. (8.13) into Eq. (8.12):

$$\left(e_{in} + \frac{e_{out}}{A}\right) R_F = -R_{in}\left(\frac{e_{out}}{A} + e_{out}\right) \tag{8.14}$$

$$e_{out} = -\left\{\frac{e_{in}(R_F/R_{in})}{1 - [(1 + R_F/R_{in})/A]}\right\} \tag{8.15}$$

If the inequality shown in Eq. (8.16) is valid, Eq. (8.15) can be simplified to Eq. (8.17):

$$A \gg 1 + \frac{R_F}{R_{in}} \tag{8.16}$$

$$e_{out} = -e_{in}\frac{R_F}{R_{in}} \tag{8.17}$$

Again, a very simple relationship for effective gain is obtained. If Eq. (8.17) holds, the gain is a function only of the ratio of the resistors and not of

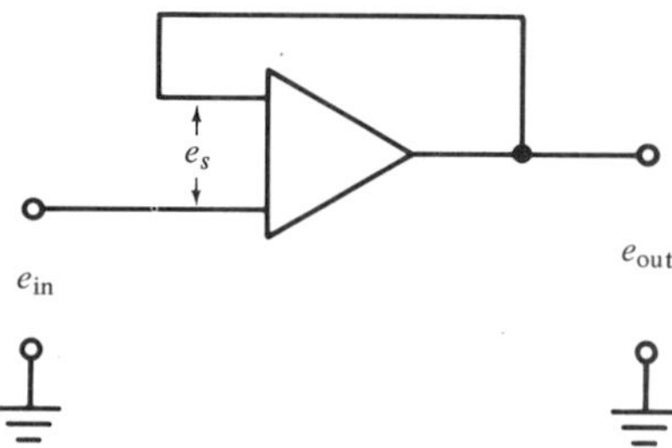

Figure 8.5 / *Operational amplifier voltage follower.*

the amplifier characteristics. For this reason, operational feedback is widely used. It must be noted, however, that the restriction shown in Eq. (8.16) imposes limits on what can be done. Signal frequency must not fall in the rolloff range of the amplifier, and the effective gain is ordinarily held to a value 40 dB less than the open-loop gain. This limitation stems not only from Eq. (8.16), but also from the fact that R_{in} is the effective input resistance, which must not be made too small lest the signal source be loaded.

8.2.4 / The voltage follower

Difficulties in achieving accurate voltage transfer may be avoided by using an operational amplifier as a follower, as shown in Fig. 8.5. Here the inverting input is shorted to the output, and the signal is imposed on the noninverting input. In an ideal amplifier, a positive value of e_{in} would tend to drive the output positive. However, e_{out} could not actually deviate from e_{in}, because the output is connected to the inverting input. Any deviation between input and output would amount to a nonzero value of e_s, which would be corrected by the inverting input. In a real amplifier, e_s is not zero, but ordinarily amounts to a few millivolts, as implied by Eq. (8.13).

This configuration does provide for amplification with voltage gain very nearly unity and with input impedance equal to the open-loop value for the amplifier which, with a FET input, can be very large. The operational amplifier follower is used to match a high-impedance source with a low-impedance receiver when the voltage level must be accurately preserved. If small offset errors and some variation of gain can be tolerated, however, a simple emitter or source follower is more economical.

8.2.5 / Offset and drift

For DC amplifiers, offset and drift constitute important problems. To the greatest practical degree, they are minimized by the choice of the amplifier design and of the components. Especially, difference amplification is used, as was pointed out in Section 4.12. The use of integrated circuit fabrication techniques is also very important, because differences between the two components of each stage of difference amplification can be minimized. Amplifiers are available that will reliably exhibit errors due to offset and

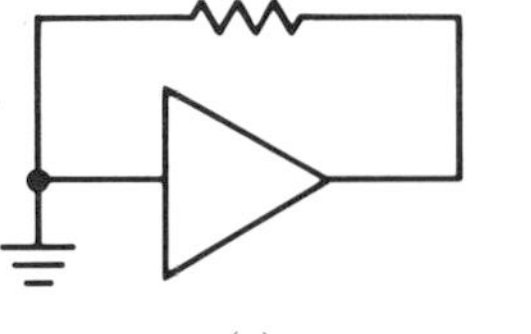

(a)

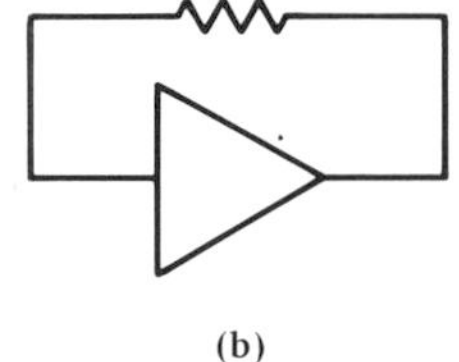

(b)

Figure 8.6 / *Effect of operational feedback on amplifier offset.*

drift amounting to less than 0.1 percent of full-scale output swing. It is possible, however, to obtain this level of performance from amplifiers that do exhibit appreciable offset and drift, by making use of chopper stabilization.

Before considering stabilization in detail, it is appropriate to note the influence of offset when operational feedback is used. Suppose that with the summing point of Fig. 8.6(a) grounded (which effectively opens the feedback loop), some value, B, of offset voltage is noted at the output. In normal use, the output is given by Eq. (8.18). If the input is opened, without a signal imposed, as in Fig. 8.6(b), the feedback current will be nearly zero because of the high open-loop input resistance of the operational amplifier.

$$e_{\text{out}} = -Ae_s + B \tag{8.18}$$

The output and summing-point voltages are therefore about equal. If we assume that $e_{\text{out}} = e_s$ under these conditions, then Eq. (8.18) can be restated as,

$$e_{\text{out}} = \frac{B}{1 + A} \simeq \frac{B}{A} \tag{8.19}$$

which indicates that the use of operational feedback reduces open-loop offset by a factor equal to the open-loop gain. An amplifier that exhibits a 100-dB open-loop gain and a 100-V offset will show a 10-mV offset with the loop closed. In the usual case, open-loop offset cannot be observed directly because it exceeds the full-scale swing of the amplifier; it is measured by the process just described.

Offset, greatly reduced by establishing the feedback loop, can be adjusted to a very small value if a suitably placed bias voltage is provided. However, this requires periodic resetting and is subject to drift. By means of automatic balance control using a chopper-modulated preamplifier, offset and drift may be reduced to very small values in the same operation.

The necessary arrangement is shown in Fig. 8.7. The feedback loop is placed around the operational amplifier with a blocking capacitor between the summing point and the inverting input. The chopper amplifier is placed between the summing point and the noninverting input. In operation, the high-frequency components of the input signal are transmitted by the blocking capacitor and amplified without being affected by the chopper-stabilized amplifier. The low-frequency components, drift and offset, are

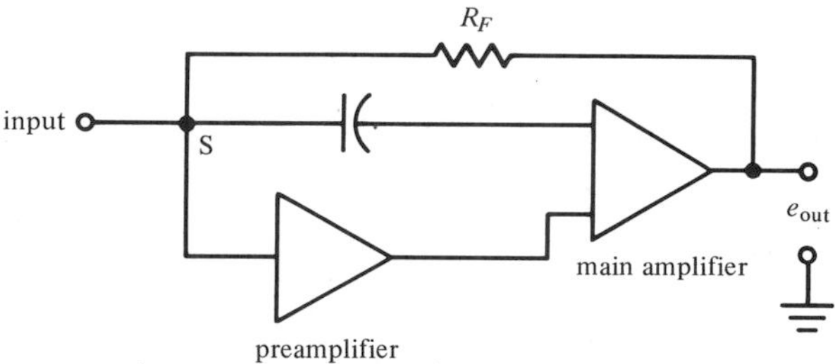

Figure 8.7 / *Automatic balance control by chopper stabilization.*

blocked from the inverting input but are seen by the chopper-stabilized amplifier.

The preamplifier is two-stage AC-coupled with a chopper, ordinarily mechanical, arranged to ground alternately the input and the output signals. The action is described in Fig. 8.8. The amplified signal is in phase with the chopped input since the preamplifier has two stages, and will be symmetrical

Figure 8.8 / *Waveforms of a chopper-stabilized amplifier.*

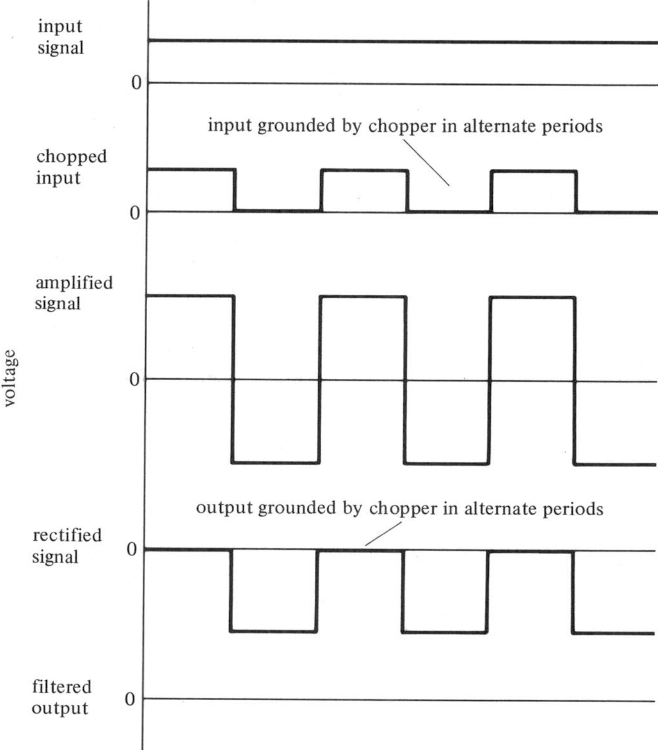

with respect to ground voltage since the amplifier is AC coupled. If the input signal is shorted at the times marked on the figure, the output signal will be shorted in alternate half-cycles, eliminating the positive segments of the output waveform in this example. This is rectification, since all signal components of one polarity are eliminated. A positive input signal will therefore produce a negative output after rectification and filtering to a smooth DC signal.

This process constitutes modulation, amplification of the modulated signal, and demodulation by phase-sensitive detection, and as in the function of the servo mechanism, it provides important benefits, which are discussed in Sections 6.4.3 and 7.2.4.

Looking at the system as a whole, high-frequency signal components are handled by the main amplifier only through its inverting input. Low-frequency components, rejected by the blocking capacitor, are accepted by the preamplifier as a modulated carrier. The chopper amplifier automatically minimizes the drift and offset of the main amplifier. That this occurs can be seen by considering the effect of introducing a positive error signal at the summing point. The effect of the preamplifier would be to amplify and invert this and to pass it to the main amplifier for further amplification. This negative voltage would be fed back to the summing point, where it would be subtracted from the positive error. Ideally, this would eliminate the error. Actually, since the relationship of Eq. (8.13) is applicable here as well, the preamplifier reduces drift and offset by a factor equal to its own gain, usually at least 60 dB. It is therefore possible to achieve reliable performance with drift and offset held automatically to the microvolt level.

8.3 / SIGNAL PROCESSING BY OPERATIONAL AMPLIFIERS

The great utility of modern instrumentation is, in part, due to the ability of the instrument to perform computations on the raw data. This is done for the convenience of the user and also to allow the instrument to serve a control function in automatic operation. For this reason, a discussion of some of the basic operations of analog computers is appropriate before considering automatic control by analog instruments.

8.3.1 / Summing

In the examples presented above, the output is a function of only one current or voltage signal; however, more than one signal could be applied. The amplifier will then respond to the sum of the input currents.

In Fig. 8.9(a), current sources are connected to the summing point. The current drawn by the feedback resistor will be the sum of the input currents if the open-loop input impedance is high. If the assumption implied in Eq. (8.9) is valid, the output will be given by

$$e_{out} = -(i_1 + i_2 + i_3)R_F \tag{8.20}$$

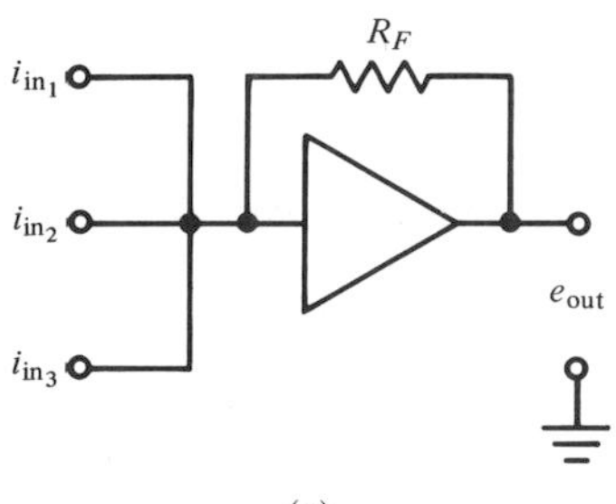

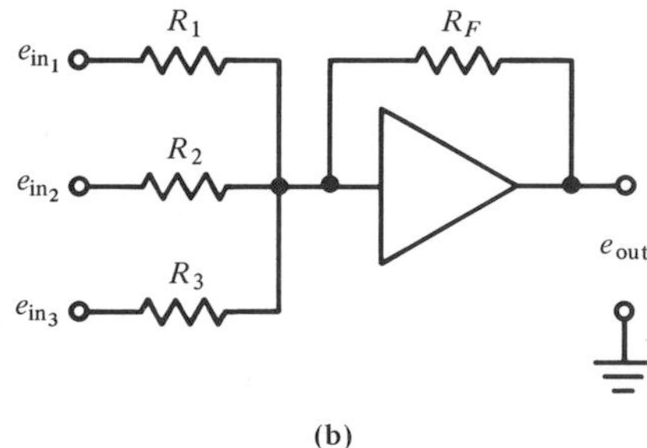

Figure 8.9 / *Analog addition.*
(a) *Current summing.*
(b) *Voltage summing.*

The same result can obviously be obtained simply by connecting the sources to a measuring resistor. The amplifier serves to hold the source end of the resistor at virtual ground, thereby providing a low-input resistance. This is the desirable condition for an accurate current measurement, and it serves to prevent interaction between the sources. With the summing point at virtual ground, none of the sources is aware of the others.

The voltage summing network is shown in Fig. 8.9(b). This provides for analogous decoupling of voltage sources. If the restrictions regarding open-loop gain are observed, Eq. (8.21) is valid:

$$e_{out} = -e_1\left(\frac{R_F}{R_1}\right) - e_2\left(\frac{R_F}{R_2}\right) - \cdots - e_n\left(\frac{R_F}{R_n}\right) \tag{8.21}$$

8.3.2 / Magnitude scaling

Equation (8.17) shows that the output from a voltage amplifier is multiplied by the ratio (R_F/R_{in}). If the two resistors have the same size, the result is inversion (multiplication by -1). By suitable adjustment of resistor sizes, the scaling constant can be varied from -0.01 to -10 or -100, the upper limit restricted as described in Section 8.2.3. In many cases where a scaling factor less than unity is needed, a passive voltage divider can be used. If this is done, the signal is not inverted, of course, and advantage cannot be taken of the low-output impedance of the operational amplifier to drive a load.

A weighted analog adder is the active component of digital-to-analog converters, as discussed in Section 9.4.2. For example, conversion of a three-digit number to analog form would require three input resistors, each inversely proportional to the desired weighting of the digit.

8.3.3 / Subtraction

It is possible to carry out subtraction by utilizing the noninverting input as shown in Fig. 8.10. If all resistors have the same value, then the output will be the difference of the inputs,

$$e_{out} = e_2 - e_1 \tag{8.22}$$

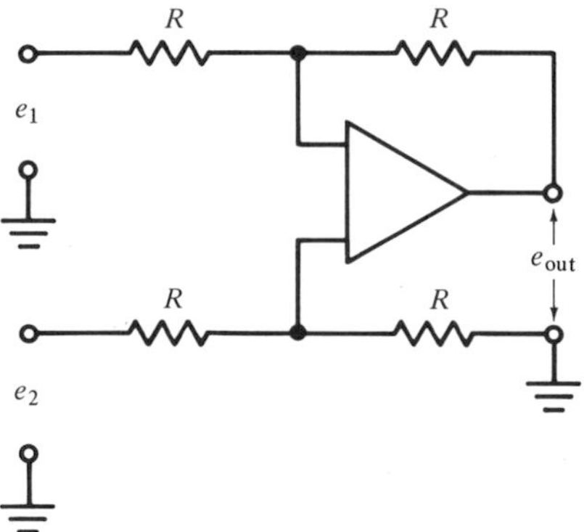

Figure 8.10 / *Analog subtraction of two voltages.*

Scaling can be accomplished by changing the relative sizes of input and output resistors. In fact, as indicated in Section 8.2.4, this arrangement is often troublesome because it ties up the noninverting input, making it impossible to stabilize the amplifier. More often, therefore, subtraction is carried out by inverting the subtrahend and adding.

8.3.4 / Integration

In the discussion of the behavior of reactive components in Chapter 1, the differential relationship between charging current and voltage across a capacitor was given by Eq. (1.43). Solving that equation for voltage gives Eq. (8.23),

$$e_{out} = \frac{1}{C} \int i_{in} \, dt \tag{8.23}$$

which shows that a capacitor, connected as in Fig. 8.11(a), acts as a current integrator. Actually, integration of voltage with respect to time is more often needed, and this can be accomplished by placing a measuring resistor in series with the capacitor, as in Fig. 8.11(b). The capacitor integrates the current that the voltage source drives through the resistor. Assuming that the capacitor is initially discharged at the instant e_{in} is applied,

$$i_{in} = \frac{e_{in}}{R} \tag{8.24}$$

Therefore,

$$e_{out} = \frac{1}{RC} \int e_{in} \, dt \tag{8.25}$$

The practical application of the relationship of Eq. (8.25) through the circuit in Fig. 8.11(b) is quite limited because, as the integration occurs and e_{out}

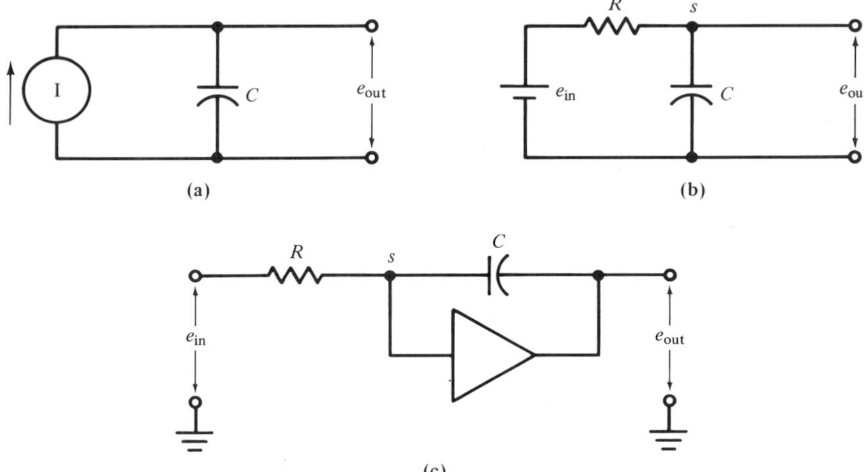

Figure 8.11 / *Analog integration.*
(a) *Current integration by a passive circuit.*
(b) *Voltage integration by a passive circuit.*
(c) *Voltage integration with an operational amplifier.*

becomes finite, the voltage that is effective in driving current into the integrator is not e_{in}, but $e_{in} - e_{out}$. For a constant value of e_{in}, therefore, e_{out} does not show a linear increase with time, but instead approaches e_{in} asymptotically, as shown in Fig. 3.3.

This difficulty arises from the fact that, in the simple circuit of Fig. 8.11(b), the capacitor and the input voltage source are both connected directly to the common point of the circuit. The problem would be avoided if it could be arranged for the input side, *s* in Fig. 8.11(b), to have a constant voltage. The other plate would therefore have to be uncoupled from the common point and be allowed to change. This can be done by placing the capacitor in the feedback loop of an operational amplifier with *s* the summing point and the other plate connected to the amplifier output, as in Fig. 8.11(c).

If the amplifier is able to hold *s* at virtual ground, that is, if *A* in Eq. (8.13) is large under the conditions of operation, the relationship between input and output is that of Eq. (8.26), up to the amplifier output voltage limit:

$$e_{out} = -\frac{1}{RC}\int e_{in}\,dt \tag{8.26}$$

The amplifier compensates for the back emf built up across the capacitor, inverts the signal and, incidentally, drives current up to its output specifications into an external load.

The operational amplifier integrator is encountered in a variety of applications, many of which are not associated with analog computers. Accordingly, it is worthwhile to consider its performance briefly. The integrator will not

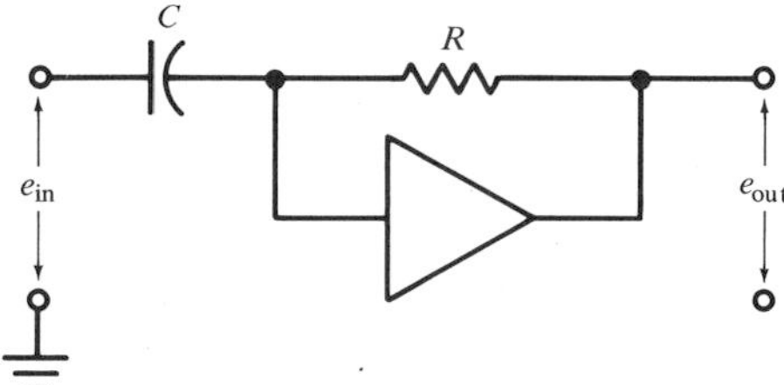

Figure 8.12 / *Differentiating with an operational amplifier.*

be affected by high-frequency noise, but it is very prone to low-frequency noise because it integrates its own offset. Therefore, it is most useful for operations that last for limited periods of time. The signal being integrated must be large as compared with the integrator offset.

From Eq. (8.26), it can be seen that if a constant input voltage is applied, the output will increase linearly with respect to time until the amplifier limit is reached. This application is used to generate scanning signals (ramp signals) that can be used to control instruments. A ramp signal, operating into a voltage-level detector, can also be used as an accurate timing device, since it will require an accurately known time to charge to a specified level.

An integrator can be used to improve the quality of DC voltage measurements on noisy signals by integrating the unknown voltage for a fixed period of time. The output voltage will be proportional to the average input voltage. It is used in various measuring operations that require integration, for example, measurement of chromatographic peaks, spectrophotometric peaks, and coulometry. The operational amplifier integrator is a basic component used in the solution of differential equations by analog computers.

8.3.5 / Differentiation

It is evidently true that differentiation can be accomplished by reversing the role of input and output in the circuits in Fig. 8.10. Such a differentiating network is shown in Fig. 8.12. The relationship between input and output is

$$e_{out} = -RC \frac{de_{in}}{dt} \tag{8.27}$$

While many possible uses for differentiation can be imagined, the analog differentiator, unlike the integrator, is not widely useful.

If the behavior of operational feedback is generalized by using impedances for resistances in Eq. (8.17), Eq. (8.28) is obtained:

$$e_{out} = -e_{in} \frac{Z_F}{Z_{in}} \tag{8.28}$$

An integrator has a capacitor in the feedback loop; therefore, Z_F decreases at high frequencies, causing decreased gain. A differentiator has the capacitor in the input; therefore, Z_{in} is reduced at high frequencies, causing increased gain. For this reason, the differentiator is very sensitive to high-frequency noise and accordingly has limited utility.

8.3.6 / Multiplication and division

Multiplication and division of variables are operations that cannot be easily carried out by analog techniques. Various approaches have been developed that can attain 0.1 percent relative accuracy and that are compatible in response time with amplifiers. Although these are available at reasonable cost, there appears to be little use of them at present, other than in analog computers.

8.3.7 / Function generation

The need often arises for conversion of a variable to some function other than those mentioned above. It may involve a logarithm, a trigonometric function, or some arbitrary function. Function generation is especially used to linearize transducer outputs. Of the several techniques available, the use of special gears, nonlinear potentiometers, devices that exhibit nonlinear behavior on being energized, and networks with adjustable variable resistance will be cited.

The spectrophotometer is a familiar example. The transducer provides a signal that is proportional to transmittance, but for many purposes, absorbance readout is preferred, necessitating the generation of a negative logarithm. If a mechanical drive is used in the instrument, a specially designed gear train can provide for logarithmic conversion of shaft rotation.

To use a nonlinear potentiometer, a constant voltage is imposed across a slidewire that exhibits nonuniform resistance along its length. The potentiometer wiper is positioned by the drive mechanism as above, or by a servo mechanism, making the voltage readout by it the specified function of the positioning variable. Potentiometers are more versatile than gear trains, because they can be fabricated with a wider variety of functions.

Various circuit components show reproducible nonlinear transfer functions. For example, within a fairly narrow range, the voltage across certain diodes is approximately proportional to the logarithm of the current. A logarithmic function generator can therefore be constructed by placing the diode within the feedback loop of an operational amplifier. The current drawn by the device will be controlled by the input current or voltage, and the output voltage imposed by the amplifier will be that appearing across the nonlinear component. This approach has the advantage of being all-electronic and therefore capable of rapid response.

This method can be made very flexible by using a piecewise linear approximation. The transfer curve is a series of linear segments that can approximate any desired curved function. The principle of the operation of this type of device can be understood by examination of the circuit in Fig. 8.13. A voltage divider is arranged with a Zener diode, which shows a characteristic voltage,

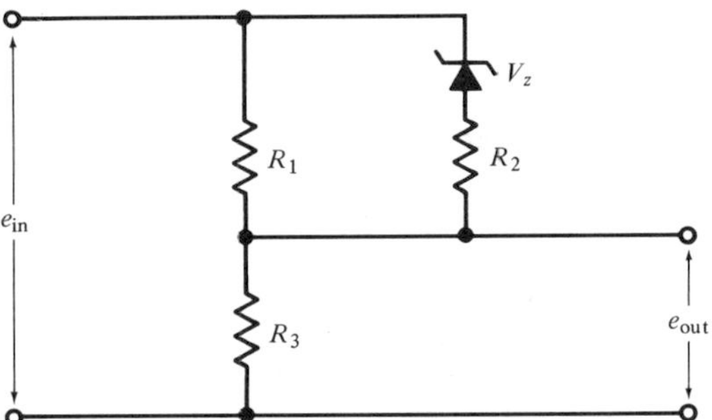

Figure 8.13 / *Function generation by piecewise linear approximation.*

V_z, above which reverse conduction occurs, in parallel with one section. If e_{in} starts at a small value and increases, the initial transfer function for this network is given by the constant term of Eq. (8.29).

$$e_{out} = e_{in}\left(\frac{R_3}{R_1 + R_3}\right) \tag{8.29}$$

If e_{in} increases sufficiently, the voltage drop across R_1 exceeds V_z and the diode starts conducting. When this happens, the transfer function for the network is given by the constant term of Eq. (8.30).

$$e_{out} = e_{in}\frac{R_3}{[R_1R_2/(R_1 + R_2)] + R_3} \tag{8.30}$$

If e_{out} is plotted against e_{in}, the slope of the curve will be greater in the second segment than in the first. This basic concept can be elaborated upon to construct function generators that provide either fixed functions—for example, logarithmic functions—or adjustable ones. For example, this can be used to produce the control signal to program the temperature of a gas chromatograph oven.

8.4 / AUTOMATIC CONTROL BY ANALOG INSTRUMENTS

A most useful aspect of analog design is the ease with which the instrument may be automated. Very often, partial automation and continuous automatic readout are combined. Details of a particular design obviously are determined by the nature of the experiment. Important factors include the type of control device used and the time scale of the measurement. In most cases, control is achieved by comparing a variable signal, derived from the quantity being controlled, with a standard. Any difference between them is amplified and applied to the control device. We encounter control devices that require both electrical and mechanical stimulus.

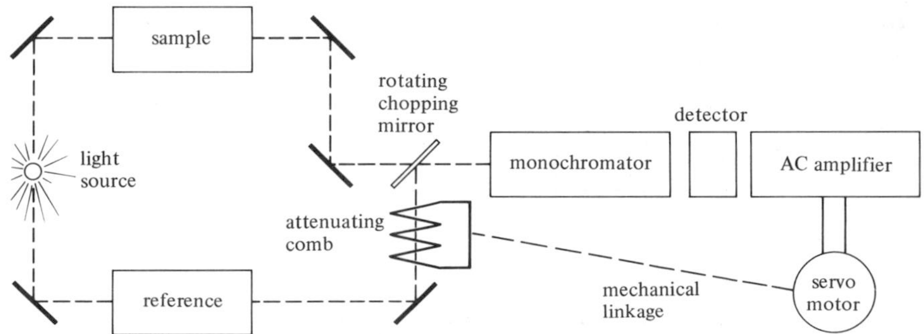

Figure 8.14 / *Double-beam spectrophotometer.*

8.4.1 / Electromechanical systems

When the control operation involves moving parts, a servo mechanism is ordinarily used. A discussion of the principle of operation of a low-power servo of the kind often used in instruments is given in Section 7.2.5. The double-beam infrared spectrophotometer is an example of an instrument that very often utilizes a servo to control an analog signal.

A simplified diagram of an optical nulling double-beam spectrophotometer is shown in Fig. 8.14. The purpose of this instrument is to measure the extent of absorption of infrared radiation by a sample. In the usual case, two measurements are made, one on the sample and one on a reference blank. The blank is usually a cell identical with that containing the sample, filled with the same solvent as is used for the sample. The radiant power transmitted through the reference cell is assumed to be that characteristic of a sample of zero concentration, and the servo-driven device compares the intensity of the beam transmitted through the sample cell with it. The source produces a wide range of frequencies, which is restricted by the monochromator before the beam reaches the detector. The detector may be a bolometer or a thermocouple (see p. 184).

Light from the source is directed simultaneously through the sample and reference cells. The two beams are recombined by a rotating chopping mirror, which alternately passes the sample and the reference beams through the monochromator to the detector. If both beams are of equal intensity, the detector is illuminated steadily and produces an invarying signal to which the AC amplifier cannot respond.

If the reference beam is more intense than the sample beam, radiation reaching the detector will fluctuate in intensity at the chopping frequency. The resulting electrical signal might be analogous to that of curve 2 in Fig. 7.6. If the sample beam is the more intense, the detector will also produce a fluctuating signal, but with the phase shifted 180°, as indicated in curve 4 in Fig. 7.6.

The rotating chopping mirror in the optical system serves the same function as the chopping switch in Fig. 7.5(b). It produces a signal at the chopping frequency that is modulated by the error signal and it furnishes, via a set of electrical contacts on its shaft, a signal that is synchronized with the modulated beam which can be used to drive the servo motor [corresponding to the line coil signal in Fig. 7.5(b)], and thereby effect phase-sensitive detection of the amplified error signal.

The servo motor is mechanically connected to a beam attenuator, which may be the metal comb indicated in Fig. 8.14. If the reference beam is more intense than the sample beam, the comb is moved to reduce the reference intensity and vice versa. The same linkage drives a pen to record the spectrum.

This is an example of a double-beam design that incorporates phase-sensitive detection and null indication, a combination that is very effective in minimizing errors from various sources. The signal is modulated by the rotating mirror and then detected by the servo motor. The system therefore rejects noise and offset that might be introduced by the monochromator, detector, or amplifier. Double-beam design involves a comparison of the sample and reference beams, not an absolute measurement of either. Therefore, fluctuations introduced by the light source, which have frequencies substantially lower than the chopping frequency, are canceled out. The rotating mirror may sample each beam 15 times per second, for example. Slow variations in light intensity will not have any effect, providing that a reasonable level of energy reaches the detector. High-frequency fluctuations will be transmitted; however, the inertia of the motor will greatly restrict the frequency range of noise actually observed.

8.4.2 / Electronic systems

If the control device can be actuated by an electrical signal, it is possible to make use of an all-electronic system, an advantage because there are no moving parts to limit response time or to wear out. Very commonly occurring examples involve control of temperature, electrode potential, and microwave frequency in esr spectroscopy. One illustration from electrochemistry will be cited.

The problem is to maintain the potential of a working electrode at some designated value relative to the bulk of the solution into which it is immersed. This is done by placing a reference electrode in the solution and measuring the potential drop between the two electrodes. Suppose that it is desired to maintain the working electrode at +1.00 V vs. the reference. The system, shown schematically in Fig. 8.15, consists of two current-carrying electrodes, with a reference electrode physically close to the working electrode. The working electrode–reference electrode potential difference is monitored and compared with a standard. Any difference is amplified and used to control the power supply that furnishes current. In operation, imposition of a voltage across the cell with the working electrode made positive will cause it to take a positive voltage relative to both the auxiliary and reference electrodes. If no component present in the solution is capable of reacting at

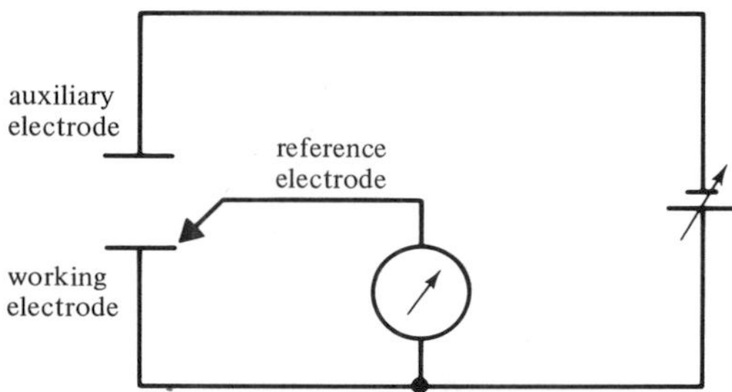

Figure 8.15 / *Electrolysis at controlled potential.*

the potential imposed, no current will flow; the solution in contact with the auxiliary electrode will be at the same potential as that in contact with the reference, ignoring any possible discontinuities of solution composition. If a current does flow, then there will be a resistance drop in the solution and the solution at the auxiliary electrode will be at a more or less positive voltage than that at the working electrode, depending on the direction of current flow.

For purposes of illustration, suppose that an oxidation takes place at this potential at the working electrode, making it the anode. Current can be controlled by controlling the voltage across the cell. An increase in voltage will cause the anode potential to assume a more positive value relative to both the reference and the cathode. There is, of course, resistance between anode and cathode, so the increased cell voltage is dropped partly across this resistance and partly across the boundary layer between the anode and the solution. Ideally, there should be no resistance in the solution between the reference and anode; a change of the anode–reference potential would therefore reflect only a change in the voltage drop between the anode and the solution surrounding it. In the usual case, then, to effect a small change in potential difference between working electrode and reference, a much larger voltage difference must be imposed between the auxiliary and the working electrode if a current is flowing.

One form of potentiostat is shown in Fig. 8.16. In understanding the operation of this circuit, the significant point is that the apparatus is connected so that the voltage of the working electrode is the same as that of the reference level (noninverting input) of the amplifier. The amplifier will try to maintain its inverting input (s) at this same value. Therefore, it must be arranged so that the reference–anode potential must be at the desired value for the amplifier to be able to balance itself. In Fig. 8.16, equilibrium will be established, going around the loop that includes the two electrodes and the amplifier, if the anode-reference voltage is equal to the negative of the bias voltage.

We can see that this will be a stable condition by a qualitative examination of the effect of a perturbation on the system. Suppose that, with the system

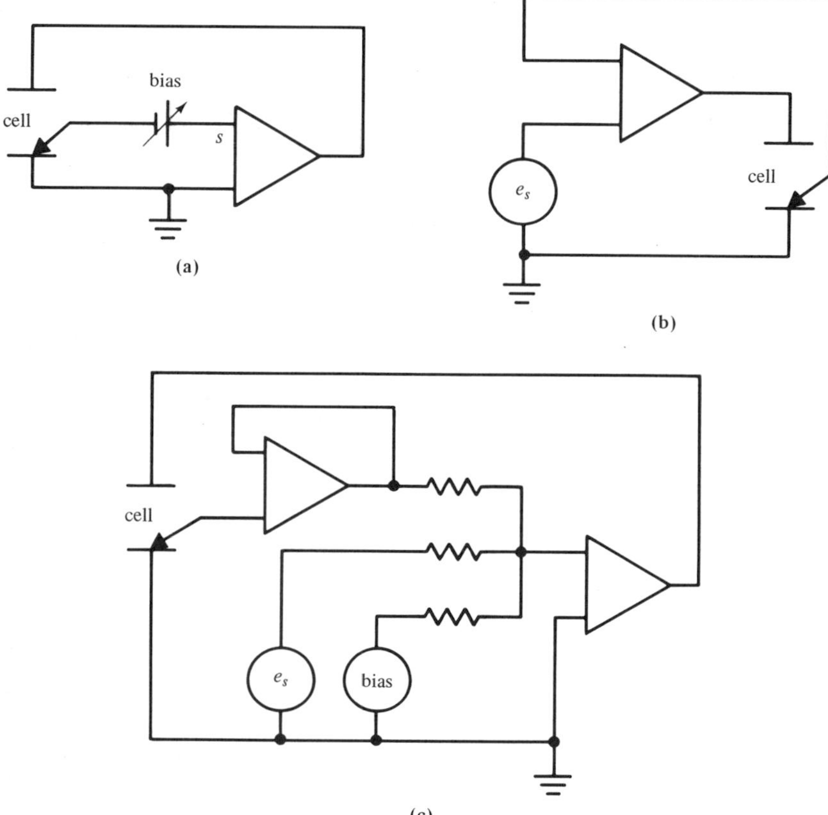

Figure 8.16 / *Potentiostat designs.*

operating, the reference is shifted 0.01 V positively from the desired value. The reference floats and the anode is grounded, so this is the same as the anode moving from +1.00 V to +0.99 V relative to the reference. A shift in this direction will tend to slow or stop the reaction. It will also cause the inverting input of the amplifier to move to a positive voltage and thereby impose an increased negative voltage at the cathode, which will tend to start the reaction. The system can therefore be stable. There are, of course, other conditions that must be met. This is a negative feedback loop and is therefore subject to oscillation if the phase shift around the loop is too large; in addition, the amplifier must be capable of furnishing the current demanded by the cell.

For this type of system to operate, the control device, in this case the cell, must be placed within the feedback loop. There are usually a number of ways to do this, all of which are not equally advantageous. In Fig. 8.16(a), the bias and the reference electrode are in series with the amplifier input. High open-loop impedance ensures that little current will be drawn from either.

This is necessary because drawing current from the reference would change its value; also, it is convenient not to load the bias because a smaller supply can be used. However, it may be very inconvenient if a signal other than a constant DC bias value is needed, because with this arrangement, the bias source must float. This is easy to do with a battery, but very awkward with the type of signal generator that might be used for polarography or voltammetry.

Another possible arrangement is shown in Fig. 8.16(b). Here the signal e_s, to be imposed on the cell, is applied to the noninverting input. This provides high-impedance connections for both components and allows single-ended operation of the signal source. It has the disadvantage of making the noninverting input unavailable for stabilization. This will be satisfactory only if the amplifier exhibits a negligibly low offset or if use of an adjustable floating bias is feasible.

An additional possibility is shown in Fig. 8.16(c). This arrangement provides for single-ended operation of the source for e_s. It also allows the use of other signals, imposing the sum of them on the cell. The main amplifier can be stabilized, but if a drain of at least several microamperes from the reference is to be avoided, a follower must be inserted as shown. Again, the question of balance of the follower must be answered.

SUGGESTED READINGS

J. G. Graeme, G. E. Tobe, and L. P. Huelsman, *Operational Amplifiers, Design and Applications*, McGraw-Hill (New York, 1971).

H. V. Malmstadt, C. G. Enke, and E. C. Toren, Jr., *Electronics for Scientists*, W. A. Benjamin (Menlo Park, Calif., 1963), Chapters 7 and 8.

R. B. Benedict, *Electronics for Scientists and Engineers*, Prentice-Hall (Englewood Cliffs, N.J., 1967), Chapter 14.

A. J. Diefenderfer, *Principles of Electronic Instrumentation*, W. B. Saunders (Philadelphia, 1972), Chapter 9.

chapter 9 / Digital instruments

chapter 9

The increasing availability of digital computers, coupled with growing awareness of their potential applications, has led to rapid changes in various aspects of our lives. This has, of course, been felt in the area of chemical instrumentation, and the use of digital equipment will undoubtedly continue to increase. The digital logic machine has had such a revolutionary impact because it is capable of performing extremely large numbers of logic operations at very high rates of speed with a high degree of reliability. Its design is based on identifiable developments in the areas of mathematics, physics, and engineering which have been accumulated over a period of many years. The computer came into existence in its present form fairly promptly following the most recent of the necessary advances in engineering, the development of reasonably priced microelectronic circuits.

In digital instruments, information to be transferred is encoded as recognizable voltage or current pulses. To be accepted by an instrument, an input must therefore first be properly encoded. For analog signals, this involves repetitive sampling, with the value of each sample being converted to coded form. The instrument will ordinarily be capable of counting the pulses, of storing the signals in a way that will preserve their information content, and of calling them up at a future time. Depending on the function of the machine, it may be capable of doing arithmetic on the numbers that are encoded and of performing various logic operations on the stored information.

In the sections below, some concepts that are basic to the operation of digital instruments are discussed. These are Boolean algebra, digital coding, electronic switching, the binary device, and data storage. Following this, brief descriptions are given of specific types of components used to implement these concepts. Finally, some examples are given of the types of uses to which digital equipment may be put.

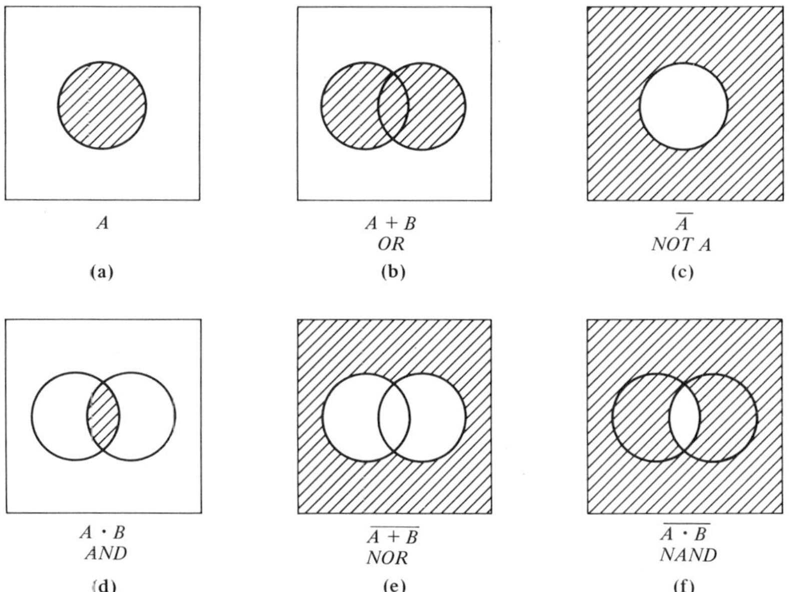

Figure 9.1 / *Venn representations of simple logic operations.*

9.1 / FUNDAMENTAL CONCEPTS

9.1.1 / Boolean algebra

The conceptual basis for logic machines has been developed from the system of algebra originated by George Boole in the middle of the nineteenth century. Boolean algebra is a symbolic method for studying logical relationships. It does not directly utilize any number system; however, quantities are expressed in terms of two possible conditions, which may be designated, true or false, **0** or **1**, or by any other convenient binary notation. The binary number system is therefore convenient to use in connection with Boolean algebra. For the purposes of this discussion, a few basic logic operations will be examined at the outset. In Section 9.2.1, the circuitry by which they are implemented is discussed.

The symbols $+$, $^{-}$, and $\cdot$ are used in Boolean algebra with meanings that are somewhat different from the conventional ones. These can be understood from the Venn diagrams in Fig. 9.1. The whole square area symbolizes the system in question, while the shaded circle in Fig. 9.1(a) represents the one element A. In Fig. 9.1(b), the *logical OR* operation on two elements, $A + B$ is symbolized by the area that includes *either A or B*. The expression $A + B$ is read "*A* or *B*." The bar symbol, for example, $\overline{A}$, indicates the complement and is read "not *A*." In Fig. 9.1(c), the circle indicates A and everything outside the circle is $\overline{A}$.

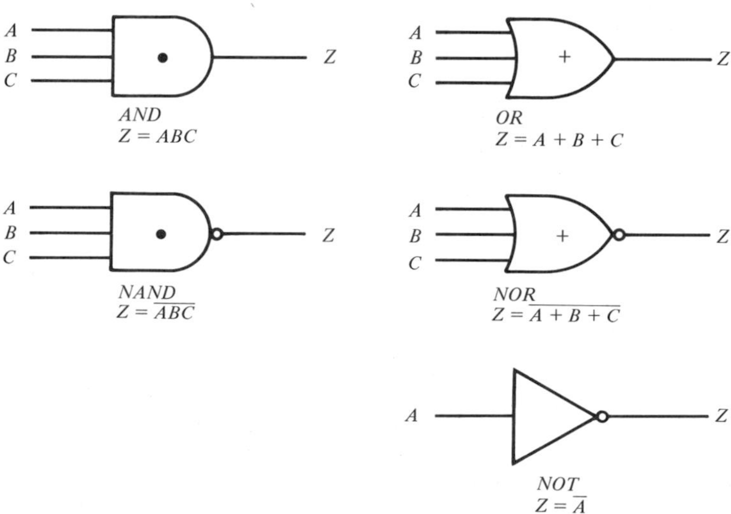

Figure 9.2 / *Logic operation symbols.*

Logical *AND*, shown by $A \cdot B$, or simply *AB*, which means "*A* and *B*," is that portion of the system which pertains to both *A* and to *B*, as indicated by the shading in Fig. 9.1(d). *NOT A OR B*, $\overline{A + B}$, *NOR*, is represented by Fig. 9.1(e) and *NOT A AND B*, $\overline{AB}$, *NAND*, by Fig. 9.1(f). Another frequently occurring symbol is $\oplus$, *EXCLUSIVE OR*. An output may occur when either *A* or *B* is true, but not if both or neither is true.

Boolean notation provides a concise and broadly accepted convention for describing the functions of circuits, without reference to the techniques used to perform them. The practice has been adopted of representing the most basic operations, *AND*, *OR*, and *NOT*, by symbols that can be used in schematic diagrams. These are illustrated in Fig. 9.2. As the name implies, the major purpose of Boolean algebra is to provide a conceptual basis for the evaluation of complex sequences of logical operations by the application of theorems. It is used in the design of digital computers; however, a detailed discussion is beyond the scope of this book. The reader is referred to the Suggested Readings cited at the end of the chapter.

9.1.2 / Digital coding of information

It is general practice to use binary codes for information handling in digital equipment. Several factors are responsible for this choice. When information is encoded in terms of only two signal levels, the maximum freedom from interference by noise, for a given signal range, is assured. A binary coding system is compatible with available switching devices that are suited to fabrication on a small scale in very large numbers and that are capable of the very fast and reliable operation required. The structure of Boolean algebra

TABLE 9.1 / *Binary–decimal equivalents*

Decimal	*Binary*	*Decimal*	*Binary*
0	0	10	1010
1	1	11	1011
2	10	12	1100
3	11	13	1101
4	100	14	1110
5	101	15	1111
6	110		
7	111		
8	1000		
9	1001		

is compatible with coding systems in which quantities are represented by a signal variable in one of two possible states.

Using the binary number system, quantities must be expressed by using only two numbers, 0 and 1. Accordingly, each digit in a binary number represents 2 raised to a power, or zero, as indicated:

digit number	8	7	6	5	4	3	2	1	0
value	2^8	2^7	2^6	2^5	2^4	2^3	2^2	2^1	2^0

Some binary numbers with decimal equivalents are given in Table 9.1. From this series, it can be seen that four binary digits, or *bits*, are required to express the 10 numbers of the decimal system and that, with four bits, a total of 16 (2^4) can be represented. In establishing a code to express decimal quantities in the binary arithmetic system, it is desired to achieve both efficiency and tolerable convenience. The most efficient system would be a simple conversion from decimal to binary; however, where human comprehension is important, this scheme becomes unsatisfactory. For example,[1]

$$5184_{10} = 1011010110111_2 \tag{9.1}$$

Accordingly, more convenient and less efficient codes are used.

Very often it is arranged to encode each decimal digit by a sequence of a fixed number of binary digits. A widely used one is the four-bit 1, 2, 4, 8, or natural code. Decimal digits are represented by blocks of four binary digits, each involving a direct conversion of a decimal digit to its binary equivalent. Thus:

$$5184_{10} = 0101\ 0001\ 1000\ 0100_2 \tag{9.2}$$

The coded version of 5184_{10} requires 16 bits in Eq. (9.2) as compared with only 13 in Eq. (9.1); however, the longer version is obviously easier to work with.

[1] Radix indicated by subscript when necessary for clarity.

TABLE 9.2 / *Binary-coded octal numbers*

Decimal	*Octal*	*Binary code*
0	00	000 000
1	01	000 001
2	02	000 010
3	03	000 011
4	04	000 100
5	05	000 101
6	06	000 110
7	07	000 111
8	10	001 000
9	11	001 001
10	12	001 010

The fact that three bits just suffice to encode eight numbers makes binary-coded octal especially attractive. It is efficient and not too cumbersome to handle. It is therefore used in small computers where space is at a premium. It can be seen from Table 9.2 that the binary–octal conversion can be performed by inspection.

Decimal–octal conversion can be performed in various ways. One involves dividing the decimal number repeatedly by 8. In each step, the remainder is taken as the least significant octal digit and the quotient is again divided by 8.

	Quotient	*Remainder*
$1520/8 =$	190	0
$190/8 =$	23	6
$23/8 =$	2	7
$2/8 =$	0	2

$$1520_{10} = 2760_8$$

To convert from octal to decimal, each digit of the octal number can be multiplied by its weighted value.

$$\begin{aligned} 0 \times 8^0 &= 0 \\ 6 \times 8^1 &= 48 \\ 7 \times 8^2 &= 448 \\ 2 \times 8^3 &= 1024 \\ \hline & \quad 1520_{10} \end{aligned}$$

Whatever code is chosen, it must be transmitted as an electrical signal. This takes the form of either a train of pulses or of a series of transitions between defined signal levels within a time frame established by a master oscillator or *clock*. This is illustrated in Fig. 9.3. Waveforms are drawn to

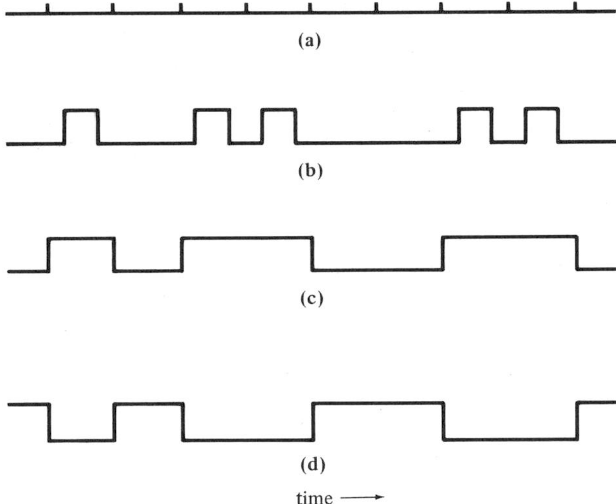

Figure 9.3 / *Digital code waveforms.*
(a) *Clock pulses.*
(b) *Positive pulses.*
(c) *Positive DC level.*
(d) *Negative DC level.*

represent the number 11001101, least significant bit first. Clock pulses are shown in Fig. 9.3(a). These delimit the period in which a bit can appear. The length of a unit signal or *word* is fixed by the design of the machine. To avoid loss of information, it is necessary to know not just the number of pulses in the signal shown in Fig. 9.3(b), but also their spacing. It would therefore be necessary to provide eight different storage locations, which could be "filled" in the proper order to receive that signal correctly. In the case of the signal shown in Fig. 9.3(c), occurrence of a positive voltage during the first period after appearance of the start of the word should cause the storage unit to indicate a **1** in the least significant bit, and so on. If the machine is designed with *negative logic*, then it would recognize a negative voltage between clock pulses as **1**, as in Fig. 9.3(d).

9.1.3 / Electronic switching

An ideal switch exhibits no resistance when closed and infinite resistance when open. In addition, it moves from one condition to the other with no time delay on receiving an actuating signal. Mechanical switches rather closely approximate the ideal so far as resistances are concerned, but in speed of operation, they of course fall far short of the ideal. Electronic switches offer the advantage of very fast response, but show much smaller differences between their open and closed resistances. Moreover, certain types cause distortion of the input signal.

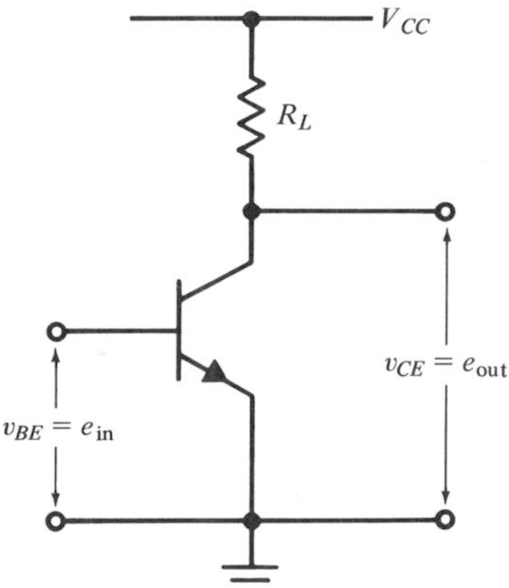

Figure 9.4 / *Common-emitter transistor as a switch.*

The demands placed on electronic switches in transmitting information encoded in binary form are fortunately congruent with their capabilities. The type of switch in question, termed a *logic gate*, provides only an indication of the presence or absence of an input signal without attempting to transmit any detail. It is actuated by an electrical signal and can be made to operate within a fraction of a microsecond.

Electronic switching is ordinarily accomplished by diodes or by amplifying devices, transistors, FETs, or tubes. Diode switches show a gain less than unity; transistors can provide gain and thus avoid degradation of the signal if it is to be processed several times, but at the cost of the use of power to activate them. Complete gate designs very often incorporate both diode and transistor switches. A switching action involves the interaction of an activating signal, which is analogous to motion of a mechanical switch, upon a second signal. The fact that in the conventional application of a transistor, there are two identifiable signal loops, usually base emitter and collector emitter, makes it easy to visualize the operation of transistorized switches.

The common-emitter inverter, shown in Fig. 9.4, is generally used. The information to be transmitted is simply the presence or absence of a positive input which will result in the transistor being in saturation or cutoff, respectively. This design can provide gain and is capable of furnishing power to an external load. Moreover, it is very simple in design; the essential requirement is that the transistor shift as rapidly as possible from cutoff to saturation or the reverse. Thus, the elaborate biasing required to utilize the active region in a conventional amplifier is entirely unnecessary. For a silicon transistor, any value of v_{BE} (notation is identified in Section 4.13.3) larger than 0.7–0.8 V can be depended on to cause saturation. A value less than 0.4–0.5 V will cause cutoff.

The switching action of a diode is not as easy to visualize because it has only two terminals. The actuating and information-bearing signals must be

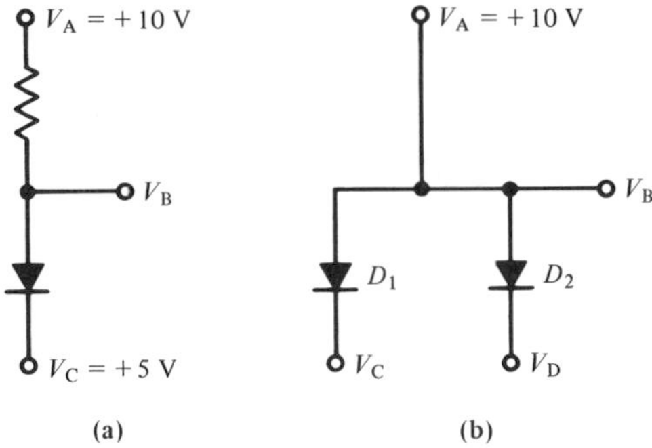

Figure 9.5 / *Diode switches.*

combined outside the diode. The diode functions as the basic element of a switch because it can be caused to show either a high or a low resistance by reverse or forward biasing it. The circuits in Fig. 9.5 illustrate this point.

That in Fig. 9.5(a) is essentially a voltage divider with the voltage V_{AC} divided between the resistor and the diode. With the specified voltage levels at A and C, the diode is forward biased, giving it a very small resistance. V_B is therefore close to V_C; specifically, the diode shows its characteristic forward-biased voltage, $\simeq 0.6$ V for a silicon diode. Therefore, V_B, which has a value of about 5.6 V, is said to be *clamped* to the 5-V level. This means that the action of the diode is to cause any value of V_A larger than $(V_B + 0.6)$ V to be dropped across the resistor. Changes in V_A have little effect on V_B. If more than one diode is used, as in Fig. 9.5(b), and V_A is more positive than either V_C or V_D by at least 0.6 V, then V_B will be clamped to the *lower* of the two, V_C or V_D. For example, if V_A is $+10$ V, V_C is 8 V, and V_D is 5 V, V_B will be clamped to $+5$ V and D_1 will be reverse biased with its cathode 3 V more positive than its anode.

Suppose that V_C is an information-bearing signal for which logical **1** is $+5$ V and logical **0** is 0 V and that V_D is a switching signal. The switch will be open for values of $V_D \leq 0$. Irrespective of whether V_C shows **0** or **1**, D_2 will be forward biased and V_B will be clamped to V_D. If $V_D \geq +5$ V, the switch will be closed. V_B will be clamped to V_C for either **0** or **1**.

9.1.4 / Binary devices

Some of the reasons for the universality of the binary design of digital instruments were given in Sections 9.1.1 and 9.1.2. Very important among these is the existence of a number of simple, reliable, and exceedingly fast binary devices, which can be used efficiently to transfer and store information in binary code. These are devices that possess two stable operating conditions and that can be switched very rapidly on command from one to the other. The following discussion is limited to transistor binaries; however, they can be constructed with tubes and FETs, as well as with other types of components.

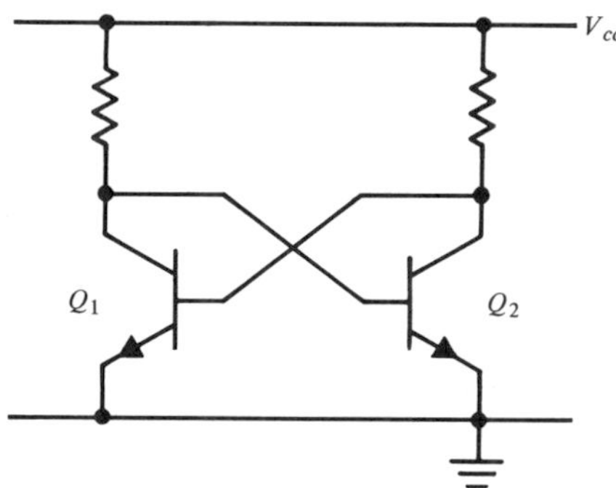

Figure 9.6 / *Transistor binary.*

A transistor binary is, in its simplest form, two common-emitter stages cross-connected so that the output of one is the input of the other and vice versa, as indicated in Fig. 9.6. The operation of this circuit can be understood qualitatively by following a signal through it. The collector of Q_1 is connected to the base of Q_2, and the collector of Q_2 is connected to the base of Q_1. If Q_1 were OFF (no collector current), its collector voltage V_{C_1} would be high, very nearly equal to V_{cc}. This would cause V_{B_2} to be high and would, given a large enough value of V_{cc}, cause Q_2 to be in saturation. V_{C_2} and V_{B_1} would therefore be small ($\simeq$0.2 V), which would indeed cause Q_1 to be cut off. Another stable state would be possible with Q_2 OFF and Q_1 ON. Once placed in one of these states, the binary would remain indefinitely in the absence of a switching signal, if the power stayed on.

With Q_1 ON and Q_2 OFF, the device could be made to shift from one state to the other either by application of a positive voltage to the base of Q_2 or by application of a negative voltage to the collector of Q_2. If V_{B_2} is made positive, V_{C_2} will drop, bringing down V_{B_1}, which will cause V_{C_2} to increase, augmenting the externally applied signal at V_{B_2}. A similar sequence of events will be initiated by application of a negative signal at the collector of Q_2. The switching action, once started, is self-sustaining because the connection of the transistors sets up a positive feedback loop and the binary will flip from one condition to the other.

In order to place a binary in operation, it is necessary to introduce a command signal to cause the binary to change states. This is termed *triggering*. In Fig. 9.7, provision is made for triggering by adding two *pullover* transistors, Q_3 and Q_4. If Q_1 were ON and Q_2 OFF, application of a positive signal to Q_4 would cause a transition because Q_4 would be turned ON, lowering its collector, and that of Q_2, almost to ground voltage. This would turn Q_1 OFF. Application of a positive trigger to the ON side, for example, to Q_3 in the illustration above, would have no effect. A description of the behavior of this binary is given by Table 9.3, which is called a *truth*

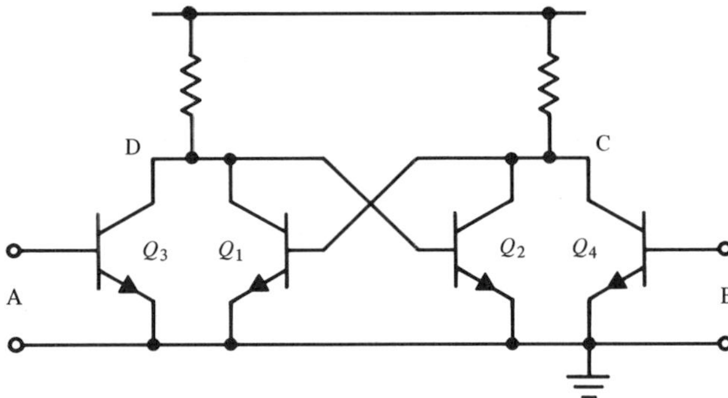

Figure 9.7 / *Transistor binary with provision for triggering.*

table. Here logical **1** is a positive signal, the exact voltage of which is determined by the design of the system. Logical **0** would be very nearly 0 V. When both inputs are **0**, the output condition is determined by which input last received a **1**. Thus, the device is capable, under these conditions, of remembering its last input.

In the literature of logic circuitry, one finds a variety of binary designs, most of which are more complex than that in Fig. 9.6. They do, however, share with it the essential feature of consisting of amplifiers, usually transistors, cross-coupled in a positive feedback arrangement. It is beyond the scope of this text to dwell at length on the details of binary design. In addition, devices constructed of discrete components are likely to be less common in the future, being increasingly supplanted by integrated circuits which come to the user as a sealed unit to which he does not have access, save through the terminals provided.

9.1.5 / Data storage

The fact that a digital instrument can readily be given the ability to store and automatically call up information is a very important factor in the great usefulness of this type of equipment. Several different types of storage devices are currently in use, and further developments in this area are subjects of continuing intense interest on the part of computer manufacturers. A

TABLE 9.3 / *Truth table for binary of Fig. 9.7*

Input		*Output*	
A	B	C	D
0	**1**	**0**	**1**
1	**0**	**1**	**0**
0 (last **1**)	**0**	**1**	**0**
0	**0** (last **1**)	**0**	**1**
1	**1**	**0**	**0**

very large number of storage locations is necessary, and the result is that the memory represents a significant fraction of the cost of a computer. Accordingly, the two major desiderata, low cost and rapid access, are counterbalanced so that we find relatively high-priced, rapid-access and low-cost, slow-access devices in the same piece of equipment.

Types of devices commonly encountered, listed in order of increasing speed of access, are as follows: punched paper tape, punched cards, magnetic tape, magnetic discs and drums, magnetic cores, and data registers. These tend to fall into two categories, devices that permit random access to stored information and those that must be scanned seriatim for a given piece of information. Registers and cores permit random access; the others must be scanned.

The least expensive and slowest methods for storage are punched paper tape and cards. Information is encoded in terms of the position of holes punched; the use of cards in connection with accounting services has made them familiar to all. This is a convenient form for long-term storage of data which can be automatically fed into a machine. In using magnetic tape, data that are previously converted to binary or binary decimal code form are recorded as small magnetized areas, with magnetization in one direction representing logical **1** and magnetization in the opposite direction, logical **0**. This provides for more compact storage and for more rapid storage and retrieval than is possible with punched cards or tape.

Magnetic discs and drums are similar in principle. They may take the form of thin discs rotated at high speeds between the jaws of the recording heads. Specific data items that are stored can be recalled without having to read out the whole content. This is done by directing the reading head to produce the information stored on a given track in a specified sector of the disc. The process by which the location of a word is specified is termed *addressing*. As data are stored, the address of each word, which itself would be a binary-coded number, must be retained. It is the task of the user, often by procedures so automatic that he may be almost unaware of them, to keep track of the addresses. Discs are rotated at several thousand revolutions per minute, so that although the average access time is imperceptibly short on the human time scale, it is very long on the microsecond time scale of the computer. Much more rapid access can be obtained by the use of a magnetic core memory.

Magnetic cores are small ($\simeq 1$ mm) toroids of ferrite (a ceramic containing magnetic iron oxide) or of a magnetic alloy. Their distinguishing characteristic is a "square" hysteresis loop. This means that a characteristic minimum magnetic field, H_c, is required to establish a fixed permanent magnetization, B_r. Imposition of $-H_c$ will reverse the magnetization to produce $-B_r$ flux density. Imposed fields smaller than $\pm H_c$ have no permanent effect.

Information storage is made possible by defining magnetization in one direction as logical **1** (*set*) and that in the other as logical **0** (*reset*). Storage is effected by applying a pulse of current of the appropriate polarity to wires passing through the hole in the core. There is a characteristic current level $\pm I_c$ required to produce the critical field $\pm H_c$. To read out information, an additional *sense* wire is passed through the core. When the core magnetization is reversed, a substantial current pulse is induced in the sense wire. If a current pulse is applied that is either too small to switch the core magnetization or that would have the effect of establishing the preexisting condition—for example, a reset command to a core that is already reset—only a very much smaller pulse is observed at the sense wire. Information readout may be accomplished by sending a reset signal to all cores of interest. Those that contained a **1** will produce a sense pulse; those that were in the **0** condition will not.

The description above represents a considerable oversimplification because it implies that each individual core has two wires uniquely associated with it. Actually, each core contains several wires, and each wire threads several cores. To read or write, only part of the minimum I_c is imposed on more than one wire; therefore, only that core threaded by wires carrying currents the sum of which equals I_c will respond. All other cores see currents less than I_c. This affords a great simplification in wiring. In addition, it should be noted that the process of reading destroys the information in the core. To avoid this, the reading process must consist of transfer of a word from memory to a buffer, followed by rewriting the word back into memory and transfer of the word to its desired destination.

The memory device offering highest speed and versatility is the data register. This consists of a number of binaries, one for each bit in the word. Conduction by one side of the binary is defined as logical **0**, by the other as logical **1**. Registers are used as temporary storage whenever data are moved about or processed. The buffer referred to above would be a register. Registers are also used to provide the interface between instruments that work on different time scales. Suppose that an analog signal is being sampled, encoded, and transferred to a digital instrument. The sampling rate would be determined by the capabilities of the encoding device and by the demands of the signal. The sampling must be done often enough not to lose the details of the input signal, and this is usually not synchronized with the operation of the digital instrument. Accordingly, the converter would be connected to transfer its sample, coded as one binary word to a register, where the information would remain until the processing unit of the digital instrument was ready for it. Similarly, human contact with a digital instrument via a keyboard, or input of information from punched tape, cards, magnetic tape, discs, or drums would involve intermediate transfer to a register.

9.2 / LOGIC CIRCUITRY

The components in which data are processed and by which decisions are made are collectively termed *logic circuits.* They are the components that implement the concepts discussed in the preceding sections. The design and

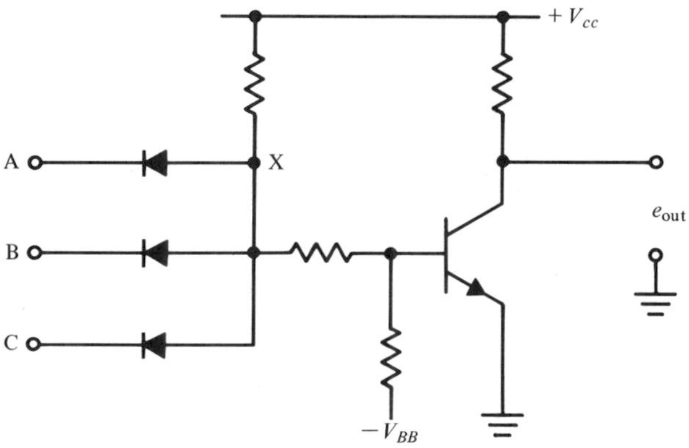

Figure 9.8 / *NAND gate (positive logic).*

construction of logic components has received much attention in recent years. The information available on this subject is very extensive. In the following sections, some of the most important types are discussed with the idea of indicating how they work and in what context they are used.

9.2.1 / Gates

The gate may be considered to be a basic component, since most of the others can be constructed by a suitable combination of gates. A gate is a combination of electronic switches that performs one of the basic logic operations, *OR*, *AND*, *NOR*, or *NAND*, and may be represented as shown in Fig. 9.2. For a specific example, the circuit in Fig. 9.8 will be cited. This is a *NAND* gate for positive logic (logical **1** is defined as a positive signal) or a *NOR* gate for negative logic (logical **1** is defined as the negative signal level). If at least one input is at logical **0**, point X is clamped to 0 V, which would cause the transistor to be OFF, producing a logical **1** output. With **1** at all inputs, X is high and the transistor can be turned ON, producing a **0** output. The circuit can be seen to produce the complement of *AND*, or *NAND*.

An attractive feature of this circuit is the fact that several gates of the same design can be connected together. Suppose that the outputs of three gates are connected to the inputs of a fourth, as shown in Fig. 9.9(a). Here the entire circuit of Fig. 9.8 is represented by a *NAND* symbol. A logical **1** at an output is caused by cutoff of the output transistor and must hold an input diode reverse biased. A logical **0** output, caused by saturation of the transistor, must hold the input diode forward biased. Thus, the two possible conditions are compatible. When the output is at the high level, it exhibits a large output impedance, but is not called upon to deliver current. In the lower level, it must take current, which it can do very well, since it exhibits a very low output impedance.

With either *NAND* or *NOR* gates, all of the logic operations can be performed without having to use other components. The advent of integrated

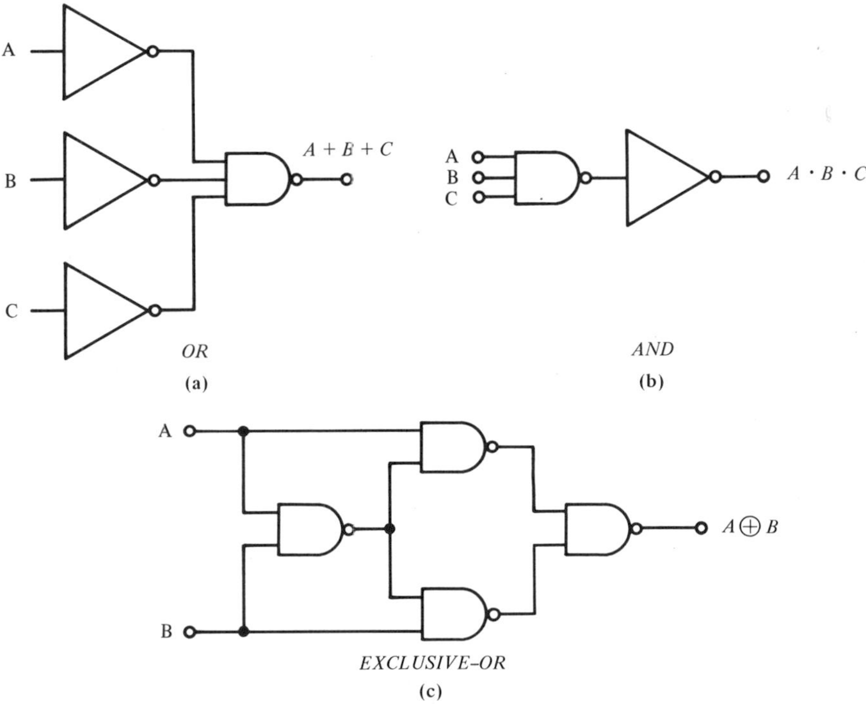

Figure 9.9 / *Logic functions with NAND gates.*

circuits has made it feasible to standardize one gate design, for example, the one shown, and to use numbers of these to build up more complex circuits, very often leaving many of the available outputs and inputs unused. This affords important simplification and economy of design. Thus, the overall operation of Fig. 9.9(a) is that of an *OR* gate. A logical **1** at A, B, and/or C produces at least one **0** at the input of the last gate, which causes it to produce a logical **1** at the output. Similarly, in Fig. 9.9(b), *NOT* following *NAND* gives *AND*. *NOR* could be obtained by adding one additional gate to Fig. 9.9(a). A similar set of circuits can be constructed, using *NOR* gates. For example, just as *OR* is *NOT NAND*, using *NAND* gates, *AND* (Fig. 9.9(b)) can be performed as *NOT NOR* with *NOR* gates. *EXCLUSIVE-OR* can be performed by replacing the *NAND* gates of Fig. 9.9(c) with *NOR* gates and adding an additional *NOR* at the output.

It is pointed out in Section 9.1.4 that a binary is basically two cross-coupled inverters. A binary can be constructed by cross-coupling *NAND* gates as shown in Fig. 9.10. The truth table for this circuit is given in Table 9.4. If the stable input condition was logical **1** at each input, the output would be decided by which input had most recently received a **0**. Thus, the binary remembers its last instruction. The circuit in Fig. 9.7 behaves in a similar fashion, except that it would remember the last input **1** (higher level)

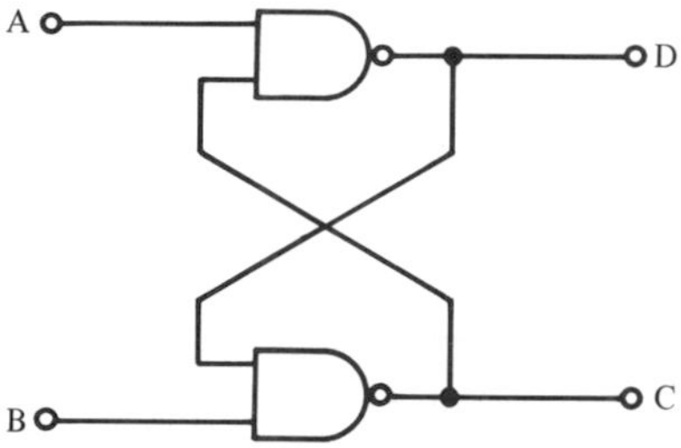

Figure 9.10 / *NAND gate binary.*

if the stable inputs were **0** (lower level). This is the effect of inversion by the pullover transistors in Fig. 9.7.

9.2.2 / Voltage-to-frequency converters

One of the basic operations encountered in digital equipment design is generation of an AC signal that has its frequency proportional to the magnitude of an analog signal. This conversion is the basic function of a digital voltmeter and of analog-to-digital converters in general. A number of types of voltage-to-frequency (V/F) converters are used. One of these, intended to be used only for unipolar signals, is illustrated in Fig. 9.11. It operates by sampling the analog signal at regular intervals. During each of these intervals, it produces a burst of pulses, the number of which is proportional to the magnitude of the analog input.

When a sample is to be taken, the control unit transmits a pulse simultaneously to the ramp generator and to the clock oscillator gate. This initiates the formation of a ramp signal which shows an accurately known increasing value with the passage of time. The gate permits a train of accurately spaced pulses to be received by the counter. When the ramp voltage equals the analog voltage, the comparator produces a pulse that opens the gate to terminate the count of clock pulses and resets the ramp. Comparison of a linear ramp with the analog signal ensures that the time from start to completion, and hence the count recorded, is proportional to the input signal voltage. By correct choice of oscillator pulse spacing and ramp charging rate, the count can be calibrated in terms of input voltage.

TABLE 9.4 / *Truth table for NAND gate binary*

Input		*Output*	
A	B	C	D
0	**1**	**0**	**1**
1	**0**	**1**	**0**
0	**0**	**1**	**1**
1	**1** (last **0**)	**1**	**0**
1 (last **0**)	**1**	**0**	**1**

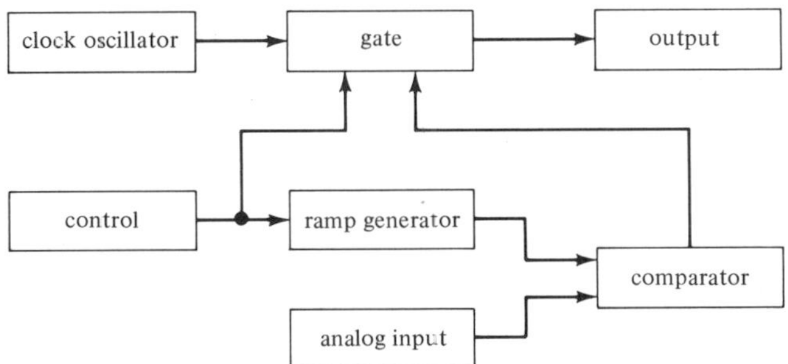

Figure 9.11 / *Block diagram of voltage-to-frequency converter.*

The ramp generator may be an operational amplifier integrator. The comparator may be a high-gain difference amplifier, similar to an operational amplifier, but with no feedback loop. Given a gain of 60 dB, the amplifier will be at one limit as long as the ramp is at least a few millivolts less than the input. As the ramp voltage approaches that of the input, the amplifier goes very suddenly from one limit to the other, producing a step that can be differentiated to yield a pulse that can control the gates.

The accuracy of this type of converter is at best comparable to that of analog techniques (0.1 percent), since the limiting factor is the accuracy of ramp generation. While this may suffice for many purposes, it evidently cannot allow full advantage to be taken of the accuracy that can be obtained with digital instruments. The reader is referred to the Suggested Readings at the end of the chapter for discussions of other types of voltage-to-frequency converters.

9.2.3 / Counters

Pulse counting is an operation of considerable utility in digital instrumentation. A counter consists basically of a number of binaries connected together. Two segments from a counter are illustrated in Fig. 9.12. These binaries differ from that in Fig. 9.7 in several ways. The inputs, A and B (Fig. 9.7), are tied to the respective collectors of Q_3 and Q_4. Capacitors are placed across the binary transistors and an additional transistor, Q_5, is provided to accept the trigger.

These changes are made in order to permit the binary to *toggle*, that is, to complement itself automatically on imposition of a trigger. From Table 9.3, it is seen that with **1** at both inputs, both outputs are **0**. All four transistors simply go into saturation. On removal of the trigger, one side will turn OFF, but there is no necessary relationship between the new state and that existing before triggering. The device must be made to remember its previous condition for a short time with **1** at both inputs, as it does with **0** at both. This is the function of the capacitors C_1 and C_2 in Fig. 9.12.

Suppose that Q_1 is ON and no trigger is present. V_{C_2} and V_{C_4} are high and C_2 is charged. It may appear inconsistent for V_{C_4} to be high because

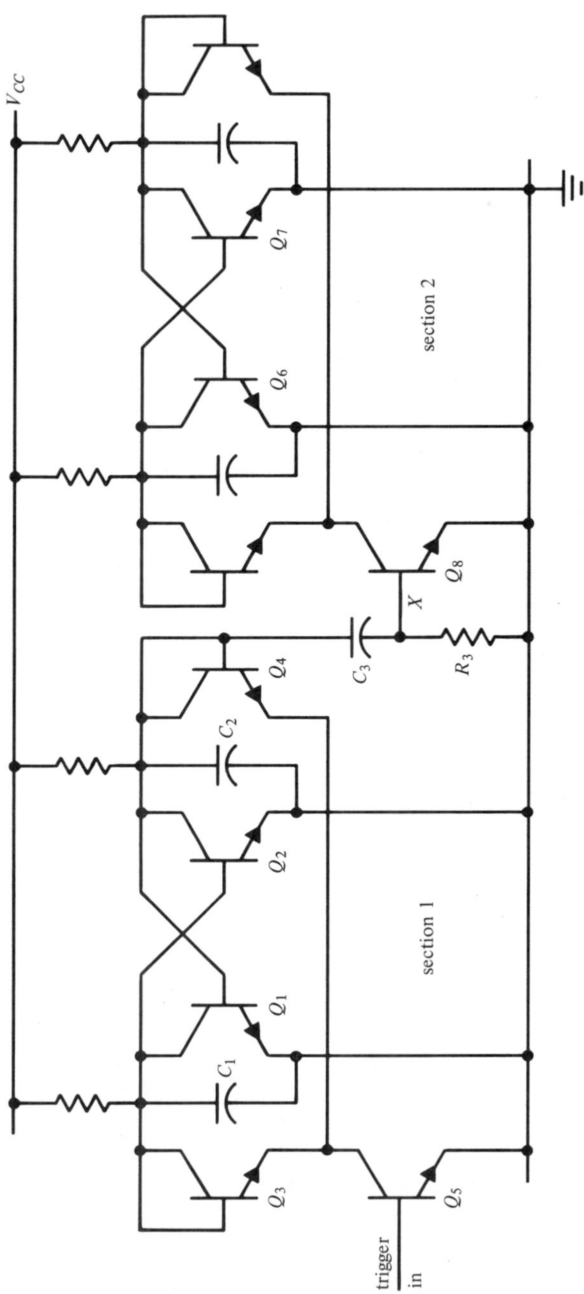

Figure 9.12 / *Two-stage counter.*

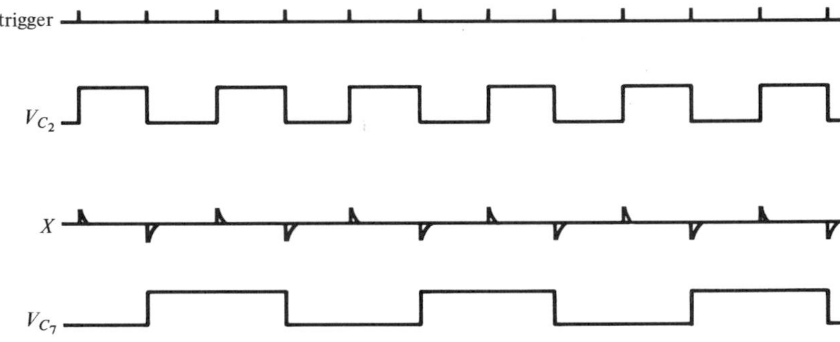

Figure 9.13 / *Waveforms from counter of Fig. 9.12.*

with the base of Q_4 connected to its collector, Q_4 should be turned ON. This would be true, except that in the absence of a trigger, Q_5 is OFF and Q_4 is thereby prevented from conducting. On imposition of a trigger, Q_5 turns ON, lowering V_{E_3} and V_{E_4} almost to ground. This has no effect on Q_3, but it turns Q_4 ON.

If the trigger persists, V_{C_2} and V_{C_4} will drop to zero, C_2 will discharge, and the binary will be in the same condition as that in Fig. 9.7 with both inputs at **1**. On removal of the trigger, one side will turn OFF, but it will not necessarily be the side most recently ON. For this reason, the trigger must not persist long enough for complete discharge of C_2. When the trigger is removed, both Q_3 and Q_4 become incapable of conducting. The transition that has started will continue only if during the time the trigger is applied, V_{C_2} drops sufficiently to bring Q_1 out of saturation. If so, with Q_1 in its active region, the positive feedback loop is established.

For this reason, the time duration of the triggering signal is critical. Either a suitable trigger pulse must be available, or it must be provided. This is the function of the high-pass filter R_3–C_3 which serves to couple the two binaries in Fig. 9.12. It differentiates the transition that occurs at Q_2 (see Fig. 3.17) to produce pulses of the correct duration. Note that useful trigger pulses will be produced only by positive transitions of V_{C_2}. Negative transitions produce pulses that drive Q_8 farther into cutoff.

As a result, while section 1 triggers on each input pulse, section 2 triggers coincident with every other input to section 1, and a third section would trigger on every eighth pulse to section 1. Waveforms are illustrated in Fig. 9.13.

The counter serves as a frequency divider if the output waveform of the last section is observed. It can be a counter, if some arrangement is made to read the condition of all binaries at the termination of a count. Suppose that in a three-stage counter, a light is lit by a high voltage at one collector in each binary. We shall define the light ON condition as logical **1** for each binary. Provision is made to set all binaries to **0** before a count is taken. The output is shown in Table 9.5 to be just a binary representation of the number

TABLE 9.5 / *Input–output relationships for a three-stage counter*

Total number of pulses	*Output*
0	0 0 0
1	0 0 1
2	0 1 0
3	0 1 1
4	1 0 0
5	1 0 1
6	1 1 0
7	1 1 1

of pulses counted. A three-stage counter has a capacity of $2^3 = 8$ counts and shows the same reading for pulse 0 and for pulse 8.

9.2.4 / Registers

The function of the data register as one form of memory device is discussed in Section 9.1.5. A register consists of an array of binaries, one for each bit to be stored. They can be arranged to accept data by parallel transfer, as indicated in Fig. 9.14(a), where each box stands for a binary, or by serial transfer as in Fig. 9.14(b). Parallel transfer is more rapid, since all bits of a word are handled simultaneously; it is, of course, much more complex. For serial transfer, a shift register is needed. This is one that is capable of transferring the contents of each binary to the next in line. In practice, there are two registers, and a shift involves a transfer of information from all stages of the main register to the appropriate stages of the temporary register, followed by transfer of the information to the receiving binaries of the main register. This avoids a simultaneous transfer into and out of a binary, which would cause a loss of information.

As an illustration of this type of information transfer, suppose that a train of pulses is to be counted for set periods at intervals and the results stored. The pulses could be counted by a binary counter of the type described above and indicated in Fig. 9.15. At the end of the count, the results might be transferred in parallel to the buffer register, leaving the counter ready for a new sample. The buffer could then discharge into memory.

The buffer register could be made up of a series of gated memory circuits constructed of *NAND* gates, as shown in Fig. 9.16. Here gates 1 and 2 constitute the binary, while gates 3 and 4 make up a control that allows the

Figure 9.14 / *Data transfer into and out of registers.*

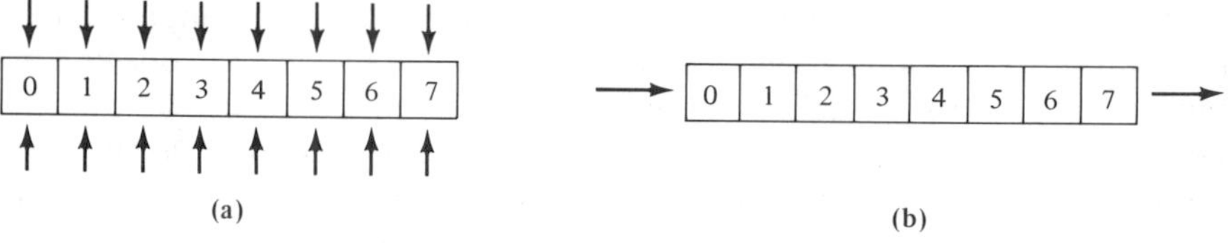

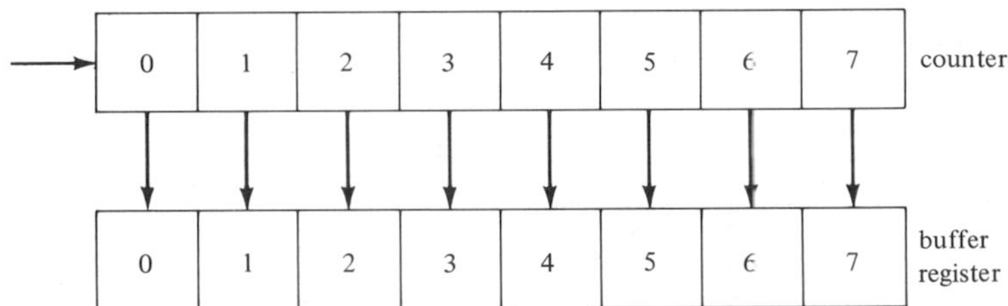

Figure 9.15 / *Storage of digitally coded data.*

binary to be unaffected by changes at its inputs A and B except when a control pulse opens 3 and 4. The action of this circuit is the same as that of Fig. 9.7, since 3 and 4 are inverters, except that three inputs, (A) (control) and (B) (control), rather than just A and B, must be present. In practice, inputs A and B are connected to the two sides of one binary in the counter. On completion of the counting period, a control pulse occurs, to apply the setting of each counter binary to the corresponding register binary. On termination of the clock pulse, the counter can be reset, the previous count safely stored in the register.

9.2.5 / Switching matrices

In designing equipment containing logic circuitry, the need continually arises to perform elaborate switching operations to transfer information along predetermined paths. For an example, the conversion of binary to octal numbers will be cited. As pointed out in Section 9.1.2, three bits are required to handle one octal digit. Let them be represented by A, B, and C in Fig. 9.17. When bit A is **1**, then in the figure, A would exhibit the voltage corresponding to logical **1** and $\bar{A}$ would have the voltage corresponding to logical **0**. If bit A were **0**, then A would be **0** and $\bar{A}$ would be **1**. The signals A and $\bar{A}$ are normally the outputs from a binary in a register.

The necessary circuitry is shown in symbolic from in Fig. 9.17. If, for example, the number being transmitted is 110_2 or 6_8, $\bar{A}$, B, and C would be **1** and A, $\bar{B}$, and $\bar{C}$ would be **0**. This would produce **0** at the outputs of all

Figure 9.16 / *Gated memory circuit.*

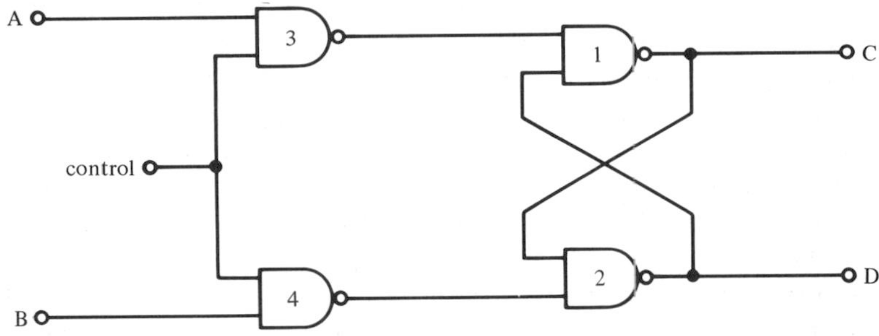

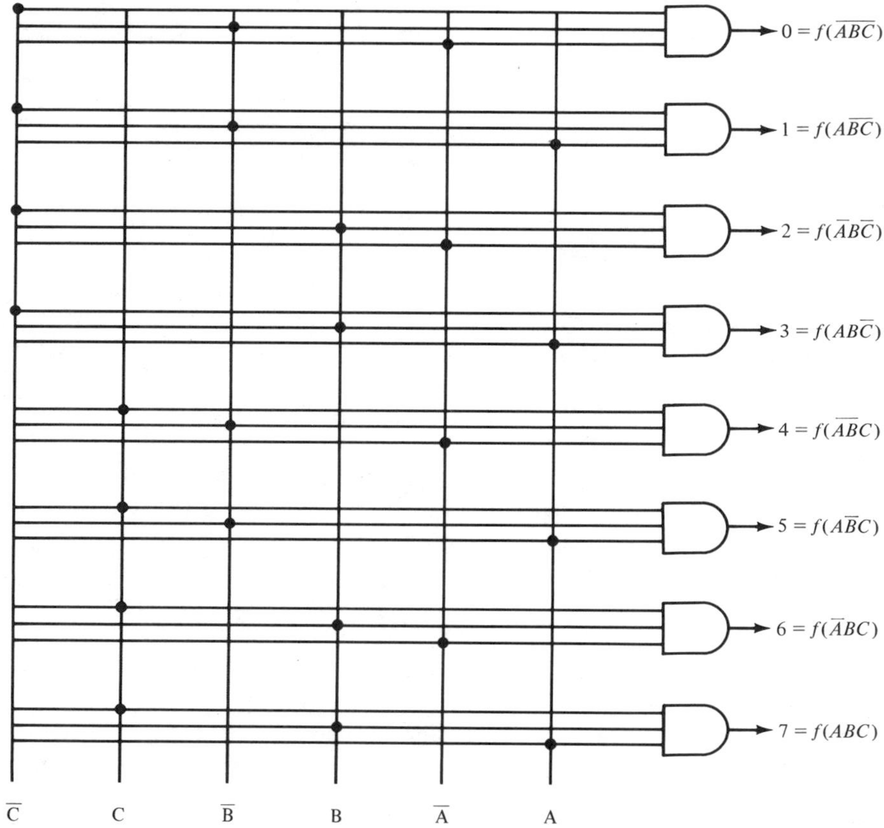

Figure 9.17 / *Binary–octal decoder, logic diagram.*

AND gates except number 6. The actual schematic of a circuit that uses diode gates is shown in Fig. 9.18. In operation, appearance of logical **0** at an input will forward bias all diodes connected to it. Logical **1** at an output is possible only if all three diodes attached to it are reverse biased.

The matrix shown in Fig. 9.17 could evidently be expanded to serve as a binary to decimal decoder. Provision would have to be made for enough inputs to accommodate the code being used, for example, eight inputs for a four-bit 1, 2, 4, 8 code; and there would be ten, rather than eight, outputs. A decoder of this type is the basic component in the readout scheme for a digital instrument. Readout is ordinarily taken to a printer or to a visual display. In either case, the matrix output can be used as the control signal. If an analog readout is desired, the matrix output can be taken to a weighted analog summing circuit. This would be similar to that in Fig. 8.8(b), but with ten inputs.

Switching matrices have many uses. For example, the adder in Fig. 9.19 can be constructed as a diode matrix. Often matrices function as digital

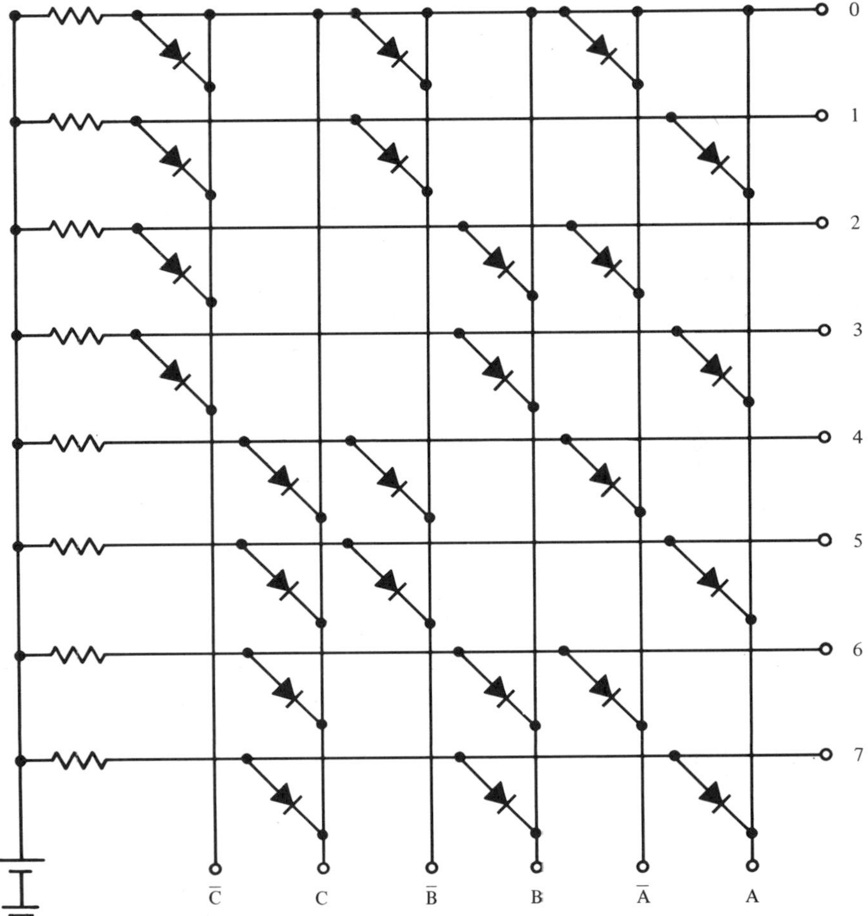

Figure 9.18 / *Binary–octal decoder, schematic diagram.*

electronic analogs of mechanical devices. Commutating action, which consists of switching an input, in sequence, to an array of outputs, can be obtained by the use of a matrix. If a signal is to be directed seriatim to six locations, it would be connected through dual input *AND* gates to each of these. The other *AND* gate input would in each case be one output from a matrix which would be driven by a counter. Matrices are used to perform the switching operations needed to produce alpha-numeric characters for visual readouts. These may consist of neon tubes with an array of electrodes each in the shape of a character ("nixie" tubes). The matrix would direct the control signal to the appropriate electrode. Another approach is to use a series of segments that can be lighted. With correct choice from a small number of these, numbers and letters can be displayed.

Matrices can be used to generate binary versions of constants. For example, if it is desired to be able to call up the constant π fairly often, this can be done with any desired number of significant figures. A binary counter is set to accept a fixed number of pulses, the condition of its registers after each pulse serving as the input to the matrix. The matrix output is the binary version of 3 after one pulse, 1 after the second, 4 after the third, and so on.

9.3 / ARITHMETIC OPERATIONS

One of the very significant differences between analog and digital instruments is in the way arithmetic operations are performed. In a digital instrument, the individual steps of addition, subtraction, multiplication, and division of numbers are carried out. Since there is, in principle, no limit on the number of bits in a word, the precision of this operation can be made very much greater than is possible with an analog device. Since digital arithmetic must be performed by multiple applications of rather simple logic operations, it is based on *algorithms*, which are rules for performance of fixed operations.

The great importance of digital instrumentation makes digital arithmetic a very significant topic; however, even the most minimal treatment would go beyond the reasonable space limitations of this book, and the reader is referred to the Suggested Readings at the end of the chapter. In the following paragraphs, an attempt is made to provide some insight into the implementation of one operation of digital arithmetic because it is felt that this is helpful in placing the strengths and limitations of digital computers in perspective. They are superbly suited to the performance of simple repetitive operations, but they do not think.

Let us consider the process of addition. Numbers are added a bit at a time, just as the reader would do in performing the sums indicated below:

$$\begin{array}{ll} 4_{10} & 0100_2 \\ 5_{10} & 0101_2 \\ \hline 9_{10} & 1001_2 \end{array} \qquad \begin{array}{ll} 2_{10} & 0010_2 \\ 5_{10} & 0101_2 \\ \hline 7_{10} & 0111_2 \end{array} \qquad \begin{array}{rr} 10_{10} & 1010_2 \\ 7_{10} & 0111_2 \\ \hline 17_{10} & 10001_2 \end{array}$$

The machine must also take into account the carry. In the following, the augend is represented by X, the addend by Y, and the sum by S. A carry from a less significant bit is C_i, and a carry to a more significant bit is C_o. Table 9.6 is a generalized statement in truth table form for one step in the addition of binary numbers.

Figure 9.19 is a symbolic representation of a full adder using *NAND* gates. It consists of two *EXCLUSIVE-OR* operations [see Fig. 9.9(c)], 1, 2, 3, 4 and 5, 6, 7, 8, and one *NAND* operation, 9. The relationship between S and the input, given in Eq. (9.3), is evident by inspection of Table 9.6.

$$S = (X \oplus Y) \oplus C_i \tag{9.3}$$

This is the relationship implied in Fig. 9.17.

TABLE 9.6 / *Truth table for addition of binary numbers*

X	Y	C_i	S	C_o
0	0	0	0	0
1	0	0	1	0
0	1	0	1	0
0	0	1	1	0
1	1	0	0	1
1	0	1	0	1
0	1	1	0	1
1	1	1	1	1

The relationship between the inputs and C_o is not so obvious; however, it can readily be shown that the operations implied by Fig. 9.19 are correct. The signal at point A of Fig. 9.19 represents $(\overline{XY})$. That at point B is $\overline{C_i(X + Y)}$. Therefore,

$$C_o = \overline{(\overline{XY})[\overline{C_i(X + Y)}]} \tag{9.4}$$

Table 9.7 demonstrates the validity of Eq. (9.4) because the last column is congruent with the C_o column of Table 9.6.

To carry out addition by a machine, registers must be provided for both augend and addend. The process may involve sequential transfer of augend and addend, a bit at a time to the adder in the same sequence that a person would use in performing the addition. It is also possible, with logic circuitry, to take all bits simultaneously (with a delay built in to permit handling the carry bits) and thereby speed up the process.

Figure 9.19 / *Full adder, logic diagram.*

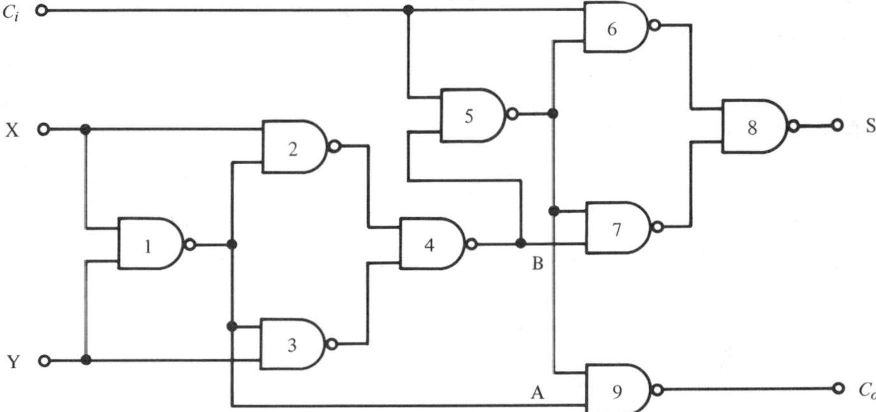

TABLE 9.7 / *Truth table for full adder*

X	Y	C_i	$(\overline{XY})$	$(\overline{X + Y})$	$[\overline{(X + Y)C_i}]$	$\overline{(\overline{XY})[\overline{(X + Y)C_i}]}$
0	0	0	1	0	1	0
1	0	0	1	1	1	0
0	1	0	1	1	1	0
0	0	1	1	0	1	0
1	1	0	0	0	1	1
1	0	1	1	1	0	1
0	1	1	1	1	0	1
1	1	1	0	0	1	1

9.4 / APPLICATIONS

9.4.1 / Use of a computer

Very often, digital equipment is included in experiments by connecting in a digital computer. This may involve the collection of data in some conventional fashion, to be recast by the experimenter into a form, such as punched cards, that is compatible with the computer input. While this procedure allows one to do computations on data, it may be more attractive to *interface* the computer directly to the experiment. This allows the process of data collection to be performed by the computer, utilizing its fast response and its patience. Moreover, when a computer monitors experimental data continuously, its capacity for logical decision making may be utilized to automate the experiment.

A computer basically consists of input and output units, control, storage, and arithmetic units. To use it, a *program* is read into the storage unit through the input. To operate, the first step of the program, which is a very detailed set of instructions arranged in logical order, is transferred to the control unit, which is adjusted to run automatically through the program, step by step. The computer is capable of receiving data from one or more transducers. It may simply store them in its memory for future use, or it may perform computations on data samples as they come in. A digital machine can be connected to activate control devices, as was described for analog instruments. Moreover, a computer can be arranged to receive several inputs and to provide several outputs essentially simultaneously, so that several aspects of an experiment may be monitored, with perhaps some data being stored and others being used for control purposes.

The direct connection of a digital instrument into an experiment requires that the signal being produced be digitized. Certain types of signals, such as those observed in radiochemistry, X-ray analysis, mass spectroscopy, and optical spectroscopic photon counting ordinarily occur as a train of pulses. These can be handled by accumulating the signal in a binary counter, as indicated in Section 9.2.4.

However, most of the signals encountered are analog in nature and must be converted to the digital form. This involves sampling and encoding as

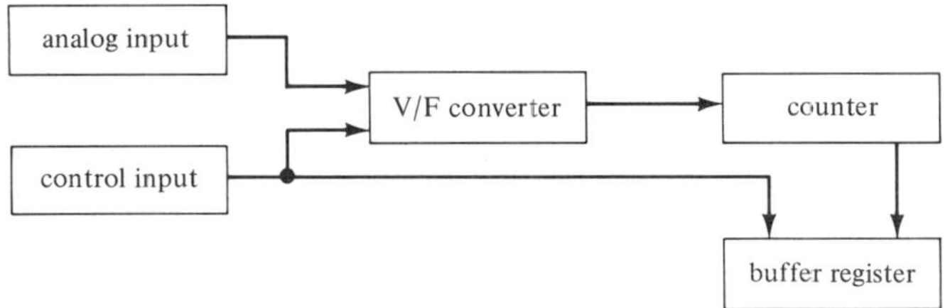

Figure 9.20 / *Block diagram of an A/D converter.*

discussed in Section 9.1.2 and is termed analog-to-digital conversion (A/D) Very often, the analog signal source and the A/D converter represent the primary error sources in an experiment. The ability to perform elaborate operations on data over extended periods of time without degradation of accuracy or precision is the major strength of digital design. One important reason is that once data are digitized, they are handled by binary devices that can be made substantially immune to the introduction of errors through interference by circuit noise. In addition, very high precision can be achieved through the use of long machine words. For example, a 12-bit word provides a resolution of 1 part in 4096 (0.025 percent), and 16 bits give 1 part in 65536 (0.002 percent).

An A/D converter consists of a voltage-to-frequency converter with the necessary logic circuitry to control it. A simplified diagram is presented in Fig. 9.20.

In operation the sample rate may be set at a fixed value, or it may be controlled by a computer. However, on receiving a command signal at the control input, the sampling cycle of the V/F converter, Fig. 9.11, is initiated. The V/F output is counted and then transferred to a buffer register. The V/F converter can then be reset and is ready for the next sample. The design of the counter, with the V/F converter, is arranged so that the count constitutes a binary-coded number which can be transferred to memory or subjected to arithmetical operations by the computer.

In addition to data acquisition, a computer can be used as a control unit, if provision is made for it to transmit information to other components of the apparatus. The use of a computer with an nmr spectrometer is an example of this type. There are occasions when the ^{13}C spectrum may be extraordinarily useful in structure elucidation. However, ^{13}C occurs in such low abundance that in a conventional experiment, its spectrum is lost in noise. By utilizing signal averaging, in which the spectrum is scanned repeatedly with the signal being sampled at intervals, as described in Section 6.4.2, the spectrum can be obtained.

This experiment presents difficulties because it must be carried out over an extended period of time with a high degree of repeatability. For example, if 1000 samples are being taken on each scan, then the 91st sample of a particular pass must coincide precisely with the 91st sample of all other passes.

Furthermore, because of the nature of nmr spectroscopy, it is necessary to maintain extraordinary stability and uniformity of the magnetic field strength throughout the experiment. Changes in field homogeneity will cause observed peak shapes to change on different passes, which would entirely vitiate the signal-averaging operation.

The capabilities of a computer are especially well utilized in this type of experiment. It can be programmed to initiate the scans and to take samples at very precisely determined times over a period of hours. It can keep track of individual samples, average them with corresponding samples in successive scans, and monitor the output of a transducer that measures field homogeneity to make adjustments as necessary.

In order to have a computer perform the control function, a transducer must be provided to monitor the variable being controlled. The signal produced is digitized and is continually compared by the computer with the desired value of the variable. The experimental setup must also include provision whereby the computer can apply this error signal, perhaps after D/A conversion, to adjust the variable.

In one approach, the height of an absorption peak of a reference compound is monitored. With field homogeneity optimized, the peak shows a maximum height. Less than optimal homogeneity reduces the observed peak height. To adjust the homogeneity, the instrument is fitted with shim coils positioned so that magnetic fields generated by them add to or subtract from the main field over limited areas. When the instrument is placed in operation, the field is optimized by adjusting the currents fed to the shims. One of these is used for automatically controlled "touch-up" adjustments by the computer during operation. The signal fed to the control shim coil is modulated by a low-frequency signal of small amplitude. If the shim current is oscillating symmetrically about its optimum value, the small fluctuations in the magnetic field have little or no effect. If, however, the shim current is not optimal, the observed peak heights will fluctuate at the modulating frequency. If shim current is smaller than optimal, fluctuations in peak height will be in phase with the modulating signal; increasing shim current will cause an increased peak height. If the average shim current is too large, peak height fluctuations will be 180° out of phase with the modulating signal; decreasing shim current results in increasing peak height. If the reference peak signal is subjected to phase-sensitive detection, using the shim-modulating signal as reference, optimal shim current corresponds to no DC output; too large a shim current can be made to result in a finite detected signal of given polarity, and too small a current can produce the opposite polarity. This detected signal can serve as the error signal to be used to control the shim current. This control function can be performed by either an analog or a digital system. When a computer is being used for data acquisition and instrument control in a signal-averaging experiment, it may also be used for this additional control function.

9.4.2 / Use of logic circuitry

While the use of a computer is an elegant and powerful application of digital instrumentation, there are many applications that do not require the capabilities of a computer. Very often, some of the basic building blocks, described in Section 9.2, may be combined to provide the capabilities of digital techniques for specific purposes at modest cost. The ready availability of integrated circuit components has made this especially practical.

This approach to instrument design is most easily applied to those experiments with outputs that can be accepted by digital circuitry without A/D conversion. For example, measurements of ionizing radiation can be performed with binary or decimal counters. Frequency measurements can be made by counting the maxima of the unknown signal through a gate controlled by an accurate timing signal. If the frequency is quite low, the period can be measured more accurately. This can be done by counting a clock pulse and using the input signal to control the gate. The V/F converter described in Section 9.2.2 is a variation of this. A detailed discussion of the potentialities of this type of design has been given by Malmstadt and Enke in the Suggested Readings listed at the end of this chapter.

9.4.3 / Hybrid instruments

In this and the preceding chapter, the point has been made that both analog and digital designs offer characteristic advantages. In particular, use of analog design is likely to simplify the input and output functions. Digital design facilitates the inclusion of information storage, decision-making capabilities and, in larger instruments, powerful data-processing capabilities. Hybrid instruments that utilize both types of design offer the possibility of taking advantage of the best points of both. It is, of course, true that most digital equipment is hybridized because of the necessity to accept analog inputs and the desirability of analog outputs. However, with the availability of convenient and reasonably priced logic circuit components, they are being increasingly incorporated into smaller instruments which have been all analog. This provides a greater degree of automation in instruments that are sufficiently inexpensive to be very widely used.

It should certainly be possible to utilize logic circuitry to overcome some of the limitations inherent in otherwise highly desirable analog instruments. For example, the analog integrator is well suited to a variety of applications; however, it is troublesome if the integrations must be prolonged because of drift. Where a signal is to be accumulated over a period of time, logic circuitry may be used to discharge the integrator periodically, and to store the information in a counter. The input can then be arranged to ensure that the integration rate remains large relative to the drift rate.

Logic circuitry and electronic switching may tend to displace mechanical switches and relays from many types of equipment. Even in applications for which speed is not of concern, electronic switches are silent, have very long life expectancies, and can be very compact and inexpensive. Automatic variable-speed scanning, automatic reset, and automatic sample changing for spectrophotometers are examples of the types of familiar laboratory operations that are at present ordinarily carried out manually or by mechanical switches, but for which electronic switching can be applied, even in inexpensive instruments. Where sequences of operations are to be performed, electronic timing circuits can supplant cams and gear trains, eliminating noise and reducing maintenance and offering the possibility of increased flexibility owing to the introduction of decision making by the instrument.

SUGGESTED READINGS

Y. Chu, *Digital Computer Design Fundamentals*, McGraw-Hill (New York, 1962).

J. Millman and H. Taub, *Pulse, Digital and Switching Waveforms*, McGraw-Hill (New York, 1965).

H. V. Malmstadt and C. G. Enke, *Digital Electronics for Scientists*, W. A. Benjamin (Menlo Park, Calif., 1969).

chapter 10 / Electromagnetic radiation and its interaction with matter

chapter 10

10.1 / INTRODUCTION

By far the largest class of instrumental methods of chemical analysis depends on the use of electromagnetic radiation as a probe to determine the identity, structure, and quantity of chemical species. The most obvious example of electromagnetic radiation is light (visible electromagnetic radiation), but light occupies only a small region in a spectrum of electromagnetic radiation that covers a range of at least 15 orders of magnitude in energy.

Instrumentation and measurement techniques differ widely in the various regions of the electromagnetic spectrum employed for chemical analysis, and at first glance the various spectroscopic methods of analysis appear to have little in common. In fact, the properties of the radiation and the types of experiments performed are basically the same whatever the region of the spectrum, and it is therefore to our advantage to attempt to understand, at least in a qualitative fashion, the properties of electromagnetic radiation and how it interacts with matter before attempting to examine the individual spectroscopic methods of analysis.

Light has been studied for centuries and much is known about its properties, but there is no straightforward answer to the simple question: What is light? In some of its aspects, as, for example, in the focusing of light by lenses and its reflection by mirrors, light is best described as a wave phenomenon. In other aspects, such as the emission or absorption of light by atoms or molecules, light can only be considered as a stream of discrete energy packets. Thus, light is said to have a dual nature, exhibiting both wave and particle characteristics. This duality is not confined to the visible portion of the electromagnetic spectrum, but can be demonstrated for all electromagnetic radiation; nor is it limited only to radiation: Wave properties can be demonstrated for all subatomic particles. The diffraction of electrons by crystals is an obvious example of the wave properties of particles.

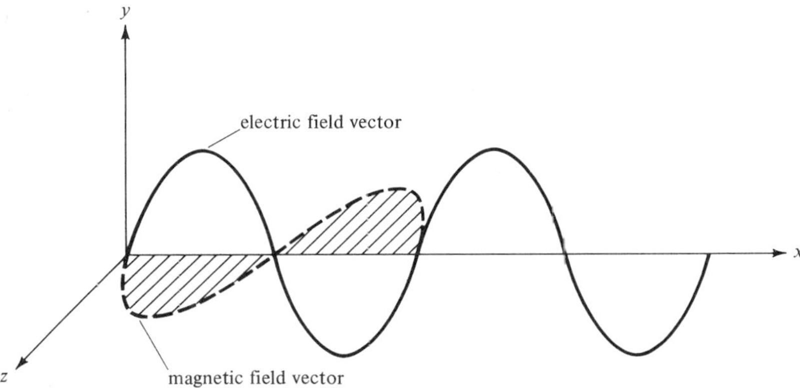

Figure 10.1 / *Representation of an electromagnetic wave.*

Electromagnetic radiation differs from other particles in that the energy packets, the photons, have zero rest mass.

It is beyond the scope of this book to consider current theories of the nature of electromagnetic radiation. In this chapter, we shall briefly review phenomena pertinent to the application of electromagnetic radiation for analytical spectroscopy.

10.2 / RADIATION AS A WAVE PHENOMENON

From the wave point of view due to Maxwell, an electromagnetic wave of frequency ν is an alternating current with an associated magnetic effect. The interaction of the wave with its surroundings can be discussed in terms of electric and magnetic vectors representing the force fields. (Compare with the discussion of the force field surrounding an isolated charge in Section 1.1.1, and the use of vector representation for electrical signals in Sections 1.3 and 1.4.) Figure 10.1 is a vector representation of a ray of light moving along the x axis. The electromagnetic field varies periodically perpendicular to the direction of propagation. Thus, the electric field varies in the direction of the y axis, and the corresponding magnetic field varies in the direction of the z axis. For most purposes, the properties of the radiation can be described by considering only the electric vector, and this will generally be our practice; but in a few cases, notably in the discussion of magnetic resonance (Chapter 20), it is more convenient to deal with the magnetic vector.

A transverse wave, such as that of Fig. 10.1, is characterized by the following quantities:

1. The *frequency*, ν, of the wave is the number of times per second that the electric field reaches its maximum positive value. Thus, frequency has

units of reciprocal time (sec^{-1}), and this quantity has been given the name hertz, which is abbreviated Hz.

2. The *wavelength*, λ, is the distance between successive maxima. The usual units of wavelength vary with the region of the electromagnetic spectrum. Some of the more commonly employed units are centimeter (cm), micrometer (μm, 10^{-6} m), nanometer (nm, 10^{-9} m), and angstrom (Å, 10^{-10} m).

3. The wavenumber, $\bar{\nu}$, is the reciprocal of the wavelength, and thus the wavenumber is the number of maxima per unit distance. Wavenumber has units of reciprocal distance (cm^{-1}), and this quantity has been given the name kayser, which is abbreviated K.

4. Since there are ν oscillations per second and each oscillation covers a distance λ, the product

$$\lambda\nu = v \tag{10.1}$$

is the distance traveled by the wave in 1 sec; that is, v is the *velocity of propagation*. The frequency, ν, is an invariant characteristic of a wave, but λ and v depend on the medium in which the wave is propagating. Wavelengths are usually reported for waves traveling in a vacuum. The velocity of propagation of all electromagnetic radiation in a vacuum is 2.998×10^{10} cm-sec^{-1}, and this quantity is given the symbol c.

5. The velocity, and hence the wavelength, of radiation in a medium is described by the *refractive index* of the medium, n, where

$$n = \frac{c}{v} \tag{10.2}$$

Thus, Eq. (10.1) can be rewritten as

$$n\lambda\nu = c \tag{10.3}$$

and this expression can be used to calculate the wavelength of radiation of a given frequency in a medium of known refractive index. For example, it is instructive to consider the difference involved in reporting the wavelength of a spectral line in air instead of in a vacuum. The refractive index of air for the yellow line due to Na is reported to be 1.0003, and the wavelength of the line measured in air is 5889.95 Å. Thus, the vacuum wavelength is given by

$$\lambda_{vac} = n\lambda_{air} = 5891.72 \text{ Å}$$

6. Electromagnetic waves, like other transverse waves, can be polarized. The plane determined by the electric vector is said to be the plane of polarization. In Fig. 10.1, the xy plane is the plane of polarization. The various waves that together constitute the beam of light generally have different directions of polarization; this is called *unpolarized light*. If all the electric vectors are confined to a single plane, the light is said to be *plane polarized.* A plane-polarized beam can be further decomposed into left and right *circularly polarized* beams. In a circularly polarized wave, the electric

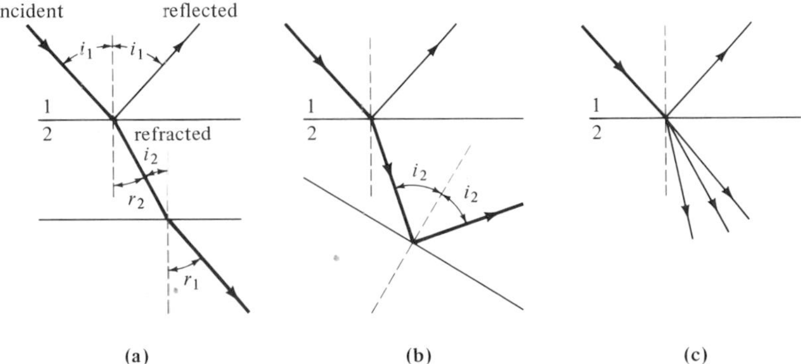

Figure 10.2 / *Some optical phenomena at the boundary of media of different refractive indices.*
(a) *Refraction and reflection of a monochromatic ray at a plane-parallel plate.*
(b) *Total internal reflection.*
(c) *Dispersion of polychromatic radiation.*

vector has a constant amplitude but rotates about the axis of propagation as the wave moves, so that to an observer the vector appears to sweep out a helix. If the direction of rotation is counterclockwise as the wave propagates toward the observer, the wave is said to be left circularly polarized; for the opposite sense of rotation, the wave is said to be right circularly polarized. Polarization phenomena are important in the consideration of such techniques as optical rotatory dispersion and circular dichroism, but will not be considered in this book.

7. The maximum value of the electric field strength is called the *amplitude*. The strength or *intensity* of the light wave is proportional to the square of the amplitude.

Some properties of electromagnetic radiation that are best understood by considering its wave aspect are illustrated in Fig. 10.2, which shows the behavior of light rays at a boundary between media of refractive indices n_1 and n_2, where $n_1 = 1.00$ (vacuum or air), and the material of refractive index n_2 is assumed to be an optically transparent material, such as glass or NaCl, with a smooth surface.

If a ray of light of a single frequency (monochromatic radiation) strikes the surface at an angle of incidence i_1, measured from a normal to the face of the material, two rays, the reflected and the refracted rays, will be produced, as shown in Fig. 10.2(a). The reflected ray will be found to propagate away from the surface at an angle of reflection equal to the angle of incidence, but on the other side of the normal. Thus, the *law of reflection* states that the angle of reflection equals the angle of incidence. The refracted ray enters the material at an angle r_2, which depends on the refractive index of the material, n_2, in the following way:

$$n_2 \sin r_2 = n_1 \sin i_1 \tag{10.4}$$

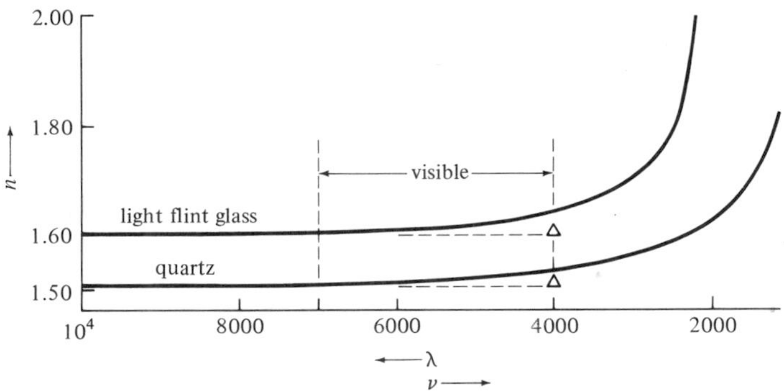

Figure 10.3 / *Dependence of refractive index on frequency in the ultraviolet and visible regions for two optical materials.*

Equation (10.4) is known as *Snell's law of refraction.* Note that if the incident ray is traveling through a vacuum, for which n_1 is by definition exactly one, measurement of i_1 and r_2 allows calculation of the refractive index of the material. It should also be noted from Eq. (10.4) that in going from a rare medium to a dense medium (i.e., from a medium of lower refractive index to a medium of higher refractive index) a light ray is bent toward the normal ($r_2 < i_1$), and in going from a dense medium to a rare medium a ray is bent away from the normal.

Reflection and refraction also occur when the ray of light reaches the second boundary of the material. If the two boundaries are parallel, it is apparent that $i_2 = r_2$ and $r_1 = i_1$. Thus, passage of a light ray through a plane-parallel plate results in a small displacement of the ray but no alteration in its direction. If the angle of incidence at the second boundary exceeds the *critical angle*, that is, the angle at which the refracted ray lies in the plane of the boundary ($r_1 = 90°$, $\sin r_1 = 1$), total internal reflection of the ray occurs, and the normal law of reflection is obeyed. Such an effect is shown in Fig. 10.2(b). From Snell's law and the definition of the critical angle, we can see that total reflection will occur whenever $\sin i_2 > n_1/n_2$. Total reflection can occur only if the incident ray is traveling in the medium of higher refractive index.

If the incident radiation is polychromatic, the angle of refraction varies with frequency, as illustrated in Fig. 10.2(c). As a general phenomenon, the refractive index of a material is found to vary with frequency. This effect is known as *dispersion* and is the basis of the use of prisms for the separation of light by frequency. Figure 10.3 shows the variation of n with frequency over the visible and near ultraviolet region of the spectrum for two kinds of optical material. The behavior exhibited in Fig. 10.3, in which n increases as ν increases, is said to be *normal dispersion.* If the plot of n is extended to still higher frequencies, a region is found in which n decreases as ν increases; such behavior is said to be *anomalous dispersion.* Regions of anomalous

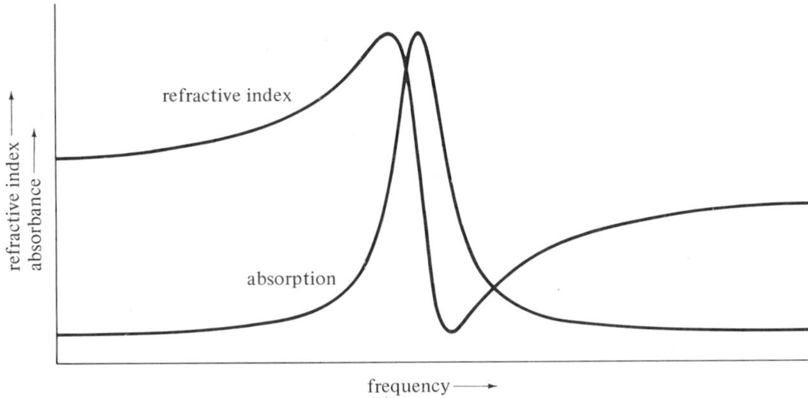

Figure 10.4 / *Idealized plot of refractive index in the region of an isolated absorption band.*

dispersion are always associated with absorption bands in the material, as illustrated in Fig. 10.4. The variation of n in the region of an absorption band is not, in fact, anomalous but typical; the names "normal" and "anomalous" merely indicate that historically "normal" dispersion was observed first and later observations in the region of absorption bands appeared anomalous until the phenomenon was better understood. Note from Fig. 10.4 that rapid change of n with ν ($dn/d\nu$ is large) occurs as the absorption band is approached from the low-frequency side, and at frequencies distant from the absorption band $dn/d\nu$ is small. Useful prisms can be made only from materials for which $dn/d\nu$ is large in the frequency region of interest.

The wave theory of electromagnetic radiation predicts that when a beam of radiation passes through an aperture, the waves will spread in all directions. This phenomenon of *diffraction* can be observed experimentally if the dimensions of the aperture are of the same order of magnitude as the wavelength of the radiation.

Diffraction of light occurs at each point of the narrow aperture, and waves spreading from different points on the aperture may at a given point overlap and give rise to a phenomenon known as *interference*. Consider two waves of equal amplitude arising from the same source but diffracted from different points, so that the two waves may travel different distances to reach a point beyond the aperture. If the distances traveled by the two waves are equal or differ by an integer multiple of the wavelength, the waves will be in phase and the electric field strength for each will reach its maximum value at exactly the same time. Thus, at the point of observation, the resultant wave amplitude is $A = A_1 + A_2$, where A_1 and A_2 are the amplitudes of the two waves. If the distances traveled by the two waves differ by half a wavelength or an odd integer multiple of half a wavelength, the waves will be out of phase so that when one wave has its most positive value the other wave has its most negative value and the resultant amplitude is zero at all times.

Intermediate cases also occur, and for a real system the resultant intensity at a given point must be derived from an addition of all contributing waves, each with its proper phase. The concepts of diffraction and interference are essential to an understanding of the operation of interference filters and diffraction gratings, both of which are discussed in Chapter 11. We should take note that interference is observed only for *coherent* waves, that is, waves generated at the same time and from the same point in a source.

10.3 / QUANTIZED ENERGY CHANGES

Every elementary system—nucleus, atom, molecule—when reasonably isolated, appears to have a number of discrete energy levels. To change from one to another, a system must be exposed to radiation or bombarded with particles having energies at least equal to the energy differences of the states. Emission and absorption of radiation arise from changes in the energy of such systems. There are two important exceptions to this type of behavior: (1) the translational motion of atoms, molecules, and ions in free space, and (2) the "blackbody" emission and absorption of radiation exhibited by solids. Blackbody radiation is discussed in Section 10.4.

10.3.1 / Energy changes in elementary systems

The energy levels of a reasonably isolated nucleus, atom, or molecule are uniquely characteristic, and the interest of the chemist in electromagnetic radiation is based chiefly on his ability to obtain information about the energy levels by measurement of the emission and absorption of electromagnetic radiation. If the intensity of emission or absorption is measured as a function of the energy of the electromagnetic radiation (or some quantity related to the energy, such as frequency, wave number, or wavelength), then one obtains a unique signature of the elementary system, which we call its spectrum. Further, with proper calibration, it is possible to relate the intensity of a particular energy transition to the concentration of interacting species.

The instrumentation required to carry out the measurements described above varies greatly as one varies the frequency of the electromagnetic radiation, but basically there are only two types of experiments—measurement of the intensity of either the emission or absorption of the electromagnetic radiation. Figure 10.5 suggests some of the possibilities.

The connection between the wave and particle characteristics of radiation was suggested by Planck; that is, when a transition occurs between energy states of a system such that a photon of energy E is emitted or absorbed, the frequency, ν, of the emitted or absorbed radiation is given by

$$E = E_i - E_j = h\nu \qquad (10.5)$$

where E_i and E_j are the energies of the states involved and h is Planck's constant. Note that in elementary systems such as are described here,

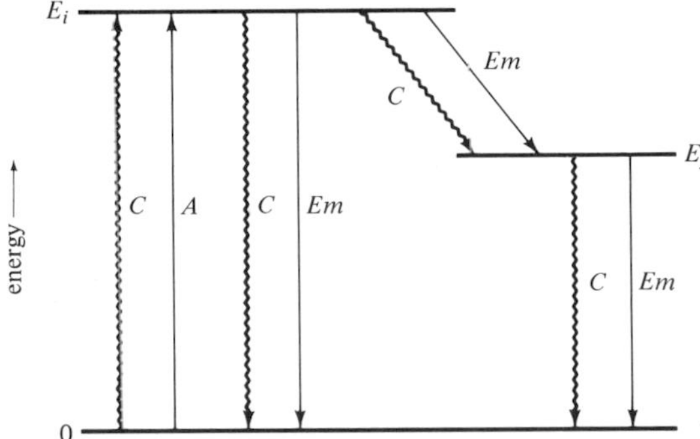

Figure 10.5 / *Generalized energy changes in an elementary system. C = collisional activation or deactivation, A = absorption of radiation, Em = emission of radiation.*

energies are measured with the ground, or most stable, state taken as the zero level. Figure 10.6 summarizes the types of transitions observed in the various regions of the electromagnetic spectrum and the approximate ranges of energy involved.

Recalling the definitions of frequency, wavelength, and wavenumber of the preceding section, one can write alternative formulations of Eq. (10.5) that relate photon energy to the wavelength or wavenumber of the radiation:

$$E = hc\bar{\nu} = \frac{hc}{\lambda} \tag{10.6}$$

In some regions of the electromagnetic spectrum, it is customery to report energy in electron volts (eV). An electron volt is defined as the energy acquired by an electron in falling through a potential of 1 V. Since an electron has a charge of 1.6×10^{-19} C, 1 eV corresponds to an energy of 1.6×10^{-19} V-C, and since a volt-coulomb is a joule, 1 eV = 1.6×10^{-19} J.

10.3.2 / Linewidths

Contrary to the implication of Fig. 10.5, photons originating from a given transition, for example, from E_i to the ground state in Fig. 10.5, are not all of the same energy, but rather there is a spread of energies which depends on the transition and the conditions under which the measurement is made. Figure 10.7 shows in an idealized fashion the type of energy distribution that is likely to be obtained. The spread of energies is usually described by the width of the line at one-half the peak amplitude, which is frequently referred to simply as the half-width of the line. It is customary to report this width in frequency rather than energy units. There are a variety of factors that contribute to the linewidth, the most important of which are described below.

The most fundamental sort of limitation on the extent to which radiation

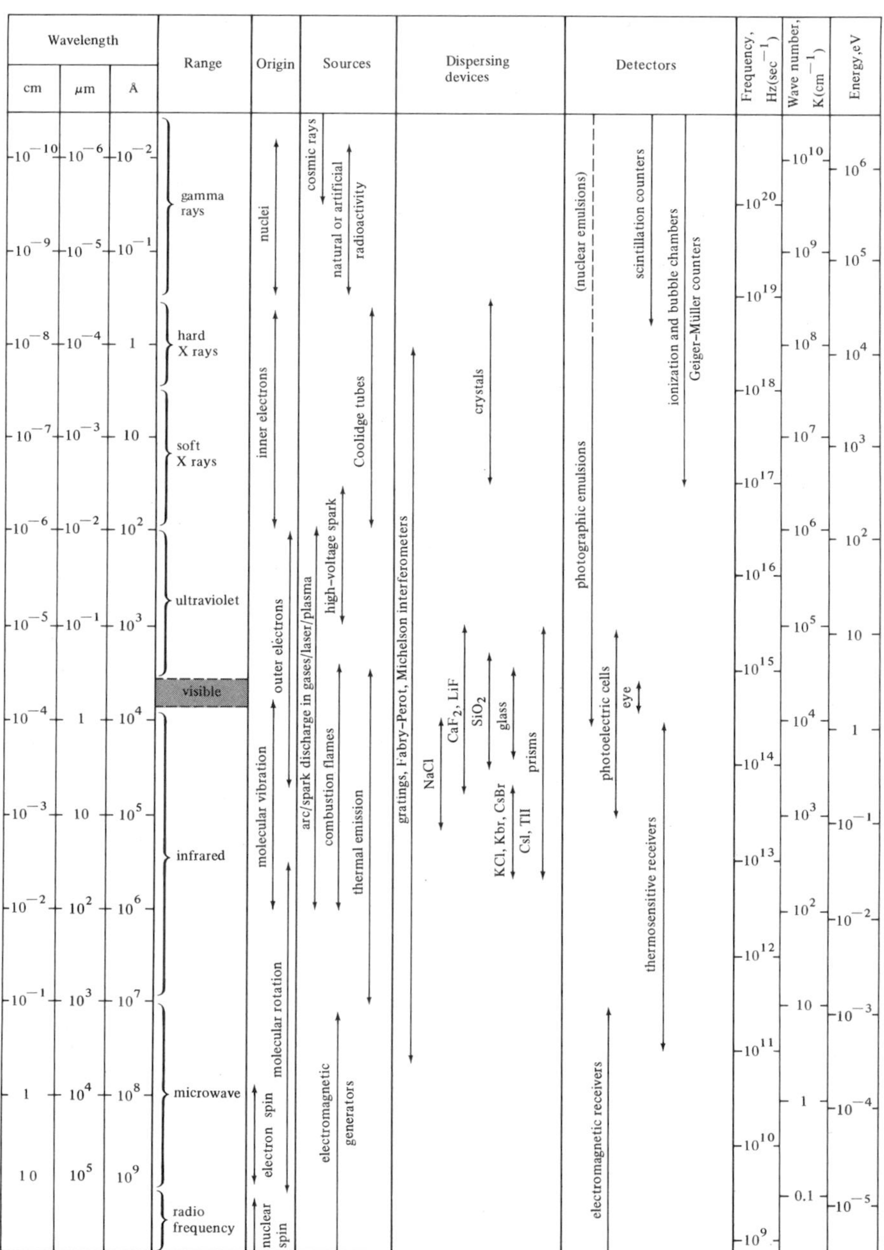

Figure 10.6 / *Survey of the electromagnetic spectrum.*
(SOURCE: R. Mavrodineanu and H. Boiteux, *Flame Spectroscopy*, Wiley, New York, 1965, p. 218)

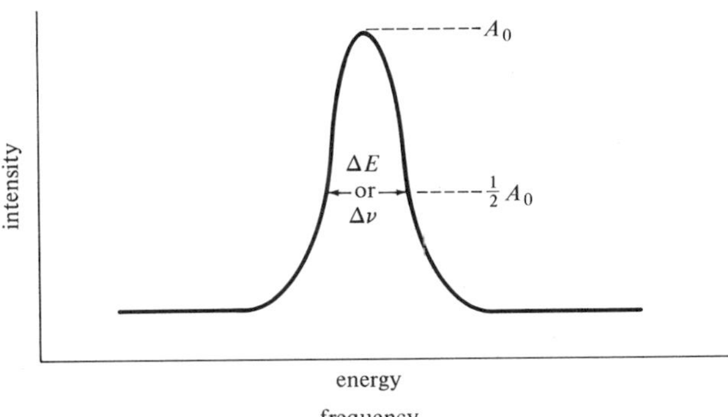

Figure 10.7 / *Typical line shape.*

may approach monochromaticity is that imposed by the Heisenberg uncertainty principle, which may be stated as

$$\Delta E\ \Delta t \simeq \frac{h}{2\pi} \tag{10.7}$$

where ΔE is the uncertainty in the energy of a given state, and Δt is the time required for the measurement. Since the measurement must be carried out in a time comparable to the lifetime of the state, ΔE can be estimated from the expression

$$\Delta E = \frac{h}{2\pi\tau} \tag{10.8}$$

where τ is the lifetime of the state. If one of the states involved in a transition is the ground state, then for most cases Eq. (10.8) gives the energy spread of the transition, since the lifetime of the ground state is usually very long. The half-width of the line in terms of frequency is then given by

$$\Delta\nu = \frac{\Delta E}{h} = \frac{1}{2\pi\tau} \tag{10.9}$$

Note that a short lifetime implies a large uncertainty in the measured frequency.

If the emitting or absorbing species are in random thermal motion with respect to the observer, the Doppler effect will contribute to the linewidth. The *Doppler effect* is the name given to the phenomenon in which the apparent frequency of radiation is increased if the emitter is moving toward the observer and decreased if moving away from the observer. Species in random motion have radial velocities relative to the observer of all possible values between the limits set by those species moving directly toward and those moving directly away from the observer. The half-width of a purely

Doppler-broadened line, if a Maxwellian distribution of velocities is assumed, is given by

$$\Delta\nu = 2\nu_0 \left(\frac{2kT \ln 2}{mc^2}\right)^{1/2} \tag{10.10}$$

where ν_0 is the frequency at the line center, k is the Boltzmann constant, T is the absolute temperature, and m is the mass of the emitting or absorbing species.

If the emitting or absorbing species are subjected to collisions, with the time between collisions short compared to the lifetime of the energy states, then there will be a collisional broadening contribution to the linewidth. The uncertainty in energy can be estimated from the expression

$$\Delta E = \frac{h}{2\pi\tau_c} \tag{10.11}$$

where τ_c is the mean time between successive collisions, and, if one of the states involved in the transition is the ground state,

$$\Delta\nu = \frac{1}{2\pi\tau_c} \tag{10.12}$$

A discussion of the relative contributions of the various types of broadening must be deferred until specific systems are considered.

10.3.3 / Relative populations of energy levels

If a system such as that of Fig. 10.5 has been equilibrated at some absolute temperature T, then the fraction of the interacting species that is in a level i is given by

$$\frac{N_i}{N} = \frac{g_i \exp(-E_i/kT)}{\sum_{m=1}^{n} g_m \exp(-E_m/kT)} \tag{10.13}$$

where N_i is the concentration of species in level i, N is the total concentration of interacting species, and E is the energy of the level identified by the subscript. The term g is called the *statistical weight* of the level and describes the degeneracy of the level, that is, the number of levels of identical energy. The denominator of Eq. (10.13) is sometimes referred to as the *partition function.*

It is instructive to consider the relative population of the excited level of a simplified system consisting of only one excited level and the ground level. The denominator of Eq. (10.13) thus becomes $g_1 + g_i \exp(-E_i/kT)$, since the ground-level energy is by definition zero, and if it is further assumed that $g_1 = g_i$, Eq. (10.13) can be rewritten as

$$\frac{N_i}{N} = \frac{\exp(-E_i/kT)}{1 + \exp(-E_i/kT)} \tag{10.14}$$

TABLE 10.1

Energy (eV)	N_i/N
10	10^{-163}
1	5×10^{-17}
10^{-1}	2.3×10^{-2}
10^{-2}	0.41
10^{-3}	0.49
10^{-4}	0.50
10^{-5}	0.50

Table 10.1 reports the results of application of Eq. (10.14) for the energies noted, where T has been taken as 300°K, or approximately room temperature. It is apparent from Table 10.1 that for excited level energies of 1 eV or more, nearly all of the interacting species will be in the ground level at room temperature, but for energies less than 0.1 eV an appreciable fraction of the species will be in the excited level, and for energies less than 0.001 eV the fraction N_i/N is approximately $\frac{1}{2}$—that is, the interacting species are approximately evenly distributed between the ground and excited level. By comparison with Fig. 10.6, these energies can be related to the various regions of the electromagnetic spectrum and the types of transitions involved.

10.3.4 / Transition probabilities

In an isolated elementary system, there are generally a large number of energy levels and hence a large number of possible transitions. However, all the possible transitions are not equally probable, and some may never be observed.

Consider the energy system of Fig. 10.8, in which there are three levels of energy, E_i, E_j, and E_k, and the concentration of species in each level is, respectively, N_i, N_j, and N_k. By spontaneous emission of a photon of energy $E_i - E_j$, a species in level i can undergo a transition to level j. It is also clear that a species in level i could undergo a similar spontaneous emission

Figure 10.8 / *A simple energy system.*

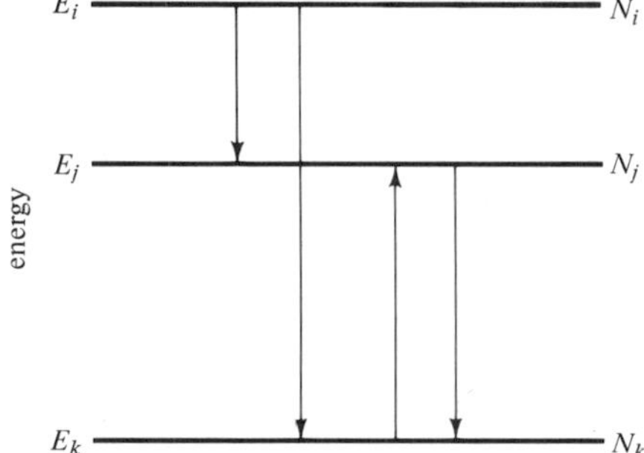

transition to level k, with emission of a photon of energy $E_i - E_k$. The number of transitions occurring per unit time depends on the concentration of species in the excited state and on the transition probability of the particular transition. Thus, for the transition $i \to j$,

$$\Delta N_{i \to j} = A_{ij} N_i \tag{10.15}$$

where $\Delta N_{i \to j}$ is the number of transitions $i \to j$ occurring per unit time, and A_{ij} is the Einstein transition probability for spontaneous emission for the transition $i \to j$ and has units of reciprocal seconds. Similarly,

$$\Delta N_{i \to k} = A_{ik} N_i \tag{10.16}$$

and it should be clear that the relative strength of the two emission lines of frequencies $(E_i - E_j)/h$ and $(E_i - E_k)/h$ is simply the ratio of the transition probabilities A_{ij} and A_{ik}.

When radiation of frequency $\nu = (E_j - E_k)/h$ is passed through the species having the energy system of Fig. 10.8, the number of species per unit time undergoing the absorption transition from level k to level j is directly proportional to the energy density of the radiation, $\rho(\nu)$, and to the concentration of species in the state k:

$$\Delta N_{k \to j} = B_{kj} \rho(\nu) N_k \tag{10.17}$$

The coefficient B_{kj} is the Einstein transition probability for absorption, and the product $B_{kj}\rho(\nu)$ has units of reciprocal seconds.

Theory requires a third process, stimulated emission, which is the reverse of the absorption process, and the rate of which depends on the radiation energy density $\rho(\nu)$ and the concentration of the excited level species:

$$\Delta N'_{j \to k} = B_{jk} \rho(\nu) N_j \tag{10.18}$$

The number B_{jk} is the Einstein transition probability of stimulated emission and, as above, the product $B_{jk}\rho(\nu)$ has units of reciprocal seconds. Stimulated emission is independent of spontaneous emission and differs from it not only in its dependence on $\rho(\nu)$ but also in its spatial distribution. Spontaneous emission is isotropic, that is, uniform in all directions, but stimulated emission occurs only in the direction of travel of the incident radiation.

The transition probabilities are related to each other in the following ways:

$$g_i B_{ij} = g_j B_{ji} \tag{10.19}$$

$$g_i A_{ij} = \frac{8\pi h \nu^3}{c^3} g_j B_{ji} \tag{10.20}$$

$$A_{ij} = \frac{8\pi h \nu^3}{c^3} B_{ij} \tag{10.21}$$

It is worthwhile to inquire at this point as to the practical significance of the preceding discussion in emission and absorption measurements. In emission measurements the radiation source is by some means (chemical combustion, electrical discharge, etc.) elevated in temperature compared to its surroundings, and thus there is a net emission of radiation. In most cases only spontaneous emission need be considered, since the radiation density is usually low in emission measurements. Note, however, the dependence of the ratio A_{ij}/B_{ij} on ν^3—the lower the frequency, the greater the significance of stimulated emission. The most outstanding exception to the predominance of spontaneous emission occurs with masers and lasers,[1] in which the experimental arrangement leads to a very high radiation density.

In absorption measurements the external radiation source is usually at a higher temperature than the absorbing medium, so that a net absorption of radiation occurs. The observed absorption must be the difference between the absorbed energy and the energy emitted in such a direction that it reaches the detector. Since spontaneous emission is isotropic, it is usually possible, by limiting the field of view of the detector, to arrange an absorption experiment so that spontaneous emission makes a negligible contribution to the measurement. In the optical region (ultraviolet, visible, and infrared) of the spectrum the number of species in excited states is small (see Table 10.1), and the radiation density is usually sufficiently small that stimulated emission is negligible. However, taking note of the ν^3 dependence of Eq. (10.21), at low frequencies stimulated emission becomes, relative to spontaneous emission, a more important pathway for depopulation of excited states and may significantly reduce the measured absorption.

A term that is closely related to the transition probability for spontaneous emission is the *oscillator strength*, usually symbolized f. This quantity arose from a classical approach to radiation theory, in which f was defined as the number of electrons per atom responsible for emission of a given radiation. This definition lost all meaning after the introduction of the quantum theory of radiation, but the term is still used, and oscillator strength is now defined by

$$f_{ij} = A_{ij}\frac{\lambda^2 mc}{8\pi^2 e^2} \tag{10.22}$$

where m and e are, respectively, the mass and charge of an electron. The absorption oscillator strength, f_{ji}, is related to the emission oscillator strength, f_{ij}, by

$$g_j f_{ji} = g_i f_{ij} \tag{10.23}$$

10.3.5 / Selection rules

Selection rules are simply attempts to present in easily remembered form information as to which of the many possible transitions of an elementary system have large transition probabilities. Formal selection rules give no

[1] Maser is an acronym for *m*icrowave *a*mplification by *s*timulated *e*mission of *r*adiation, and laser is an acronym for *l*ight *a*mplification by *s*timulated *e*mission of *r*adiation.

information as to how strong a line or band will be if the corresponding transition satisfies certain requirements, that is, if the transition is "allowed," but they do say that certain lines or bands should have zero intensity because the corresponding transition is "forbidden." Selection rules are based on idealized systems, such as the harmonic oscillator, which are not always accurate descriptions of physical reality, and hence selection rules sometimes fail, and forbidden transitions are, in fact, observed. It will be convenient to defer presentation of specific selection rules until subsequent chapters in which various types of transitions are discussed individually.

10.3.6 / Types of transitions

Figure 10.5 is a generalized energy-level diagram that might be applicable to transitions in any region of the electromagnetic spectrum. So far we have dealt with considerations applicable to all types of transitions, but before concluding this discussion, it is appropriate to consider individually the types of transitions that occur in each of the regions of the electromagnetic spectrum.

A / *Gamma-ray region*

Gamma radiation is emitted or absorbed as a result of transitions between nuclear energy levels. A nucleus may be produced in an excited state in the course of radioactive decay or by capture of a particle. Gamma radiation may then be emitted as the nucleus deactivates to the ground state. The measured energy spectrum of gamma rays, emitted in transitions between the energy states of a nucleus, is used to determine the energies of these states—the connection being made by the same trial-and-error procedures as were used in the very early days of the study of atomic spectra and energy states. The energies of the nuclear states give important information about nuclear structure, but from the point of view of the chemist a more interesting type of information is obtained in a gamma-ray absorption experiment of a type that has come to be known as Mössbauer spectroscopy.

If the gamma ray emitted by an excited nucleus is allowed to interact with a ground-state nucleus of the same element, the gamma ray may be re-absorbed. Although this process was a well-recognized possibility, it was believed to be a rather improbable process for the following reason: Because of the very high energy of the gamma ray, a small but appreciable fraction of the energy provided by the nuclear transition should go into recoil of the nucleus upon emission, and thus the energy of the gamma ray should be somewhat less than the energy change of the transition. Moreover, in the absorption process the gamma ray should impart some of its energy to the nucleus as kinetic energy. Consequently, for identical emitting and absorbing nuclei, it would be expected that the emitted gamma ray would fall short of supplying the energy required for the absorption transition by an amount equal to the kinetic energy imparted to the emitting and absorbing nuclei.

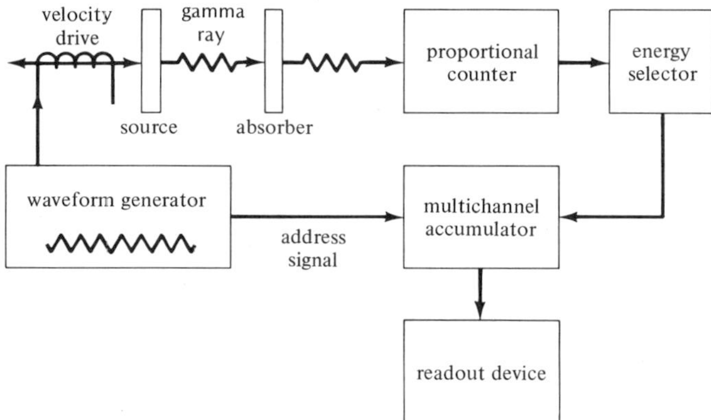

Figure 10.9 / *Block diagram of apparatus for Mössbauer spectroscopy.*

This expectation was disproved in 1958 by Mössbauer, who showed that gamma-ray resonance absorption could occur with moderate efficiency if the source and absorber were both contained in solid lattices. This is possible because in a sizable fraction of the emission and absorption processes the momentum is taken up by the entire crystal rather than by the individual nucleus. Since the kinetic energy is inversely proportional to the mass for a given momentum, the energy lost to the gamma ray is nearly zero when the recoiling mass becomes very large.

The energies of the gamma rays from such recoilless emissions may be very sharply defined, since they approach the spread of energies set by the Heisenberg uncertainty principle. Thus, if the absorbing nucleus is in a different chemical environment than the emitting nucleus, the perturbation of the nuclear energy levels in the two environments is sufficiently different to destroy resonance; that is, the energies of emission and absorption no longer match. Resonance can be restored by altering the energy of the gamma ray or the energy required for absorption. The requisite small energy shift can be obtained by moving the emitter with respect to the absorber to take advantage of the Doppler shift. If the emitter is in motion with a relative velocity v with respect to the absorber, then the energy of the gamma ray in the reference frame in which the absorber is at rest appears to be shifted by an amount $\Delta E = Ev/c$. If the velocity can be varied in a known fashion, then the absorption spectrum can be scanned. The spectrum is observed as the relative number of gamma rays per second passing the absorbing sample as a function of the relative velocity of the source and sample. The shift of the absorption band with respect to some reference absorber can be interpreted in terms of the chemical environment of the nucleus. Figure 10.9 indicates the principal components of the Mössbauer absorption experiment.

B / *X-ray region*

X-ray radiation arises from transitions between the innermost electronic levels of atoms. If a sufficiently high-speed electron is directed into a metal target, it is possible to eject an electron from one of the inner shells of the atoms constituting the target material. The place of the ejected electron will then promptly be filled by an electron from an outer shell, whose place in turn will be taken by an electron from still farther out. The ionized atom thus returns to its ground state by several steps, in each of which an X-ray photon of definite energy is emitted. In addition, some continuous radiation is produced.

The spectrum obtained by the electron-bombardment process is characteristic of the target material, and this direct excitation process is the basis of the very sensitive electron-probe microanalysis technique. The X-ray radiation generated by electron bombardment can also be used to irradiate a sample material, and if the photons correspond to a characteristic transition energy of the sample material, the photons may be absorbed and the sample material excited. There are analytical methods based on the measurement of both the absorption of X-ray radiation and the intensity of X-ray radiation reemitted by the sample material.

C / *Ultraviolet-visible region*

Spectra in the ultraviolet-visible region are due to transitions between outer electronic levels of atoms and molecules. Simultaneous changes in vibrational and rotational energies may be superimposed in the case of molecular species. Spectra can be excited by collisions with atomic and molecular species at high temperatures, by absorption of radiation, and by chemical reactions, in addition to electron impact. Flame atomic emission spectra (Chapter 13) are due to excitation by collisional processes and, in some cases, by chemical reactions. Arc and spark atomic emission spectra (Chapter 14) are due to collisional and electron-impact excitation. Atomic absorption and atomic fluorescence spectrometry (Chapter 13) depend on the absorption of radiation at characteristic frequencies and measurement, respectively, of the intensity of radiation absorbed and reemitted. Molecular spectra are largely observed as absorption spectra (Chapter 16), but molecular fluorescence and phosphorescence (Chapter 17) are based on measurement of the reemission of absorbed radiation.

D / *Infrared region*

Spectra in the infrared region are mostly associated with changes in vibrational energy levels of molecules, although some pure rotational energy transitions for heavier molecules are found in the low-energy end of this region. Changes in vibrational energy may also be accompanied by changes in rotational energy. Note from Table 10.1 and Fig. 10.6 that lower vibrational energy levels may be appreciably populated at room temperature.

Almost all infrared spectra are observed as absorption spectra. The spectra obtained in this region are highly characteristic, and infrared spectroscopy is today an almost indispensable tool in the characterization of organic compounds. Infrared spectroscopy is described in Chapter 18.

E / *Microwave region*

Spectra in the microwave region arise from two distinct sources: changes in rotational energies of molecules, and changes in the energies associated with the alignment of unpaired electrons in an external magnetic field. In both cases excited levels may be appreciably populated, and in both cases absorption measurements are experimentally more convenient in most situations. The analytical applications of molecular rotational spectra are described briefly in Chapter 19. The electron spin resonance technique is described in Chapter 20.

F / *Radio frequency region*

Spectra in the radio frequency region are associated with changes in nuclear orientation in an external magnetic field. At room temperature, all levels are almost equally populated. Again, absorption measurements prove experimentally most convenient in the majority of cases.

The energy required for the nuclear orientational transition is dependent on the molecular environment of the nucleus, and the technique of nuclear magnetic resonance has become an indispensable tool in the characterization of complex chemical species. This technique is described in detail in Chapter 20.

10.4 / BLACKBODY RADIATION

Radiation from a heated solid does not exhibit discrete lines or bands but rather is continuous; that is, the intensity of the radiation varies smoothly with wavelength over an extended range of wavelengths. Moreover, to a large extent the distribution of radiation with wavelength does not depend on the material of which the emitter is composed, but depends only on the absolute temperature of the source. If the solid has the property of absorbing all the radiation incident on it, then the source is said to be an *ideal blackbody*. A close approximation to an ideal blackbody can be achieved by an arrangement such as that shown in Fig. 10.10. An evacuated hollow inside any solid may be maintained at equilibrium with respect to both emission and absorption. If, as shown in Fig. 10.10, the only radiation that escapes from the hollow comes through an opening that is small compared to the total surface area of the cavity, then the spectrum obtained is that of an ideal blackbody and does not depend on the nature, size, or shape of the solid, but only on the temperature.

A blackbody source obeys the following laws:

1. The Stefan-Boltzmann law,

$$E_{tot} = \sigma T^4 \tag{10.24}$$

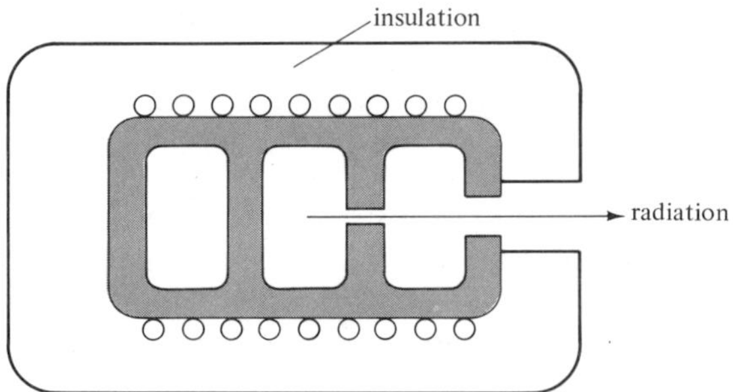

Figure 10.10 / *Cross-sectional view of a wire-wound electrical furnace constructed to behave as a blackbody.*

states that the total energy, E_{tot} (all wavelengths), emitted by a blackbody increases as the fourth power of the absolute temperature, T. If the total energy is given in units of ergs per centimeter^3, then the constant σ is 7.56×10^{-15} erg-cm^{-3}-deg^{-4}. The radiation flux in units of ergs per centimeter^2-second can be obtained by multiplying the right side of Eq. (10.24) by $c/4$.

2. The Wien displacement law,

$$\lambda_{max} T = b \tag{10.25}$$

states that the product of the wavelength of maximum blackbody emission and the absolute temperature of the source is a constant, or, expressed in different form, the wavelength of maximum blackbody emission varies inversely with the absolute temperature of the source. If a spectrometer is used to find the distribution of energy over various wavelengths for a blackbody source, a smooth plot of energy vs. wavelength is obtained. Figure 10.11 shows the expected result of such a measurement for a blackbody source at three temperatures. Note that as the temperature increases, the total energy increases and the wavelength of maximum emission decreases. The constant b has a value of 2898 μm-°K, and a straightforward calculation shows that at a temperature of 6000°K, λ_{max} falls in the visible region of the spectrum at 4830 Å.

3. The Planck radiation law is

$$E_\lambda = \frac{2\pi hc^2}{\lambda^5[\exp(hc/\lambda kT) - 1]} \tag{10.26}$$

where E_λ has units of energy per unit area unit time unit wavelength interval, typically ergs per centimeter^2-second-angstrom, h is Planck's constant, and k is the Boltzmann constant. The curves of Fig. 10.11 are actually plots of Eq. (10.26), and both Eqs. (10.24) and (10.25) can be obtained from (10.26).

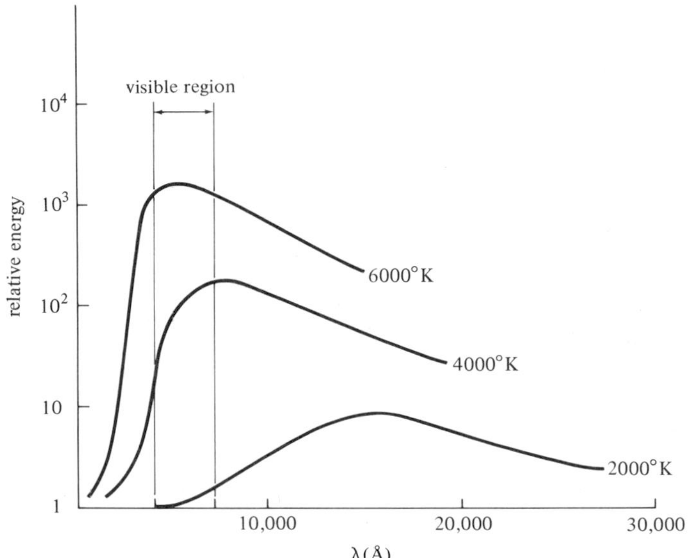

Figure 10.11 / *Energy of blackbody radiation as a function of wavelength.*

SUGGESTED READINGS

G. Feinberg, "Light," *Sci. Amer.* **219** (3), 50 (1968).
V. F. Weisskopf, "How Light Interacts with Matter," *Sci. Amer.* **219** (3), 60 (1968).
R. W. Wood, *Physical Optics*, Dover (New York, 1967).
E. J. Bowen, *The Chemical Aspects of Light*, Clarendon Press (Oxford, 1946).
D. H. Whiffen, *Spectroscopy*, Wiley (New York, 1966).

PROBLEMS

Convert each of the following quantities used to describe electromagnetic waves to wavelength (in vacuum) in meters.

1 / 5000 Å (in vacuum)

2 / 5000 Å in a medium with $n = 1.4$

3 / 1000 K

4 / 10^{15} Hz

Calculate the angle of reflection and/or refraction for each of the following conditions. Assume that the refractive index of air is 1.00.

5 / A monochromatic ray in air is incident on a medium of refractive index 1.6 at an angle of 30°.

6 / A monochromatic ray in a medium of refractive index 1.6 is incident on an air medium at an angle of 30°.

7 / A monochromatic ray in a medium of refractive index 1.6 is incident on an air medium at an angle of 45°.

Complete the following table.

	Frequency	Transition energy: Joules	Transition energy: Electron volts	Spectral region
8 /			0.5	
9 /	3×10^{10}			
10 /		1.6×10^{-19}		

11 / The 589-nm radiation emitted by Na atoms in a flame is due to a transition between the first excited state and the ground state. The lifetime of the first excited state is approximately 2.2×10^{-8} sec.
(*a*) Calculate the approximate natural half-width (i.e., that imposed by the Heisenberg uncertainty principle) of the line in hertz and angstroms.
(*b*) Calculate the Doppler half-width of the line in hertz and angstroms for a flame temperature of 2500°K.

Calculate the relative populations of the energy levels for a system consisting of three levels if the statistical weights are identical for all three levels, the system is at equilibrium at a temperature of 300°K, and the relative energies of the levels are as follows:

12 / 0, 0.01 eV, 0.02 eV

13 / 0, 0.1 eV, 0.2 eV

14 / 0, 1 eV, 2 eV

15 / For a blackbody emitter, calculate the factor increase in total emitted energy and the shift in the wavelength of maximum emission resulting from an increase in temperature from 1500° to 2000°K. In what region of the spectrum does the wavelength of maximum emission occur for a blackbody of 2000°K?

chapter 11 / Instrumentation for optical spectroscopy

chapter 11

11.1 / INTRODUCTION

In Chapter 10, we emphasized the general aspects of the interaction of electromagnetic radiation with matter. In this and succeeding chapters, we shall deal separately with the various regions of the electromagnetic spectrum and discuss the instrumentation required and the chemical information obtained in each region. Because it is the region of most common experience, we shall begin our discussion with the so-called optical region of the spectrum, that is, the regions of the spectrum that we have previously labeled as ultraviolet (UV), visible, and infrared (IR). In this chapter, we shall describe the types of instrumentation required for optical spectroscopic measurements.

Instruments used to study the absorption or emission of electromagnetic radiation as a function of wavelength are known as spectrometers, spectrophotometers, or spectrographs. An optical spectrometer is an instrument with an entrance slit, a dispersing device, and one or more exit slits, with which measurements are made at selected wavelengths or by scanning over a range of wavelengths. A spectrophotometer is a spectrometer equipped with some additional components so that it furnishes the ratio, or a function of the ratio, of the radiant power of two beams as a function of spectral wavelength. The two beams may be separated in time, space, or both. A spectrograph is an instrument with an entrance slit and dispersing device that uses photography to obtain a record of the spectrum. There are three essential components in all three devices: (1) a stable source of radiant energy, (2) a monochromator to resolve the source radiation into its component "monochromatic" elements, and (3) a radiation detector and appropriate readout device. One additional component is usually required if absorption measurements are to be made: a transparent container for the sample. Figure 11.1 shows the essential components of a spectrometer.

Despite the basic similarity of all optical spectrometers, in practice no single instrument is suited to cover the whole optical region. The arrange-

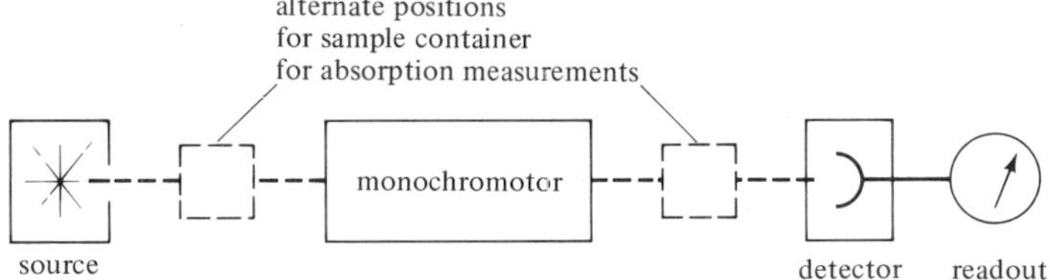

Figure 11.1 / *Components of a spectrometer.*

ment found most often in a chemical laboratory is to have one instrument for the near UV and visible range (approximately 190–800 nm) and another for the IR region (2–15 μm, 5000–667 cm^{-1}). The range of the instruments is determined by the transparency of optical materials, the spectral output of the source, and the spectral response of the detector, as well as by the characteristics of the monochromator. More sophisticated instruments extend the range of wavelengths covered by a single instrument, but the requirements for a UV-visible instrument are sufficiently different from those of an IR instrument that in practice a minimum of two instruments is required; in this chapter, we shall discuss instrumentation for the UV-visible region in Section 11.2 and instrumentation for the IR region in Section 11.3.

Although spectrometers are the most widely employed devices for the measurement of optical spectra, they are not the only devices for such measurements. Interferometers provide an alternative approach to the same measurements, and recent developments in technology have made their use feasible in a variety of spectroscopic experiments. The principles of interferometric measurements and the associated technique of Fourier transform spectroscopy will be discussed briefly in Section 11.6.

11.2 / INSTRUMENTATION FOR THE ULTRAVIOLET-VISIBLE REGION

11.2.1 / Radiation sources

An ideal radiation source would be of high intensity and monochromatic at the wavelength of interest. Such properties are rarely approached in real sources. Real sources may be broadly classified as *line sources*—those emitting chiefly discrete atomic spectral lines—and *continuum sources*—those emitting chiefly radiation with intensity that varies smoothly over an extended range of wavelengths.

A frequently encountered example of a line source is a flame containing a volatile element such as sodium. When a volatile compound of sodium, such as NaCl, is introduced to a flame, the presence of sodium is indicated by the bright yellow color imparted to the flame. The sodium atoms produced by the dissociation of NaCl in the flame can be activated to an excited state by collision with sufficiently energetic particles in the flame, and upon return to the ground state they emit radiation at a wavelength of 589 nm. This wavelength is in the yellow region of the visible spectrum, and it is this radiation the eye detects when we observe the characteristic color sodium gives to the flame.

A variety of line sources exist in which the excitation of the atoms is produced by an electrical discharge. These do not differ in principle from the flame source: A dilute atomic vapor is produced, the atoms are excited by collisions with energetic particles, and some of the atoms deactivate by emission of photons of characteristic wavelengths. Some examples of line sources will be discussed in detail in Chapters 13 and 14.

The laser is a type of line source that is finding increasing application in chemistry. A discussion of the principles of laser action is beyond the scope of this chapter, but it is pertinent to note that as a spectral source the laser is capable of exhibiting very high intensity and a high degree of monochromaticity. The laser is also a valuable tool in the alignment of optical components because the laser radiation is highly collimated—that is, the beam exhibits a very low angle of divergence.

In some situations, line sources approach rather closely our stated characteristics of an ideal source. The individual lines may be of rather high intensity, and the half-intensity linewidth is typically measured in fractions of an angstrom, Å. A variety of analytical determinations are based on the direct excitation of characteristic atomic emission spectra (see Chapters 13 and 14), and Raman spectroscopy (Chapter 18) also makes use of electrical discharge and laser atomic line sources. Line sources are also employed in the measurement of absorption by dilute atomic vapors (see Chapter 13) and have been employed in certain types of measurements of absorption by molecular species.

For molecular absorption measurements, it is usually desirable to be able to vary continuously the measurement wavelength, that is, to scan the spectrum. The ideal source for such absorption measurements is a monochromatic source that is "tunable," that is, its wavelength can be varied at will. The line sources described above are, of course, not tunable. Certain types of lasers, the so-called dye lasers, show promise of providing a close approximation to a tunable, monochromatic source, but at present the only practicable approach to the measurement of absorption spectra is to employ a continuum source and a monochromator. The continuum source provides high-intensity radiation over an extended range of wavelengths, and the monochromator allows resolution of small-wavelength intervals of the radiation and smooth variation of the central wavelength of the resolution element.

An ideal continuum source for absorption measurements would have uniform intensity over the range of wavelengths of interest. The intensity of actual sources shows a strong dependence on wavelength. Most familiar continuum sources are of the hot-body type, and, although most are not true blackbody sources, many of their properties can be described in terms of blackbody behavior as discussed in Section 10.4.

Reference to the discussion of blackbody radiation shows that inconveniently high temperatures are required for the production of appreciable blackbody radiation in the ultraviolet by a hot-body source, and hence a different approach has been taken in fabricating sources of continuum UV radiation. As noted above, an electrical discharge through a low-pressure gas gives rise to line spectra that are characteristic of the gas. However, as the gas pressure is increased the lines broaden, and at sufficiently high pressure the lines coalesce to produce a usable continuum with, perhaps, some superposed line structure. Two common examples are the high-pressure mercury arc and the xenon arc discharge lamps. These lamps are operated at electrical powers ranging from about 100 W to in excess of 1000 W and are capable of producing a continuum of usable intensity that extends from about 2000 Å into the infrared.

The hydrogen or deuterium lamp represents a somewhat different approach to the production of an ultraviolet continuum. A relatively low-power (less than 100 W) electrical discharge through a low pressure (about 10 torr) of hydrogen or deuterium leads to the production of a continuum extending from the cutoff of the envelope material (as low as 1600 Å) to about 3500 Å, with a line spectrum superposed on a continuous background at longer wavelengths. The continuum arises from a transition from a stable excited electronic state of the hydrogen molecule to a lower state in which the molecule dissociates. A continuum, rather than a discrete band spectrum, is observed because a portion of the energy is taken up in translational motion (unquantized) of the atoms resulting from dissociation of the molecule. Hydrogen and deuterium lamps operate in the same fashion, but deuterium, while more costly, produces a two or three times increase in intensity and lamp life.

To the extent that comparisons can be made without specifying actual lamps and power supplies, xenon and mercury lamps are more intense and less stable, except at the shortest wavelengths, than are the hydrogen and deuterium lamps. We have noted above that the xenon and mercury lamps have usable continua extending from the ultraviolet to the near infrared. This characteristic may be desirable in some situations, but must be considered a disadvantage when measurements are to be made in the ultraviolet, since the very intense radiation in the visible range greatly increases the difficulty of preventing stray radiation (i.e., radiation at wavelengths other than the desired measurement wavelength) from reaching the detector.

In the near infrared and visible region, the source most commonly employed is the tungsten filament lamp. The output of this lamp closely resembles that of a blackbody operated at about 2000°K. The life of a tungsten filament lamp can be greatly extended by inclusion of a low pressure of iodine or bromine vapor in the lamp. With this addition and a fused silica envelope, it is known as a quartz-halogen lamp. The halogen reacts with the tungsten to form a volatile compound, which pyrolyzes when it comes in contact with the hot filament and redeposits the tungsten instead

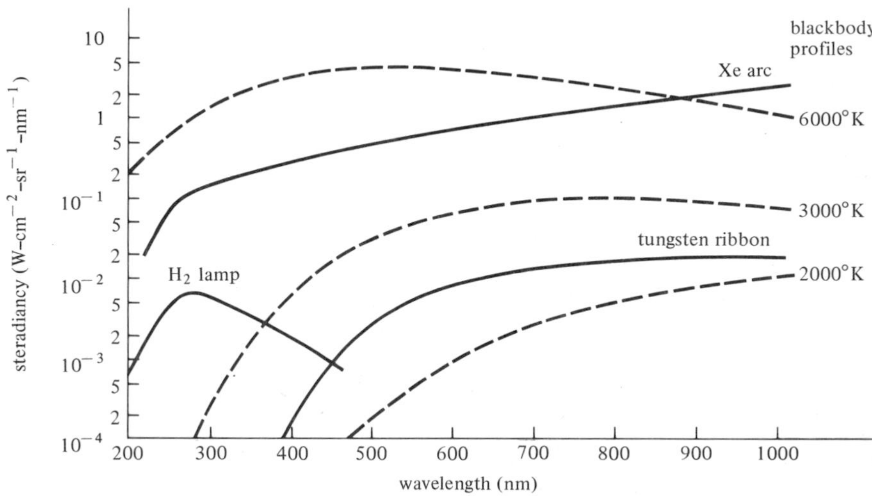

Figure 11.2 / *Typical outputs for some continuum sources.*

of allowing tungsten volatilized from the filament to collect on the cooler envelope of the lamp.

Typical output characteristics of several continuum sources are compared in Fig. 11.2.

11.2.2 / Optical materials

An obvious requirement for a spectrosopic measurement is that all materials in the optical path be transparent to the radiation of interest. Table 11.1 reports the range of transparency for a number of optical materials.

For measurements in the visible range, glass, because of its strength and low cost, is the material of choice. The range of transparency of glass is quite dependent on its composition. As noted in Table 11.1, the ultraviolet limit for most ordinary glasses is about 3600 Å, but some special types of glasses are usable down to approximately 2200 Å. The better transparency of fused silica at short wavelengths makes it the material of choice for most UV applications.

For measurements below 2000 Å, the optical properties of air must also be considered. The O_2 molecule absorbs strongly at wavelengths shorter than 1950 Å, and must be excluded if measurements are to be made in this region. In many spectrometers, it is possible to purge the optical system with dry nitrogen to eliminate O_2 absorption and thus extend the range of the instrument to approximately 1600 Å. At still shorter wavelengths, it is necessary to evacuate the entire optical system, and this usually requires an instrument specifically designed for vacuum UV spectroscopy.

TABLE 11.1 / *Useful transmission range[a] for optical materials*

Material	*Range*
fused silica	1700 Å–3.6 μm
glass	3600 Å–2.5 μm
sodium chloride	2000 Å–15 μm
potassium bromide	2300 Å–25 μm
potassium chloride	2000 Å–18 μm
thallium bromide–thallium iodide	5000 Å–35 μm
cesium iodide	2300 Å–50 μm
calcium fluoride	1250 Å–9 μm
barium fluoride	1300 Å–12 μm
lithium fluoride	1040 Å–7 μm
sodium fluoride	1950 Å–10.5 μm
cadmium fluoride	2000 Å–10 μm
lead fluoride	2900 Å–11.6 μm
manganese fluoride	1200 Å–8.4 μm
strontium fluoride	1200 Å–11 μm
lanthanum fluoride	4000 Å–9 μm
magnesium fluoride	1100 Å–7.5 μm

[a] Limits are taken as wavelengths where percent transmittance falls to 60 percent for a 1-cm thickness.

11.2.3 / Filters

An optical filter is a device that allows the transmission of a fixed range of wavelengths and eliminates all others. There are many situations in which wavelength separation is required, but the high degree of spectral purity achievable with a monochromator is unnecessary. It may be desirable to isolate a band of wavelengths (bandpass filter), or it may be sufficient to ensure that all radiation below a given wavelength or above a given wavelength is eliminated (cutoff filter). Several examples can be cited:

1. In absorption measurements where the absorbing species offers a wide band absorption or in emission measurements where the spectral features are widely separated, a bandpass filter may provide sufficient monochromaticity for the measurements.
2. A filter may be used to reduce stray light in a spectrometer by eliminating a large wavelength region of unwanted radiation.
3. In a grating spectrometer, a filter may be used to eliminate unwanted "orders," that is, radiation of different wavelengths dispersed at the same angle as the desired radiation.

A / *Absorption filters*

All of the optical materials described in the preceding section sometimes serve as filters. For example, in certain applications it is desirable to prevent UV radiation from impinging upon the sample or from entering the spectrometer. As may be seen from Table 11.1, ordinary soft glass serves as a very effective, and readily available, filter for this purpose.

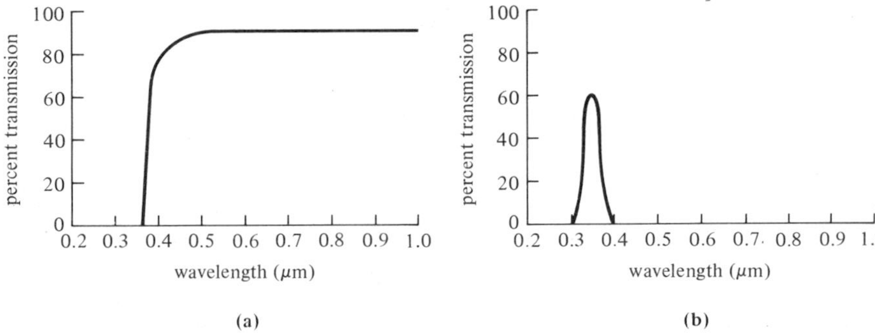

Figure 11.3 / *Typical glass absorption filters.*
(a) *Cutoff (long pass).*
(b) *Bandpass.*

The most widely employed filters for the visible region are of colored glass. Filters of dyed gelatin between glass plates serve the same purpose and are less expensive, but they are also less permanent than the colored glass filters. Figure 11.3 shows transmission curves for some glass filters. Sometimes a narrow bandpass filter can be obtained by using two filters, one of which has a short-wavelength cutoff just below the desired wavelength and the other of which has a long-wavelength cutoff just above the desired wavelength.

In some situations, a filter may conveniently be made from gases or solutions. Raman spectroscopy provides an example of a situation in which a solution filter is particularly advantageous because of the large size and unusual shape of the required filter. Before the advent of the laser, the most common experimental arrangement for excitation of Raman spectra was to place the sample cell inside a helical mercury line source. Selection of an exciting mercury line was accomplished by placing a filter solution of the proper composition in an annular space between the source and the Raman cell.

B / *Interference filters*

Isolation of narrow spectral regions can also be accomplished by means of interference filters of both the Fabry-Perot and multilayer dielectric type. A Fabry-Perot interference filter consists of two transparent plates on which have been evaporated partially reflecting metal films, separated by a transparent dielectric material. Such an interference filter is represented in Fig. 11.4, which also shows in schematic fashion the mode of operation of the filter. A portion of the entering beam (A) is transmitted directly (beam B), and a portion is reflected at surface 2 (beam C). A portion of beam C is again reflected at surface 1 to form beam D, which exits with beam B. Beam D has a longer path length than beam B by twice the thickness of the layer separating surfaces 1 and 2. If the beam of radiation is incident at

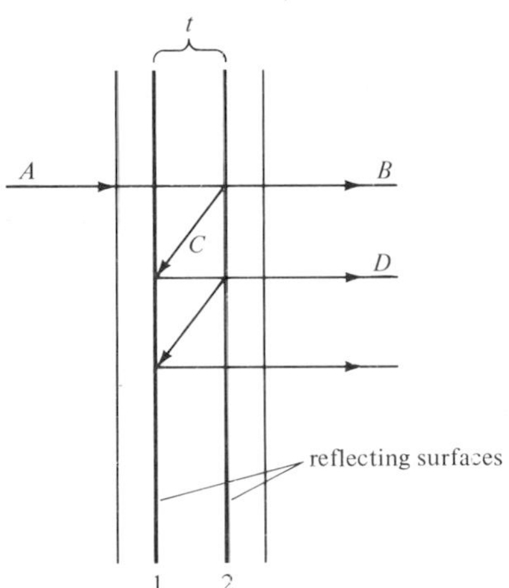

Figure 11.4 / *Fabry-Perot interference filter. Reflections are shown at an angle for clarity.*

right angles to the face of the filter (normal incidence) and if the thickness of the dielectric layer is denoted by t, then the emergent beams will be in phase and will reinforce one another for wavelengths that are integer multiples of $2t$; that is, the equation for constructive interference is

$$m \frac{\lambda}{n} = 2t \tag{11.1}$$

where m is an integer known as the order number, n is the refractive index of the dielectric layer, λ is the wavelength of the radiation in a vacuum, and λ/n is thus the wavelength of the radiation in the dielectric layer. The intensity of radiation at wavelengths that do not satisfy Eq. (11.1) will be reduced by destructive interference of the emergent beams.

An interference filter is characterized optically by the wavelength of peak transmittance, by the transmittance (ratio of intensity transmitted to intensity incident on the filter) at the peak wavelength, by the width of the band of radiation passed (bandpass) at one-half peak transmittance, and by the background transmittance. The significance of these terms is illustrated in Fig. 11.5. Fabry-Perot interference filters are available with peak wavelengths ranging across the visible region. Peak transmittances are typically 35 to 45 percent. Half-widths are typically 100 Å or greater, depending on the order employed and somewhat on the peak wavelength. Background transmittances are typically 0.5 percent or less.

A filter in which the optical spacing of the reflecting metal surfaces is one-half the wavelength at the peak of the band for which it was designed ($\lambda/n = 2t$) is called a first-order filter ($m = 1$); one in which the spacing is equal to the wavelength ($\lambda/n = t$) is called a second-order filter ($m = 2$). Higher-order filters are designated in the same fashion. The width of the

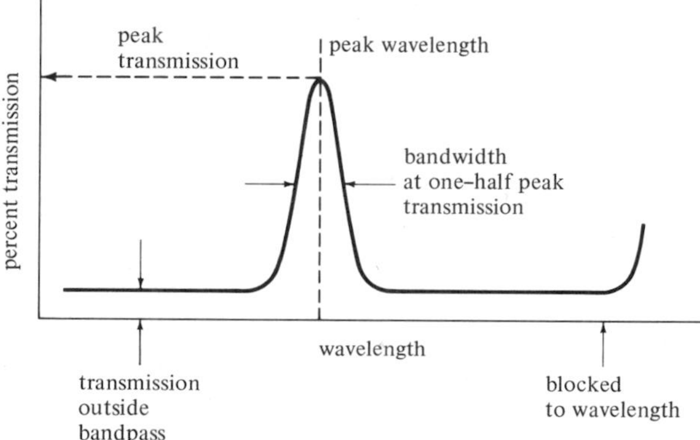

Figure 11.5 / *Significance of terms used to characterize interference filters.*

band passed as expressed by the half-width depends on the order of interference of the filter and also on its transmittance. Transmittances being equal, a first-order filter inherently has a wider bandpass than a second-order filter. For a given order, the greater the peak transmittance, the greater the half-width of the band. In many cases, interference filters are constructed with an auxiliary colored glass filter as one of the cover plates to eliminate unwanted orders. In manufacturers' literature, it is customary to specify the range of wavelengths for which unwanted radiation has been blocked.

Multilayer interference filters are prepared by successive evaporation of 5 to 25 layers of high- and low-refractive index dielectric materials in alternating layers. Each interface of high- and low-refractive index material behaves as a partially reflecting surface, and the interference effects described above for the Fabry-Perot filter occur at each interface. Thus, the multilayer filter gives an effect similar to that of placing several Fabry-Perot filters in series. By proper choice of layer materials, thicknesses, number of layers, and substrate materials, filters can be designed with a wide choice of central wavelengths, half-widths, and background transmission characteristics.

Multilayer filters are available with peak wavelengths ranging from about 4000 Å to 1.2 μm, with peak transmittances of about 50 percent and half-widths of 100 Å. Typical background transmittance is less than 0.01 percent. Special filters are available with half-widths as low as 10 Å and peak transmittances of 40 percent or better. Filters are also available for the UV region, but half-widths typically are must greater and peak transmittances much less than in the visible.

11.2.4 / Monochromators

A monochromator is a device for producing a beam of radiation of high spectral purity, the wavelength of which can be varied at will over a wide range. Thus, monochromators and filters differ in the half-width of the band of radiation produced, although there is some overlap in resolution achieved

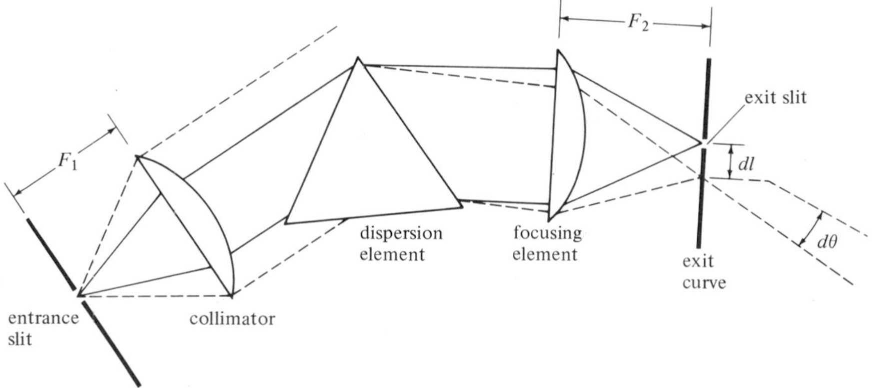

Figure 11.6 / *Components of a monochromator.*

with the best filters and the lowest power monochromators, and they differ in their ability to be tuned to a selected wavelength.

The essential components of a monochromator are: (1) an entrance slit; (2) a collimating device, that is, a lens or mirror that causes the beam of radiation to travel as parallel rays; (3) a dispersion device, either a prism or grating, which causes different wavelengths of radiation to travel at different angles; (4) a focusing lens or mirror, which images each of the monochromatic beams in the exit curve of the monochromator; and (5) an exit slit. Figure 11.6 shows schematically the arrangement of components in a monochromator. A prism is shown as the dispersion device, but a grating could as well have been shown.

A / *Dispersion devices*

The dispersion device is of central importance in the operation of a monochromator, and we shall therefore consider this component first. Two types of dispersion devices are in wide use: prisms and gratings.

Prisms. The utility of a prism as a dispersion device arises from the phenomenon of refraction. In Section 10.2, it was pointed out that a collimated beam of radiation incident on a block of dielectric material follows a path through the material in accordance with Snell's law, that is,

$$n_1 \sin i_1 = n_2 \sin r_2 \tag{11.2}$$

Thus, when a beam of radiation passes from air into a denser medium, it is refracted toward the perpendicular; and when it passes from a denser medium into air, it is refracted away from the perpendicular. If the faces of the dielectric material are parallel, the net result is a small displacement of the beam without angular deviation. If the beam of radiation consists of two

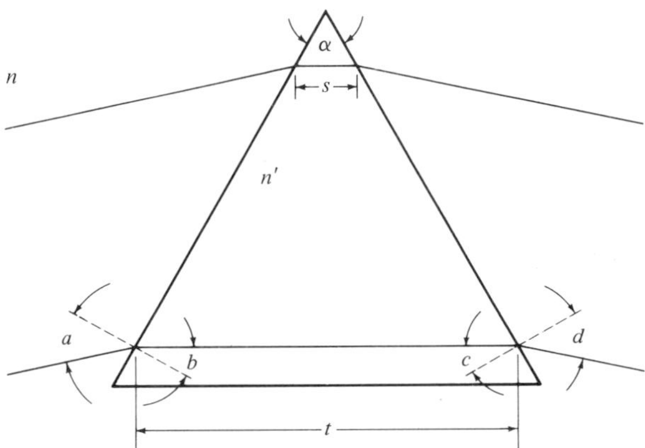

Figure 11.7 / *Path followed by a beam of monochromatic radiation through a prism.*

wavelengths of radiation for which n_2 differs appreciably, their paths through the block differ, but no useful wavelength separation is achieved. If the faces of the block are not parallel but are arranged as in a prism, as shown in Fig. 11.7, a useful separation according to wavelength is achieved. This can be demonstrated by a straightforward application of Snell's law. Assume the prism apex angle, α, is 60°, the angle of incidence of the beam, a, is 60°, the refractive index of air is 1.00, the refractive index of the prism material is 1.70 for λ_1, the shorter wavelength radiation, and 1.60 for λ_2, the longer wavelength radiation. If the subscripts 1 and 2 are associated with λ_1 and λ_2, respectively, for the angles b, c, and d shown in Fig. 11.7, then the calculation of the angles proceeds as follows:

$$\sin b_1 = \frac{\sin 60^\circ}{1.70} = 0.51 \tag{11.3}$$

$$b_1 = 30.7^\circ \tag{11.4}$$

$$\sin b_2 = \frac{\sin 60^\circ}{1.60} = 0.54 \tag{11.5}$$

$$b_2 = 32.7^\circ \tag{11.6}$$

$$c_1 = \alpha - b_1 = 29.3^\circ \tag{11.7}$$

$$c_2 = \alpha - b_2 = 27.3^\circ \tag{11.8}$$

$$\sin d_1 = 1.70 \sin 29.3^\circ = 0.83 \tag{11.9}$$

$$d_1 = 56.1^\circ \tag{11.10}$$

$$\sin d_2 = 1.60 \sin 27.3^\circ = 0.74 \tag{11.11}$$

$$d_2 = 47.7^\circ \tag{11.12}$$

It should be apparent from the preceding calculation that the ability of a prism to resolve one wavelength of radiation from another depends both on the geometry of the prism and the difference in refractive indices of the material at the two wavelengths. This latter factor in turn depends on the dispersion of the material, that is, on $dn/d\lambda$, in the wavelength region of interest.

It is usual to use a prism at minimum deviation, that is, in such a fashion that the beam to be observed traverses the prism parallel to the base, as shown in Fig. 11.7. With this condition, it can be shown that the angular dispersion, $d\theta/d\lambda$, where $d\theta$ is defined in Fig. 11.6, is given by

$$\frac{d\theta}{d\lambda} = \frac{d\theta}{dn}\frac{dn}{d\lambda} \tag{11.13}$$

where the first term depends only on the geometry of the prism and the second is the dispersion of the prism material. Further, assuming minimum deviation, the resolving power, R, of the prism is given by

$$R = (t - s)\frac{dn}{d\lambda} \tag{11.14}$$

where resolving power is a measure of the ability of the prism to separate radiation of one wavelength from other radiation of nearly the same wavelength. The concept of resolving power is described more fully in Section 11.2.4.E. If the whole face of the prism is illuminated, $(t - s)$ is just the width of the base of the prism. Thus, resolving power is improved by a large prism angle and a large dispersion. The prism angle is limited by reflection losses, which become larger as the angle is increased, and an apex angle of 60° is the usual compromise. Recalling the discussion of Section 10.2, the dispersion of a dielectric material increases sharply as an absorption band is approached from the long-wavelength end. Thus, it is desirable to use a prism material close to its absorption edge. When both transparency and dispersion are considered, fused silica is the material of choice for the UV region, but glass is a better choice for measurements in the visible. It should also be noted that because of the nature of the relationship of n to λ for dielectric materials, a prism gives a nonlinear distribution of wavelengths—the red, or long-wavelength, end of the spectrum is compressed as compared to the blue, or short-wavelength, end of the spectrum.

Gratings. The action of a diffraction grating can be described in terms of a light wave interference effect. Interference effects were discussed in Section 10.2, and it was pointed out that passage of a light beam through a narrow aperture results in a spreading of the beam by diffraction. Such an effect is shown in Fig. 11.8. When collimated monochromatic radiation of wavelength λ is incident on a slit of width b and the transmitted radiation is focused in a plane by a lens of focal length f, the distribution of intensity in the focal plane is that shown to the right of the figure: a bright central

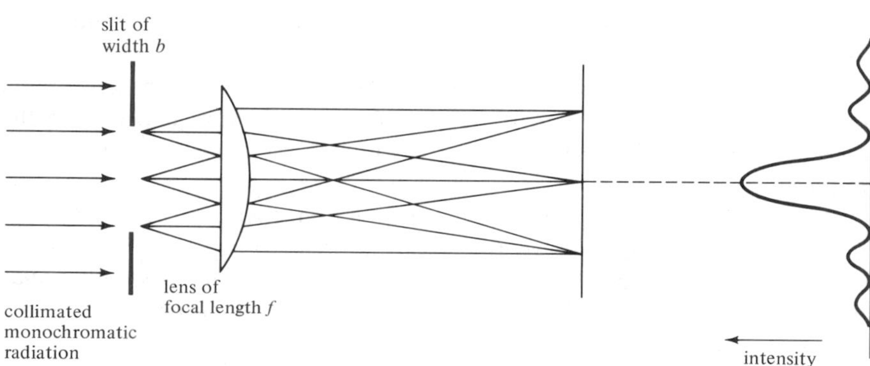

Figure 11.8 / *Fraunhofer diffraction of a single slit.*

maximum with a series of intensity minima and maxima to either side. Thus, the image observed at the focal plane is not truly a replica of the slit, but rather it is a diffraction pattern. The width of the central maximum is given approximately by $2f\lambda/b$, and is then inversely proportional to b. If b is of the order of magnitude of λ, the maximum may become so wide as to cover the whole field of view.

A diffraction grating is made up of a large array of parallel, equally spaced, closely spaced slits which are of approximately the width of the wavelength of light to be diffracted. In a transmission grating, light passes through this array of slits; in a reflection grating, light is reflected from a surface that has been "ruled" with a large number of parallel grooves or lines such that each groove acts as an individual slit. In either a transmission or a reflection grating, the fundamental phenomenon is the same; and since the geometry of the transmission grating is more readily visualized, it will be used in the subsequent discussion.

If the individual slits of a grating are so narrow that the central maximum fills the field of view, then new maxima and minima will be observed due to interference between beams of radiation passing through the individual slits. With a very large number of slits, these new maxima become very sharp; and they are narrow compared to the distances between the maxima. It is these maxima that constitute the spectra of a grating. The spacing of maxima due to a monochromatic beam incident on a grating can be derived in the following fashion, employing the geometrical arrangement shown in Fig. 11.9. In Fig. 11.9, $ABCDE$ represents a section of a plane transmission grating, where BC and DE are two adjacent openings and AB and CD are opaque sections between the openings. The collimated rays L_1, L_2, and L_3 are incident on the grating from the left at an angle α measured from a perpendicular to the face of the grating. The openings are assumed to be of a width comparable to the wavelength of the incident light, and hence to an observer at the right of the grating, each opening would appear as a light source emitting in all directions. The condition for constructive interference

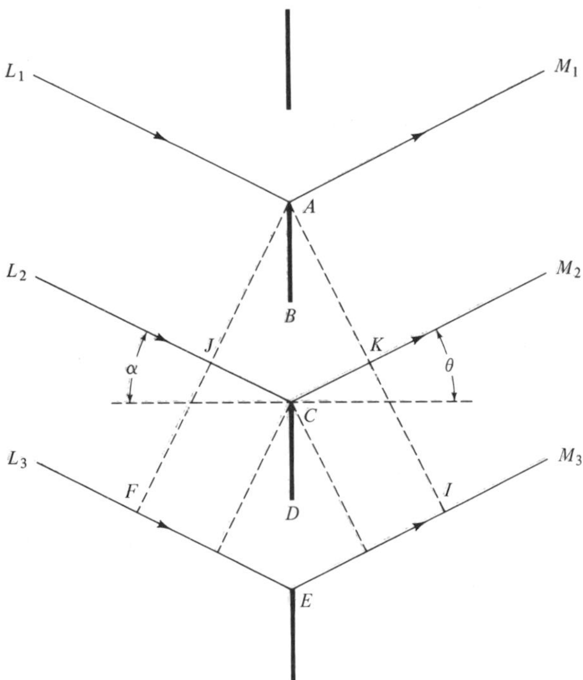

Figure 11.9 / *Diffraction by a plane transmission grating.*

between light emanating from the various openings can be obtained by considering the set of parallel rays labeled M_1, M_2, and M_3.

In the incident beam, the rays L_1, L_2, and L_3 are in phase at *AJF*, where *AJF* is perpendicular to the direction of travel of the rays. *AKI* is perpendicular to the direction of travel of the diffracted rays, and it is apparent that the ray L_2M_2 has traveled a distance *JCK* greater than ray L_1M_1 to reach the *AKI* plane, and, in the same fashion, ray L_3M_3 has traveled a distance *FEI* greater than L_1M_1. It should be noted that *FEI* = 2*JCK*, and that the path difference for succeeding openings will increase in arithmetical progression; that is, for succeeding rays the path difference will be 3*JCK*, 4*JCK*, and so on.

When the rays M_1, M_2, M_3 are brought to a focus, they will reinforce and produce a bright image of the source if they are in phase along the plane front *AKI*. This condition will be satisfied only if *JCK* is an integral multiple of the wavelength of the light. Therefore, if the spacing is $d = AC = CE$, then bright maxima will occur for those conditions that satisfy the relationship

$$JC + CK = d(\sin \alpha + \sin \theta) = m\lambda \tag{11.15}$$

Equation (11.15) is not a complete solution, since the condition for

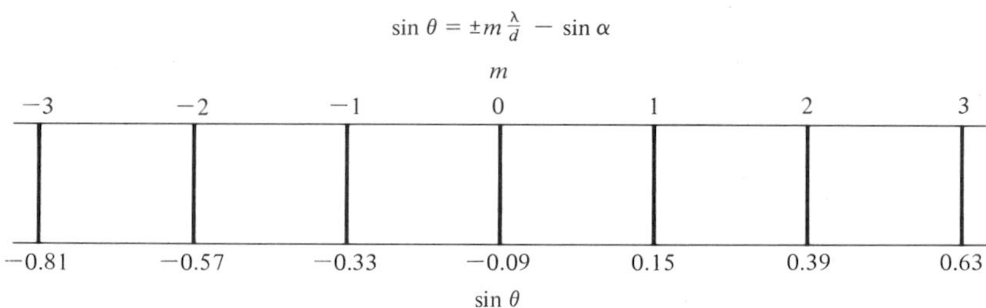

Figure 11.10 / *Distribution of* 2000-Å *radiation in the exit plane of a grating monochromator.* $d = 8.3 \times 10^{-5}$ cm, $\alpha = 5°$.

reinforcement can be met by rays diffracted to the opposite side of the direct ray, that is, rays for which θ is negative and $|\theta| > \alpha$. Under these conditions, the expression $d(\sin \alpha + \sin \theta)$ is negative. Since d, λ, and α can always be defined as positive, m is taken as negative for spectra on the other side of the direct beam, and the complete expression is accordingly

$$\pm m\lambda = d(\sin \alpha + \sin \theta) \tag{11.16}$$

This equation holds for all values of α and θ and for both reflection and transmission gratings. The integer m is known as the *order number*, and d is known as the *grating constant*. As noted above, d is the distance between corresponding points on adjacent grooves or openings, and hence d is the reciprocal of the number of grooves or openings per unit distance. Note that d and λ must be in the same units. For reflection gratings, θ is taken as negative if α and θ are on opposite sides of the perpendicular, and m is taken as negative if α and θ are on opposite sides of the direct reflection.

The significance of Eq. (11.16) can perhaps best be understood by consideration of some sample calculations. For a plane grating for the UV and visible, a typical value of the number of lines per centimeter is 12,000 (approximately 30,000 lines/in.), and thus $d = 8.3 \times 10^{-5}$ cm. We shall assume that $\alpha = 5°$ ($\sin \alpha = 0.09$) and calculate the angles at which 2000-Å (2.0×10^{-5} cm) radiation will be diffracted. The result is shown in Fig. 11.10. The solution obtained with $m = 0$ corresponds to the direct ray for a transmission grating and the direct reflection ray for a reflection grating. This "zero-order" radiation always occurs at $-\alpha$. It does not depend on the wavelength, and hence all wavelengths represented in the incident beam will be found at this position. The spectrum is symmetrically disposed about the zero-order position. For radiation of a given wavelength, the angular separation from the zero-order radiation increases as the order number increases.

Figure 11.11 shows the result of calculating the diffraction angle θ in the first order as a function of wavelength, other conditions being the same as described above. Note that the wavelength increases linearly with $\sin \theta$, and that the shorter wavelength radiation occurs at a smaller angle of diffraction

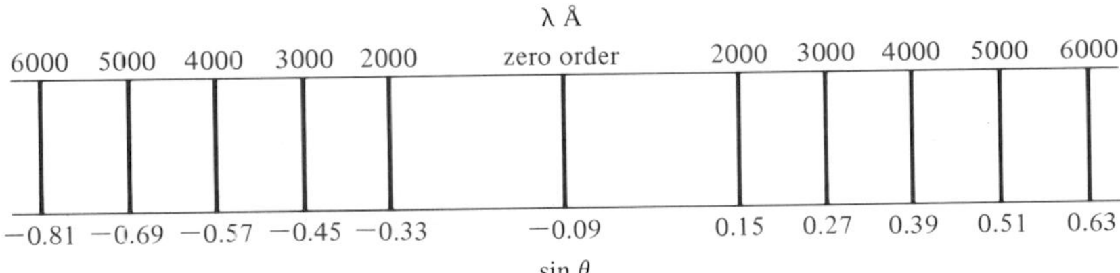

Figure 11.11 / *Diffraction angle as a function of wavelength for a plane grating monochromator in the first order.* $d = 8.3 \times 10^{-5}$ cm, $\alpha = 5°$.

than the longer wavelength radiation. Note by comparison of Fig. 11.10 and Fig. 11.11 that 2000-Å radiation in the second order falls at the same position as 4000-Å radiation in the first order, and third-order 2000-Å radiation falls at the same position as first-order 6000-Å radiation. This overlapping of orders is a characteristic of gratings, and care must be taken to eliminate this difficulty during use. A variety of means are available for elimination of unwanted orders, but perhaps the most common solution is to place a filter of appropriate cutoff in the light path.

The distribution of available intensity of a given wavelength among the various orders results in an undesirable attenuation of the intensity available at any single measurement position. By proper shaping of the grooves in a reflection grating, it is possible to concentrate the available intensity at a particular diffraction angle. Figure 11.12 shows a section of the face of such a plane grating. A grating of this type, in which the face is made up of plane facets lying at an angle β to the surface of the grating, is known as an *echelette* grating, and the angle β is known as the *blaze angle*. The effect of shaping the grooves in this fashion is to concentrate the diffracted energy around the diffraction angle for which the ordinary law of reflection from the individual facets is most nearly obeyed. From Fig. 11.12, it can be seen that

Figure 11.12 / *Section of the face of a plane echelette grating.*

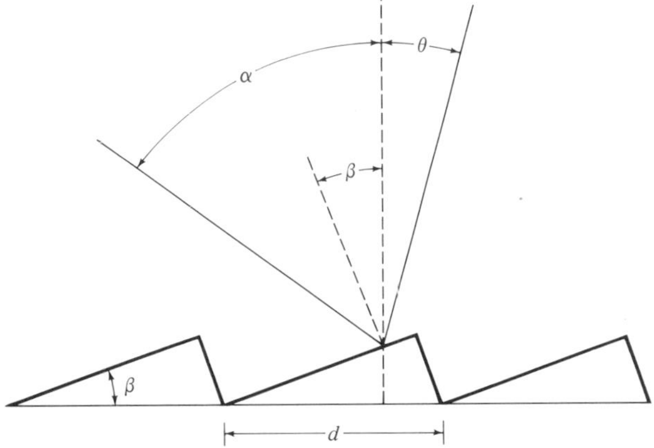

this condition is met when $\alpha - \beta = \beta - \theta$. (Recall that θ is negative if the incident and diffracted rays are on opposite sides of the perpendicular.) Thus, if $\beta = 15°$ and the other conditions are as described above, that is, $\alpha = 5°$ and $d = 8.3 \times 10^{-5}$ cm, then maximum efficiency of diffraction will occur at a diffraction angle of 25°, and this corresponds to a wavelength in the first order of approximately 4200 Å. For the conditions described, 4200-Å radiation will be concentrated in the +1 order, and 2100-Å radiation will be concentrated in the +2 order. By optimization of the blaze angle, as much as 90 percent of the available energy of a particular wavelength can be made to fall in a single order.

An expression for the angular dispersion of a grating is readily obtained by assuming that α is constant and differentiating Eq. (11.16) with respect to λ. The result is

$$\frac{d\theta}{d\lambda} = \frac{m}{d \cos \theta} \tag{11.17}$$

Thus, it is apparent that the angular separation of adjacent wavelengths increases as the order number increases and as the grating constant decreases. In most spectrometers the range of values of θ is small, so that the angular dispersion remains nearly constant as the wavelength is varied.

The theoretical resolving power of a grating is given by the expression

$$R = mN \tag{11.18}$$

where N is the total number of grooves illuminated. Thus, the larger the ruled area and the higher the order number, the greater the resolving power of a plane diffraction grating. Note that a 10-cm grating with 1200 grooves/cm, when employed in the tenth order, provides the same theoretical resolving power as a 10-cm grating with 12,000 grooves/cm employed in the first order, that is, 120,000. Provided the two gratings have the same blaze angle, the only drawback to the use of the coarser grating is the greater difficulty in eliminating overlapping orders. Since coarse gratings are more easily ruled than fine gratings, some very high resolution measurements have been made with relatively coarse gratings employed in very high orders. With such a system, order sorting is usually performed by employing a low resolving power prism to limit the range of wavelengths incident on the grating.

B / *Slits*

The entrance slit of a monochromator acts as a virtual source for the optical system of the monochromator. If the entrance slit is illuminated with monochromatic radiation, the optical system forms an image of the entrance slit on the exit curve of the monochromator. If a source emitting several spectral lines is used to illuminate the slit, an image of the slit corresponding to each of the spectral lines will be formed in the exit curve. It should be apparent that the ability to distinguish one line from another depends not

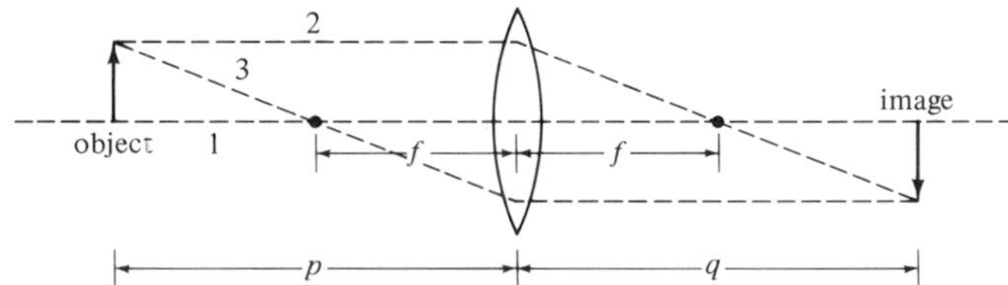

Figure 11.13 / *Formation of an image in accordance with the thin-lens equation. Ray 2, which travels parallel to the optic axis, 1, on the object side of the lens, must pass through the focal point on the image side of the lens. Ray 3, which passes through the focal point on the object side of the lens, must travel parallel to the optic axis on the image side of the lens.*

only on the dispersion of the prism or grating but also on the size of the image. Hence, it is desirable to use a narrow entrance slit. In spectrographs a photographic plate is placed on the exit curve, and all the images are recorded simultaneously. In spectrometers there is an exit slit, and the images are brought successively across the exit slit to be detected, usually by rotation of the dispersing element.

C / *Collimating and focusing devices*

In the discussion of prisms and gratings as dispersion devices, it was assumed that collimated radiation was incident on the dispersion device and that some means was provided for focusing the dispersed rays. Either a spherical lens or a mirror may be used to collimate the beam of radiation from the entrance slit. The action of both lenses and mirrors is described approximately by the thin-lens equation,

$$\frac{1}{p} + \frac{1}{q} = \frac{1}{f} \tag{11.19}$$

where p and q are the source-to-lens and lens-to-image distances, respectively, as shown in Fig. 11.13, and f is the focal length, that is, the distance to the point at which the lens will cause incident parallel rays to converge. Parallel rays are considered to arise from a source at infinity so that, in terms of Eq. (11.19), $p = \infty$ and $q = f$. In like fashion, the image of a source placed at the focal distance ($p = f$) from a lens will be formed at infinity ($q = \infty$). Thus, the light beam originating at the entrance slit of a monochromator may be collimated by placing a lens or a mirror at its focal distance from the slit. In the same way, the dispersed beams may be imaged by placing a lens or mirror in the path of the parallel rays at its focal distance from the exit curve.

Most spectrometers today employ mirrors rather than lenses for the collimating and focusing elements. The mirrors are usually of the front surface type; that is, a blank of glass or silica is ground and polished to the desired curvature and a highly reflecting metal film is then vaporized on the

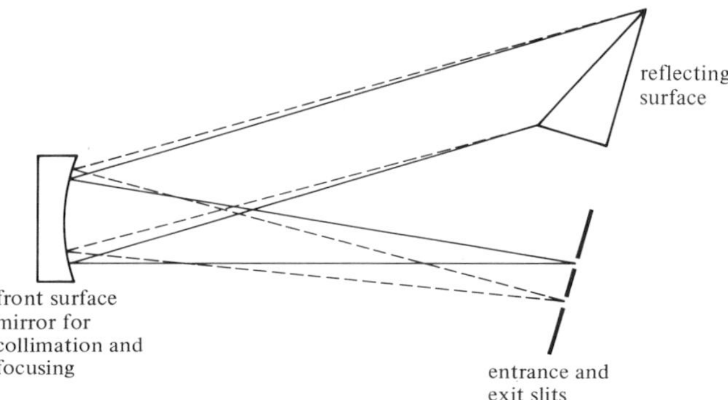

Figure 11.14 / *Littrow mount.*

face. For use in the UV and visible region, aluminum is the preferred metal because of its durability and high reflectivity.

D / *Some typical monochromator systems*

Three of the more common optical arrangements used in small spectrometers are shown in Figs. 11.14–11.16. Figure 11.14 shows one version of the Littrow mount, which is used with both prisms and gratings. It is both economical of material and compact. The entrance and exit slits are different portions of the same mechanism, and one mirror serves both for collimation and focusing. Light passes through the prism, is reflected from the back surface, and travels back through the prism so that the 30° prism acts as though it were a 60° prism. Moreover, the highly symmetrical arrangement tends to reduce the effect of aberrations in the system.

Figures 11.15 and 11.16 show two rather similar arrangements that are widely used with plane gratings: the Ebert mount and the Czerny-Turner

Figure 11.15 / *Ebert mount.*

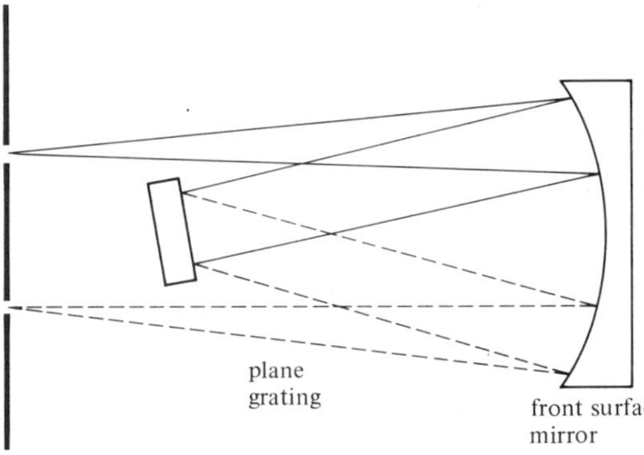

plane
grating

front surface
mirror

entrance
and exit slits

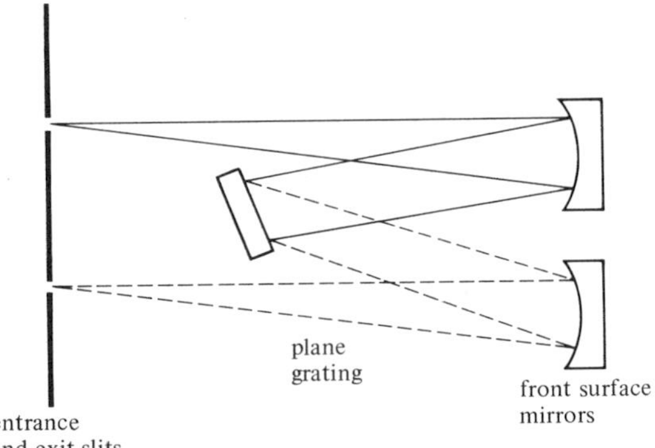

Figure 11.16 / *Czerny-Turner mount.*

mount. The Ebert mount uses a single large mirror for collimating and focusing. The center portion of the mirror is usually masked to reduce stray light. The Czerny-Turner mount uses individual smaller mirrors for collimating and focusing. One common variation with both types is to place the entrance and exit slits in line, as shown in Fig. 11.17. This allows a great deal more freedom in the arrangement of components at the entrance and exit slits.

E / *Figures of merit for monochromators*

The characteristics of a monochromator of most interest to the experimenter are its dispersion, resolving power, resolution, and speed. The meaning of these terms will be discussed in terms of the optical diagram of Fig. 11.6.

In employing the term "dispersion," careful distinction must be made between the related quantities angular dispersion, linear dispersion, and reciprocal linear dispersion. The angular dispersion depends solely on the

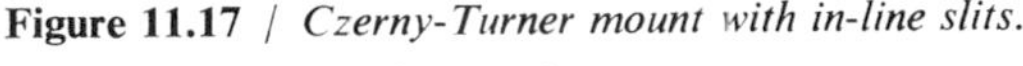

Figure 11.17 / *Czerny-Turner mount with in-line slits.*

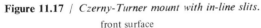

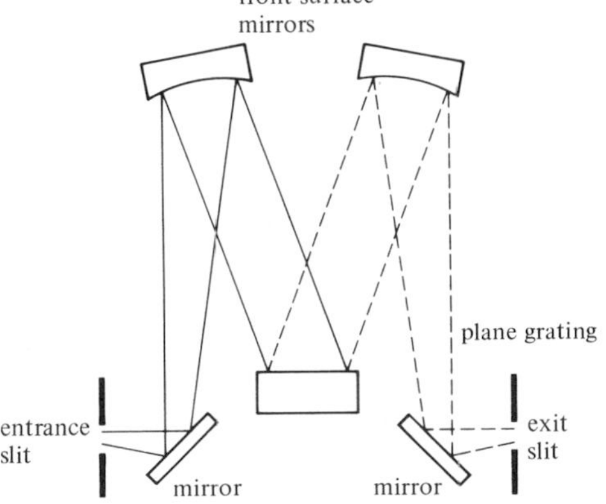

qualities of the dispersing element, but the linear and reciprocal linear dispersions depend also on the focal length of the instrument. The angular dispersion, $d\theta/d\lambda$, has been defined for a prism in Eq. (11.13) and for a grating in Eq. (11.17). Customary units for angular dispersion are radians per angstrom and radians per centimeter. The linear dispersion is the linear separation of the two images at the focal curve of the monochromator divided by $d\lambda$, that is, $dl/d\lambda$, where dl is as shown in Fig. 11.6. It follows that the linear dispersion depends on the focal length of the focusing element since, if $d\theta$ is in radians, $dl = F_2\, d\theta$, and hence

$$\frac{dl}{d\lambda} = F_2 \frac{d\theta}{d\lambda} \tag{11.20}$$

Customary units are millimeters per angstrom. The most widely reported "dispersion" is the reciprocal linear dispersion, which is simply the inverse of Eq. (11.20), that is,

$$\frac{d\lambda}{dl} = \frac{d\lambda}{F_2\, d\theta} \tag{11.21}$$

Typical values for table-top grating monochromators are in the range of 10–40 Å/mm, and the value for a given grating monochromator is nearly constant with wavelength. Figure 11.18 reports the reciprocal linear dispersion for a typical table-top prism monochromator.

Expressions for the theoretical resolving power of prisms and gratings have been given in Eqs. (11.14) and (11.18). A working definition of resolving power is provided by the expression

$$R = \frac{\bar{\lambda}}{\Delta\lambda} \tag{11.22}$$

where $\Delta\lambda$ is the wavelength separation of two lines that can just be distinguished as two rather than one, and $\bar{\lambda}$ is the mean wavelength of the pair. In practice, a subjective decision must be made as to whether two lines are distinguishable; Eqs. (11.14) and (11.18) are based on an objective definition for the distinguishability of lines—the Rayleigh criterion. The Rayleigh criterion assumes that the spectral line image widths are diffraction limited and states that two lines are just distinguishable as two rather than one when the central maximum of one falls at the same position on the exit curve as the first minimum in the diffraction pattern of the other. Theoretical resolving powers of 100,000 or more are not uncommon with grating monochromators, and actual values closely approach the theoretical limit. The resolving power of prism monochromators is usually considerably less than 100,000 even in the most favorable region of the dispersion curve. Note that Eq. (11.22) provides a convenient means of calculating the approximate resolving power required to accomplish a particular line separation. For example, if one wishes to just resolve the sodium doublet at 5890 and 5896 Å,

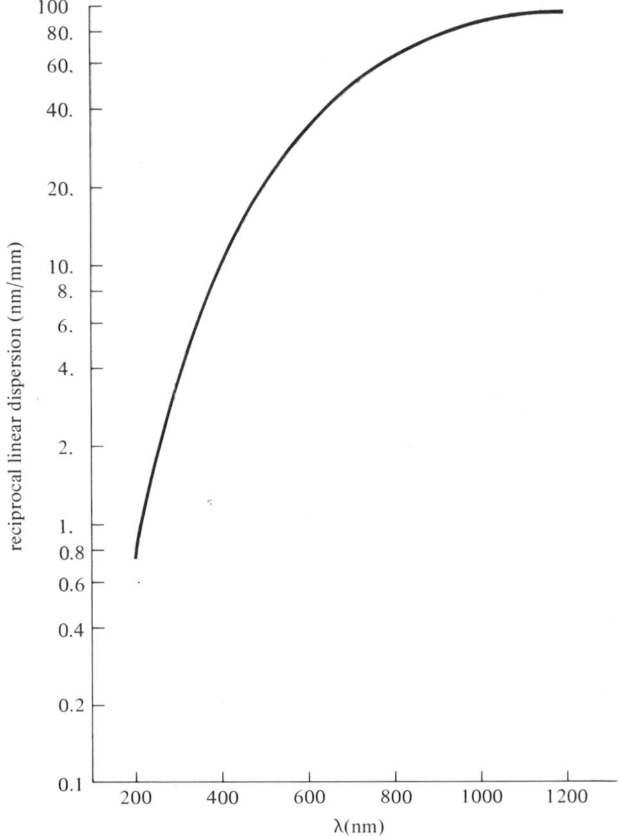

Figure 11.18 / *Reciprocal linear dispersion for a small prism monochromator.*

a monochromator with a resolving power of only $R = 5893/6 = 980$ is required.

The resolution of a monochromator is the smallest wavelength interval that it is capable of isolating. The size of the wavelength interval depends on the resolving power, the slit width, and the perfection of the optical system, and is usually reported as the half-intensity bandwidth. The significance of the term "half-intensity bandwidth" may be understood by considering the following experiment: Assume that a monochromator with equal rectangular entrance and exit slits is illuminated by a monochromatic source. If the optical system is perfect and arranged for unit magnification, an image of the entrance slit of the same dimensions as the exit slit will be formed on the exit curve. If a detector is placed behind the exit slit to respond to the intensity of radiation passing through the exit slit, as the dispersing element is rotated to bring the image across the exit slit the response of the detector will be as shown in Fig. 11.19; the intensity will increase until the image and exit slit coincide and then decrease again. If the slit widths were

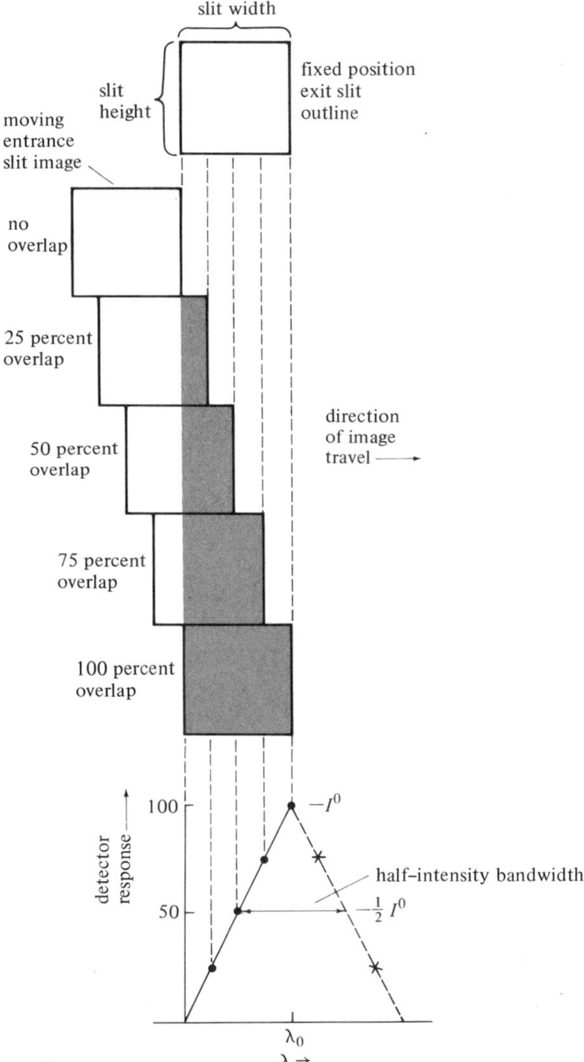

Figure 11.19 / *Illustration of the significance of "half-intensity bandwidth."*

decreased, the width of the curve in Fig. 11.19 decreases. If a continuum source replaced the monochromatic source, one could visualize the exit curve as containing an infinite number of monochromatic images of the entrance slit, and if in some fashion the radiation passing through the exit slit for a particular setting of the dispersion element could be analyzed, then one would find that a plot of intensity vs. wavelength would be identical to that of Fig. 11.19. In real optical systems the image is somewhat widened by aberrations, and the distribution of intensity with wavelength tends to be

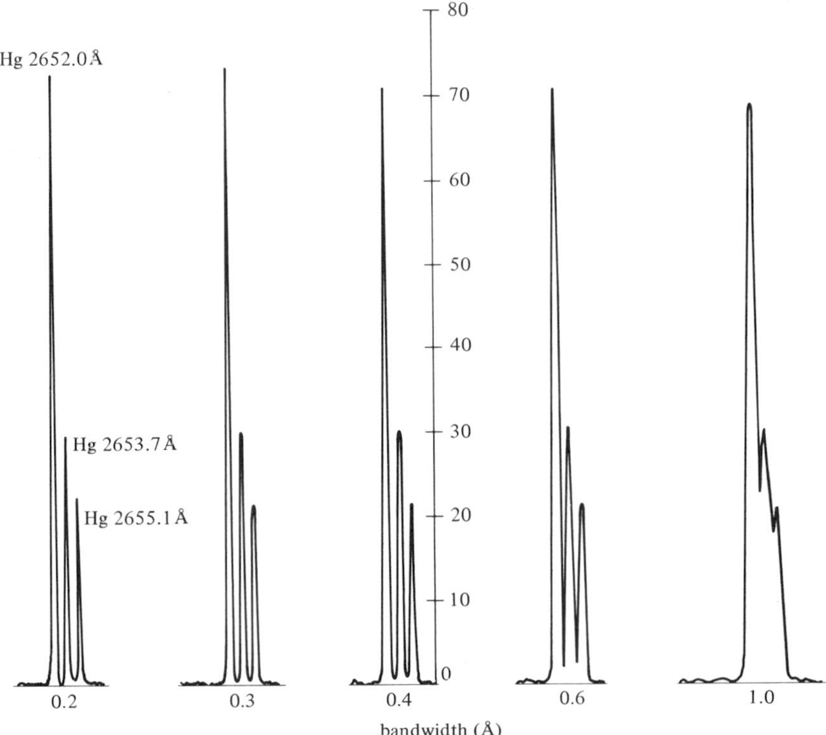

Figure 11.20 / *Effect of half-intensity bandwidth on the measured spectrum.*

Gaussian rather than triangular. The half-intensity bandwidth is the width of the distribution curve at one-half peak intensity. Note that the approximate half-intensity spectral bandwidth of a monochromator may be calculated by multiplying the reciprocal linear dispersion by the slit width:

$$s \simeq D_R W \tag{11.23}$$

where s is the half-intensity spectral bandwidth, D_R is the reciprocal linear dispersion, and W is the mechanical slit width. As W decreases, s approaches a minimum determined by aberrations and the effect of diffraction at the slit. The effect of variation of the half-intensity bandwidth on a measured spectrum is shown in Fig. 11.20.

The term "speed" as applied to a monochromator originated with spectrographic work, in which the length of time required for proper exposure of the film was directly related to the light-gathering power of the optical system. A high-speed monochromator is one with a large light-gathering ability. Light-gathering ability is described quantitatively by the *f*/ number of the optical system, which is given by

$$f/ = \frac{F_1}{A} \tag{11.24}$$

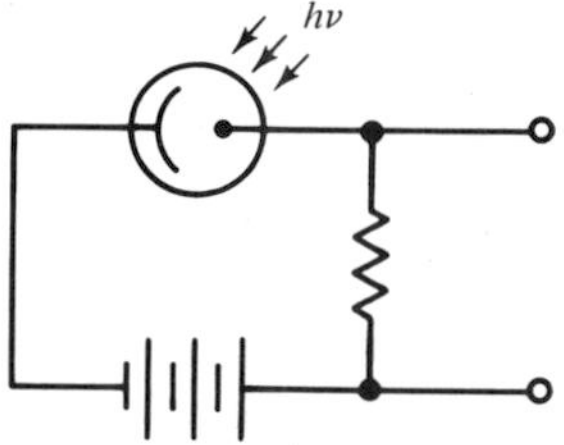

Figure 11.21 / *Vacuum photodiode.*

where F_1 is the focal length of the collimator and A is the diameter of a circular aperture of the same area as the limiting aperture of the monochromator. The limiting aperture is usually set by the dispersing element, since this is the most expensive item in the optical system. Note in Fig. 11.6 that if F_1 were decreased, a larger fraction of the radiation could be collimated within the limiting aperture. In the same fashion, as indicated by the dashed lines in Fig. 11.6, an increase in A would result in collection of a larger fraction of the radiation. Thus, the speed increases as the $f/$ number decreases.

11.2.5 / Detectors

Although a number of devices may serve as radiation detectors in the UV-visible range, we shall be concerned only with the two types that are of greatest practical interest: photographic detectors and detectors based on the photoelectric effect.

A / *Photoelectric detectors*

Some features of photoelectric detectors have been described in Section 7.3.1. In a photoelectric detector, photons falling on a specially prepared

Figure 11.22 / *Current–voltage characteristics of a vacuum photodiode.*

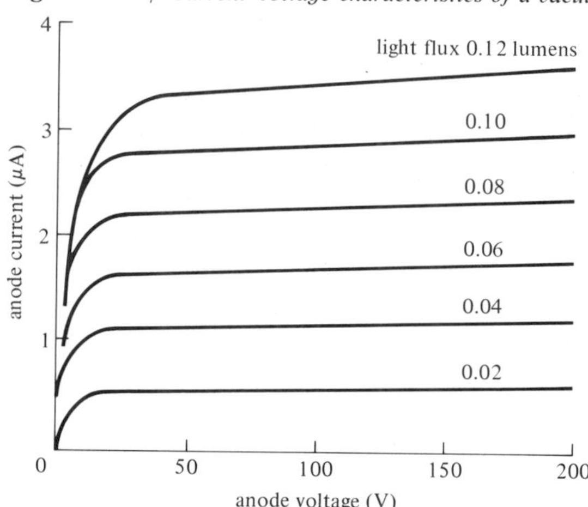

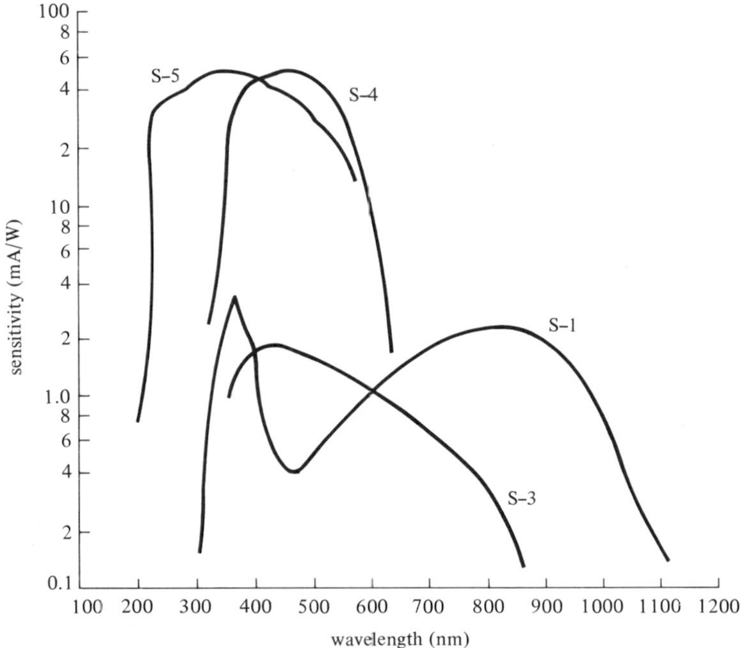

Figure 11.23 / *Typical spectral response for some photoemissive surfaces.*

surface are absorbed and, provided the photon energy exceeds a certain threshold, electrons are emitted from the surface. The number of electrons emitted per unit time is directly proportional to the radiant power, and hence a photoelectric detector is a radiant power-to-electrical current transducer.

The form of a typical photodiode and its associated circuit are shown in Fig. 11.21. For spectroscopic applications, the electrodes are sealed in an evacuated envelope that is transparent or that contains a window that is transparent to the radiation of interest. In operation, the electrode with the emitting surface, the photocathode, is held at a negative potential with respect to the collecter electrode, the anode. Figure 11.22 shows the current–voltage characteristics of a vacuum photodiode at several radiation intensity levels. At low positive voltages of the anode with respect to the cathode, the number of electrons collected increases as the voltage increases, but at anode voltages of 50 V or more, every electron emitted is collected and the current becomes nearly independent of the anode voltage. The tube is usually operated in this saturation region, and with this condition the current increases linearly with radiation intensity over a wide range of intensities.

The electron-emitting layer and the window material determine the response of the tube to various wavelengths of radiation. Figure 11.23 shows spectral response curves for some typical photodiodes. The long-wavelength cutoff is determined chiefly by the emitting layer, and the short-wavelength cutoff is determined chiefly by the window material. The S-3 curve of Fig.

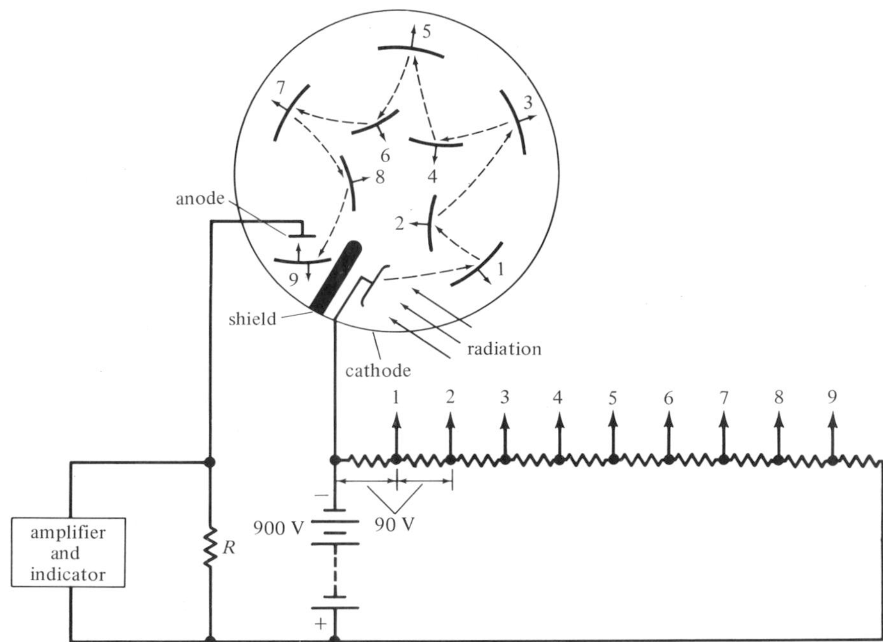

Figure 11.24 / *A multiplier phototube.*

11.23 is for a surface containing Ag, Rb, and their oxides with a lime-glass envelope. The S-4 and S-5 curves are both for a Cs and Sb surface, S-4 with a lime-glass envelope and S-5 with an envelope of UV-transmitting glass. Photodiodes with S-4 and S-5 surfaces typically show sensitivities in the range 0.03–0.05 A/W at the wavelength of maximum response; typical peak values for S-3 surfaces are an order of magnitude lower. Photoelectric surfaces are available for wavelengths to approximately 1 μm.

With a multiplier phototube, the photocurrent can be amplified by a factor of 10^6 to 10^8. This amplification is obtained by incorporating in the evacuated envelope a string of electrodes, called *dynodes*, which when struck by a sufficiently energetic electron emit a number of secondary electrons, thus leading to a cascade effect. The process is indicated schematically in Fig. 11.24. Each dynode consists of a plate coated with a layer that, like the photocathode, readily emits electrons. The tube is constructed so that each primary photoelectron is accelerated toward the first dynode and strikes it with considerable force, each liberating two to five secondary electrons. Each of these in turn is accelerated to the following dynode, and so on for a total of 9 to 16 stages. The potential increases progressively from photocathode to anode in steps of typically 30–100 V. In the usual arrangement, the photocathode is held at a large negative potential and the anode is nearly at ground potential.

Unlike the photodiode, the output current of the multiplier phototube increases sharply as the applied voltage increases, since the gain achieved at each dynode is dependent on the accelerating potential. Since the overall gain is therefore extremely sensitive to changes in the applied voltage, the power supply for a multiplier phototube must be very well regulated. As a rough guide, the stability of the power supply should be 10 times better than the stability required in the overall measurement. The spectral response depends on the photocathode coating and the window material, just as in the case of the photodiode.

Both the photodiode and multiplier phototube are high output-impedance current sources, and the output of either is readily amplified. Both exhibit very short response times, typically of the order of 10^{-8} sec, and both show an appreciable dark current, that is, a residual current in the absence of illumination. Dark current arises from several causes, the most important of which is thermionic emission of electrons from the photocathode and, in the case of multiplier phototubes, from the dynodes. This component of dark current can be eliminated almost entirely by cooling the phototube to approximately $-40°C$. Most of the remainder of dark current is due to decay of radioactive materials occurring naturally in the envelope material and to cosmic radiation. Typical dark current values at room temperature are of the order of 5×10^{-9} A for photodiodes and about an order of magnitude higher for multiplier phototubes (measured at the anode).

The lowest level of light that can be measured with a phototube is determined by the statistical fluctuation in the rate of emission of electrons from the photocathode and dynodes. This statistical noise in phototubes arises from random fluctuations in (1) the electron current ("shot" noise), and (2) thermal motion of the conducting electrons in the load resistor (Johnson noise).

Both shot noise and Johnson noise have been discussed in Section 6.2. Equation (6.2) gives the mean-square Johnson noise voltage as

$$\bar{e}_R^{\,2} = 4kTR\,\Delta f$$

where k is the Boltzmann constant, T is the absolute temperature, Δf is the frequency-response bandwidth of the measurement equipment, and, for phototubes, R is the value in ohms of the load resistor. The mean-square noise current at the photocathode due to shot noise is given by the expression

$$\bar{i}_s^{\,2} = 2ei\,\Delta f \qquad (11.25)$$

where e is the charge on an electron in coulombs and i is the cathode current in amperes. The mean-square noise voltage due to shot noise is thus

$$\bar{e}_s^{\,2} = 2ei\,\Delta f R^2 \qquad (11.26)$$

where R is the load resistance.

Since Johnson and shot noise occur in random phase to each other as well as to other signal components, the overall mean-square noise voltage for a *photodiode* is given by

$$\begin{aligned}\bar{e}_T &= (2ei\,\Delta f R^2 + 4kTR\,\Delta f)^{1/2} \\ &= \left[eR\,\Delta f\left(2iR + \frac{4kT}{e}\right)\right]^{1/2}\end{aligned} \tag{11.27}$$

Equation (11.27) contains the information required for optimization of the signal-to-noise ratio for a photodiode. At an absolute temperature of 300°K, $4kT/e = 0.1$ to a good approximation. It is desirable to choose the load resistance to have a value such that the shot-noise component (iR) is larger than $4kT/e$, since both the signal and shot-noise voltages increase as R increases but $4kT/e$ is invariant; thus, if R were chosen so that iR was small compared to $4kT/e$, the signal would be reduced without effecting a reduction in noise. At low light levels, the cathode current, i, is largely due to dark current, and the minimum shot noise is thus set by the dark current. If 5×10^{-9} A is taken to be a typical value of dark current for a photodiode, then the shot-noise component will equal the Johnson-noise component if a load resistance of 10 MΩ is used. Once shot noise becomes large compared to Johnson noise, no further improvement in signal-to-noise ratio is obtained by increasing the load resistance, and thus it is the shot noise due to dark current (not the dark current itself) that sets a limit to the lowest light level detectable with a photodiode. Assuming a dark current of 5×10^{-9} A, the rms shot-noise current for a photodiode is approximately $4 \times 10^{-14}\sqrt{\Delta f}$ A, which implies that photocurrents of this order of magnitude may be detected with a photodiode. Lower current levels require a multiplier phototube.

With a multiplier phototube, the overall shot noise is increased by the amplification factor, G, of the tube and by shot-noise contributions from the dynodes. The overall mean-square shot-noise voltage is given by

$$\bar{e}_s^{\,2} = 2aGei_a\,\Delta f R^2 \tag{11.28}$$

where i_a is the current at the *anode*, and a is a factor to account for the dynode contribution that typically has a value of about 1.2. The overall rms noise voltage for a multiplier phototube is thus

$$\bar{e}_T = \left[eR\,\Delta f\left(2aGi_aR + \frac{4kT}{e}\right)\right]^{1/2} \tag{11.29}$$

Since G is typically about 10^6, it is apparent that the condition that shot noise exceeds Johnson noise is readily achieved. For typical values of a and G, if the anode dark current is approximately 10^{-8} A, then the dark-current-limited rms noise current is approximately $6 \times 10^{-11}\sqrt{\Delta f}$ A. This implies that photo*cathode* currents of the order of $6 \times 10^{-11}\sqrt{\Delta f}/10^6 = 6 \times 10^{-17}\sqrt{\Delta f}$ A can be detected with a multiplier phototube.

Note that it is characteristic of photoelectric detectors that the noise increases as the current increases. At low light levels, the dark current is significant and determines the minimum noise; but as the level of illumination increases, the dark current becomes negligible and the noise increases as the square root of the intensity of the radiation incident on the photocathode.

B / *Photographic detectors*

The basis of photographic detection is the chemical change engendered in silver halides when they are exposed to light. When light is incident on a silver halide crystal, it produces a latent effect that makes it easier to liberate silver from the crystal that has been exposed than from one that has not. The liberation is performed by a suitable reducing agent or "developer," which deposits metallic silver from only those crystals that have previously been exposed to light; the unreduced silver halide is removed by a second solution, the "fixer."

In practice, the light-sensitive silver salts are used in the form of an emulsion—a dispersion of small silver halide crystals in gelatin. The emulsion is spread in a thin layer on a support—most commonly glass plate or plastic film. When the emulsion is exposed in a spectrograph, a latent image of the spectrum is formed which becomes an actual image after the emulsion is developed and fixed. A photographic detector, unlike a photoelectric detector, simultaneously detects a large portion of the spectrum. The "blackening" of the plate for a particular image is roughly proportional to the product of the intensity of the radiation and the exposure time. This is the basis of quantitative spectrographic measurements. Note that the photographic detector response is proportional to the total incident energy, unlike the response of the photoelectric detector which is proportional to incident power.

The amount of blackening of a photographic plate at any point can be measured with an instrument called a *densitometer* or *microphotometer*, the principle of which is shown in Fig. 11.25. The radiation from an incandescent lamp is focused as a narrow slit of light on the photographic plate and again on a phototube. The blackening is reported as the optical density of the image on the plate,

$$D = \log \frac{P_0}{P} \qquad (11.30)$$

where P_0 is the power, as detected by the phototube, transmitted through a clear portion of the plate, and P is the power transmitted through the spectral image. A typical plot of plate blackening against the log of the product of intensity and time (this product is usually called exposure) is shown in Fig. 11.26. Note that there is a considerable range in which the density varies in a nearly linear fashion with log exposure. The preparation of plate calibration curves such as that of Fig. 11.26 is considered in Chapter 14.

Photographic detectors can be made for all wavelengths shorter than 1.3 μm. A wide variety of emulsions has been developed for various spectral

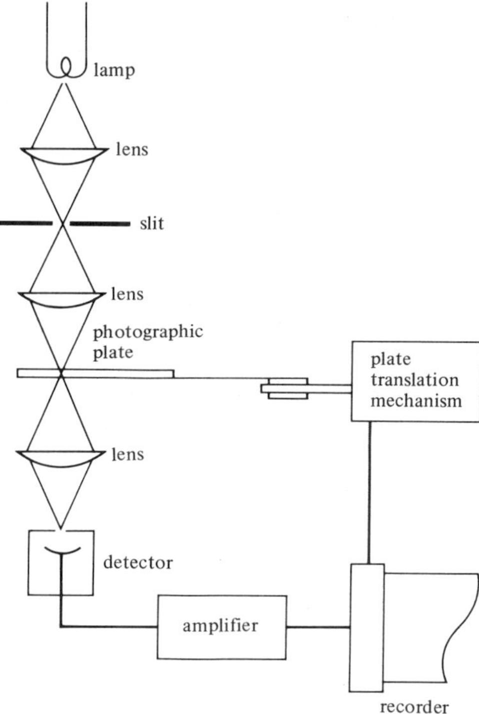

Figure 11.25 / *Recording microphotometer.*

regions. Sensitivity (sometimes called speed) and resolution (graininess) are also variable and can be selected to suit the problem. The sensitivity of an emulsion is wavelength dependent—the spectral response is determined by gelatin absorption at short wavelength, by the silver salt used, and by the presence of various sensitizing agents for short- and long-wavelength

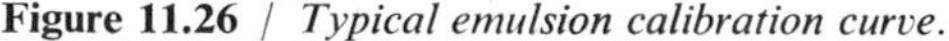

Figure 11.26 / *Typical emulsion calibration curve.*

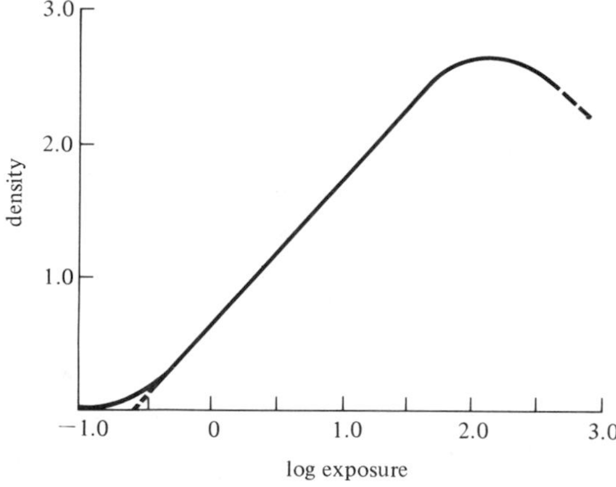

measurements. Sensitivity and graininess are not entirely independent qualities. A small number of photons, probably less than 10, is enough to permit chemical reduction of a whole silver halide grain to produce a dark speck on the plate. As the size of the grain is increased, a larger dark speck is produced for the same number of photons; thus, sensitivity is increased, but the increased graininess may result in some loss of detail in the image.

C / *Comparison of photoelectric and photographic detectors*

The ability of a photographic plate to detect and record a large portion of a spectrum simultaneously makes it a logical choice for qualitative-analysis applications, particularly where the spectral source may be changing rapidly with time or when only a limited amount of sample is available. It should also be noted that a photographic detector is an integrating device, and thus it is useful when signal averaging is required (see Section 6.4.2). The large number of variables that must be controlled in the manufacture, exposure, and development of the photographic detector make it a rather undesirable choice for quantitative measurements. Photoelectric detectors have largely supplanted photographic ones for this purpose.

11.3 / INSTRUMENTATION FOR THE INFRARED REGION

11.3.1 / Radiation sources

Sources for the infrared region are, for the most part, of the hot-body type. Typical spectral outputs for some IR sources are shown in Fig. 11.27. A simple coil of tungsten or nichrome wire may be used as a source for the IR region if the required wavelength range and intensity are not too great, but the most widely employed sources are the Nernst glower and the Globar.

The Nernst glower is a rod or hollow tube about 2 cm long by 0.1 cm in diameter made by sintering together oxides of such elements as cerium, zirconium, thorium, and yttrium. The rod is nonconducting when cold, but when heated the resistance drops and sufficient current flows when a voltage is applied across the heated rod to maintain the rod at a high temperature. Resistors in the circuit limit the current flow, so that a steady temperature in the range of 1500° to 2000°K is maintained. The Globar is a rod of silicon carbide, usually about 5 cm long and 0.5 cm in diameter, which, like the Nernst glower, can be heated by passage of an electrical current. It is usually operated at a somewhat lower temperature, about 1400°K maximum.

11.3.2 / Optical materials

Table 11.1 lists the useful ranges of various materials. Sodium chloride has been the material most widely employed in analytical IR instruments, and thus most measurements reported are for the 2–15 μm range (5000–667 cm^{-1}). In addition to transparency, physical strength, ability to take an optical polish, and reaction to atmospheric moisture are factors in the choice of

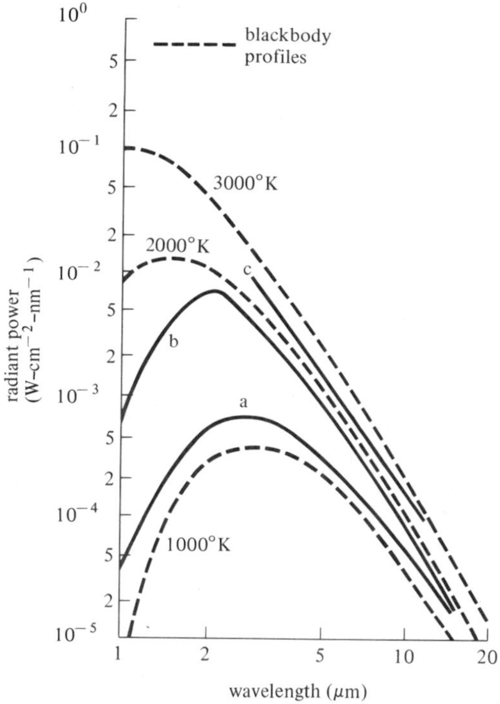

Figure 11.27 / *Typical spectral outputs for some IR sources. a, Globar; b, Nernst glower; c, tungsten glower.*

material. Because the alkali chlorides, bromides, and iodides are hygroscopic and the polished surfaces are readily etched by moisture, optical components made of these materials must be protected from moist air. Although lenses have been made from the alkali halides, most IR spectrometers use mirrors for collimating and focusing.

11.3.3 / Filters

The problems of order sorting and stray light removal are more severe in the IR- than in the UV-visible range, and hence the need for good filters is correspondingly greater. Examples of each of the types of filters discussed in Section 11.2.3 may also be cited for the IR region.

A / *Absorption filters*

In the infrared, absorption filters are principally used to remove short-wavelength radiation. Filters that are opaque out to 1–5 μm may be made by controlled treatment of potassium chloride or bromide by X rays and heat. Sharp, short-wavelength cutoff filters may also be made from high-purity semiconductor materials such as selenium and germanium. If a photon of greater energy than the gap between the valence and conduction bands is incident, it is absorbed and an electron–hole pair created. If the photon energy is less than the gap, there is no absorption. The absorption edge is very sharp and is temperature sensitive. The cutoff wavelength varies with material, being in the region of 5900 Å for selenium and 7.5 μm for indium antimonide. Because of the high refractive indices of semiconductors, it is necessary to use antireflection coatings to obtain high transmission in the transparent region, and the coating may limit the range for which the filter is usable.

A somewhat different approach to the use of a filter for the elimination of an unwanted signal due to short-wavelength radiation is to use a chopper that is transparent to the unwanted (stray) radiation but opaque to the radiation of interest. Thus, a sectored disc of fused silica rotated in the radiation beam will periodically interrupt radiation in the 4–50 μm range but allow shorter-wavelength radiation to pass unmodulated. An amplifier tuned to the chopping frequency may be used to discriminate against the signal due to the unwanted radiation.

B / *Interference filters*

Multilayer dielectric filters are now available with peak wavelengths out to at least 20 μm. Bandwidths are typically about 2 percent of the peak wavelength. A circular filter is commercially available in which the coating has been deposited on the substrate in such a manner that the peak transmission wavelength varies as the filter is rotated. Wedge-shaped filters of this type have been employed in place of prism or grating dispersing elements in a low-cost IR spectrophotometer and in an instrument for rapid scanning of the effluent from a gas chromatograph.

11.3.4 / Monochromators

A monochromator for the IR region consists of the same basic components as one for the UV-visible region, but there are some important differences in detail that require mention.

A / *Dispersion devices*

The principal virtue of prisms as dispersive elements for the IR region is the relative simplicity in design of the monochromator. The entire spectral region over which the material is transparent can be scanned in one continuous sweep of the scanning arm with no auxiliary spectral separating devices. Recommended prism materials for various regions of the infrared are given in Table 11.2. Although lithium fluoride and calcium fluoride give better dispersion in the short-wavelength region, sodium chloride is the usual prism material for instruments covering the 2–15 μm range.

The use of reflection gratings as dispersion elements in the IR region offers the advantages of better dispersion and nearly constant dispersion over a considerable range of wavelengths. Moreover, for wavelengths longer than about 50 μm, gratings offer the only feasible approach to construction of a

TABLE 11.2 / *Recommended prism materials*

Region (μm)	*Material*
2.7–5	LiF
5–9	CaF_2
9–15	NaCl
15–25	KBr
25–50	CsI

dispersion device, since no suitable materials have been found for construction of prisms in this range. However, the use of gratings introduces some additional complications in instrument design because of the necessity of eliminating radiation in unwanted orders. This problem is somewhat more severe in the infrared than in the UV-visible because of the greater range of wavelengths examined and because, for maximum efficiency, the groove spacing must be of the same order of magnitude as the wavelength of radiation to be dispersed; thus, IR gratings are more coarsely ruled than those for the UV-visible range. If we assume an optical arrangement in which the angle of incidence and the angle of diffraction are equal (Littrow arrangement), then Eq. (11.16) may be rearranged to the form

$$\sin \theta = \frac{m\lambda}{2d} \tag{11.31}$$

In this form it can readily be seen that the larger the value of d, the smaller the spacing between orders for a given wavelength of radiation. A grating for the IR region is typically ruled with 1200 lines/cm ($d = 8.3 \times 10^{-4}$ cm), whereas a value of 12,000 lines/cm is typical for the UV-visible region. The required order separation in IR instruments is most commonly achieved by use of either a low-resolution foreprism or a long-wavelength pass filter to reject short wavelengths.

The efficiency of a grating falls off on either side of the blaze wavelength. A useful rule-of-thumb for predicting the range in which the efficiency will be better than 50 percent is given in Eq. (11.32), where λ_b is the blaze wavelength in the first order:

$$\left(\frac{2}{2m + 1}\right) \lambda_b < \lambda < \left(\frac{2}{2m - 1}\right) \lambda_b \tag{11.32}$$

Two approaches have been used to circumvent this loss of efficiency in IR instruments designed to cover a wide range of wavelengths. One approach is to use a single grating in several orders. For example, a grating blazed for 12 μm has been used to cover the range from 2.2 to 18 μm in five orders. The range becomes successively shorter in the higher orders, with increasing difficulty of separation. A second approach is to use more than one grating, each limited to use in the first and second orders. Most commercially available instruments provide for automatic interposition of filters at the appropriate wavelengths and automatic interchange of gratings.

B / *Collimating and focusing devices*

Mirrors are used almost exclusively in IR instrumentation. Gold film has a somewhat higher reflectance than aluminum for IR radiation, and hence it is used when maximum performance is required.

C / *Some typical monochromator systems*

The Littrow, Ebert, and Czerny-Turner mounts employed in UV-visible monochromators are also widely used in IR systems. The Littrow prism mount usually consists of a full 60° prism with a separate mirror to reflect the beam back through the prism.

11.3.5 / Detectors

For purposes of our discussion, IR detectors will be divided into two main types: thermal detectors and photon detectors. A convenient term for comparison of IR detectors is the detectivity, D^*, where

$$D^* = \frac{S/N}{P_D}\left(\frac{\Delta f}{A}\right)^{1/2} \tag{11.33}$$

where the units of D^* are centimeter-hertz$^{1/2}$-watt^{-1}, S and N are signal and noise voltage, P_D is the radiant power per unit detector area incident on the detector, A is the detector area, and Δf is the frequency-response bandwidth of the measurement system. It should be clear from Eq. (11.33) that a large value of D^* is a desirable quality in a detector.

A / *Thermal detectors*

Thermal detectors absorb radiation and measure the power of the incident radiation in terms of the heating effect resulting from the absorption. Characteristics of some detectors of this type are shown in Table 11.3. They may be classified by the method in which the temperature rise is sensed.

Bolometers. As described in Section 7.3.2, the operating principle of the *bolometer* is the change of electrical resistance of a material due to the temperature change caused by the absorbed radiation. Certain semiconductor

TABLE 11.3 / *Characteristics of some thermal detectors*

Detector	*Operating temperature*	*Detectivity* (cm-Hz$^{1/2}$-W^{-1})	*Response time* (msec)
bolometer	ambient	2×10^8	1.5
thermocouple	ambient	1.5×10^9	15
Golay	ambient	1.5×10^9	15
pyroelectric	ambient	10^8	0.002

materials exhibit a large change in resistance with change of temperature, and bolometers made from one of these materials are known as *thermistors*. Contrary to the behavior of metals, the resistance of a thermistor falls as the temperature rises.

Thermistors are made in the form of a semiconductor flake, approximately 10 μm thick, mounted on a heat-dissipating substrate. The surface of the flake is blackened to act as an efficient radiation absorber. The temperature of the flake rises as radiation falls on its surface and returns to its initial value when the radiation is blocked. The time constant of the detector depends on the thermal coupling between the flake and the substrate and on the radiation interchange with the surroundings. The change of resistance is detected by making the flake one arm of a bridge circuit. A second flake, shielded from radiation, is incorporated in the other arm of the bridge to compensate for changes in ambient temperature. Resistances are typically in the megohm range.

Thermocouples. As described in Section 7.3.3, *thermocouples* are made by welding together two wires of different thermoelectric properties at two points. A potential is developed between the two junctions when the junctions are at different temperatures. When used as IR detectors, one of the junctions is kept at constant temperature while the other, mounted on a blackened radiation receiver, is placed in the radiation beam. To achieve reasonably rapid response, the thermocouple must be of low thermal mass. To reduce heat loss by convection and thus improve detectivity, the thermocouple is mounted in an evacuated enclosure with a suitable window to admit the radiation. The signal developed at a thermocouple detector is typically 0.1–2.0 μV with an output impedance of about 10 Ω.

Pneumatic detectors. In the *pneumatic*, or *Golay*, *detector*, the IR radiation is detected by the pressure rise it causes by heating an enclosed gas. One form of Golay detector is shown in Fig. 11.28. A thin metal coating on a plastic film acts as a very broad-band absorber; it has nearly uniform absorptivity for radiation from the UV to the microwave range. The pressure change in the small enclosure behind the plastic film causes a distortion of the diaphragm that serves as one wall of the enclosure. A variety of means have been employed to sense the change in the diaphragm. In the example shown, the diaphragm is the reflector in an optical lever arrangement. Movement of the diaphragm alters the position of the light beam in its passage through a grid, and, if the grid is properly constructed, the intensity of light falling on the photocell varies directly with the position of the beam. The characteristics listed for the Golay detector in Table 11.3 are not significantly better than those of the thermocouple or bolometer detector, but because of the wide radiation-receiving area of the Golay detector, it finds use in instruments where wide slits are required, and in the far infrared (beyond 50 μm) it has better detectivity because its metalized

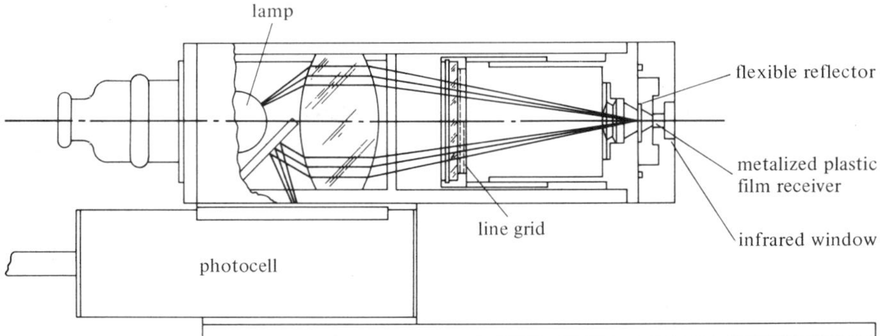

Figure 11.28 / *Golay detector.*

plastic film receiver is a far better absorber in this region than the usual black coating of thermocouple or bolometer detectors.

Pyroelectric detectors. Certain acentric crystals have a permanent electric dipole moment. Normally, the field produced by this is neutralized by surface charges attracted to the faces of the crystal. A temperature change will cause a change of lattice dimensions in the crystal, and hence a change in the permanent dipole moment. If the material of the crystal is an insulator, a relatively long time is required for the surface charges to rearrange themselves, and an electric field appears. A pair of electrodes normal to the polar axis of the crystal may be used to measure the voltage generated within the crystal. Pyroelectric detectors are most commonly made from single-crystal triglycine sulfate (TGS). The detector is fabricated in the same manner as a capacitor. Two electrodes, one of them transparent, are formed on opposite sides of a TGS slice. Radiation is received through the transparent electrode. The voltage generated within the crystal is usually applied directly to a field effect transistor which is an integral part of the detector package.

Pyroelectric materials absorb quite strongly in the far infrared and, unlike the bolometer and thermocouple detectors, the TGS detector has essentially flat wavelength response from the near infrared through the far infrared. As indicated in Table 11.3, the detectivity of the TGS detector is somewhat less than that of the Golay detector, but because of its very much faster response time, it has replaced the Golay detector for some far infrared applications.

Noise in thermal detectors. The limiting noise in thermal detectors arises from random arrival of photons from the background. Thus, unlike photoelectric detectors, the detector noise is largely independent of the intensity of signal radiation incident on the detector.

B / *Photon detectors*

No materials are known that have sufficiently small work functions to allow construction for the IR region of electron-emissive detectors of the type used

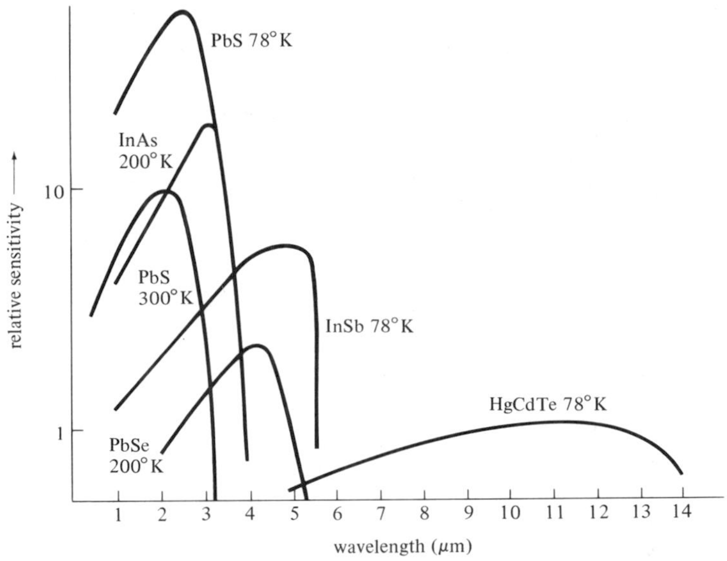

Figure 11.29 / *Spectral response of typical intrinsic detectors.*
SOURCE: H. Levinstein, *Anal. Chem.* **41**(14), 81A(1969).

in the UV-visible region. However, certain semiconductor materials exhibit what might be called an "internal photoelectric effect," which can be used for detection of IR radiation. As discussed in Section 7.3.1, the effect consists of excitation of a charge carrier in a semiconductor from the valence band (instrinsic detector) or from an impurity band (extrinsic detector) into the conduction band. The energy required for this process is frequently very much less than that required for photoelectric emission. The magnitude of the energy required depends, of course, on the semiconductor material and any added impurities.

Semiconductor detectors are operated most often in the photoconductive mode. A photoconductive detector is a resistor (with a value of about 200,000 Ω in the dark) that decreases in value as the intensity of radiation striking its surface increases.

Detectors of PbS were the earliest example of this class. However, PbS detectors do not respond to radiation at wavelengths greater than 3 μm. At present, semiconductor detectors are available for wavelengths up to about 100 μm. Since the excitation energy required is very small in the long-wavelength devices, the detectors must be cooled to prevent background thermal excitation from obscuring observation of the IR radiation. The spectral response of semiconductor detectors is, like that of photoelectric detectors for the UV-visible, sharply peaked. Figure 11.29 shows the characteristics of some detectors of the intrinsic type. Note the high detectivity at peak response and the temperatures required for proper operation of the long-wavelength detectors. Response times range from several hundreds of microseconds for some of the intrinsic detectors to nanoseconds for some of the extrinsic detectors.

The limiting noise of semiconductor detectors is associated with fluctuations in the density of free charge carriers, produced either by vibrations of the crystal lattice or by random arrival of photons from the background. The lattice vibration contribution may be eliminated by adequate cooling, and the detector is then said to be *background limited.* As in the case of thermal detectors, the noise is largely independent of the intensity of the signal radiation incident on the detector.

11.4 / MEASUREMENT OPERATIONS IN SPECTROMETERS

The principles of measurement operations were described in Chapter 7. In this section, we shall deal specifically with the application of these principles in instrumentation for optical spectroscopy.

11.4.1 / Characteristics of spectrometric signals

Each of the types of signals described in Chapter 7 are represented in optical spectroscopic measurements.

A / *Current sources*

Photoelectric detectors are current sources. The current typically is in the range of microamperes. The output impedance of photoelectric detectors is very high, hundreds of megohms, and thus a photoelectric detector approaches the characteristics of an ideal current source, and its output signal is readily amplified. Frequently, the output signal is read as the voltage across a resistor in series with the detector, as shown in Fig. 11.21, or in multiplier phototubes, as the voltage across a resistor between the anode and the last dynode, as shown in Fig. 11.24. Figure 11.30 shows a circuit in which an operational amplifier is used as an inverting current-to-voltage transducer at the output of a multiplier phototube. This arrangement is particularly convenient because of the ease with which dark current offset can be provided at the summing point, *S*. To minimize noise and drift, the source radiation is often modulated (e.g., by periodic mechanical interruption of the light beam) and AC (usually phase-sensitive) amplification employed. The advantages of modulation have been discussed in Sections 6.3 and 6.4. Because of the fast response time of photoelectric detectors, the frequency at which the radiation is modulated is determined by considerations pertaining to the electronic system rather than the detector.

B / *Voltage sources*

Thermocouple detectors are voltage sources that typically give an output in the range of a few microvolts and have output impedances in the range of a few ohms. Because of the inherent low impedance of a thermocouple, its

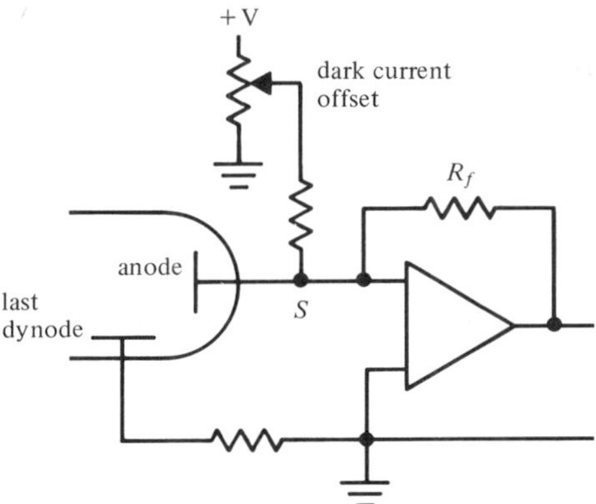

Figure 11.30 / *System for measuring current output from a multiplier phototube.*

Johnson noise is so low that noise in the amplifier can limit the signal-to-noise ratio. The signal-to-noise ratio may be improved by modulating the radiation incident on the detector and using a transformer to step up the signal before amplification. The pertinent considerations in optimization of the signal-to-noise ratio are that (1) the signal increases directly as the transformer turns ratio; (2) apparent source impedance, and Johnson noise, increase as the square of the turns ratio; and (3) the noise due to the amplifier is, of course, invariant with the turns ratio. Thus, one can see that nearly optimum conditions are achieved when the transformer turns ratio is such that Johnson noise is approximately the same magnitude as amplifier noise. If the turns ratio were decreased, the signal would decrease without a proportionate decrease in the noise, since amplifier noise would predominate; if the turns ratio were increased, Johnson noise would predominate and noise would increase as the square of the turns ratio, but signal would increase only as the first power of the turns ratio. Turns ratios of about 300:1 prove optimal. The modulation frequency is usually around 10 Hz because of the relatively long response time of thermocouple detectors.

C / *Resistance*

Bolometer detectors and most semiconductor detectors are essentially resistors. In both cases, the quantity of interest is not the magnitude of the resistance but the *change* in resistance that occurs as radiation is incident on the detector. Figure 11.31 shows a simple circuit for a bolometer detector. If the two thermistors are well matched, changes in ambient temperature result in equal changes in resistance of each and no change in potential at point *S*. When radiation falls on the active thermistor, its temperature rises, its resistance falls, and the voltage at point *S* changes by an amount proportional to the intensity of the radiation. Both bolometer and semiconductor detectors have resistances in the range of megohms. The voltages to be measured as a result of resistance changes in the detector are usually

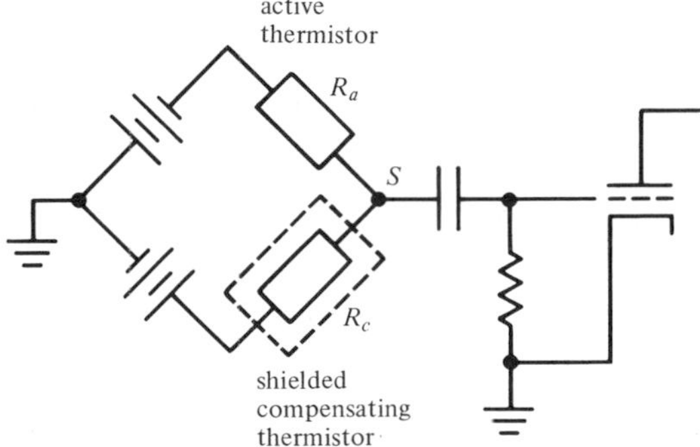

Figure 11.31 / *Bolometer detector circuit.*

of the order of microvolts. With bolometers, the long response time limits the modulation frequency to about 10 Hz; detector response time is usually not a factor in selection of the modulation frequency with semiconductor detectors.

11.4.2 / Single-beam and double-beam measurements

In spectroscopic measurements, frequent calibration checks are necessary to ensure the validity of results. As a minimum, this means that it is necessary to compare the reading for the analytical sample with the reading for a blank or reference. If the spectrometer is arranged in such a way that a relatively long time (several seconds or longer) elapses between the measurements of sample and blank, then the instrument is single beam in concept. Almost all emission spectrometers are single-beam instruments, and manual interchange of sample and blank is required. Figure 11.32 shows schematically a single-beam absorption spectrophotometer. To obtain one absorption measurement with this instrument requires the following steps:

1. With the shutter closed, the instrument is adjusted to give a zero meter reading.
2. With blank in the optical path, the shutter is opened and the instrument is set to give a meter reading of 100.
3. The sample is placed in the optical path and the percent transmittance is read directly from the meter.

A single-beam instrument is thus characterized by *slow, sequential* measurements.

The name "double beam" implies simultaneous measurement of sample and reference, but this is almost never the case. A double-beam spectrometer differs from a single-beam spectrometer chiefly in that comparison of sample and blank is caused to occur rapidly (many times a second), and thus double beam usually means *rapid sequential* measurements.

Figure 11.33 illustrates the principles of double-beam measurement. The mirror assembly vibrates between positions A–A and B–B, alternately

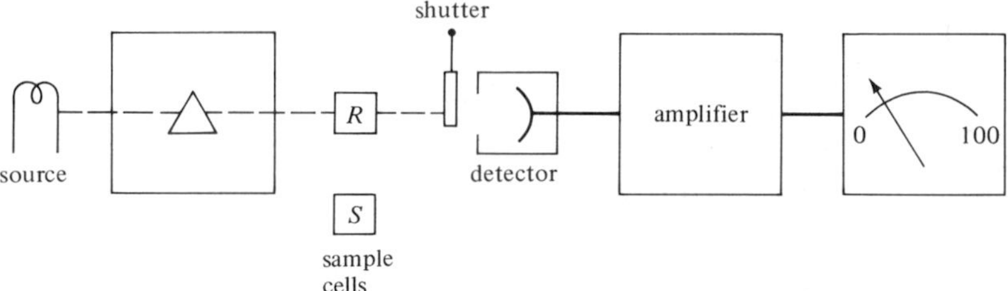

Figure 11.32 / *A single-beam absorption spectrophotometer.*

directing the radiation through the sample, *S*, and reference, *R*, solutions. A switch, *Sw*, synchronized with the motion of the mirrors, switches the output signal of the phototube to the appropriate electronic channel. If arrangements are made to check the zero level when the mirror assembly lies between positions A–A and B–B, the instrument diagrammed in Fig. 11.33 will carry out automatically the three steps listed above as required in making a single-beam absorption measurement.

There are two chief advantages to double-beam operation as compared to single-beam operation: (1) Because sample and reference are compared with greater frequency, effects such as source intensity drift, variation in amplifier gain, and other changes in optical and electrical components that are common to the two channels have less effect on the measurements; and (2) automatic recording of absorption spectra is made possible.

11.4.3 / Direct presentation and null balancing

The single-beam absorption spectrophotometer of Fig. 11.32 is an example of a direct-presentation measurement system: The amplified output of the detector is measured directly in terms of meter deflection. With such a system, the accuracy of the absorption measurement is directly dependent on the linearity, accuracy, and reproducibility of the amplifier and meter used. The principles of null-balancing measurements are described in Section

Figure 11.33 / *A double-beam absorption spectrophotometer.*

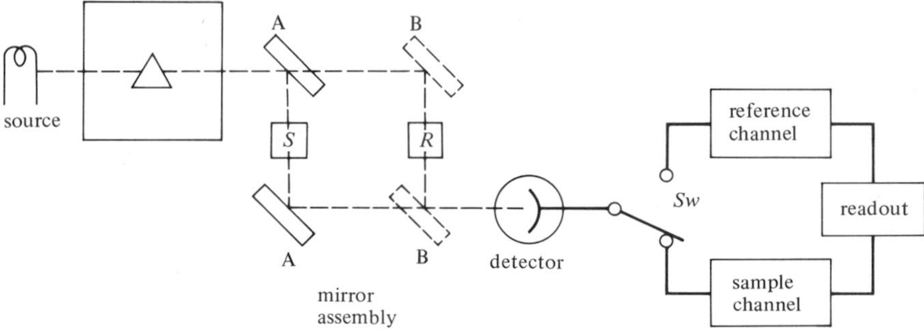

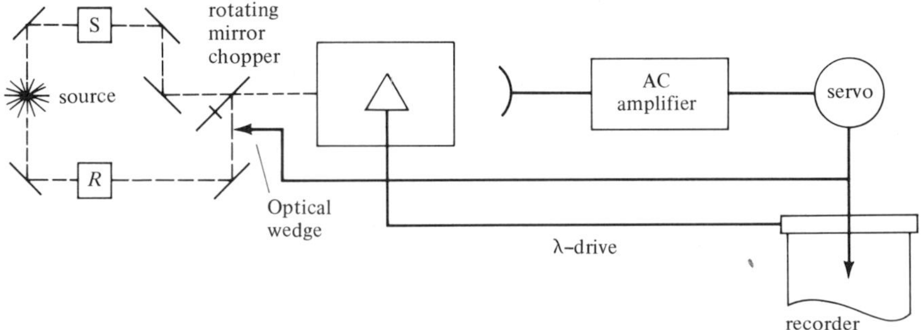

Figure 11.34 / *An optical-nulling spectrophotometer.*

7.2.4. When null balancing is used in the measurement system of a single-beam spectrophotometer, the following steps are required to make an absorption measurement:

1. With the shutter closed, one end of a linear potentiometer is established as 0 percent transmittance (%T) by adjusting the indicating meter to its null condition with the slidewire contactor at the 0%T position.
2. With blank in the optical path, the shutter is opened and the opposite end of the slidewire is established as 100%T by restoring the indicating meter to its null condition with the slidewire contactor at the 100%T position.
3. The sample is placed in the optical path, the contactor is moved on the slidewire until the indicating meter is returned to the null condition, and the percent transmittance is read directly from the potentiometer scale.

Since measurements are always made with the signal level adjusted to the same value, linearity is not required in the amplifier and meter used in the null-indicating circuit. The slidewire must, of course, be linear, but this is rather readily achieved.

Because manual null balancing is both tedious and time consuming and because present-day electronics are capable of achieving the required degree of linearity for direct-presentation measurement, manual null-balancing spectrophotometers are rarely used at present. However, most double-beam instruments do make use of a null-balancing measurement system. The electromechanical systems used in double-beam spectrophotometers have been discussed in Sections 7.2.5 and 8.4.1. Double-beam null-balancing spectrophotometers are of two types: optical nulling and electronic nulling. Infrared instruments are almost invariably optical nulling, and UV-visible instruments are usually electronic nulling. Electronic nulling is preferred, but the long response time of IR detectors makes this approach difficult to apply to IR measurements and dictates the use of optical nulling.

An optical-nulling spectrophotometer is diagrammed in Fig. 11.34. As the chopper rotates, radiation from the sample and reference beams is alternately passed through the monochromator and allowed to fall on the detector. If

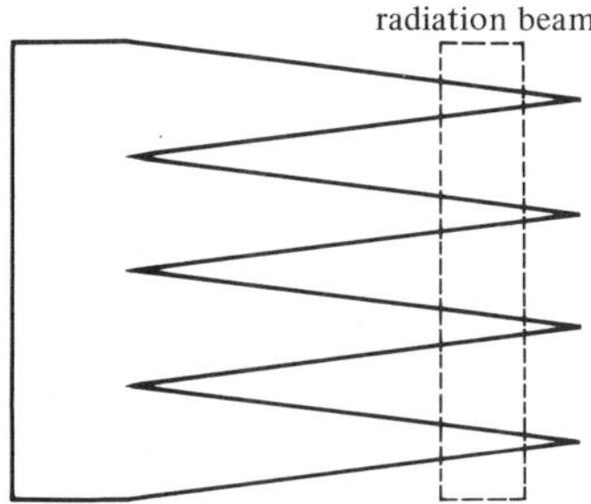

Figure 11.35 / *Comb-type beam attenuator.*

the intensity is the same in both beams, then the output of the detector is DC and no signal is present to drive the servo. If absorption occurs in the sample beam so that the reference beam is of greater intensity, then an AC signal will be presented to the amplifier and the servo will drive the optical wedge into the reference beam until the intensities again balance. (The operation of servo mechanisms is described in Section 7.2.5.) If the sample beam becomes more intense than the reference beam, an AC signal will again be produced, but it will be opposite in phase to that produced when the reference beam is of greater intensity. The servo will respond to this signal by withdrawing the optical wedge until the beam intensities match. The optical wedge is linked to the recorder pen, so that the position of the wedge is recorded. The wedge position is directly related to the ratio of transmitted intensities of sample and reference. An optical wedge usually takes a comb-like form such as that shown in Fig. 11.35. The wedge is constructed of several evenly tapered fingers to minimize inhomogeneity in the radiation beam. It must be carefully constructed so that the degree of attenuation of the light beam is directly proportional to the distance the wedge is inserted in the beam.

An electronic-nulling spectrophotometer is diagrammed in Fig. 11.36. The chopper is so constructed that reference and sample beams, separated by a dark interval, are alternately passed to the detector. When reference-beam radiation is incident on the detector, relay K1 is closed so that a voltage proportional to the reference signal, as attentuated by the measuring potentiometer, is stored on the capacitor. When sample-beam radiation is incident on the detector, relay K2 is closed so that a voltage proportional to the sample signal is stored on the capacitor. Relay K3 is used to sample alternately the reference and sample signals. If they are unequal, an AC signal is produced, which drives the servo mechanism to readjust the measuring potentiometer and the recorder pen until the signals are equal. The synchronization of the operation of K1 and K2 with the chopper can readily be achieved by having the switches that control the operation of the relays operated by the chopper or another disc on the same shaft.

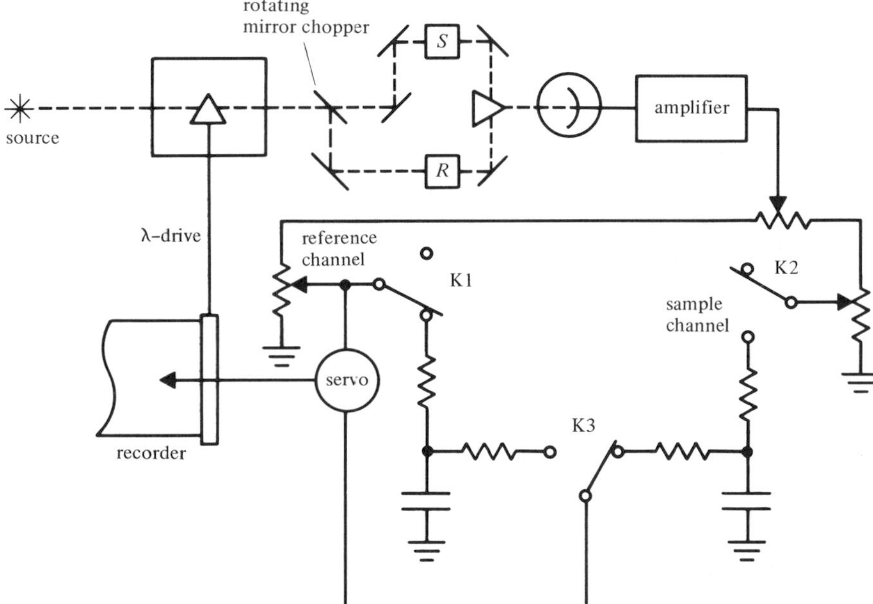

Figure 11.36 / *An electronic-nulling spectrophotometer.*

11.5 / SPECTROMETER PERFORMANCE

In spectrometry, as in all other instrumental methods of analysis, frequent tests of instrument performance are required to ensure valid measurements. In this section, we shall consider means of testing several aspects of spectrometer performance, that is, wavelength calibration, accuracy of the intensity or transmittance scale, resolution, and stray light.

11.5.1 / Wavelength calibration

For most chemical-analysis applications, highly accurate wavelength calibration is not required. In the UV-visible region, almost any source providing a known line spectrum will suffice for wavelength calibration. Perhaps the most widely used such source is the low-pressure mercury arc, which is available in a variety of inexpensive and convenient forms. Table 11.4 lists the wavelengths of the more useful mercury lines. The several visible lines are convenient as a visual check on wavelength. For any spectrometer that makes use of a hydrogen lamp source, the hydrogen line spectrum available from this source is a convenient means of wavelength calibration. The three well-separated lines at 3799, 4861, and 6563 Å should be detectable without great difficulty. A deuterium lamp gives the same lines at slightly longer wavelengths (about 1 part in 2000), but the difference is usually not significant for routine analytical purposes. In double-beam absorption spectrophotometers, solid standards containing a rare earth oxide

TABLE 11.4 / *Wavelengths (in air) of mercury lines useful in the calibration of ultraviolet-visible spectrometers*

λ_{air}(Å)	λ_{air}(Å)	λ_{air}(Å)
2536.52	3131.83	4077.83
2967.28	3650.15	4358.35 (blue)
3021.50	3654.83	5460.74 (green)
3125.66	3663.28	5769.59 (yellow)
3131.55	4046.46 (purple)	5790.65

or mixture of oxides fused into silica are generally more convenient. Holmium oxide and didymium (a mixture of rare earth oxides) filters are usually available from spectrophotometer manufacturers in a form that allows insertion directly into the sample holder. The absorption lines are reasonably sharp and distributed across the UV-visible region. In the IR region, standard absorption spectra must be used for wavelength calibration. Figure 11.37 shows the absorption spectrum of a polystyrene film with wavelengths marked. In using wavelength standards, it is usually sufficient to compare each standard wavelength, λ_s, with the instrument reading, λ_0, at which the standard line appears, and to plot $\Delta\lambda = \lambda_s - \lambda_0$ against λ_0. If there is a systematic error, a smooth curve drawn through the points may be used as a correction curve.

11.5.2 / Intensity and transmittance calibration

The measurement of absolute intensities is seldom required in spectroscopy for analytical purposes. When required, it is usually accomplished by calibration of the spectrometer intensity scale with a tungsten lamp source that has been standardized against a carefully designed blackbody source. Blackbody intensity standards are maintained by the National Bureau of Standards.

In the more usual case in which relative intensity measurements are required for either emission or absorption measurements, a variety of means are available for calibration. All the calibration techniques to be described depend on attenuation of the source intensity by a known amount, and before describing the techniques it is appropriate to recall some pertinent facts concerning the absorption of radiation. When radiation is incident on an absorbing solution in which the concentration of absorbers is c, in grams per liter, and the path length through the solution is b, in centimeters, then the incident and transmitted intensities are related by the expression

$$I = I^0 10^{-abc} \tag{11.34}$$

where a is a constant characteristic of the absorber, for a given wavelength, and is known as the *absorptivity*. The ratio I/I^0 is the transmittance, and the quantity $\log I^0/I$ is the absorbance. The relationship between absorbance,

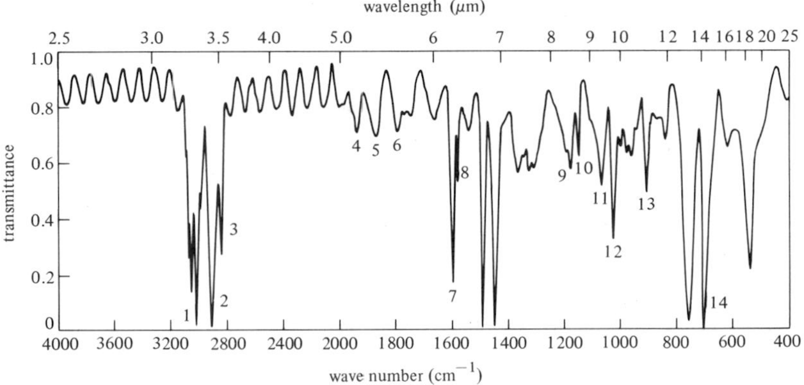

Instrument: Beckman IR-9

Scanning Speed: 100 cm^{-1}/min from 400 to 2000 cm^{-1}; 200 cm^{-1}/min from 2000 to 4000 cm^{-1}

Slit Width: Select (3 X Standard)

Sprectral Slit Width: 1.7 to 2.7 cm^{-1}

Gain: 2 percent

Period: 2

Supression: None

Sample Thickness: 40 μm

Band Number	Wavelength, Air (μm)	Wave number, Vacuum (cm^{-1})	Band Number	Wavelength, Air (μm)	Wave number, Vacuum (cm^{-1})
1	3.3026	3027.1	8	6.3150	1583.1
2	3.4190	2924	9	8.4622	1181.4
3	3.5070	2850.7	10	8.6609	1154.3
4	5.1426	1944.0	11	9.3511	1069.1
5	5.3433	1871.0	12	9.7250	1028.0
6	5.5491	1801.6	13	11.026	906.7
7	6.2428	1601.4	14	14.304	698.9

Figure 11.37 / *Spectrum of polystyrene showing reference wavelengths in the IR region.* SOURCE: "Tables of Wavenumbers for the Calibration of Infrared Spectrometers," *Pure and Applied Chemistry* **1**, 684(1961).

path length, and concentration (Beer's law) usually takes the form

$$A = abc \tag{11.35}$$

From Eq. (11.35), it can be seen that in the simplest situation, in which only a test of spectrometer readout linearity is required, any convenient absorbing solution may be employed in cells of several thicknesses. A plot of absorbance against cell thickness will be linear if the readout system is linear. It is better to vary the path length rather than the concentration of the absorbing solution, since changes in concentration frequently lead to equilibria shifts which complicate interpretation of the measurements.

The usefulness of the above measurement depends on an accurate knowledge of the path length, b. The cuvettes employed for UV-visible measurements are usually of sufficient size that the measurement of b presents no difficulty, and one can accept the stated value for the path length. However, in most measurements two cuvettes are required, one for sample

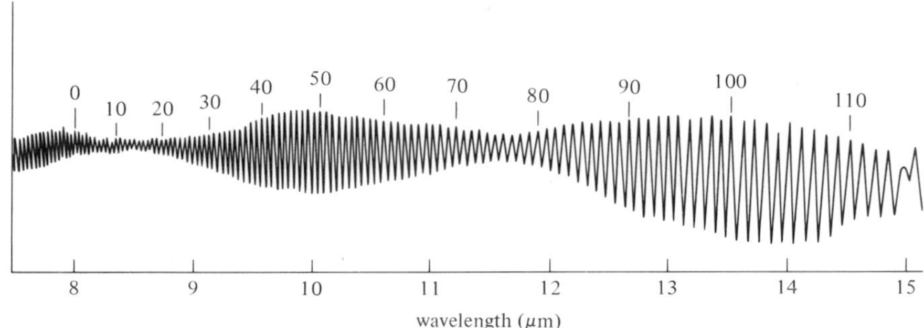

Figure 11.38 / *Interference pattern for measurement of cell thickness. In this case,* 105 *fringes between* 8 μm *and* 14 μm (1250 cm^{-1} *and* 714 cm^{-1}) *gives* b = 0.098 cm.

and one for blank, and it is not unusual to find that cells of nearly identical dimensions give noticeably different absorption readings. There are several possible reasons for this, including minor optical imperfections that lead to the light beam striking at different points on the detector surface (which is seldom perfectly uniform in response across the whole surface) and visually imperceptible "UV dirt" on the face of the cuvettes. The simplest remedy is to interchange sample and blank and their respective cuvettes after the first reading, repeat the reading, and take the arithmetic mean of the two readings.

For absorption cells of thicknesses less than 1 mm, such as those frequently employed in IR measurements, the accuracy of direct measurement is low, and the best approach is to calculate the cell thickness from the interference effect observed when the spectrum is scanned with an empty cell. Figure 11.38 shows a typical interference pattern. The path length, b, through the cell is given by

$$2b = \frac{m}{\Delta\bar{\nu}} \tag{11.36}$$

where m is the number of peaks in the wavenumber interval $\Delta\bar{\nu}$.

In the UV-visible region, several solutions can serve as acceptable standards when calibration of a spectrometer readout in terms of actual absorbance or transmittance values is required. The two most widely accepted are K_2CrO_4 and $K_2Cr_2O_7$. The K_2CrO_4 solution is prepared by dissolving approximately 35 mg of accurately weighed K_2CrO_4 and approximately 2 g of KOH in water and diluting to 1 liter. The accepted value of the absorptivity at 373 nm is 24.85 $cm^{-1}(g/liter)^{-1}$. The $K_2Cr_2O_7$ solution is prepared by dissolving 50–60 mg of accurately weighed dichromate, adding 0.3–0.4 ml of concentrated sulfuric acid, and diluting to 1 liter. The accepted values of absorptivity at 235, 257, 313, and 350 nm are, respectively, 12.5, 14.5, 4.9, and 10.7 $cm^{-1}(g/liter)^{-1}$.

The best and most convenient absorbance standards for the UV-visible region are neutral density filters. Neutral-density filters are prepared by vacuum evaporation of a thin metal film on a substrate of glass or fused silica. A neutral-density filter shows approximately the same transmittance at all wavelengths within a wide range, the range determined in most cases by the optical properties of the substrate. The transmittance of a neutral-density filter is determined by the coverage achieved by the metal film, and filters are commercially available in transmittances ranging from approximately 0.01 percent to 50 percent.

11.5.3 / Stray light measurement

Stray light is radiation at wavelengths other than the intended wavelengths that reaches the detector in a spectrometer. We do not intend to include as stray that radiation which lies within the normal spread of wavelengths due to the finite bandpass of the monochromator. Stray light arises from scattering and random reflections within the monochromator and usually manifests itself only at the extremes of the overall spectral response of the instrument. For example, when a tungsten source is used for measurements in the near ultraviolet, stray light may strongly affect the measurements because the intensity of the source at the measurement wavelength is so low compared to the intensity of potentially stray radiation available at longer wavelengths. In the same fashion, when measurements are carried out in the near infrared with a photoelectric detector, the likelihood of an appreciable effect due to stray light is greatly increased due to the very much greater sensitivity of the detector for radiation of wavelengths shorter than the measurement wavelength. The effect of stray light on absorption measurements is described in greater detail in Chapter 16.

The amount of stray light present at a given wavelength is usually expressed as a percentage of the available intensity at the measurement wavelength. Thus, 1 percent stray light indicates that $I_s = 0.01I^0$, where I_s is the intensity of the stray radiation and I^0 is the intensity of the radiation transmitted through the reference at the measurement wavelength.

The basis of the measurement of stray light in a spectrophotometer may be understood by considering the effect of stray light on a measured transmittance. The measured transmittance, T_m, in the presence of stray radiation is given by

$$T_m = \frac{I + I_s}{I^0 + I_s} \tag{11.37}$$

If it is assumed that the sample is transparent to stray radiation, and further that $I^0 \gg I_s$ (greater than 1 percent stray light would be unusual), then

$$T_m = \frac{I + SI^0}{I^0} \tag{11.38}$$

TABLE 11.5 / *Filters for measurement of stray light*

Filter	*Approximate measurement wavelength* (nm)
H_2O, 1 mm	170
KCl, 2 *M*, 1 mm	190
NaBr, 10^{-3} *M*, 10 mm	200
NaBr, 0.1 *M*, 10 mm	210
NaI, 1 *M*, 10 mm	240
Corning 0–54	285
acetone, 10 mm	310
Corning 0–52	330
Corning 3–73	380
Chance OY–18	410
Jena OG–1	500
Jena RG–2	580
Jena RG–8	650
Corning 4–96	680
silicon	900
Chance ON–22	1100
germanium	1600

SOURCE: W. Slavin, *Anal. Chem.* **35,** 561 (1963).

where S is the fraction of stray radiation. If a sample is chosen that is completely opaque at the measurement wavelength, that is, $I = 0$, then the measured transmittance is equal to S, and this is indeed a suitable technique for measurement of stray radiation. Scale expansion, as described in Chapter 16, is usually required for stray light measurements. Filters (solid and solution) suited to stray light measurement for UV-visible spectrophotometers are listed in Table 11.5.

Stray light in excess of 1 percent is unacceptably high for almost any purpose and calls for remedial action, which usually is the use of a different source–detector combination or insertion of a filter to eliminate a large region of potentially stray radiation.

In the IR region, practically all the stray radiation is at wavelengths shorter than the measurement wavelength, because IR measurements are made on the long-wavelength side of the intensity peak of the hot-body emitter sources (see Fig. 11.28). LiF, CaF_2, NaCl, and KBr will pass all frequencies from their IR cutoff (see Table 11.1) through the visible and ultraviolet and thus are suitable filters for stray light measurements. Compounds with very strong absorption bands may also be used for stray light measurements. For example, measurement of the amount of energy passed in the 700-cm^{-1} absorption band of polystyrene film will give a reasonable estimate of stray radiation. This band transmits less than 0.01 percent in a film of 0.05-mm thickness.

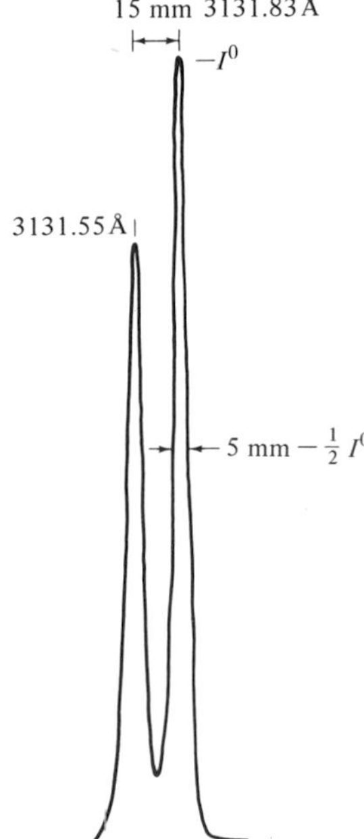

Figure 11.39 / *Recorder tracing obtained in the second order of a* 0.5-m *plane grating spectrometer for* Hg 3131.55 *and* Hg 3131.83 Å *lines.*

11.5.4 / Measurement of half-intensity spectral bandwidth

A method of measuring the half-intensity spectral bandwidth of a monochromator is suggested by the discussion of the term in Section 11.2.4. If the monochromator is illuminated with monochromatic radiation and the intensity as a function of the wavelength setting of the monochromator is measured, then the half-intensity spectral bandwidth may be measured from the resulting I vs. λ curve, as shown in Fig. 11.22. No real source produces monochromatic radiation, but many sources produce atomic lines that have half-intensity widths much less than the bandwidth of a monochromator, and such lines may be considered monochromatic for the purposes of the bandwidth measurement.

The procedure for bandwidth measurement is illustrated in Fig. 11.39, which shows the recorder tracing obtained in the second order of a 0.5-m plane grating spectrometer for a pair of Hg lines from a low-pressure source. The half-width of the larger peak is 5 mm. The separation of the two lines provides a means of converting the measured half-width to wavelength units. The two lines have a wavelength separation of 0.28 Å and show a separation of 15 mm on the chart. Thus, the conversion factor is 0.28/15 = 0.019 Å/mm. The measured half-intensity bandwidth is thus $5 \times 0.019 \simeq$ 0.10 Å.

11.6 / INTERFEROMETERS AND FOURIER TRANSFORM SPECTROSCOPY

Interferometers provide an alternative to dispersion spectrometers in the measurement of emission and absorption spectra. One type of interferometer has already been described, that is, the Fabry-Perot interference filter. The Fabry-Perot interferometer is similar in design and operation to the interference filter, but provision is made for varying the optical spacing between the reflecting surfaces, either by moving the plates mechanically or by varying the refractive index of the dielectric material between the plates. Interferometers of this type have for many years been used in very high resolution studies of very limited portions of atomic spectra, principally in the study of isotope effects and linewidth measurements. The discussion of specialized measurements of this type is not appropriate to a general text and will not be presented here. However, a second type of interferometer has in recent years come into widespread use in the technique known as *Fourier transform spectroscopy*, and in this section we shall briefly describe the instrumentation required for Fourier transform spectroscopy in the optical region of the electromagnetic spectrum. To date, nearly all optical Fourier transform spectroscopic measurements have been confined to the IR region, and Fourier transform spectrometers have almost exclusively made use of Michelson interferometers. Hence our attention will be devoted chiefly to the IR region and to instruments based on the Michelson interferometer. The basis of Fourier transform spectroscopy has been described in Section 6.4.4.

11.6.1 / The Michelson interferometer

The arrangement of the basic components of a Michelson interferometer is shown in Fig. 11.40. Radiation that enters the interferometer is split into two beams. Beam A follows a path of fixed distance before returning to the beam splitter, but the distance traveled by beam B before recombination with A at the beam splitter can be varied. When beams A and B are recombined, an interference pattern is produced. The detector is placed so that radiation in the central image of the interference pattern will be incident upon it; with this condition, maximum intensity will reach the detector when the radiation in the two beams is in phase at the beam splitter, and minimum intensity will reach the detector when radiation in the two beams differ in phase by 180° at the beam splitter.

Imagine that monochromatic radiation of wavelength λ enters the interferometer. If the path distances of beams A and B are initially identical, then A and B will be in phase at the beam splitter and the detector output will be at its maximum value. If mirror $M2$ is displaced by a distance $\lambda/4$, then beam B will travel a distance that differs by $\lambda/2$ from the distance traveled by beam A, and on recombination of A and B at the beam splitter destructive interference will occur and the detector output will be at its minimum value.

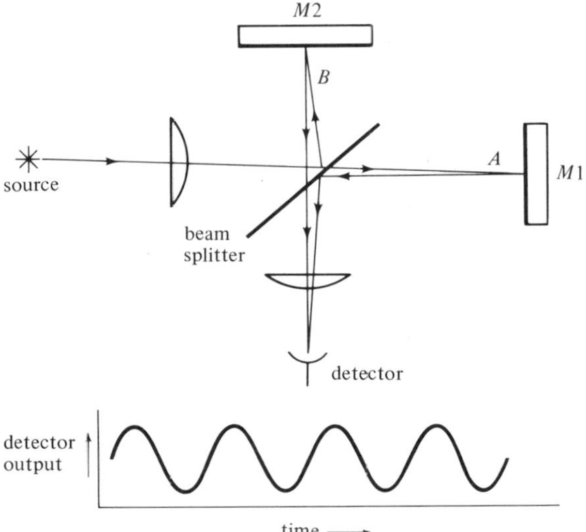

Figure 11.40 / *Basic components of a Michelson interferometer. Light paths are shown off axis for clarity.*

If $M2$ is further displaced so that the paths of A and B differ by λ, a maximum detector signal will again be noted. Such a maximum will be observed whenever the paths differ by an integer multiple of λ. If provision is made for uniform translation of $M2$, the detector output as a function of time will be as shown in Fig. 11.40. The frequency of the detector signal is determined by the translational velocity of $M2$ and the wavelength of the monochromatic radiation, that is,

$$f = \frac{v}{(\lambda/2)} = 2v\bar{\nu} \tag{11.39}$$

A straightforward calculation shows that convenient signal frequencies are obtained if the mirror velocity is in the range of 0.1 cm-sec^{-1}. If $v = 0.1$ cm-sec^{-1}, $f = 200$ Hz for radiation of 10-μm wavelength ($\nu = 3 \times 10^{13}$ Hz, $\bar{\nu} = 1000$ cm^{-1}), and $f = 400$ Hz for radiation of 5-μm wavelength ($\nu = 6 \times 10^{13}$ Hz, $\bar{\nu} = 2000$ cm^{-1}). The amplitude of the detector signal is proportional to the intensity of the incoming monochromatic radiation.

If the incoming radiation is polychromatic, it should be apparent from Eq. (11.39) that the interferometer has the effect of transforming each frequency component so that a detector output wave of unique frequency is produced for each component. The overall detector output as a function of time, the interferogram, will be the sum of the waves for each frequency component and will have a form such as that shown in Fig. 11.41. The central spike occurs when $M1$ and $M2$ are equidistant from the beam splitter, since with this condition all components will be in phase at the beam splitter.

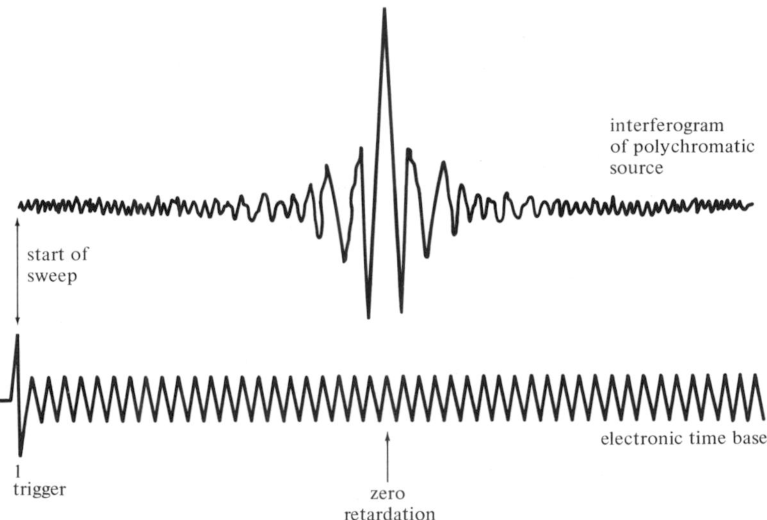

Figure 11.41 / *Interferogram of polychromatic source.*

The interferogram contains all the information associated with a conventional spectrum, but it is the Fourier transform of the spectrum. All the spectral information can be extracted by performing the appropriate inverse Fourier transformation, as described in Section 6.4.4. Such a transformation is always carried out with the assistance of a digital computer.

11.6.2 / The Fourier transform spectrometer

The components of a typical Fourier transform spectrometric system are shown in Fig. 11.42. A glower source is the usual choice for the IR region. The pyroelectric detector is shown, although other detectors have been employed, because its wide spectral response and rapid response time make it particularly well suited to the requirements of Fourier transform spectroscopy. Most systems include a small computer directly interfaced to the spectrometer to control the scan system and carry out the mathematical transformation.

Perhaps the most critical part of the Fourier transform system is the mirror scan mechanism. In order faithfully to reproduce the spectrum from the interferogram, the detector output must be accurately known as a function of mirror displacement. It is easy to see that the difficulty of accurately measuring mirror displacement increases as the wavelength of interest decreases. In the far IR region, the radiation wavelengths are of the order of hundreds of micrometers, and mirror movements of fractions of a wavelength are readily obtained with conventional mechanical devices. At shorter wavelengths, another approach has been taken to achieve sufficient accuracy, that is, the fringe-referencing technique shown schematically in

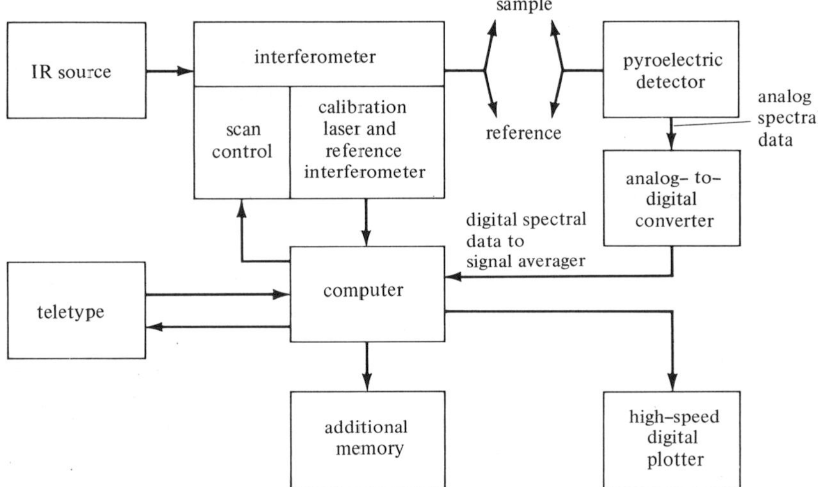

Figure 11.42 / *Schematic of a typical Fourier transform spectrometer.*

Fig. 11.43. One transducer is used to drive simultaneously the movable mirrors of two interferometers through identical displacements. Sample radiation is passed through one interferometer and produces the sample interferogram. Monochromatic light from a laser is passed through the second interferometer and produces a periodic wave like that of Fig. 11.40,

Figure 11.43 / *Fringe-referencing system.*

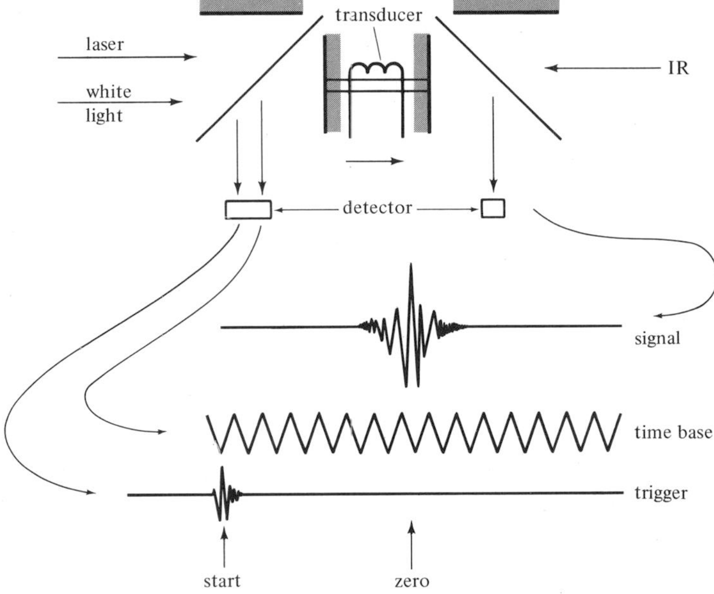

the laser interferogram. White light passed through the laser interferometer will produce a detector output spike when the light paths of the two beams in the reference interferometer are identical. The two interferometers are aligned so that the white light spike occurs at the beginning of a scan. Thus, the mirror displacement in the sample interferometer may be determined by counting the number of peaks in the laser interferogram since the white light detector spike; a strictly linear mirror drive is therefore not required.

11.6.3 / Resolution

The resolution obtained in an interferometer depends on (1) the optical quality of the components, and, especially, the planarity of the mirrors; (2) the degree of collimation of the radiation beam; and (3) the length of the mirror sweep. Of the three, the last is most important in determining the attainable resolution. Truly monochromatic radiation passed through an interferometer would produce an infinite cosine wave if the mirror travel were infinitely long. The Fourier transform of such a wave would be a spike of zero width. Since only finite mirror travel is possible, the duration of the cosine wave is finite, and the waveform is truncated. The Fourier transform of the truncated waveform will be a spike centered at the same frequency but with nonzero width. The width of the spike is inversely related to the length of the truncated waveform and hence to the length of mirror travel. Some commercial instruments employing fringe-referencing provide for mirror travel of up to 5 cm and resolution better than 0.1 cm^{-1}. Further improvement in resolution seems likely.

11.6.4 / Advantages

Since an interferometer does not require narrow slits but rather has an entrance aperture that is typically 1 cm or more in diameter, the energy throughput of a Fourier transform spectrometer is quite large compared to a dispersion spectrometer. This can be quite advantageous in energy-limited situations and probably accounts in part for the early application of Fourier transform spectroscopy to far IR measurements and spectroscopic measurements of distant stellar sources.

A somewhat more subtle advantage of Fourier transform spectroscopy is that known as *Fellgett's advantage.* In a dispersion spectrometer, the resolution elements are scanned sequentially across the detector, and, if there are m resolution elements, the intensity of each is measured for only the fraction t/m of the total time, t, required to scan the spectrum. With an interferometer, radiation from all regions of the spectrum is incident on the detector throughout the whole scan time, and it can be shown that this results in an improvement in the signal-to-noise ratio by a factor of $m^{1/2}$ compared to that obtained with a dispersion spectrometer, provided that a background-limited detector is employed and both instruments have the same throughput, employ the same detector, and operate for the same length of time. It should be evident that a corollary of the preceding statement is that the measurement time with an interferometric instrument may be reduced by a factor of $1/m^{1/2}$, compared to that required with a dispersion spectrometer, without reduction of the signal-to-noise ratio.

Infrared thermal detectors are background limited, that is, the noise does not increase as the intensity of incident radiation increases, and Fellgett's advantage is realized in IR Fourier transform spectrometers. With photoemissive detectors, except at very low intensities, the noise increases as the square root of the incident intensity, and if m resolution elements reach the detector instead of only one, the noise increases as $m^{1/2}$ and Fellgett's advantage is not obtained.

To obtain a spectrum by the Fourier transform technique requires a great deal more effort than the conventional dispersion technique. For this reason, one would not undertake Fourier transform spectroscopy unless a very significant advantage could be realized. At present, the chief areas of application are in the far infrared, in IR emission studies of weak emitters, in absorption studies with remote sources, and in rapid-scanning measurements.

SUGGESTED READINGS

J. F. James and R. S. Sternberg, *The Design of Optical Spectrometers*, Chapman and Hall (London, 1969).

R. A. Sawyer, *Experimental Spectroscopy*, 3rd ed., Dover (New York, 1963).

R. P. Bauman, *Absorption Spectroscopy*, Wiley (New York, 1962).

J. R. Edisbury, *Practical Hints on Absorption Spectrometry*, Plenum (New York, 1967).

ASTM Committee E-13, *Manual on Recommended Practices in Spectrophotometry*, 3rd ed., American Society for Testing and Materials (Philadelphia, 1969).

M. J. D. Low, "Fourier Transform Spectrometers," *J. Chem. Ed.* **47**, A163, A255, A349, A415 (1970).

G. Horlick, "Fourier Transform Approaches to Spectroscopy," *Anal. Chem.* **43** (8), 61A (1971).

PROBLEMS

1 / An interference filter was constructed with a dielectric layer of exactly 0.500-μm thickness and a refractive index of 1.500. What visible and UV wavelengths would be transmitted by this filter?

2 / For the prism shown in Fig. 11.44, calculate the angular separation of the refracted rays if the incident beam is made up of radiation of two wavelengths for which the prism material exhibits refractive indices of 1.5 and 1.6, respectively.

Radiation is incident at an angle of 5° on a grating of 6000 grooves/cm.

3 / Calculate the angles at which 300-nm radiation will be diffracted in the first three orders.

4 / Calculate the angles at which 200-, 300-, 400-, 500-, and 600-nm radiation will be diffracted in the first order.

A certain grating employed in a Littrow arrangement has 750 grooves/cm and a blaze angle of 26°45′.

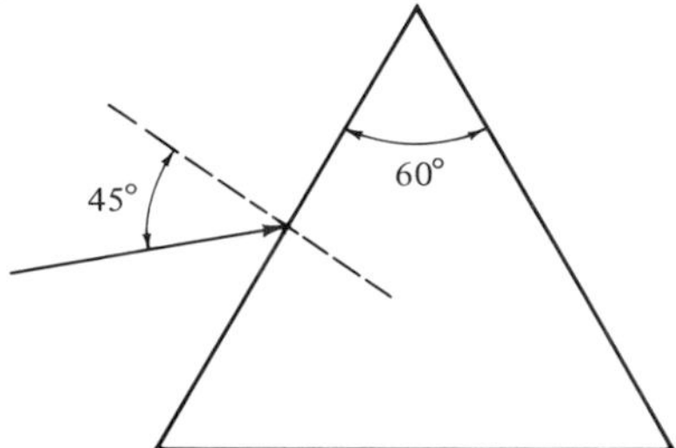

Figure 11.44

5 / Calculate the wavelengths that will be observed at the blaze angle in the first five orders.

6 / In what order will the mercury green line (546.1 nm) be observed closest to the blaze angle?

A Czerny-Turner spectrometer having the optical arrangement shown in Fig. 11.45 is equipped with a 1200-grooves/mm grating.

7 / What wavelengths will appear at the exit slit in the first two orders if the light beam is incident on the grating at an angle of 15°?

8 / If the grating has a blaze angle of 22°, for what wavelength in the first order is it blazed?

The manufacturer's description of a monochromator reads as follows:

type of mount—Littrow
aperture ratio—*f*/6.8
focal length—350 mm
reciprocal dispersion—20 Å/mm at exit slit
grating—precision plane grating replica, 48 mm × 48 mm ruled area, 1180 lines/mm, blazed for 6000 Å in first order

9 / Indicate on a diagram of the optical system the dimension that corresponds to the focal length.

10 / Describe the significance of the term "aperture ratio" and calculate the size of the limiting aperture. How is this size related to the grating size?

11 / Calculate the theoretical half-intensity spectral bandpass of this instrument with slits of 50-μm width.

Figure 11.45

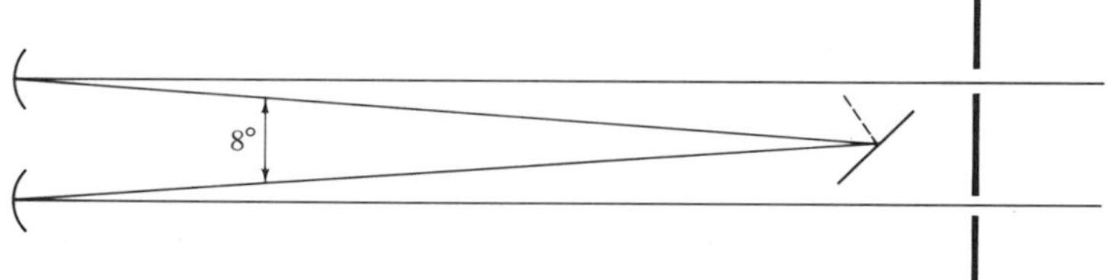

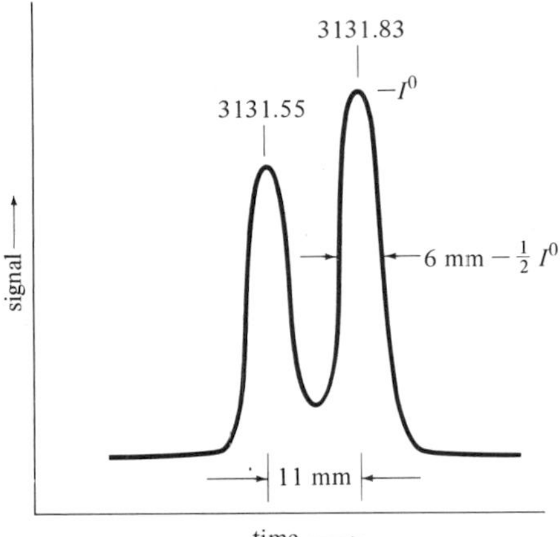

Figure 11.46

12 / Calculate the theoretical resolving power of the grating in the first order.

13 / In what order will Hg 2537 Å radiation be diffracted with greatest efficiency? Justify your answer.

14 / What is the blaze angle of the grating?

15 / If a grating is blazed for 750 nm in the first order, at what first-order wavelengths should it be set in order to transmit each of the following lines with maximum efficiency: 750, 375, 250 nm?

16 / If a grating is blazed for 450 nm in the first order, at what first-order wavelengths should it be set in order to transmit each of the following lines with maximum intensity: 750, 375, 250 nm?

17 / If a grating spectrometer has a resolving power of 20,000 at 400 nm, what must be the minimum separation between two lines in order that the lines may be just resolved?

18 / A recorder tracing of the Hg 3131.55, 3131.83 Å doublet obtained with a 0.5-m, Ebert mount, plane grating spectrometer with 10-μm slits, operated in the second order, is shown in Fig. 11.46. If the doublet is said to be just resolved, calculate:
(*a*) the resolving power of the spectrometer in the second order;
(*b*) the spectral bandwidth of the spectrometer in the second order;
(*c*) the approximate reciprocal linear dispersion of the spectrometer in the first order.

19 / What is the theoretical spectral bandpass of a monochromator for which the reciprocal linear dispersion is stated to be 15 Å/mm when slits of 20 μm are used? What is the actual bandpass if lines at wavelengths 4040.10 and 4040.60 Å are found on a recorder output to have a separation between centers of 10 mm and a half-width of 7 mm?

20 / What is the minimum resolving power required to resolve two lines that differ by 0.1 Å and have an approximate wavelength of 4000 Å? Can these lines be resolved with a grating that has 500 grooves/mm and a ruled width of 50 mm?

A certain multiplier phototube has a maximum radiant sensitivity of 120,000 A/W at the anode for a supply voltage of 1000 V. Under the same operating conditions, the dark current is 5 nA.

21 / What is meant by the term "dark current," and to what phenomena can it be attributed?

22 / What is meant by the term "shot noise" as applied to multiplier phototubes?

23 / Calculate the factor by which the signal-to-noise ratio of the output of the multiplier phototube changes if the incident radiant power is increased from 5×10^{-11} W to 1×10^{-10} W.

24 / Calculate the percent stray light necessary to account for an absorbance reading of 1.5 when the true absorbance is 1.7.

chapter 12 / Introduction to atomic spectroscopy

chapter 12

12.1 / INTRODUCTION

Atomic spectra originate in transitions between outer electronic levels of atoms. Thus, as was pointed out in Section 10.5, atomic spectra occur in the UV and visible portions of the electromagnetic spectrum. Since there are no vibrational levels associated with atoms, electronic transitions within atoms give rise to sharp "line" spectra rather than the "band" spectra of molecules.

Atomic spectroscopy is the most ancient of the various branches of spectroscopy. Kirchhoff and Bunsen had by 1860 established the basis of qualitative atomic spectral analysis and elegantly demonstrated its power by the discovery of two previously unknown elements, rubidium and cesium. Moreover, atomic spectroscopy played a central role in the development of our understanding of the electronic structure of the elements, and formed the basis for later developments in molecular spectroscopy. Because of its historical position, many of the concepts and much of the terminology of atomic spectroscopy have been carried forward into other areas, and it is appropriate to begin our consideration of spectroscopic methods with atomic spectroscopy so that we may have the benefit of seeing the origin of the concepts and terms that will recur throughout our discussions.

This chapter develops the principles common to all types of atomic spectroscopy. In the two subsequent chapters, atomic spectroscopy is rather arbitrarily divided according to the method used for production of atoms and excitation of spectra. Chapter 13 deals principally with flame spectroscopic methods, and Chapter 14 with arc/spark emission spectroscopy. This division is justified by the rather considerable differences in instrumentation employed in the two methods, which in turn has led them to develop as somewhat separate disciplines. We should note, however, that the two disciplines have tended to grow toward each other, so that there has been a considerable blurring of the dividing line in recent years.

12.2 / ORIGIN OF ATOMIC SPECTRA

12.2.1 / Historical introduction

During the earliest period in the development of atomic spectroscopy, the wavelengths of lines associated with various elements were measured and both qualitative and quantitative analysis techniques were developed, but attempts at systematic classification of the spectral lines were unsuccessful until Balmer, in 1885, succeeded in describing the wavelengths of the visible lines of the hydrogen spectrum by an equation that is written in modern form as

$$\frac{1}{\lambda} = \bar{\nu} = R\left(\frac{1}{2^2} - \frac{1}{n^2}\right) \tag{12.1}$$

where R is a constant (Rydberg's constant) and n assumes the values 3, 4, 5, 6 for the four visible lines. This formula served as a first example of two important generalizations. Rydberg, in the last decade of the nineteenth century, showed how the simple Balmer formula might be generalized to give an account of series of lines in other elements. Ritz pointed out that when written in the form

$$\bar{\nu} = \frac{R}{{n_1}^2} - \frac{R}{{n_2}^2} \tag{12.2}$$

where n_1 and n_2 are both integers, the Balmer formula states that the wavenumber of any spectral line may be written as the difference of two terms, T' and T'',

$$\bar{\nu} = T'' - T' \tag{12.3}$$

The allowed electronic energy levels, or terms, of an atom, and the lines arising from transitions between these levels can be summarized in a diagram of the type due to Grotrian. The Grotrian diagram for hydrogen is shown in Fig. 12.1. The ordinates are energy scales, and energy levels are drawn as horizontal lines. A transition between two terms is represented by a vertical line, the length of which is a measure of the wavenumber of the emitted line. The energy scale to the right has its zero at the top of the diagram and is in wavenumber units to emphasize the relationship of the diagram to Eqs. (12.1) and (12.2). The H_α line represented by the equation

$$\bar{\nu} = R\left(\frac{1}{2^2} - \frac{1}{3^2}\right)$$

may be used to illustrate this relationship. The H_α line appears in Fig. 12.1 as the difference of two lengths marked A and B, which are in fact the quantities $R/4$ and $R/9$, both measured from a line that is the limit of the term sequence, that is, the value of R/n^2 when $n \to \infty$. This limit represents ionization—the removal of an electron to an indefinitely large distance. The

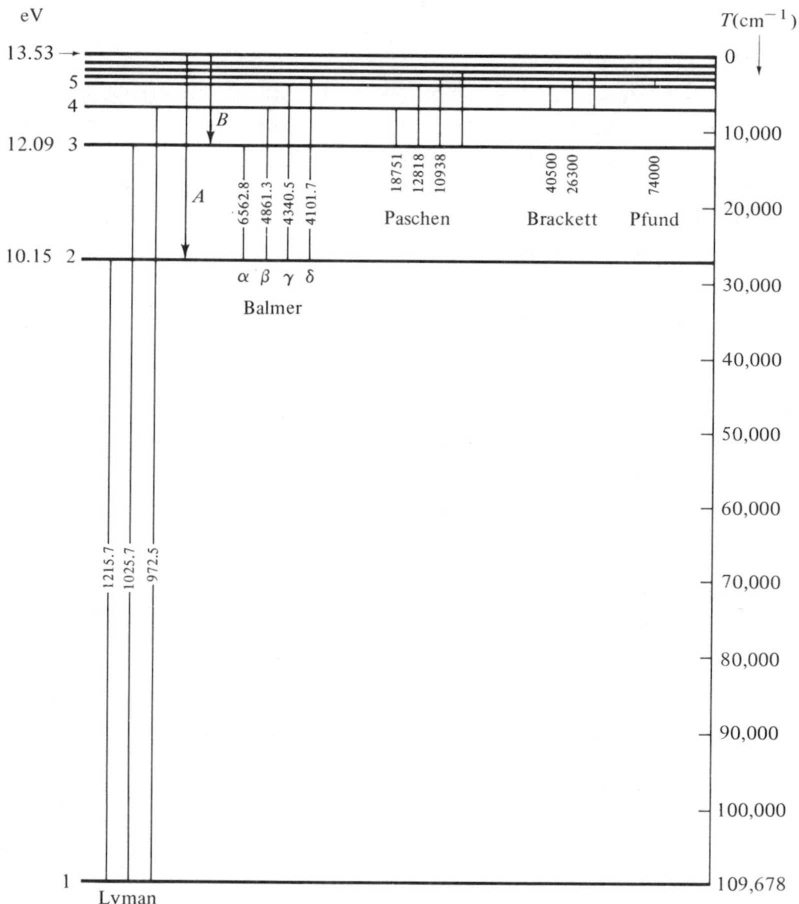

Figure 12.1 / *Grotrian diagram for hydrogen.*
SOURCE: H. G. Kuhn, *Atomic Spectra*, 2nd ed., Academic (New York, 1969), p. 87.

energy scale on the left has as its zero the lowest-lying term, or ground state, and is given in units of electron volts (eV). This electron volt scale may be used directly to obtain the excitation potential of a line, which is a measure of the energy required to excite an atom to emit the line and is customarily given in electron volts. Thus, 10.15 eV must be provided to excite the first line of the Lyman series, the line of lowest excitation potential (sometimes called the resonance line), and 12.09 eV is required to produce the H_α line.

12.2.2 / Quantum numbers

The number n in Eq. (12.1) is a quantum number. It is not, however, the only quantum number associated with electrons, but because it is the most important in describing the energy levels in the hydrogen atom, it is called the *principal quantum number*. There are three other quantum numbers required to characterize a particular electron in an atom. The four quantum numbers may be defined in various ways, but the various definitions all lead to the same result. The most commonly employed set of quantum numbers is n, l, m, and s.

The quantum number l is called the *orbital* or *azimuthal quantum number*. It takes integral values in the range between 0 and $n - 1$. The energy of the electron varies as l varies, and hence the orbital quantum number adds a term to the energy defined through the principal quantum number.

The quantum number m is called the *magnetic quantum number*. It also takes on integral values, and its values can range from $-l$ to $+l$. In the presence of an applied external field, different values of m correspond to different energy contributions to the energy defined by n and l. In the absence of an external field, the various values of m are equivalent in energy, and the states corresponding to the various values are said to be *degenerate*.

The quantum number s is called the *spin quantum number*. It can have either of two values, $+\frac{1}{2}$ or $-\frac{1}{2}$, corresponding to the two possible orientations of the electron spin.

12.2.3 / Electron shells and subshells

Since the electronic energy levels within an atom depend primarily on the principal quantum number n, the electrons are first classified according to the values of this number. We say that electrons with the same value of n belong to the same *shell*, and the shells are designated with a capital letter as shown in Table 12.1. Within a shell, the electrons are further classified according to the value of the orbital quantum number l, as shown in Table 12.2. An electron that belongs to an s subshell is said to be an s electron, an electron in a p subshell, a p electron, and so on.

12.2.4 / The Pauli exclusion principle

The Pauli exclusion principle states that no two or more electrons within an atom can have all quantum numbers the same. Thus, when one applies this principle and the previously enunciated restrictions on the values taken by the quantum numbers, there is only a limited number of electrons that can

TABLE 12.1 / *Symbols of the electron shells*

n =	1	2	3	4	5	6	7
shell =	K	L	M	N	O	P	Q

TABLE 12.2 / *Symbols of the subshells*

l =	0	1	2	3
subshell =	s	p	d	f

TABLE 12.3 / *Building up of K, L, and M shells*

Shell	*Quantum numbers* n	l	m	s	*Max. no. of electrons in subshell*
K	1	0	0	$\pm\frac{1}{2}$	2 *s* electrons
L	2	0	0	$\pm\frac{1}{2}$	2 *s* electrons
		1	−1	$\pm\frac{1}{2}$	6 *p* electrons
			0	$\pm\frac{1}{2}$	
			1	$\pm\frac{1}{2}$	
M	3	0	0	$\pm\frac{1}{2}$	2 *s* electrons
		1	−1	$\pm\frac{1}{2}$	6 *p* electrons
			0	$\pm\frac{1}{2}$	
			1	$\pm\frac{1}{2}$	
		2	−2	$\pm\frac{1}{2}$	10 *d* electrons
			−1	$\pm\frac{1}{2}$	
			0	$\pm\frac{1}{2}$	
			1	$\pm\frac{1}{2}$	
			2	$\pm\frac{1}{2}$	

occupy each shell and subshell. This can be illustrated by considering the first several values of n as shown in Table 12.3.

From a combination of the Pauli principle with the energy characteristics of electrons as determined by their quantum numbers, the ground-state (lowest-energy) configuration of an atom or ion of any number of electrons can be understood. Ground-state configurations for the first 20 elements are shown in Table 12.4. If we recall that in atomic spectroscopy we are principally concerned with transitions of the outer electrons, then there are two important principles that are evident from a consideration of Table 12.4. First, there is a periodicity of outer-electron configuration. For example, Li, Na, and K all exhibit closed shells with one *s* electron in the next outer shell, and Be, Mg, and Ca all exhibit closed shells with two *s* electrons in the next outer shell. The periodicity of electron structure, of course, accounts for the periodicity of chemical properties, and also exhibits itself in a periodicity of spectral properties. Compare, for example, the Grotrian diagrams for Na and K shown in Fig. 12.2. Except for the spacing between levels, the two are clearly identical. (As the number of outer electrons increases, the term diagrams become more complex and similarities in the spectra within chemical families become more difficult to detect.) Second, atoms and ions with the same number of electrons (i.e., isoelectronic species) give closely similar spectra except for a scale factor. Thus, H, He^{+}, and Li^{2+} give similar spectra.

12.2.5 / Spectral terms

The Rydberg-Ritz combination principle [Eq. (12.3)] states that every spectral line arises from a transition between two terms. Modern theory

TABLE 12.4 / *Ground-state electron configuration of the elements*

Nuclear charge	Element	K: 1s	L: 2s	L: 2p	M: 3s	M: 3p	M: 3d	N: 4s	N: 4p	Term symbol
1	H	1								$^2S_{1/2}$
2	He	2								$^1S_0$
3	Li	2	1							$^2S_{1/2}$
4	Be	2	2							$^1S_0$
5	B	2	2	1						$^2P_{1/2}$
6	C	2	2	2						$^3P_0$
7	N	2	2	3						$^4S_{3/2}$
8	O	2	2	4						$^3P_2$
9	F	2	2	5						$^2P_{3/2}$
10	Ne	2	2	6						$^1S_0$
11	Na	2	2	6	1					$^2S_{1/2}$
12	Mg	2	2	6	2					$^1S_0$
13	Al	2	2	6	2	1				$^2P_{1/2}$
14	Si	2	2	6	2	2				$^3P_0$
15	P	2	2	6	2	3				$^4S_{3/2}$
16	S	2	2	6	2	4				$^3P_2$
17	Cl	2	2	6	2	5				$^2P_{3/2}$
18	Ar	2	2	6	2	6				$^1S_0$
19	K	2	2	6	2	6		1		$^2S_{1/2}$
20	Ca	2	2	6	2	6		2		$^1S_0$

associates these terms with particular configurations of the electrons in an atom, that is, particular distributions of the electrons within the available shells and subshells of the atom. In optical spectroscopy, only the electrons in incomplete shells, the outer or valence electrons, need be considered. For example, the ground term of Na corresponds to an electron configuration in which the single outer electron is in the 3*s* subshell; the terms in the vicinity of 2.1 eV correspond to an electron configuration in which the single outer electron is in the 3*p* subshell.

The spectroscopist's shorthand notation for the various terms is the *term symbol.* Term symbols are shown in the last column of Table 12.4 and at the head of each column in the Grotrian diagrams (Fig. 12.2). A term symbol contains the following information:

$$^{2S+1}L_J$$

where L is a quantum number that is arrived at by considering the orbital quantum numbers l_i of all the outer electrons, S is a quantum number derived from the spin quantum numbers s_i of all the outer electrons, and J is a quantum number arrived at by considering the resultant of the orbital angular momentum described by L and the spin angular momentum described by S.

The L quantum number is an integral positive number or zero. It is customary to characterize the values of L by a capital letter, or term symbol,

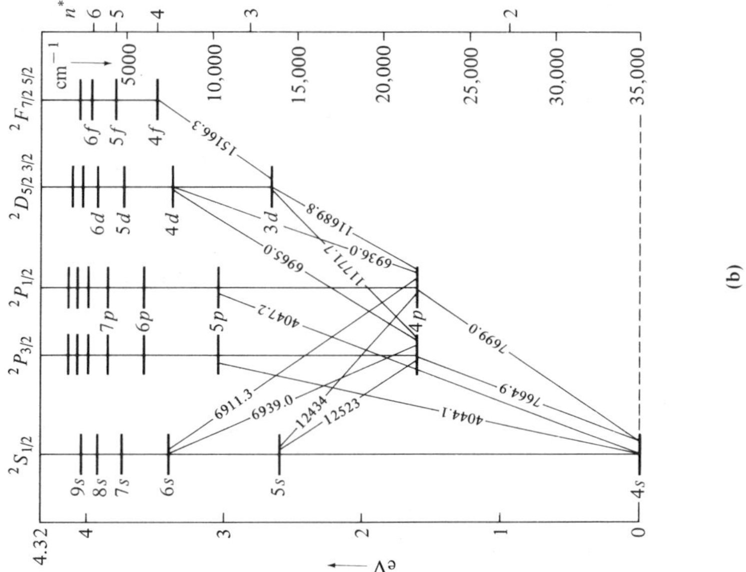

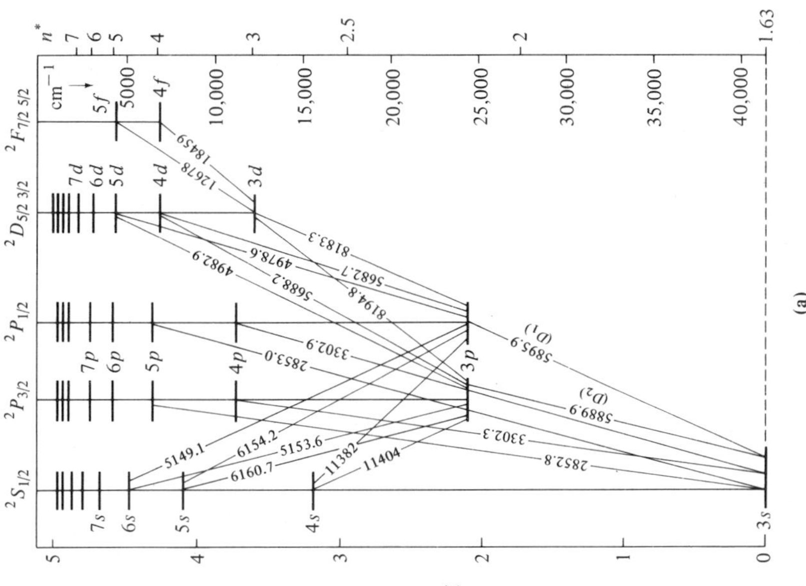

Figure 12.2 / *Grotrian diagrams for* (a) Na *and* (b) K.
SOURCE: H. G. Kuhn, *Atomic Spectra*, 2nd ed., Academic (New York, 1969), pp. 154, 155.

TABLE 12.5 / *Term symbols*

L =	0	1	2	3	4	5	6	7	8
symbol =	S^a	P	D	F	G	H	I	K	L^a

[a] These symbols should not be confused with the resultant orbital quantum number L and resultant spin quantum number S.

as shown in Table 12.5. The quantity $2S + 1$ is called the *multiplicity* of the term, and if $S \leq L$ the number of levels associated with the term is given by the multiplicity. For example, the first excited term of Na is ^2P, and the term diagram does in fact show two levels of slightly different energy in the vicinity of 2.1 eV. The Na ground term, however, is ^2S ($L = 0$, $S = \frac{1}{2}$) and is single. Whenever $L < S$, the number of levels is given by $2L + 1$, but the name multiplicity is reserved for the quantity $2S + 1$. The S quantum number is an integral or half-integral positive number. The quantum number J takes values of $L + S$, $L + S - 1$, $L + S - 2, \ldots, |L - S|$. Thus, J takes integral or half-integral positive values.

12.2.6 / Selection rules

In section 10.2.5, we introduced the concept of selection rules. We wish now to consider the rules applicable to atomic spectra. If we consider a set of energy terms, such as that for Mg in Fig. 12.3, it becomes obvious that not all possible transitions give rise to spectral lines—some transitions are "forbidden." Selection rules, which have been derived from observation and from quantum mechanical calculations, enable one to make a selection of the pairs of states that combine (the "allowed" transitions), and the pairs that do not combine ("forbidden" transitions). Selection rules usually take the form of a statement of the changes in quantum numbers associated with allowed transitions. We shall describe one simple set of selection rules, but before doing so let us hasten to point out that not all selection rules are rigorously valid—"forbidden" transitions sometimes occur.

In terms of the quantum numbers L, S, and J, the selection rules governing allowed transitions are

$$\Delta L = \pm 1 \tag{12.4}$$

$$\Delta S = 0 \tag{12.5}$$

$$\Delta J = \pm 1, 0 \qquad J = 0 \rightarrow J = 0 \text{ excluded} \tag{12.6}$$

Equation (12.4) states that only those transitions are allowed for which the value of L changes by ± 1. Stated in terms of the Grotrian diagram, this means that transitions occur only between adjacent columns. Thus, in Fig. 12.3 we see $S \rightarrow P$ ($L = 0$ to $L = 1$), $P \rightarrow S$ ($L = 1$ to $L = 0$), $D \rightarrow P$

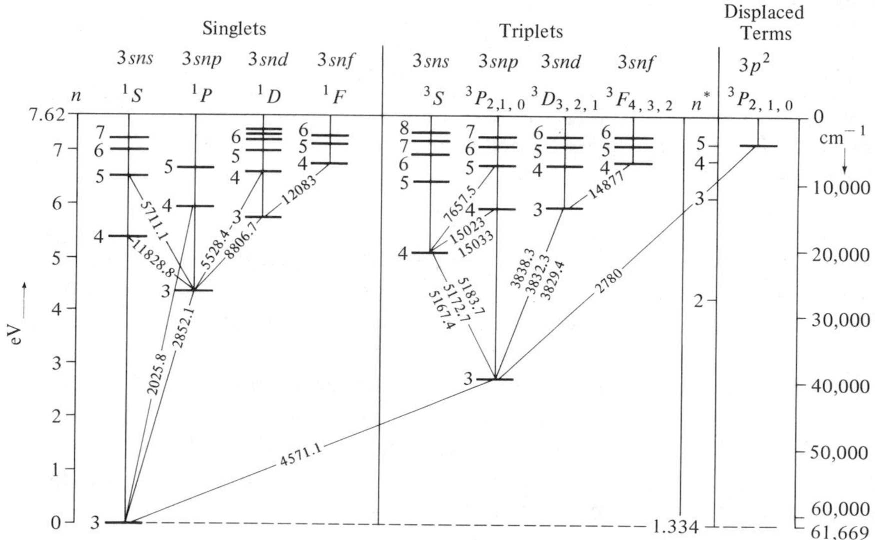

Figure 12.3 / *Grotrian diagram for* Mg.
SOURCE: H. G. Kuhn, *Atomic Spectra*, 2nd ed., Academic (New York, 1969), p. 186.

(L = 2 to L = 1) but no $S \rightarrow D$ ($L = 0$ to $L = 2$) or $F \rightarrow P$ ($L = 3$ to $L = 1$) transitions. This selection rule is applicable only to one-electron transitions. Where two-electron transitions occur, the selection rule must be stated in terms of l values for the individual electrons. The so-called displaced terms of Fig. 12.3 arise from two-electron activity; note that the 2780-Å line occurs in violation of Eq. (12.4), since $\Delta L = 0$ for the transition. Equation (12.5) states that only those transitions are allowed for which no change in S occurs. Thus, transitions are forbidden between terms of different multiplicity. This rule is fairly often violated, the "intercombination" line 4571.1 Å of Fig. 12.3 being an example ($S = 1$ to $S = 0$). The final selection rule states that those transitions are allowed for which J changes by ± 1 or not at all, but that a $J = 0$ to $J = 0$ transition is forbidden. The application of this rule can be seen in the transitions giving rise to the "diffuse series"[1] of the alkali metals as shown in Fig. 12.4. The four states give rise only to a triplet of lines corresponding to ${}^2D_{3/2} \rightarrow {}^2P_{3/2}$ ($\Delta J = 0$), ${}^2D_{3/2} \rightarrow {}^2P_{1/2}$ ($\Delta J = -1$), and ${}^2D_{5/2} \rightarrow {}^2P_{3/2}$ ($\Delta J = -1$). The transition ${}^2D_{5/2} \rightarrow {}^2P_{1/2}$ ($\Delta J = -2$) is not seen, and thus the selection rule is observed. This last selection rule is the most rigorously observed.

[1] The series was called "diffuse" because the triplet structure was unresolved in early measurements, and thus the lines of this series had a "diffuse" appearance. The early names of the Na series—*s*harp ($S \rightarrow P$), *p*rincipal ($P \rightarrow S$), *d*iffuse ($D \rightarrow P$), and *f*undamental ($F \rightarrow D$)—are responsible for the *s, p, d, f* nomenclature for subshells.

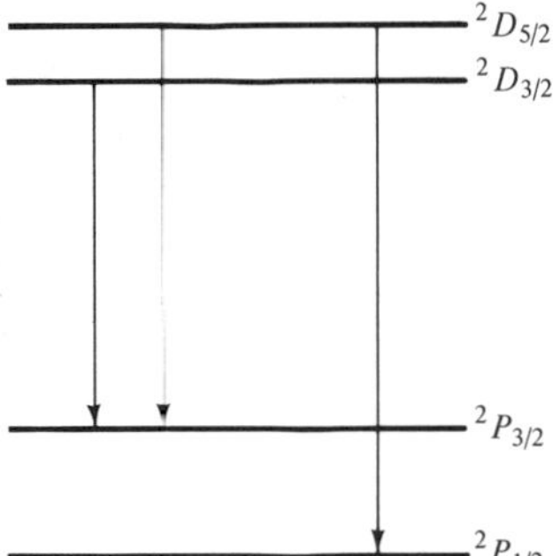

Figure 12.4 / *Origin of "diffuse series" of alkali metals.*

12.3 / FUNDAMENTALS OF ANALYTICAL APPLICATIONS

12.3.1 / Emission, absorption, and fluorescence

In the preceding sections, we have seen that even the simplest atoms are capable of emitting a large number of lines of distinctive wavelength. It should be apparent from this that, in principle, atomic spectroscopy can serve as a convenient means of qualitative elemental analysis, even for rather complex mixtures. In subsequent chapters, we shall describe the practice of qualitative analysis by atomic spectroscopy, but at this point we wish to describe some of the fundamental aspects of *quantitative* analysis by atomic spectroscopy.

There are three types of experiments we can use in analytical atomic spectroscopy—emission, absorption, and fluorescence. The three experiments are indicated in schematic fashion in Fig. 12.5.

In the emission experiment, excited atoms are produced principally by collisions of the ground-state atoms with energetic species—electrons, other atoms, or molecules. This is the type of excitation that occurs in flames and in various types of electrical discharges. In this case, the number of atoms in a given excited state at any time is described primarily by the temperature that characterizes the excitation medium and the excitation energy of the state and does not depend in any way on the selection rules. The length of time that a given atom remains in an excited state before returning to the ground state by emission of a photon is very short, typically 10^{-8} sec for an allowed transition. The selection rules are applicable to the emission process. Analytical information is obtained by detecting the emitted radiation at selected wavelengths.

In the absorption experiment, radiation from an external source is passed through the atomic vapor, and the attenuation of the intensity by absorption of photons of the proper energy is measured. Selection rules apply to the absorption process, and the resonance lines are of chief interest, since under the conditions of most experiments only the ground state is appreciably populated, and the transition to the first excited state usually has the largest

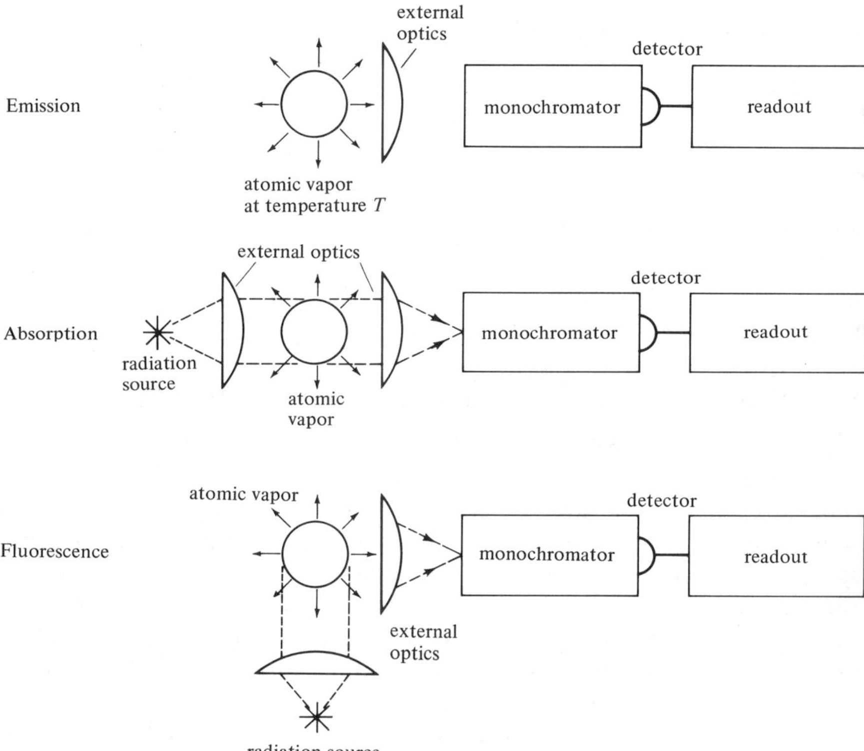

Figure 12.5 / *Schematic representation of emission, absorption, and fluorescence experiments.*

transition probability (is most allowed). Some fraction of the absorbed radiation will be reemitted at the same wavelength, but since it is emitted isotropically, only a small fraction will be collected by the optical system and it will not interfere in the absorption measurement. The attenuation of the incident intensity is usually expressed as a function of the ratio of the transmitted to incident radiation intensity.

In the fluorescence experiment, radiation from an external source is passed through the atomic vapor, photons of the proper energy are absorbed, and some fraction of the absorbed radiation is isotropically reemitted. The intensity of the reemitted radiation is measured. The basic difference between the emission and fluorescence experiments is the mode of excitation. In the fluorescence experiment, selection rules apply to both the absorption and emission processes, which makes its mode of excitation considerably more selective than that of the emission experiment.

12.3.2 / Intensity relationships

Quantitative analysis by atomic spectroscopy depends on measurement of radiation intensity. The basic assumption of spectroscopic methods is that the intensity of radiation emitted or absorbed is proportional to the atomic concentration, or, in equation form,

$$I = KN^x \tag{12.7}$$

where I is the intensity, K is a proportionality constant, N is the atomic concentration, and x is an exponent that is usually near unity. Most analytical applications of spectroscopy do not require actual evaluation of K and x in terms of fundamental processes, but rely instead on comparison of relative intensity values for standards and unknowns to derive the concentrations of the unknowns. A comparative measurement of this type requires only that K and x be identical for the two sets of measurements, standards and unknowns. It is, however, useful to attempt to evaluate K and x, if only for an idealized situation, so that we may understand their dependence on experimental variables.

The term "intensity" is useful as a qualitative concept, but for a quantitative discussion it lacks precision. Table 12.6 lists some of the quantities for which the term "intensity" has been used. In our subsequent discussion, we shall adhere to the nomenclature of Table 12.6 when it is necessary to distinguish among the various measures of intensity.

A / *Emission*

To derive an expression for the intensity of spontaneous emission, let us return to a situation first proposed in Section 10.3.4. Given a unit volume of material containing N_i emitting atoms in an excited state i of energy E_i above the ground state, a number of atoms per unit time, $\Delta N_{i \to j}$, undergo spontaneous transitions to a lower state j of energy E_j through emission of an equal number of photons of frequency $\nu = (E_i - E_j)/h$. The number of

TABLE 12.6 / *Measures of radiant intensity*

Name	*Meaning*	*Typical units*
radiant flux	radiant power, 1 W = 1 J sec^{-1} = 10^7 ergs-sec^{-1}	ergs-sec^{-1}
radiance	radiant power (unit area)$^{-1}$ (unit solid angle)$^{-1}$; solid angles are measured in steradians[a] (abbreviated sr)	ergs-sec^{-1}-cm^{-2}-sr^{-1}
spectral radiance	radiant power (unit area)$^{-1}$ (unit solid angle)$^{-1}$ (unit wavelength interval)$^{-1}$	ergs-sec^{-1}-cm^{-2}-sr^{-1}-nm^{-1}
irradiance	radiant power (unit area)$^{-1}$	ergs-sec^{-1}-cm^{-2}
spectral irradiance	radiant power (unit area)$^{-1}$ (unit wavelength interval)$^{-1}$	ergs-sec^{-1}-cm^{-2}-nm^{-1}

[a] A steradian is defined as the solid angle that encloses a surface on a sphere equivalent to the square of the radius of the sphere. A steradian is thus related to a sphere in the same way that a radian is related to a circle. There are 4π steradians in a sphere.

photons emitted per unit time and unit volume is given by

$$\Delta N_{i\to j} = A_{ij}N_i \tag{12.8}$$

where A_{ij} is the Einstein transition probability of spontaneous emission. If we allow the lower state j to be the ground state ($E_j = 0$), then each photon is of energy $h\nu$, and the irradiance of the emitting volume is given by

$$I_{em} = h\nu L\, \Delta N_{i\to j} = h\nu L A_{ij}N_i \tag{12.9}$$

where L is the thickness of the emitting volume in the direction of observation. If the emitting volume has been equilibrated at temperature T, then application of Eq. (10.13) allows the intensity to be expressed in terms of the total concentration of emitting atoms N, and

$$I_{em} = h\nu L A_{ij}N \frac{g_i}{B(T)} \exp\left(\frac{-E_i}{kT}\right) \tag{12.10}$$

where g_i is the statistical weight of state i, and $B(T)$ is the partition function, which is given by $B(T) = g_j + g_k \exp(-E_k/kT) + g_l \exp(-E_l/kT) + \cdots$, and for many cases may be replaced by g_j, the statistical weight of the ground state j. The statistical weight simply describes the degeneracy of a state. Recall from Section 12.2 that in the absence of an external field there are degenerate levels signified by the quantum number m. When the interaction of the orbital and spin angular momenta is considered, it can be shown that the degeneracy of the resultant states is given by $2J + 1$. Thus, for the ${}^2P_{3/2}$ and ${}^2P_{1/2}$ states of Na, $g_i = 4$ and 2, respectively, and, from Eq. (12.10), the intensities of the two resonance lines, 5890 Å and 5896 Å, will be approximately in the ratio of 2:1.

In Section 10.3.2, it was pointed out that there is always a range of energies associated with a given transition, due principally in the case of atomic spectroscopy to Doppler and collisional broadening effects. As a consequence of our assumption that all the emitted photons are of a single energy $h\nu$, the intensity represented by Eq. (12.10) is an integrated intensity; that is, it corresponds to simultaneous measurement of the entire atomic line. Since atomic lines are typically narrow, a few hundredths to a few tenths of an angstrom, compared to the bandpass of monochromators, the integrated intensity is the quantity most often measured.

Comparison of Eqs. (12.7) and (12.10) shows that $x = 1$ and that K depends on a number of factors, some of which may vary with experimental conditions. The quantities h and k are fundamental constants; ν, A_{ij}, g_i, and E_i are determined by the atomic system; and L and T vary with the conditions of the experiment. The partition function, $B(T)$, is characteristic of the atomic system, but may show some dependence on T.

We have so far neglected to consider one further factor that determines the intensity of atomic emission lines—self-absorption. *Self-absorption* is the phenomenon in which photons originating within the emitting volume are reabsorbed before escaping. A quantitative description of self-absorption is

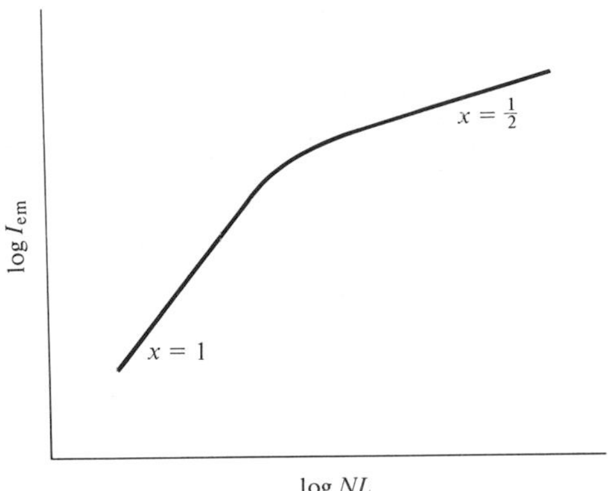

Figure 12.6 / *Theoretical dependence of integrated emission intensity on NL.*

beyond the scope of this discussion, but we can note that self-absorption increases as NL increases, and for large values of NL the intensity of emission is given by

$$I_{em} = K'(NL)^{1/2} \exp\left(\frac{-E_i}{kT}\right) \tag{12.11}$$

where K' depends in a complex fashion on the profile of the atomic line. Note that for self-absorption to occur, the photon must have an energy corresponding to an allowed absorption transition from a populated level. Since most of the atoms will be in the ground state, self-absorption is important chiefly for resonance lines. Figure 12.6 shows the expected dependence of the integrated emission intensity on NL. For a log-log plot, x equals the slope of the curve.

B / *Absorption*

To derive an expression for the intensity of radiation absorbed, let us consider a uniform volume of material containing a concentration of absorbing atoms N on which is incident a parallel beam of radiation of spectral irradiance I^0. As the radiation passes through the absorbing medium, its intensity decreases in exponential fashion according to the expression

$$I_\lambda = I_\lambda{}^0 \exp(-k_\lambda L) \tag{12.12}$$

where $I_\lambda{}^0$ and I_λ are the incident and transmitted spectral irradiances, respectively, at wavelength λ, and k_λ is the atomic absorption coefficient of the absorbing atoms at that wavelength. The intensity of radiation absorbed at λ, $I_{A\lambda}$, is thus

$$I_{A\lambda} = I_\lambda{}^0 - I_\lambda = I_\lambda{}^0[1 - \exp(-k_\lambda L)] \tag{12.13}$$

Since the absorption line cannot be truly monochromatic, it is necessary to

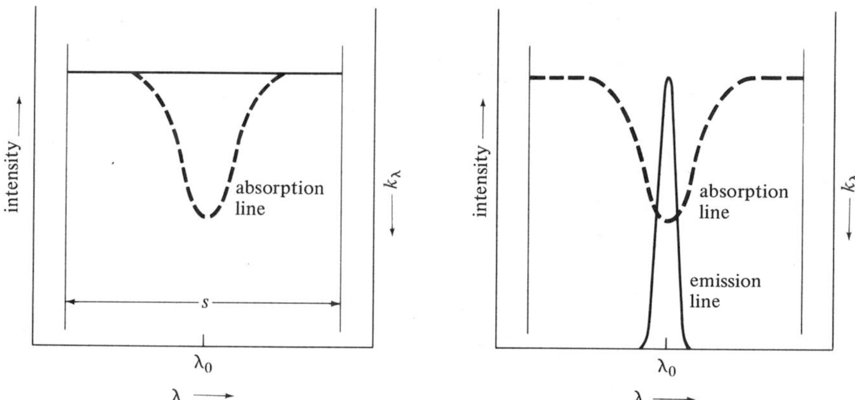

Figure 12.7 / *Representation of absorption with a continuum and line source.*

integrate the absorption over the range of wavelength for which absorption occurs. Thus,

$$I_A = \int I_\lambda{}^0[1 - \exp(-k_\lambda L)]\, d\lambda \tag{12.14}$$

where I_A is the integrated intensity of radiation absorbed.

In absorption measurements, we usually wish to express the intensity of radiation absorbed as a fraction of the incident intensity:

$$\alpha = \frac{I_A}{\int I_\lambda{}^0 d\lambda} \tag{12.15}$$

where α is the fraction of incident radiation absorbed, or, in brief form, the absorption. The absorbance, A, is related to α by

$$A = -\log(1 - \alpha) \tag{12.16}$$

Figure 12.7 presents schematically the two limiting cases we can conveniently consider in evaluating Eq. (12.15). If a continuum source is used, we can assume that $I_\lambda{}^0$ is constant over the range of wavelengths passed by the monochromator, and write

$$I_A = I_C{}^0 \int_0^\infty [1 - \exp(-k_\lambda L)]\, d\lambda \tag{12.17}$$

where $I_C{}^0$ is the spectral irradiance of the source at the wavelength λ_0. The only case that will interest us is the one for which $k_\lambda L$ is sufficiently small that we can make use of the approximation

$$\int_0^\infty [1 - \exp(-k_\lambda L)]\, d\lambda = \int_0^\infty k_\lambda L\, d\lambda \tag{12.18}$$

The right-hand portion of Eq. (12.18) has been evaluated and is given by

$$\int_0^\infty k_\lambda L\, d\lambda = \frac{\pi e^2}{mc} N_j f_{ji} L \tag{12.19}$$

where e and m are the charge and mass of an electron, N_j is the atomic concentration in the lower level of the transition, and f_{ji} is the oscillator strength of the absorption transition (see Section 10.3.4). For most cases of interest to us, the lower level of the transition is the ground state, and we can assume that $N_j \simeq N$, since at any given time almost all the atoms are in the ground state. Thus, the intensity of radiation absorbed is

$$I_A = I_C{}^0 \frac{\pi e^2}{mc} N f_{ji} L \tag{12.20}$$

Comparing Eqs. (12.10) and (12.20), it can be seen that the expressions for I_{em} and I_A are similar in form in that both I_{em} and I_A are directly proportional to NL. I_A, however, does not depend on T.

The integration in the denominator of Eq. (12.15) must be carried out over the bandpass of the monochromator:

$$\int_s I_\lambda \, d\lambda = I_C{}^0 s \tag{12.21}$$

Thus, the fraction of radiation absorbed is given by

$$\alpha = \frac{\pi e^2}{mcs} N f_{ji} L \tag{12.22}$$

From Eq. (12.22) we can see that when a continuum source is used, the measured absorption is dependent on the resolution of the monochromator —as the bandpass of the monochromator increases, the measured absorption decreases.

Let us consider briefly the other limiting case presented in Fig. 12.7. In this case we imagine that a source of line radiation is used, that the line produced is effectively monochromatic with an integrated intensity (irradiance) I_L, and that both the emission and absorption lines peak at λ_0. Under these conditions, Eq. (12.14) becomes

$$I_A = [1 - \exp(-k^0 L)] \int I_\lambda{}^0 \, d\lambda = [1 - \exp(-k^0 L)] I_L \tag{12.23}$$

where k^0 is the atomic absorption coefficient at λ_0. If we again make the assumption that $k^0 L$ is small, then we may write

$$I_A = I_L k^0 L = I_L \frac{2(\ln 2)^{1/2} \pi^{1/2} e^2 \lambda_0{}^2}{mc^2 \, \Delta\lambda_D} N_j f_{ji} L \tag{12.24}$$

where $\Delta\lambda_D$ is the half-intensity width of the absorption line as determined by Doppler broadening (see Section 10.3.2). We should note again the direct proportionality between I_A and NL.

For the case of a monochromatic line source, the absorption is given by

$$\alpha = \frac{2(\ln 2)^{1/2} \pi^{1/2} e^2 \lambda_0{}^2}{mc^2 \, \Delta\lambda_D} N_j f_{ji} L \tag{12.25}$$

There are two differences between Eqs. (12.22) and (12.25) that are worth noting. (1) In the case of a line source, α shows no dependence on the characteristics of the monochromator—the effective resolution of the system is the half-width of the source line. (2) In the case of a line source, α depends on the broadening of the absorption line. If the temperature of the absorbing medium increases, $\Delta\lambda_D$ increases, and α decreases. The dependence on linewidth arises because we are attempting to measure the peak absorption, and as the absorption line is broadened its area remains constant, but its peak value decreases.

Before concluding, let us note that the continuum source case, as represented by Eq. (12.22), is readily achieved experimentally, but that the monochromatic line case, as represented by Eq. (12.25), is seldom approached in practice. In the usual case, the width of the source line is comparable to the width of the absorption line. This situation does not alter the dependence of α on $Nf_{ji}L$, but the proportionality factor depends in a complex fashion on the shape of both the emission and absorption line.

C / *Fluorescence*

To derive an expression for the intensity of atomic fluorescence, let us consider a uniform cube of material with concentration N of absorbing atoms on which is incident a parallel beam of radiation of spectral irradiance I^0. The intensity of fluorescence, I_F, is given by

$$I_F = \Phi I_A f_s \tag{12.26}$$

where I_A is the intensity of radiation absorbed, f_s is a self-absorption factor that accounts for the reabsorption of photons, and Φ is the power efficiency, that is, the ratio of emitted radiant power to absorbed radiant power. For the low-concentration case, which is of greatest interest to us, $f_s = 1$, and $I_F = \Phi I_A$. Substitution of Eqs. (12.20) and (12.24) leads to the appropriate expressions for excitation of the fluorescence by continuum and line sources. Thus, the intensity of fluorescence is also directly proportional to NL for the low-concentration case. At larger values of NL, the self-absorption factor, f_s, is less than one and dependent on NL. The effect of f_s on I_F is illustrated in Fig. 12.8. A thorough, but readable, account of atomic fluorescence intensity expressions is available in the review article by Winefordner and Elser [*Anal. Chem.* **43** (4), 24A (1971)].

The power efficiency, Φ, of Eq. (12.26), is closely related to the quantum efficiency, ϕ, which is the ratio of the number of photons emitted per unit time to the number of photons absorbed per unit time:

$$\Phi = \phi \frac{\nu_F}{\nu_A} \tag{12.27}$$

where ν_F is the frequency of the fluorescence radiation and ν_A is the frequency of the absorbed radiation. Consider the term diagram shown in Fig. 12.9. The quantum efficiency of the fluorescence transition $i \rightarrow j$ is simply the rate

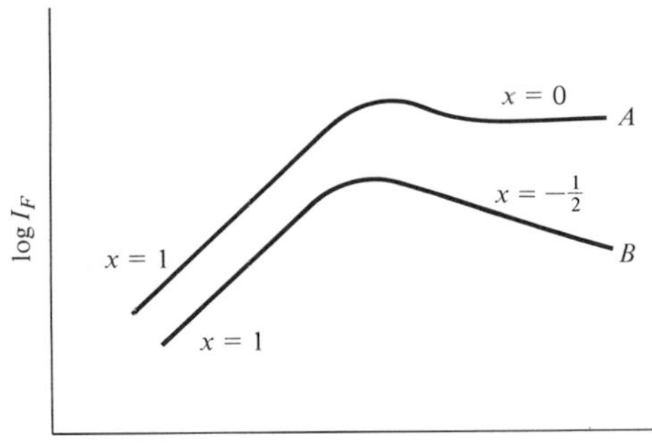

Figure 12.8 / *Dependence of I_F on NL for continuum, A, and narrow-line, B, source. For small values of NL, I_F is directly proportional to NL, but at large values the self-absorption factor, f_s, is proportional to $1/NL$ for the continuum case and to $1/(NL)^{3/2}$ for the line case.*

of the transition $i \rightarrow j$ divided by the total rate at which the state i is depopulated. Thus,

$$\phi_1 = \frac{A_{ij}}{A_{ij} + A_{ik} + K_i} \tag{12.28}$$

where A_{ij} is the Einstein transition probability for spontaneous emission for the transition $i \rightarrow j$, A_{ik} is the same quantity for the transition $i \rightarrow k$, and K_i is the rate constant for radiationless deactivation (quenching) of the state i. The maximum value of the quantum efficiency is thus one, and for the transition $i \rightarrow j$ the maximum value of the power efficiency is less than one since $\nu_{F1} < \nu_A$. The quantum efficiency for the transition $i \rightarrow k$ is given by

$$\phi_2 = \frac{A_{ik}}{A_{ij} + A_{ik} + K_i} \tag{12.29}$$

and $\Phi_2 = \phi_2$ since $\nu_{F2} = \nu_A$.

Figure 12.9 / *A simplified term diagram.*

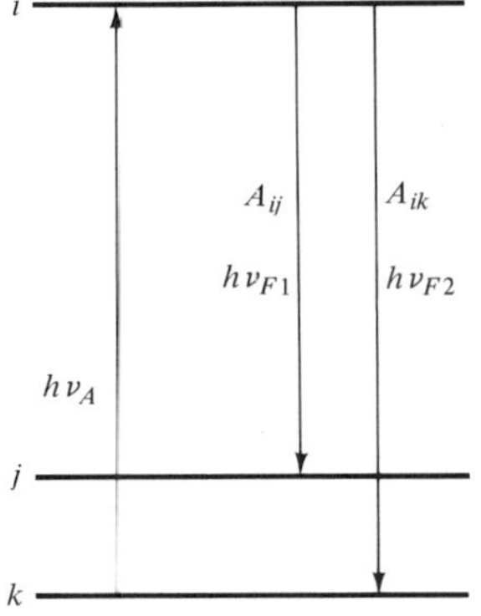

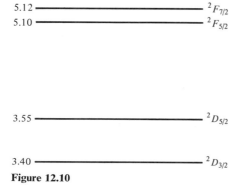

Figure 12.10

SUGGESTED READINGS

Chris Candler, *Atomic Spectra*, Van Nostrand Reinhold (New York, 1964).
H. G. Kuhn, *Atomic Spectra*, 2nd ed., Academic (New York, 1969).
R. Mavrodineanu and H. Boiteux, *Flame Spectroscopy*, Wiley (New York, 1965), Chapters 15 and 16.

PROBLEMS

1 / Figure 12.10 shows a portion of an atomic energy-level diagram with the term symbol for each of the states represented shown at the right and the energy (in electron volts) of the state shown at the left.
(*a*) State the values of L, S, and J for each of the states.
(*b*) State the selection rules governing ΔL, ΔS, and ΔJ and show how they apply to transitions for the energy levels represented.
(*c*) Which of the allowed transitions will give rise to the lowest-energy radiation? Calculate the wavelength (in centimeters) of the emitted radiation.

2 / What are the values of L, S, and J for a state with the term symbol ^2F?

3 / How many lines are predicted by the selection rules for a $^3D_{3,2,1}$–$^3P_{2,1,0}$ transition? Justify your answer by reference to the selection rules.

4 / A portion of a term diagram for indium is shown in Fig. 12.11. Refer to this diagram as required to answer the following questions.
(*a*) List all the allowed transitions for the states shown.
(*b*) Pick the line or group of lines that is likely to have the greatest emission intensity by flame atomic emission, and justify your choice by reference to the expression describing the intensity of atomic (thermal) emission.
(*c*) If an external continuum source is used to observe the absorption of In atoms in a flame, list the lines that will be seen in absorption.
(*d*) Write an expression describing the ratio of the absorption, α_1, of the longest wavelength absorption line to the absorption, α_2, of the next shorter wavelength absorption line.
(*e*) If In atoms in a flame are irradiated only with radiation of energy corresponding to the transition $5^2P_{1/2}$–$^2S_{1/2}$, at what wavelengths, if any, would your expect to see atomic fluorescence?

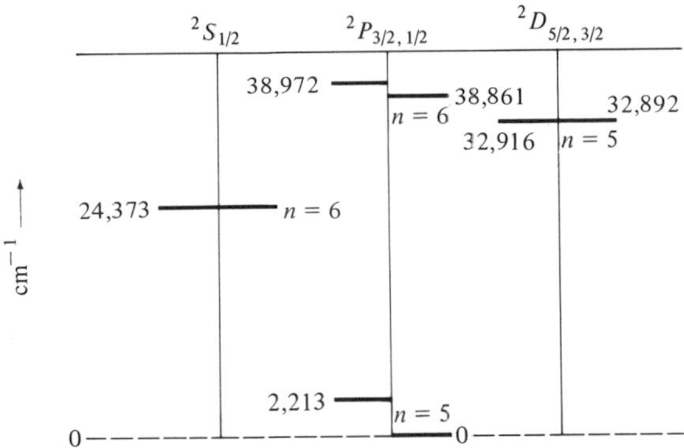

Figure 12.11 / *Partial term diagram for* In.

5 / Figure 12.12 is a partial term diagram for Hg. Use the information provided on this diagram to answer the following questions.

(*a*) Assuming that the selection rules are strictly obeyed, list the allowed transitions.

(*b*) Assuming that the selection rule governing ΔS is not operative, list any additional allowed transitions to the ground state.

(*c*) If an external continuum source is used to observe the absorption of Hg atoms in a room-temperature vapor, list the transitions that will be seen if all the selection rules are obeyed.

6 / If the $^3P_{2,1,0}$ states of Hg are thermally populated, calculate the ratio of the populations of the $^3P_2$ and $^3P_0$ levels in a flame of 2700°K. The splitting is 6398 cm^{-1}, and $k = 0.697$ cm^{-1}/°K.

Figure 12.12 / *Partial term diagram for* Hg.

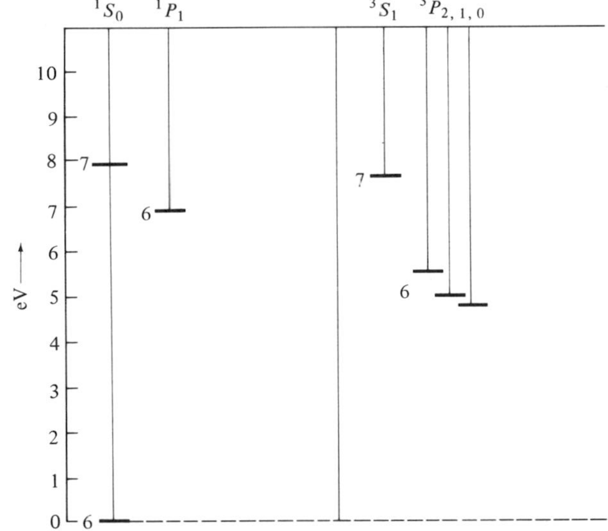

chapter 13 / Flame spectrometry

chapter 13

13.1 / INTRODUCTION

In this chapter, we shall describe three techniques that have in common the use of a flame for production of the atomic vapor. The three techniques are flame atomic emission spectrometry (AE), flame atomic absorption spectrometry (AA), and flame atomic fluorescence spectrometry (AF), and, as their names suggest, they are examples of the three types of experiments described in Chapter 12. These techniques constitute an important portion of the field of trace-element analysis,[1] and many of the determinations of this type, particularly in the analysis of biological materials, are today performed by these techniques.

The material of this chapter is presented in five main sections: The first considers the processes by which atoms are produced in flames, the succeeding three describe in turn the three techniques, and the final section compares the three techniques.

13.2 / ATOMIZATION IN FLAMES

13.2.1 / The atomization process

In Section 12.3.2, we derived intensity relationships applicable to AE, AA, and AF, but in all cases we began by assuming a volume of material with a uniform concentration of atoms. We must now enter the real world, in which we seldom encounter a sample as a uniform atomic vapor, and consider the processes by which such an atomic vapor is produced. In our subsequent discussion we shall use the term *atomization* to mean the production of atoms; for the more colloquial sense of "atomization," which is understood to mean the reduction of a liquid to a fine spray, we shall use the term *nebulization*.

[1] Trace analysis has been defined as the field concerned with the determination of species occurring at concentrations of 100 ppm or less.

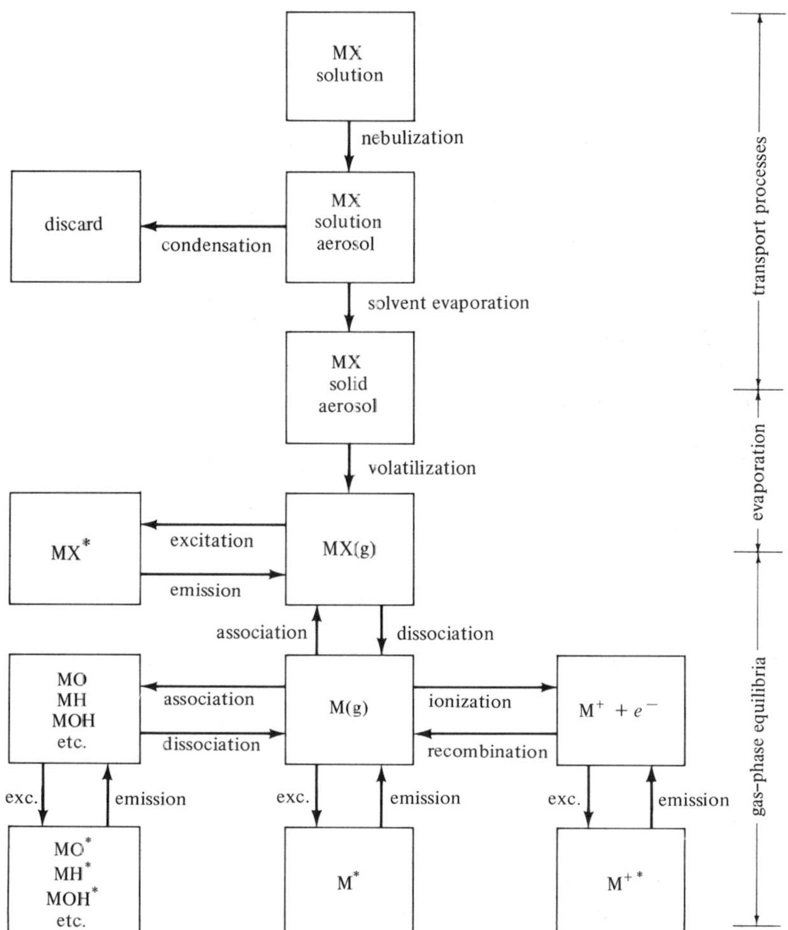

Figure 13.1 / *Flame atomization processes.*

We begin by assuming that we are interested in determining the concentration of a metal M, which occurs in solution as a salt MX. Thus, we must consider the conversion of MX in solution to an atomic vapor of M. (It is reasonable to choose a solution sample as our starting place, since very many real samples are in fact solutions or may readily be obtained in solution form.) The conversion from MX (solution) to M (atomic) may in many cases be effected quite simply by spraying the solution into a flame. The processes involved are illustrated in Fig. 13.1. The solution is broken up into fine droplets, which may be injected directly into the flame or may pass through a nebulization chamber in which the larger droplets are separated and discarded and only the smaller droplets enter the flame. Solid aerosol particles are produced as solvent evaporates from the droplets. When a nebulization chamber is used, some solvent evaporation occurs even before

TABLE 13.1 / *Properties of some fuel-oxidant mixtures*

Mixture	*Ignition temp.* (°C)	*Burning velocity* (cm sec^{-1})	*Approximate max. flame temp.* (°C)
H_2–air	530	440	2045
H_2–O_2	450	3680	2660
H_2–N_2O	—	390	2650
C_2H_2–air	350	160	2125
C_2H_2–O_2	335	2480	3100
C_2H_2–N_2O	400	285	2955

the particles enter the flame. We shall refer to the processes leading to the actual introduction of the sample into the flame as *transport processes*. As the aerosol enters the flame, the remaining solvent is removed and the solid particles are vaporized. Processes at the gas–solid interface are complex and not yet well understood, but one possible further path is indicated in Fig. 13.1—production of molecular species, which may then dissociate to give free atoms. The atoms produced undergo further reactions such as ionization or association with flame gas products to give other molecular species.

13.2.2 / Flames and burners

It is customary to distinguish two types of flames: premixed and diffusion. The flame produced on a Bunsen burner is of the premixed type—the fuel and oxidant are thoroughly mixed before ignition, the flow of gases is nonturbulent, and the flame exhibits distinct zones of different properties. The rate of combustion in such a flame is determined by the burning velocity of the mixture. When a combustible mixture is ignited at a point in the mixture, the flame front propagates through the whole mixture at a uniform rate. The rate at which the flame front propagates is the *burning velocity* of the mixture. If we ignite a combustible mixture that is issuing from a tube, a stable flame will be produced only if the rate of flow of the mixture at some point just equals the velocity with which the flame front propagates in the opposite direction. In a diffusion flame, the pure fuel gas flows from a tube, either into the open atmosphere, where it burns in the oxygen of the air, or into a stream of oxidant gas flowing from an adjacent or concentric orifice. When the rate at which a gas issues from an orifice exceeds a certain critical velocity, it becomes a turbulent jet that entrains the surrounding gas and broadens out in the form of a cone. The flame in such a jet consists of irregular vortexes of gas and oxidant mixture forming random patches of combustion waves throughout the jet, but no cohesive combustion surface. This is a turbulent diffusion flame and is the type of diffusion flame most often encountered in flame spectrometry. The rate of combustion in such a flame is determined by the rate of mixing of the fuel and oxidant gases.

Table 13.1 lists properties of some of the fuel–oxidant combinations employed in flame spectrometry. Turbulent diffusion flames can readily be produced from any of the mixtures, since no mixing of fuel and oxidant

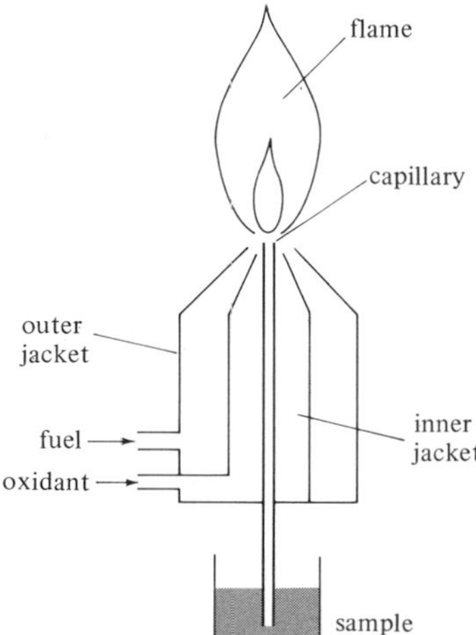

Figure 13.2 / *One version of the total-consumption burner.*

occurs prior to the actual combustion zone. Premixed flames, however, are limited by the properties of some of the mixtures. The rate of issue of the gas from the burner orifice must exceed the burning velocity. Thus, it is difficult to produce premixed flames when oxygen is used as the oxidant, because the burning velocity is very high. The ignition temperature determines the temperature the burner head should not exceed, and the flame temperature characterizes the possibilities of the flame for evaporating and dissociating compounds.

As one might expect, burners are classified into two types according to the type of flame produced. The *total-consumption burner*, shown in Fig. 13.2, produces a turbulent diffusion flame. It consists of three concentric chambers —a central capillary, an inner jacket, and an outer jacket. In the usual arrangement, oxidant flows through the inner jacket and fuel through the outer jacket. The flow of oxidant over the capillary tip causes a pressure drop at the tip and aspiration of solution through the capillary. The velocity of the oxidant stream at the capillary tip is sufficiently high that the solution stream is broken into fine droplets as it issues from the capillary. In this arrangement, all of the solution is sprayed into the flame, and thus the name "total consumption." The name should not, however, be understood to mean that all the sample material is atomized in the flame. A rather large range of droplet sizes is produced with this arrangement, and some pass through the flame incompletely vaporized.

In the *premix burner*, shown in Fig. 13.3, the sample is aspirated by the oxidant stream into a large chamber. The larger droplets are collected and discarded, and the fuel, oxidant, and remaining aerosol are mixed and then forced to the burner opening. The burner orifice may be a narrow slot, up to 15 cm long, or an array of small-diameter holes. The slot-type burner is

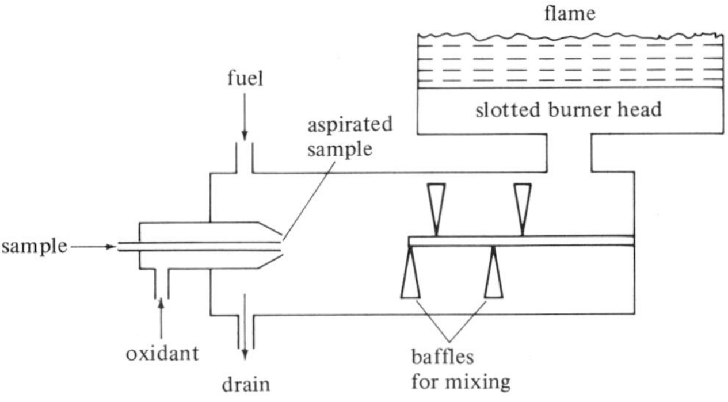

Figure 13.3 / *A premix burner.*

readily constructed and provides a long path length suitable for emission and absorption measurements. A burner head with a square or circular array of holes is better suited to atomic fluorescence measurements.

Total-consumption burners produce noisy, poorly characterized flames, with a rather short path length, but they achieved widespread use in flame spectrometry because they were able safely to produce high-temperature flames with oxygen as the oxidant. The superiority of total-consumption burners in this regard has been challenged in recent years by the use of nitrous oxide as an oxidant. The C_2H_2–N_2O mixture has a burning velocity comparable to that of C_2H_2–air, and thus may safely be burned in premix burners, and the flame temperature is only slightly less than that of C_2H_2–O_2 flames. Most flame spectrometric measurements today are made with premix burners designed to burn mixtures with either air or nitrous oxide as the oxidant.

13.2.3 / Interferences

In Section 12.3, we discussed the proportionality between the intensity of emission, absorption, or fluorescence and the atomic concentration N. In Section 13.2.1, we have described some of the processes that affect the proportionality between N and the solution concentration C. The basic assumption of flame spectrometric measurements is that there is a constant proportionality between the measured quantity and C for all the measurements required for a given determination. In the usual case, a determination requires measurements for one or more standards in addition to the unknown. An interference occurs when the proportionality between the measured quantity and C varies within a group of measurements. Effects that alter the proportionality between N and C constitute interferences in all three flame techniques and are conveniently considered together. These interferences may be classified (see Fig. 13.1) as transport interferences, evaporation interferences, or gas-phase equilibria interferences.

To transport interferences belong all factors affecting the amount of sample entering the flame. The most important factors are viscosity (influencing the sample aspiration rate), surface tension (influencing the size of the aerosol droplets), and the vapor pressure of the solvent (influencing solvent evaporation rate and condensation losses). Because transport effects affect all elements equally, they constitute interferences not specific to particular elements.

Evaporation interferences arise from alterations in the rate of evaporation of solid aerosol particles in the flame. These may be either specific to particular elements or nonspecific. A specific interference of this type occurs when a chemical reaction between the analyte and a concomitant in the solution leads to the formation of a compound that vaporizes at a different rate. An example of this type of behavior is found in the effect of phosphate anion on calcium. In the presence of phosphate, a relatively involatile Ca—P—O compound is formed, and a marked depression of the Ca atomic concentration occurs in the lower regions of the flame. A nonspecific evaporation interference can occur when the analyte occurs in a solution of high total solids. After solvent evaporation, the test element may be embedded in a large salt particle from which it cannot be vaporized quickly.

Gas-phase equilibria interferences arise from shifts in the dissociation and ionization equilibria. These equilibria may be considered in much the same fashion as solution equilibria. Thus, for the ionization reaction,

$$\mathrm{M} \rightarrow \mathrm{M}^+ + e$$

the equilibrium constant customarily takes the form

$$K_i = \frac{p_{\mathrm{M}^+} p_e}{p_\mathrm{M}} \tag{13.1}$$

where p_{M^+}, p_e, and p_M are the partial pressures of ions, electrons, and atoms, respectively. The partial pressure of a species may be related to its concentration by the ideal gas law. For example, the partial pressure of M, p_M, is related to the concentration of M in atoms per centimeter^3 of flame gases, N_M, by

$$p_\mathrm{M} = N_\mathrm{M} kT \tag{13.2}$$

From Eq. (13.1), it can be seen that if the partial pressure of electrons in the flame is increased—for example, by the addition of another easily ionized element to the flame—that the partial pressure of M (and N_M) is increased. This type of effect may be observed by comparing the intensity of Na flame emission for two solutions containing identical concentrations of Na, one of which also contains K. The solution containing K will give a larger Na emission signal because of the repression of the Na ionization by ionization of K.

13.2.4 / Elimination of interferences and optimization of atomization

The detection and elimination of interferences depends largely on the ingenuity and experience of the analyst, but we can note some common approaches that have proved useful.

One very general approach is to prepare standards that imitate the sample composition. This method is reliable but laborious. If the overall composition of the samples is not known or varies from sample to sample, then the standard additions approach described in Section 13.3.4 may be applicable. It is usually possible, by either the imitation or standard-additions approach, to ensure that physical properties of samples and standards are sufficiently similar that transport interferences are avoided.

Interferences may sometimes be eliminated by addition of a relatively large concentration of the interferant species to standards and samples so that the natural variation is "buffered" out. Thus, it is customary to add a large excess of an easily ionized species to repress ionization. The buffer approach has sometimes been adopted to eliminate evaporation interferences such as the effect of phosphate on calcium, but it is less useful in this case because of the rather considerable depression of the signal that results. A better approach in such cases is the use of "releasing agents," which prevent the depression by setting up a competitive equilibrium in the aerosol droplet. For example, La is sometimes used as a releasing agent for Ca since it effectively "competes" with Ca for the phosphate anion, and the involatile Ca—P—O compound is avoided.

The choice of flame, flame region, and flame gas composition may also have a profound effect on the efficiency of atom production. Higher temperature flames reduce the effect of slow evaporation steps and reduce compound formation in the flame. The effect of slow evaporation steps may also be reduced by making observations higher in the flame, but compound formation may be greater in this region because of greater entrainment of ambient air and cooling of the flame. The flame gas composition affects the efficiency of atomization by its effect on gas-phase equilibria. Many elements form stable monoxides in flames, and a substantial gain in atomic concentration can sometimes be obtained by increasing the amount of fuel in the flame mixture, thereby reducing the free oxygen concentration in the flame and shifting the equilibrium reaction,

$$MO \rightarrow M + O$$

to favor formation of M.

Figure 13.4 helps to illustrate the effect of flame gas composition and temperature on atomic concentration. Sodium is representative of the elements that form readily vaporized and easily dissociated compounds. There are rather broad regions of constant atomic concentration, the maximum concentration occurs low in the flame, and the atomic concentration is not strongly affected by the change from lean to rich conditions. Calcium represents an intermediate case in which the evaporation proceeds more slowly, and compound formation becomes important in the cooler and less fuel-rich regions of the flame. Thus, the maximum atomic concentration is rather sharply localized, occurs at some distance from the burner top, and

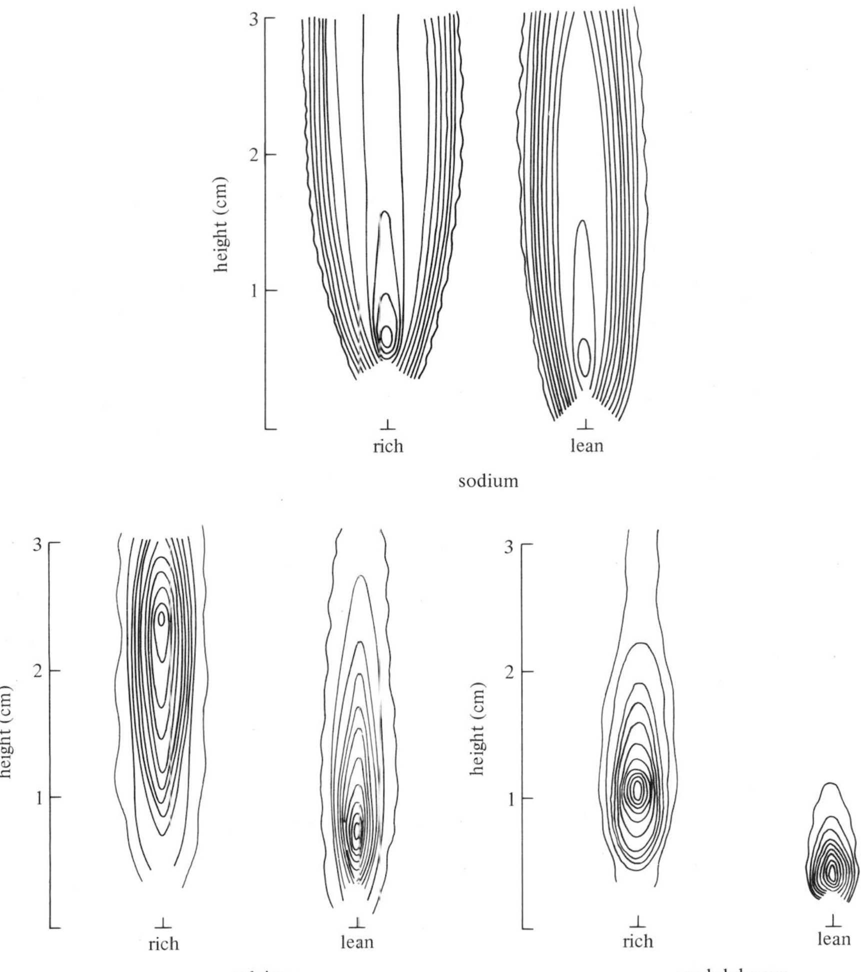

Figure 13.4 / *Distributions of atoms in* 10-cm *slot-type premix acetylene–air flame. View is end on. Contours are drawn at intervals of* 0.1 *absorbance units, with maximum absorbance at center.*
SOURCE: C. S. Rann and A. N. Hambly, *Anal. Chem.* **37**, 879 (1965).

its position is strongly altered by the change from fuel-lean to fuel-rich conditions. Molybdenum represents the extreme case of an element that forms a very stable monoxide in the flame. The atomic concentration at a given point is very strongly dependent on flame gas composition, and fuel-rich conditions are essential to useful measurements.

For most elements, the atomic concentration produced in the flame can be increased if an organic solvent is used for the sample. The principal effect of the organic solvent is on the transport processes—the solution aspiration

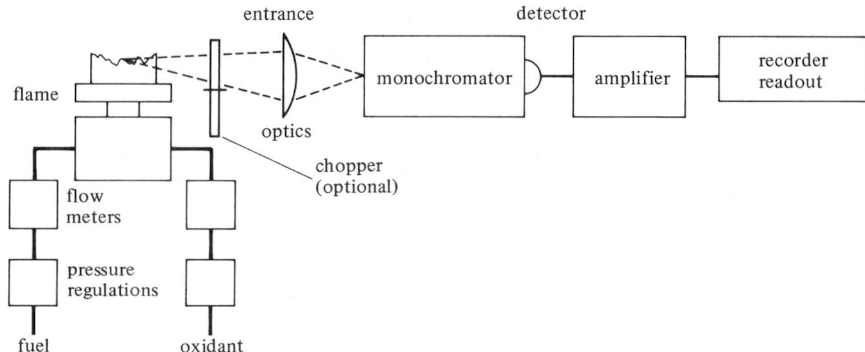

Figure 13.5 / *Basic components for a flame emission measurement.*

rate is increased, smaller droplets are formed, and solvent evaporation occurs more readily with many organic solvents. If the analyte element in an aqueous sample is extracted into an immiscible organic solvent, there are advantages in addition to the increased transport rate: (1) A concentration of analyte is effected if the volume of the organic phase is less than the volume of the aqueous phase, and (2) the analyte may be separated from interferant species.

13.3 / FLAME ATOMIC EMISSION SPECTROMETRY

13.3.1 / The flame emission experiment

The basic components of a flame emission measurement are shown in Fig. 13.5. Typical features are as follows. The burner is of the premix slot type, capable of burning both air-supported and nitrous oxide-supported flames. Pressure regulators and flow meters are used to maintain constant flame operating conditions. A lens or mirror is used to form an image of the flame on the entrance slit of the monochromator. The monochromator is capable of providing a resolution of about 0.2 Å and has variable slits and provision for wavelength scanning. (These features are commonly met in 0.5-m focal length plane grating spectrometers.) The detector is a multiplier phototube with spectral response appropriate to the wavelength region to be investigated. The electronic measurement system consists of an amplifier and recorder readout. The amplifier may be of the DC type, but more usually a mechanical chopper modulates the flame emission signal and a tuned AC or phase-sensitive amplifier is used.

13.3.2 / The flame as an excitation source

Flames are not homogeneous excitation sources; the temperature and chemical environment vary from area to area. In turbulent diffusion flames this variation is not so readily apparent, since there are no well-defined zones, but in premixed flames there are four zones that can be distinguished, as

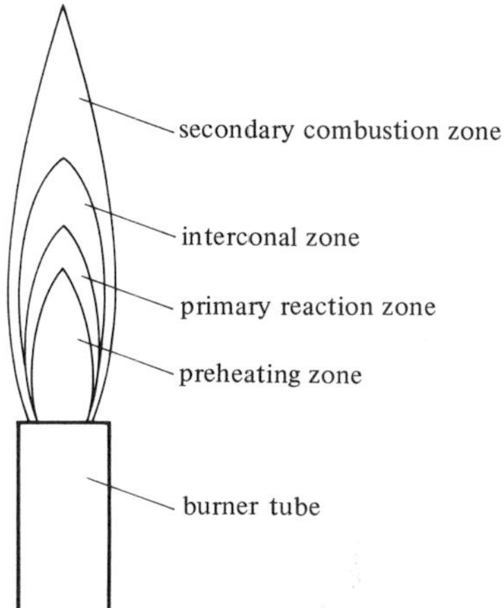

Figure 13.6 / *Structure of a premixed flame.*

shown in Fig. 13.6. The first is the *preheating zone*, where the combustion mixture is heated to the ignition temperature. The second is the *primary reaction zone*, which is characteristically a rather thin blue cone surrounding the preheating zone. The complex processes of combustion occur in the primary reaction zone. This region is rich in ions and radical species that have not had sufficient time to reach thermodynamic equilibrium. For our purposes, this means simply that excitation occurring in this region cannot be described by a single temperature [Eq. (10.25)], but only in terms of specific excitation mechanisms. Thermodynamic equilibrium is closely approached in the third and fourth zones, and these zones are of greatest interest in flame spectrometry. The fourth, or outermost, zone is the *region of secondary combustion* within which the products of the primary combustion are burned to stable molecular species by air entrained from the atmosphere. The third zone lies between the two reaction zones and is called the *interconal zone*. In flames produced from a stoichiometric mixture of fuel and oxidant, that is, where the amounts of each are in the ratio of their coefficients in the balanced equation that best describes the primary combustion reaction, the interconal zone is extremely small; but in fuel-rich hydrocarbon flames, this zone increases greatly in size and may extend to a height several centimeters beyond the primary reaction zone. It is within this zone that the chemical environment is most conducive to the production of free atoms from those elements that form stable monoxide molecules. The maximum flame temperature usually occurs just above the primary reaction zone, and thus the interconal zone is also the region of maximum excitation for atomic emission spectrometry. It is well to recall that the point of maximum emission intensity does not necessarily coincide with the point of maximum temperature, since the intensity depends on both the temperature *and* the atomic concentration [Eq. (12.15)].

If we confine our attention to the third and fourth flame zones, then the ability of the flame to excite emission from a given concentration of free

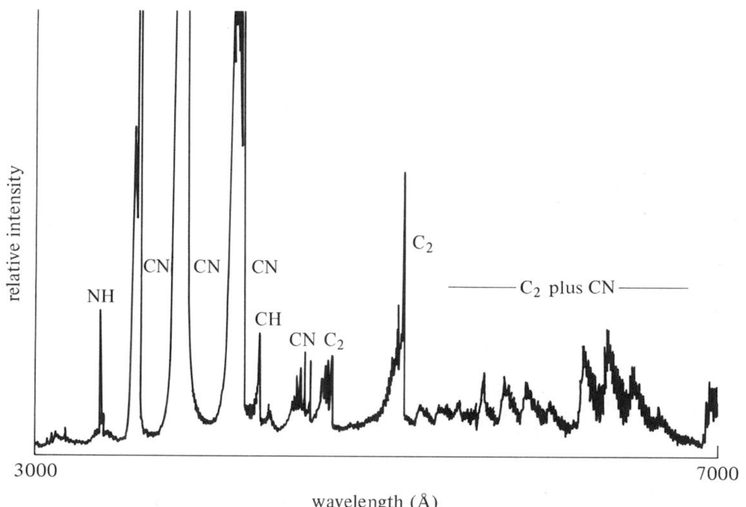

Figure 13.7 / *Emission spectrum of a premixed fuel-rich nitrous oxide–acetylene flame.* SOURCE: E. E. Pickett and S. R. Koirtyohann, *Anal. Chem.* **41**(14), 28A(1969).

atoms is described entirely by the flame temperature, approximate maximum values of which for various mixtures are given in Table 13.1. With only a few exceptions, the premixed, fuel-rich nitrous oxide–acetylene flame is the flame of choice for atomic emission spectrometry. The chief exceptions are the elements with extremely high excitation potentials, such as arsenic, tin, and antimony, for which improved sensitivities have been obtained by taking advantage of nonequilibrium excitation in the primary reaction zone of special flames.

One disadvantage of the nitrous oxide–acetylene flame is the relatively intense background that is emitted within certain spectral regions. This background spectrum is the result of radicals and molecules that are present in the flame gases in large quantities. The strongest emission arises from CH, OH, CN, and C_2 molecules. The emission spectrum of a premixed, fuel-rich nitrous oxide–acetylene flame is shown in Fig. 13.7.

13.3.3 / Emission measurements

Flame atomic emission spectrometry is primarily a method of quantitative analysis. Qualitative analysis is possible but not entirely acceptable, because the flame is not sufficiently energetic to guarantee high sensitivity for a large number of elements for one choice of flame composition and region of observation.

Quantitative analysis requires measurement of relative values of the intensity of emission of an atomic line of the element of interest. This task is complicated by the possible occurrence of background emission within the monochromator bandpass—that due to concomitants in the sample and the flame background emission. Flame spectra are relatively simple because

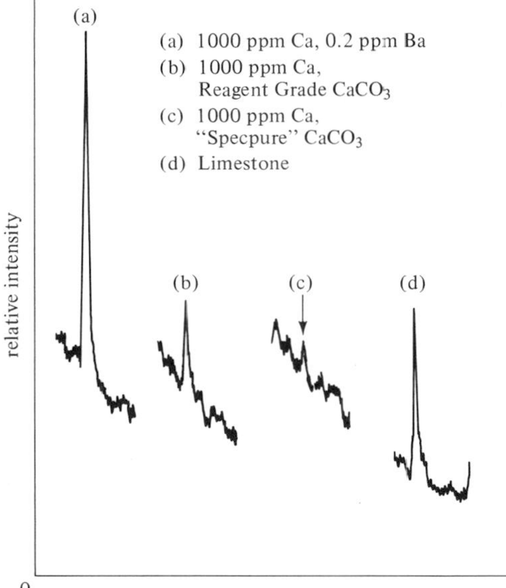

Figure 13.8 / *Background correction by means of a wavelength scan:* Ba 5535 Å *line with strong* CaOH *molecular emission background.*
SOURCE: E. E. Pickett and S. R. Koirtyohann, *Anal. Chem.* **41**(14), 28A(1969).

the excitation energy of the flame is relatively low, and the resolution of the small monochromators employed is sufficiently high that correction for background does not, in most cases, present great difficulty. Two approaches are commonly used.

In the first, the wavelength vicinity of the analytical line is slowly scanned and the spectrum recorded. Figure 13.8 shows scans of the Ba 5535 Å line in the presence of calcium. The base line of molecular emission due to CaOH and the net Ba intensity are both easily established, and the practicality of determining a trace of Ba in a Ca matrix is apparent. This approach is slow and consumes a rather large quantity of sample. Moreover, the peak line intensity is observed only briefly, thereby reducing the precision of the measurement.

The second approach is less wasteful of sample and gives better precision. This approach is illustrated in Fig. 13.9. The line-plus-background intensity is first measured for a group of samples at the wavelength of the line. The wavelength is then changed by $1\frac{1}{2}$ to 2 times the monochromator bandpass, and the adjacent background contributed by the matrix is measured and subtracted for each sample. If the background is complex, it should also be measured on the opposite side of the line and the average reading subtracted. The base line should be established by aspirating pure solvent before and after each reading, so that the background contribution from the flame itself is not included.

Both of these background-correction methods fail if unresolved line

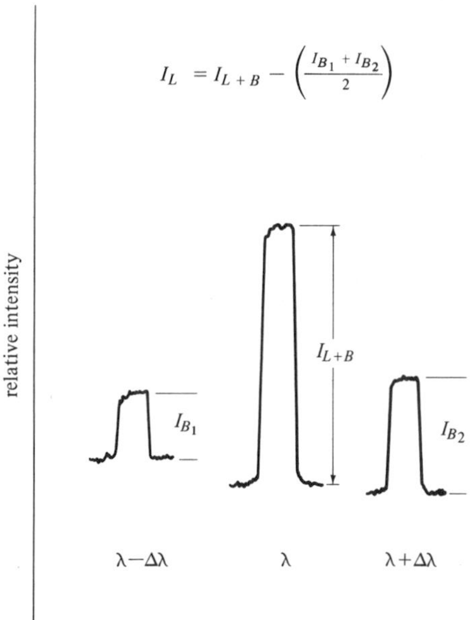

Figure 13.9 / *Illustration of the intensity contributions involved in nonscanning background correction for most general case. Each of the three recorder tracings would be made by aspirating solvent, then sample solution, then solvent again.*
SOURCE: E. E. Pickett and S. R. Koirtyohann, *Anal. Chem.* **41**(14), 28A(1969).

structure is present, such as, for example, that due to atomic emission of a concomitant in the sample. Such emission constitutes a spectral interference, and if the monochromator is incapable of resolving the two lines, then either another analyte line must be chosen for measurement or a prior separation of the analyte and concomitant must be carried out.

13.3.4 / Methods of quantitative analysis

The two principal methods of quantitative analysis are the analytical curve and the standard addition method.

For the first, a series of standards is prepared to contain known concentrations of the element of interest. These standards should have compositions as nearly like the samples as possible, and the range of concentrations should include the concentration of the analyte in the samples. Emission measurements for samples and standards are carried out as described in the preceding section, and the background-corrected readings for the standards are plotted against their analyte concentrations to construct the analytical curve. The concentration of analyte in the samples is then read from the analytical curve employing the background-corrected readings for each of the samples.

Sometimes the exact composition of the sample matrix is not known, the matrix is too complicated to imitate with standard solutions, or only a small number of samples are to be analyzed, and the labor of preparing complicated

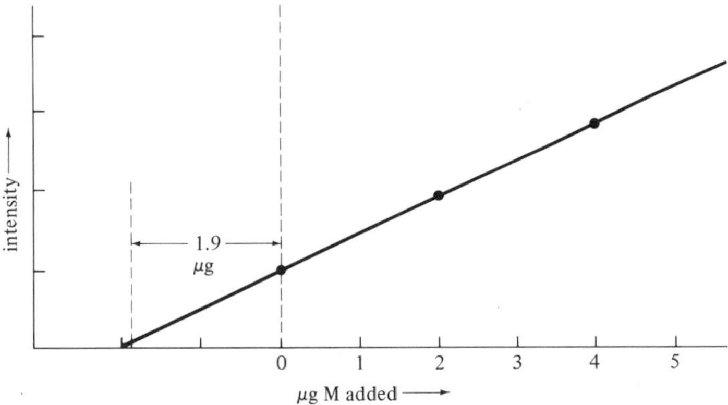

Figure 13.10 / *Graphical evaluation using the standard additions method.*

standard solutions is excessive. In these cases it may be advantageous to use the standard-addition method. There are several variations of the standard-addition method, but one of the simplest is as follows. Three separate, equal aliquots of the sample solution are taken, and to two of them are added definite amounts of the element to be determined. If the amount of the element in the sample is x, then suitable additions would be x and $2x$. The three fractions are then diluted to a definite volume, and the emission measurements made. The background-corrected readings are plotted, as shown in Fig. 13.10, against the quantity of analyte element added, and a line is drawn through the three points, extending to an emission reading of zero. The point of intersection on the negative concentration axis indicates the amount of the element present in one aliquot of the sample, and the analyte concentration in the sample is the amount found divided by the aliquot volume.

Several conditions must be met if the standard-additions method is to give an accurate result: (1) A zero reading will be obtained if the concentration of analyte in the sample is zero, (2) the chemical form of the added standard must be the same as that of the analyte or must produce an equivalent analytical signal, and (3) the analytical curve must be linear.

13.4 / FLAME ATOMIC ABSORPTION SPECTROMETRY

13.4.1 / The flame atomic absorption experiment

The basic components of a flame atomic absorption experiment are shown in Fig. 13.11. Typical instrumentation is similar to that for flame emission except for the obvious addition of a source of line radiation of the element of interest and a lens or mirror for forming an image of this source in the center of the flame. In all commercial atomic absorption instruments the source radiation is modulated, either electronically or mechanically, and some type of tuned amplifier system is employed. Since radiation from the

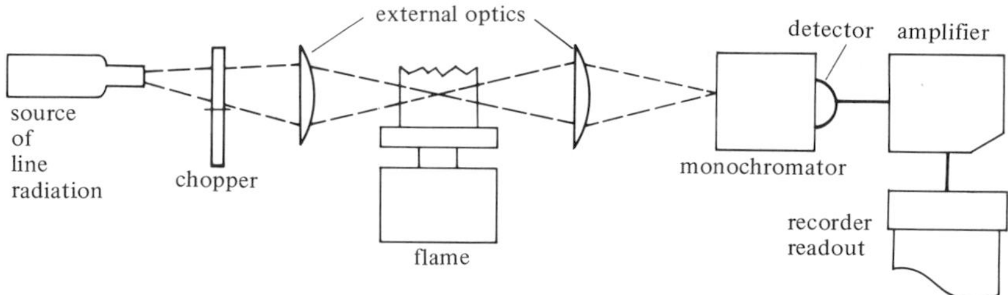

Figure 13.11 / *Basic components for a flame atomic absorption measurement.*

flame is not modulated, the measurement system discriminates against flame emission. In some instruments, the measurement system is designed to perform some computation so that a quantity proportional to absorbance or absorption is read out directly. Double-beam systems are also used, in which the source radiation alternately passes through and around the flame.

The choice of a radiation source and its operating conditions can strongly affect the atomic absorption measurements. In most cases, the radiation source is a hollow cathode discharge tube, one version of which is illustrated in Fig. 13.12. The hollow cathode discharge tube is usually formed from glass tubing. Electrodes are sealed in one end, a window of glass or silica is sealed on the other, and the internal pressure is adjusted to 1–10 torr with a fill gas of, in most cases, neon or argon. The cathode is a hollow cylinder with an internal diameter of less than 10 mm. The position and form of the anode are variable and largely irrelevant to our discussion. If sufficient voltage is applied between the two electrodes, a glow discharge starts and, at a suitable pressure, the discharge concentrates in the hollow of the cathode.

The mechanism of the discharge is usually explained as follows. Electrons leaving the cathode due to the applied voltage collide with atoms of the fill gas and cause their ionization. The positively charged ions gain kinetic energy from the potential difference, strike the cathode surface, and eject atoms from its crystal lattice. The sputtered atoms are excited in the hollow, where a relatively large concentration of atoms may be maintained in a region of favorable excitation conditions. At present, the excitation mechanism is not well understood. It is generally assumed that the excitation proceeds by collisions with excited atoms or ions of the fill gas or by direct electron impact.

The cathode must contain the element whose spectrum the lamp is to

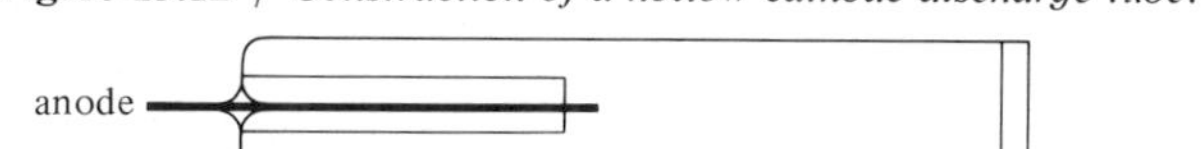

Figure 13.12 / *Construction of a hollow cathode discharge tube.*

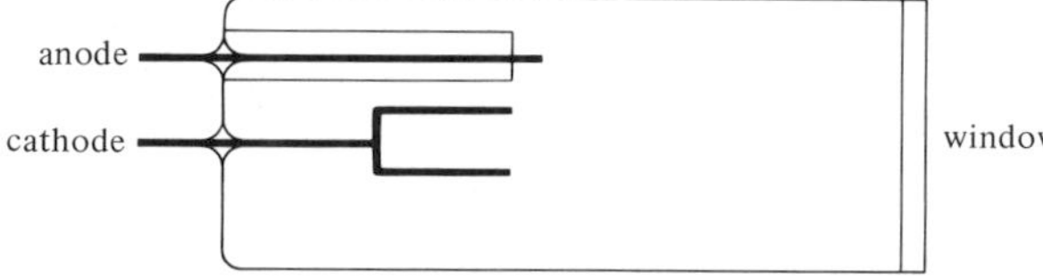

emit. In many cases, the cathode is fabricated from the pure metal. For precious metals, a small quantity of the metal in the form of a foil is placed in a cathode of a more common metal, such as copper or aluminum. For still other metals, the mechanical properties necessary for fabrication of a hollow cathode are obtained by using an alloy of the element of interest with some other metal.

The ideal source for atomic absorption measurements would emit a single, very sharp line of the element of interest. A hollow cathode discharge tube emits moderately sharp lines of all the elements present in the cathode and of the fill gas. The monochromator, as illustrated in Fig. 13.13, serves to isolate the one line of interest from all the lines emitted from the hollow cathode discharge tube.

The reason for employing a line source is discussed in Section 12.3.2. Atomic lines are considerably narrower than the bandpass of small monochromators. Thus, if a continuum source is employed, much of the radiation that reaches the detector lies at wavelengths outside the absorption line, and the effect of absorption is "diluted" by this unwanted radiation. If an infinitely narrow isolated source line is used, then all the radiation lies at the wavelength of maximum absorption, and the intensity of radiation reaching the detector will be most sensitive to changes in the atomic concentration in the absorbing medium. The width of a line isolated from a hollow cathode discharge tube is sufficiently great that the line cannot be considered "infinitely narrow" or "monochromatic" compared to the width of the absorption line. Thus, not all the source line radiation falls at the wavelength of maximum absorption, but in the usual case it all falls within the absorption line. From the preceding considerations, it can be seen that as the width of the source line increases, the sensitivity of the absorption measurement to changes in atomic concentration decreases until the continuum-source case is approached.

In hollow cathode discharge tubes, the width of the emitted lines is a function of lamp current, and it is desirable that the current remain as low as possible consistent with adequate intensity. Optimum operating conditions vary from element to element and with the construction of the lamp. Figure 13.14 illustrates the effect of lamp current on measured absorption as observed with a Ca hollow cathode discharge tube.

13.4.2 / The flame as an absorption medium

For atomic absorption measurements, the chief function of the flame is to produce atoms of the element of interest. It must be regarded as a disadvantage of the flame that the atomic emission spectrum of many elements is excited at the temperatures required for atomization. Although the measurement system of an atomic absorption spectrophotometer can be designed to discriminate against the flame emission signal, the emitted radiation at the wavelength of interest falls on the detector and contributes noise to the measurement. Thus, optimization of the flame for absorption

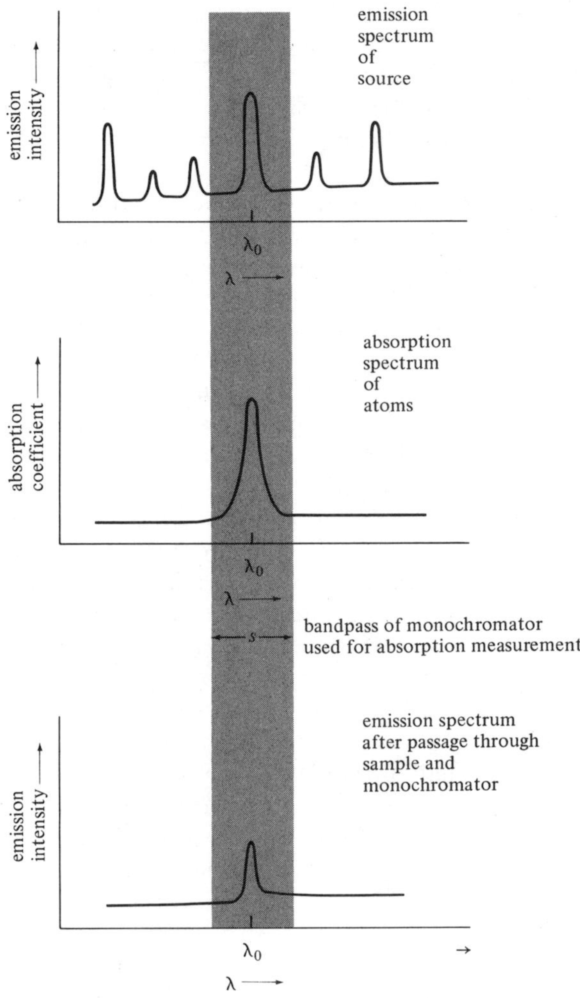

Figure 13.13 / *Isolation of a source line by the monochromator and absorption of the line by atoms in the flame.*

measurements requires a compromise between maximum atom production and minimum evaporation interferences in high-temperature flames and low flame emission in cooler flames. The dilemma is clearly illustrated in the atomic absorption determination of Ca. In the acetylene–air flame, the Ca absorption may be markedly depressed in the presence of phosphate anion. This interference can be eliminated by use of the nitrous oxide–acetylene flame, but in this flame the intensity of Ca flame atomic emission reaching the detector may be comparable to the intensity of radiation from the hollow cathode discharge tube.

For most elements, either the air–acetylene or the nitrous oxide–acetylene

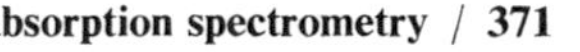

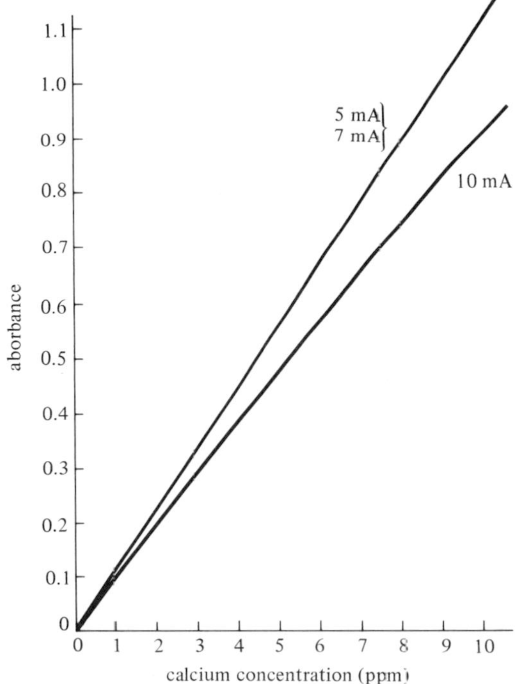

Figure 13.14 / *Effect of lamp current on measured absorption for* Ca *at* 4227 Å *with a* Ca *hollow cathode discharge tube.*

flame produced on a premix slot-type burner is used for atomic absorption measurements. The premixed air–hydrogen flame proves somewhat better for a few elements, principally those, such as As and Se, that have resonance lines in the vicinity of 2000 Å, where the transparency of hydrocarbon flame gases is low compared to that of hydrogen flames.

13.4.3 / Absorption measurements

Clearly, atomic absorption is not well suited for qualitative analysis, since in the typical arrangement a different radiation source must be substituted for each element. In quantitative analysis, a direct proportionality between absorbance and concentration, as in the Beer's-law relationship of solution spectrophotometry, is assumed to exist. The absorbance is given by

$$A = \log \frac{I^0}{I} \tag{13.3}$$

where I^0 and I are defined, respectively, as the readings obtained when a blank solution (or solvent) is aspirated and when the sample solution is aspirated. As in the flame emission technique, two types of spectral effects may interfere in the measurement: incomplete resolution of the source line and attenuation of the source radiation by concomitants of the analysis element.

By reference to Fig. 13.13, it can be seen that the monochromator serves to isolate a single source line from the emission spectrum of the hollow cathode discharge tube. Usually, the emission spectrum is sufficiently simple that the resolution of the monochromator employed is adequate to isolate a resonance line, but two types of difficulty may occur. (1) Some lamps exhibit a continuum that arises principally from hydrogen impurity in the fill gas. (2) The lamp exhibits a nonabsorbed line that is not resolved from the resonance line. This line may be due to fill-gas atoms or ions or to atoms or ions of any other elements present in the lamp as well as nonabsorbed atomic or ionic lines of the analysis element. In either of the above cases, the effect is the addition of an intensity term to both the numerator and denominator of Eq. (13.3), so that the measured absorbance can be represented by

$$A = \log \frac{I_1{}^0 + I_2{}^0}{I_1 + I_2} \tag{13.4}$$

where $I_1{}^0$ and $I_2{}^0$ are the signals due to the resonance line and the nonabsorbed radiation, respectively, when blank is aspirated, and I_1 and I_2 are the readings due to the resonance line and nonabsorbed radiation, respectively, when sample is aspirated. From Eq. (13.4), it can be seen that the effect of the nonabsorbed radiation is an undesirable nonlinearity in the relationship between measured absorbance and concentration.

If the nonabsorbed radiation is due to a continuum emission, the usual remedy is to replace the lamp. The line-to-continuum intensity ratio can be improved up to the point at which the limiting resolution of the monochromator is approached by decreasing the slit width of the monochromator —the intensity of continuum radiation reaching the detector is proportional to the square of the slit width, and the line radiation intensity is proportional to the first power of the slit width. A decrease in slit width may also be used in some cases to eliminate an unwanted source line. If this remedy fails, then it may be necessary to seek an alternative absorption line.

Attenuation of the source radiation by concomitants in the analysis solution can occur by background molecular absorption or scattering of the source line or actual overlap of atomic lines. These difficulties are analogs of the concomitant emission problem discussed with respect to the flame emission method. This type of interference is less likely to occur in AA than in AE, because the effective resolution is considerably better in AA (the width of the source line) than it is in AE (the monochromator bandpass). Thus, the ratio of line-to-background absorption is usually high because all the radiation to be absorbed falls within the absorption line; and, since atomic lines are narrow, the occurrence of overlap is rare.

Although spectral interferences of this type occur less often in AA than in AE, they are more difficult to detect and eliminate, because the use of a line source eliminates the possibility of scanning to observe the background on either side of the absorption line. An efficient correction for broad-band background may be made by using a continuum source in addition to the hollow cathode source. The line source reads line-plus-background absorption; the continuum source reads only background absorption. Background-compensation devices of this type are now included on some commercial instruments. When a line-overlap interference is detected, the only remedies are to seek another absorption line or to perform a separation of the analyte and interferant elements before measurement.

When very small absorbance values are to be measured, it is desirable to use scale-expansion techniques. It is convenient in such cases to make use of a recorder readout and transmittance or absorption values rather than absorbance. The three quantities are related as follows:

$$T = \frac{I}{I^0} \tag{13.5}$$

$$\alpha = \frac{I^0 - I}{I^0} = 1 - T \tag{13.6}$$

$$T = 10^{-A} = \exp(-2.3A) \tag{13.7}$$

where T and α are, respectively, transmittance and absorption. If A is sufficiently small (less than about 0.08, corresponding to 80–100 %T) then we can make use of the approximations

$$\exp(-2.3A) = 1 - 2.3A = T \tag{13.8}$$

$$\alpha = 2.3A \tag{13.9}$$

Thus, for low concentrations, α should be directly proportional to concentration in the same way that absorbance is directly proportional to concentration.

The scale-expansion technique is illustrated in Fig. 13.15. The first frame shows an unexpanded 10 percent absorption reading. The second frame shows a scale expansion of 2. To achieve this scale expansion, the sensitivity of the measurement system is increased (amplifier gain increased or multiplier phototube voltage increased) and the zero level is offset until the pen returns to its original 100 percent level. For 2× scale expansion, the true zero reading now lies one full chart span off scale, as indicated by the dashed line. Still greater scale expansion may be obtained by further increasing the sensitivity and offsetting the zero level as indicated for a scale expansion of 5. The amount of scale expansion that can be achieved may be limited by the amount of zero offset available, or, more fundamentally, by the signal-to-noise ratio. Scale expansion does not improve the signal-to-noise ratio; it only makes the signal easier to read, and may therefore improve the precision of the measurement by reducing uncertainty in the reading.

The line chosen for the atomic absorption measurement is generally the

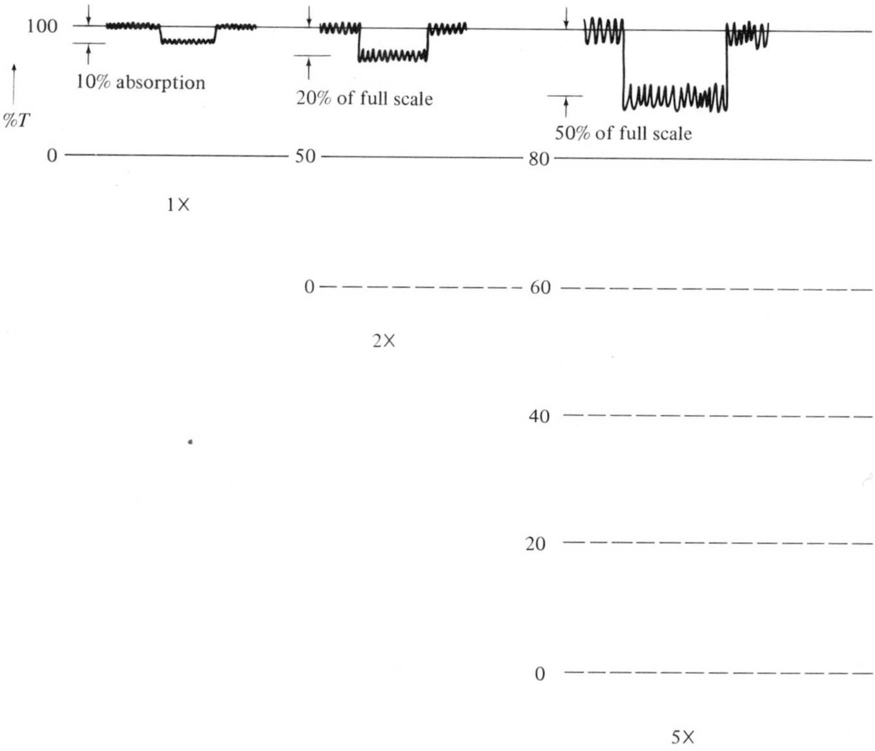

Figure 13.15 / *Illustration of the technique of scale expansion for absorption measurements.*

line with the greatest oscillator strength that corresponds to a transition from the ground state to some excited state. The reason for this choice is clearly shown in Eq. (12.25), which states that the absorption is directly proportional to the concentration of atoms in the lower state and to the oscillator strength of the transition. A line different from this may be chosen for any of the following reasons.

If the ground state is a multiplet with only slightly differing energies, the concentration of atoms in a level of slightly higher energy but greater statistical weight may be higher than the concentration of atoms in the lowest energy level. From the Boltzmann equation [Eq. (10.25)], it follows that

$$N_i = N_j \frac{g_i}{g_j} \tag{13.10}$$

where the subscripts i and j denote the upper and lower level, respectively. Aluminum provides an example of this situation. Its line at 3082 Å, with an oscillator strength of 0.22, corresponds to a transition from the ground state, which has a statistical weight of 2; but the 3082-Å line has a lower sensitivity in flame absorption than the 3093-Å line, with an oscillator

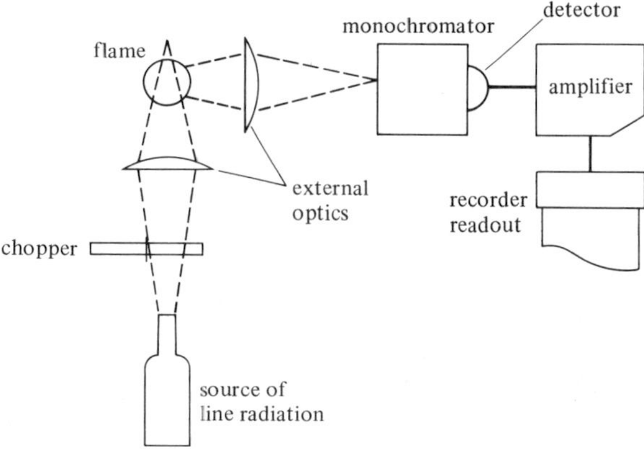

Figure 13.16 / *Basic components for a flame atomic fluorescence measurement.*

strength of 0.23, which corresponds to a transition from a level with a statistical weight of 4 lying 0.13 eV above the ground state.

The transparency of the flame gases may dictate the choice of an analytical line for elements such as As, Se, and Hg, for which the resonance line of greatest oscillator strength lies at very short wavelengths. For Hg, the strongest line lies at 1850 Å and is completely inaccessible for flame absorption.

Spectral interferences associated with either the source line or the absorption line may also influence the choice of an analytical line.

13.4.4 / Methods of quantitative analysis

Both the analytical-curve and standard-additions methods are widely used in AA measurements. Absorbance, or, for low concentrations, absorption readings, are used rather than emission readings, but otherwise the application of the two methods to AE and AA measurements is identical.

13.5 / FLAME ATOMIC FLUORESCENCE SPECTROMETRY

13.5.1 / The flame atomic fluorescence experiment

Atomic fluorescence is the most recent of the three techniques to be applied to chemical analysis, and, for this reason, the theory and practice of flame fluorescence are less well defined than that of the other two techniques. The basic components of the flame atomic fluorescence experiment are, as shown in Fig. 13.16, similar to those of AA, but the radiation source is arranged so that it is not viewed directly by the monochromator. In the usual case, the source–flame axis has been approximately at right angles to the flame–monochromator axis, but this is not an essential feature of the experiment, and other arrangements have been employed.

In addition to the difference in source arrangement, there are less obvious differences in the AA and AF experiments, principally in the flame configuration and the external optics. A square or circular cross-section flame can be more readily irradiated for AF than the slot-type flame used for AA, and an optical system of high light-gathering ability is much more important for AF than for AA.

The characteristics sought in a source for AF measurements are also somewhat different from those for AA measurements. Line sources, rather than continuum sources, usually are employed because they provide greater intensity over the absorption linewidth of the analyte atoms and greater selectivity (the effective resolution is determined by the width of the exciting line rather than by the monochromator bandpass). The importance of high intensity can be understood by consideration of Eqs. (12.24) and (12.26), which show that the intensity of the fluorescence is directly proportional to the source intensity. Equation (12.25) shows that the absorption, α, does not depend on source intensity. Since the monochromator does not "see" the source in AF measurements, the width of the source line is unimportant in AF but is, of course, very important in AA measurements.

The two types of spectral sources most widely employed for AF measurements are hollow cathode discharge tubes and electrodeless discharge tubes. The properties of hollow cathode discharge tubes have been described in Section 12.4.1. When employed for AF measurements, they have usually been operated at somewhat higher current than proves optimum for AA, thus gaining intensity at the expense of linewidth.

Electrodeless discharge tubes have been most successfully employed for the more volatile elements. In favorable cases, they produce intense narrow lines and have been used in place of hollow cathode discharge tubes for AA determination of elements such as As and Se. They are usually constructed in the form shown in Fig. 13.17 from silica tubing. The bulb is typically 1 cm in diameter and several centimeters long. A small amount of the element of interest and some inert gas (Ar, Kr, Xe) at a pressure of 0.5–10 torr are sealed in the bulb.

Most electrodeless discharge tubes used for AF measurements have been excited by microwave radiation at 2450 MHz. Power supplies capable of producing up to 100 W of radiant power are typically employed. The discharge tube is immersed in the microwave field in a cavity, and ionization of the fill gas is initiated by use of a spark coil of the type used for vacuum testing (Tesla coil). The mechanism by which the spectrum of the element of interest is excited is usually described as follows. Free electrons in the tube move under the influence of the field and transfer energy to the contents of the tube by collisions with fill-gas atoms and ions. Further ionization results, and the discharge becomes self-sustaining. Initially, only the spectrum of the fill gas is produced, but as the temperature of the tube rises, the element of interest is volatilized and its atoms are also excited. Eventually, the fill-gas spectrum is almost entirely replaced by that of the element of interest, and, as the tube temperature reaches a steady state, the intensity of the line emission stabilizes. To obtain a stable output, the tube walls must

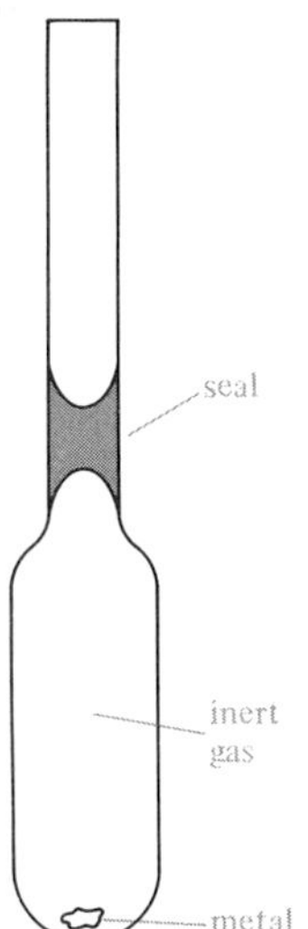

Figure 13.17 / *Illustration of usual form of an electrodeless discharge tube line source.*

be at sufficiently high temperature to prevent condensation of the element. It is usually necessary to provide some kind of shielding to keep the temperature of the tube walls constant.

Electrodeless discharge tubes can be prepared for elements for which a vapor pressure in the range 10^{-3}–1 torr can be produced at the operating temperature of the tube. For volatile elements, such as Zn, Cd, Hg, Se, Te, the tubes are prepared from the pure element; for less volatile elements, the tubes are prepared from volatile compounds, usually the halides.

13.5.2 / The flame as a medium for atomic fluorescence

The requirements of a flame for AF are identical to those for AA with one exception. The flame composition affects the power efficiency [the Φ term of Eq. (12.26)], and it is desirable to choose a flame that has a low concentration of "quenchers," that is, species that are efficient in deexcitation of the element of interest. The most important quenching species in flames are CO, CO_2, and N_2. The CO and CO_2 concentrations may be minimized by the use of hydrogen flames instead of hydrocarbon flames. The N_2 concentration may be minimized by preventing air entrainment and by use of $Ar–O_2$ mixtures as the oxidant instead of air. Air entrainment is minimized by the use of burners that give a flame of minimum turbulence and that make provision for surrounding the analytical flame with a second flame or a flow of inert gas.

13.5.3 / Fluorescence measurements

AF, like AA, is not well suited to qualitative analysis. Quantitative analysis requires measurement of relative values of the intensity of emission of an

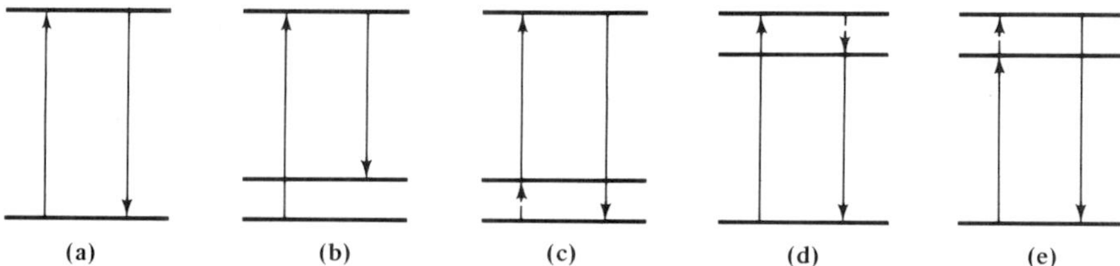

Figure 13.18 / *Some types of atomic fluorescence.*
(a) *Resonance fluorescence.*
(b) *Stokes direct-line fluorescence.*
(c) *Anti-Stokes direct-line fluorescence.*
(d) *Stokes stepwise-line fluorescence.*
(e) *Thermally assisted anti-Stokes stepwise-line fluorescence.*
The solid lines represent radiational processes, and the dashed lines represent nonradiational processes.

atomic line of the element of interest. In making such measurements, AF combines good features of the AE and AA techniques in that it is an emission measurement like AE, but it has the selectivity of AA since the excitation depends on the absorption of atomic line radiation. The principal difficulty encountered in the measurement of fluorescence intensities is due to the scatter of source radiation by concomitants of the analyte. Scattered radiation that reaches the detector is read as fluorescence intensity but is unrelated to the analyte concentration. Like the background absorption of AA, this type of interference is difficult to detect and eliminate when line sources are employed for excitation. Nearby nonresonance lines and nearby lines of elements not present in the sample have been used to detect the interference. Correction should be possible by use of a technique similar to that employed in AA for background absorption, in which the signals due to a continuum and line source are read alternately.

The factors that govern the choice of the measurement line for AF are somewhat more complex than for AA. In the most important case for analytical atomic fluorescence, the wavelength of the fluorescence emission is the same as the wavelength of the exciting radiation; but there are several processes by which the fluorescence radiation may occur at either longer or shorter wavelengths than the exciting radiation, and some of these are of analytical significance. The processes are illustrated in Fig. 13.18. A complete discussion of factors affecting the relative intensity of fluorescence lines is inappropriate to this text, but some of the more important considerations are apparent from the intensity expressions derived in Section 12.3.2. Equation (12.26) states that the fluorescence intensity is directly proportional to the intensity of radiation absorbed, which, in turn, as shown in Eq. (12.24), depends on the source intensity, the concentration of atoms in the lower term of the transition, and the oscillator strength of the transition.

Consider, for example, the term diagram for Mg in Fig. 12.3. The most intense line in a Mg line source is usually the 2852.1-Å line, only the ground

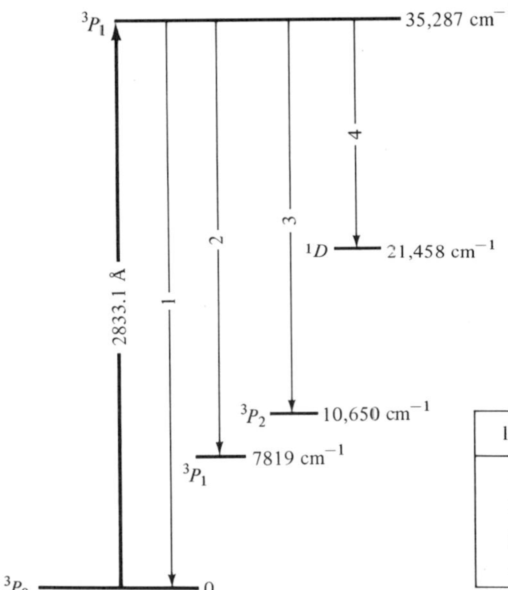

line	λ (Å)	A_i (sec^{-1})
1	2833.1	0.6 × 10^8
2	3639.6	0.4 × 10^8
3	4057.8	3.1 × 10^8
4	7229.0	0.02 × 10^8

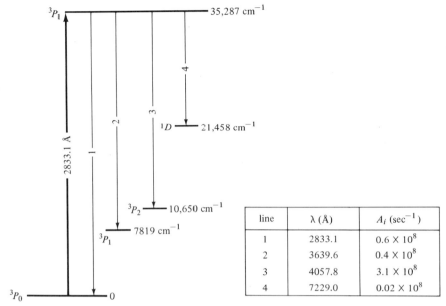

Figure 13.19 / *A partial term diagram for* Pb.

term is appreciably populated at flame temperatures, and the 2852.1-Å line has the largest oscillator strength. Clearly, the 2852.1-Å line is expected to be the strongest line in the fluorescence spectrum of Mg.

A partial term diagram for Pb is shown in Fig. 13.19. Only the ground term $(6s^2 6p^2)\ {}^3P_0$ is appreciably populated at flame temperatures, and the 2833.1-Å line is the strongest Pb absorption line. Unlike the Mg case, however, there are four possible routes for radiational deactivation of the excited state, and, neglecting reabsorption, the four fluorescence lines will, from Eq. (12.26), have intensities in the ratio

$$I_1 : I_2 : I_3 : I_4 = \Phi_1 : \Phi_2 : \Phi_3 : \Phi_4$$

From the definition of Φ, the power efficiency [Eqs. (12.27) and (12.28)], it can be shown that

$$I_1 : I_2 : I_3 : I_4 = A_1 : A_2 \frac{\nu_2}{\nu_1} : A_3 \frac{\nu_3}{\nu_1} : A_4 \frac{\nu_4}{\nu_1}$$

where A_i is the Einstein transition probability for spontaneous emission for the transition designated by the subscript. The transition probability for the 4057.8-Å line is by far the largest, and the 4057.8-Å line is the strongest line in the fluorescence spectrum of Pb.

Thus, in selecting a line for fluorescence measurements, attention must be given to both the absorption and emission processes. Instrumental factors or a need to avoid interferences may also affect the line choice.

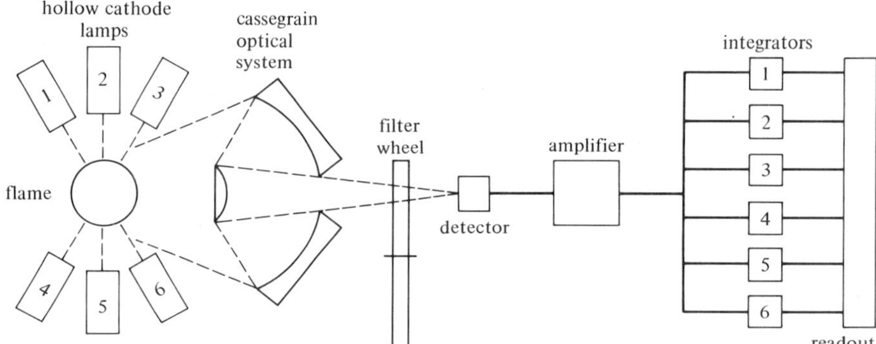

Figure 13.20 / *Schematic diagram of instrument for simultaneous multielement analysis by atomic fluorescence.*

In AF measurements, the role of the monochromator is somewhat different than in either AE or AA measurements. Since the monochromator does not view the source and since flame emission may be discriminated electronically, we might very well ask if it serves any function. When a flame atomizer and a detector with a wide spectral response are used, the monochromator does serve the very useful and necessary function of sharply limiting the amount of radiation falling on the detector. However, nondispersive systems can be used to considerably greater advantage for AF than for either AE or AA measurements, and such systems have been described in which interference filters and multiplier phototubes of limited spectral response are used.

Of the three flame techniques, AF is best suited to simultaneous multielement analysis, that is, the simultaneous determination of the concentration of several elements in a single portion of sample. This capability arises primarily from the effectiveness of nondispersive atomic fluorescence measurements. One approach that has been taken is to use individual sources for each element, a single flame, and a single detector. The radiation sources may be arranged in any convenient fashion about the flame with only one restriction—the detector must not view the source directly. An optical filter for each element is mounted on a filter wheel such that the filters can be brought sequentially into position between the flame and detector, and there is a separate electronic channel for each element. An arrangement of this type for simultaneous determination of six elements is illustrated in Fig. 13.20. Logic circuitry controls the operation of the instrument so that as sample is aspirated into the flame, the measurement operation might proceed as follows.

1. A filter is brought into position.
2. An electronic pulse is applied to the appropriate lamp so that a pulse of radiation is emitted, exciting the fluorescence of the selected element.

3. The fluorescence signal produced at the detector is amplified and applied to the appropriate integrator.
4. The next filter is brought into position and steps 2 and 3 are repeated.

The process might continue until a number of signals have been integrated for each of the six elements.

13.5.4 / Methods of quantitative analysis

The analytical curve and standard additions methods are commonly used. They have been described for AE in Section 13.3.4, and the approach for AF is identical.

13.6 / COMPARISON OF FLAME TECHNIQUES

In the preceding sections, we have described three rather similar techniques for quantitative determination of elements at the trace level. Inevitably, the question arises as to which of the three techniques is "best." Thoughtful consideration of the preceding material provides the answer: There is no "best" technique; there are only "best" solutions to particular problems. The three techniques should be regarded as complementary, and the one used for a given analysis should be determined on the basis of the types of interferences to be encountered, available instrumentation, and so on. Some general guidelines may be summarized as follows.

A comparison of experimental limits of detection for the three techniques has led to the following rule of thumb: For elements with resonance lines at wavelengths longer than 4000 Å, best detection limits are obtained by AE; for elements with resonance lines at wavelengths shorter than 3000 Å, either AA or AF will give the best limits. In the intermediate wavelength range, rather similar limits will be obtained by all three techniques. Table 13.2 compares experimental detection limits for the three techniques for 32 elements.

The three techniques suffer in approximately equal measure from interferences. All interferences that pertain to the atomization process affect all three techniques in identical fashion. Spectral interferences have been described in the appropriate sections for each technique. Spectral interferences occur to a greater extent in AE than in AA or AF, but are perhaps easier to correct for.

The precision of measurement for a given sample should be about the same for all three methods—the relative standard deviation should be 1–2 percent. The accuracy of measurement depends more on proper choice of standards, elimination of interferences, and sample preparation than on the measurement technique.

The instrumentation required for the three methods is similar in cost and complexity. That for AE is simplest, since no external radiation source is required. We have noted in describing AF measurements the suitability of

TABLE 13.2 / *Comparison of experimental limits of detection* (μg/ml) *in atomic flame spectrometry*

	AE		AA		AF	
Element	λ(Å)	*Detection limit*	λ(Å)	*Detection limit*	λ(Å)	*Detection limit*
Ag	3281	0.02	3281	0.0005	3281	0.0001
Al	3962	0.005	3962	0.04	3962	0.1
As	2350	50	1937	0.1	1937	0.1
Au	2676	4	2428	0.01	2676	0.005
Be	2349	0.1	2349	0.002	2349	0.1
Bi	2231	2	2231	0.05	2231	0.005
Ca	4227	0.0001	4227	0.0005	4227	0.02
Cd	3261	2	2288	0.0006	2288	0.000001
Co	3454	0.05	2497	0.005	2497	0.005
Cr	4254	0.005	3579	0.005	3579	0.05
Cu	3274	0.01	3247	0.003	3247	0.001
Fe	3720	0.05	2483	0.005	2483	0.008
Ga	4172	0.01	2874	0.07	4172	0.01
Ge	2652	0.5	2652	0.1	2652	0.1
Hg	2537	40	2537	0.2	2537	0.0002
In	4511	0.005	3039	0.05	4511	0.1
Mg	2852	0.005	2852	0.0003	2852	0.001
Mn	4031	0.005	2795	0.002	2795	0.006
Mo	3903	0.1	3133	0.03	3133	0.5
Ni	3415	0.6	2320	0.005	2320	0.003
Pb	4058	0.2	2833	0.01	4058	0.01
Pd	3635	0.05	2746	0.02	3405	0.04
Rh	3692	0.3	3435	0.03	3692	3
Sb	2598	20	2175	0.07	2311	0.05
Si	2516	5	2516	0.1	2040	0.6
Se	—	—	1960	0.1	1960	0.04
Sn	2840	0.3	2246	0.03	3034	0.05
Sr	4607	0.0002	4607	0.004	4607	0.03
Te	2383	200	2143	0.1	2143	0.005
Tl	3776	0.02	2768	0.02	3776	0.008
V	4379	0.01	3184	0.02	3184	0.07
Zn	2139	50	2139	0.002	2139	0.00002

SOURCE: J. D. Winefordner and R. C. Elser, *Anal. Chem.* **43**(4), 24A(1971).

this technique for nondispersive measurements and simultaneous multi-element analysis.

As indicated in Table 13.2, a large number of elements can be determined by flame spectrometry. In fact, a survey of the literature of the field seems to indicate that only a few elements have not been determined by at least one of the three flame techniques. Elements not determined directly by flame spectrometry are the short-lived radioactive elements, and the elements at the upper right and extreme right of the periodic table, that is, C, N, P, O, S, H, the halogens, and the noble gases. The principal atomic lines of these latter elements fall in the far UV region of the spectrum, and thus their direct determination by flame techniques is precluded; but for many of these, indirect methods have been developed in which the concentration of the element is determined by flame spectrometric measurement of another element bearing a known stoichiometric relationship to the element of interest.

The range of materials analyzed by flame techniques is also impressively large. Trace elements may have profound effects on the health of plants and animals and on the physical and chemical properties of materials. Thus, flame techniques are routinely applied to the determination of trace elements in biological materials, such as blood, serum, urine, tissue, plant materials, soils and soil extracts, fertilizers, and foodstuffs, and to the determination of trace elements in such other materials as metals, ores, glass, plating solutions, petroleum products, and cement. Slavin (see the Suggested Readings at the end of the chapter) has given an interesting survey of applications of the atomic absorption technique. Two examples may serve to illustrate the utility of flame spectrometry in elemental analysis.

Sodium and potassium are important constituents of human body fluids. They possess no single specific function, but they must be maintained within narrow concentration ranges for proper functioning of the body. Consequently, monitoring of the sodium and potassium concentrations in blood or serum is often required in medical treatment. The determination of sodium and potassium, and indeed of all the alkali metals, is of more than average difficulty by classical techniques, but these elements are readily determined by flame emission, and the flame emission determination of Na and K is today a very common procedure in clinical laboratories.

Flame atomic absorption has achieved widespread use in the diagnosis of heavy metal poisioning. Lead, mercury, thallium, and cadmium are all readily determined in blood or urine by the atomic absorption technique. Cases of lead poisoning often occur among young children living in older urban areas. Many of these cases appear to be due to the ingestion of paint chips containing large amounts of lead. In an attempt to detect evidence of lead poisoning before the victim suffers irreversible damage, screening programs have been set up in many areas, employing flame atomic absorption to measure the lead content of blood samples of children with no overt symptoms of lead poisoning.

SUGGESTED READINGS

R. Mavrodineanu and H. Boiteux, *Flame Spectroscopy*, Wiley (New York, 1965).

R. Herrmann and C. T. J. Alkemade, *Chemical Analysis by Flame Photometry*, P. T. Gilbert, Trans., Wiley-Interscience (New York, 1963).

W. J. Price, *Analytical Atomic Absorption Spectrometry*, Heyden (London, 1972).

W. Slavin, *Atomic Absorption Spectroscopy*, Wiley-Interscience (New York, 1968).

I. Rubeska and B. Moldan, *Atomic Absorption Spectrophotometry*, P. T. Woods, Trans., Ed., CRC Press (Cleveland, Ohio, 1969).

J. D. Winefordner and R. C. Elser, "Atomic Fluorescence Spectrometry," *Anal. Chem.* **43** (4), 24A (1971).

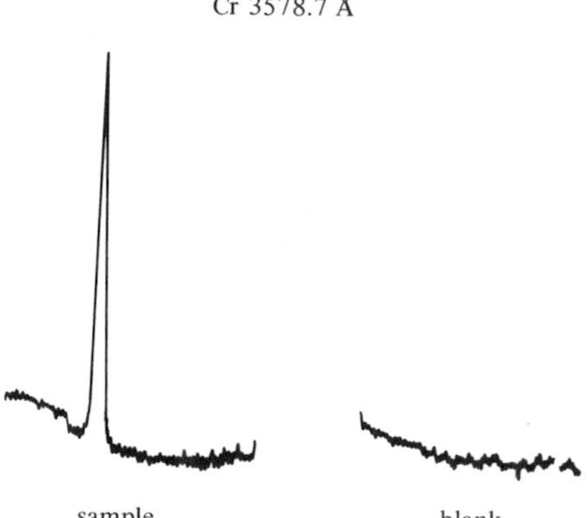

Figure 13.21 / *Wavelength scan of the emission line for determination of* Cr *in an* Fe *alloy.* SOURCE: V. A. Fassel, R. W. Slack, and R. N. Kniseley, *Anal. Chem.* **43**, 186 (1971).

PROBLEMS

1 / Two solutions contain identical amounts of the element M. Solution 1 is to be considered the standard and solution 2 the sample. For each of the measurement conditions noted below, compare the atomic concentration produced with solutions 1 and 2, using the following descriptive terms: (i) atomic concentrations *identical*, (ii) atomic concentration for solution 2 *higher* than for solution 1, (iii) atomic concentration for solution 2 *lower* than for solution 1, (iv) atomic concentration varies in *indeterminate* fashion due to two or more possibly compensating effects. Briefly justify your answers. Assume that a premix burner is used.

(*a*) The flow rate of air (the gas used for aspirating the solutions) is higher when solution 2 is sprayed.
(*b*) The viscosity of solution 2 is greater than that of solution 1.
(*c*) Solution 2 contains a surface-active agent that results in a smaller mean drop size when solution 2 is sprayed.
(*d*) M is present in solution 1 as the hydrated cation and in solution 2 as the metal acetyl acetonate.
(*e*) Solution 2 contains 10 ppm KCl in addition to M.
(*f*) Solution 2 is 1 *M* in NaCl.
(*g*) The flow rate of fuel is higher when solution 2 is sprayed.

2 / If the total pressure of Na (atoms plus ions) in a flame is 10^{-8} atm and the equilibrium constant for the ionization reaction, K_i, is 7.4×10^{-8} atm at the temperature of the flame (2800°K), calculate the fraction of Na present as un-ionized atoms for the following conditions:

(*a*) All the electrons in the flame result from the ionization of Na.
(*b*) The partial pressure of electrons in the flame is buffered at 10^{-7} atm by ionization of K.

3 / The wavelength scan of the emission line used for the determination of Cr in an Fe alloy is shown in Fig. 13.21. The background recording was made while aspirating a blank solution containing 2.5 mg/ml of Fe (corresponding to a sample containing 50 wt percent of Fe). Relative intensity values were obtained by setting the monochromator to the peak of the Cr 3578.7-Å line, alternately aspirating sample and blank solutions, and recording the relative intensity reading for the sample as the difference between the sample and blank readings. The relative intensity values reported below are average values for five replicate measurements.

(*a*) Comment on the validity of the above-described approach to measurement of the relative intensity values, basing your comments on the recorder tracing shown.

(*b*) Determine the concentration of Cr in the Fe alloy marked X.

Sample	*Wt* % Cr	*Relative intensity at* Cr *3578.7* Å
1	0.016	5.3
2	0.046	16
3	0.063	21
4	0.074	24
5	0.143	49
X		18

4 / Determine the concentration of Ca in a sample for which the following data were obtained by the standard addition method. The Ca standard solution contained 1 μg/ml Ca.

Solution	*Relative intensity*
20-ml sample, dil. to 25 ml	10
20-ml sample + 1-ml std., dil. to 25 ml	13.3
20-ml sample + 2-ml std., dil. to 25 ml	16.5

5 / Determination of the concentration, C_x, of the element M in a sample is to be carried out in the following way: The background-corrected relative intensity, I_x, is read while aspirating a portion of the sample. To a portion of the sample of known volume V_x is added a known volume V_s of standard solution of known concentration C_s, and the background-corrected relative intensity, I_{x+s}, is read while aspirating the resulting solution. Derive an expression for calculating C_x from the known quantities C_s, V_s, V_x, I_x, I_{x+s}.

6 / In the atomic absorption determination of the element M, an aliquot of the unknown gave an absorbance reading of 0.435. A solution obtained by adding 1 ml of 100 ppm M to a 9-ml aliquot of the unknown gave an absorbance reading of 0.835. What was the concentration of M in the unknown?

7 / Assume that Fig. 13.22 describes the radiation output of a spectral source for three sets of operating conditions. Predict the effect on the measurement of atomic absorption and atomic fluorescence at λ_0 as the source output changes from 1 to 2 to 3.

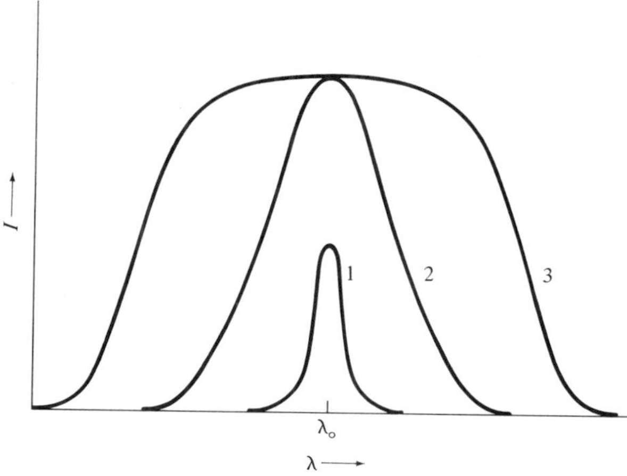

Figure 13.22

8 / Calculate the ratio of absorption expected for the Al lines at 3082 Å and 3093 Å in a flame at 2800°K. The 3082-Å line arises from a transition between the $^2P_{1/2}$ ground state and a $^2D_{3/2}$ excited state. The oscillator strength of the transition is 0.22. The 3093-Å line arises from a transition between a $^2P_{3/2}$ state that is 0.13 eV above the ground state and a $^2D_{5/2}$ excited state. The oscillator strength is 0.23. ($k = 8.8 \times 10^{-5}$ eV/°K.)

9 / Given an atomic absorption spectrophotometer with a slot-type premixed burner and a supply of common fuels and oxidants, suggest remedies to the following problems encountered in the use of the atomic absorption technique. The sample may not be altered, but you may change any of the settings normally accessible on a spectrophotometer.

(*a*) The sensitivity is poor and oxide formation in the flame is suspected.
(*b*) A chemical interference, such as the effect of phosphate on Ca, is suspected.
(*c*) The signal due to the source line is so low that amplifier noise is unacceptably high.
(*d*) The source line intensity is high and there is very little noise associated with the measurement, but the absorption is so low that it is difficult to read.

10 / For an atomic absorption measurement:

(*a*) What are the values of the transmittance, *T*, the absorption, α, and the absorbance, *A*, if a scale reading of 95 is obtained when solvent is aspirated and a reading of 90 is obtained when sample is aspirated?
(*b*) If the hollow cathode discharge tube used for the measurement above emitted appreciable continuum radiation in the region of the line of interest, how would the measured absorption differ from that obtained with a lamp that did not emit the continuum radiation?
(*c*) If a concomitant in the sample produced species in the flame that scattered the source radiation, in what way would the measured absorption differ from that for a solution with an identical concentration of the element of interest but no concomitants?

11 / In the flame emission determination of Mn, the following data were obtained. Calculate the concentration of Mn in the original sample solution.

Solution	*Reading*
original solution	43.2
10-ml aliquot of sample + 1 ml of 100-μg/ml Mn std.	75.6

12 / What values of transmittance, T, and absorbance, A, correspond to a measured absorption, α, of 0.30?

13 / The following emission data were obtained by the standard addition method, using a standard solution containing 2 μg/ml of the element of interest:

Solution	*Relative intensity*
20-ml sample, dil. to 25 ml	5
20-ml sample, 1-ml std., dil. to 25 ml	11.6

Calculate the concentration of the element of interest in the original sample.

chapter 14 / Arc/spark emission spectroscopy

chapter 14

14.1 / INTRODUCTION

In this chapter, we consider techniques of elemental analysis that employ electrical discharges for excitation of atomic line spectra. These techniques find application for both trace and micro analysis, and are particularly well suited to the rapid analysis of solid samples (e.g., mineral ores, metal alloys) for multiple elements.

The basic components of an arc/spark emission measurement are shown in Fig. 14.1. Typical features are as follows. The electrical discharge strikes between two electrodes, one of which is the sample or holds the sample. Several types of discharges are in common use, and these will be described in Section 14.2. Electrical discharges are typically more energetic than the flame excitation sources described in Chapter 13; as a consequence, the spectra produced by arc or spark excitation are considerably more complex, and dispersing elements of high resolving power are required for adequate resolution. Most spectrographs manufactured today for analytical purposes are grating instruments of 1.5–3.5 m focal length with plane gratings of 600–1200 grooves/mm, but many large quartz prism instruments of similar focal length are still in use. Instruments of this type typically provide a reciprocal linear dispersion of 5 Å/mm or better at the detector. Photographic and photoelectric detectors are both widely used. Photographic detectors are most commonly plates (i.e., the photographic emulsion is mounted on one side of a thin glass plate), with 4 × 10 in. plates perhaps the most commonly used size. Multiplier phototubes with good UV spectral response are the most commonly used photoelectric detectors. If a photographic detector is used, readout of the spectral information is an entirely distinct operation, and may be separated from the other aspects of the measurement by a considerable distance and lapse of time. Typical readout devices are described in Section 14.3 for both photographic and photoelectric detectors.

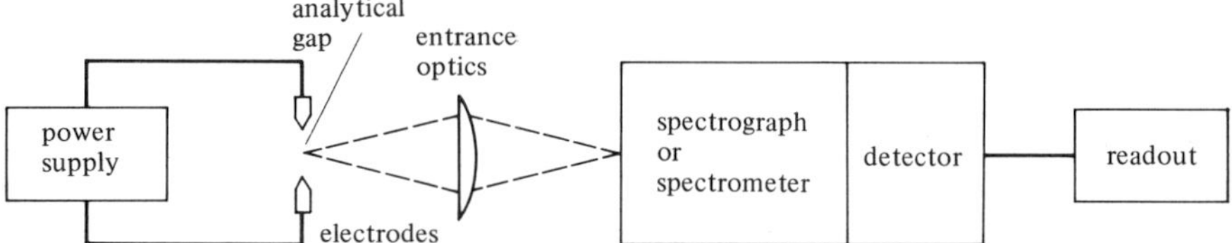

Figure 14.1 / *Basic components of an arc/spark emission measurement.*

14.2 / EXCITATION

14.2.1 / The free-burning DC arc

As usually employed for spectrochemical analysis, the arc is a steady, unidirectional flow of current of 1–30 A between a pair of metal or graphite electrodes spaced 1–20 mm apart. The voltage drop in such an arc is from 10 to 100 V, depending on the composition of the arc plasma,[1] the current, and the gap size. The arc does not obey Ohm's law, that is, the voltage across the gap decreases with increasing current; and the voltage drop is not linear from one electrode to the other but changes more rapidly near the electrodes than over most of the arc column, as shown in Fig. 14.2.

Figure 14.3 shows a simple rectifier circuit for production of a DC arc. Because the resistance of the arc decreases with increasing current, the current must be controlled in the supply circuit. A resistor could function

Figure 14.2 / *Variation of voltage between the electrodes of a DC arc at two levels of current.*

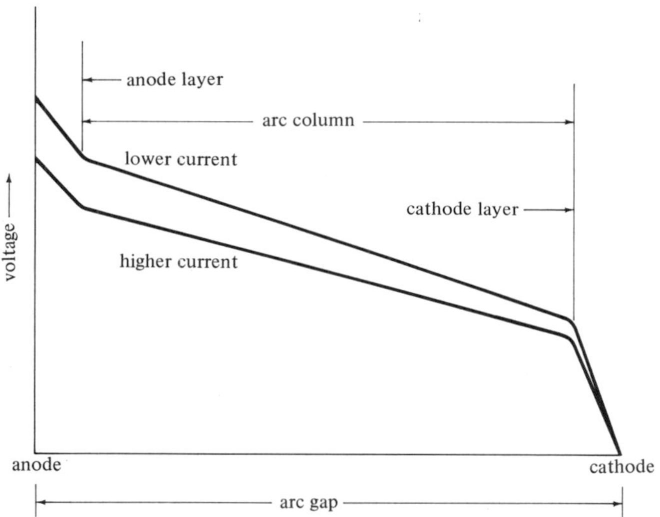

[1] Plasma in this context means simply an ionized gas that is approximately electrically neutral.

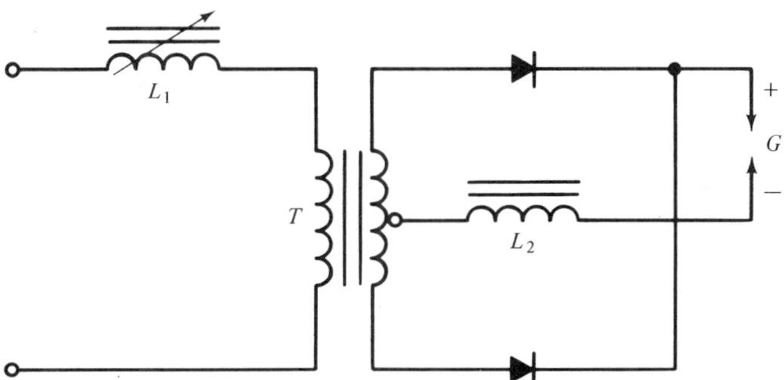

Figure 14.3 / *Rectifier power supply for a DC arc. Variable inductance L_1 provides control of the arc current, and L_2 filters the rectified current. T is a center tap transformer, and G is the analytical gap.*

as a simple current control, but the use of a resistor for this purpose entails the conversion of considerable electrical power into heat. A variable inductance in the input to the rectifier, as shown in Fig. 14.3, is usually a more satisfactory solution to the problem of current control. For stable operation of the arc, the voltage available from the power supply when no current is being drawn should be at least twice the voltage drop in the arc column. Most DC arc power supplies are designed for an open-circuit voltage of 200–250 V.

The arc is ignited by bringing the two electrodes into momentary contact or by striking a low-current spark to provide initial ionization of the arc gap. The arc is then maintained by thermal ionization of the plasma and by a supply of electrons from the cathode and positive ions from the anode. Some of the principal features of the free-burning DC arc are as follows:

A / *Temperature*

There cannot be exact thermodynamic equilibrium in the arc because energy and matter are continually flowing into and out of the arc, but in any small element of volume the conditions are close enough to equilibrium to be described by a temperature; that is, Eqs. (10.13) and (12.10) are applicable. The temperature in the arc column varies across its diameter, being highest in the center and decreasing toward the edges. The temperature may also be higher in the regions of larger voltage drop near the electrodes. Typical arc temperatures along the axis of the arc column range from about 4000° to 8000°K. The temperature achieved is dependent on the composition of the arc plasma and especially on the ionization energy of material in the gap region. If the elements in the plasma are easily ionized, the electron concentration is higher and the resistance is lower. At a given arc current, the energy dissipation is proportional to the resistance, so that the energy input to the arc and the temperature of the discharge are lower in the presence of elements of low ionization energy. The temperature of the arc is not

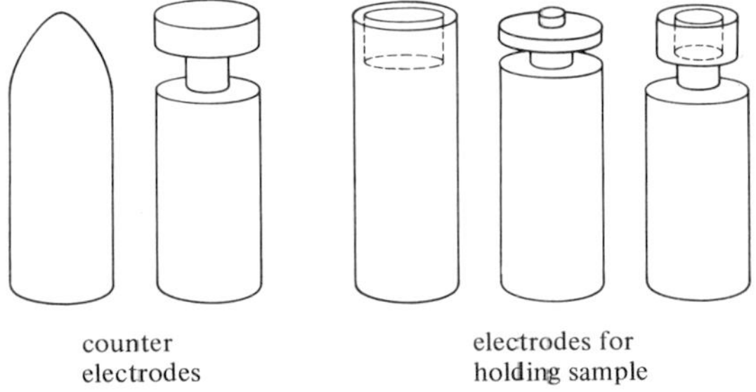

Figure 14.4 / *Some typical graphite electrode shapes. Narrow necks reduce thermal conductivity and thereby raise electrode operating temperature.*

strongly dependent on the arc current. Increasing the arc current increases the diameter of the arc column and decreases the voltage drop, so that the energy dissipation per unit volume of the arc plasma remains relatively constant.

B / *Sample introduction*

The material to be analyzed may be made one or both of the electrodes of the arc if it is suitable in form, is conductive, and does not melt excessively at the arc operating conditions; or the sample may be a powder, chips, or solution residue placed in or on one or both of the electrodes. The electrodes are heated by the passage of current, by radiative and conductive transfer of energy from the arc plasma, and by chemical reactions with the atmosphere. When the electrodes become sufficiently hot, both sample and electrode material vaporize and enter the arc plasma. Rods of carbon or graphite are most often used for the electrodes, because these materials are good conductors of electricity, remain solid at temperatures as high as 4000°K, and can be obtained in a highly purified form. Some of the more typical electrode shapes are shown in Fig. 14.4.

As the electrodes are heated by the arc, the most volatile components of the sample vaporize first, followed by those with higher boiling points. Figure 14.5 indicates the course of vaporization of the components of a sample. As the individual components vaporize, the composition of the arc plasma changes with time, and, since the temperature of the arc column depends on the composition, the excitation conditions also change with time. Since the intensity of emission of a spectral line depends on the concentration of the emitting atom and the arc temperature [Eq. (12.10)], it can be seen that this "selective volatilization" effect may greatly complicate the establishment of a quantitative relationship between emission intensity and sample composition.

Selective volatilization is not an unmitigated evil; benefit may sometimes be derived from it. It is apparent in Fig. 14.5 that volatilization of elements such as Cd and Pb is essentially complete before appreciable volatilization of elements such as Sc and Zr begins. Thus, spectral interferences may

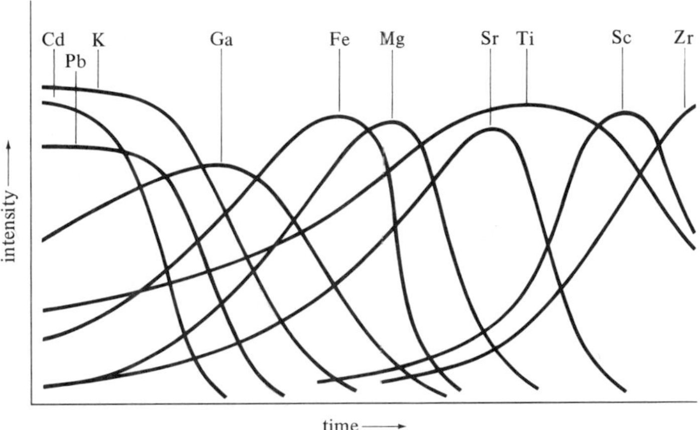

Figure 14.5 / *Idealized selective volatilization curves.*

sometimes be reduced and line-to-background contrast improved by proper selection of the exposure period: If the more volatile elements are sought, the exposure may be limited, for example, to the first 30 sec of the arc burn; if more refractory elements are sought, the arc might be allowed to burn for 30 sec before the spectrograph shutter is opened.

C / *Background emission*

In addition to the atomic line radiation sought in emission spectroscopy, the DC arc emits both continuum and band background radiation. Continuum radiation arises principally from the blackbodylike radiation of the electrode tips and incadescent particles within the arc column and from radiative recombination reactions. Band emission arises from some simple species that are stable in the cooler regions of the arc, the most notable of which is the CN radical. This species is formed when carbon electrodes are burned in air, and its intense band emission from this source makes most of the region between 3600 and 4200 Å unusable.

D / *Arc wander*

The arc column does not cover the electrode ends, but contacts them only at small spots. These contact spots tend to wander erratically over the electrode surface, sampling different portions of the electrode material as they move. In addition, the thermal gas currents generated by the arc can alter the position of the arc column. This wandering of the arc alters the fraction of emitted radiation that reaches the detector, so a detector that responded to instantaneous values of intensity would produce a signal that varied erratically with time.

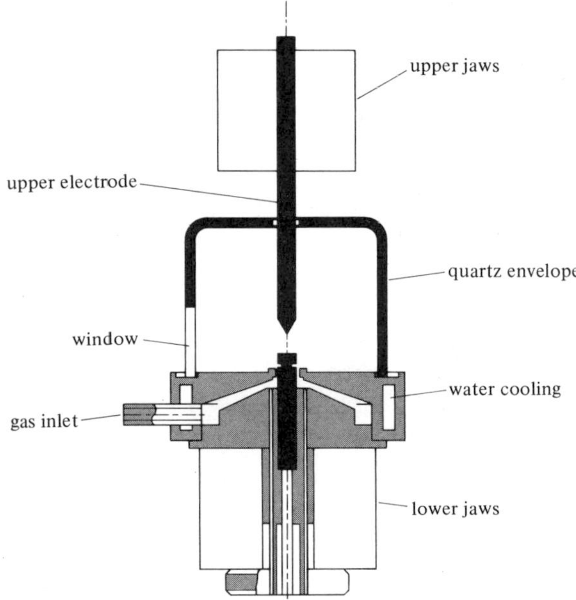

Figure 14.6 / *Vertical cross-section view of Stallwood jet.*

14.2.2 / The stabilized DC arc

Various approaches have been proposed for eliminating the shortcomings of the free-burning DC arc described above. It will suit our purposes to describe only one which appears to have gained rather wide acceptance, the Stallwood jet illustrated in Fig. 14.6. With the Stallwood jet, the sample electrode is placed in a water-cooled holder, and an argon–oxygen mixture is caused to swirl upward around the electrode as it burns. Both electrodes are enclosed in a quartz envelope so that a slight positive pressure is developed by the Ar–O_2 gas flow, and nitrogen is excluded from the arc region.

The Stallwood jet functions in at least four ways to improve the performance of the DC arc:

1. The annular stream of gas reduces arc wander and thereby improves the accuracy and precision of intensity measurements.
2. The exclusion of nitrogen prevents the formation of CN and thereby opens for use the 3600–4200 Å region obscured by CN emission with the free-burning arc.
3. Selective volatilization is reduced. The Stallwood jet may be arranged so that the sample electrode can be adjusted upward as it burns away. Material in the bottom of a deep-cratered electrode is kept cool by the forced flow of gas, and samples tend to volatilize unselectively in layers.
4. The gas composition establishes the excitation conditions, and, by varying the ratio of argon to oxygen, excitation conditions favorable to particular elements can be attained.

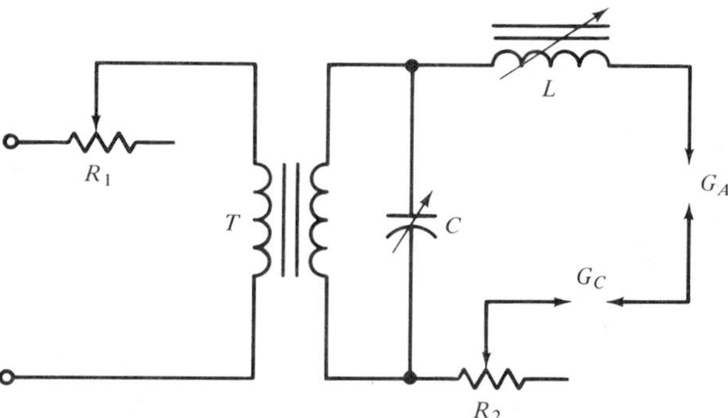

Figure 14.7 / *Power supply for the high-voltage spark. T is a high-voltage transformer; R_1 is for current control; R_2, L, and C control the electrical characteristics of the spark; G_C is the control gap; and G_A is the analytical gap.*

14.2.3 / Other arc devices

In the AC arc, the direction of current flow changes regularly with time, periodically reaching zero current. Thus, the arc must be restruck on each half-cycle. The intermittent arc is similar to the AC arc in that the current periodically goes to zero, but it differs in that the polarity of the discharge does not change. The advantage of the AC or intermittent arc is that with each reignition, the contact spots on the electrodes tend to move to new points. Thus, the effect of arc wander is minimized, since for the slow random motion of the unstabilized DC arc there is substituted a more rapid random motion, resulting in more uniform sampling and statistically more uniform illumination of the spectrograph.

14.2.4 / The high-voltage spark

A representative circuit for a high-voltage spark is shown in Fig. 14.7. The transformer, *T*, with an output of 15,000–40,000 V, charges the capacitor, *C*, until enough voltage is developed to break down the gaps. Such breakdowns typically occur 2–20 times during each half-cycle of the 60-Hz line voltage. The capacitor discharges through the inductance and resistance of the discharge circuit as well as through the gaps. The form of the discharge current and voltage pulse is controlled by the values of *C*, *L*, and R_2. These have values, respectively, of the order of a few thousandths to a few hundredths of a microfarad, a few ohms, and tens to hundreds of microhenries.

As in the case of arc discharges, at least one electrode forming the analytical gap contains the sample material. The size of the gaps is adjusted so that the breakdown voltage of the control gap determines the start of current flow through the circuit, and thereby ensures reproducible breakdown conditions. If the analytical gap determined the start of current flow, the breakdown conditions might vary with each spark because of the changing composition of the gap region. Various mechanical and electronic devices have been

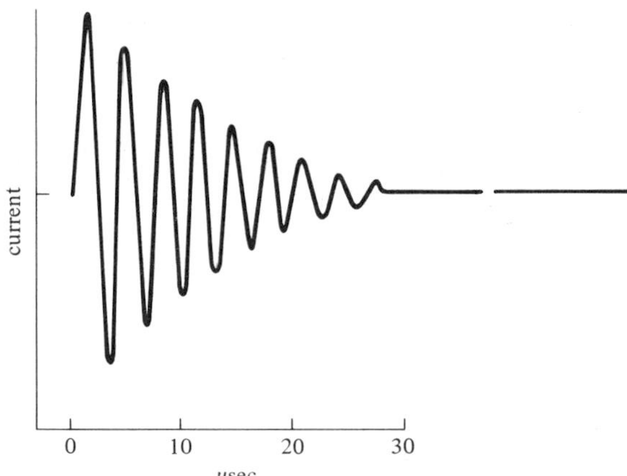

Figure 14.8 / *Variation of current with time for a single discharge of a high-voltage spark.*

employed to ensure reproducible breakdown of the control gap. A fuller discussion of these is to be found in the Suggested Readings at the end of the chapter.

Some of the principal features of the spark discharge are as follows.

A / *Excitation conditions*

With each breakdown of the gap, the current flow through the discharge circuit is a damped oscillatory discharge, as shown in Fig. 14.8. The natural frequency of the oscillation in hertz is equal to $[(1/LC) - (R^2/4L^2)]^{1/2}$, where L is in henries, C is in farads, and R is in ohms. For typical operating conditions, the frequency is of the order of 10^6 Hz. After the initial breakdown, the voltage drop across the gap is similar to that of the DC arc, about 50 V, and the train of discharge pulses continues until the voltage across the capacitor falls below the level required to sustain the breakdown. There then follows a relatively long off period, during which the capacitor is again charged to the breakdown voltage of the control gap.

The total energy available in a train of discharge pulses is determined by the size of C and the initial voltage on C. The peak current is determined by the energy stored on C and the values of L, C, and R. The peak current of the initial pulse may be as high as 1000–2000 A, but the average current of a spark discharge is usually from a few tenths of an ampere to a few amperes (reflecting the long off periods).

Excitation in the spark discharge varies as the current varies. At the maximum of the first current pulse, the temperature in the core of the spark may be as high as 40,000°K, and the intensities of ion lines are high compared to the intensities of the lines of neutral atoms.[2] Excitation

[2] This is the basis of the terminology of the older spectroscopic literature in which lines due to neutral atoms are called "arc" lines and lines due to atomic ions are called "spark" lines.

conditions can be altered by varying the discharge circuit parameters. For example, if L is increased, the period of the oscillations becomes longer, the peak current decreases, the maximum temperature decreases, and the emission intensity decreases. The emission intensity decreases at different rates for lines of different excitation energies and for the background. Thus, interferences may be minimized and the line-to-background ratio improved by a proper choice of L, C, and R.

B / *Sample introduction*

As each spark contacts the electrodes, some material at the contact point is vaporized. The mechanism of the vaporization step is not clear. Local heating of some material to its boiling point is the most probable mechanism, since the current density is very large at the contact point. Another possible mechanism of sample vaporization for which some evidence exists is the sputtering of material from the electrode surface by impact of ions accelerated in the large voltage gradients of the spark.

Individual sparks strike the electrode at various points, and, except for very small samples, the relatively long off periods of the spark ensure that very little heating of the bulk of the sample occurs. Thus, problems due to selective volatilization are largely avoided with the spark. Since each spark may strike a different portion of the sample, the intensity measured for each spark may vary drastically due to sample inhomogeneities and the position of the spark discharge with respect to the entrance slit of the spectrograph (analogous to the arc wander described for the DC arc). However, as already noted, the repetition rate of the sparks may be from 2 to 20 per half-cycle, with 120 half-cycles per second. Since an exposure period of 30 sec or more is usually required, the intensity measured is usually integrated over many thousands of sparks, and the fluctuations tend to be smoothed statistically. Precisions of the order of 1 percent of the amount present can usually be achieved with spark excitation.

In the course of an exposure with spark excitation, only some 1–5 mg of material may be volatilized from the electrodes, compared to 10–50 mg in an arc exposure. In addition, the concentration of neutral atoms is reduced by the rather high degree of ionization in the spark. As a result, spark spectra tend to be more complex and less intense than arc spectra, and the sensitivity of determinations with the spark is usually considerably less than with the arc.

14.2.5 / The laser microprobe

Certainly the most advanced application of the laser to analytical spectroscopy is the use of the laser microprobe for emission spectroscopy of micro samples. Certain large, pulsed lasers are capable of emitting very intense pulses of radiation which are well collimated and nearly monochromatic and hence can be focused to a small spot (as small as 5–50 μm in diameter). The radiation density under such conditions is so high that the selected portion of sample material is instantaneously vaporized. A pair of electrodes with a voltage of 1–2 kV applied across the gap is placed just above the sample, and movement of the vaporized sample material into the gap triggers a discharge through the vapor, exciting the emission spectrum of the material.

In addition to its ability to examine directly very small samples, the microprobe has the advantage of being applicable to the direct volatilization of materials that are not electrically conductive. Its advantages have been rather spectacularly demonstrated in the analysis of selected portions of individual blood cells.

14.3 / DETECTION

The characteristics of the detection and readout systems are determined largely by the characteristics of the excitation source. With most excitation sources for emission spectroscopy, there are large fluctuations in the intensity of the emitted radiation due to (1) fluctuations in the excitation conditions and (2) fluctuations in the atomic concentration of the element of interest as the sample is consumed. Thus, to obtain adequate measurement precision it is usually necessary to provide for integration of the signal over a relatively long time (of the order of tens of seconds to minutes), and, if more than one element is to be detected in a single sample, arrangement must be made for simultaneous detection at several wavelengths. Both of these conditions can be met with either photographic or photoelectric detectors.

14.3.1 / Photographic detectors

The chief features of photographic detectors have been described in Section 11.2.5, where it was pointed out that a photographic emulsion is an integrating detector—it responds to the product of intensity and time—and it simultaneously detects a large region of the spectrum. Thus, the photographic detector has some obvious advantages for emission spectroscopy.

Figure 14.9 illustrates the appearance of photographically detected emission spectra. On this plate, two separate exposures have been made at a single wavelength setting of the spectrograph. The vertical separation of the two spectra has been obtained by moving the plate holder between exposures. Most spectrographs have a mechanism that allows the plate to be moved, or "racked," so that several exposures may be made on a single plate without altering any other settings of the spectrograph. The iron arc spectrum, in which a number of lines have been identified, serves as a wavelength scale for the plate.

In Fig. 14.9, there are quite obvious differences in the "blackness" of spectral lines, and it is intuitively obvious that the degree of "blackness" is in some fashion related to the inherent intensity of the lines in the excitation source. The blackening of an exposed area in a photographic plate is a function of the exposure E, which is defined by

$$E = It \tag{14.1}$$

where I is the intensity and t is the exposure time. The blackness of a spectral image on the plate is measured by the use of a microphotometer, as described in Section 11.2.5. The blackening is reported as the optical density, D, where

$$D = -\log T \tag{14.2}$$

$$T = \frac{P}{P_0} \tag{14.3}$$

T is the transmittance of the image, P_0 is the radiant power transmitted through a clear portion of the plate, and P is the power transmitted through the spectral image.

To determine line intensities from the microphotometer readings, the plate must be calibrated so that the optical density produced for a given exposure is known quantitatively. This calibration is greatly simplified by the fact that in spectrochemical analysis we are concerned only with relative intensity values, and hence we need not calibrate the emulsion on an absolute scale. One type of plate calibration curve consists of a plot of optical density as a function of the logarithm of the exposure as shown in Fig. 14.10. This type of plot is known as an H and D curve (Hurter-Driffield curve). The tangent of the angle between the linear segment and the exposure axis is called the *gamma* of the emulsion and is a measure of the contrast of the emulsion. The intercept of the linear portion of the curve with the abscissa is a measure of the inertia of the emulsion, or the smallest exposure needed to give a detectable image. The most sensitive plates have low inertias.

The relationship between optical density and exposure is dependent on the wavelength of radiation that caused the blackening, the emulsion type, and the development procedure. Unfortunately, the relationship varies with different batches of the same emulsion type and with the age of the chemicals used in the development procedure. Thus, an emulsion characteristic curve must be prepared periodically at each wavelength of interest if accurate intensity measurements are to be made.

A number of techniques exist for relating optical density to relative exposure values. We shall describe only one: the rotating step sector technique. Figure 14.11 shows a step sector. When the sector is rotated at high speed between the entrance slit of the spectrograph and a source of line radiation, such as an iron arc, each spectral line in the developed plate consists of segments that are related to one another in known ratios of

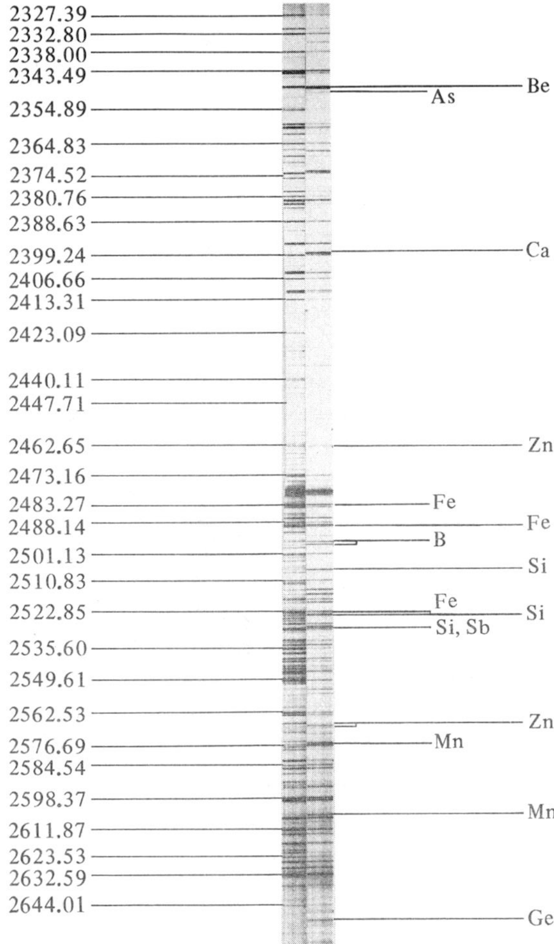

Figure 14.9 / *Example of photographically detected emission spectra. These spectra were recorded on a* 254-mm *plate with a* 1.8-m *focal length quartz prism spectrograph. The reciprocal linear dispersion at the short-wavelength end of the plate is approximately* 1.7 Å/mm. *The upper spectrum is for* Fe *and the numbers are the wavelength in* Å *of the indicated* Fe *lines. The lower spectrum is for* R.U. *powder, and the prominent lines of a number of elements are identified.* (R.U. *powder contains a mixture of elements in such proportions that at least several lines of each element will appear when the powder is burned in an arc.*)

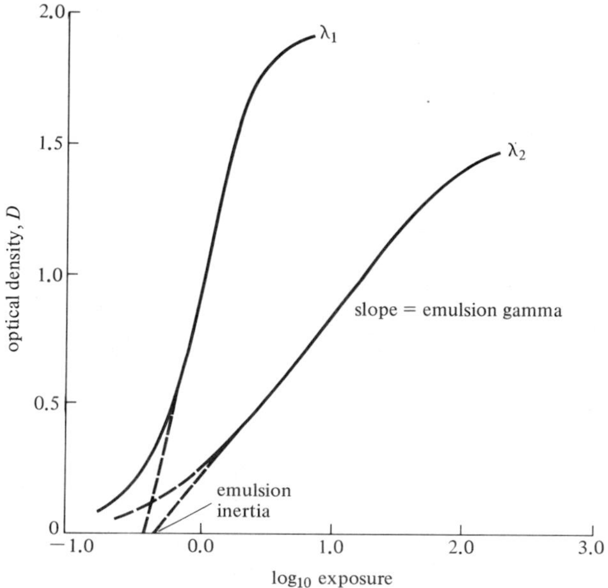

Figure 14.10 / *Emulsion calibration curves.*

exposure. It is assumed that the average line intensity remains constant and that the exposure time for each segment varies in accord with the geometry of the sector. For example, for the five-step sector of Fig. 14.11, the portion of the line image that corresponds to the top of the entrance slit is "on" during the whole exposure [2(90° + 45° + 22.5° + 11.25° + 11.25°) = 360°]; the next segment is "on" one-half the time [2(45° + 22.5° + 11.25° + 11.25°) = 180°]; the next, one-fourth the time [90°]; the next, one-eighth [45°]; and the last, one-sixteenth [22.5°]. Thus, if the least dense portion

Figure 14.11 / *Step sector disc with a step ratio of 2.*

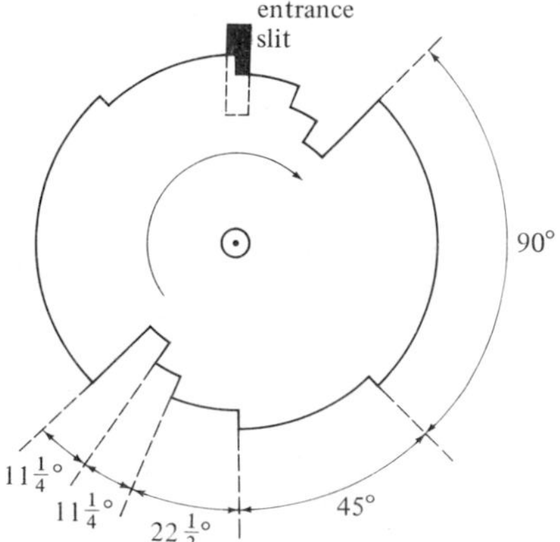

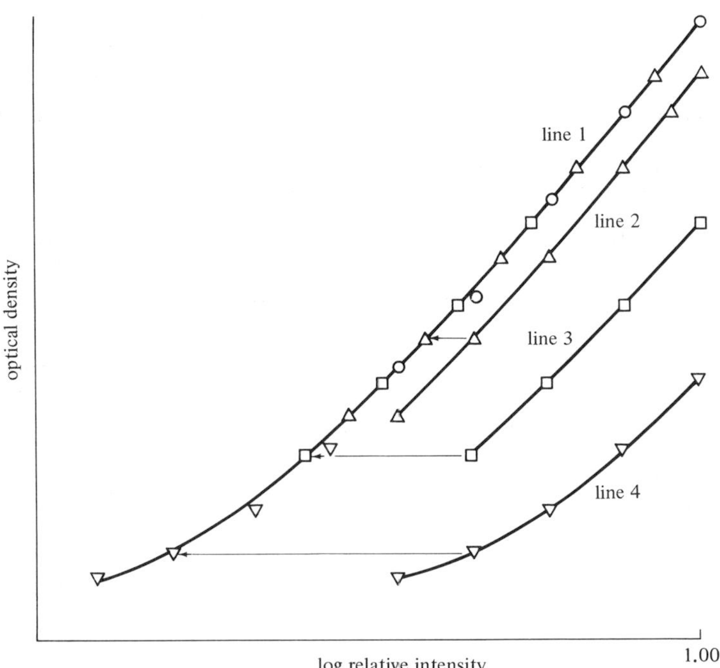

Figure 14.12 / *Preparation of an emulsion calibration curve. Lines 2, 3, and 4 are merged with line 1 by application of arbitrary constants, as indicated by the arrows.*

of the image is arbitrarily assigned a relative exposure value of one, the various segments of each line have relative exposures in the ratio 1:2:4:8:16.

To prepare a characteristic curve covering a sufficient range of optical density values, it is customary to measure the optical densities of the steps for several lines in the spectral region of interest. Plots of optical density vs. the logarithm of the relative exposure then yield a family of parallel curves displaced from one another along the ordinate (provided that the gamma of the emulsion is constant over the wavelength region in question). These are combined into a single curve by applying arbitrary shifts along the abscissa, as shown in Fig. 14.12.

A considerably more linear emulsion characteristic curve may be drawn by plotting the logarithm of the relative exposure against the Seidel function $\Delta = \log(1/T - 1)$, where T is the transmittance of the spectral image defined in Eq. (14.3). With most emulsions, this curve remains linear over the range of relative transmittances that are ordinarily measured in emission spectroscopy.

14.3.2 / Photoelectric detectors

Multiplier phototubes have been described in some detail in Section 11.2.5. Unlike a photographic plate, a multiplier phototube is not inherently an integrating device (it responds to radiant energy per unit time), and it does not possess the photographic plate's ability to record simultaneously a large portion of the spectrum. Integration is achieved by storing the signal from a multiplier phototube on a capacitor. The net charge on the capacitor at the end of the exposure is proportional to the time-integrated intensity reaching the detector. Simultaneous detection of a number of elements of the spectrum is obtained by providing an exit slit and multiplier phototube at each position on the exit curve corresponding to a line or background feature to be recorded. Photoelectric spectrometers (so-called direct readers) have been built with positions for as many as 60–70 detectors.

Photoelectric spectrometers are obviously expensive and, because of the considerable time and effort associated with exit-slit alignment, relatively inflexible. Their principal application is to routine analytical tasks, such as in the control of the production of metals, where the need for rapid analysis outweighs other factors.

14.4 / SAMPLE HANDLING

14.4.1 / Solid samples

A solid metal sample in the form of a plate, rod, or disc can be analyzed with the sample as one or both of the electrodes. Either arc or spark excitation can be applied. Arc excitation is possible when the metal is able to withstand the heat of the arc, and this will vaporize more of the sample and allow better sensitivity of detection than spark excitation. Some samples are not homogeneous enough or are not suitable in form for excitation, and these must be sampled by taking chips, drillings, or filings, according to the nature of the material. Care must be excercised to avoid contamination by the cutting tool and to obtain a representative sample.

Some samples, such as minerals, cement, chemicals, and so on, are often obtained as powders. The powdered sample, if not already finely divided, is ground or crushed to produce a material of more uniform characteristics which can be more evenly volatilized. The powdered sample may be analyzed directly, or it may first be mixed with some other materials. Common additives include internal standards (see Section 14.6), spectroscopic buffers (materials added to samples of widely varying composition so that like excitation conditions are obtained with all samples), and graphite powder to promote a more uniform burn of nonconductive materials. The material is usually placed in a cup electrode for excitation, and arc excitation is usually employed.

14.4.2 / Solution samples

Solution samples are advantageous in that they are homogeneous and standards are readily prepared. Two chief methods exist for dealing with solution samples: residue excitation and direct excitation. In the residue technique, a small portion of the solution is placed on a suitable electrode, solvent is evaporated, and the electrode is burned in an arc or spark discharge. Residue techniques are valuable for the analysis of samples that are available in limited quantity, since only a small volume of solution is needed. They are not particularly applicable to very dilute solutions, since the amount of sample that can be evaporated on an electrode is limited.

Most techniques for direct solution analysis employ spark excitation. The high electrode temperatures characteristic of arc excitation would produce quick vaporization of solutions. The spark rotating-disc technique is perhaps the most widely used method. In this technique, a small graphite disc is driven by a motor so that it rotates in a vertical plane, and, as it rotates, its lower surface passes continuously through the sample solution. The counter electrode is located above the disc, so that each spark volatilizes a small volume of the solution from the upper surface of the disc. A second approach to direct solution analysis makes use of the porous cup electrode. The porous cup is a graphite tube, typically $\frac{1}{4}$ in. in diameter and 1–1.5 in. long, with a hole drilled along its axis to leave only a thin (0.025–0.040 in.) bottom. The solution is placed in this tube, which serves as the upper electrode for a spark discharge. The sample seeps through the bottom of the electrode, forming a thin film that is evaporated by the spark discharge.

14.5 / QUALITATIVE ANALYSIS

14.5.1 / Excitation and exposure

The virtues of emission spectrography are most clearly apparent in qualitative elemental analysis. Some 70 elements can be simultaneously detected in concentrations ranging from the trace level to 100 percent by a rapid procedure that also allows an estimate of the concentration to be made and that requires only a few milligrams of sample.

The high sensitivity obtained with the DC arc makes it the excitation method of choice for qualitative analysis. To ensure that all of the constituents of the sample will be detected, the current is chosen to evaporate the entire sample within a period from 15 sec to 1 or 2 min, and the spectrum is photographed during the entire excitation period. The exposure may be adjusted in several ways to provide suitable line intensities. Use of a step sector or step filter (a neutral-density filter with several regions corresponding to different optical densities) is perhaps the most convenient, since varied exposures are provided with a single arcing of the sample.

14.5.2 / Line identification

The wavelengths of the lines are determined from their positions on the photographed spectrum. This task is facilitated by recording a standard spectrum (ordinarily that of an iron arc) on each plate or film. The identification of wavelengths is most easily done with the aid of a microphotometer equipped to project the spectrum on a screen along with a comparison spectrum having wavelength markings.

The wavelengths of the lines may be determined by measuring their distance from lines of known wavelength (the Fe lines, for example) and applying the known reciprocal linear dispersion of the spectrograph. This is easily accomplished with grating spectrographs, for which the dispersion is very nearly independent of wavelength. The task is more complicated with prism spectrographs, but can be accomplished by use of interpolation formulas, the best known of which is that due to Hartmann. Interpolation procedures are described in the Suggested Readings at the end of the chapter. Once the wavelengths of the lines have been obtained, the elements can be identified from an atlas of wavelengths such as the *M.I.T. Wavelength Tables*. The presence of one line at the expected position in the spectrum cannot be taken as positive identification of an element, since it is entirely possible that another element in the sample may have a line at very nearly the same wavelength. At least two or three lines should be identified for each element. The relative intensity of the lines is a valuable clue as to whether they actually arise from the same element.

A second approach to qualitative analysis is that of direct comparison of positions without specific measurement of wavelengths. This is facilitated by development of a "master plate" for the specific instrument and measurement conditions to be employed. A master plate is shown in Fig. 14.9 for a large quartz spectrograph. It contains an Fe arc spectrum for indexing and the spectrum produced from a burn of R.U. (*raies ultimes*) powder. R.U. powder contains a mixture of elements in such proportions that at least several lines of each of the more important elements will appear when the powder is burned in an arc. The elements from which the stronger lines arise are identified on the master plate, and qualitative analysis is accomplished by projecting the master plate spectra and sample plate spectra on a screen, shifting one plate with respect to the other until the lines of the Fe arc spectra coincide, and then identifying lines in the sample spectrum that coincide with marked lines in the R.U. powder spectrum.

14.6 / QUANTITATIVE ANALYSIS

14.6.1 / General principles

For quantitative analysis, the samples must be prepared and excited and the photographic plates exposed and developed in a carefully controlled fashion. The analytical curves, showing the relationship between the line intensities and element concentrations, must be prepared by exciting standards in the same way as samples, and the standards must conform, as far as possible, to the samples in both chemical and physical form.

14.6.2 / The internal standard

Because of the many factors that can affect the line intensities in arc or spark discharges, an internal standard is necessary in order to achieve acceptable precision in quantitative measurements. The internal standard is an element that is present in each sample and standard at a fixed concentration. The intensities of a line of the analyte and a line of the internal standard are determined, and the ratio of the two intensities is calculated. If the factors affecting the measured intensities are well controlled, the intensity of the internal standard line will not change, since the concentration of the internal standard element is fixed. However, if one or more of the factors affecting measured line intensity varies, the internal standard line and analyte line should be affected equally, so that the intensity ratio will depend only on the concentration of the analyte. A plot of the intensity ratio (analyte to internal standard) as a function of the concentration of the analyte forms the analytical curve.

In choosing an internal standard element, the following criteria must be considered if adequate compensation for the factors that can affect intensity is to be achieved:

1. The internal standard element must occur in all samples and standards at the same concentration, or it must be available as a pure material that can be added to the samples and standards without introducing contaminants.
2. The internal standard should have physical and chemical properties that are similar to those of the analyte, so that the two elements are volatilized at the same rate and undergo the same reactions.
3. The excitation energies of the two lines should be identical, so that both lines will be affected identically by changes in source temperature.
4. The two elements should have similar ionization energies, so that changes in source electron concentration and temperature will affect the two lines in the same fashion.
5. The two lines should be close in wavelength, so that the same emulsion calibration curve is applicable to both.
6. The two lines should be similar in intensity, to reduce the effect of errors in the emulsion calibration curve.
7. The spectrum of the internal standard element should be simple, so that it does not cause spectral interferences.

Obviously, it is difficult in practice to meet all these criteria, and compromises must be made in the selection of internal standards. Very often, a major matrix element that does not vary significantly from one sample to the next is chosen as the internal standard element, even if it does not meet the criteria very well, because of the great convenience of such a choice.

14.6.3 / Standard additions

The method of standard additions has been described in connection with flame spectrometry. For photographic emission spectroscopy, it differs only in that the intensity *ratio* (analyte to internal standard) is plotted vs. the added concentration of analyte. In order for the additions method to be applicable, the sample must be in a form to which varying amounts of the analyte can be added.

SUGGESTED READINGS

G. R. Harrison, R. C. Lord, and J. R. Loofbourow, *Practical Spectroscopy*, Prentice-Hall (Englewood Cliffs, N.J., 1948).

N. H. Nachtrieb, *Principles and Practice of Spectrochemical Analysis*, McGraw-Hill (New York, 1950).

L. H. Ahrens and S. R. Taylor, *Spectrochemical Analysis*, 2nd ed., Addison-Wesley (Reading, Mass., 1961).

B. F. Scribner and M. Margoshes, "Emission Spectroscopy," in *Treatise on Analytical Chemistry*, I. M. Kolthoff and P. J. Elving, Eds., Part 1, Vol. 6, Wiley-Interscience (New York, 1965).

ASTM, *Methods for Emission Spectrochemical Analysis*, 5th ed., American Society for Testing and Materials (Philadelphia, 1968).

M. Slavin, *Emission Spectrochemical Analysis*, Wiley (New York, 1971).

PROBLEMS

1 / Determine the percent Sb in a tin-base alloy X from the following DC arc emission data. Six exposures were made under identical conditions. Five of the exposures were for tin-base alloys containing known amounts of Sb, and the sixth was for the alloy X. The optical densities of the selected lines on the photographic plate were measured and gave the results shown in columns 3 and 5. The calibration curve of the photographic emulsion was used to convert the density readings to the relative intensity values shown in columns 4 and 6.

		Sb *2878* Å		Sn *2785* Å	
Sample	% Sb	*D*	*I*	*D*	*I*
1	0.126	0.26	2.15	1.57	16.4
2	0.316	1.01	6.45	1.57	16.4
3	0.708	1.55	15.1	1.44	12.5
4	1.334	1.43	12.0	0.83	5.05
5	2.512	1.58	16.4	0.45	2.95
X		0.98	6.24	0.92	5.75

2 / Use the following arc emission data to calculate the percent Hf in a powdered zirconium oxide sample. Ignore any effect of dilution.

Sample	*Relative I* Hf *2773.36* Å	*Relative I* Zr *2783.56* Å
unknown	79.4	275.4
unknown + 0.4% Hf	263.0	218.8

3 / Use the data provided below to prepare emulsion calibration curves, plotting D vs. log E (H and D curve) and the Seidel function Δ vs. log E. The step factor was 2 for the sector employed.

Microphotometer readings for emulsion calibration

	%*T*			
Line	*Step 1*	*Step 2*	*Step 3*	*Step 4*
A	—	—	89.6	64.5
B	92.0	66.9	27.0	6.5
C	—	85.3	51.0	17.0
D	19.5	5.2	—	—
E	37.5	10.0	3.0	—

4 / Use either of the emulsion calibration curves prepared above and the following spark emission data to determine the percent Pb in the tin alloy sample X.

Sample	%Pb	%*T* Pb *2833* Å	%*T* Sn *2761* Å
1	0.126	54.9	2.71
2	0.316	9.71	2.69
3	0.708	2.84	3.61
4	1.334	3.74	14.9
5	2.512	2.63	35.7
X		21.4	12.0

chapter 15 / Introduction to molecular spectroscopy

chapter 15

15.1 / INTRODUCTION

Molecules, even diatomic species, exhibit spectra that are much more complex than atomic spectra. The additional complexity occurs because the molecular energy is composed of quantized components associated with rotation and vibration of the molecule in addition to the electronic transitions observed in atoms. In this chapter, we shall consider some general principles important to our later consideration of the analytical applications of molecular spectroscopy.

15.2 / GENERAL NATURE OF ROTATIONAL, VIBRATIONAL, AND ELECTRONIC SPECTRA

The energy of a molecule may be treated as though it consisted of additive quantized components due to (1) rotation of the molecule, (2) vibration of constituents of the molecule with respect to one another, and (3) intramolecular electronic transitions. The three types of motion can be separated because of the great differences in their frequency of occurrence. The time required for an electronic transition is of the order of 10^{-15} or 10^{-16} sec. In this time, the components of a molecule, vibrating with a frequency of approximately 10^{12} sec^{-1}, will have covered a distance of only one-thousandth or less of the amplitude of the vibration and can therefore be regarded as fixed during the time required for the electronic transition. In the same fashion, rotations, with a frequency of the order of 10^{10} sec^{-1}, are much slower than vibrations.

The energy levels for the rotation of a molecule are relatively close to each other and, if a rotational transition is permitted, the radiation associated with the transition is of long wavelength. Pure rotational spectra are observed in the region of the electromagnetic spectrum extending from the far infrared into the microwave, with wavelengths of the order of 1 cm. The

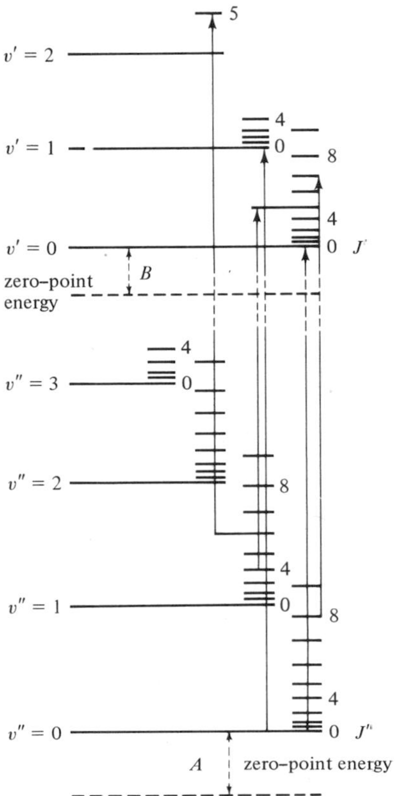

Figure 15.1 / *Rotational and vibrational levels associated with electronic levels of a diatomic molecule, and some corresponding rotational and vibrational changes accompanying electronic excitation.*

separation of the vibrational levels is greater, and the corresponding spectra are of shorter wavelength. Vibrational spectra are most often observed in the IR region between 1 and 30 μm. The spacing of electronic levels is usually large, and electronic transitions are, of course, most frequently observed in the UV and visible portions of the electromagnetic spectrum.

In gases, vibrational transitions are generally accompanied by changes in the rotational state, so that the spectrum corresponding to a transition between two vibrational levels, instead of being a single line, contains a number of lines close together representing the rotational transitions superposed on the vibrational transition. Similarly, electronic spectra of gaseous molecules may assume a very complicated structure because of the superposition of vibrational and rotational changes on the electronic transition. Figure 15.1 illustrates this state of affairs for transitions occurring between two electronic levels, *A* and *B*, of a diatomic molecule. The energy increases upward in the figure; the relative spacings of the rotational levels—indicated by the short horizontal lines—and the vibrational levels—represented by the longer horizontal lines—are realistic, but, for economy of space, the electronic levels have been drawn much closer together than is reasonable for transitions occurring in the ultraviolet or visible. The vertical lines, representing some of the possible transitions in this system, are broken to indicate the compression of the electronic energy scale.

The numbers 0, 1, 2, and so on, assigned to the rotational levels in Fig. 15.1 indicate the values of the rotational quantum number[1] J and, similarly, the vibrational levels are characterized by a vibrational quantum number v. The spacing between rotational levels tends to increase with increasing J, while the spacing between vibrational levels tends to decrease with increasing v. The dashed horizontal lines associated with the electronic levels A and B indicate the energy that the molecule would possess at the absolute zero of temperature if it were possible to suppress all vibratory motion within the molecule. The energy represented by the separation between the dashed line and the level with $v = 0$ and $J = 0$ is the residual energy of the molecule at the absolute zero by virtue of the vibratory motion. This quantity is called the *zero-point energy*.

There are a great many permitted transitions between the vibrational and rotational levels of A and B, a few of which are indicated as absorption transitions by the vertical arrows. Thus, it is clear that a very complicated pattern may arise in the absorption spectrum corresponding to an electronic transition in a molecule, whereas in an atom an electronic transition gives rise to a single line.

When observed with a spectrometer of sufficiently high resolving power, the visible and UV spectra of diatomic molecules and of some relatively simple polyatomic molecules are indeed found to consist of a large number of fine lines constituting the rotational structure of the spectrum. The rotational structure is, however, seldom observed in analytical spectroscopy. Various circumstances are responsible for this failure to see the rotational structure:

1. The lines may be so close together as to be unresolvable with the available instrumentation.
2. Most spectra are measured for liquid samples, and molecular collisions occur before a rotation can be completed. When this occurs, the energy levels are perturbed to varying degrees, and the concept of quantized rotations breaks down.
3. The spacing between rotational levels for large molecules may be so small that even in the gas phase the rotational structure is lost, because the width of the lines imposed by Doppler and collisional broadening (see Section 10.3.2) may be greater than the spacing between consecutive lines.

[1] The rotational quantum number unfortunately bears the same symbol as the total angular momentum quantum number of atomic spectroscopy, but the two are usually easy to distinguish by context.

4. Recognizable structure in a spectrum requires the existence of discretely quantized states in both the initial and final electronic levels. If a molecule is stable, the ground electronic state will always be quantized, but on excitation, the molecule may dissociate. The dissociated state corresponds to a continuum of energy levels separated by arbitrarily small intervals representing infinitesimally small increases in the kinetic energy of the dissociated fragments, and transitions to such a state give rise to continuous spectra.

Unlike rotational structure, vibrational structure of electronic spectra is often seen. The vibrational spectrum associated with an electronic transition consists of a series of more or less broad bands, usually with an asymmetric distribution of intensity determined by the concentration of unresolved rotational lines. To illustrate, the UV absorption spectrum of benzene shown in Fig. 15.2 is associated with a single electronic transition, and the band structure is determined by the vibrational changes which accompany the electronic transition. Vibrational bands persist in the spectra of many species even for liquid and solution samples, as may be seen in Fig. 15.2(b).

15.3 / PURE ROTATIONAL SPECTRA

If a diatomic molecule is regarded as a system of two particles separated by a fixed distance R_0, rotating about an axis through the center of gravity and perpendicular to the molecular axis, its energy of rotation is given by

$$E_J = \frac{J(J + 1)h^2}{8\pi^2 I} \tag{15.1}$$

where J is the rotational quantum number and I is the moment of inertia of the molecule. The moment of inertia plays the same role in rotational motion as the mass in rectilinear motion. Its value for the rigid diatomic rotator is

$$I = \mu R_0^2 \tag{15.2}$$

where μ is a quantity known as the reduced mass and is given by

$$\mu = \frac{m_1 m_2}{m_1 + m_2} \tag{15.3}$$

in which m_1 and m_2 are the respective masses of the atoms in the molecule. Thus, for the diatomic molecule CO:

$$m_1 = \frac{12.00 \text{ g/mole}}{6.02 \times 10^{23} \text{ atoms/mole}} = 1.99 \times 10^{-23} \text{ g}$$

$$m_2 = \frac{16.00 \text{ g/mole}}{6.02 \times 10^{23} \text{ atoms/mole}} = 2.66 \times 10^{-23} \text{ g}$$

$$\mu = 1.14 \times 10^{-23} \text{ g}$$

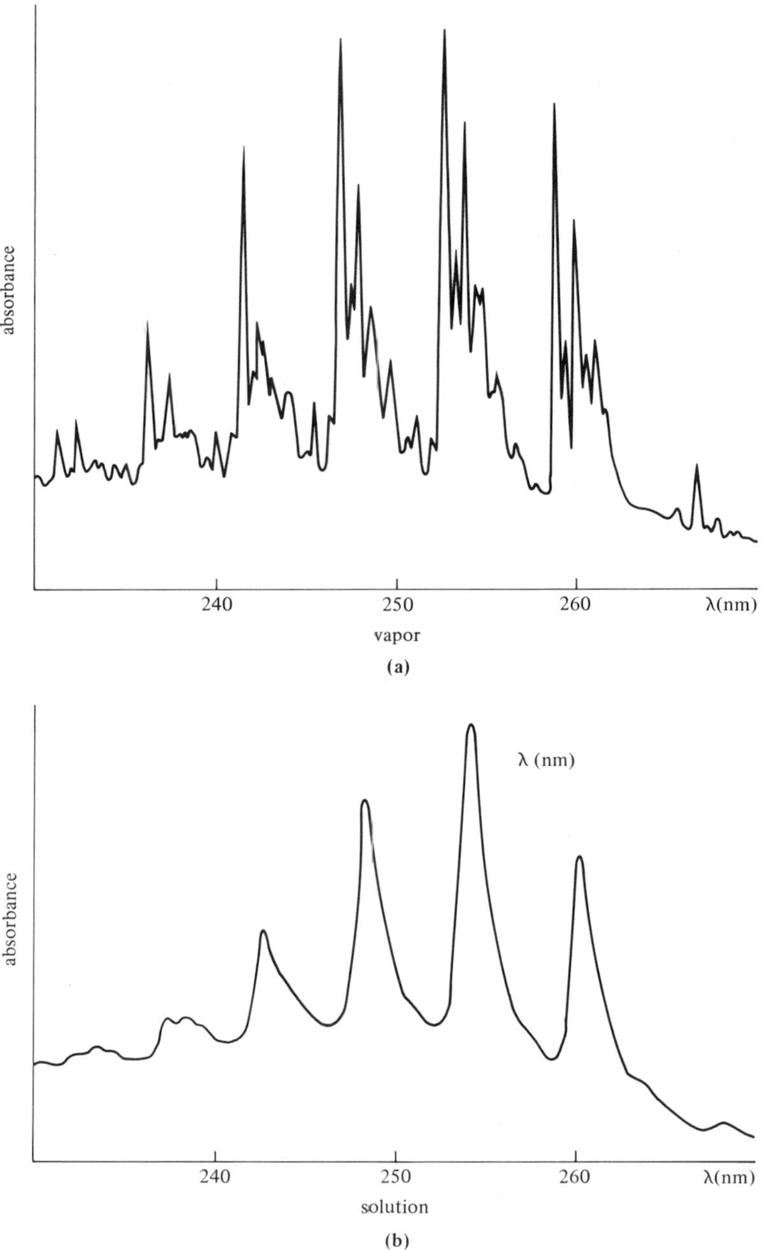

Figure 15.2 / *Ultraviolet absorption spectrum of benzene.*
(a) *Vapor.*
(b) *Solution in ethanol.*
SOURCE: W. West, "Introductory Survey of Molecular Spectra," in *Chemical Applications of Spectroscopy*, W. West, Ed., 2nd ed., Part 1, Wiley-Interscience (New York, 1968), p. 23.

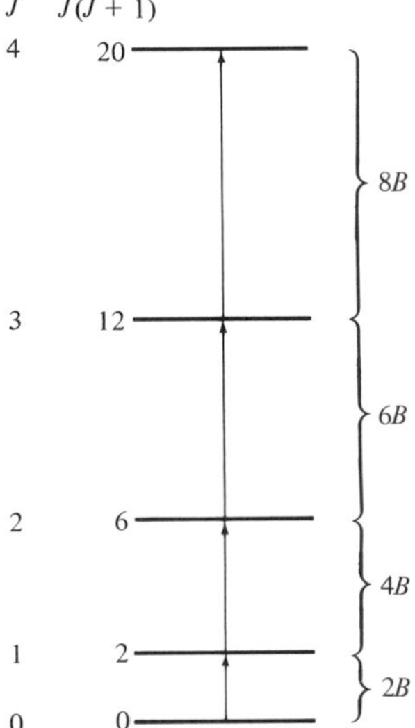

Figure 15.3 / *Energies associated with allowed changes for purely rotational transitions.*

The internuclear separation,[2] R_0, is 1.13×10^{-8} cm at equilibrium, and the moment of inertia is therefore 14.5×10^{-40} g-cm^2.

The permitted changes in J for a diatomic molecule undergoing purely rotational changes are $\Delta J = \pm 1$. Thus, for the rotational transition $J = 0$ to $J = 1$, $E_0 = 0$, $E_1 = 2h^2/8\pi^2 I$, and the wavenumber of the corresponding radiation is obtained by application of the Bohr condition

$$E_1 - E_0 = hc\bar{\nu} \tag{15.4}$$

such that

$$\bar{\nu} = \frac{2h}{8\pi^2 Ic} = 2B \tag{15.5}$$

where

$$B = \frac{h}{8\pi^2 Ic} = \frac{27.99 \times 10^{-40}}{I} \tag{15.6}$$

In the same fashion, the wavenumbers for other allowed transitions can be calculated to show that the pure rotational spectrum for a diatomic molecule (with fixed internuclear distance) consists of a series of lines at wavenumbers $2B$, $4B$, $6B$, and so on, as indicated in Fig. 15.3. The spacing is inversely proportional to the moment of inertia, and the pure rotational spectrum

[2] One of the ways in which the internuclear separation is determined is by calculation from Eq. (15.2) after determination of I from rotational spectra. Thus, we have reversed the usual fashion of carrying out this calculation.

therefore furnishes a means of measuring the moment of inertia and hence the internuclear separation of a diatomic molecule. For CO, the value of B calculated from Eq. (15.6) is 1.93 cm^{-1}, and thus the $1 \leftarrow 0$ transition gives rise to radiation of wavenumber slightly less than 4 cm^{-1} and a wavelength of about 0.25 cm.

Diatomic molecules do not, in fact, behave entirely as rigid rotators. As the rotational energy increases, the centrifugal force causes a stretching of the molecule, the moment of inertia increases, and the separation of the lines of the rotational spectrum decreases with increasing wavenumber.

Only molecules with a permanent electrical moment exhibit pure rotational spectra, since only then will there be a periodic variation of the electrical distribution along directions fixed in space. Homonuclear diatomic molecules, such as O_2, H_2, N_2, and so on, thus have no pure rotational spectra. Polyatomic molecules, with the exception of a few linear molecules such as $HC{\equiv}CH$ or HCN, have three moments of inertia about mutually perpendicular axes and give rise to more complex spectra than do diatomic molecules.

15.4 / VIBRATION-ROTATION SPECTRA

The atoms in a molecule carry out periodic vibrations about equilibrium positions with amplitudes that are usually small compared with the internuclear separation. Under the proper conditions, the vibrational energy of the molecule can change, in quantized steps, by interaction of the molecule with electromagnetic radiation. Such an interaction is most easily described for a diatomic molecule. A diatomic molecule has a single vibration along the line of centers. During the vibration, the electrical distribution in the bond changes. If the two atoms are not identical, the electrical center of the molecule oscillates in synchronism with the nuclei, and the molecule behaves as an oscillating electrical dipole. Such a dipole can, of course, interact with the electric field of electromagnetic radiation so that radiation is absorbed from the field and the vibrational energy of the molecule is increased (an absorption process), or the molecule may lose a discrete quantity of vibrational energy and emit electromagnetic radiation.

A net change in dipole moment of the molecule is essential to the direct observation of a vibrational spectrum. Thus, homonuclear molecules, such as H_2, N_2, O_2, and so on, do not exhibit a vibrational spectrum in the IR region. Symmetrical polyatomic molecules such as $H_3C{-}CH_3$ do not possess a permanent dipole moment but can execute periodic motions other than the symmetric stretching vibration in which a periodic motion of the electrical center of the molecule occurs, and for which a vibration-rotation spectrum appears.

If the molecule is in the gaseous state and therefore freely rotating, changes in the vibrational energy may be accompanied by changes in the rotational

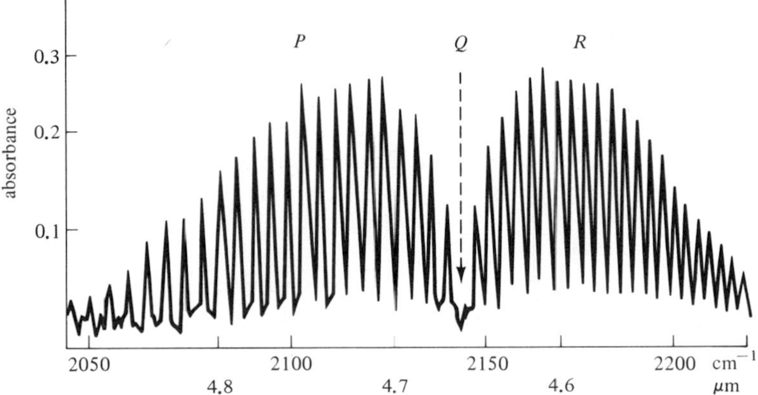

Figure 15.4 / *Rotation-vibration absorption band of* CO. *Rotation lines result from transitions between rotational energy levels of the two lowest vibrational levels* ($v = 0$ *and* $v = 1$).

energy. The spacing of vibrational levels is sufficiently large that at room temperature most of the molecules are in the lowest vibrational state ($v = 0$), and the chief vibrational transition observed by absorption of IR radiation is to the $v = 1$ state. At room temperature, there is a distribution of molecules over a considerable number of rotational states, so that it is possible for the increase in vibrational energy induced by IR radiation to be accompanied by both decreases and increases in rotational energy. The IR spectrum of a gas thus contains bands for which the position is fixed by the vibrational change but which, because of the accompanying rotational changes, consists of a series of lines at longer and shorter wavelengths than the pure vibrational change. For most diatomic molecules in the gaseous state, the selection rule for the rotational change accompanying the vibrational change is $\Delta J = \pm 1$, and thus the pure vibrational change (i.e., without rotation change) is prohibited. Hence, the typical vibration-rotation spectrum of a diatomic molecule consists of two branches, one toward longer wavelength, the *P* or negative branch (ΔJ negative), the other toward shorter wavelength, the *R* or positive branch (ΔJ positive), about a missing line. The spectrum of CO shown in Fig. 15.4 is typical.

If the vibrational motion of the molecule involves a component of the vibrating electrical moment perpendicular to the line joining the atoms, $\Delta J = 0$ also becomes permitted, and there then appears between the *P* and *R* branches a zero or *Q* branch. This situation is rare among diatomic molecules but common among polyatomic molecules.

For quantitative treatment of the vibrational behavior of diatomic molecules, the simplest model is the harmonic oscillator. In this model, the atoms oscillate in unison about equilibrium positions, each acting under the influence of a restoring force proportional to the displacement from its equilibrium position. The frequency of oscillation is given by

$$\nu = \frac{1}{2\pi}\left(\frac{k}{\mu}\right)^{1/2} \tag{15.7}$$

where k, the force constant, is the restoring force per unit displacement from the equilibrium position and thus has units of dynes per centimeter. The expression for the energy of the simple harmonic oscillator is

$$E_v = (v + \tfrac{1}{2})h\nu \tag{15.8}$$

where the vibration quantum number v assumes values 0, 1, 2, ... The lowest vibrational state, $v = 0$, is associated with the zero-point energy $(\frac{1}{2})h\nu$, and the highest states are arranged equidistant above the lowest state at a separation of $h\nu$. The selection rule for the absorption or emission of radiation by the simple harmonic oscillator is $\Delta v = \pm 1$; hence, a molecule in the lowest vibrational state could gain only one quantum by absorption, and the absorption spectrum would consist of a single vibrational band of frequency ν equal to the frequency of mechanical oscillation.

Actual molecules do not conform to the harmonic oscillator model in detail. As a consequence, the spacing of vibrational levels is not uniform but decreases with increasing values of v, and transitions greater than $\Delta v = \pm 1$ are allowed. By analogy with acoustical terminology, the $1 \leftarrow 0$ vibrational band is often called the fundamental or first harmonic, and the bands at higher frequency corresponding to the $2 \leftarrow 0$ and $3 \leftarrow 0$ transitions are the first and second overtones, or the second and third harmonics.

15.5 / ELECTRONIC SPECTRA

Each electronic state in a molecule has its own pattern of vibrational and rotational levels, as illustrated in Fig. 15.1 for two electronic states of a diatomic molecule. The electronic transition is generally accompanied by vibrational and rotational changes, and since in polyatomic molecules a great many possibilities may exist for these latter changes, the spectrum becomes a complicated assembly of lines spread over a considerable spectral region. Figure 15.5 illustrates the vibrational transitions accompanying an electronic transition of a hypothetical diatomic molecule. Typically, the vibrational spacing differs in the various electronic states. In the diagram, the interval between levels in the upper electronic state is less than that in the lower electronic state, indicating a decrease in the force constant accompanying the electronic excitation.

The absorption spectrum consists of a series of vibrational subbands, representing the $0 \leftarrow 0$, $1 \leftarrow 0$, $2 \leftarrow 0$, . . . , transitions, the numbers representing the value of v in the excited and ground states, respectively. Such a series of bands associated with transitions from a common vibrational level in one electronic state to successive levels in another constitutes a band *progression*. Bands corresponding to transitions with a given value

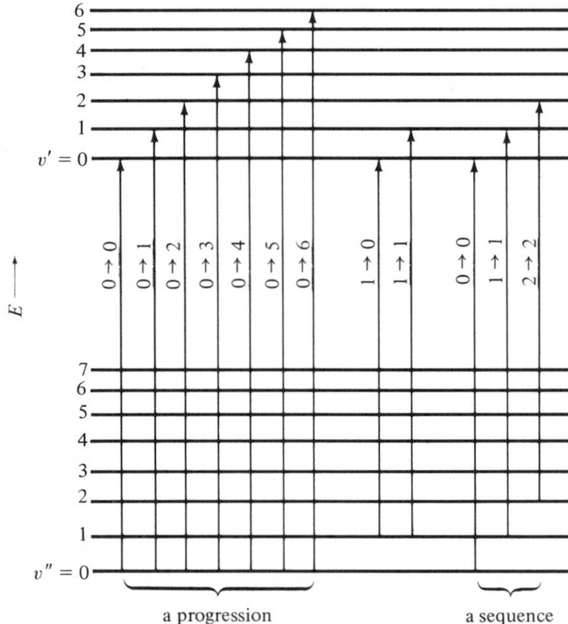

Figure 15.5 / *Vibrational transitions accompanying an electronic transition of a diatomic molecule.*

of Δv, as the $0 \leftarrow 0$, $1 \leftarrow 1$, $2 \leftarrow 2$, . . . , bands are near each other and form a band *sequence*. Sequences are often prominent in the emission spectra of thermally or electrically excited vapors, containing a population of electronically excited molecules distributed over a number of vibrational states, but, since in normal absorption measurements at room temperature the initial state of most of the molecules is the level $v = 0$ of the ground electronic state, sequences are not usually pronounced in absorption spectra.

15.6 / POTENTIAL ENERGY CURVES

Figure 15.6(a) is a schematic drawing of the potential energy of a diatomic molecule in its lowest electronic state as a function of the distance between the two atoms. One of the nuclei is regarded as fixed, and the other vibrates about an equilibrium point on the line joining the nuclei, under the influence of restoring forces that hold the atoms together when their separation exceeds the equilibrium value and prevents their collapse into each other when the molecule is compressed. The total vibrational energy, kinetic plus potential, must have one of the quantized values appropriate to the vibrational motion.

During a vibration, as the atoms are separated to greater than the equilibrium distance, a constantly increasing restoring force is exerted, and the atoms are ultimately brought to rest and begin to move toward each other

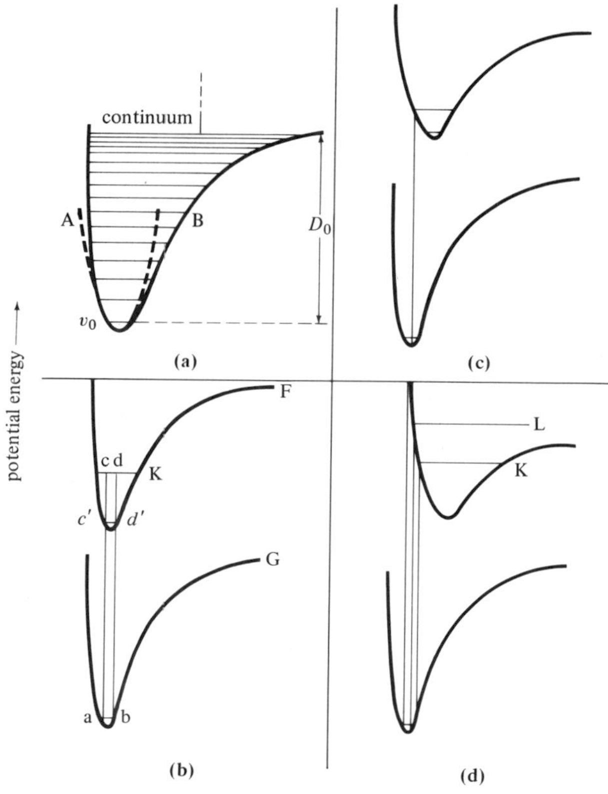

Figure 15.6 / *Potential energy curves.*
(a) *For the ground state.*
(b), (c), *and* (d) *For the ground state and various excited states.*
SOURCE: W. West, "Introductory Survey of Molecular Spectra," in *Chemical Applications of Spectroscopy*, W. West, Ed., 2nd ed., Part 1, Wiley-Interscience (New York, 1968), p. 29.

under an attractive force. As they approach each other at distances less than the equilibrium distance, increasing repulsive forces are brought into play, the atoms are again brought to rest and again begin to separate. The motion of the atoms then consists of an oscillatory motion about an equilibrium distance of separation at which the attractive and repulsive forces balance.

The points on the continuous curve represent the variation of potential energy with internuclear distance, while a horizontal line such as AB represents a state of constant total vibrational energy. Since the total vibrational energy is quantized, only discrete values are possible, and these are represented by the horizontal lines for the different possible values of the vibrational energy corresponding to vibrational quantum numbers 0, 1, 2, ...

The information contained in Fig. 15.6(a) can be more fully appreciated by considering a particular vibrational state, such as that represented by AB. The total energy, kinetic and potential, is constant, but the relative amounts of kinetic and potential energy change over the vibration. At the turning points A and B, the kinetic energy is zero and all the energy is potential; while at the equilibrium distance, the velocity of the atoms and the kinetic energy are at a maximum. For any point on the horizontal line representing the total vibrational energy, the kinetic energy is represented by the vertical distance between the point and the potential energy curve.

If the atomic oscillations were simple harmonic, the potential energy–distance curve would be a parabola symmetrical about a vertical line through the equilibrium distance. Such a parabola is shown as a dashed curve in Fig. 15.6(a). For small oscillations, there is close correspondence between the actual potential energy curve and the parabola, and thus it is apparent that the actual motion is approximately simple harmonic for small oscillations. A simple harmonic oscillator, however, would possess infinite stability and therefore does not correspond to any real molecule. Real molecules may be dissociated and are subject to smaller and smaller restoring forces as the separation of the atoms from the equilibrium position is increased. The true potential energy curve, corresponding to smaller attractive forces than for the harmonic oscillator for large separations and to greater repulsive forces for small separations, therefore deviates from a parabola in the manner shown in Fig. 15.6(a). At sufficiently great internuclear separation the restoring force vanishes, the two atoms separate entirely, that is, the molecule dissociates, and the potential energy curve becomes a horizontal line. The vertical distance D_0 between the lowest vibrational energy ($v = 0$) and the horizontal portion of the potential energy curve is the dissociation energy of the molecule in the lowest electronic state. The lowest vibrational energy is above the minimum in the potential energy curve by an amount equal to the zero-point energy.

In Fig. 15.6(b) a potential curve similar to that of Fig. 15.6(a) is also shown for an excited electronic state. For the case shown, the equilibrium separation between the atomic nuclei is approximately the same for the ground and excited states; that is, the minimum of the potential energy curve for the excited state is approximately vertically above the minimum for the ground state. This situation implies that for the excited state the force constant for oscillations of small amplitude, and therefore the vibration frequency and the shape of the potential energy curve near the minimum will be about the same as for the ground state.

As we have seen in Section 15.2, during the time required for an electronic transition the position of the atomic nuclei scarcely changes. This is an important principle of molecular spectroscopy, which has come to bear the name *Franck-Condon principle.* At room temperature, most of the molecules are in the lowest vibrational state of the ground electronic state (for the

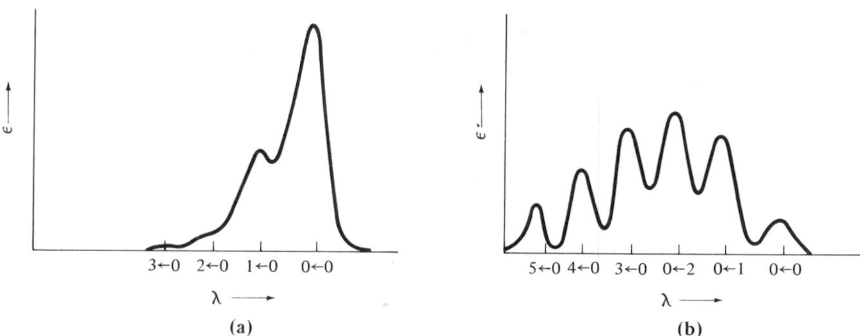

Figure 15.7 / *Distribution of intensity among vibrational bands in an electronic spectrum as determined by the Franck-Condon principle.*
SOURCE: W. West, "Introductory Survey of Molecular Spectra," in *Chemical Applications of Spectroscopy*, W. West, Ed., 2nd ed., Part 1, Wiley-Interscience (New York, 1968), p. 32.

case represented, ab), and, according to the Franck-Condon principle, an absorption transition to the excited state without change in the position and momentum of the atomic nuclei (a vertical transition) will find the molecule represented in Fig. 15.6(b) in the lowest vibrational state of the excited electronic state ($c'd'$). The corresponding absorption band (the $0 \leftarrow 0$ band) will be intense. Consider the probability of a transition to a high vibrational state K of the excited electronic state represented in Fig. 15.6(b). According to the Franck-Condon principle, the excited molecule will have a configuration defined by the coordinates corresponding to points between c and d. These configurations, however, are associated with large amounts of kinetic energy, given by the vertical distances cc' and dd', whereas the kinetic energy for this configuration in the ground state is zero at a and b and small at intermediate points. The transition to level K is therefore improbable, and the corresponding absorption band is of low intensity. Figure 15.7(a) shows schematically the shape of the absorption band associated with this type of relation between the potential energy curves for an assembly of molecules in solution, for which the vibrational bands are broadened by a variety of intermolecular interactions. The $0 \leftarrow 0$ band, at the longest wavelength, is the most intense, the $1 \leftarrow 0$ band appears as a shoulder at shorter wavelength, and higher vibrational bands are of very low intensity.

In Fig. 15.6(c) the equilibrium separation between the atoms in the diatomic molecule is greater in the excited electronic state than in the ground state. The shallower and broader minimum in the potential energy curve for the excited state indicates a loosening of the binding forces. The Franck-Condon principle now predicts the most probable transition to be associated with a vibrationally excited upper electronic state, and the most intense vibrational band in the absorption spectrum is not the $0 \leftarrow 0$ band, but the

$2 \leftarrow 0$ for the situation represented. Figure 15.7(b) shows the corresponding absorption spectrum for the substance in solution.

Figure 15.6(d) represents the case in which the potential energy curve for the excited electronic state is so much displaced to greater distances with respect to the curve for the ground state that, although transitions to states such as K are possible in the part of the diagram corresponding to a stable, discretely quantized molecule, so also are transitions to states such as L, in the part of the diagram corresponding to a dissociated molecule with a continuum of energy values. The absorption spectrum in this case consists, therefore, of a discrete band series at longer wavelengths, converging on a continuum that extends to the shorter wavelengths.

Two-dimensional potential energy diagrams, such as that of Fig. 15.6, are strictly applicable only to diatomic molecules, but qualitative insight into the behavior of polyatomic molecules may often be derived from such a two-dimensional representation.

SUGGESTED READINGS

W. West, "Introductory Survey of Molecular Spectra," in *Chemical Applications of Spectroscopy*, W. West, Ed., 2nd ed., Part 1, Wiley-Interscience (New York, 1968).

L. de Galan, *Analytical Spectrometry*, Adam Hilger Ltd. (London, 1971), Chapter 2.

D. H. Whiffen, *Spectroscopy*, Wiley (New York, 1966), Chapter 13.

chapter 16 / Ultraviolet-visible absorption spectroscopy

chapter 16

16.1 / INTRODUCTION

In previous chapters, we have described instrumentation for UV-visible spectroscopy (Chapter 11) and the origins of molecular spectra (Chapter 15). In this chapter, we are principally concerned with applications of molecular spectroscopy in the UV-visible region for (1) quantitative determination of one or more species in a mixture; and (2) qualitative analysis, in which the goal is the identification of a pure compound, the determination of the presence or absence of a particular species in a mixture, or the identification of certain functional groups in a compound under structural investigation.

Quantitative analysis may be carried out with a relatively simple and inexpensive single-beam spectrophotometer, but useful information for qualitative analysis usually requires a double-beam spectrophotometer with a monochromator of moderately good resolution (1 nm or better) and provision for automatic recording. For measurements in the visible region, a tungsten filament lamp is commonly used, while a hydrogen or deuterium arc lamp is required for the UV region. Cells for holding the absorbing solution are of glass for the visible and quartz for the UV region. The most commonly encountered cell provides a path length of approximately 1 cm, but a variety of special-purpose cells are available which, for example, provide longer path lengths or minimize the volume of sample required. Multiplier phototube detectors are employed in all but the simplest single-beam instruments. Because some compounds decompose under the influence of UV radiation, the absorption cells are placed behind the monochromator. In this way, only a small fraction of the radiation emitted by the source falls on the absorption cell. A variety of solvents are suitable for UV-visible spectroscopy. Table 16.1 lists some common solvents and the approximate wavelength below which they cannot be used because of absorption. The solvent, of course, must be transparent in the region of

TABLE 16.1 / *Transmission limits of some solvents*

	Approximate wavelength at which transmittance is 50% (nm)	
Solvent	0.1 cm	1.0 cm
water		180
n-hexane	<200	210
cyclohexane	<200	220
2, 2, 4-trimethylpentane	<200	210
n-heptane	<200	210
methanol	199	216
ethanol	199	215
diethylether	203	245
acetonitrile	210	215
chloroform		250
carbon tetrachloride		270
benzene		280
acetone		310

interest, but in addition its possible effects on the absorbing system must be considered. Polar solvents, such as water, alcohols, esters, and ketones, often obliterate the vibrational structure of absorption bands. Spectra obtained with nonpolar hydrocarbon solvents more closely resemble gas-phase spectra.

16.2 / QUANTITATIVE ANALYSIS

16.2.1 / The Beer-Lambert law

Information as to the identity of the absorbing species and the details of the absorption process is not a prerequisite to application of absorption spectroscopy to quantitative analysis. Indeed, the utility of absorption spectroscopy for quantitative analysis was recognized many years before the absorption process was understood. In 1730, Lambert and Bouguer postulated the relationship between radiant power and path length through an absorbing medium that would be stated in modern terms as: "The decrease in radiant energy of a beam of monochromatic radiation is proportional to the intensity of the beam and the length of absorbing medium through which the beam passes." In 1852, Beer added a postulate concerning the concentration of absorbers in the absorbing medium, and the revised statement is thus: "The decrease in radiant energy of a beam of monochromatic radiation is proportional to the intensity of the beam and the number of absorbers in its path."

The mathematical statement that has come to bear the name *Beer-Lambert law*, or simply *Beer's law*, may be derived as follows. Consider, as shown in Fig. 16.1, a beam of monochromatic radiation of intensity I_0 that strikes an absorption cell with an optical path length b, containing a solution with

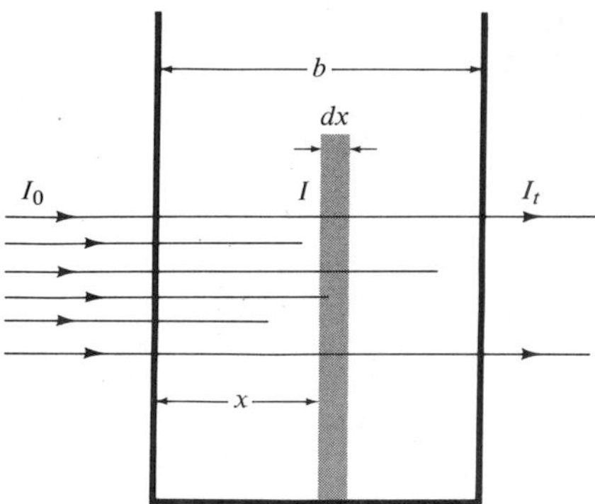

Figure 16.1 / *Absorption of radiation in a cell.*

a concentration c of absorbing molecules. After absorption over a distance x in the cell, the intensity of the beam is diminished to I. The decrease of the intensity in the adjacent layer dx is, according to Beer's hypothesis, proportional to the intensity I and to the number of absorbing molecules $c\,dx$ in the layer:

$$-dI = kIc\,dx \tag{16.1}$$

where k is a proportionality constant. Because absorption of radiation results from a spectral transition from the ground state, c is, strictly speaking, the concentration of molecules in the ground state. At room temperature, however, almost all the molecules are in the ground electronic state, and c can safely be taken equal to the total concentration.

With the boundary condition $I = I_0$ for $x = 0$, the solution of the differential equation (16.1) is found to be

$$I = I_0 e^{-kcx} \tag{16.2}$$

When the absorption cell has been traversed completely, the intensity is reduced to I_t, and the transmittance T of the solution is given by

$$T = \frac{I_t}{I_0} = e^{-kcb} = 10^{-0.43kbc} = 10^{-abc} \tag{16.3}$$

The proportionality constant a is related to the transition probability between two energy levels and indicates the strength of the absorption band. In older literature, this constant is designated by a variety of names, the most popular of which is absorption coefficient. The name *absorptivity* is presently recommended, and the units are centimeters^{-1} (grams/liter)$^{-1}$, where b has units of centimeters and c has units of grams per liter. An alternative representation that is widely used is

$$T = 10^{-\varepsilon bC} \tag{16.4}$$

where ε is called the *molar absorptivity* and has units of centimeters^{-1} (moles/liter)$^{-1}$ and C is the molar concentration of the absorbing species.

Molar absorptivities of 10^4–10^5 represent strong absorption bands. Weaker bands, which usually arise from formally forbidden transitions, have molar absorptivities of the order of 100–1000 $\text{cm}^{-1}(\text{moles/liter})^{-1}$. The value of the absorptivity varies with wavelength and for each different species, so the Beer-Lambert law is strictly valid only for monochromatic radiation and applies separately to each component of a solution.

The detector of a spectrometer measures the intensity of the radiation transmitted by the absorption cell, and, in the usual case, the measurement is carried out in such a way that I_0 is the intensity transmitted through an absorption cell containing pure solvent or some other blank solution. Thus, the transmittance of the solution can readily be obtained, but the inverse exponential relationship between transmittance and concentration is inconvenient for quantitative analysis. Therefore, another quantity, the *absorbance*, A, of the solution is introduced:

$$A = -\log T = abc = \varepsilon bC \tag{16.5}$$

16.2.2 / Deviations from the Beer-Lambert law

Equation (16.5) predicts a straight-line relationship between absorbance and concentration, but it is often observed that at larger concentrations the measured absorbance deviates from the predicted value. This is known as a deviation from the Beer-Lambert law. Strictly speaking, true deviations can occur only if the absorptivity varies with concentration. In molecular absorption spectrometry, this happens only at very large concentrations; most deviations observed are due to a failure to meet the conditions for which the Beer-Lambert law was derived. Such deviations would most appropriately be called *apparent* deviations from the Beer-Lambert law, the most important causes of which are as follows.

A / *The effect of spectral bandwidth*

In deriving the Beer-Lambert law, it was assumed that monochromatic radiation was employed for the absorption measurement. Actually, however, the monochromator isolates from the continuum radiation of the source a small-wavelength range given by the spectral bandwidth, s. The intensity striking the detector in the absence of an absorbing substance is then given by

$$I_0 = \int_s I_{0,\lambda}\, d\lambda = I_{0,\lambda} s \tag{16.6}$$

where $I_{0,\lambda}$ is the intensity per unit wavelength interval. Over the small range covered by the spectral bandwidth, $I_{0,\lambda}$ is practically independent of wavelength.

When the absorption cell contains an absorbing substance, the intensity striking the detector is given by

$$I_t = \int_s I_{0,\lambda} 10^{-a(\lambda)bc}\, d\lambda = I_{0,\lambda} \int_s 10^{-a(\lambda)bc}\, d\lambda \tag{16.7}$$

and the measured absorbance is

$$A_m = -\log \frac{I_t}{I_0} = -\log \frac{\int 10^{-a(\lambda)bc}\, d\lambda}{s} \tag{16.8}$$

If the absorptivity does not change over the spectral bandwidth, the exponential term can be removed from the integral, and the measured absorbance will be equal to the absorbance predicted by Eq. (16.5).

The spectral bandwidth of UV spectrometers is typically of the order of 1 nm. Molecular absorption bands are generally smooth and much broader than 1 nm, so that the influence of spectral bandwidth is in most cases negligible, especially when the absorbance is measured at the maximum of the absorption band. If the absorption band is sharp, however, or measurements are made on a steep slope of the absorption band, the absorptivity may well vary over the spectral bandwidth, and apparent deviations from the Beer-Lambert law will be observed. Figure 16.2 provides an example of the effect of spectral bandwidth. With increasing slit width (and hence increasing spectral bandwidth), the recorded vibration bands gradually blend together. Simultaneously, the absorbance at the band maximum decreases, and the slope of the analytical curve is reduced.

B / *The effect of stray light*

Ideally, a monochromator transmits radiation only within the spectral band around the central wavelength selected. However, some radiation outside this spectral band—stray light—invariably emerges from the exit slit. Stray light may alter the measured absorbance in a variety of ways, but if we confine our attention to the case in which the stray light is not absorbed by the sample, then the measured absorbance is given by

$$A_m = -\log \frac{I_t + I_s}{I_0 + I_s} \tag{16.9}$$

Stray light is usually reported as a fraction of I_0, and if S represents the fraction of stray light, then Eq. (16.9) may be written as

$$A_m = -\log \frac{I_t + SI_0}{I_0 + SI_0} = -\log \frac{T + S}{1 + S} \tag{16.10}$$

and the effect of stray light can be seen from the calculated measured absorbances for various values of S, as shown in Table 16.2.

In a good UV spectrometer, the amount of stray light is usually less than 0.1 percent over most of the range, and thus the effect of stray light is negligible in most cases. The effect of stray light inevitably becomes noticeable, however, if the transmittance of the sample is made sufficiently small, and it may become noticeable even for samples of moderate transmittance as the UV limit of the monochromator is approached. For most spectrometers, the stray light becomes an increasingly larger fraction of I_0 as the

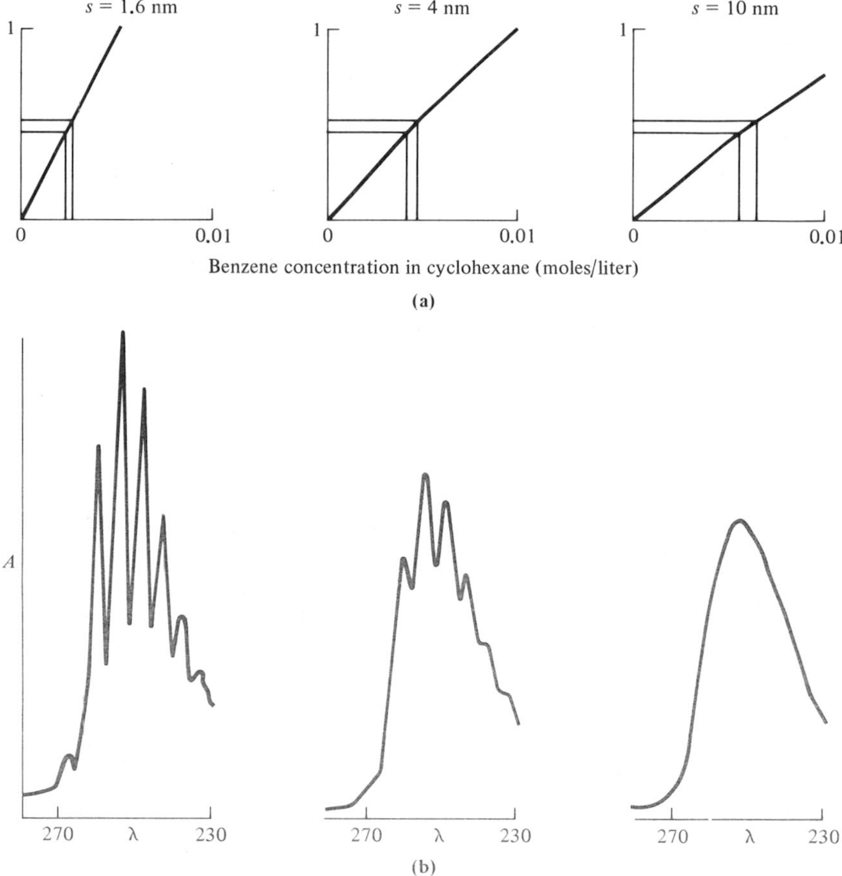

Figure 16.2 / (a) *Analytical curves and* (b) *absorption spectra for benzene at increasing spectral bandwidth.*

UV limit is approached, because both the source intensity and detector response decrease at shorter wavelengths, and thus the signal due to I_0 decreases. Visible radiation usually presents the most serious stray light problem for UV-visible spectrometers, because the spectral radiance of most sources is high and the spectral response of most detectors is high for visible radiation.

TABLE 16.2 / *Effect of stray light on measured absorbance for a true absorbance of 1.00*

A_m	% *Stray light*
1.00	0.01
0.996	0.10
0.963	1.0

C / *The effect of chemical equilibria*

In Eq. (16.5), c represents the true concentration of the absorbing substance. Practical analysis often deals with substances that are in equilibrium with other components of the solution. Any shift of this equilibrium as the concentration of the compound of interest is varied leads to an apparent deviation from the Beer-Lambert law. If, for example, the absorbing species is the anion of a weak acid, the degree of dissociation of the acid decreases with increasing concentration, and the measured absorbance will be less than the predicted value. It should also be clear that deviations may occur as a result of changes in other components of the solution. For example, for the case of the anion of a weak acid, a change in solution pH will alter the concentration of the absorbing species.

16.2.3 / The analytical curve

Equation (16.5) might seem to provide a method for calculating the concentration of an absorbing species, but in practice this method is seldom used because the apparent value of the absorptivity is, as we have seen in the preceding section, sensitive to the conditions of measurement. Quantitative analysis is usually carried out by employing an analytical curve prepared from a series of standard solutions of known concentration.

Some examples of analytical curves are contained in Fig. 16.2, and these may be used to illustrate some of the considerations that affect the choice of conditions for preparation of the analytical curve. The slope of the analytical curve should be as large as possible to obtain the best sensitivity (i.e., the ability to detect small changes in concentration) and precision. By comparing the three analytical curves, it can be seen that an uncertainty of a given magnitude in the measured absorbance, represented by the pair of horizontal lines on the three curves, leads to an uncertainty in the measured concentration, represented by the pair of vertical lines on the three curves, which is smallest for the curve of steepest slope. To obtain a curve of large slope, measurements should be made at the wavelength corresponding to the maximum value of the absorptivity for the species of interest. The choice of a band maximum has another advantage as well: The band is often relatively flat at the maximum, and the measurement will be relatively insensitive to errors in reproducing the wavelength setting.

16.2.4 / Precision

An expression for the relative error in the determination of the concentration can be derived from Eq. (16.5):

$$\frac{dc}{dA} = \frac{1}{ab} = \frac{c}{A} \tag{16.11}$$

$$\frac{dc}{c} = \frac{dA}{A} = \frac{d\log T}{\log T} = \frac{0.43}{T\log T}\,dT \tag{16.12}$$

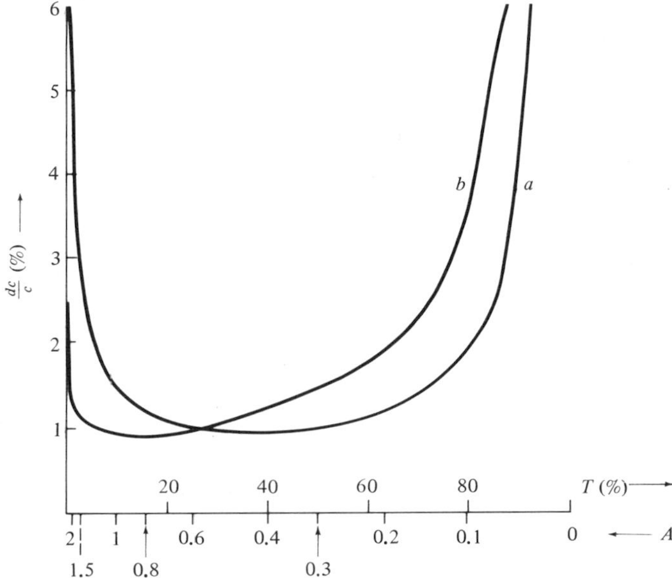

Figure 16.3 / *Relative error in the measurement of transmittance. Curve a, constant error in the transmittance. Curve b, error proportional to the square root of the transmittance.*

where dc/c is the relative error in the concentration. Thus, it is apparent that the relative error in concentration depends on the error in measuring the transmittance, dT, and on the magnitude of the transmittance. By plotting dc/c against T, it is possible to examine the dependence of precision on the magnitude of T and thereby determine the optimum range of transmittance values, but before doing so it is necessary to make an assumption as to the behavior of dT as T varies. The assumption most commonly presented is that dT is invariant with T and lies in the range 0.002–0.01 for commercial spectrophotometers. Curve *a* of Fig. 16.3 was obtained by assuming a constant error in the transmittance of 0.0035. The discussion of noise associated with photoelectric detectors (Section 11.2.5) suggests that the assumption of a constant error in transmittance is somewhat unreasonable. The shot noise of a photoelectric detector increases as the square root of the incident intensity, and thus we may expect dT to be proportional to $T^{1/2}$, and Eq. (16.12) can be rewritten, replacing dT with $kT^{1/2}$, as

$$\frac{dc}{c} = \frac{0.43k}{T^{1/2}\log T} \tag{16.13}$$

where k is the proportionality constant relating dT and $T^{1/2}$. Curve *b* of Fig. 16.3 was obtained from Eq. (16.13), using a value of k such that the relative error is 1 percent at an absorbance of 0.6.

For either assumption about dT, it is seen that dc/c changes little in the absorbance range between 0.1 and 1.5 but increases rapidly for higher and lower absorbances. In practice, the error curves for most spectrophotometers lie somewhere between curves *a* and *b*, and it is preferable to prepare solutions such that the absorbance falls within the range 0.1–1.5. In this optimum range, the precision of absorption measurements is found to be about 0.5 percent.

16.2.5 / The limit of detection

It is instructive to estimate, from Eq. (16.5), the range of concentrations determinable by the absorption spectrophotometric method. For a favorable case the molar absorptivity may be as high as 10^5 cm^{-1} (mole/liter)$^{-1}$, and for a 1-cm path length cell, absorbances in the range 0.1–1.5 will be obtained for concentrations in the range 1×10^{-6} to 1.5×10^{-5} *M*. Still lower concentrations may be detectable, although the precision of measurement may be inadequate for a quantitative determination. The limit of detection depends not only on the value of the absorptivity but also on the noise level inherent in the instrument employed. With most commercial instruments, an absorption of 1 percent ($T = 0.99$, $A = 0.004$) is close to the limiting detectable signal. For a compound with a molar absorptivity of 10^5, if a 10-cm path length cell is used, the limit of detection will be approximately 4×10^{-9} *M*.

16.2.6 / Multicomponent analysis

To this point, it has been assumed that only one absorbing component was present in any sample. Since many substances absorb in the UV-visible region, it is often necessary to carry out a number of chemical manipulations prior to the final absorption measurement in order to achieve this condition and avoid serious overlap of absorption bands. Sometimes, instead of resorting to tedious and time-consuming manipulations, it is possible to determine two or more absorbing components simultaneously.

Absorbances are additive, and hence the absorbance of a mixture of absorbing substances is equal to the sum of the separate absorbances. Consider a mixture of two components, X and Y. The following equations can be written to describe the absorbance of the solution measured at two wavelengths λ_1 and λ_2:

$$A(\lambda_1) = a_{X,\lambda_1} b c_X + a_{Y,\lambda_1} b c_Y \tag{16.14}$$

$$A(\lambda_2) = a_{X,\lambda_2} b c_X + a_{Y,\lambda_2} b c_Y \tag{16.15}$$

where a_{X,λ_1} is the absorptivity of the component X at λ_1, and the other subscripts similarly indicate the component and wavelength of measurement. The four absorptivity values can be determined by absorption measurements on solutions of the pure components, and, after measurement of $A(\lambda_1)$ and $A(\lambda_2)$ for the mixture, the two simultaneous equations can be solved for the unknown concentrations c_X and c_Y.

For best results, λ_1 and λ_2 are chosen so that the absorbance at one wavelength is due chiefly to component X and the absorbance at the other wavelength is due chiefly to component Y. As the complexity of the mixture increases, the difficulty of finding suitable wavelengths for measurements increases rapidly, and, in practice, the multicomponent technique is seldom useful for mixtures with more than two components with overlapping spectra.

16.2.7 / Spectrophotometric titrations

In spectrophotometric titrations, a spectrophotometer is used to facilitate observation of the endpoint of a titration. The measured quantity of interest is the volume of titrant required to cause the absorbance of a solution to change in a specified fashion.

The endpoint indicator color change in titrations of metals with EDTA (ethylenediaminetetraacetic acid) is notoriously difficult to detect visually, and provides an example of a situation in which the spectrophometric method may be used to advantage. If the indicator changes from red to blue, the spectrometer may be set to a wavelength of 650 nm. Under favorable conditions, the absorbance of the titrated solution then shows a sudden increase from a very low value to a rather high value at the endpoint of the titration.

For titrations in which one of the reactants can serve as the indicator, the titration may be carried out in a somewhat different fashion. The titration of ferrous ion by permanganate provides a good example of this type of titration. The visual endpoint consists of the first appearance of the pink color indicating unreacted permanganate ion. In the spectrophotometric mode, if the absorbance of the solution is monitored at 525 nm, the absorbance will remain zero until the equivalence point and then increase linearly as excess permanganate is added. Therefore, a plot of permanganate added yields two straight lines, as shown in Fig. 16.4, and the point of intersection corresponds to the endpoint of the titration.

16.3 / QUALITATIVE ANALYSIS

16.3.1 / The concept of a chromophore

The occurrence of absorption bands in the spectrum of a molecule is often associated with the presence of specific groups in the molecule and is approximately independent of the rest of the molecule. In the visible and UV spectral regions, the primary effect of the absorption of a quantum of radiation by a molecule is usually to excite one electron. The energy required, and hence the wavelength of the effective radiation, is determined by the nature of specific atoms and atomic groupings in the molecule and is influenced to only a secondary degree by other atoms and groups for which the electronic binding is very different. Atomic groups in a molecule and their associated chemical bonds that determine an electronic absorption

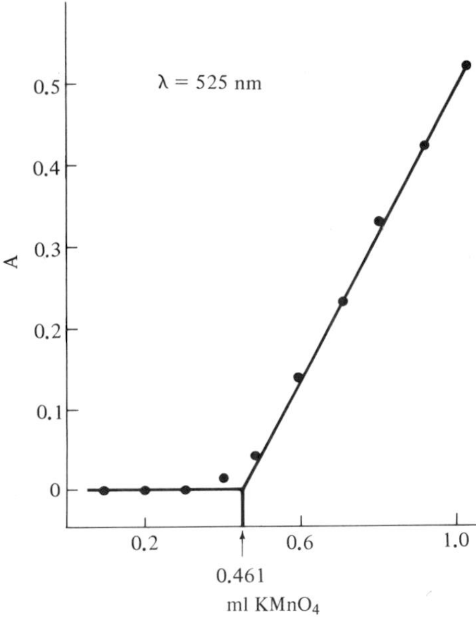

Figure 16.4 / *Photometric titration of* Fe(II) *by* $KMnO_4$.

band are called *chromophores.* In the visible and UV regions, the wavelength, intensity, and general appearance of the electronic band spectra, along with the effect of environmental conditions, such as pH and solvent, are the most important factors for the characterization of compounds. For example, aldehydes and ketones containing the unconjugated carbonyl group show a characteristic absorption band of relatively low molar absorptivity with a maximum at about 280–290 nm. The band lacks notable vibrational structure in solution, and the maximum is displaced to shorter wavelengths with increasing polarity of the solvent. An unknown compound exhibiting an absorption spectrum with all these characteristics would be judged to contain an unconjugated aldehydic or ketonic carbonyl group, while a deviation of the spectrum from any of the characteristics of this carbonyl chromophore would throw serious doubt on this identification.

16.3.2 / Absorption by organic compounds

The spectra of organic compounds in the visible and UV regions arise from valence electrons, which participate in holding the atoms in the molecule together, or from nonbonding electrons, localized on specific atoms such as nitrogen, oxygen, sulfur, or halogen atoms. The "lone-pair" electrons on nitrogen are such nonbonding electrons. The spectroscopic designation of a nonbonding electron or the associated orbital is *n*.

Since the valence electrons are associated with the bonding between atoms, they move in the field of at least two atomic centers; their state is

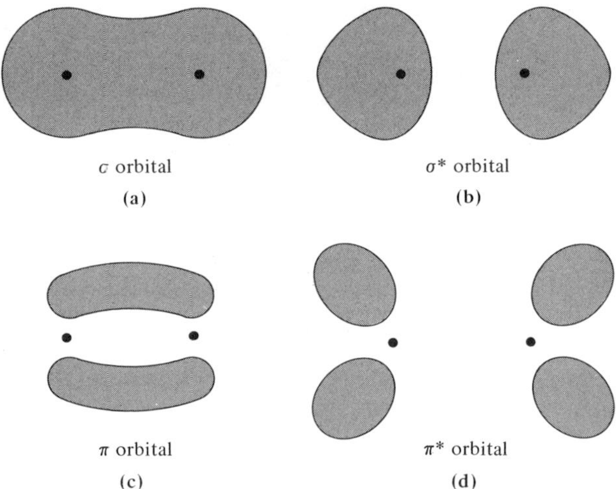

Figure 16.5 / *Electron distribution in σ and π molecular orbitals.*

thus described by a molecular orbital about the two or more atomic nuclei, in contrast to the atomic orbital, which describes the state of an electron localized about the nucleus of a single atom. In saturated organic compounds containing only single bonds, the two electrons associated with the bond are for the most part localized about the two bonded atomic nuclei, and are influenced to only a minor degree by atoms outside that bond. The molecular orbitals occupied by such localized bonding electrons are called *σ orbitals*, the electrons are *σ electrons*, and the bond arising from the presence of the two electrons shared between two atomic centers is a *σ bond.* As illustrated in Fig. 16.5(a), the probability distribution of the σ electrons is rotationally symmetrical about a line connecting the two atomic centers.

The absorption of radiation of the appropriate frequency can excite the σ electron to an orbital designated σ^*, for which the probability distribution is still rotationally symmetrical about the line connecting the atomic centers but which has a node at a point between the atomic centers, as illustrated in Fig. 16.5(b). The σ^* state is said to be *antibonding*, because the electron probability is at a minimum between the atomic centers. Excitation of an electron to an antibonding state does not, however, imply dissociation of the molecule, since only one of the two originally bonding electrons is excited by the absorption and some bonding energy remains in the excited state.

In the simplest unsaturated stable molecule, ethylene, three of the valence electrons of each carbon atom occupy σ orbitals, two of them embracing the carbon atom and one of the attached hydrogen atoms and the third embracing the two carbon atoms. The remaining valence electron of each carbon atom occupies a molecular orbital of a different type, a *π orbital*, formed by the overlap of *p* atomic orbitals of each of the carbon atoms.

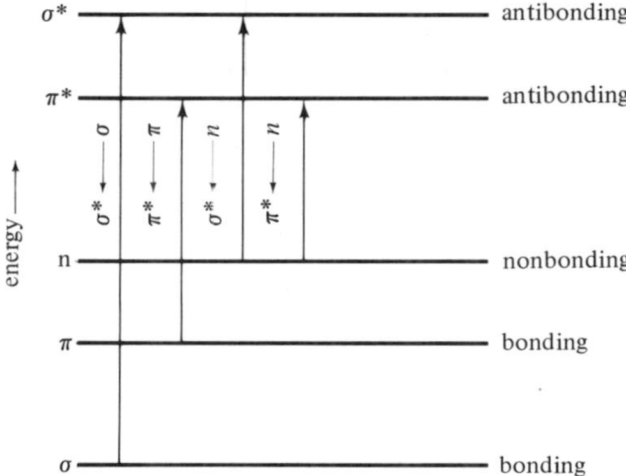

Figure 16.6 / *Electronic molecular energy levels.*

The π orbital, like the p atomic orbital from which it is derived, has a node in the plane containing the carbon and hydrogen nuclei, as shown in Fig. 16.5(c), and the maximum electron probability therefore occurs in two regions above and below the molecular plane.

The absorption of radiation of the proper frequency can excite a π electron to a state designated π^*. Like the σ^* state, the π^* state is antibonding. The probability distribution for π^* is shown in Fig. 16.5(d).

Figure 16.6 illustrates the usual arrangement by energy of the various electronic states. Absorption of radiation can bring about transitions among certain of the states. As shown in Fig. 16.6, the commonly encountered transitions are $\sigma^* \leftarrow \sigma$, $\sigma^* \leftarrow n$, $\pi^* \leftarrow n$, and $\pi^* \leftarrow \pi$.

A / $\sigma^* \leftarrow \sigma$ *transitions*

The σ electrons are the most tightly bound of the valence electrons; that is, they are of lowest energy. The σ^* state is the highest-energy excited state, and thus a $\sigma^* \leftarrow \sigma$ transition requires quite high-energy radiation. In the usual case, $\sigma^* \leftarrow \sigma$ transitions are observed only in the region below 200 nm. The valence electrons of saturated hydrocarbons are all of the σ type; hence, their absorption spectra can be observed only with spectrometers that can be purged of atmospheric oxygen, since O_2 absorbs strongly in the region below 200 nm.

B / $\sigma^* \leftarrow n$ *transitions*

Since n electrons are of considerably higher energy than σ electrons, a number of saturated compounds containing atoms with n-type electrons absorb in a conveniently accessible region of the ultraviolet due to $\sigma^* \leftarrow n$ transitions. Table 16.3 contains some examples of compounds showing absorption bands due to $\sigma^* \leftarrow n$ transitions. The energy requirement for a $\sigma^* \leftarrow n$ transition depends on the properties of the atom with the unshared pair and only to a lesser extent on the structure of the molecule.

TABLE 16.3 / *Some examples of absorption due to $\sigma^* \leftarrow n$ transitions*

Compound	λ_{max} (nm)	ε_{max}	*Compound*	λ_{max} (nm)	ε_{max}
H_2O	167	1480	$(CH_3)_2S$	229	140
CH_3OH	184	150	$(CH_3)_2O$	184	2520
CH_3Cl	173	200	CH_3NH_2	215	600
CH_3Br	204	200	$(CH_3)_2NH$	220	100
CH_3I	258	365	$(CH_3)_3N$	227	900

For example, the electrons in the sulfur atom are more loosely bound than those in the oxygen atom, and the $\sigma^* \leftarrow n$ absorption band for $(CH_3)_2S$ occurs at a longer wavelength than the corresponding band for $(CH_3)_2O$. A similar trend is apparent in the series CH_3Cl, CH_3Br, CH_3I. Note that molar absorptivities for $\sigma^* \leftarrow n$ transitions are typically rather small, falling in the range 100–3000 cm^{-1} $(mole/liter)^{-1}$.

C / $\pi^* \leftarrow n$, $\pi^* \leftarrow \pi$ *transitions*

Most applications of absorption spectroscopy to organic compounds are based on $\pi^* \leftarrow n$ or $\pi^* \leftarrow \pi$ transitions, because the energies required for these processes are such that the absorption bands for many compounds lie at wavelengths greater than 200 nm. Both types of transitions require the presence of an unsaturated site in the molecule to provide the π^* orbitals.

Some examples of $\pi^* \leftarrow n$ and $\pi^* \leftarrow \pi$ transitions are shown in Table 16.4. Note that molar absorptivities for $\pi^* \leftarrow n$ transitions are even lower than for $\sigma^* \leftarrow n$ transitions, falling typically into the range 10–100 cm^{-1} (mole/liter)$^{-1}$. Molar absorptivities for $\pi^* \leftarrow \pi$ transitions, on the other hand, are large, typically about 10^4 for a single unsaturated site. Another characteristic difference of the two types of transitions occurs in the effect of solvent

TABLE 16.4 / *Absorption characteristics of some common chromophores*

Chromophore	*Example*	*Solvent*	λ_{max} (nm)	ε_{max}	*Type*
alkene	$C_6H_{13}CH{=}CH_2$	*n*-heptane	177	13,000	$\pi^* \leftarrow \pi$
alkyne	$C_5H_{11}C{\equiv}CCH_3$	*n*-heptane	178	10,000	$\pi^* \leftarrow \pi$
carbonyl	$CH_3\overset{\overset{O}{\|}}{C}CH_3$	*n*-hexane	186	1,000	$\sigma^* \leftarrow n$
			280	16	$\pi^* \leftarrow n$
	$CH_3\overset{\overset{O}{\|}}{C}H$	*n*-hexane	180	large	$\sigma^* \leftarrow n$
			293	12	$\pi^* \leftarrow n$
carboxyl	CH_3COOH	ethanol	204	41	$\pi^* \leftarrow n$
amido	CH_3CONH_2	water	214	60	$\pi^* \leftarrow n$
azo	$CH_3N{=}NCH_3$	ethanol	339	5	$\pi^* \leftarrow n$
nitro	CH_3NO_2	isooctane	280	22	$\pi^* \leftarrow n$
nitroso	C_4H_9NO	ethylether	300	100	—
			665	20	$\pi^* \leftarrow n$
nitrate	$C_2H_5ONO_2$	dioxane	270	12	$\pi^* \leftarrow n$

polarity on the wavelength of the bands. Peaks arising from $\pi^* \leftarrow n$ transitions are generally shifted to shorter wavelengths (*hypsochromic* or *blue shift*) as the polarity of the solvent is increased. Usually, but not always, a reverse trend (*bathochromic* or *red shift*) is observed for $\pi^* \leftarrow \pi$ transitions. The hypsochromic effect apparently arises from the increased solvation of the nonbonded electron pair, which lowers the energy of the n orbital. The most dramatic effects of this kind (blue shifts of 30 nm or more) are seen with polar hydrolytic solvents, such as water or alcohols, in which hydrogen bond formation between the solvent protons and the nonbonded electron pair is extensive.

D / *Effect of multiple chromophores*

If a molecule contains more than one chromophore and the chromophores are separated from one another by more than one single bond, the chromophores absorb independently, and the absorption spectrum is a superposition of the absorption bands due to the various chromophores. If the isolated chromophores are identical, the wavelength of the absorption will be approximately that for a single chromophore but the molar absorptivity will increase approximately as the number of identical chromophores. For example, the absorption maximum of 1,5-hexadiene,

$$CH_2{=}CHCH_2CH_2CH{=}CH_2,$$

occurs at 178 nm, and the absorption maximum of 1-hexene,

$$CH_2{=}CHCH_2CH_2CH_2CH_3,$$

occurs at 177 nm, but the molar absorptivity for the diene (26,000) is slightly more than double that of the monoene (11,800).

The absorption spectrum of compounds containing conjugated ethylenic bonds is not merely an intensified version of the simple olefinic spectrum. In the molecular orbital treatment, π electrons are considered to be further delocalized by the conjugation process; for example, in 1,3-butadiene, $CH_2{=}CHCH{=}CH_2$, the π orbital involves four atomic centers. The effect of this delocalization is to give more bonding character to the π^* orbital, lower its energy, and thereby shift the absorption maximum corresponding to the $\pi^* \leftarrow \pi$ transition to longer wavelengths. For butadiene, the absorption band occurs at 210 nm and the molar absorptivity is approximately 21,000. Thus, the absorption maximum is shifted approximately 30 nm compared to that for the unconjugated ethylenic chromophore in 1,5-hexadiene and 1-hexene, and the molar absorptivity corresponds to that of a diene. A further red shift occurs if the conjugation involves additional ethylenic bonds. Table 16.5 illustrates the trend in wavelength and molar absorptivity.

Conjugation between the doubly bonded oxygen of aldehydes, ketones, and carboxylic acids and an olefinic double bond gives rise to a similar bathochromic shift of absorption bands. Since the effect of conjugation

TABLE 16.5 / *Absorption of conjugated polyenes*

Compound $H(CH{=}CH)_nH$		*Solvent*	λ_{max} (nm)	ε_{max}
$n = 1$	ethylene	vapor	162	~10,000
$n = 2$	1, 3-butadiene	vapor	210	
		hexane	217	20,900
$n = 3$	1, 3, 5-hexatriene	isooctane	268	42,700
$n = 4$	1, 3, 5, 7-octatetraene	cyclohexane	304	
$n = 5$	1, 3, 5, 7, 9-decapentaene	isooctane	334	121,000
$n = 11$	all-*trans*-β-carotene	hexane	480	139,000

is to lower the energy of the π^* orbital, both the short-wavelength band due to the $\pi^* \leftarrow \pi$ transition and the long-wavelength band due to the $\pi^* \leftarrow n$ transition are shifted. For example, the absorption spectrum of crotonaldehyde, $CH_3CH{=}CHCHO$, in ethanol, shows a band at 220 nm with a molar absorptivity of 15,000 and a band at 322 nm with a molar absorptivity of 28. The low molar absorptivity of the longer wavelength band is typical of that for the $\pi^* \leftarrow n$ transition of the carbonyl chromophore, but the absorption band for an unconjugated carbonyl typically occurs at about 280 nm.

E / *Aromatic systems*

The UV spectra of aromatic hydrocarbons are characterized by three sets of bands that originate from $\pi^* \leftarrow \pi$ transitions. For example, benzene has a strong absorption band at 184 nm ($\varepsilon_{max} = 68{,}000$), a weaker band at 204 nm ($\varepsilon_{max} = 8800$), and a still weaker band at 254 nm ($\varepsilon_{max} = 250$). When measured in solution with nonpolar solvents, the 184-nm band has no vibrational structure, the 204-nm band has poorly resolved vibrational structure, and the 254-nm band has well-developed vibrational structure.

All three of the characteristic bands for benzene are strongly affected by ring substitution. The effects on the two longer wavelength bands are of particular interest because these two bands are readily observed, and changes in the appearance of the bands may be used to identify substituents that do not themselves absorb in an accessible region of the UV spectrum. Table 16.6 lists the absorption characteristics of some benzene derivatives.

By definition, an *auxochrome* is a functional group that does not absorb in the UV region, but has the effect of shifting chromophore peaks to longer wavelengths, as well as increasing their intensities. In Table 16.6, —OH and $—NH_2$ can be seen to be examples of auxochromes. Auxochromic substituents have at least one pair of n electrons capable of interacting with the π electrons of the ring. This interaction apparently has the effect of lowering the energy of the π^* state, and thus a bathochromic shift of the absorption band occurs. Note that the auxochromic effect is more pronounced for the phenolate anion than for phenol itself, probably because the anion has an extra pair of unshared electrons to contribute to the inter-

TABLE 16.6 / *Absorption of some benzene derivatives*

Compound	*Formula*	*Solvent*	λ_{max} (nm)	ε_{max}	λ_{max} (nm)	ε_{max}
benzene	C_6H_6	hydrocarbon	254	250	204	8800
toluene	$C_6H_5CH_3$	hydrocarbon	262	260	208	7900
hexamethyl-benzene	$C_6(CH_3)_6$	hydrocarbon	271	230	221	10000
chlorobenzene	C_6H_5Cl	hydrocarbon	267	200	210	7400
iodobenzene	C_6H_5I	hydrocarbon	258	660	207	7000
phenol	C_6H_5OH	hydrocarbon	271	1260	213	
phenolate ion	$C_6H_5O^-$	dil. NaOH	286	2400	235	9400
benzoic acid	C_6H_5COOH	ethanol	272	855	226	9800
aniline	$C_6H_5NH_2$	methanol	280	1320	230	7000
anilinium ion	$C_6H_5NH_3^+$	dil. acid	254	160	203	7500

action. With aniline, on the other hand, the nonbonding electrons are lost by formation of the anilinium cation, and, as a consequence, the auxochromic effect disappears.

16.3.3 / Absorption by inorganic systems

Although UV-visible absorption spectroscopy has long been applied to the quantitative determination of inorganic and metal-organic species, development in understanding the transition processes observed has been slow, and hence absorption spectroscopy for qualitative analysis of inorganic species in solution has not progressed to the same extent as for qualitative analysis of organic species. The principal absorption processes presently recognized for inorganic species are (1) transitions within incompletely filled *f* orbitals of the metal ion, (2) transitions within incompletely filled *d* orbitals of the metal ion, and (3) charge-transfer transitions.

A / *f-electron transitions*

The ions of most of the lanthanide and actinide elements absorb in the UV-visible region. In distinct contrast to spectra observed for most organic and inorganic species in solution, lanthanide and actinide ions show rather sharp line spectra, which are relatively unaffected by changes in the solvent or other species associated with the metal ion in solution. The line spectra for these two groups of ions appear to be due to transitions among the *f* levels in the ion (4*f* for the lanthanides, 5*f* for the actinides), which are partially filled. The independence of the spectra from external influence is presumably due to the screening of the *f* orbitals by the outer filled orbitals. For example, the electron configuration for Ce is $[Kr]4d^{10}4f^{2}5s^{2}5p^{6}6s^{2}$.

B / *d-electron transitions*

The transition metal ions have partially filled *d* orbitals, and their spectra seem largely due to transitions within the *d* levels. Unlike the lanthanide and actinide ion spectra, however, the transition metal spectra are broad

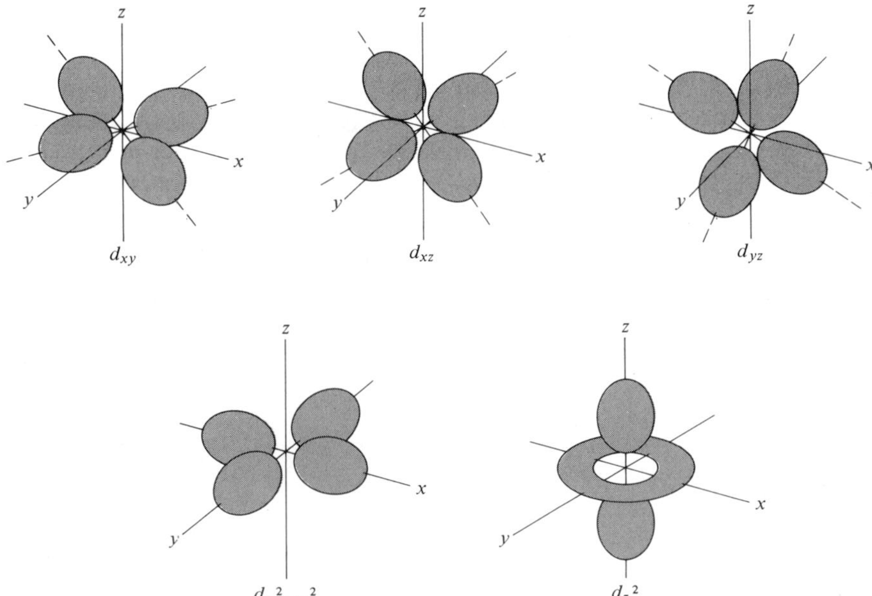

Figure 16.7 / *Electron density distribution in various d orbitals.*

and strongly influenced by the nature of the species (the ligand) bonded to the metal ion in solution. Explanation of the transition metal spectra is based on the premise that the energies of the *d* orbitals, which are degenerate in the absence of an external electric or magnetic field, are split by the influence of the ligand. In the simplest treatment, that provided by the so-called crystal field theory, the splitting is presumed due to the mutual electrostatic repulsion of electrons in the ligands for electrons in the *d* orbitals.

Consider the application of the crystal field approach to the simplest case, that of the Ti(III) ion, which has a single *d* electron. In the absence of a perturbing field, the five *d* orbitals have the electron distributions shown in Fig. 16.7, and the five orbitals are all of the same energy; that is, the single *d* electron of Ti(III) may be found in any of the orbitals with equal probability. The degeneracy of the five orbitals is removed, however, if the Ti(III) ion finds itself in a complex such as $[Ti(H_2O)_6]^{3+}$. The hexaaquo complex is octahedral; that is, the six ligands are evenly distributed about the central ion, one ligand at each end of the three axes shown in Fig. 16.7. In this arrangement, the negative ends of the water dipoles are pointed toward the metal ion; the electric fields from these dipoles will thus have a repulsive effect on electrons in the *d* orbitals, so the energy of the *d* orbitals is increased. The effect of the ligands, however, is not the same for all of the orbitals. The d_{xy}, d_{xz}, and d_{yz} orbitals have their maximum electron density between the axes and hence experience less repulsive effect from the ligands

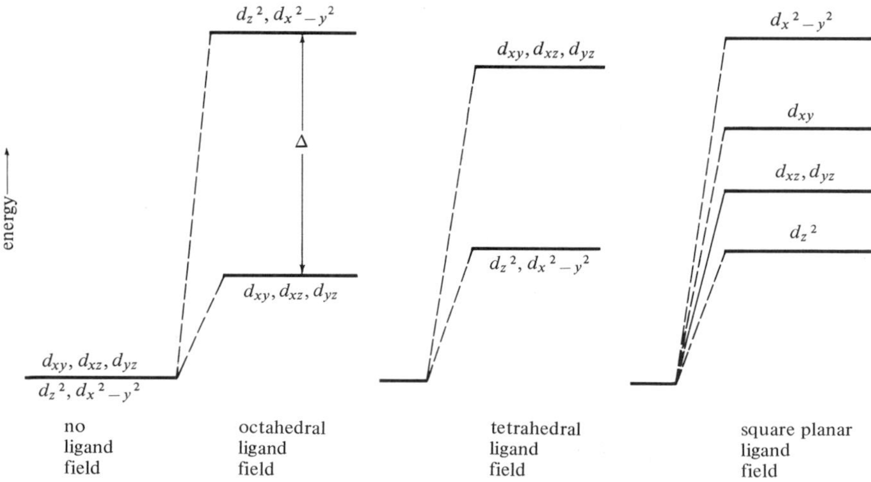

Figure 16.8 / *Effect of ligand field on d-orbital energies.*

than the $d_{x^2-y^2}$ and d_{z^2} orbitals, which are oriented along the axes. The overall effect of the field exerted by the water dipoles is to split the d orbitals into two groups, as shown in Fig. 16.8. Also shown are energy diagrams derived through similar considerations for complexes involving four ligands in either a tetrahedral or square planar arrangement.

We are now in a position to understand some of the spectra of complex ions. We should expect relatively simple spectra from ions, such as $[Ti(H_2O)_6]^{3+}$, with a d^1 configuration, and, in fact, this is the case. The absorption spectrum of $[Ti(H_2O)_6]^{3+}$, as shown in Fig. 16.9, consists of a single band with a maximum at about 490 nm. This band has been identified

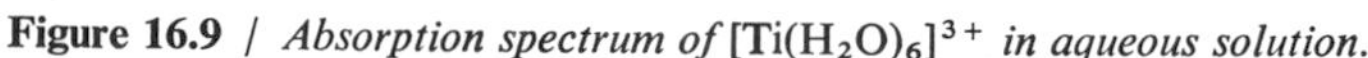

Figure 16.9 / *Absorption spectrum of* $[Ti(H_2O)_6]^{3+}$ *in aqueous solution.*

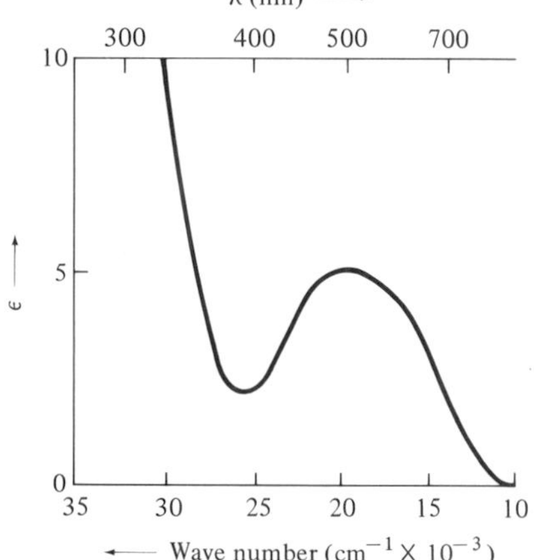

as arising from a transition between the threefold degenerate level and the twofold degenerate level, and this implies that the splitting energy, Δ, is approximately 20,400 cm^{-1} ($1/4.9 \times 10^{-5}$ cm). For configurations other than d^1, the arrangement of energy levels is more complex and spectra may be more difficult to interpret.

The magnitude of Δ is dependent on a number of factors, including the valence state of the metal ion and the position of the parent ion in the periodic table. An important variable attributable to the ligand is the so-called ligand field strength, which is a measure of the extent to which a complexing group will split the energies of the *d* orbitals; that is, a complexing agent with a high ligand field strength will cause Δ to be large. It is possible to arrange ligands in a series according to their capacity to cause *d*-orbital splittings. For the more common ligands, the series is $I^- < Br^- < Cl^- < F^- < OH^- < C_2O_4^{2-} \simeq H_2O < SCN^- <$ pyridine $\simeq NH_3 <$ ethylenediamine $<$ dipyridyl $<$ *o*-phenanthroline $< NO_2^- < CN^-$. The utility of this series is that the *d*-orbital splittings and hence the relative frequencies of visible absorption bands for two complexes containing the same metal ion but different ligands can be predicted from the above series whatever the particular metal ion may be. For a few cases, inversions of the order of adjacent or nearly adjacent members of the series are found.

Absorption spectra arising from the *d–d* transitions have provided much of the information on which current theories of bonding in inorganic complexes are based and provide information as to the structure of complexes. The *d–d* absorption bands are not particularly useful for quantitative analysis because the molar absorptivities are low (typically around 50).

C / *Charge-transfer spectra*

For quantitative analytical purposes, the most important type of absorption by inorganic species is charge-transfer absorption, because the molar absorptivity is very high (typically greater than 10,000). Charge-transfer absorption is not limited to transition metal ions but occurs for the majority of ions in solution. In order for a complex to exhibit a charge-transfer spectrum, it is necessary that one of its components have electron-donor characteristics and the other, electron-acceptor properties. A charge-transfer absorption involves transition of an electron of the donor to an orbital that is largely associated with the acceptor. As a consequence, the excited state is the product of a kind of internal oxidation–reduction process. This behavior differs from that of an organic chromophore, where the electron in the excited state is in the molecular orbital formed by two or more atoms.

A well-known example of a charge-transfer absorption is observed in the Fe(III) thiocyanate ion complex. Absorption of radiation results in the transition of an electron from the thiocyanate ion to an orbital state that is largely associated with the Fe(III) ion. The product is thus an excited species involving predominantly Fe(II) and the thiocyanate radical, SCN. Under

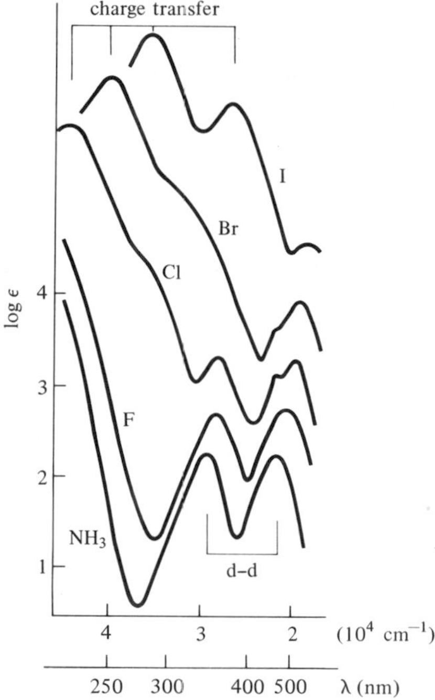

Figure 16.10 / *Absorption spectra of* $[Co(NH_3)_5X]^{n+}$, *where* X = NH_3 (n = 3) *and* X = F, Cl, Br, I (n = 2). *Each spectrum is raised half a unit of log ε above the preceding one.*

ordinary circumstances, after a brief time the excitation energy is lost and the complex returns to the ground state. Occasionally, however, dissociation of the excited complex may occur to produce a photochemical oxidation–reduction.

The energy required for the charge-transfer absorption can be related to the electron affinity of the ligand. As the electron affinity of the ligand decreases, the energy required for excitation decreases, and the charge-transfer absorption band moves to longer wavelengths. This type of behavior is illustrated in Fig. 16.10 for the complex ion $[Co(NH_3)_5X]^{n+}$, where X = NH_3, F, Cl, Br, or I. It may also be noted that as the number of ligands of the same kind increases around a central atom, the charge-transfer absorption moves to longer wavelength. Thus, as the concentration of chloride ions increases, the color of a copper solution, first blue because of slight absorption in the red and green from a *d–d* band of $[Cu(H_2O)_4]^{2+}$, gives way to green, as replacement of H_2O by Cl^- progressively moves a charge-transfer band into the visible from the UV side.

The strong UV-visible absorption spectra of anions such as sulfate, chromate, perchlorate, permanganate, nitrate, and so on are also classified as charge-transfer spectra. They are considered as arising from electron transfer from molecular π orbitals to a higher state more closely associated with the central atom.

16.3.4 / Applications of ultraviolet-visible absorption spectroscopy to qualitative analysis

The appearance of absorption spectra may be strongly affected by the conditions of measurement, and compound identification can best be carried out by direct comparison of the spectrum of the unknown with the spectrum of an authentic sample of the suspected compound. It is usually easier to compare spectra that have been plotted as log A vs. wavelength. If the Lambert-Beer law is written in the form

$$\log A = \log \varepsilon + \log bC \tag{16.16}$$

it can be seen that only ε, of the terms on the right, varies with wavelength, and variations in b or C result only in a linear displacement of log A at every wavelength. Thus, the appearance of a log A vs. wavelength spectrum is unaltered by changes in b or C.

The certainty with which a compound can be identified from spectral data depends on the number of separate spectral features (peaks, minima, inflection points) that are found to coincide between the unknown and an authentic reference spectrum. In the UV-visible region the absorption bands, especially if measured in solution, tend to be few in number, broad, and rather featureless. Further, the absorption peaks for many chromophores are not greatly influenced by nonabsorbing groups attached to the molecule, and as a consequence, only the absorbing functional group can be identified. Thus, unambiguous identification from UV-visible spectral data alone rarely occurs, even when the information available from additional spectra of the unknown and reference in a variety of solvents, at different pH, and after suitable chemical treatment is considered.

Even though UV-visible absorption spectra may not provide unambiguous identification of a compound, they are nevertheless useful for detecting the presence of certain functional groups that act as chromophores. For example, a weak absorption band in the region of 280–290 nm that is displaced toward shorter wavelengths with increased solvent polarity is a strong indication of the presence of the carbonyl group. A weak absorption band at about 260 nm with indications of vibrational structure constitutes evidence for the presence of an aromatic ring. Ultraviolet-visible absorption spectra may also make possible an unambiguous choice between two possible isomeric arrangements of a known molecular formula. Thus, the two compounds

$$\underset{\text{I}}{H_3C-\overset{\overset{\displaystyle O}{\|}}{C}-CH_2-CH_2-\overset{\overset{\displaystyle O}{\|}}{C}-CH_3} \qquad \underset{\text{II}}{H_3C-CH_2-\overset{\overset{\displaystyle O}{\|}}{C}-\overset{\overset{\displaystyle O}{\|}}{C}-CH_2-CH_3}$$

are readily distinguished by their UV spectra. Compound I, acetonylacetone, absorbs at about 270 nm, as does acetone, but with a molar absorptivity approximately twice that of acetone. Compound II, 3,4-hexadione, absorbs at about 400 nm.

SUGGESTED READINGS

W. West, "Introductory Survey of Molecular Spectra," in *Chemical Applications of Spectroscopy*, W. West, Ed., 2nd ed., Part 1, Wiley-Interscience (New York, 1968).

L. de Galan, *Analytical Spectrometry*, Adam Hilger Ltd. (London, 1971), Chapter 4.

R. E. Dodd, *Chemical Spectroscopy*, Elsevier (New York, 1962), Chapter 4.

C. N. R. Rao, *Ultraviolet and Visible Spectroscopy*, 2nd ed., Plenum (New York, 1967).

L. E. Orgel, *An Introduction to Transition-Metal Chemistry: Ligand-Field Theory*, Methuen (London, 1960).

PROBLEMS

1 / A solution showed a transmittance of 0.100 at the measurement wavelength.
(*a*) What was the absorbance of the solution?
(*b*) If the solution had a concentration of 0.02 g/liter and the transmittance was measured for a 1-cm cell, what was the absorptivity of the compound at the measurement wavelength?
(*c*) If the compound had a molecular weight of 100, what was its molar absorptivity?
(*d*) Calculate the expected transmittance of the solution in a 5-cm cell.
(*e*) Calculate the expected transmittance of a solution measured in a 5-cm cell that is half as concentrated as the original.

2 / From the data provided below, calculate the weight of the compound that should be dissolved and diluted to 100 ml to obtain a transmittance of 50 percent in a 1-cm absorption cell.

Compound	*Measurement wavelength* (nm)	ε cm^{-1} (moles/liter)$^{-1}$
(a) C_6H_5COOH	226	9800
(b) $C_6H_5CH_3$	208	7900
(c) $C_8H_{10}N_4O_2$ (caffeine)	278	9200

3 / By what factor should the solutions prepared above be diluted to provide 50 percent transmittance in a 5-cm absorption cell?

4 / At 257 nm, the following absorbances are measured for solutions of potassium dichromate:

(*a*) Draw the analytical curve and from it determine the absorptivity of the dichromate anion.
(*b*) Determine the concentration of dichromate in solutions for which the absorbances are, respectively, 0.156, 0.729, and 1.20.

Concentration (mg/liter)	*Absorbance in 1*-cm *cell*
10	0.142
20	0.284
40	0.565
60	0.840
80	1.105
100	1.37

5 / At 210 nm, a certain monochromator exhibits 1 percent stray light. If a sample shows an absorbance of 1.1 at this wavelength, what is the true absorbance? Assume that the sample is transparent to the stray light.

6 / If the true absorbance of a sample is 2.00, what absorbance will be measured with a spectrometer that exhibits 1 percent stray light at the measurement wavelength?

7 / A certain acid has a dissociation constant of 5×10^{-7}. At 440 nm, the absorbance (1.00-cm cell) for a 5×10^{-4} M solution of this acid is 0.401 at pH = 1 and 0.067 at pH = 13. What will be the absorbance of the solution at pH = 7?

8 / A 5.00×10^{-4} M solution of A gave absorbance readings (in a 1-cm cell) of 0.683 at 440 nm and 0.139 at 590 nm. A 8.00×10^{-5} M solution of B gave absorbance readings (in a 1-cm cell) of 0.106 at 440 nm and 0.470 at 590 nm. Calculate the concentrations of A and B in a solution that gave absorbance readings (in a 1-cm cell) of 1.022 at 440 nm and 0.414 at 590 nm.

9 / Assuming the titration reaction is A + B → C, where B is the titrant, draw curves showing the course of a photometric titration for each of the following cases:
(*a*) A and C are colorless, B absorbs.
(*b*) A and B absorb, C is colorless.
(*c*) A absorbs, B and C are colorless.

10 / What conclusion would you draw from the presence of a very weak absorption band at 250 nm in the spectrum of cyclohexane?

11 / Mesityl oxide exists in two isomeric forms:

$CH_3—C(CH_3)=CH—CO—CH_3$ $\quad$ $CH_2=C(CH_3)—CH_2—CO—CH_3$

One exhibits an absorption maximum at 235 nm with ε = 12,000; the other shows no high-intensity absorption beyond 220 nm. Identify the isomers.

12 / Assign the structures shown to the respective isomers on the basis of this information: The α-isomer shows a peak at 228 nm (ε = 14,000) while the β-isomer has a band at 296 nm (ε = 11,000).

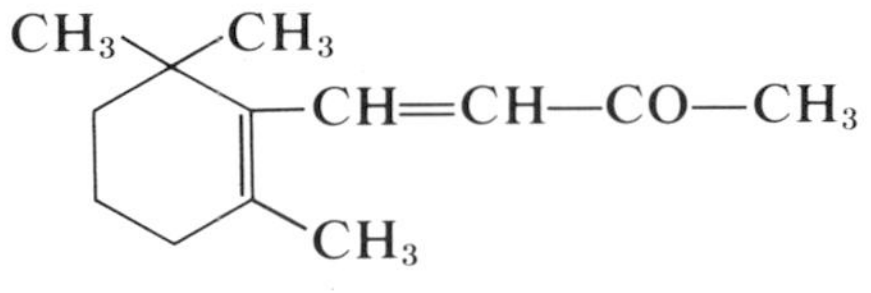

CH_3 CH_3
$—CH=CH—CO—CH_3$
CH_3

I

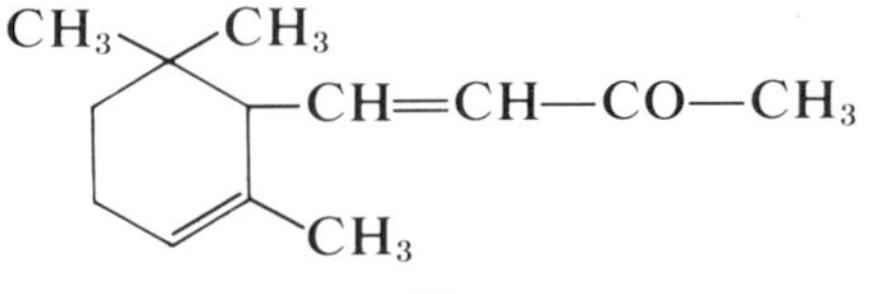

CH_3 CH_3
$—CH=CH—CO—CH_3$
CH_3

II

13 / The spectrum of compound A (in cyclohexane) shows maxima at 210 nm (ε_{max} = 15,500) and at 330 nm (ε_{max} = 37). Compound B (in cyclohexane) shows maxima at 190 nm (ε_{max} = 4000) and 290 nm (ε_{max} = 10). Assign the structures shown to compounds A and B and justify your assignments by identifying the most likely type of transistion and the chromophore responsible for each of the bands.

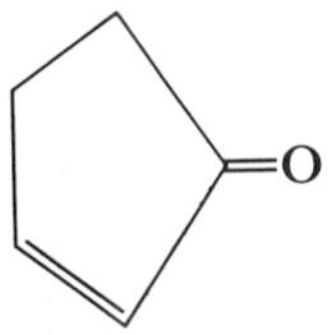

I

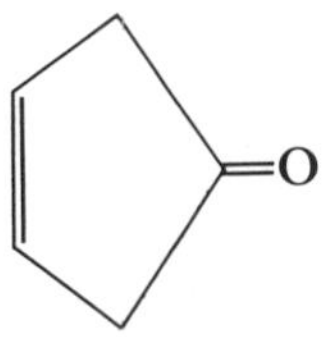

II

chapter 17 / Molecular luminescence spectrometry

chapter 17

17.1 / INTRODUCTION

In the preceding chapter, we considered the analytical utility of the absorption process in which molecules absorb electromagnetic radiation and undergo a transition to an excited electronic state. If a molecule excited by absorption of radiation returns to the ground electronic state with emission of radiation, it is said to luminesce, and the emitted radiation is termed luminescence. There are two types of luminescence processes that will be considered in this chapter: fluorescence and phosphorescence. The two processes are often distinguished phenomenologically—fluorescence (in the usual case) occurs promptly after excitation (delays in the range of nanoseconds to microseconds), but phosphorescence occurs after an appreciable delay (milliseconds or longer)—however, as we shall see, there is a more fundamental distinction between the two processes.

17.2 / FUNDAMENTAL CONSIDERATIONS

17.2.1 / Luminescence processes

The absorption processes of greatest interest in exciting luminescence are, as described in Section 16.3.2, $\pi^* \leftarrow \pi$ and $\pi^* \leftarrow n$. A simplified energy-level diagram for a molecule with both n- and π-type electrons is shown in Fig. 17.1, where arrows represent the electrons and indicate the spin states. In the usual case, a molecule in the ground state has its electron spins paired, the net spin is zero, and the state is said to be a *singlet*. (The name "singlet" derives from the considerations of atomic spectroscopy described in Chapter 12. In the terms used to describe atomic spectra, we would say that $S = 0$ and the multiplicity of the state is $2S + 1 = 1$.) When a ground-state molecule absorbs radiation, the transition usually occurs without change of spin ($\Delta S = 0$), and the excited state is also a singlet. Two such excited singlet states are indicated in Fig. 17.1 If the electron spins become un-

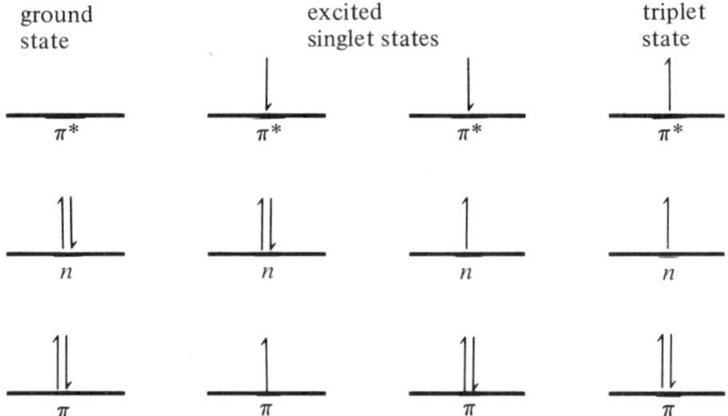

Figure 17.1 / *Some electronic states for a molecule with n- and π-type electrons.*

paired, $S = 1$ for the excited state, the multiplicity of the state is 3, and the state is said to be a *triplet*. One such triplet excited state is shown in Fig. 17.1. The triplet state is always of somewhat lower energy than the corresponding singlet state. Thus, the energy required for the $\pi^* \leftarrow n$ transition is less for the singlet-to-triplet transition than for the singlet-to-singlet transition, but the molar absorptivity is usually much less for the singlet-to-triplet transition because the change of spin is formally forbidden. Typically, the probability of a singlet–triplet process is about 10^{-6} that of a corresponding singlet–singlet process. Thus, if direct excitation of triplet states from the ground state by absorption of radiation were the only mechanism by which these states could be populated, they would be of little significance in analytical spectroscopy. There is, however, another mechanism for population of the triplet state.

Consider the potential energy curve representation (strictly applicable only to a diatomic molecule) for a molecule with a singlet ground state and one singlet and one triplet excited state shown in Fig. 17.2. (The potential energy curve representation has been described in Section 15.6.) The potential energy curves of the singlet and triplet excited states intersect in the region of higher vibrational energy, and in this region a radiationless transition is possible from the singlet excited state to the triplet excited state. As indicated by the potential energy curves, the molecule undergoes a transition from a condition of low vibrational energy in an excited electronic state to a condition of higher vibrational energy in a lower-energy excited electronic state. An alternative representation for such a process is given in Fig. 17.3, which shows the electronic energy levels of benzene. We shall employ this representation to consider some of the processes that are important to an understanding of molecular luminescence.

A / *Vibrational relaxation*

Molecules in solution at room temperature may be assumed to be in the lowest vibrational level of the ground state. Absorption occurs, therefore, from the $v = 0$ vibrational level of the ground state to various vibrational

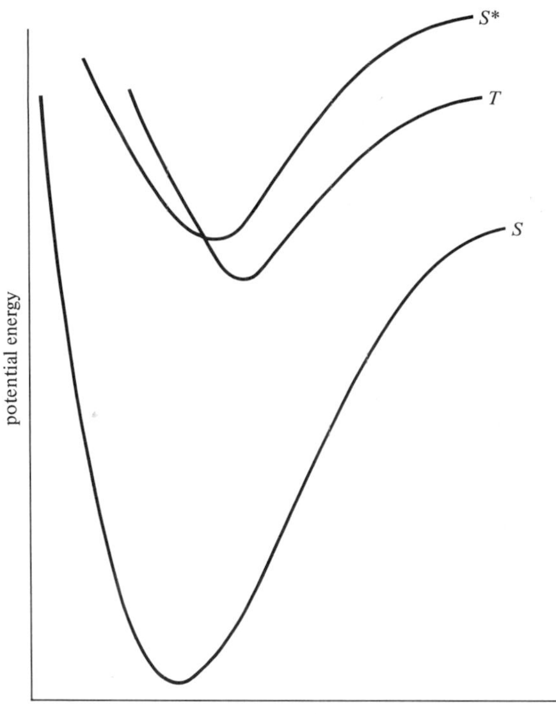

Figure 17.2 / *Potential energy curves for ground state, and for singlet and triplet excited states.*

levels of an excited state. The time required for an absorption transition is short (10^{-15} sec) compared to the period of molecular vibrations, and immediately after excitation a molecule has the same geometry and is in the same environment as it was in the ground state (Franck-Condon principle, Section 15.2). In this situation, it can do one of two things: it can emit a photon from the same vibrational level to which it was excited initially, or it can undergo changes in vibrational levels prior to emission of radiation. In solution, relaxation of a molecule through transfer of excess vibrational energy from the solute molecule to the solvent is quite rapid, and the molecule undergoes a radiationless transition to the lowest vibrational level of the excited state in a period of 10^{-13}–10^{-11} sec. As a consequence of the rapidity of vibrational relaxation for molecules in solution, photon emission always occurs from the lowest vibrational level of an excited state.

B / *Fluorescence*

Once a molecule arrives at the lowest vibrational level of an excited state, it can do several things, one of which is to return to the ground state by emission of a photon. This process is called flourescence. Note that the fluorescent emission is of lower energy than the absorbed radiation because

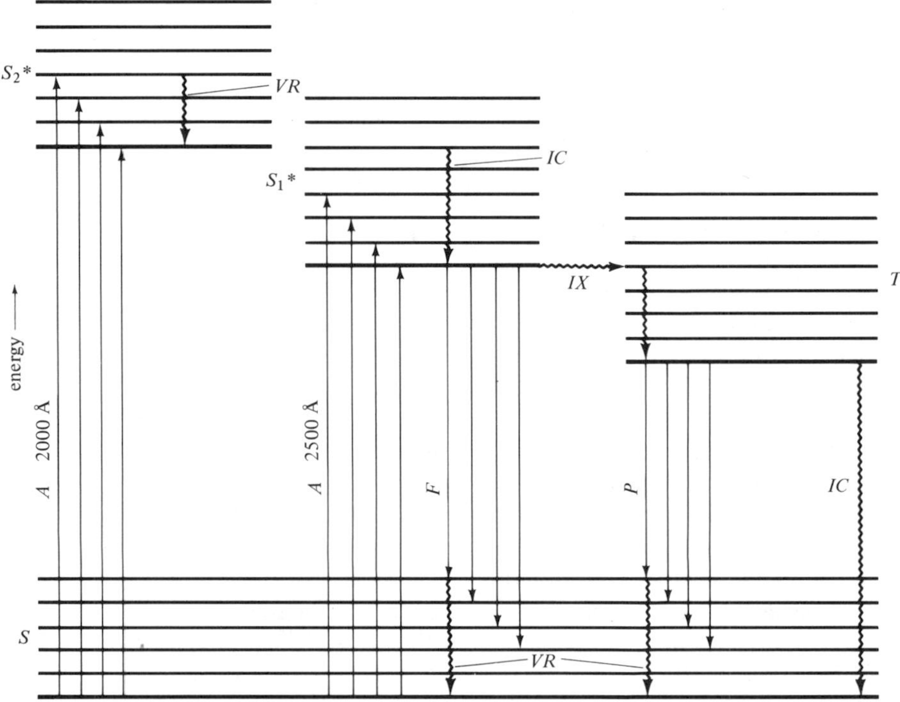

Figure 17.3 / *Energy-level diagram of the electronic states of benzene: A, absorption; F, fluorescence; P, phosphorescence; IC, internal conversion; IX, intersystem crossing; S, singlet ground state; S*, singlet excited state; T, triplet state; VR, vibrational relaxation.*

of vibrational relaxation both before and after the emission. Thus, for molecules in solution, the fluorescence spectrum always occurs at longer wavelengths than the absorption spectrum.

The lifetime of an excited singlet state is of the order of 10^{-9}–10^{-7} sec, and, if no other processes compete with fluorescence, all the excited molecules will return to the ground state by fluorescence in this period of time. If, in fact, all the excited molecules return to the ground state by fluorescence, the *quantum efficiency* of fluorescence for the system is said to be unity. The quantum efficiency of fluorescence is defined as the fraction of excited molecules that return to the ground state by fluorescence, and obviously takes values ranging from zero to one.

C / *Internal conversion*

One of the processes that may compete with fluorescence is a return of the excited molecule to the ground state by conversion of all the excitation energy into heat. Such a process is termed *internal conversion.* Internal conversion between an excited singlet state and the ground state is thought to be an inefficient process. Internal conversion between two excited singlet states is a much more probable process, especially if the energy separation of the states is small so that a considerable coupling of the vibrational states may occur. Such an internal conversion process is shown in Fig. 17.3. As a general rule, a molecule may be considered to undergo internal conversion to the lowest vibrational level of its lowest excited singlet state in a time (10^{-13}–10^{-11} sec) that is short relative to photon emission, regardless of the singlet state to which it was excited initially.

D / *Phosphorescence and intersystem crossing*

Although population of triplet states by direct absorption from the ground state, as discussed in connection with Fig. 17.2, is inefficient, there is another process for population of triplet states that may be very efficient: the internal conversion from the lowest-energy singlet state to a lower-energy excited triplet state. Such a process is termed an *intersystem crossing.* Recalling that singlet–triplet processes are typically 10^{-6} times less probable than singlet–singlet processes, it may seem unlikely that intersystem crossing can occur during the lifetime of an excited singlet state (10^{-9}–10^{-7} sec). Recall, however, that vibrational relaxation processes such as those involved in internal conversion occur in a time (approximately 10^{-13} sec) that is short relative to photon emission, and thus it may be anticipated that times of the order of $10^{-13}/10^{-6} = 10^{-7}$ sec, that is, comparable to the lifetime of the excited singlet state, are required for intersystem crossing. Therefore, intersystem crossing can compete with fluorescence emission in depopulating the $v = 0$ vibrational level of the lowest excited singlet state. In general, the smaller the energy gap between the lowest-energy excited singlet state and the triplet state and the longer the lifetime of the lowest-energy excited singlet state, the greater the probability of intersystem crossing.

Once intersystem crossing has occurred, the molecule undergoes the usual rapid vibrational relaxation and falls to the $v = 0$ level of the triplet state. In the usual case, the energy difference between the $v = 0$ level of the triplet state and the $v = 0$ level of the excited singlet state is sufficiently great compared to thermal energy that repopulation of the singlet state from the triplet state is improbable. The transition from the triplet state to the singlet ground state is formally forbidden, and thus, in the absence of competing processes, the lifetime of the triplet state is long (10^{-3}–10 sec); ultimately, however, a radiative transition can occur. Radiation arising from the triplet-to-singlet transition is termed *phosphorescence.* Thus, the fundamental distinction between fluorescence and phosphorescence is that fluorescence arises from a singlet–singlet transition and phosphorescence arises from a triplet–singlet transition.

The likelihood of a radiationless transition from the triplet state to the ground state is much greater than for a singlet–singlet radiationless transition. Two factors foster this condition: (1) The energy difference between the triplet state and the ground state is less than the separation between the

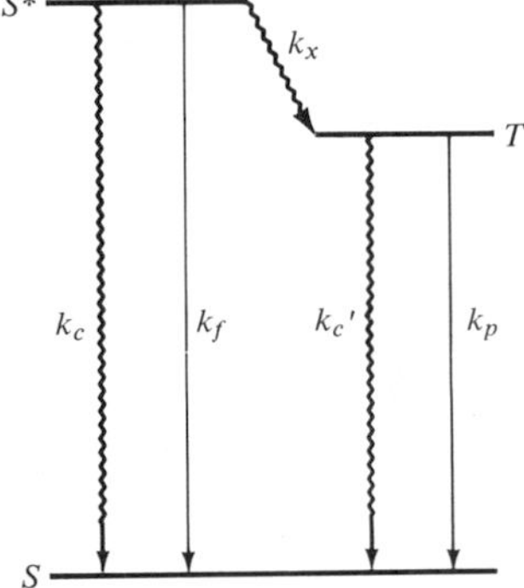

Figure 17.4 / *Summary of rate constants for excited-state processes.*

lowest excited singlet state and the ground state, which tends to enhance vibrational coupling between the triplet and ground states, and therefore to enhance internal conversion. (2) The lifetime of the triplet state is much longer than that of an excited singlet state, and therefore loss of excitation energy by collisional transfer to solvent molecules is greatly enhanced. In fact, this second process is so important that phosphorescence is almost never observed in solution at room temperature. Generally, phosphorescence is observed at the temperature of liquid nitrogen in solvents that freeze as rigid glasses at this temperature.

17.2.2 / Factors affecting luminescence

The factors affecting luminescence can best be discussed in terms of the rate constants for excited-state processes, as outlined in Fig. 17.4. The symbols S, S^*, and T represent the ground state, the first excited singlet state, and the triplet state, respectively. The rate constants k_c and k_c' are for radiationless energy loss from the first excited singlet state and the triplet state, respectively; k_f and k_p are for fluorescence and phosphorescence, respectively; and k_x is for intersystem crossing.

The quantum efficiency for fluorescence is given by

$$\phi_f = \frac{k_f}{k_f + k_c + k_x} \tag{17.1}$$

If k_f is much greater than k_c and k_x, ϕ_f will approach unity, while if k_c or k_x is large compared to k_f, ϕ_f will approach zero. The quantum efficiency for phosphorescence depends on the rate of intersystem crossing and on the competition between phosphorescence and radiationless transition from the triplet state:

$$\phi_p = \frac{k_p}{k_p + k_c'} \times \frac{k_x}{k_f + k_c + k_x} \tag{17.2}$$

In general, k_f and k_p are chiefly dependent on molecular structure and are affected only slightly by the molecular environment. The value of k_x also depends on molecular structure but is influenced somewhat by environment. At the other extreme, k_c and k_c' are strongly affected by the molecular environment and depend to a small extent on molecular structure.

Consider the factors that affect k_f. This rate constant is a measure of the probability of an electronic transition between the lowest excited singlet state and the ground state. This means that the factors that affect the intensity of the long-wavelength absorption band in a molecule will also affect the rate of spontaneous emission from the lowest excited singlet state. Therefore, the maximum molar absorptivity, ε_{max}, of the long-wavelength absorption band can serve as a qualitative measure of k_f. As ε_{max} decreases, one expects to see a corresponding decrease in k_f (and a corresponding increase in the lifetime of the excited singlet state). Recall from Chapter 16 that $\pi^* \leftarrow \pi$ absorption bands have molar absorptivities about 10^3 times larger than those for $\pi^* \leftarrow n$ absorption bands. Thus, if the nature of the molecule is such that the transition from the lowest excited singlet to the ground state is $\pi^* \rightarrow n$, the rate constant for fluorescence will be much less than for a molecule in which the transition is $\pi^* \rightarrow \pi$. In fact, it is found that for $\pi^* \rightarrow n$ transitions the rate of intersystem crossing, k_x, is greater than k_f, but that for $\pi^* \rightarrow \pi$ transitions the two rates are approximately the same. An interesting demonstration of this behavior is provided by the fluorescence of quinoline in solvents of varying polarity. In nonpolar solvents, such as benzene, the lowest-energy transition is $\pi^* \rightarrow n$, but in more polar solvents, as solvent interaction with the n-type electrons increases, the lowest-energy transition becomes $\pi^* \rightarrow \pi$. Thus, the quantum efficiency for fluorescence of quinoline shows marked dependence on the solvent polarity. For example, the relative ϕ_f values of quinoline in benzene, ethanol, and water are, respectively, 1, 30, and 1000.

The rate of intersystem crossing, k_x, depends strongly on the energy split between the singlet and triplet states. The smaller the difference in energy between the singlet and the triplet, the greater the probability of intersystem crossing. This behavior occurs because intersystem crossing is brought about by coupling between a vibrational level of the triplet and the lowest vibrational level of the lowest excited singlet. For a molecule to go from the singlet state to a very high vibrational level of a triplet state is unlikely because of the large change required in nuclear configuration. However, if the singlet–triplet split is small, intersystem crossing will require only a small change in nuclear configuration and the probability of the crossing will be increased.

Another factor affecting intersystem crossing is the so-called heavy-atom effect. This effect is similar to the progressive failure of the $\Delta S = 0$ selection rule of atomic spectroscopy as the atomic number of the atom increases. Substitution of a heavy atom into a π-electron system increases the rate of intersystem crossing and enhances phosphorescence at the expense of fluorescence. For example, in the halobenzene series, the fluorescence efficiency of fluorobenzene is approximately 0.16, chlorobenzene is 0.05, bromobenzene is approximately 0.01, and iodobenzene shows no fluorescence.

Species with unpaired spins (paramagnetic species) are also found to increase the rate of intersystem crossing. For example, the mesoporphyrin dimethyl ester chelate of zinc (atomic number 30) shows fluorescence of moderate intensity and weak phosphorescence; however, if copper (atomic number 29) is substituted for zinc in this compound, no fluorescence is observed but very strong phosphorescence occurs. This decrease in fluorescence and increase in phosphorescence cannot be attributed to an atomic-number effect, since copper has an atomic number one less than zinc, but must be attributed to the paramagnetic character of the Cu(II) ion.

The rates of radiationless deactivation, k_c and k_c', are most strongly affected by environmental factors, one of the most important of which is temperature. During the lifetime of an excited state, a molecule in solution will undergo many collisions with other molecules. Some of these collisions will be such that the excitation energy can be transferred and lost as heat. As the temperature of the solution increases, the collision rate increases and the probability of radiationless deactivation increases. Thus, it is found that ϕ_f for most molecules decreases with increasing temperature. Because of the longer lifetime of the triplet state, the effect of temperature is even greater on ϕ_p, and phosphorescence, with only a few exceptions, can be observed only when the solution is frozen into a rigid glass.

17.3 / EXPERIMENTAL CONSIDERATIONS

The basic instrumental components for molecular luminescence are not unlike those used in other UV-visible methods, but the arrangement of components is somewhat different. Figure 17.5 illustrates the basic arrangement of components for fluorescence and phosphorescence. The chemical system to be investigated is held in a glass or quartz container. For fluorescence measurements, the container is usually similar to the 1-cm-square cross-section cell employed for absorption spectrometry, but all four sides are polished. For phosphorescence, the container is usually a small-diameter (about 1 mm i.d.) tube, which is placed in a silica Dewar flask filled with liquid nitrogen. The sample is excited by UV or visible light from the excitation source, which is usually a high-pressure xenon or mercury arc lamp. The desired excitation wavelength region is selected by means of the excitation monochromator (or filter). The luminescence is emitted in all directions, and a small portion of the luminescence is collected by the emission monochromator, usually arranged at right angles to the excitation axis. The emission monochromator selects the wavelength region of luminescence to be measured, the radiation is detected by a multiplier phototube, and the resulting electrical signal is amplified and displayed. To obtain an accurate representation of the luminescence spectrum, the signal must be corrected

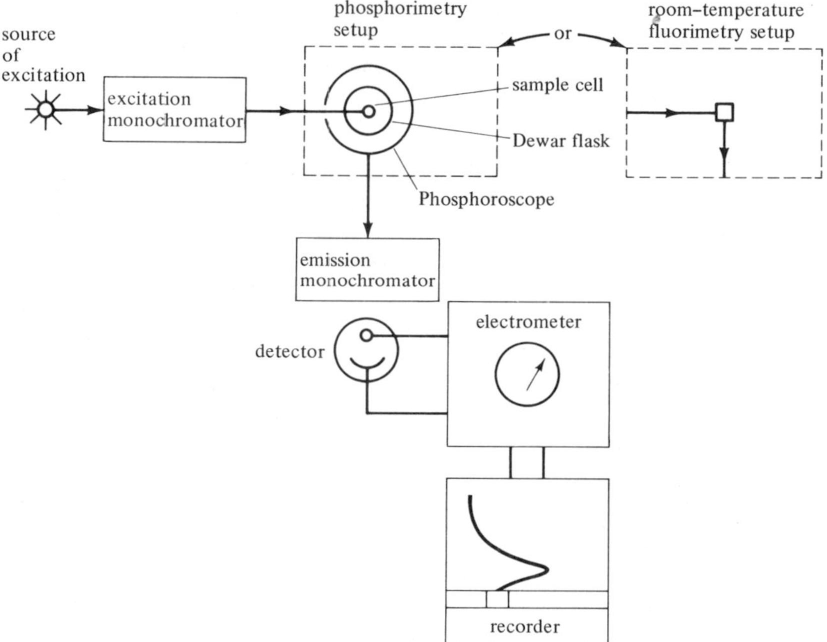

Figure 17.5 / *General arrangement of components for luminescence spectrometry.* SOURCE: J. D. Winefordner, S. G. Schulman, and T. C. O'Haver, *Luminescence Spectrometry in Analytical Chemistry*, Wiley (New York, 1972), p. 140.

for variation of the source intensity, monochromator transmission efficiency, and detector sensitivity with wavelength. Modern instruments provide means of automatic compensation. Correction of the luminescence spectrum is basically similar to the use of a double-beam instrument for absorption spectrometric measurements.

In the measurement of phosphorescence, a mechanical light chopping device, the *phosphoroscope*, is employed to separate the phosphorescence from stray source radiation and rapidly decaying fluorescence. Perhaps the simplest phosphoroscope is that illustrated in Fig. 17.5: an opaque cylinder with one or more apertures cut into it so that as the cylinder is rotated the sample cell alternately "sees" the exit slit of the excitation monochromator and the entrance slit of the emission monochromator. By controlling the rate of rotation of the cylinder, the time between excitation and observation can be varied. Thus, the phosphorescence lifetime is readily measured, and by proper adjustment of the phosphoroscope speed it may be possible to differentiate between phosphorescent species with similar spectra but different decay times.

17.4 / QUANTITATIVE ANALYSIS BY FLUORESCENCE

17.4.1 / Relationship between intensity and concentration

Fluorescence is necessarily preceded by the absorption of radiation at the excitation wavelength, and an expression to describe the intensity of fluorescence is readily derived by beginning with the statement that the intensity of fluorescence, I_f, is equal to the intensity of radiation absorbed, I_a, times the quantum efficiency, ϕ_f:

$$I_f = \phi_f I_a \tag{17.3}$$

The intensity of radiation absorbed can be derived in the fashion described in Section 16.2:

$$\begin{aligned} I_a &= I_0 - I_t \\ &= I_0(1 - 10^{-\varepsilon bC}) \end{aligned} \tag{17.4}$$

Thus, the intensity of fluorescence is given by

$$I_f = \phi_f I_0(1 - 10^{-\varepsilon bC}) \tag{17.5}$$

The exponential term can be rewritten as a series expansion, and if the concentration of absorbers is sufficiently low that all terms raised to powers larger than one can be ignored, Eq. (17.5) can be written

$$I_f = 2.3\phi_f I_0 \varepsilon bC \tag{17.6}$$

Thus, it is clear that I_f increases linearly with C for small values of C. Also, it should be noted that I_f depends directly on I_0, and thus the sensitivity of a determination may be improved by increasing I_0.

Figure 17.6 shows a typical analytical curve for fluorescence measurements. Two factors not considered in the derivation of Eq. (17.5) are responsible for the reversal of slope at high concentrations: self-quenching and self-absorption. The former is the result of collisions between fluorescing molecules, and results in a radiationless deactivation. Self-quenching can be expected to increase with concentration. Self-absorption occurs when the emitted radiation occurs at a wavelength at which absorption can also occur.

17.4.2 / Inorganic analysis

With a few exceptions, such as uranium salts, inorganic ions do not exhibit fluorescence. However, many inorganic ions, metal or nonmetal, have the ability to form complexes with a variety of organic molecules, and many of these complexes are highly fluorescent. Some of the elements that can be determined as fluorescent complexes are Al, Au, B, Be, Ca, Cd, Cu, Eu, Ga, Gd, Ge, Hf, Mg, Nb, Pd, Rh, Ru, S, Sb, Se, Si, Sm, Sn, Ta, Tb, Th, Te, W, Zn, and Zr. In addition, some inorganic ions, such as fluoride or cyanide, can quench the fluorescence of other species, and this effect can be made the basis of methods for determining the concentration of the quenching ions.

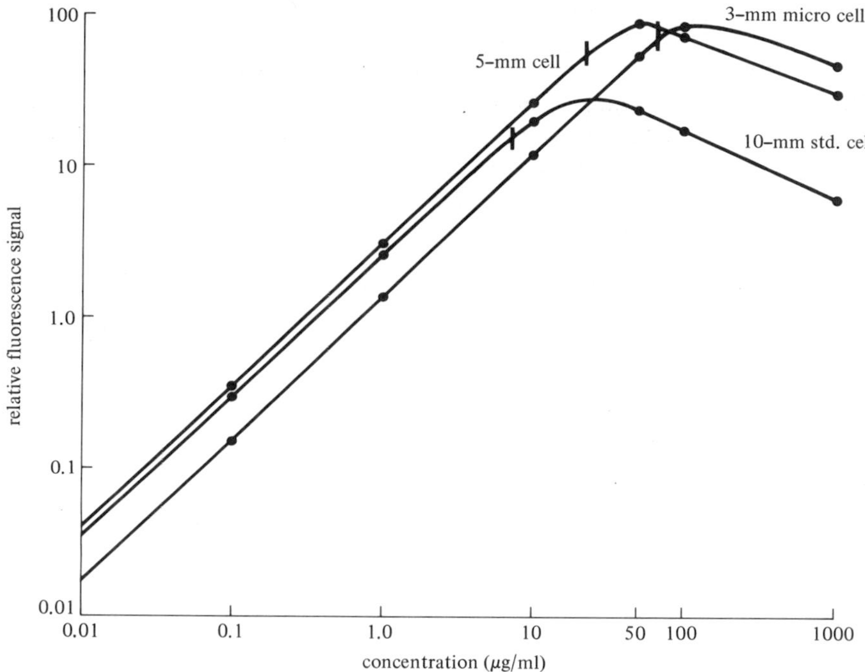

Figure 17.6 / *Analytical curves for the fluorescence of quinine sulfate.*
SOURCE: J. D. Winefordner, S. G. Schulman, and T. C. O'Haver, *Luminescence Spectrometry in Analytical Chemistry*, Wiley (New York, 1972), p. 293.

A wide variety of organic compounds form chelates that are useful for fluorescence measurements; some of the best known of these are shown in Fig. 17.7. A π-electron system is usually present in the organic ligand, and the excitation and emission are associated with $\pi^* \leftrightarrow \pi$ transitions. These transitions usually result in broad fluorescence peaks with little or no structure. Some of the lanthanide ions are exceptions to this behavior. Their fluorescence spectrum consists of narrow bands, presumably arising from transitions within the f orbitals of the metal ion.

Many transition metal ions fail to form fluorescent complexes because they are paramagnetic and their presence favors intersystem crossing at the expense of fluorescence. This behavior may be viewed as an advantage in avoiding interferences in the determination of other metal ions present in solution with the paramagnetic ions.

A limitation of chelate formation for inorganic ion analysis is that it frequently lacks specificity; that is, the same ligand will complex with a number of ions, and the emission maxima of these complexes are very similar in wavelength. On the other hand, fluorescence determinations, when applicable, are extremely sensitive, allowing some elements to be determined in nanogram quantities.

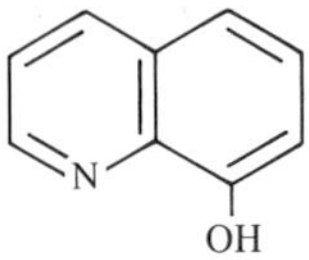

(a) 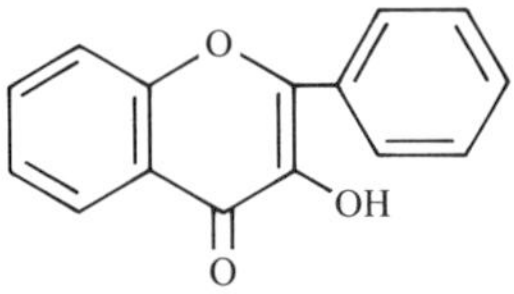 (b)

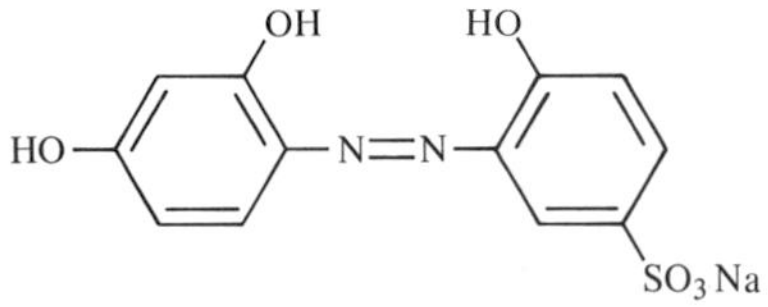

(c) 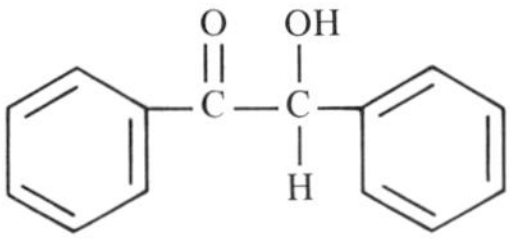 (d)

Figure 17.7 / *Some selected reagents used in the fluorescence determination of inorganic ions.*
(a) 8-*Hydroxyquinoline* (Al, Be, *and others*).
(b) *Flavanol* (Zr, Sn).
(c) *Alizarin garnet* R (Al, F^-).
(d) *Benzoin* (B, Zn, Ge, Si).

17.4.3 / Organic analysis

The number of applications of fluorescence to the determination of organic species is impressively large. The most important applications of the technique appear to be in the analysis of food products, pharmaceuticals, clinical samples, and natural products. The sensitivity and selectivity of the method make it particularly valuable in these fields. Many more compounds absorb than fluoresce, and thus fluorescence often offers a useful alternative to absorption spectrometry in attaining the selectivity required for a determination. In addition, fluorescence provides some additional experimental flexibility in achieving selectivity among similar compounds in that both the wavelength of excitation and the wavelength of observation can be selected.

17.5 / QUANTITATIVE ANALYSIS BY PHOSPHORESCENCE

17.5.1 / Relationship between intensity and concentration

Derivation of the expression describing the dependence of the intensity of phosphorescence on the concentration of the absorbing species proceeds in the same fashion as that employed for Eq. (17.4). The result is

$$I_p = \phi_p I_0(1 - 10^{-\varepsilon bC}) \tag{17.7}$$

and for sufficiently small concentrations,

$$I_p = 2.3\phi_p I_0 \varepsilon bC \tag{17.8}$$

Figure 17.8 shows a typical analytical curve for phosphorescence. Note the

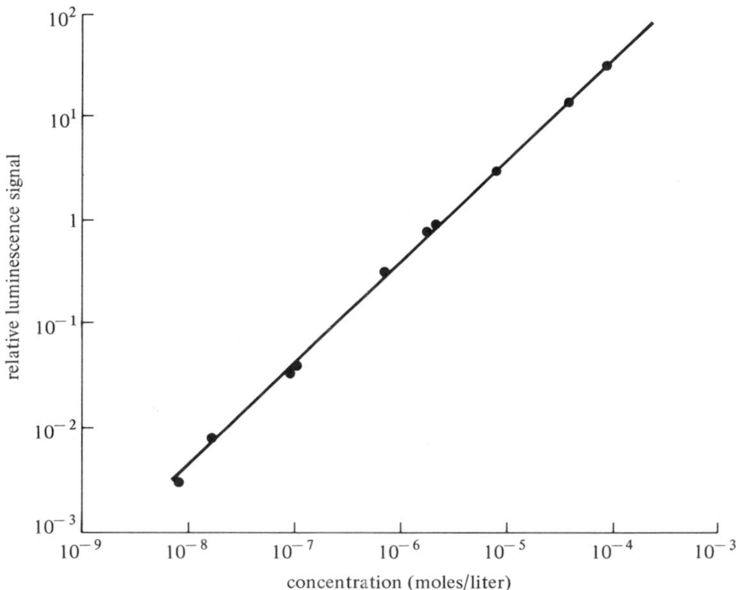

Figure 17.8 / *Analytical curve for the phosphorescence of indole butyric acid at* 77° K. SOURCE: J. D. Winefordner, S. G. Schulman, and T. C. O'Haver, *Luminescence Spectrometry in Analytical Chemistry*, Wiley (New York, 1972), p. 295.

high sensitivity and wide dynamic range of the phosphorescence method for this analyte.

17.5.2 / Applications

Applications of phosphorescence to the determination of inorganic compounds have been very rare, and the greatest number of applications of the phosphorescence technique have been for the determination of trace concentrations of organic compounds in samples of biological or medical interest. For example, Table 17.1 describes the phosphorescence characteristics of a number of biologically significant compounds (so-called antimetabolites, which inhibit enzyme activity). The detection limits obtained are comparable to those achieved by absorption spectrophotometric and other more specialized methods, such as those involving enzyme reactions. In many of the cases cited, the sensitivity of the phosphorescence technique can be enhanced by preliminary chemical treatment.

Many compounds that absorb UV radiation but do not fluoresce may show strong phosphorescence, and thus fluorescence and phosphorescence are complementary in many respects. Because many compounds that absorb do not show either appreciable fluorescence or phosphorescence, the breadth of application of the luminescence techniques is not as great as absorption spectrophotometry, but the luminescence techniques may have

Table 17.1 / *Phosphorescence characteristics of some antimetabolites*[a]

Compound	*Wavelength of excitation* (nm)	*Wavelength of emission* (nm)	*Detection limit* (μg/ml)	*Decay time* (sec)	*Range of utility*[b] (*decades*)
adenosine	280	422	0.80	3.2	>1
2-amino-4-methyl pyrimidine	302	438	0.033	2.1	>3
8-azaquanine	282	442	0.30	1.8	>2
2-chloro-4-amino-benzoic acid	312	447	0.069	1.0	2
2,6-diaminopurine sulfate	294	424	1.2	1.7	>2
oxythiamine HCl	272	460	3.4		>2
α-picolinic acid HCl	278	400	0.014	5.2	3
quercetin	343	480	0.30	2.1	>3
2-thiouracil	312	432	0.0038		>4

[a] All measurements taken at 77°K in ethanolic solution.
[b] Concentration range over which the analytical curve is nearly linear.

SOURCE: W. J. McCarthy, "Phosphorescence Spectrometry," in *Spectrochemical Methods of Analysis*, J. D. Winefordner, Ed., Wiley (New York, 1971), p. 487.

significant advantages of selectivity for many analysis problems. With the phosphorescence technique, one additional variable can be exploited to enhance selectivity, that is, the delay time between excitation and observation.

17.6 / QUALITATIVE ANALYSIS BY LUMINESCENCE

A full discussion of the relationship between molecular structure and luminescence does not fall within the scope of this book, but we can take note of some general principles concerning the utility of luminescence spectroscopy for qualitative analysis. Fluorescence and phosphorescence are not broadly useful methods for determining molecular structure in the same way that, for example, IR and nuclear magnetic resonance spectroscopy are. Nonetheless, significant qualitative information concerning molecular structure can be obtained from luminescence studies. Of the large number of known organic compounds, relatively few exhibit intense luminescence. Therefore, the mere fact that an organic molecule does luminesce can constitute significant information regarding its structure.

In order for a compound to luminesce, it first must absorb, and hence much of the material of Section 16.3 is pertinent to the present discussion. Luminescence techniques are applicable chiefly to compounds exhibiting $\pi^* \rightarrow \pi$ and $\pi^* \rightarrow n$ transitions. For the reasons discussed in Section 17.2, if the lowest-energy emission transition is $\pi^* \rightarrow n$, phosphorescence is likely to be the dominant mode of luminescence, while fluorescence usually dominates for compounds in which the lowest-energy transition is $\pi^* \rightarrow \pi$.

The certainty of a qualitative analysis increases as the number of features coincident in the sample and reference compound increases. If the luminescence techniques are considered together, a large number of features can be provided for comparison for many compounds. These are, in addition to the spectral features of the absorption spectrum, (1) the spectral features of a fluorescence and/or phosphorescence spectrum, (2) measured quantum efficiencies of fluorescence and phosphorescence, (3) the phosphorescence lifetime, and (4) changes in any or all of the above due to environmental and structural factors such as the heavy-atom effect, the effect of paramagnetic species, and solvent effects.

SUGGESTED READINGS

J. D. Winefordner, S. G. Schulman, and T. C. O'Haver, *Luminescence Spectrometry in Analytical Chemistry*, Wiley (New York, 1972).

J. D. Winefordner, Ed., *Spectrochemical Methods of Analysis*, Wiley (New York, 1971).

R. S. Becker, "Fluorescence and Phosphorescence Spectroscopy," in *Chemical Applications of Spectroscopy*, W. West, Ed., 2nd ed., Part 1, Wiley-Interscience (New York, 1968).

D. M. Hercules, Ed., *Fluorescence and Phosphorescence Analysis*, Wiley-Interscience (New York, 1966).

M. Zander, *Phosphorimetry*, T. H. Goodwin, Trans., Academic (New York, 1968).

chapter 18 / Vibrational spectroscopy

chapter 18

18.1 / INTRODUCTION

Spectroscopy based on transitions between the vibrational levels of molecules and polyatomic ions forms a powerful and well-developed tool for identification, structure determination, and quantitative estimation of these species. Two rather different experimental approaches to vibrational spectroscopy exist. These are the direct observation of the absorption of radiant energy in the IR region of the spectrum, and the indirect observation of vibrational transitions using a scattering phenomenon known as the Raman effect. Both IR and Raman spectroscopy will be discussed in this chapter.

Vibrational transitions are, in general, of much greater energy than the thermal energy ($kT = 4 \times 10^{-14}$ erg/molecule at 298°K). A typical (5-μm) vibrational transition requires 4×10^{-13} erg; as a result, only the ground vibrational level is appreciably populated at normal temperatures as noted previously in Section 15.4. There will, therefore, usually be only one absorption band corresponding to each mode of vibration of a molecule. The vibrational transitions of organic molecules that are most often used in analysis lie in the wavelength region 2–15 μm. This region is referred to as the "ordinary" infrared. The spectral region between 2 μm and the visible is known as the *near infrared* (near to the visible), while the region from approximately 50 μm to the microwave region is termed the *far infrared* (far from the visible). These regions are utilized, but less so than the ordinary infrared, and some of those applications will be discussed in this chapter.

Infrared radiation was discovered by William Herschel in 1800, but because of the difficulty in producing suitable detectors, it was not until the experiments of W. W. Coblentz in the period 1900–1910 that the chemical value of IR radiation even began to be recognized. Although Coblentz actually measured the IR absorption spectra of a number of organic liquids, the structure–spectra correlations known now were not perceived and IR spectroscopy remained confined to the realm of the research of academic physicists.

In the 1930s, organic chemists began to consider IR as a possible analytical method, and some of the larger chemical companies undertook the construction of spectrometers. It was World War II, however, that provided the stimulus for the commercial development of IR. At that time, it was realized that IR provided a rapid, accurate method of analyzing the C_4 hydrocarbon fraction required for the production of much-needed synthetic rubber. The combination of instrumental availability with the recognition of the chemical potentialities of the method has since made IR a ubiquitous and well-exploited field.

Although the Raman effect was known during this period of explosive development in IR, Raman spectrometry has lagged behind IR in its chemical applications because of instrumental difficulties that are only now being overcome. This "exploitation gap" seems certain to narrow in the future.

18.2 / MOLECULAR VIBRATIONS

In this section, the energies and number of vibrations in molecules will be discussed. Since these are properties of the molecule, the discussion is common to both IR and Raman spectroscopy although, as we shall see later, the selection rules for transitions are different for the two techniques.

18.2.1 / Normal modes of vibration

A system consisting of N particles can be described completely by specification of the positions and velocities of the particles in a three-dimensional Cartesian coordinate system. Any arbitrary displacement or velocity can be expressed in terms of combinations of suitable displacements or velocities along the three principal directions. Each particle, therefore, is said to possess three *degrees of freedom*, and a system of N such particles has $3N$ degrees of freedom, or requires $3N$ sets of coordinates for its complete specification. If the N particles are bound together as the atoms of a molecule, they are constrained to move together in certain ways. Thus, the molecule as a whole has three translational degrees of freedom and, if it is a nonlinear molecule, three rotational degrees of freedom. The remaining motions of the constituent atoms, the vibrations under the restraints of the chemical bonds, constitute $3N - 6$ degrees of freedom. In the case of linear molecules, no rotation about the molecular axis is of significance, so $3N - 5$ vibrational degrees of freedom remain. All vibrational motions of the atoms, no matter how complex, can be described completely in terms of these $3N - 6$ (or $3N - 5$) fundamental vibrations, which are termed the *normal modes* of vibration for the molecule.

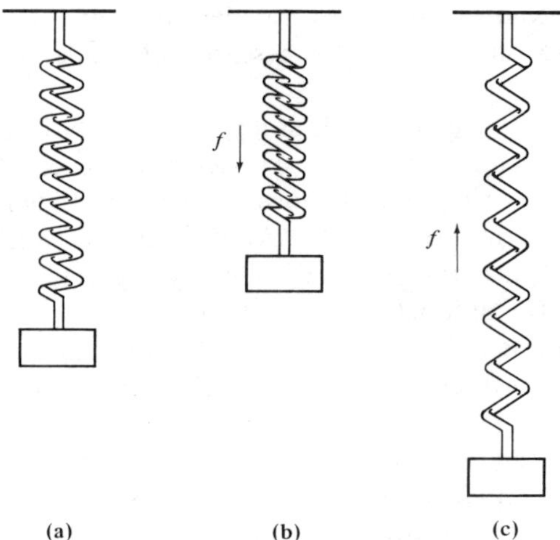

Figure 18.1 / *A mass, m, attached to a fixed object by a spring.*
(a) *Equilibrium extension.*
(b) *Spring compressed and force shown acting to resist compression.*
(c) *Spring stretched and force shown acting to resist stretching.*

18.2.2 / Harmonic oscillators

For small amplitudes of vibration, the atoms in a molecule behave as if they were point masses attached together by springs that obey Hooke's law, that is, the force that tends to restore the atom to its equilibrium position is directly proportional to the extent of the displacement and in the opposite direction. This relation is stated in Eq. (18.1),

$$f = -kx \tag{18.1}$$

where x is the displacement and k is the restoring force constant of the bond. Systems that obey Eq. (18.1) are termed *harmonic oscillators*. For a single mass, m, attached to a fixed object by a spring, we can depict the stretching and compressing of the spring as shown in Fig. 18.1.

Application of Newton's second law allows replacement of the force in Eq. (18.1) by the product of mass times acceleration. Since acceleration is the second time derivative of displacement, the net relationship is

$$m\frac{d^2x}{dt^2} = -kx \qquad \frac{d^2x}{dt^2} = -\left(\frac{k}{m}\right)x \tag{18.2}$$

We see, therefore, that x must be a function of time that differs from its second derivative by only a negative constant, thus suggesting a sinusoidal relation. If we choose $x = A\cos(2\pi\nu t)$, where A is the amplitude of vibration (the maximum value of x, or the maximum displacement) and ν is the frequency of vibration, then $d^2x/dt^2 = -4\pi^2\nu^2A\cos(2\pi\nu t)$. To obtain consistency with Eq. (18.2), the equality $4\pi^2\nu^2 = k/m$ must hold. Solving

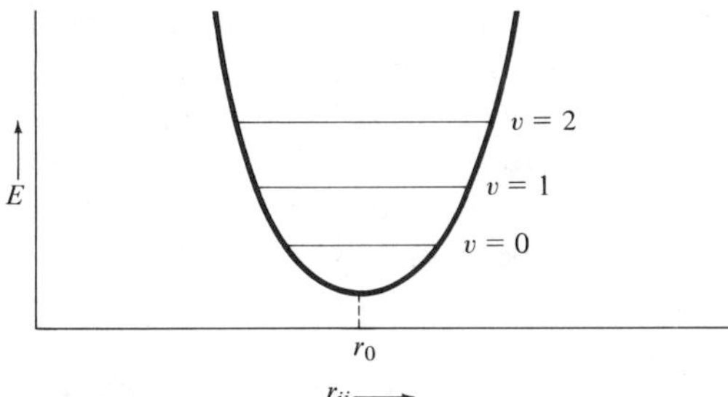

Figure 18.2 / *Potential energy as a function of internuclear separation (r_{ij}) for a harmonic diatomic vibrator. The equilibrium bond distance is r_0 and the first three quantized vibrational energy levels are indicated showing larger displacements from r_0.*

for the frequency, we find that

$$\nu = \frac{1}{2\pi}\left(\frac{k}{m}\right)^{1/2} \tag{18.3}$$

For a diatomic molecule, we may simply substitute the reduced mass of the system, μ, for m and thereby transform Eq. (18.3) into the same form as Eq. (15.7):

$$\nu = \frac{1}{2\pi}\left(\frac{k}{\mu}\right)^{1/2} \tag{15.7}$$

Classically, the potential energy of such a system is a parabolic function of the amount the bond is stretched (or compressed) from its equilibrium length, as shown in Fig. 18.2. A strictly parabolic potential well is required if the vibration is to be considered harmonic. Since with sufficient vibrational energy the molecule will dissociate, the potential energy curve is not parabolic for large-amplitude vibrations and significant deviations from the harmonic oscillator model occur; however, these are of little concern for analytical IR.

Quantum mechanics shows that the energy of a vibrational mode is given by Eq. (15.8):

$$E_v = (v + \tfrac{1}{2})h\nu \tag{15.8}$$

It is possible to insert the classical expression for the frequency into Eq. (15.8) to obtain the energy in terms of the reduced mass and bond force constant:

$$E_v = (v + \tfrac{1}{2})\left(\frac{h}{2\pi}\right)\left(\frac{k}{\mu}\right)^{1/2} \tag{18.4}$$

Recall that v is the vibrational quantum number, which may take on values of zero or any positive integer, and that the additive term $\frac{1}{2}$ represents the zero-point vibrational energy, energy possessed by a vibrational mode even at the absolute zero of temperature. The lowest three vibrational levels are indicated in the energy diagram in Fig. 18.2. As indicated in Chapter 15, the selection rule $\Delta v = \pm 1$ obtains; as a result, since for most cases only the ground level is appreciably populated, the transition from $v = 0$ to $v = 1$ is that which is observed experimentally.

18.2.3 / Observed absorption bands

The number of bands observed in an IR spectrum is usually not equal to the computed number of fundamental frequencies. Several factors may contribute to such an inequality. Some vibrations may not be active in the IR because they do not result in a change in the dipole moment of the molecule. As noted in Chapter 15, a change in dipole moment must accompany a vibration in order that it may interact with, and therefore absorb energy from, the electric field of the IR radiation. Two vibrational modes may occur at identical frequencies in a molecule (these are termed *degenerate* bands); the number may be further reduced by overlap or weak absorption, or a band may fall outside the range of the particular spectrometer being used. The number of bands may be increased by the observation of weak (forbidden) absorption in which $\Delta v > 1$, which are known as *overtones.* Two or more fundamental modes may interact to produce discrete absorptions which occur at the frequency that corresponds to the sum or the difference of the individual band frequencies. Such absorptions are known as *combination bands.*

It should be noted that even relatively simple molecules have many vibrational fundamentals; benzene, for example, has 30. Only quite small molecules have vibrational spectra simple enough for complete analysis and assignment of their normal modes. The vast majority of vibrational spectra–structure correlations are empirical and are not based on complete theoretical analysis of the spectra.

18.3 / THE RAMAN EFFECT

Several reasons that will be discussed later in this chapter suggest the desirability of having a method of examining vibrational transitions in a different way than by direct IR absorption. Perhaps the most fundamental reason is that vibrations that do not cause a change in molecular dipole moment are inactive in IR spectroscopy; information about these vibrations must be obtained in a different way. The Raman effect provides such a way.

18.3.1 / Rayleigh scattering

Consider a clear substance (solid, liquid, or gaseous) which is irradiated with monochromatic light of frequency ν_0. The incident light is normally in the visible region and is chosen to be of a frequency that does not coincide with any absorption by the sample. The majority of the light simply passes through the sample in the direction of the incident beam. A small amount, corresponding to about 1 part in 10^5, however, is scattered by the sample in all directions and can be observed by viewing the sample at right angles to the incident beam. The scattering of light at the same frequency as the incident light is a well-known phenomenon called *Rayleigh scattering* after Lord Rayleigh, who investigated it in the late nineteenth century.

Rayleigh scattering may be described as an elastic collision of a photon with a molecule of the sample substance; that is, no energy is exchanged, but the direction of travel of the photon is altered. An alternative description is that of simultaneous absorption and reemission of the photon by a sample molecule, again with no exchange of energy. While the incident beam is collimated, scattering is isotropic, thus permitting its observation. Rayleigh scattering intensity is proportional to the fourth power of the incident frequency, so short wavelengths are more strongly scattered than long wavelengths. This effect has long been recognized as the reason that the sky appears blue. The sky is illuminated by sunlight which is scattered by the atmosphere; the blue light intensity predominates because of wavelength dependence of Rayleigh scattering.

18.3.2 / Raman scattering

Spectroscopic investigation of light scattered from a sample illuminated with a monochromatic beam reveals that about 1 percent of the total scattered intensity occurs at frequencies that are different from the incident frequency. Complete investigation of the scattered radiation shows a pattern of spectral lines. These include an intense line at the incident frequency called the Rayleigh line, and much weaker lines shifted to both lower and higher frequencies. The appearance of these weak lines was predicted in 1923 by a German physicist, A. Smekal, and observed experimentally in 1928 by an Indian physicist, C. V. Raman. For this work, Raman was awarded the Nobel Prize in physics in 1930, and the effect has since borne his name in most of the scientific literature.

In a simplified way, one can describe Raman scattering as an inelastic collision (i.e., one that involves energy transfer) between a photon and a sample molecule. Two possible situations may occur. In the first, the molecule is encountered in its ground vibrational level and excited to a higher level during the collision. The photon then emerges with lower energy than the incident beam, $E = (h\nu_0 - E_v)$ and its frequency is shifted down, $\nu = \nu_0 - \Delta\nu$. Alternatively, the photon can collide with a vibrationally excited molecule which is deactivated during the collision. The photon then carries off the extra energy, $E = (h\nu + E_v)$ and its frequency is shifted up, $\nu = \nu_0 + \Delta\nu$.

The quantity $\Delta\nu$ is known as the *Raman shift*. Lines for which the shift is negative are known as *Stokes lines* after G. G. Stokes, who did much pioneering work in fluorescence spectroscopy. The name was given because of the formal similarity of the lines at diminished frequencies to fluorescence

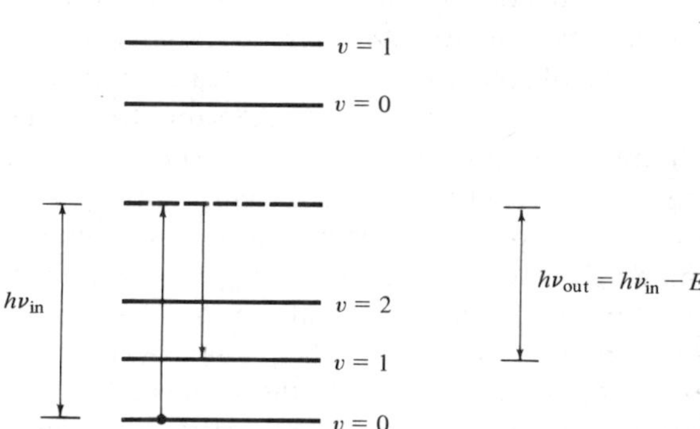

Figure 18.3 / *Origin of a Stokes Raman line in the "virtual" transition model. The first three vibrational levels of both the ground and first excited electronic levels are shown. The entering photon has insufficient energy to stimulate a transition between the electronic levels, but it can excite the molecule to the "virtual" level indicated by the dashed line. If, on deexcitation, the molecule returns to the $v = 1$ vibrational level, the energy of the exciting photon is reduced as shown.*

lines. In fact, of course, the Raman lines arise from a totally different mechanism, and the name is inappropriate. Nonetheless, the terminology extends even to those lines with positive shifts, which are called *anti-Stokes lines*.

The transition probability is the same for lines with positive and negative shifts, but, since the anti-Stokes lines arise from vibrationally excited levels which are populated to only a small extent, they are always less intense than the corresponding Stokes lines. Moreover, since the population of an excited level must decrease with greater energy separation from the ground level, the intensity of anti-Stokes lines falls off with increasing Raman shift. This frequency dependence does not occur to a discernable extent for the Stokes lines, since the energy of the first excited level has only a very small effect on the ground-level population. In the usual case, the greater intensity of the Stokes lines makes them the more useful, and they are preferred for analysis in Raman spectroscopy.

Raman scattering involves the intercession of the molecular electronic levels. In one picture of the mechanism of the phenomenon for a Stokes line, one may imagine a "virtual" transition from the $v = 0$ level of the ground electronic level to a "virtual" electronic level. When the molecule returns immediately to the ground electronic level, as indicated in Fig. 18.3, it may return to the $v = 1$ vibrational level with the reemission of a less energetic photon. Such a "virtual" transition which begins in the $v = 1$ level and ends in the $v = 0$ level produces an anti-Stokes line. Since in both this picture and the collisional model the process is very different from direct absorption of an IR photon, it should not be surprising that different selection rules apply. Although the allowed transitions are characterized by $\Delta v = \pm 1$ in both methods, Raman spectroscopy does not require that a change in dipole moment be associated with the vibration.

For a band to be observed in Raman spectroscopy, it is required that the vibration that gives rise to the band be associated with a change in the *polarizability* of the molecule. Polarizability refers to the ease with which the electron distribution of the molecule may be altered. The electric field of a photon that has *insufficient* energy to induce an electronic transition can, nonetheless, deform, or polarize, the molecule's electron cloud. If a particular molecular vibration results in a change in the polarizability, then the act of polarization can, as a second-order process, induce a vibrational transition. The molecule need not have a permanent dipole moment, since during the time it is subjected to the electric field it acquires an *induced* dipole moment with which here is associated an induced transition moment for vibrational excitation.

The information that may be derived from measurement of the Raman shifts for a particular species is of the same type as that which may be obtained from IR spectra. Different vibrations are available for study by the two techniques. In fact, if the formal selection rules were always obeyed strictly, for many molecules that possess a center of symmetry a rule of "mutual exclusion" would apply: Those vibrations that are IR active would be inactive in the Raman spectrum and vice versa. In actual practice, the same bands are frequently observed by both methods.

18.4 / INSTRUMENTATION

Many of the differences between IR and Raman spectroscopy with regard to both applications and sample-handling techniques are a direct result of the differences in the instrumentation. It is therefore convenient to discuss the instrumentation briefly before proceeding with chemical applications.

18.4.1 / Spectrophotometers

Instrumentation for the IR region has been described in Section 11.3, and the reader is referred there for a discussion of IR spectrophotometers. Since Raman spectroscopy employs visible radiation, the analyzer and detector used in a Raman spectrometer are the same as those used in high quality UV-visible spectrophotometers, and the reader is referred to Section 11.2 for details. It is necessary only to note here that the monochromator employed for Raman work should be one of large dispersion in order to observe Raman lines with small shifts in the presence of the very much stronger Rayleigh line. The additional requirement of minimizing scattered light usually limits Raman monochromators to designs that employ two successive dispersing elements, that is, double monochromators.

18.4.2 / Raman sources

The only device that differentiates a Raman spectrometer from a visible spectrophotometer is the source. This includes the sample holder, a means of

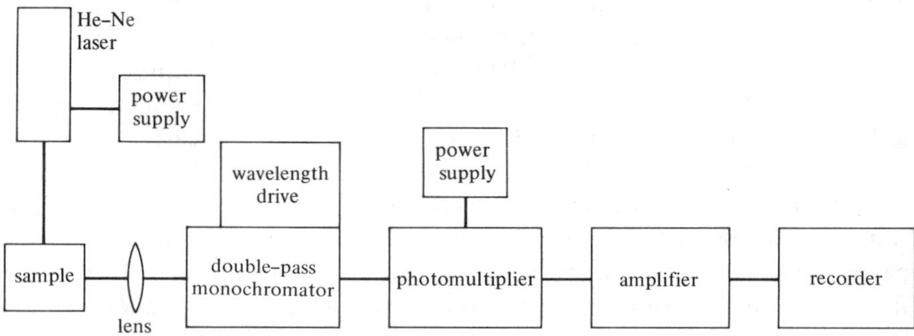

Figure 18.4 / *Block diagram of a laser Raman instrument.*

illuminating the sample with monochromatic light, and the optics to observe the scattered light and bring it efficiently into the analyzing monochromator.

Since the Raman effect is inherently a very weak one, it is necessary to have the maximum possible light intensity incident on the sample. In traditional Raman instruments, this goal is approached by the use of a cylindrical sample cell placed inside a helical mercury arc lamp. Monochromatic light of either 4358 Å or 5461 Å can be isolated from the output of such an arc lamp (known as the Toronto arc lamp after the university at which its use was developed) by jacketing the sample in an appropriate filter solution containing an organic dye. Since the two isolable lines are in the blue and green visible region, respectively, the necessity of avoiding absorption of the exciting line by the sample restricts sample solutions to those that are colorless or pale yellow. The mercury arc source is quite inefficient and produces a large amount of heat from which the sample must be protected, usually by circulating the filter solution that also serves as coolant. It can be ascertained that a 2.5-kW arc lamp radiates about 50 W at 4358 Å, of which about 1 W is actually effective in exciting the sample.

Raman spectrometers that employ laser excitation are now commercially available. The helium–neon gas laser, which emits coherent, highly monochromatic light at 6328 Å, seems to be the most widely used source. In spite of the loss of scattering intensity associated with the use of a longer-wavelength exciting light, lasers with power outputs in the range 5–35 mW give at least as good intensity Raman spectra as the older arc sources. The longer wavelength has the advantage that fewer samples have interfering absorption bands, thus permitting the use of a wider variety of samples. The fact that the laser beam is narrow and readily collimated means that small-volume, multipass cells are easier to use than with a mercury arc source. As a result, the volume of liquid samples required is reduced from the order of milliliters to a few microliters.

A block diagram of a laser Raman spectrometer is shown in Fig. 18.4; in Fig. 18.5, the details of one possible configuration of source optics are shown. The combination of sample cell and source optics must be designed

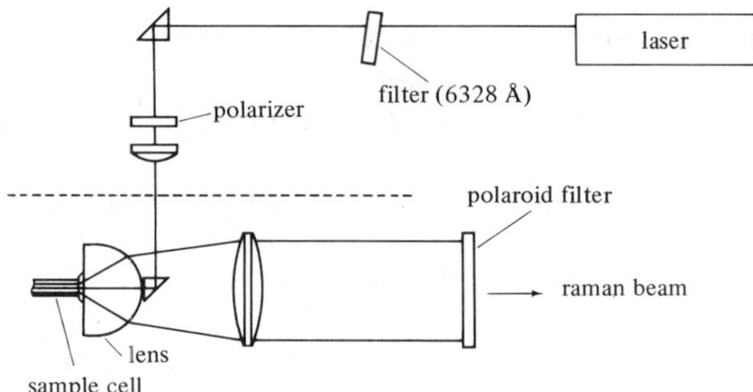

Figure 18.5 / *One arrangement of source optics for a laser Raman instrument. Note that in this geometry the scattered light is collected along the axis of the exciting radiation at* 180° *to the direction of the laser beam rather than at right angles.*

to maximize efficiency of both excitation and collection of the scattered light. A number of different approaches to the solution of these problems is possible, and the one chosen will depend on the specific design of the remainder of the instrument. In Fig. 18.5, we note the additional feature that the exciting light is plane polarized, and that a polarizing filter is provided at the entrance to the monochromator. Measurement of the polarization of the scattered lines provides information regarding the symmetry of the vibration under study. A discussion of this aspect of Raman spectroscopy is outside the scope of this text, and the reader is referred to the book by Herzberg (listed in the Suggested Readings at the end of the chapter) for details.

18.5 / SAMPLE HANDLING

In both IR and Raman spectrometry, samples may be examined as solids, liquids or solutions, or in the gas phase. The intensity of bands in both techniques is proportional to the amount of material that absorbs or scatters the radiation, with the result that gaseous samples and very dilute solutions pose special problems. Infrared is the more sensitive of the two methods, but even so, typical cells for examination of gases at low pressure range in optical path from 10 cm to >1 m. Gaseous samples give the highest-resolution spectra which may reveal (see Section 15.4) rotational fine structure. Such samples, however, are of limited applicability to routine analytical measurements, which are usually made using samples in condensed phases.

18.5.1 / Samples for infrared spectroscopy

Infrared spectra may be obtained using solutions, thin films, or solids dispersed in a transparent matrix. Since glass and quartz transmit poorly in the IR (see Chapter 11), sample cells must be fabricated from a material such a crystalline NaCl, which is transparent in the region 2–15 μm. Longer-wavelength IR studies require more exotic materials for sample cells, such as crystalline KBr or CsBr. Since these materials are readily soluble in aqueous media and water, even in 1-mm thickness, is poorly transparent throughout the IR, IR solution spectra are largely restricted to those that can be obtained in organic solvents.

The simplest sampling technique is the thin-film method, which is appropriate for medium-resolution studies of some plastic solids and of pure liquids that are not too volatile. A thin section of a solid (such as polyethylene film) may be placed directly in the radiation beam. In the case of liquids, a few drops of sample are placed on a flat, polished disc of crystalline NaCl; a second such disc is placed on top. The two discs are squeezed together until an evenly distributed liquid film is included between them; this combination is then placed into a suitable holder that fits in the radiation beam.

Powders or small crystals must be dispersed in a medium of approximately the same refractive index or dissolved in an appropriate solvent in order to avoid large radiation losses due to scattering from the surface of the particles. A method similar to the film approach for liquids is known as the *mull* technique. A few milligrams of solid is ground in a mortar; when the sample is finely divided, a few drops of heavy paraffin oil are added and grinding is continued until the mixture, or mull, is uniform. The mixture is then squeezed between salt flats as described for liquid films. Paraffin oil has only C—H and C—C absorptions, which fall near 3 and 7 μm; if it is necessary to examine these regions, the sample may be mulled with a fully halogenated oil, fluorolube. Both oils have refractive indices that make them suitable for use with a wide variety of organic samples.

As an alternative approach, about 1 mg of solid may be finely ground with about 100 mg of dry KBr. The mixture is then placed in a steel die and subjected to a pressure of approximately 15,000 psi. KBr is quite plastic, and the result of pressing is a disc, usually about 1 cm in diameter, that, when expertly prepared, appears glassy clear.[1] The sample is believed to form a solid solution in the halide matrix. The KBr disc method can be extended to microgram samples by the use of smaller diameter discs in conjunction with a condensing lens to concentrate the IR beam over the smaller area. In all of the methods so far described, no sample is placed in the reference beam of the double-beam spectrophotometer.

Infrared absorptions are moderately intense, and typical solution conditions involve the use of 10 percent weight/weight solutions in cells of 0.1–1 mm optical path. Such cells are fabricated from two flat, polished salt plates separated by a lead or Teflon gasket of the required thickness. This "sandwich" is then mounted between metal plates that have a central opening for the radiation beam to pass through. One of the more common designs has filling ports attached to the metal mounts; these ports line up

[1] Visible clarity, although reassuring to the experimenter, is not necessary in IR optics; imperfections too small to scatter IR wavelengths do scatter visible light and may cause opacity.

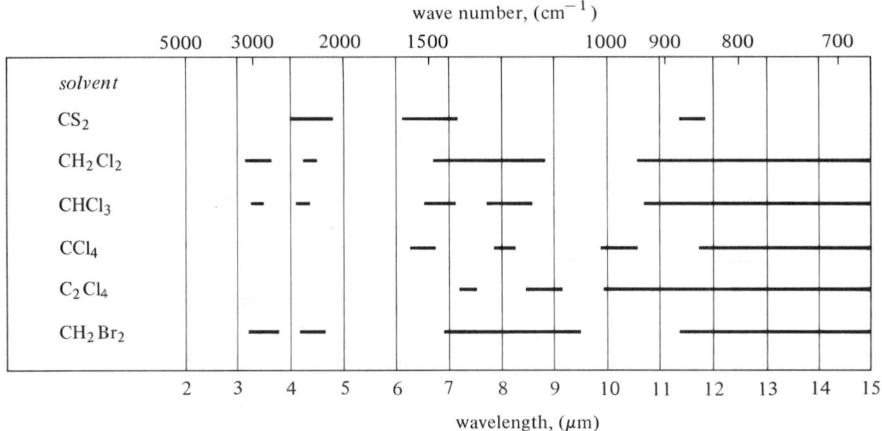

Figure 18.6 / *Transparent regions of some common solvents used for* IR *spectroscopy. Regions lined out are those in which a* 1-mm *thickness of solvent transmits* 25 *percent or less of incident radiation.*

with holes drilled through one of the salt plates. In almost no case is the solvent entirely devoid of IR absorptions, so advantage is taken of the double-beam capability of the spectrophotometer. The cell containing solvent plus sample is placed in one beam, and a second cell of accurately matched path length, containing only solvent, is placed in the reference beam. The double-beam instrument then nulls small absorptions due to the solvent.

Even using a double-beam instrument, it is not possible to make meaningful spectral measurements if the solvent transmits less than about 50 percent of the incident radiation. Unfortunately, no single solvent has the required transparency over the entire IR range; in fact, none is sufficiently transparent over the 2–15 μm region. The usefully transparent regions for some common solvents are shown in Fig. 18.6. One useful combination of solvents is CCl_4 for the wavelengths shorter than 7 μm and CS_2 for the remaining range to 15 μm.

18.5.2 / Samples for Raman spectroscopy

Since Raman spectra are obtained using visible light, glass and quartz cells are used routinely and no difficulty is encountered in handling aqueous solutions. Moreover, water is an excellent solvent for Raman spectroscopy because it has few interfering bands. It dissolves readily large numbers of inorganic materials, and these have formed a large fraction of the materials studied by Raman spectroscopy. Organic materials may be examined as neat liquids or in solution in organic solvents chosen for their lack of interfering bands. The low sensitivity of the Raman method usually indicates the use of fairly concentrated solutions, which may range from 1*M* to saturated.

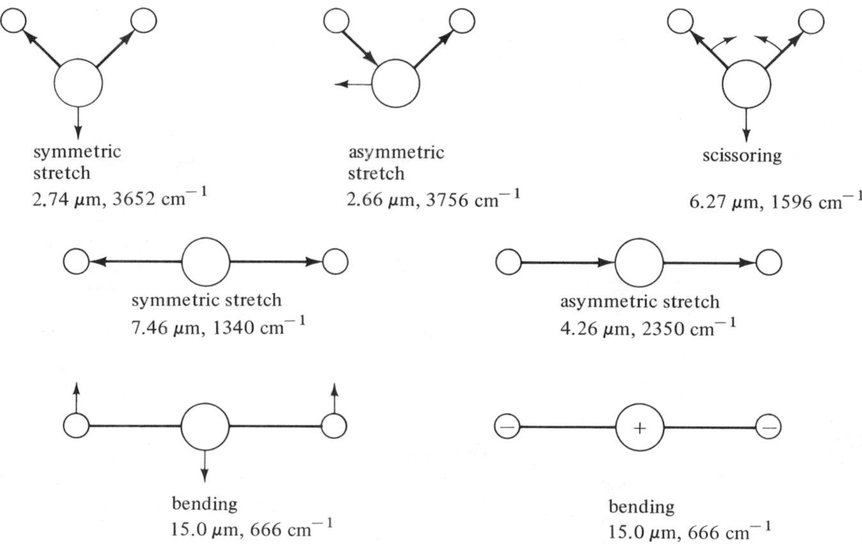

Figure 18.7 / *Normal modes of vibration for water and carbon dioxide. Arrows indicate directions of motion of the atoms; the minus and plus signs indicate, respectively, motion into and out of the plane of the page.*

Useful Raman, as well as IR, data may be obtained from single crystals and solid polycrystalline samples, but these are not routine analytical applications and the reader is referred to more specialized texts for details.

18.6 / INTERPRETATION OF INFRARED SPECTRA

As noted previously, the majority of structure–spectra correlations based on IR frequencies are empirical. Identical correlations can be made for Raman spectra, although there are currently fewer data with which to do so. The distinction between the two methods occurs in the relative intensities of observed absorptions; the frequencies will be the same. The discussion in this section is limited, therefore, to the 2–15 μm IR, the region in which the majority of data have been obtained and for which empirical correlations are most completely developed.

18.6.1 / Assigned spectra

To give the reader some feeling for the types of vibrational motions that molecules undergo, we begin this section with two examples of small molecules for which complete assignment of the vibrational spectra has proved possible. The normal modes for H_2O and CO_2 are shown in Fig. 18.7. Note that in all cases, the normal modes of vibration must be such that the center of mass of the molecule does not move under the influence of the vibration. If this were not so, the vibration would become a translational

TABLE 18.1

Bond	k (dyne/Å × 10^3)
C—H	5
C—F	6
N—H	6.3
O—H	7.7
C—Cl	3.5
C—C	4.5
C═C	10
C═O	12
C≡C	15.6
C≡N	17.7

motion. For H_2O, $3N - 6 = 3$, and all three vibrations can be observed in the IR. For CO_2, a linear molecule, $3N - 5 = 4$, but only two absorptions are observed in the IR. The two bending modes are degenerate and occur as a single band, while the symmetric stretch is inactive in the IR. The change in dipole moment caused by stretching one bond is exactly canceled by the symmetric deformation of the second C—O bond; the frequency of the vibration comes, of course, from the Raman spectrum.

18.6.2 / Generalizations from theory

Despite the fact that most vibrational spectra are not susceptible to complete theoretical analysis, some generalizations can be made on the basis of Eq. (18.4). Thus, strong chemical bonds (large force constants) will result in absorptions at high energy (short wavelength), while weaker bonds will have absorptions at lower energy. Light atoms will favor higher-energy absorptions, while heavy atoms will favor lower-energy vibrations. In the case of single, light atoms (e.g., H), it is sometimes possible to consider the remainder of the molecule as a fixed object and therefore employ the simpler form of Eq. (18.3). This approach is frequently useful in calculating the isotope shift of an IR frequency when, say, deuterium is substituted for protium. Some approximate force constants for common organic fragments are given in Table 18.1.

These generalizations rationalize a number of experimental observations. For example, NH, OH, and CH stretching vibrations are all observed in the region near 3 μm. Other groups with similar bond strengths and masses tend to occur together in narrow spectral regions, such as C═C, C═O, and N═N near 6 μm and C≡C and C≡N near 4.5 μm. In addition, molecular symmetry may be important. Thus, in trans symmetric olefins and symmetric acetylenes, the carbon–carbon multiple-bond frequency may be weak or absent in the IR because of the cancellation of change in dipole moment that can occur in molecules possessing a center of symmetry. Such cancellation cannot occur in cis olefins, and for those molecules the C═C absorption is usually strong.

TABLE 18.2 / *Approximate locations of selected functional group absorptions*

Group	*Vibration type*	$\tilde{\nu}$ (cm^{-1})	λ (μm)
$—CH_3$	asymmetric stretch	2962	3.38
	symmetric stretch	2872	3.48
	asymmetric bend	1450	6.90
	symmetric bend	1375	7.28
$—CH_2—$	asymmetric stretch	2926	3.43
	symmetric stretch	2853	3.51
	bending (scissor)	1465	6.83
—C=C—	stretching	1660–1640	6.0–6.1
=C(H)—H	out-of-plane bending	1000–650	10–15.4
—C≡C—	stretching	2260–2100	4.4–4.8
≡C—H	stretching	3300	3.03
	bending	700–610	14.3–16.4
—O—H	stretching	3650–3685	2.7–2.8
	in-plane-bending	1420–1330	7.0–7.5
$—NH_2$ (—N(H)H)	asymmetric stretch	3500	2.86
	symmetric stretch	3400	2.94
	bending (scissor)	1650–1580	6.1–6.3
$—NO_2$ (—N(O)O)	asymmetric stretch	1660–1500	6.0–6.7
	symmetric stretch	1390–1260	7.2–7.95
>C—O—	stretching	1300–1000	7.7–10.0
>C—C(=O)—O—	asymmetric stretch (combination)	1210–1160	8.3–8.6
>C—C(=O)—C—	stretching and bending combination	1230–1100	8.1–9.1

18.6.3 / Empirical correlations

Experience shows that the normal IR can be subdivided approximately into three sections. The region 2–7.7 μm ($>1300\ cm^{-1}$) is termed the *functional-group region*, since it is there that the important functional groups of organic chemistry absorb. Examples include ketones, $R_2C{=}O$, which characteristically show a strong band near 6 μm due to the carbonyl group; RNO_2 shows a strong band near 6.5 μm and a medium-intensity band between 7 and 8 μm due to the nitro group; the hydroxyl function of a primary alcohol, ROH, absorbs typically near 2.9 μm. Some selected group frequencies are collected in Table 18.2.

The 7.7–11 μm region is known as the *fingerprint region*. Absorptions there are complex, usually combinations of interacting modes. This region is probably a unique fingerprint for every molecule, which gives rise to the name. In addition, some bands in this region can be assigned and are useful in confirming conclusions drawn from the pattern in the functional group region. For example, if an OH band is observed for a phenol or alcohol, the position of the C—C—O combination band (7.9–10 μm) can frequently be used to assign a more specific structure.

General classification of molecules is possible from the pattern in the 11–15 μm region. Specifically, the absence of absorptions in this region is a good criterion for the absence of an aromatic ring, since aromatics and heteroaromatics have ring bending modes and C—H out-of-plane bending modes in this range. Substitution patterns on a benzene nucleus can also be determined from bands in this region.

When interpreting an IR spectrum, it is always desirable, if possible, to confirm a conclusion drawn from one portion of the spectrum with data from another portion. Consider a carbonyl band observed in the region 5.84–5.94 μm; this is a slightly shorter wavelength than is exhibited by typical ketones and could be an aromatic aldehyde. If such a conclusion is correct, then the aldehyde C—H band (doublet, 3.4–3.7 μm) must be present.

The summation of empirical correlations such as those discussed is usually presented as a *correlation table* such as that given in Fig. 18.8.

18.6.4 / Factors that shift bands

Just as the shift in the frequency of the carbonyl band is a clue that one has an aldehyde rather than a ketone, there are a number of effects that shift typical group frequencies; it is useful to examine a few examples. We shall look, therefore, at the effect of the following factors on the position of one particular band, the ketone carbonyl absorption: physical state of the sample, substituent effects, conjugation, and hydrogen bonding.

The absorption of a pure sample (neat) of an aliphatic ketone at 5.83 μm (1715 cm^{-1}) is customarily regarded as the "normal" position with respect to which shifts are measured. The absorption frequency is increased if the neat liquid is diluted with a nonpolar solvent, and decreased by dilution with a polar solvent. Clearly, the reason for these effects is that *inter*molecular interactions are reduced in the nonpolar medium and increased in the polar solvent. Typical solvent shifts have a range of about 25 cm^{-1}.

Upon conversion of $R_2C{=}O$ to $RXC{=}O$, the direction of the shift observed will depend on the nature of X. If X is an electronegative substituent, its effect may be primarily inductive. This causes the carbonyl carbon atom to become more electropositive, as indicated in Fig. 18.9(a). Thus, the force constant of the C═O bond is increased and the absorption is shifted to higher energy. Substituents that, by conjugation, increase the contribution of resonance structures such as that in Fig. 18.9(b) cause the carbonyl bond to take on more single-bond character and hence have a smaller force constant. The result is a shift of the absorption to lower

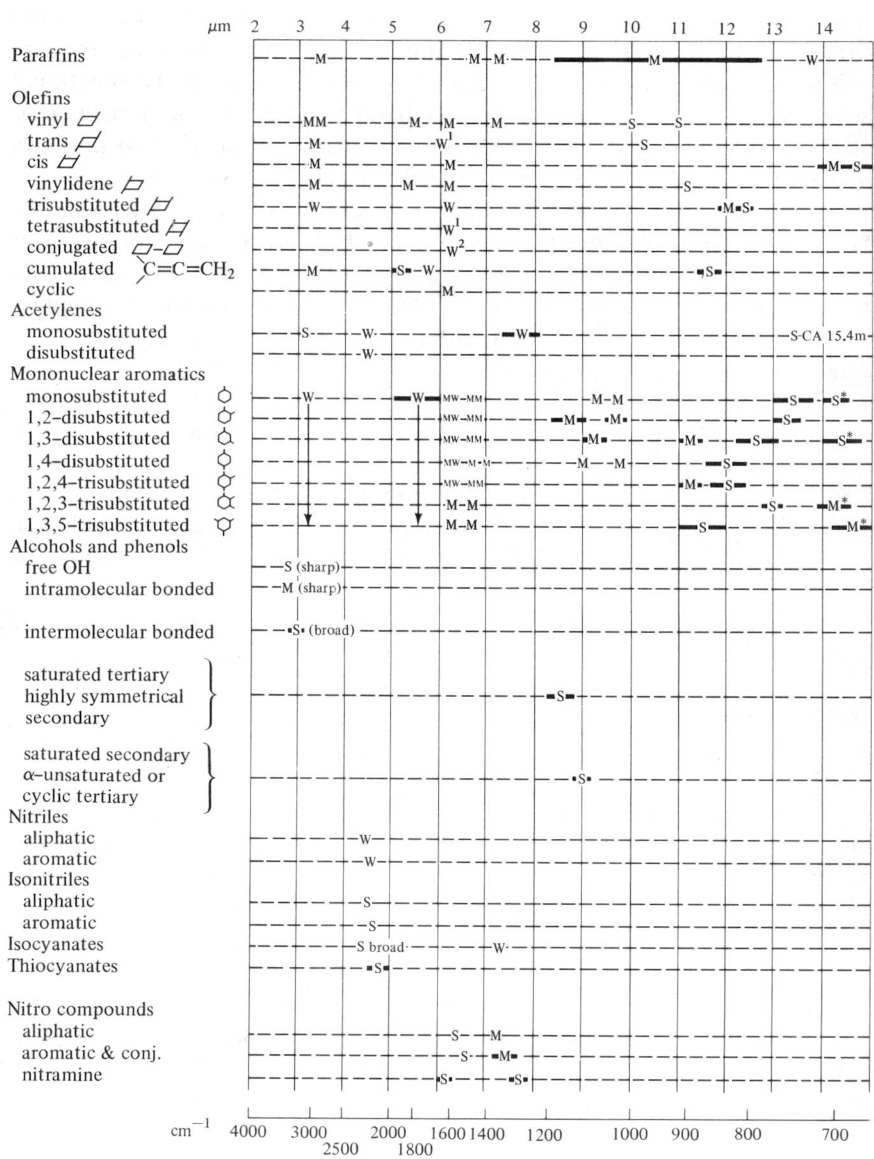

Figure 18.8 / *Characteristic* IR *frequencies for selected organic moieties. Intensities are indicated*, S = *strong*, M = *medium*, W = *weak*.
SOURCE: R. M. Silverstein and G. C. Bassler, *Spectrometric Identification of Organic Compounds*, 2nd ed., Wiley (New York, 1967), pp. 7–8, 73–77.

energy. When X = F or Br, the effect is predominately inductive, and the carbonyl absorptions are found at 1869 cm^{-1} and 1812 cm^{-1}, respectively. Resonance effects predominate when X = NH_2, C=C, or phenyl; the approximate carbonyl frequencies are 1670 cm^{-1} when X = NH_2 and 1685–1666 cm^{-1} when X is an unsaturated substituent.

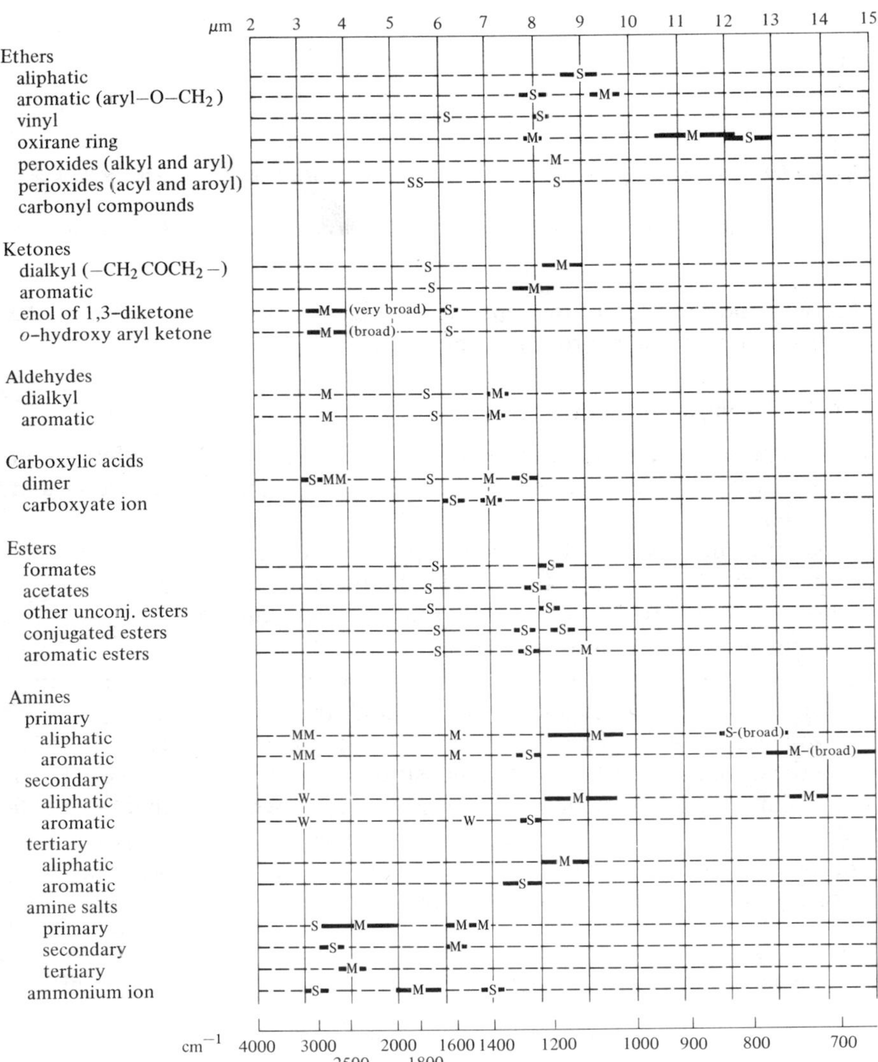

Hydrogen bonding leads to small shifts to lower frequencies, since the formation of the hydrogen bond weakens the C═O bond. Neat methylethylketone has its carbonyl absorption at 5.83 μm, while when observed as a 10 percent solution in methanol, the band occurs at 5.86 μm. Structures such as $CH_3(C_2H_5)C{=}O\cdots HOCH_3$ are formed in solution; obviously, both partners of the hydrogen bond are affected and the OH band for methanol will also be shifted to lower energy. Separate bands may sometimes be observed for hydrogen bonded and free hydroxyl groups in the same solution; in such cases the hydrogen bonding equilibria may be studied by

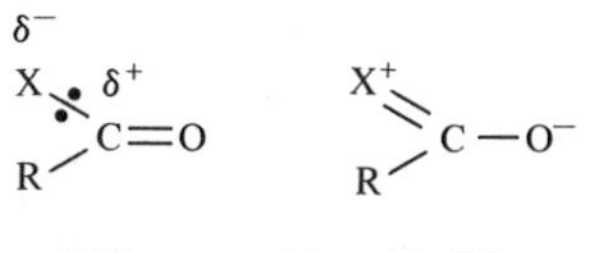

(a) (b)

Figure 18.9 / (a) *An inductive effect in which the substituent causes an increase in the* C=O *force constant.*
(b) *A resonance effect in which conjugation decreases the* C=O *force constant.*

observing the variation of the separate band intensities as a function of the concentrations of the hydrogen bonding species.

18.6.5 / Examples of spectra

With some of the foregoing ideas in mind and the use of the information in Fig. 18.8 and Table 18.2, we now give some examples of IR spectra and their interpretation.

The spectrum given in Fig. 18.10 represents the limit of a very simple IR spectrum. The bands near 6.4 and 7.3 μm are characteristic of the nitro group and arise from N—O stretching modes. We note also the characteristic doublet near 3.4 μm from the methyl stretching modes. With the elemental composition CH_3NO_2 at hand, there is little difficulty in assigning the structure of nitromethane on the basis of the IR data alone.

A somewhat more complex spectrum is shown in Fig. 18.11. This spectrum is obtained from a neat liquid of composition C_3H_6O. The prominent band at 3.3 μm suggests a methyl stretching vibration, while the bands at 7.0 and 7.3 μm can be assigned to symmetric and asymmetric methyl bending vibrations. The strong absorption at 5.8 μm suggests the presence of a carbonyl group, which is confirmed by the presence of the combination

Figure 18.10 / *The* IR *spectrum of* CH_3NO_2; *pure liquid sample run in a cell of* 0.01-mm *path.*
SOURCE: © Sadtler Research Laboratories, Inc.

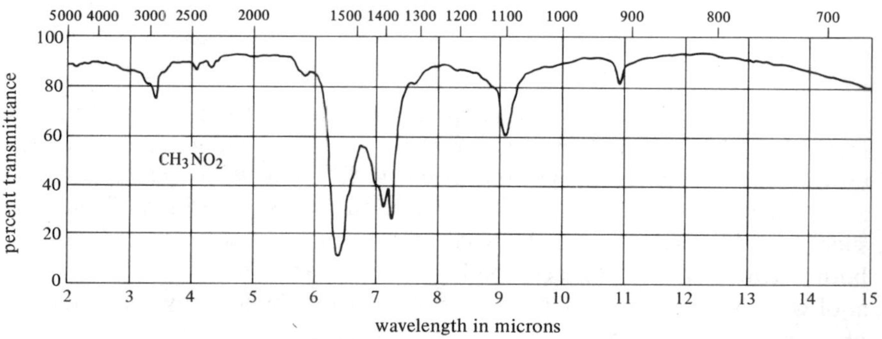

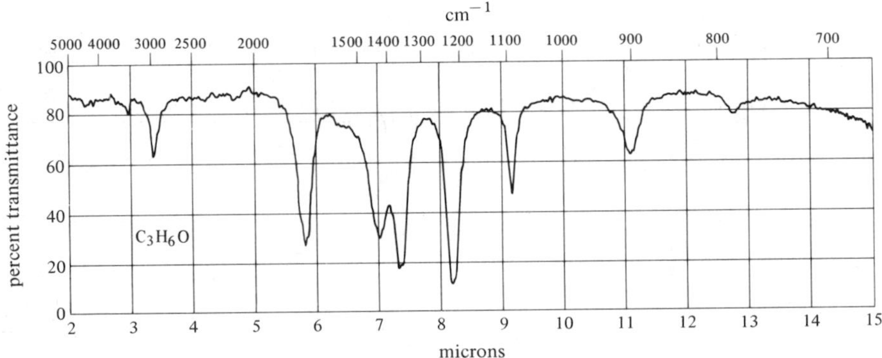

Figure 18.11 / *The* IR *spectrum of* C_3H_6O*; pure liquid sample run in a cell of* 0.01-mm *path.*
SOURCE: © Sadtler Research Laboratories, Inc.

band expected for C—C(=O)—C near 8.2 μm. It is therefore reasonably straightforward to assign this spectrum to acetone.

The spectrum in Fig. 18.12 shows new features and greater complexity than the previous two. The doublet on the short-wavelength side of 3.0 μm is characteristic of an N—H stretch associated with a primary amine. The presence of an amine group is further supported by the N—H bending band near 6.2 μm. The medium-intensity band at 7.8 μm is diagnostic for the C—N stretching vibration of aromatic primary amines. The three weak bands at 3.3, 9.7, and 10 μm, along with the two strong bands at 13.2 and 14.4 μm represent a pattern typically found for monosubstituted benzene rings. Given the empirical formula C_6H_7N, it is reasonable to suppose that the molecule can be identified as aniline from the IR data alone. Even the

Figure 18.12 / *The* IR *spectrum of* C_6H_7N*; pure liquid run between salt flats.*
SOURCE: © Sadtler Research Laboratories, Inc.

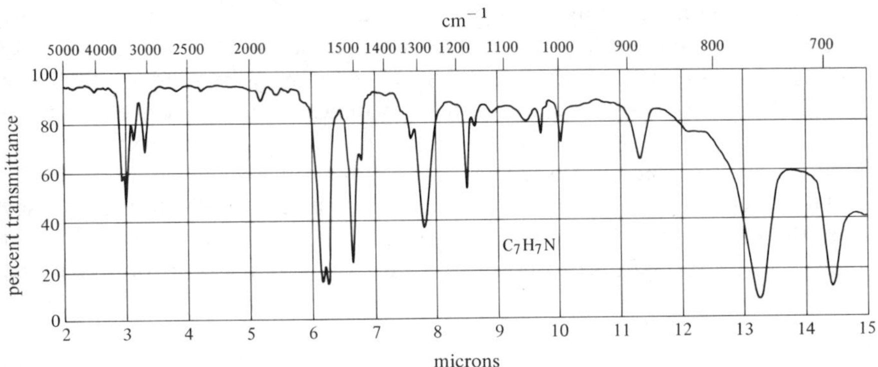

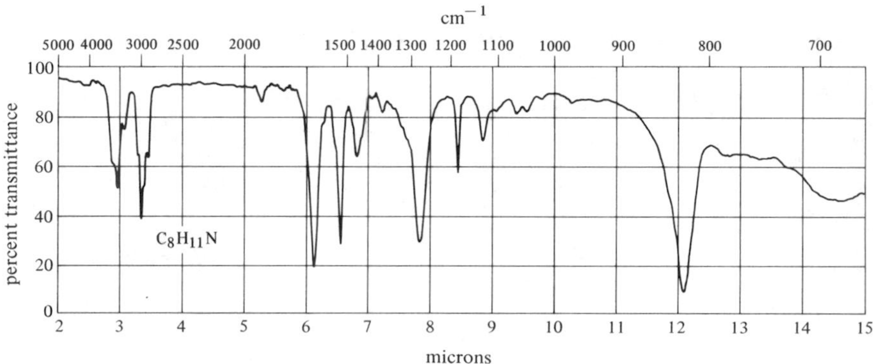

Figure 18.13 / *The* IR *spectrum of* $C_8H_{11}N$; *pure liquid run between salt flats.* SOURCE: © Sadtler Research Laboratories, Inc.

relative neophyte to IR spectroscopy has sufficient information to interpret this spectrum as belonging to an aromatic amine; comparison of these data with cataloged spectra would then provide unambiguous identification.

Many features of the spectrum in Fig. 18.13 are the same as in the preceding spectrum. The N—H stretch, N—H bend, and C—N stretch at 2.95, 6.2, and 7.8 μm, respectively, suggest a molecule quite similar to aniline. The prominent multiplet near 3.4 μm suggests the combination of methyl and methylene group stretching frequencies. Note the difference in the long-wavelength region, however, where only a single strong band at 12.1 μm is present. This single band in the aromatic region is typical for 1,4-disubstituted benzene rings, and this substitution pattern is supported by the medium and weak bands near 6.6 and 6.8 μm. Thus, given the formula $C_8H_{11}N$, we may rationally assign the structure 4-ethylaniline.

Figure 18.14 shows a spectrum that suggests an aromatic structure, but with a substitution pattern different from the preceding molecule. The intense peak at 13.5 μm can be characteristic for 1,2 disubstitution. In addition, the pattern of peaks in the 6–7 μm region and the two weak bands at 9 and 9.8 μm fit the spectral pattern expected for *ortho* substitution. At 3.4 μm we find a methyl absorption; the absorption at 7.2 μm is probably one of the methyl bending frequencies, and the second of these bands may be the shoulder on the peak near 7.8 μm, or it may be hidden under the aromatic bands. In either case, we have evidence for the presence of methyl and an *ortho*-substituted benzene ring, which, in combination with the formula C_8H_{10}, leads to the structure 1,2-dimethylbenzene.

The spectrum shown in Fig. 18.15 represents a case in which it is unlikely that the complete structure can be deduced from IR data alone. We can pick out a carbonyl absorption at 5.8 μm; the broad absorption centered near 8.5 μm is a characteristic combination band for $\mathrm{C{-}\overset{\overset{\displaystyle O}{\|}}{C}{-}O}$, suggesting

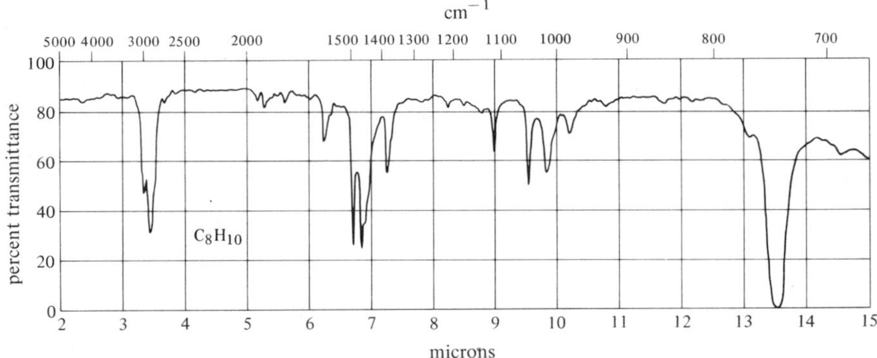

Figure 18.14 / *The* IR *spectrum of* C_8H_{10}*; pure liquid sample run in a cell of* 0.025-mm *path.* SOURCE: © Sadtler Research Laboratories, Inc.

that an ester group is present. The presence of the medium-intensity band at 6.83 μm is diagnostic for the scissoring vibration of methylene groups and, of course, the band at 3.3 μm can be assigned to methylene plus methyl absorptions. The lack of appreciable absorption beyond 11 μm suggests that no aromatic structures are present. It seems likely, however, that more information would be required in order to establish the structure cyclohexanecarboxylic acid, ethyl ester. Moreover, note that the spectrum in Fig. 18.16 is essentially indistinguishable from the one just discussed. It is not possible to distinguish unambiguously between cyclohexaneacetic acid, ethyl ester and the similar molecule described in the discussion of Fig. 18.15 on the basis of IR data alone.

18.7 / FURTHER APPLICATIONS

In this section, we shall review briefly some applications of vibrational spectroscopy beyond the identification and structure determination applica-

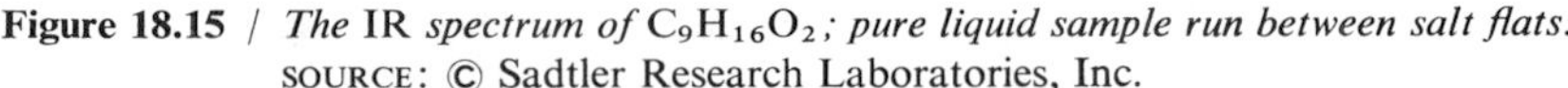

Figure 18.15 / *The* IR *spectrum of* $C_9H_{16}O_2$*; pure liquid sample run between salt flats.* SOURCE: © Sadtler Research Laboratories, Inc.

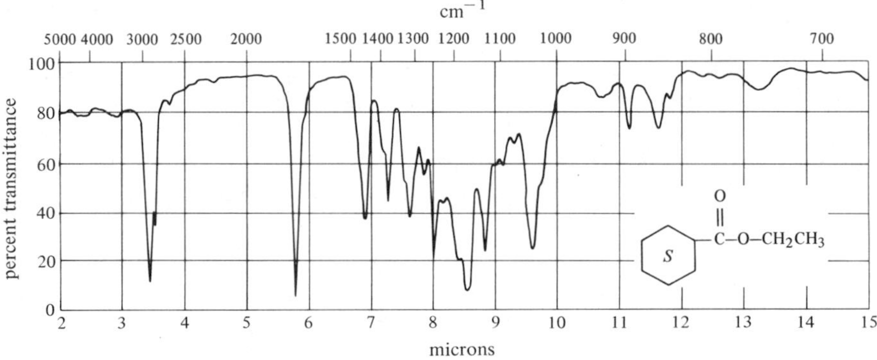

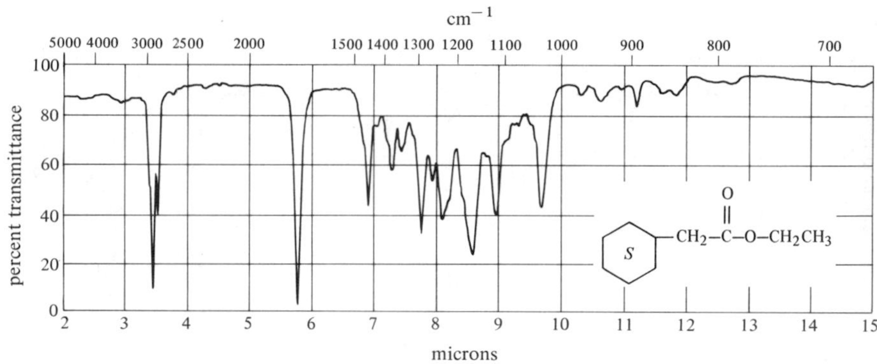

Figure 18.16 / *The* IR *spectrum of* $C_{10}H_{18}O_2$*; pure liquid sample run between salt flats.* SOURCE: © Sadtler Research Laboratories, Inc.

tions in organic chemistry emphasized in the preceding section. This review is by no means exhaustive, but rather is intended to give the reader some feeling for the potentialities of other aspects of this branch of spectroscopy.

18.7.1 / Quantitative analysis

Resolved bands in the ordinary IR region are usually found to obey Beer's law and may form the basis of rapid quantitative analyses. In the past, when the majority of IR spectrometers were of the single-beam type, more use was made of quantitative IR analysis than is done in current practice. There are two reasons for the declining use of IR in quantitative procedures. The convenient, scanning double-beam instruments now in general use are largely of the optical-null (Chapter 11) design and have lower photometric accuracy than their single-beam predecessors; and other techniques, principally gas chromatography, have replaced IR for quantitative measurements. Nonetheless, IR should not be overlooked as a quantitative tool when an accuracy of a few percent is satisfactory.

18.7.2 / Near infrared

The wavelength region between 0.7 and 2.5 μm is customarily regarded as the near infrared. The near IR is distinguished from the NaCl region by instrumental differences: Quartz is a suitable optical material for cells, and double-quartz monochromators using photoconductive detectors are employed. As a result of the differences in instrumentation, high-resolution, ratio-recording (Chapter 11) spectrometers are generally available, making the near IR suitable for quantitative measurements with precisions in the 0.004–0.5 percent range.

With few exceptions, all of the absorption bands in this region arise from overtones ($\Delta v > 1$) and combinations involving hydrogen stretching vibrations of molecular fragments X—H. In practice, the majority of spectra studied involve C—H, N—H, and O—H stretching vibrations. In addition to the fundamental spectroscopic interest in these bands, in particular the study of the interactions from which combination bands arise, useful analytical procedures have been developed for the determination of water, alcohols, amines, organic acids, and olefins. For example, the amine content of hydrocarbons can be assayed with good precision and a detectability limit of the order of 0.04 percent using a near IR absorption at 1.5 μm. Studies of hydrogen bonding are also well suited to the near IR.

18.7.3 / Far infrared

The far infrared is, as one worker in the field observed, "the region beyond where your instrument stops." While in the recent past, anything beyond 15 μm was considered "far," the routine availability of grating instruments that cover the 2–50 μm region has caused the far IR to mean the region from 50–1000 μm (200–1 cm^{-1}). This region is instrumentally difficult to study, since no source is available that provides the major fraction of its energy output at the desired frequencies. The source employed is the high-pressure mercury arc which has, of course, it greatest power output in other spectral regions. Only the Golay and pyroelectric detectors (Section 11.3.5) are suitable for this very low-energy radiation. Grating instruments function to about 700 μm, although difficulties are encountered in filtering out unwanted orders; the longer-wavelength region is accessible only by interferometry.

Far IR has been of interest to inorganic chemists for some time as a method of studying crystal lattice vibrations. More recently, it has been applied to measurement of the very low molecular vibrational frequencies. Knowledge of these frequencies is needed to complete normal-mode analyses and for the calculation of thermodynamic properties. Other applications of far IR include the study of chain and lattice vibrations in polymers, studies of hindered *intra*molecular rotations, and the detection of very weak *inter*molecular interactions. In the last category, studies of hydrogen bond vibrations in the 200–100 cm^{-1} region have provided insight on molecular interactions of the liquid phase. Bands below 90 cm^{-1} in spectra of nitriles and nitro compounds have been observed and assigned to vibrations of dipole–dipole complexes. Spectra of organic salts in benzene solution have revealed bands due to interionic vibrations and the vibrations of ion aggregates.

The far IR appears to be exclusively the province of spectroscopists and physical chemists; no routine analytical procedures in this region are known.

18.7.4 / Raman spectroscopy

Since Raman spectra consist of scattered radiation shifted in wavelength, they appear as emission, rather than absorption spectra. This is apparent in Fig. 18.17, where a portion of the Raman spectrum of neat CCl_4 is shown. In this spectrum, the intense line with zero shift is the Rayleigh line; the Raman line shifted about 200 cm^{-1} appears symmetrically on both sides of

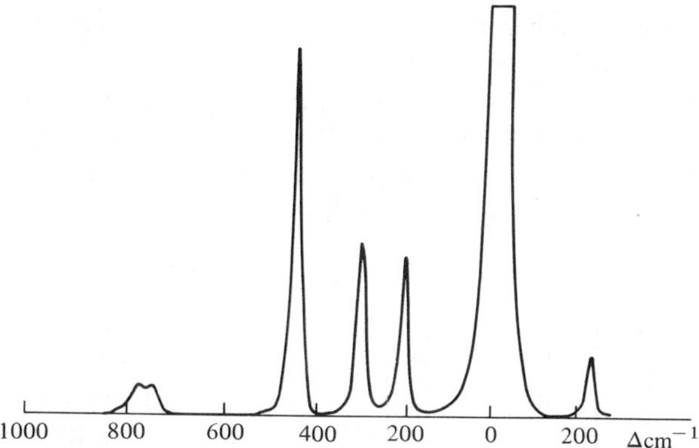

Figure 18.17 / *The laser Raman spectrum of carbon tetrachloride run as the neat liquid. Note that both the Stokes and anti-Stokes lines appear for the vibration with the smallest frequency and that these lines have the expected intensity relationship.*
SOURCE: H. A. Szymanski, *Raman Spectroscopy: Theory and Practice*, Plenum (New York, 1967), p. 55.

the Rayleigh line with the high-frequency (anti-Stokes) line noticeably less intense than the Stokes line, as anticipated.

In many instances, the magnitudes of the Raman shifts correspond exactly to the frequencies of IR absorptions. This is true for the doublet in the CCl_4 spectrum centered near 770 cm^{-1}: The most intense bands in the IR spectrum of this molecule occur at 782 and 755 cm^{-1}. If, of course, the same information were always available from both techniques, one of the two would be redundant. However, as discussed previously, the two methods represent very different fundamental processes, and some vibrational motions interact only with IR radiation while others produce only Raman effects. For such cases, the IR and Raman spectra complement each other, and the complete vibrational spectrum can be obtained only by using both techniques. From theoretical considerations, more vibrational modes should be active in the Raman than in the IR, so intrinsically more information should be contained in the Raman spectrum.

As noted above, for some structures a rule of mutual exclusion applies, and none of the IR-active modes gives rise to a Raman effect. In such cases, of which polyethylene is one example, the differences in the vibrational frequencies obtained from the two spectra may be used to great advantage in determination of symmetry and structure. A discussion of the assignment of vibrations to symmetry groups is, however, beyond the scope of this chapter.

One notes in the discussion of IR functional-group frequencies that most of the organic functions that absorb strongly are highly polar. This fact is, of course, related to the requirement that a change in molecular dipole

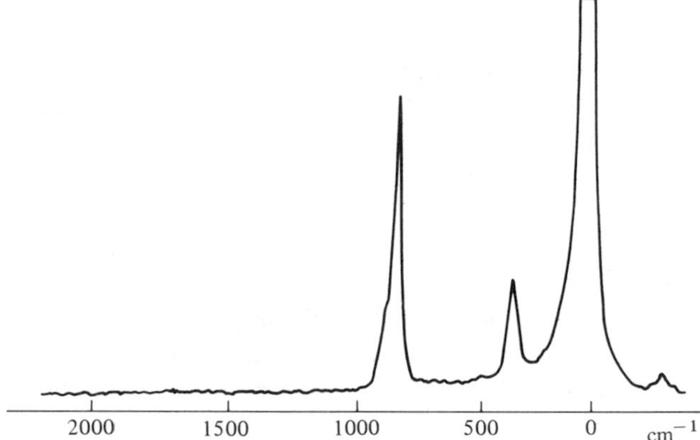

Figure 18.18 / *The laser Raman spectrum of potassium chromate run as a saturated aqueous solution.*

SOURCE: H. A. Szymanski, *Raman Spectroscopy: Theory and Practice*, Plenum (New York, 1967), p. 59.

moment is associated with an IR-active vibration. On the contrary, Raman effects are associated with a change in polarizability, and *nonpolar* groupings tend to be the most easily polarized; thus these give strong Raman scattering. Application of vibrational spectroscopy to the study of polymeric materials is divided along these lines: IR investigations usually involve study of polar substituents on the polymer chain, whereas Raman spectra give information about the carbon skeleton itself.

Perhaps the single most exploited advantage of Raman spectroscopy is its ability to deal with aqueous solutions, which are anathema to commonly used IR-transparent materials. Much use has been made of the Raman effect in studying polyatomic inorganic ions and metal complexes. An example of such a spectrum is given in Fig. 18.18 where, in addition to the Rayleigh line, three vibrational bands for the chromate ion, CrO_4^{2-}, can be seen. Information regarding the symmetry of these species, types of bonding, and the degree of covalency of metal–ligand bonds is obtainable from Raman spectra. It is also possible to make quantitative measurements of concentrations of species in solution from the integrated intensities of Raman bands. On this basis, ionic equilibria in solution can be investigated.

One can say with certainty that the perfection of laser-source Raman instruments will increase the utilization of this method. It is likely to become effective competition for IR in organic analysis and structure determination for the reasons of larger information content of the spectra, and because the entire vibrational frequency range can be examined with one instrument rather than the three separate instruments required to cover the near, medium, and far IR. It can be seen in Fig. 18.17 that it would be easy to resolve Raman bands with shifts as small as 100 cm^{-1};

the limitation on observation of low-frequency bands is simply the ability of the instrument to limit the width of the Rayleigh line. The source and detector problems of far IR are circumvented.

Biochemistry is a field in which Raman spectroscopy has obvious applications. The ability to study conformations of polymeric materials combined with the ease of handling aqueous samples means that biopolymers may be studied with relative ease under conditions approximating those in living organisms. An example is the comparison of the Raman spectrum of solid poly-L-proline with that obtained from an aqueous solution of this biopolymer. It was possible to conclude that the polymer retains its helical conformation even in solution, in spite of the fact that it lacks the *intra*-molecular hydrogen bonds usually believed to be the basis of the stability of such helical conformations.

SUGGESTED READINGS

General

D. H. Whiffen, *Spectroscopy*, Wiley (New York, 1966), Chapters 8–10 (paperbound).

G. Herzberg, *Infrared and Raman Spectra of Polyatomic Molecules*, Van Nostrand Reinhold (New York, 1945).

Infrared Spectroscopy

R. M. Silverstein and G. C. Bassler, *Spectrometric Identification of Organic Compounds*, 2nd ed., Wiley (New York, 1967), Chapter 3.

L. Bellamy, *The Infrared Spectra of Complex Molecules*, Wiley (New York, 1958).

K. Nakanishi, *Infrared Absorption Spectroscopy: Practical*, Holden-Day (San Francisco, 1962).

H. A. Szymanski, *IR Theory and Practice of Infrared Spectroscopy*, Plenum (New York, 1964).

K. Nakamoto, *Infrared Spectra of Inorganic and Coordination Compounds*, Wiley (New York, 1963).

Raman Spectroscopy

H. A. Szymanski, *Raman Spectroscopy Theory and Practice*, Plenum (New York, Vol. 1, 1967; Vol. 2, 1970).

E. I. Loader, *Basic Laser Raman Spectroscopy*, Heyden-Sadtler (New York, 1970).

PROBLEMS

1 / Calculate the frequency and wavenumber for IR radiation of wavelength 6 μm.

2 / The IR spectrum of chloroform shows a C—H stretching absorption at 3100 cm^{-1}. At what wavelength would this absorption be found for tritiated chloroform, C^3HCl_3?

3 / An instrument has been designed to detect the presence of SO_2 in smoke stack effluent from a distance. The device consists only of a telescope that focuses the image of the stack plume on an IR monochromator–detector system. How does this instrument function?

4 / The vibrational absorption of HCl occurs at 2886 cm^{-1}. Calculate the force constant of the bond. When a photon is absorbed, find the amplitude of vibration. Recall that at the extreme of the vibration all of the energy must be in the form of potential energy ($V = \int_0^x kx\ dx$). Express the amplitude of vibration as a percentage of the equilibrium bond distance, 1.27 Å.

5 / The IR spectra in Fig. 18.19 arise from the *cis* and *trans* isomers of $C_2H_2Cl_2$. Decide from the spectral data which is which.

Figure 18.19 / *Spectra for Problem 5.*

SOURCE: © Sadtler Research Laboratories, Inc.

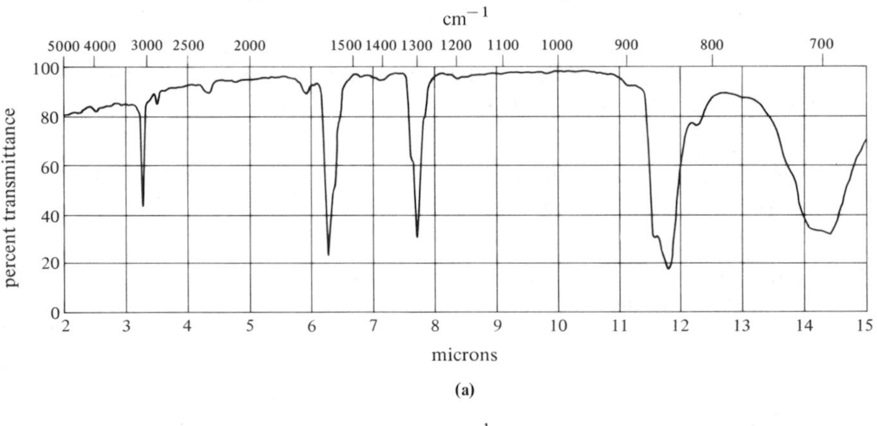

(a)

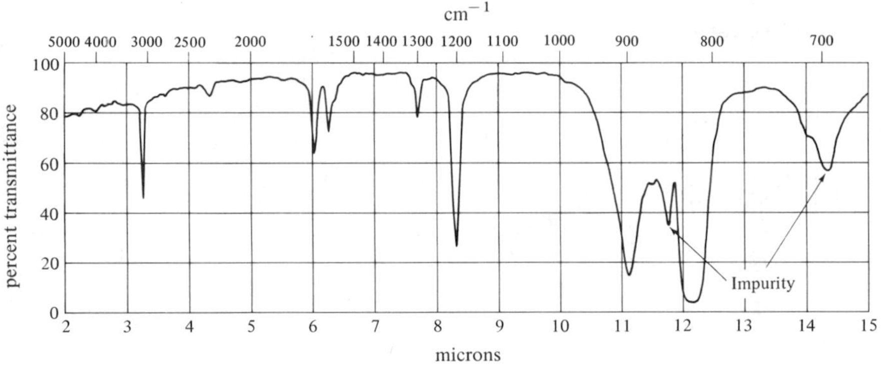

(b)

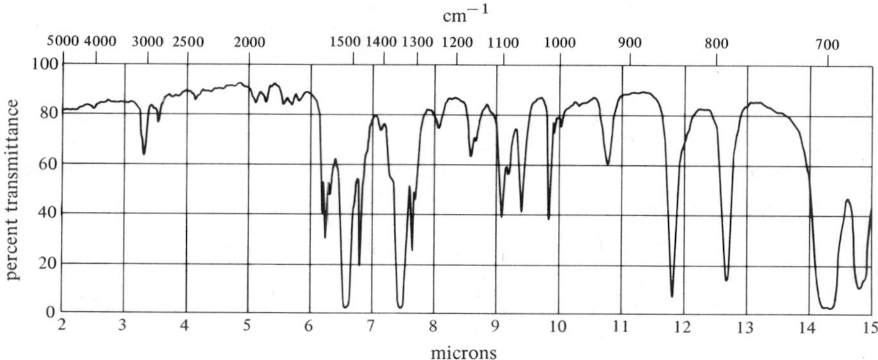

Figure 18.20 / *Spectrum for Problem 6.*
SOURCE: © Sadtler Research Laboratories, Inc.

6 / Assign a rational structure for the compound $C_6H_5NO_2$ for which the IR spectrum is given in Fig. 18.20.

7 / Assign a rational structure for the compound $C_6H_6N_2O_2$ for which the IR spectrum is given in Fig. 18.21.

8 / See how many structural features you can assign for the compound $C_{13}H_{18}O$, for which the IR spectrum is given in Fig. 18.22.

Figure 18.21 / *Spectrum for Problem 7.*
SOURCE: © Sadtler Research Laboratories, Inc.

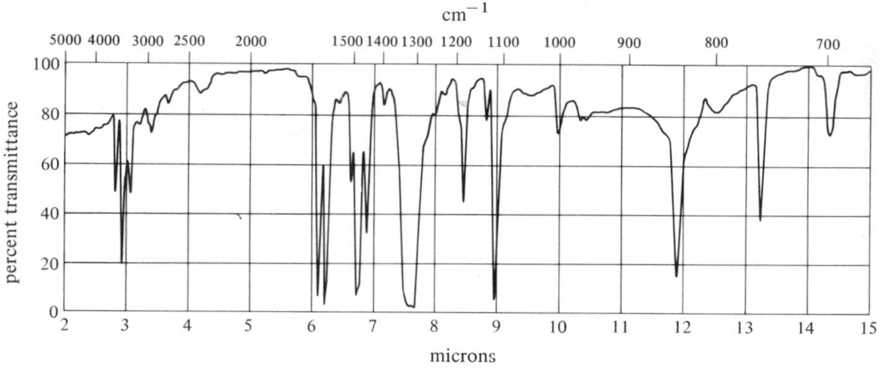

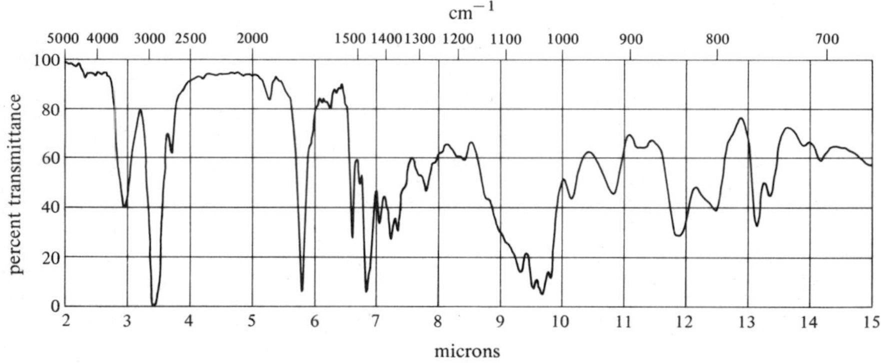

Figure 18.22 / *Spectrum for Problem 8.*
SOURCE: © Sadtler Research Laboratories, Inc.

chapter 19 / Rotational spectroscopy

chapter 19

19.1 / INTRODUCTION

As indicated in Section 10.3.6, microwave[1] radiation excites pure rotational transitions in molecules that possess a permanent dipole moment. Since there are many rotational levels even for simple molecules, and many of the higher levels are appreciably populated at ordinary temperatures, a large number of transitions is possible. Microwave spectra are obtained with gas-phase samples, so the spectral lines are typically very narrow, and the rich structure of the rotational spectra can be observed experimentally.

Since the frequencies of microwave transitions depend only on the moments of intertia of the sample molecules and not all on functional groups, the information available from a microwave spectrum is both different from, and complementary to, that obtained from IR spectroscopy (Chapter 18). Until quite recently, the principal utilization of microwave spectra lay in the determination of the moments of inertia of molecules and, thereby, highly precise determinations of bond lengths and angles; this was exclusively the province of molecular spectroscopists. As commercial spectrometers have become available, however, the potential of microwave spectroscopy as an analytical tool is beginning to be realized. Both qualitative and quantitative analyses can be performed, and closely similar molecules are easily differentiated. In fact, since only moments of inertia determine the spectrum, molecules that differ only by isotopic substitution at a single site have markedly different spectra.

In this chapter, we shall briefly outline the use of this branch of spectroscopy both for analysis and for the determination of molecular dimensions.

[1] Wavelengths in the microwave region, typically millimeters to several centimeters, are very long compared to visible or IR wavelengths, and the name may appear anomalous. The name, however, was affixed to this spectral region by radio engineers, to whom these waves appeared small indeed compared to radio wavelengths of many meters.

19.2 / GENERATION AND TRANSMISSION OF MICROWAVE ENERGY

As indicated previously in this text, the distinctions among the various regions of the electromagnetic spectrum are arbitrary human distinctions, while wavelength, frequency, and energy vary smoothly and continuously throughout the spectrum. Microwaves are commonly defined to lie in the spectral region between approximately 1 GHz (1000 MHz) and 100 GHz ($\lambda = 30 - 0.30$ cm). This definition is based, at the lower extreme of frequency, on the transition from efficient transmission of energy via cables to transmission via wave guides, and on the necessity of abandoning ordinary electronic oscillators as radiation sources. The upper limit of the region is fixed by limitations of available sources for the generation of appreciable power. Above 100 GHz, other sources become competitive with oscillators of current design, and the microwave region merges with the far infrared (see Section 18.7.3).

19.2.1 / The klystron oscillator

As higher frequencies are required, vacuum tube sources require special design; and it has been found that above 1 GHz the engineering problems become essentially insurmountable. The upper limit to the frequency range of conventional vacuum tubes is set by several factors, among which are the interelectrode capacitance and the inductance of the leads to the tube elements, neither of which can be conveniently decreased. In addition, the transit time of the electrons in the space between the electrodes approaches a significant fraction of the radio frequency period, and appreciable power losses occur because the tube structure and leads function as antennas. All of these factors necessitate small tube structures and, in the limit, reduction of size leads to the inability of the tube to dissipate heat adequately. Such limitations led to the design of several different microwave sources, initially as an outgrowth of the development of radar (*ra*dio *d*irection *a*nd *r*anging) technology during World War II. Of these sources, primarily the reflex klystron has found significant application in spectroscopic measurements requiring high stability. It has been the source of choice for both microwave spectroscopy and electron spin resonance spectroscopy (Chapter 20), although solid state sources are now becoming available.

The unique feature of the klystron is that it is designed to take advantage of the finite transit time of the electron beam, which had proved to be one of the significant difficulties in achieving high frequencies with ordinary vacuum tubes. A simplified schematic of a reflex klystron is shown in Fig. 19.1.

Electrons are emitted from the heated cathode and accelerated by the beam voltage, V_b, applied to the first cavity grid. The beam is velocity modulated by a rf voltage induced between the two cavity grids; thus, groups of electrons are alternately speeded up and slowed down. In the drift space, fast-moving electrons catch up with those that move more

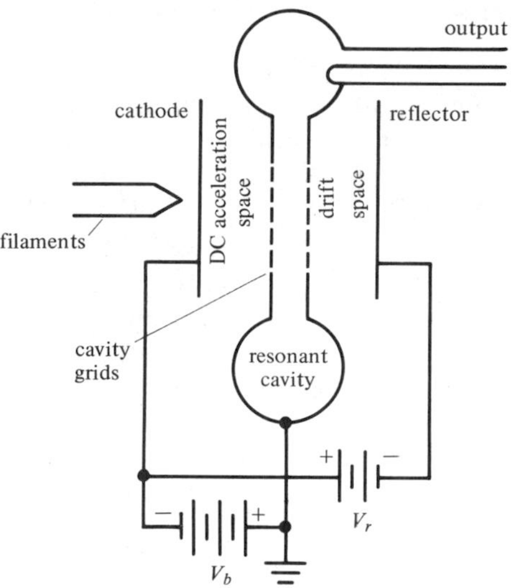

Figure 19.1 / *Schematic diagram of a reflex klystron. The term "reflex" refers to the fact that the electron beam is reflected back toward its source.*

slowly, with the result that the beam is converted into a series of groups or "bunches" of electrons. The reflector is operated at a voltage somewhat more negative than the cathode, and it serves to return the bunched beam to the cavity. When the cavity size, beam voltage, and reflector voltage are all properly adjusted, the bunched electrons return to the cavity with the proper phase and are "debunched" (accelerated), with the result that they give up microwave energy to the cavity; this process sets up a rf voltage and the tube oscillates. If the cavity is coupled to a transmission line or wave guide, energy can be extracted from the tube.

A reflex klystron has a limited range of operating frequencies defined by the construction of the tube. Typically, the mechanical tuning range (varying the cavity dimensions) is 5–50 percent of the design frequency; electronic tuning (varying the reflector voltage) covers a much smaller range—for example, 0.2–0.8 percent. For instance, the Varian X-13 klystron has a mechanical tuning range of 8.1–12.4 GHz and can be tuned electronically over 60 MHz at the lower end of the mechanical range and over 45 MHz at the upper end of the mechanical range. Invariably, both the electronic tuning range and the power output of a klystron are frequency dependent. While klystrons are available for radar applications that have very large pulsed power output (10^4 kW, intermittently) the klystrons used in spectroscopy are low power, typically providing 1 W or less of continuous power.

19.2.2 / Wave guides

A wave guide is a pipe made of conducting material through which microwave energy can be transmitted with great efficiency. The electromagnetic

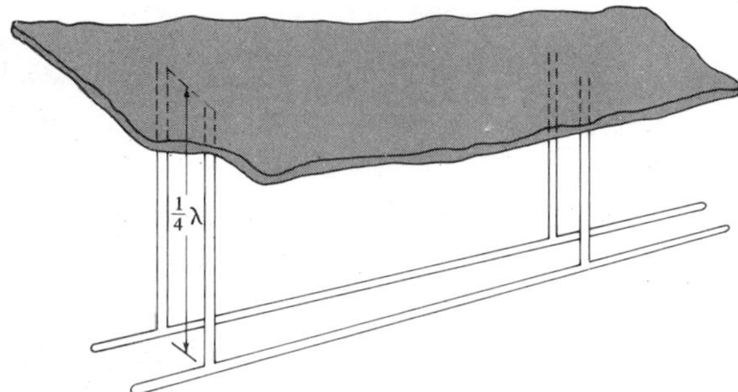

Figure 19.2 / *A two-wire power transmission line supported from a conducting overhead by metallic quarter-wavelength "insulators."*

radiation propagates in the air space inside the pipe, with the conductor serving both to confine and direct it. These pipes may be of circular or rectangular cross section, and the dimensions of the cross section will be comparable to the wavelength of the radiation that is transmitted. In the low-frequency end of the microwave region, the efficiency of coaxial cables rivals that of wave guides, and the bulk of the latter necessitated by the long wavelengths restricts their use. At frequencies above 3 GHz, however, the superior efficiency of wave guides dictates their use in preference to cables whenever the transmission distance is more than a few inches.

The function of a wave guide can be understood in the following way. Consider a pair of power transmission lines such as that shown in Fig. 19.2. These lines are suspended from a metal overhead by conducting rods chosen to be one-quarter wavelength long. Since the two lines are separated by half a wavelength, the signals are exactly out of phase and the all-metal system presents an extremely high impedance between the lines. Alternatively, these transmission lines could be supported from both above and below by quarter-wave sections of conducting material, as shown in Fig. 19.3. The impedance between the two lines is unaffected by the number of such supports, and if the number of these "insulating" frames is simply increased until they touch, a section of rectangular wave guide results. It might appear

Figure 19.3 / *A two-wire power line supported by quarter-wave sections of metallic "insulation" from both above and below.*

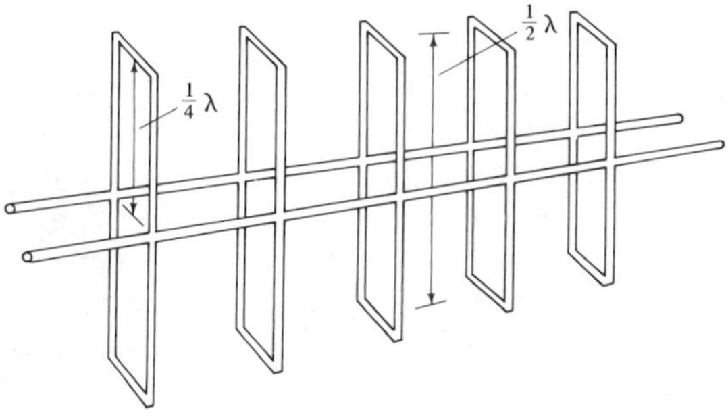

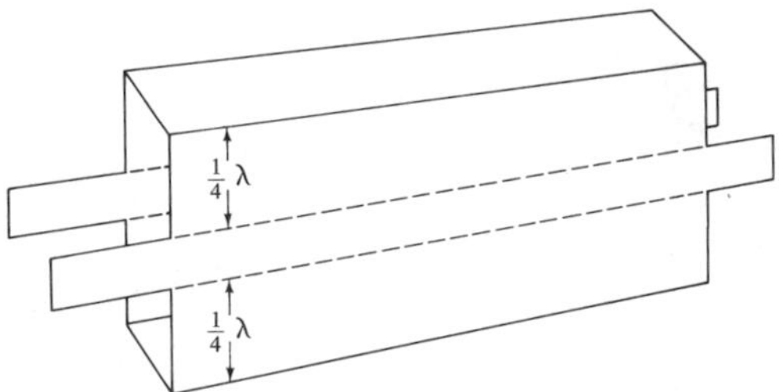

Figure 19.4 / *A section of wave guide with two imaginary bus bars shown running through its sides.*

at this point that a wave guide can carry only a single frequency. In fact, however, the wave guide can carry the anticipated frequency and also all higher frequencies. The transmission lines of Fig. 19.3 could just as easily have been metal bus bars. In Fig. 19.4, a section of wave guide is illustrated with two imaginary bus bars running through its sides. The guide can be imagined as having an upper and a lower quarter-wave section of metallic "insulation" and a center section of bus bar. When the transmitted frequency is increased, the bus bar section width can be imagined to increase also, while the quarter-wave "insulators" diminish in size. As a result, the guide will pass all frequencies above a certain minimum corresponding to that frequency at which the upper and lower quarter-wave "insulators" begin to overlap, leaving no bus bar at all.

We see, therefore, that a rectangular wave guide must have its long cross-sectional dimension greater than half a wavelength for the radiation it is designed to carry. The short dimension has little or no effect on the frequency limitation of the guide, but it does control the voltage at which the guide will arc. Thus, this dimension is not critical except in applications requiring very high powers, and for those uses it must be large. In practice, the larger inside dimension of a rectangular wave guide is usually 0.7, and the shorter dimension between 0.2 and 0.5, respectively, of the wavelength in air for the design frequency. For comparison of efficiency, the attenuation of a 10-GHz signal by a 3-in. cylindrical wave guide is less than 0.01 dB/m, while the same signal is reduced more than 2 dB/m by half-inch coaxial cable.

19.3 / MICROWAVE SPECTROMETERS

Microwave spectroscopy shares with all other branches of absorption spectroscopy the necessity of providing a source of monochromatic radiation, a sample holder, and a radiation detector. This technique differs from optical spectroscopy in two significant ways. First, the microwave oscillator source is inherently highly monochromatic, so no dispersive element is required; however, in general, a single source will not be tunable over the entire region of interest, so several sources may be required. Second, since

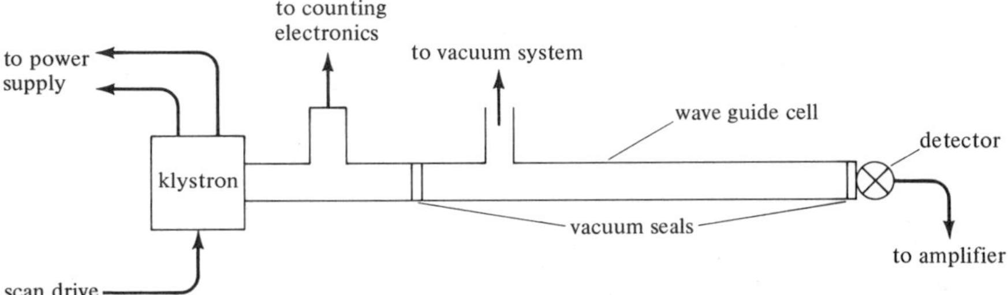

Figure 19.5 / *A simple microwave spectrometer including a klystron (source), wave guide cell (sample holder), and crystal detector. The absorption frequencies are determined by counting the source radiation using electronic techniques.*

the samples examined will be gases at low pressure, a long path length cell will be needed, as will a vacuum system and gas-handling apparatus for sample introduction and manipulation. To extend the range of samples that can be examined, it is desirable to provide a means of heating the inlet system and sample cell to obtain sufficient vapor pressure from materials that are not gaseous at ambient temperature.

19.3.1 / A simplified spectrometer

The basic components required for microwave spectroscopic measurements are shown in Fig. 19.5. The klystron oscillator provides highly monochromatic microwave radiation, which can be varied over at least part of the spectrum by mechanical tuning as described above. The sample holder is a section of vacuum-tight wave guide, usually of the order of 2 m in length and sealed at each end with mica windows. A crystal detector is placed at the far end of the sample cell to detect the intensity of transmitted radiation. The detector is a silicon–tungsten crystal that has the property of producing a DC voltage output proportional to the intensity of microwave radiation striking it.

The output of the detector crystal is amplified and observed as the vertical axis displacement of a chart recorder. The spectrum is scanned over the available frequency interval by driving the mechanical tuning screw of the klystron with a motor. The scan drive can be coupled to the horizontal axis of the display device in several ways; the simplest of these is simply to use synchronous motor drives for each with a time measurement to provide calibration. The frequency of the oscillator output is counted with an electronic frequency counter and read out directly in gigahertz; these readings may be periodically noted on the recording chart by the operator. As noted above, the tuning range of a given klystron may be insufficient to scan the entire region of interest, so it is possible to have several such sources coupled to the spectrometer and switch from one to the next as necessary without dismantling the apparatus.

A device such as that just described will produce spectra only under the most favorable circumstances, since its sensitivity is poor. High sen-

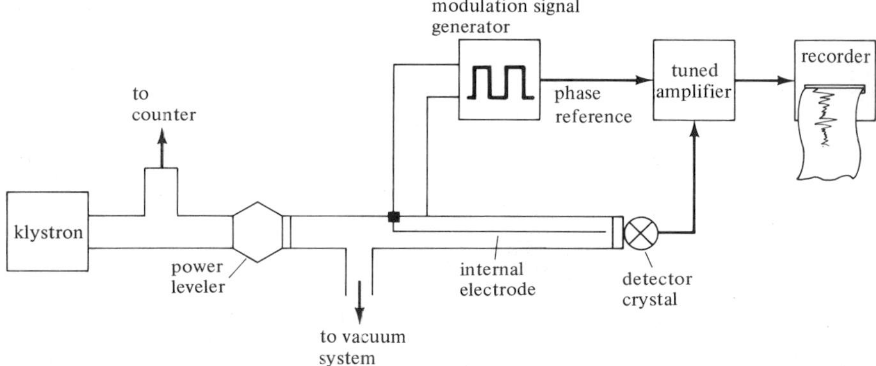

Figure 19.6 / *A schematic diagram of a microwave spectrometer that incorporates Stark-effect modulation and AC amplification of the signal to provide enhancement of the signal-to-noise ratio.*

sitivity is almost always a necessity, however, since in order to obtain optimum resolution it is necessary to work at low sample pressures to avoid line broadening due to molecular collisions. The difficulties of amplifying DC signals have been discussed in previous chapters. In addition, the noise generated by the detector crystal is found to be inversely proportional to frequency, so the signal-to-noise ratio can be improved if the spectral information can be detected as a relatively high-frequency AC signal. The fact that the power output of the source varies with frequency during a scan will prove detrimental, since such variations may produce spurious signals stronger than those due to microwave absorption.

19.3.2 / The Stark modulated spectrometer

A spectrometer that overcomes many of the difficulties mentioned in the preceding section is shown schematically in Fig. 19.6. The principal innovation in this instrument is the introduction of modulation of the spectral information at 100 kHz and the introduction of a tuned, phase-sensitive AC amplifier. The modulation is provided by use of the Stark effect, which is described more fully in Section 19.4.1. In brief, the energies of microwave transitions are shifted if the sample is exposed to a strong electric field. As shown in the figure, an electrode is added that runs the length of the sample wave guide cell; this electrode is, of course, well insulated from the wave guide, and a square-wave voltage, often at 100 kHz, is applied between the two. The AC voltage is selected to produce an electric field of the order of 1 kV/cm in the sample region. The square-wave modulation is chosen with zero base so that the electric field oscillates between zero and its maximum value on each half-cycle.

Consider the situation when the klystron frequency is such that a rotational transition would be observed in the absence of the electric field. When the electric field is applied, the energy of this transition is shifted slightly, and the spectrometer is no longer at the transition frequency; as a result, the absorption is "turned off." When the modulating field has returned to zero, the absorption is "turned on" again. Since the modulation frequency is very large compared with the rate at which the microwave spectrum is scanned, the signal is turned on and off many times while the scan passes through the line; thus, the detector receives information in the form of microwave radiation amplitude modulated at 100 kHz when an absorption is being observed, and constant microwave power when only the base line is scanned. Since the detector rectifies microwave radiation, the constant microwave power produces only a DC signal, which is rejected by the AC amplifier. The detector will, however, pass the modulation frequency, and the spectral information is thus delivered to the AC amplifier at the IF frequency. Since the amplifier is phase referenced to the modulator (see Section 6.4.3), signals with any other phase are rejected. With instruments of this type, very great sensitivity is possible.

It is convenient to insert a power-leveling device between the source and the spectrometer cell. The source output is generally a rounded convex function of frequency with its maximum located near the center of the source tuning range. The power leveler is simply an electronic attenuator that flattens this function (albeit by reducing the source power below its maximum) and provides constant power input to the spectrometer over most of the useful source range.

The effective dispersion of a microwave spectrometer is very large because the source output bandwidth is small compared with the separations between the usual absorption lines. Resolution of 0.1 MHz or better is typically obtained; at an operating frequency of 30 GHz, this corresponds to about 3 ppm. The frequency of an absorption line is readily measured to better than 0.05 MHz or about 1 ppm using electronic frequency counting techniques.

The sample cell of a typical spectrometer requires 0.25–1 liter of gas at a pressure of less than 1 torr so that very small samples can be examined. Since the method is nondestructive, the sample may be recovered if desired simply by pumping it out of the cell through a suitable cold trap. Only losses by adsorption on the walls of the system prevent complete sample recovery.

19.4 / PURE ROTATIONAL SPECTRA

The basic equation for the energy of a diatomic rotor was stated in Section 15.3. The energy of a rotational level characterized by quantum number J is given by Eq. (15.1), which is reproduced here for convenience:

$$E_J = \frac{J(J + 1)h^2}{8\pi^2 I} \tag{15.1}$$

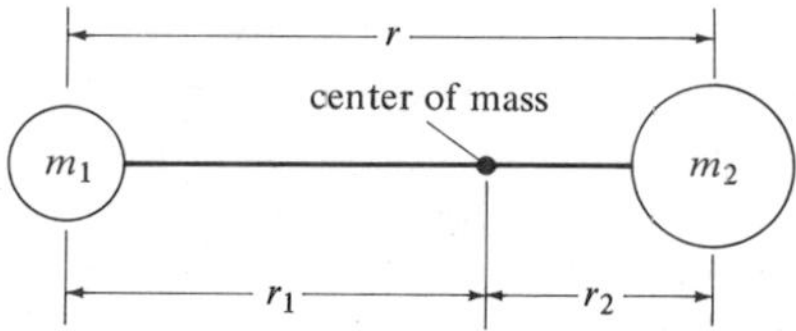

Figure 19.7 / *A diatomic rotor showing the location of the center of mass on the internuclear axis.*

As described in Chapter 15, I is the moment of inertia for rotation of the diatomic molecule about an axis that passes through the center of mass and is perpendicular to the bond joining the two atoms. For a diatomic molecule, there can be only one such moment of inertia; the small spatial extent of the nuclei precludes a moment for rotation about the bond axis, and the rotations about the two orthogonal axes are equivalent by symmetry. In this section, we shall examine in somewhat more detail the behavior of diatomic rotors and also extend the discussion to polyatomic species.

19.4.1 / Diatomic molecules

The moment of inertia of a classical rotor is defined by

$$I = \sum m_i r_i^2 \tag{19.1}$$

where r_i is the perpendicular distance from the ith particle of mass m_i to the center of mass of the system. With reference to Fig. 19.7, the center of mass is defined as that point on the internuclear axis which satisfies the relation $m_1 r_1 = m_2 r_2$. Since $r_1 + r_2 = r$, the center of mass is readily shown to be located by Eq. (19.2):

$$r_1 = \frac{m_2}{m_1 + m_2} r \qquad r_2 = \frac{m_1}{m_1 + m_2} r \tag{19.2}$$

Combination of these relations with the definition of the moment of inertia gives, upon rearrangement, the relation

$$I = \frac{m_1 m_2}{m_1 + m_2} r^2 = \mu r^2 \tag{19.3}$$

The product of the masses divided by their sum is termed the *reduced mass*, symbolized μ, as stated in Chapter 15.

In analogy to Newton's second law for linear momentum, the angular momentum of a rotating system, $I\omega$, is given by the product of the moment of inertia (the analog of mass) and the angular velocity of rotation, ω (the analog of linear velocity) expressed in radians per second. In linear motion, the kinetic energy is given by $mv^2/2$ or, in terms of momentum, by $(mv)^2/2m$. The analogous relations for rotational energy are

$$E_{rot} = \frac{I\omega^2}{2} = \frac{(I\omega)^2}{2I} \tag{19.4}$$

In contrast to linear momentum, angular momentum of molecular systems is quantized, and solution of the appropriate form of Schrödinger's equation shows that the allowed values of angular momentum are given by $\sqrt{J(J+1)}h/2\pi$, where J is the rotational quantum number and may be zero or any integer. Substitution of this expression for the angular momentum into Eq. (19.4) yields Eq. (15.1) directly. Conversion of the rotational energy to wave number and setting $B = h/8\pi^2 Ic$ as indicated in Eqs. (15.5) and (15.6) leads to the spacing of the rotational levels shown in Fig. 19.8. When the selection rule $\Delta J = +1$ is introduced for absorption, the anticipated spectrum is a series of equally spaced lines with their spacing equal to $2B$ cm^{-1}, as indicated in the figure.

For a rotational energy level characterized by a particular value of J, the angular momentum may take on any of the $2J + 1$ values of $\sqrt{J(J+1)}h/2$ obtained by inserting J, $J-1$, $J-2, \ldots, 0, -1, -2, -J+1, \ldots, -J$. The quantity $2J + 1$ specifies the degeneracy, or multiplicity, of a particular rotational level and corresponds to the number of sublevels into which that energy level will be split by application of an electric field. This splitting is indicated schematically in Fig. 19.9.

It is, of course, the presence of these sublevels, which can be revealed by an electric field, that makes the use of Stark-effect modulation possible. The Stark splitting observed for a particular level may be as large as 30 MHz (10^{-3} cm^{-1}) for levels with small values of J, but the splitting diminishes rapidly with increasing J. Nonetheless, the resolution of microwave spectroscopy is sufficient that the shifts are large enough to provide the necessary modulation of the absorption signal. Since the Stark splitting is proportional to the square[2] of the product of the applied electric field and the molecular dipole moment, measurement of these splittings serves as the most accurate method of determining dipole moments.

Any microwave spectrum contains a large number of lines that may be observed if the instrument covers a sufficient range of frequencies, because the energy separation of the various rotational levels is small compared to kT, the thermal energy at ambient temperatures. As a result, even levels with large J values are appreciably populated. As an example, the relative populations of the rotational levels of nitrous oxide at 298°K are illustrated in Fig. 19.10. An absorption line may, of course, originate in any of these populated levels. The number of lines observed may be further increased if the molecule contains nuclei that possess electric quadrupole moments. In such molecules, coupling occurs between the rotational motion of the

[2] This is a second-order Stark effect. The first-order interaction vanishes for the levels of diatomic molecules. Both first- and second-order Stark effects may be observed for nonlinear polyatomic species.

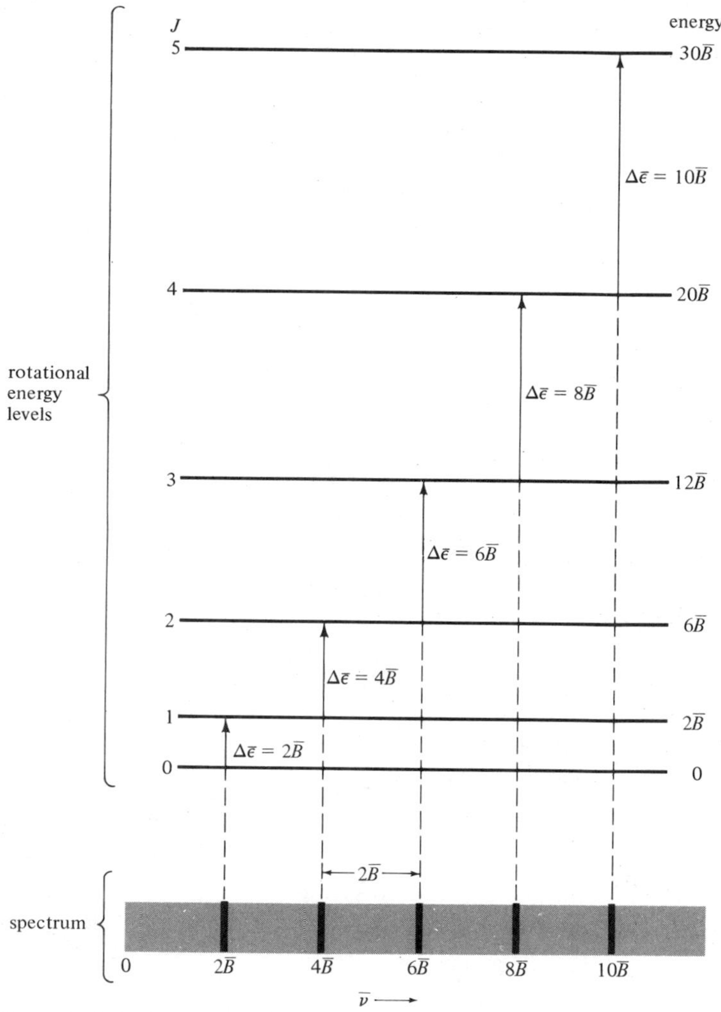

Figure 19.8 / *The rotational energy levels and absorption transitions for a diatomic rotor. Energies are in reciprocal centimeters, and the symbol* $B = h/8\pi^2 Ic$.

molecule and the nuclear spin angular momentum. This interaction splits all but the $J = 0$ level, thus increasing the number of possible transitions. Analysis of quadrupole couplings can lead to estimates of multiple-bond character, the relative amount of ionic and covalent bonding, and the type of hybridization in molecular bonds. The reader is referred to more specialized texts for further discussion of this effect.

In the case of a diatomic molecule, assignment of the rotational spectrum permits determination of the moment of inertia, and therefore the bond length, with great precision. If this measurement is to reflect the actual equilibrium internuclear separation, it must be corrected for both zero-

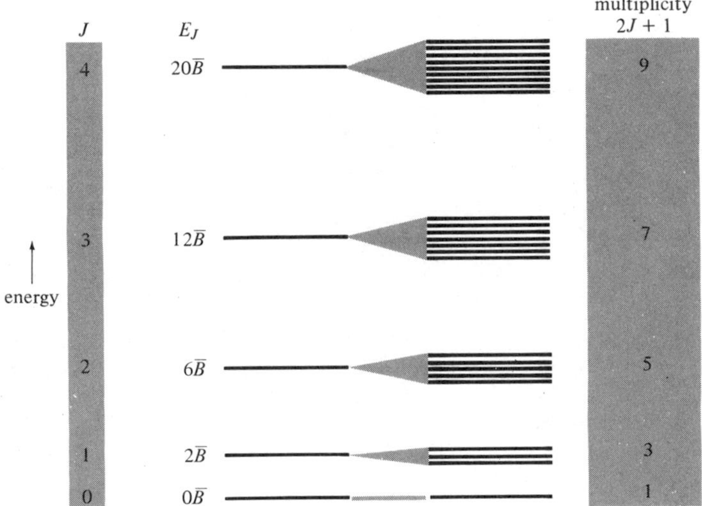

Figure 19.9 / *The rotational energy diagram showing the multiplicity of the levels. The figure indicates only the number of sublevels that will be observed in the presence of a strong electric field; the quantitative splitting will depend on the molecule and on the value of J.*

point vibration and bond stretching due to the rotational motion. Such corrections may be made from the experimental results by using the deviations from the predicted $2B$ spacing of the lines.

19.4.2 / Larger molecules

Linear polyatomic molecules are treated just as described for diatomics; however, since these molecules have a single moment of inertia but at least two bond lengths, more data are required. Consider, for example, the linear molecule OCS, in which two bond lengths must be determined. Measure-

Figure 19.10 / *The relative populations of rotational energy levels of* N_2O *at* 298°K. *Note that the most probable level at this temperature has J about 15.*

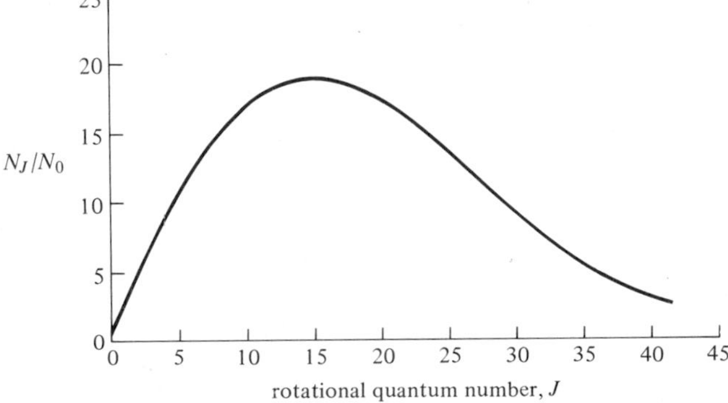

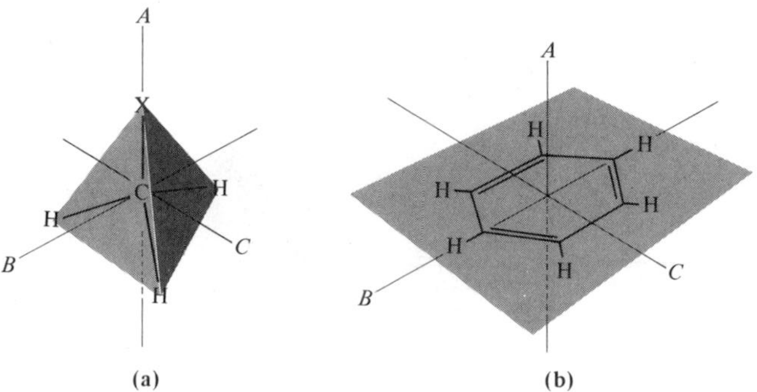

Figure 19.11 / *Two examples of symmetric top molecules for which $I_B = I_C$, but both differ from I_A.*
(a) $I_A < I_B$; *prolate symmetric top.*
(b) $I_A > I_B$; *oblate symmetric top.*

ment of the moment of inertia for two isotopic species can provide the data required to compute these structural parameters. The common species is $^{16}O^{12}C^{32}S$, but $^{16}O^{12}C^{34}S$ is present at 4.1 percent abundance and its $J = 0 \rightarrow 1$ transition is readily measured without isotopic enrichment.

The situation is more complex for a generalized polyatomic species. These molecules may be divided into two classes, symmetric rotors and asymmetric rotors. A symmetric top molecule is one in which the moments of inertia for rotation about two axes are equal, but different from the moment of inertia for rotation about the third, unique axis. In asymmetric rotors, all three moments of inertia are different.

Examples of symmetric top molecules include the methyl halides, CH_3X, and symmetric aromatic molecules such as benzene.[3] In Fig. 19.11 it should be apparent that in both examples the moments of inertia for rotation about the B and C axes, I_B and I_C, are equal, while that for rotation about A, I_A, is unique. The methyl halides represent the case in which $I_A < I_B = I_C$ and are termed *prolate* symmetric tops; the aromatics represent the opposite case, $I_A > I_B = I_C$ and are known as *oblate* symmetric tops.

As with simpler molecules, the total angular momentum of a symmetric top is quantized as discussed above, but in addition, it can be shown that the component of angular momentum about the unique axis is also quantized in accordance with Eq. (19.5),

$$K \frac{h}{2\pi} \qquad K = 0, \pm 1, \pm 2, \ldots, \pm J \tag{19.5}$$

where the positive and negative values of K correspond to clockwise or anticlockwise rotation. It can be shown that the energy of such a molecule

[3] While useful as an example of this type of symmetry, benzene has, of course, no dipole moment, and therefore no microwave spectrum.

is given in ergs by Eq. (19.6) and in reciprocal centimeters by Eq. (19.7).

$$E_{JK} = \frac{J(J+1)h^2}{8\pi^2 I_B} + \left(\frac{h^2}{8\pi^2 I_A} - \frac{h^2}{8\pi^2 I_B}\right) K^2 \tag{19.6}$$

$$E_{JK} = J(J+1)B + (A - B)K^2 \tag{19.7}$$

Here, as before, $B = h/8\pi^2 cI_B$ and $A = h/8\pi^2 cI_A$ while $J = 0, 1, 2, \ldots$ and $K = 0, \pm 1, \pm 2, \ldots, \pm J$. Equation (19.7) corresponds to a rigid rotor and, as in the case of diatomics, small correction terms must be added to account for centrifugal stretching. The selection rules for absorption are $\Delta J = +1$ and $\Delta K = 0$, so the spectrum of a symmetric top is characterized by absorptions at wave numbers $\bar{\nu} = 2B(J + 1)$, just as in the case of a diatomic molecule.

Spectra of symmetric tops may therefore be easily interpreted, and structural data have been obtained for a number of such molecules. Included as examples of dimensions determined from microwave spectral data are CH_3CN and NH_3. In the former, the HCH angle is 109°8′ and the CH, CC, and CN bond distances are, respectively, 1.092, 1.460, and 1.158 Å. In the latter case, the NH bond distance is 1.016 Å and the HNH angle is 107° ± 2°, where the relatively large uncertainty is a result of the rapid umbrellalike inversion vibration that is characteristic of ammonia.

Asymmetric rotors present a much more difficult problem. The total angular momentum of such molecules is quantized and has the same dependence on J as is found in simpler molecules; however, no general expression can be given for the energies of any but the lowest few energy levels of an asymmetric rotor. It can be shown that there will be $2J + 1$ sublevels within each energy level and that the energies of these sublevels will lie between the extremes obtained by artificially distorting the molecule first into an oblate symmetric top and then into a prolate symmetric top. In the past, an estimate of the energy-level diagram for the asymmetric rotor was obtained by interpolating between these extremes using a correlation diagram. In current practice, the energies of interest are rapidly calculated using digital computers. The reader is referred to more specialized works for further discussion of this problem.

The complexity of the energy-level schemes for asymmetric rotors leads to correspondingly more complex microwave spectra, and assignment of such spectra is a problem of some difficulty. Such assignments are facilitated by comparison of observed spectra with the pattern of lines calculated using computer techniques. Occasionally, examination of spectra of molecules that differ only in isotopic substitution is used. Each isotopically different molecule poses a complete new problem in assignment of the spectrum; nonetheless, this approach may be necessary to obtain data sufficient to determine completely the molecular structure. In these ways, the structures of a large number of asymmetric rotors have been determined.

19.5 / ANALYTICAL APPLICATIONS

Despite the task of achieving a complete assignment of the microwave spectrum of even relatively simple asymmetric molecules, the technique has considerable potential as an analytical tool for both qualitative and quantitative analysis. Just as in the case of IR spectroscopy, it proves to be unnecessary to assign the spectral transitions in order to use the data for analytical purposes.

19.5.1 / Qualitative analysis

As has been pointed out above, the microwave spectrum of a molecule is determined entirely by its moments of inertia, which are unique for every species. Such spectra may serve, therefore, as fingerprints for identification of molecules in a complementary manner to analysis by IR. For example, the IR spectra of methylethylketone and diethylketone consist (see Chapter 18) of absorptions due, predominately, to the carbonyl group and methyl and methylene carbon–hydrogen stretching vibrations. It would require some sophistication to distinguish the two spectra unambiguously. The microwave spectra of these molecules are very different, however, and prove easily distinguishable.

Microwave spectroscopy thus provides an absolute method of identifying closely similar molecules. Since the number of lines in a complete spectrum is large, the system is always overdetermined and, in fact, the identification could normally be made by recording only a small portion of the spectrum. To apply this method, all that is required is a catalog of molecular frequencies such as those that have already been prepared for IR spectroscopy. There is no need to wait for the assignment of these frequencies. Identification using microwave spectroscopy is not so well developed as the analogous procedure in IR spectroscopy, primarily because of the unavailability of commercial spectrometers until quite recently. It is a method that seems destined to undergo extensive growth.

The sensitivity of microwave spectroscopy depends on the particular molecular species studied, since the absorption intensity depends on the molecular dipole moment, the particular transition examined, and the frequency of the transition. Typical asymmetric rotors have absorption intensities that indicate that detectability limits between 0.1 and 1 percent should be routinely obtained with current instruments. In particularly favorable cases, much lower limits of detectability have been achieved. Samples required are small, being of the order of a micromole to fill a typical wave guide cell to the usual working pressure of a few hundredths to 0.1 torr. A portion of the low-resolution microwave spectrum of acetone is reproduced in Fig. 19.12 to illustrate the type of spectrum expected.

19.5.2 / Quantitative analysis

Quantitative measurements are usually made on a single line chosen from the spectrum of the analyte. The resolution of the technique is sufficient that

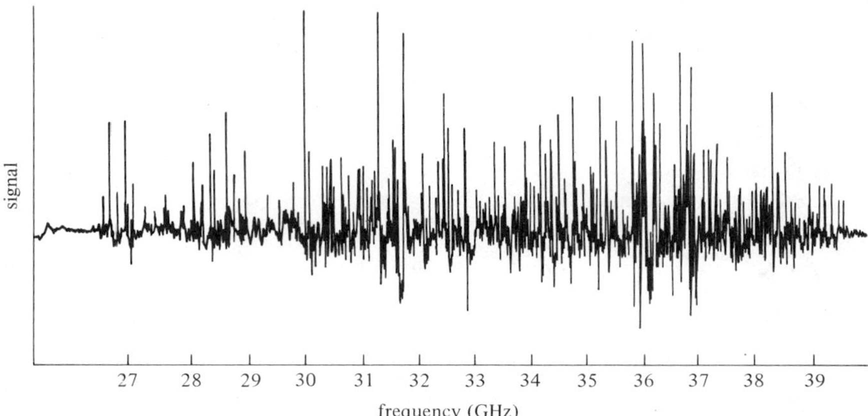

Figure 19.12 / *The microwave spectrum of acetone over the frequency range* 26–40 GHz.
SOURCE: J. T. Funkhouser, S. Armstrong, and H. W. Harrington, *Anal. Chem.* **40**(11), 22A(1968). Reprinted by permission of the American Chemical Society.

isolation of such a line is typically not difficult. A single line from the spectrum of acetone is shown uniquely resolved on an expanded frequency scale in Fig. 19.13.

Several factors suggest that peak areas rather than heights should be used for quantitative measurements. Among these is the fact that microwave lines are subject to pressure broadening. As the pressure of a sample of absorbing

Figure 19.13 / *Expanded presentation of the* 36.934-GHz *line of acetone. The line was scanned in* 100 *sec, and each interval indicated on the abscissa corresponds to* 1 kHz.
SOURCE: J. T. Funkhouser, S. Armstrong, and H. W. Harrington, *Anal. Chem.* **40**(11), 22A(1968). Reprinted by permission of the American Chemical Society.

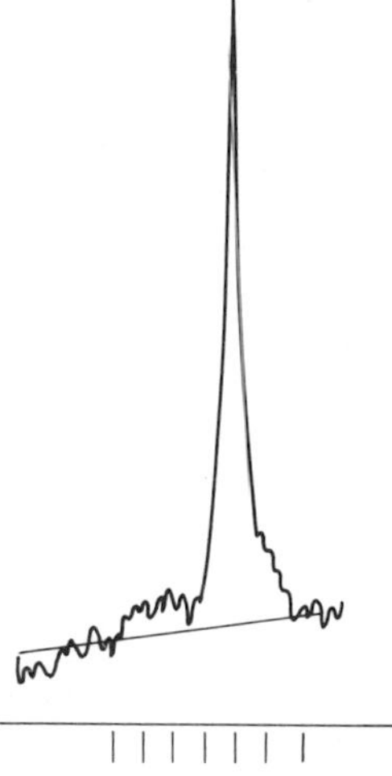

gas is increased, the width of the lines increases in proportion to the pressure over a considerable range. The integrated absorption (area) also increases in proportion to pressure, so that the peak height may remain approximately constant and not reflect the change in sample pressure. In mixtures, each constituent may have its own specific broadening power on the line selected for measurement. In addition, the small transition energies and appreciable thermal population of levels above the ground rotational level cause microwave spectroscopy to share with magnetic resonance (see Chapter 20) appreciable sensitivity to power saturation; peak heights are, as a result, not proportional to applied microwave power except at power levels sufficiently low to preclude saturation.

In the past, microwave measurements were usually made in the so-called Beer's-law region, that is, at power levels below saturation where the signal is found to be linear with the square root of the applied power. This has required working at very low power levels, which leads to very small signals and associated increases in experimental difficulty. Moreover, the Beer's-law coefficient contains the lifetime of the rotational levels as well as concentration, thus mitigating its utility as a quantitative parameter. More recently, it has been shown that quantitative data can be obtained accurately and reproducibly by working at whatever power level is found experimentally to produce the optimum signal. When peak areas are measured under these conditions of optimum power, linear working curves are obtained over a considerable range of total pressure and analyte concentration. In the case of acetone–freon–nitrogen mixtures, successful quantitative analyses for acetone have been demonstrated over a range of 10^{-4} to 10^{-2} torr total pressure with 1–10 percent acetone concentrations. As with the qualitative applications, the quantitative analytical applications of microwave spectroscopy are only now beginning to be exploited and will surely experience growth.

The method should be ideal for continuous monitoring of mixtures of gases and will, therefore, undoubtedly find applications in process control. Since microwave spectroscopy depends so heavily on the masses in the analyte, identification of specific isotopic substitution in samples, even at the natural abundance level for ^{13}C (ca. 1 percent) poses no difficulty. The location of the position of the isotopic substitution within the analyte molecule is a problem to which this technique is particularly well suited. It holds, therefore, considerable potential for the quantitative following of tracer isotopes through a chemical reaction sequence.

SUGGESTED READINGS

D. H. Whiffen, *Spectroscopy*, Wiley (New York, 1966), Chapter 7 (paperbound).

G. M. Barrow, *Introduction to Molecular Spectroscopy*, McGraw-Hill (New York, 1962), Chapters 3, 5, and 7.

W. Gordy, W. V. Smith, and R. F. Trambularo, *Microwave Spectroscopy*, Dover (New York, 1966).

C. Towers and A. Schawlow, *Microwave Spectroscopy*, McGraw-Hill (New York, 1955).
J. E. Wollrab, *Rotational Spectra and Molecular Structure*, Academic (New York, 1967).
W. Gordy and R. L. Cook, "Microwave Molecular Spectra" in *Technique of Organic Chemistry*, A. Weissberger, Ed., Vol. IX, Part 2, Wiley-Interscience (New York, 1970).

PROBLEMS

1 / Calculate the moment of inertia for $^1H^{35}Cl$ given that the equilibrium bond distance is 1.27 Å.

2 / Calculate the moments for inertia for H_2S that contains common isotopes given that the H—S bond distance is 1.34 Å and the HSH angle is 92.2°.

3 / If the equilibrium bond distance in ClF is 1.63 Å, what is the energy of the $J = 5$ rotational level? At what frequency will the transition $J = 4 \rightarrow J = 5$ be observed? Since ^{37}Cl occurs naturally in appreciable abundance, there will be two answers to the questions above. What frequency difference separates the $J = 4 \rightarrow J = 5$ transitions for the two species?

4 / Estimate the *ratio* of populations for the $J = 3$ and $J = 4$ rotational levels of $^1H^{35}Cl$. (Remember to include the degeneracies.)

5 / Given that the $J = 0 \rightarrow J = 1$ transitions for $^{16}O^{12}C^{32}S$ and $^{16}O^{12}C^{34}S$ are, respectively, 12162.98 MHz and 11865.63 MHz, calculate the equilibrium bond lengths. You may wish to consult R. H. Schwendeman, H. N. Volltraner, V. W. Laurie, and E. C. Thomas, *J. Chem. Ed.* **47,** 526 (1970).

chapter 20 / Magnetic resonance spectroscopy

chapter 20

20.1 / INTRODUCTION

Magnetic resonance spectroscopy differs from the ordinary forms of absorption spectroscopy in two fundamental ways. First, the energy levels among which transitions are observed owe, in the usual case, their energy separation entirely to the imposition of an external DC magnetic field. Without such a DC field, no spectroscopy of this type is possible. Second, the transitions observed are nearly always induced by the magnetic field associated with the incident electromagnetic radiation; that is, they are *magnetic dipole*, rather than electric dipole, transitions. Magnetic resonance spectroscopy depends for its existence on the fact that many isotopes of the elements and many subatomic particles possess *magnetogyric* properties. By this we mean that these particles behave, within the limits of a restricted analogy, as if they were tiny, spinning bar magnets. As we shall see, this inadequate analogy has led to the term "spin" to describe the magnetogyric properties. While spin is useful as a code word, it can lead to misconceptions about the physics of the experiments.

When a sample containing nuclei that exhibit spin properties is placed in a DC magnetic field of appropriate strength and is simultaneously irradiated by a much weaker radio frequency magnetic field, the nuclei reveal their presence and identity, and provide a description of their chemical environment. The availability of such information makes the nuclear magnetic resonance, or nmr, experiment of immense importance to a wide variety of scientists ranging from organic chemists to solid state and nuclear physicists.

Electron paramagnetic resonance (epr) or electron spin resonance (esr) both refer to the same experiment, which is basically similar to that just described for nuclei. However, in the case of esr, it is the population of *unpaired* electrons in the samples that is subjected simultaneously to a DC magnetic field and a radio frequency magnetic field, and from which information is then derived. Because the magnetogyric constant associated with the electron is about a thousand times larger than those associated with nuclei, the experimental apparatus for detecting esr signals appears rather different from that used for nmr. The major factor in the apparent differences lies in the use of megahertz radiation in nmr and gigahertz radiation (microwave) in esr, the latter being of the same type as that used in rotational absorption spectroscopy (Chapter 19). Beneath the apparent differences, the two techniques have many similarities, and they are therefore discussed together in the present chapter. The greater emphasis is given to nmr because of the broader scope of its application and its more extensive utilization in analytical problems.

20.2 / MAGNETOGYRIC PROPERTIES

The magnetogyric properties, or spin, of nuclei and electrons are characterized by both an intrinsic angular momentum and a magnetic dipole moment. The terminology "spin" arose from the classical picture of a charged particle spinning about an axis. Such behavior would generate both angular momentum and a magnetic dipole, since a spinning charge is equivalent to a current loop and these are well known from classical electromagnetic theory to generate magnetic fields. It should be noted that these magnetogyric properties were ascribed to nuclei and electrons only as a last resort, as Max Born put it, "under the compulsion of experiment" by Pauli (1924) and by Uhlenbeck and Goudsmit (1925) in order to account for the observed features of atomic spectra. It is, of course, quite ridiculous to conceive of an electron or nucleus as a spinning ball of charge, and such a picture is inconsistent with a number of results of both classical and quantum physics. A discussion of the origin of spin is beyond the scope of this text, and for our purposes it is sufficient to say simply that these are quantum mechanical properties for which no truly good classical analogy exists.

20.2.1 / Initial experiments

The atomic-beam experiments of Stern and Gerlach (1924), in which the magnetic deflection of lithium and silver atoms were observed, proved that spin properties were quantized and permitted measurement of the atomic magnetic moments that arise in these cases primarily from the presence of a single unpaired electron. Later, Stern and coworkers (1933) extended the beam method to H_2 molecules and were able to measure the much smaller magnetic moment of the proton to an accuracy of about 10 percent. The beam method was powerfully improved by Rabi and his group (1939) by introduction of the resonance method. In these experiments, the beam of molecules was focused on a detector by a DC magnetic field and simultaneously irradiated at radio frequency; when the rf was adjusted to a frequency that could be absorbed, the molecular beam was defocused. Thus, the resonant absorption of energy was observed as a sharp decrease in the detector current. Although these elegant molecular beam studies provided the first successful example of nuclear magnetic resonance, we shall not be further concerned with gas-phase work, but rather our attention will be directed to magnetic resonance in bulk matter.

The first successful resonance experiment in bulk matter was an electron resonance experiment reported by the Soviet physicist Zavoisky in 1945. The following year, proton nuclear magnetic resonance was reported essentially simultaneously by two American research groups, those of Edward Purcell at Harvard and Felix Bloch at Stanford. In honor of their work, Bloch and Purcell were awarded the Nobel Prize in physics in 1952. Since that time, magnetic resonance spectroscopy has developed at an almost unbelievable pace. The ubiquitous occurrence of protons in organic chemicals, in addition to the favorable nuclear properties of protons, has caused proton nuclear magnetic resonance to become the single most important field in nmr. It is, therefore, the subject of a large part of this chapter.

20.2.2 / Properties and definitions

As stated above, all the subatomic particles of chemical interest—electrons, protons, and neutrons—possess magnetogyric or spin properties, by which we mean an intrinsic angular momentum and an associated magnetic dipole moment. Both of these quantities are vector quantities; that is, they have both magnitude and direction. They are always collinear vectors, but may, depending on the particular particle, point in the same or opposite directions. The rules of vector addition are obeyed, so when more than one particle is present, the *resultant* (vector sum) angular momentum and magnetic moment are the experimentally observable properties.

Since the fundamental spin property belongs to the subatomic world, the ways in which the individual momenta and magnetic moments combine govern the properties of atoms and molecules. For example, a *pair* of electrons in a chemical bond has no resultant spin, because the Pauli principle requires that their momenta have the opposite sense and the vector sum is zero. The subatomic world is strictly segregated in the sense that this cancellation by pairing occurs only for like particles. Neutrons may cancel neutrons, protons may cancel protons, and electrons cancel electrons, but there is no interparticle cancellation.

The magnetogyric nature of a particle is specified by stating the *spin number*, I, for that species. The spin number multiplied by $\hbar$ (Planck's constant divided by 2π) gives the maximum *measurable* component of the angular momentum for a particular species. Electrons, protons, and neutrons are all spin $\frac{1}{2}$ particles; that is, $I = \frac{1}{2}$ for these species. The nature of the magnetic properties of nuclei depends on the numbers of neutrons and protons, and the following generalizations can be made. Nuclei that have even numbers of both protons and neutrons have no magnetic properties, since all the like particles are paired. Examples of common nuclei without spin are ${}^{4}_{2}He$, ${}^{12}_{6}C$, and ${}^{16}_{8}O$. If *either* the number of neutrons *or* the number of

TABLE 20.1 / *Properties of selected nuclei*

Isotope	nmr *frequency at* 10 kG (MHz)	*Natural abundance* (%)[a]	*Relative sensitivity at constant field (assuming equal numbers of nuclei)*	*Magnetic moment (in nuclear magnetons)*	*Spin, I* ($h/2\pi$)
^{1}H	42.576	99.98	1.00 (reference)	2.79	$\frac{1}{2}$
^{2}H	6.536	1.56×10^{-2}	9.6×10^{-3}	0.86	1
^{6}Li	6.265	7.43	8.5×10^{-3}	0.82	1
^{7}Li	16.547	92.57	0.29	3.25	$\frac{3}{2}$
^{10}B	4.575	18.83	2.0×10^{-2}	1.80	3
^{11}B	13.660	81.17	0.16	2.69	$\frac{3}{2}$
^{13}C	10.705	1.11	1.6×10^{-2}	0.70	$\frac{1}{2}$
^{14}N	3.076	99.64	1.0×10^{-3}	0.40	1
^{15}N	4.315	0.36	1.0×10^{-3}	−0.28	$\frac{1}{2}$
^{17}O	5.772	3.7×10^{-2}	2.9×10^{-2}	−1.89	$\frac{5}{2}$
^{19}F	40.055	100	0.83	2.63	$\frac{1}{2}$
^{31}P	17.236	100	6.6×10^{-2}	1.13	$\frac{1}{2}$

[a] Average abundance of the indicated isotope in naturally occurring materials.

protons is odd, but not both, the nuclei have magnetic properties characterized by spin numbers that are odd integral multiples of $\frac{1}{2}$; that is, $I = n(\frac{1}{2})$, where $n = 1, 3, 5, 7, 9$. Examples of such nuclei include $^{11}_{5}B$, $^{13}_{6}C$, and $^{15}_{7}N$, for which the spin numbers are, respectively, $\frac{3}{2}$, $\frac{1}{2}$, and $\frac{1}{2}$. Note that nuclei in this class, of necessity, occur at *odd* mass numbers. Nuclei that contain odd numbers of *both* protons *and* neutrons possess magnetic properties, but unlike the previous group are characterized by spin numbers that are odd integral multiples of one. Examples include $^{10}_{5}B$, $^{14}_{7}N$, and $^{2}_{1}H$, with $I = 3$, 1, and 1, respectively. These nuclei occur at *even* mass numbers. Properties of some common nuclei are summarized in Table 20.1.

In the absence of an external magnetic field, an isolated magnetic nucleus or electron finds all orientations of its angular momentum vector in space of equal energy. Under such circumstances, the nuclear moment has no mechanism for interaction with its environment, so no particular orientation is preferred or even defined. When a steady (DC) magnetic field is applied, only certain orientations of the angular momentum vector are allowed; thus, the imposed field defines a direction. Now only certain discrete values of angular momentum and magnetic moment are possible. It is conventional to denote the external magnetic field as H_0 and to associate its direction with the positive z direction of a Cartesian coordinate system. In the presence of the DC field, we assert that the total length of the angular momentum vector for a particle of spin number I is $[I(I + 1)]^{1/2}\hbar$; however, the only observable is the z component of this vector which is, of course, its projection on the z axis defined by the external field. Allowed values of this projection are given by $m_I\hbar$, where m_I is the *magnetic quantum number*. This quantum number may take on any of the $2I + 1$ values of the series

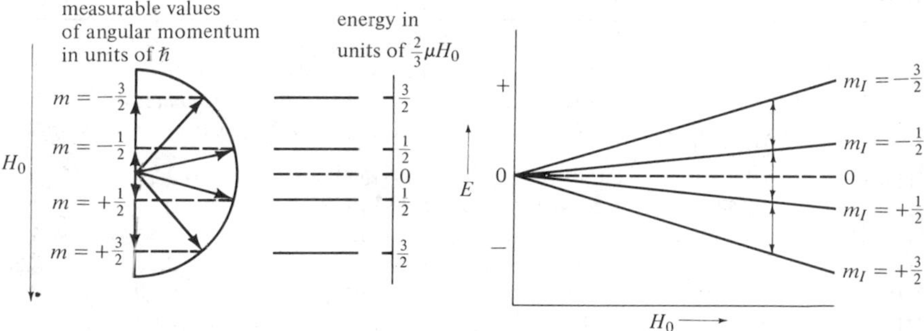

Figure 20.1 / *On the left is a diagram showing schematically the four measurable values of angular momentum for a nucleus having spin number $I = \frac{3}{2}$ in a magnetic field H_0; the corresponding energy states are also shown. On the right the dependence of the energies on H_0 is illustrated. Arrows represent allowed transitions in which $\Delta m_I = \pm 1$. The three indicated transitions all occur at the same energy.*

$I, I-1, I-2, \ldots, -(I-1), -I$. The collinear magnetic moment vector must follow a similar pattern. The maximum measurable component of magnetic moment is usually given the symbol μ. Since it is the physical observable, it is frequently called simply the magnetic moment. The actual magnetic moment vector has length $[(I+1)/I]^{1/2}\mu$. The observable components of the magnetic moment are given by the $2I + 1$ values of $m_I\mu/I$ that form the series $\mu, (I-1)\mu/I, (I-2)\mu/I, \ldots, -(I-1)\mu/I, -\mu$.

The energy of the particle in the external field differs for each state represented by a different value of m_I. The allowed energies are the $2I + 1$ values of $-m_I\mu H_0/I$. Allowed transitions are those for which $\Delta m_I = \pm 1$. As shown in Fig. 20.1, the separation of these states increases linearly with increasing applied field. Note the implications of these statements: The magnetic moment vector can never lie exactly along the direction of the imposed magnetic field; this is a consequence of the uncertainty principle. By virtue of the minus sign in the expression for energy, the lowest energy state is that in which the magnetic moment is, as we might have anticipated, *most nearly* parallel to the imposed field.

20.2.3 / Some numerical quantities

We shall need numerical evaluation of the energy spacing of the various states in magnetic resonance in order to compute the fields and frequencies required for a particular experiment. An electron can have two kinds of magnetic moment, spin and orbital. The *orbital* magnetic moment of the electron was shown to be $e\hbar/2M_ec$ on the basis of Bohr's theory (here e is the electron charge, M_e is the electron mass, and c is the speed of light, all in cgs units). It happens that the spin moment of the electron has almost an identical value, a circumstance that Dirac has shown follows from quantum electrodynamics. Therefore, the fundamental unit of magnetic moment for electrons is $e\hbar/2M_e c = \beta = -0.927 \times 10^{-20}$ erg-gauss^{-1}, which is known as the *Bohr magneton*. Since the charge on the electron is negative, the Bohr magneton is a negative quantity; the reader should be aware that β is often used without sign.

By analogy, one might expect the magnetic moment of a proton to be $e\hbar/2M_p c$, where the charge and mass of the proton are used in the same formula. Note that this quantity is of opposite sign to the electron magnetic moment. This formula does not give the actual magnetic moment of any nucleus; however, it is of the correct order of magnitude and may be conveniently used to represent nuclear moments with the aid of small, dimensionless multipliers, one for each nucleus. This quantity is known as the *nuclear magneton* and is symbolized either μ_0 or β_n. The former symbol is much to be preferred, since the latter is too easily confused with the Bohr magneton. The nuclear magneton has the value 0.505×10^{-23} erg-gauss^{-1}, which is more than three orders of magnitude smaller than the electron moment.

For convenience, at this point we shall define a useful quantity, γ, the *magnetogyric ratio*, which is the ratio of the maximum observable magnetic moment to the maximum observable angular momentum; that is, $\gamma = \mu/I\hbar$. The magnetogyric ratio of a free electron, γ_e, has the value 17.6×10^6 rad-sec^{-1}-gauss^{-1}, while the magnetogyric ratio of the proton, $\gamma_H = 2.76 \times 10^4$ rad-sec^{-1}-gauss^{-1}.

20.2.4 / Magnetic resonance transitions

For a particle of spin number I, it was indicated that there are $2I + 1$ equally spaced energy states and that the energy spacing of these states is $\Delta E = H_0\mu/I$. The frequency of radiation required to induce transitions among such levels is given by the usual relation:

$$h\nu_0 = \frac{H_0\mu}{I} \tag{20.1}$$

It is frequently more convenient to use the nuclear magneton or, for electron resonance, the Bohr magneton. For this reason, we introduce the quantity g, the splitting factor, or simply g factor. For nuclei $g = \mu/\mu_0 I$, and it is a different constant for each different magnetic nucleus. For example, the g factor for the proton is 5.58. In the case of electron resonance, the g factor is 2.00 for a "free" electron, but may vary considerably from that value when there are appreciable orbital contributions to the net magnetic moment. The resonance conditions under which energy may be absorbed by the spin system are given by

$$\begin{aligned} h\nu_0 &= \frac{H_0\mu}{I} = g\mu_0 H_0 \qquad \text{(for nuclei)} \\ h\nu_0 &= g\beta H_0 \qquad \text{(for electrons)} \end{aligned} \tag{20.2}$$

To get some idea of the values involved, we calculate the resonance frequency first for protons and then for electrons in a magnetic field of 5000 gauss.

Substitution of the numerical values for the proton gives the result indicated in Eq. (20.3), or a resonance frequency of 21.3 MHz.

$$\nu_0 = \frac{5.58 \times 0.505 \times 10^{-23} \text{ erg-gauss}^{-1} \times 5000 \text{ gauss}}{6.62 \times 10^{-27} \text{ erg-sec}}$$

$$= 21.3 \times 10^6 \text{ sec}^{-1} \tag{20.3}$$

The calculation for electrons,

$$\nu_0 = \frac{2.00 \times 0.927 \times 10^{-20} \text{ erg-gauss}^{-1} \times 5000 \text{ gauss}}{6.62 \times 10^{-27} \text{ erg-sec}}$$

$$= 14.0 \times 10^9 \text{ sec}^{-1} \tag{20.4}$$

shows that the larger magnetic moment of the electron results in the necessity for a very much higher frequency radiation, 14.0×10^3 MHz, or 14.0 GHz.

Another way of expressing the resonance condition that is frequently encountered in the literature uses the magnetogyric ratio. This is indicated in Eq. (20.5):

$$h\nu_0 = \frac{H_0\mu}{I} = \hbar\gamma H_0 = \frac{h}{2\pi}\gamma H_0 \tag{20.5}$$

Since it is often simpler to work with angular frequency in magnetic resonance problems, we recognize that Eq. (20.5) can be rearranged to

$$2\pi\nu_0 = \omega_0 = \gamma H_0 \tag{20.6}$$

which is often employed as the fundamental equation of magnetic resonance.

20.2.5 / Boltzmann distributions

Since the intensity of a magnetic resonance signal depends on the difference in population between the upper and lower states involved in the transition, it will be of interest to compute the equilibrium differences encountered in typical nuclear and electron resonance transitions. We recall from the discussion in Chapter 10 that the probabilities of absorption and induced emission are equal; and, since the probability of spontaneous emission varies as the cube of the transition frequency, spontaneous emission is a relatively improbable event at these low frequencies. The maintenance of the population difference is therefore crucial to the observation of any net absorption of energy. From Eq. (20.2), we can calculate the transition energies involved and since $kT = 4.14 \times 10^{-14}$ erg at 300°K, a simple Boltzmann equation will give the ratio of populations of upper and lower states for the particles. (The reader should not blithely accept the use of Boltzmann statistics. Spin $\frac{1}{2}$ particles obey Fermi-Dirac statistics, and the use of classical Boltzmann statistics is justified only in the case of low particle density, that is, noninteraction, as is generally the case in fluid media.)

The appropriate numerical constants give $\Delta E = 2.82 \times 10^{-23}$ erg-gauss^{-1} for protons, so the energy separation of the two spin states in a field of

5 kG is 14.1×10^{-20} erg. Since the two states are of equal degeneracy, the appropriate Boltzmann expression is given by

$$\frac{n^*}{n} = \exp\left(\frac{-\Delta E}{kT}\right) = \exp\left(-\frac{14.1}{4.14} \times 10^{-6}\right) = \exp(-3.41 \times 10^{-6}) \qquad (20.7)$$

The simplest way to evaluate such an expression is to recall that e^x may be approximated as $(1 + x)$ when x is small. Taking N as the total number of spins and noting that n^* and n are nearly equal, then, approximately, $n^* = (1 - 3.41 \times 10^{-6})N/2$ and $n = (1 + 3.41 \times 10^{-6})N/2$. Thus, at this field strength in a sample of a million spins, the lower-state population exceeds that of the upper state by only seven spins! This population difference is responsible *in toto* for any resonance signals that may be observed.

For electrons $\Delta E = 1.85 \times 10^{-20}$ erg-gauss^{-1}, and the energy separation of the two states at 5 kG is 9.27×10^{-17} erg. The population ratio is therefore $\exp(-2.24 \times 10^{-3})$ and, following the procedure used for protons, the relative populations are approximately $n^* = (1 - 2.24 \times 10^{-3})N/2$ and $n = (1 + 2.24 \times 10^{-3})N/2$. In a sample of a million electrons, therefore, the lower-state population contains an excess of about 4000.

We shall delve more deeply into the theory of magnetic resonance in Sections 20.5 and 20.6, but first we shall consider the most important topic for the ordinary chemist, chemical applications of proton nmr.

20.3 / CHEMICAL APPLICATIONS OF PROTON NUCLEAR MAGNETIC RESONANCE

In this section, we shall examine the ways in which chemists employ ^1H nuclear magnetic resonance spectroscopy for the identification and structure elucidation of organic compounds. The type of experiment we shall consider is referred to as *high-resolution* proton nmr. The meaning of this terminology should become clear as we proceed with the discussion.

The initial resonance experiments of the Purcell group revealed a single broad absorption for the protons in a block of paraffin. By contrast, a high-resolution spectrum of 1-nitropropane is shown in Fig. 20.2. The three groups of closely spaced peaks arise from protons in the sample, while the single peaks at the left and right ends of the trace arise, respectively, from the solvent and a standard compound added as a reference. The initial part of our discussion will consider the origin of such a spectrum as this one.

20.3.1 / The chemical shift

In 1950 two Stanford physicists, Proctor and Yu, were attempting to make precise measurements of the magnetic moment of ^{14}N using aqueous NH_4NO_3 as the source of nitrogen nuclei. They found, to their surprise, two distinct resonance signals, and thereby accidentally discovered that nuclei can absorb energy under slightly different resonance conditions depending on their chemical environment. Almost at the same time, the

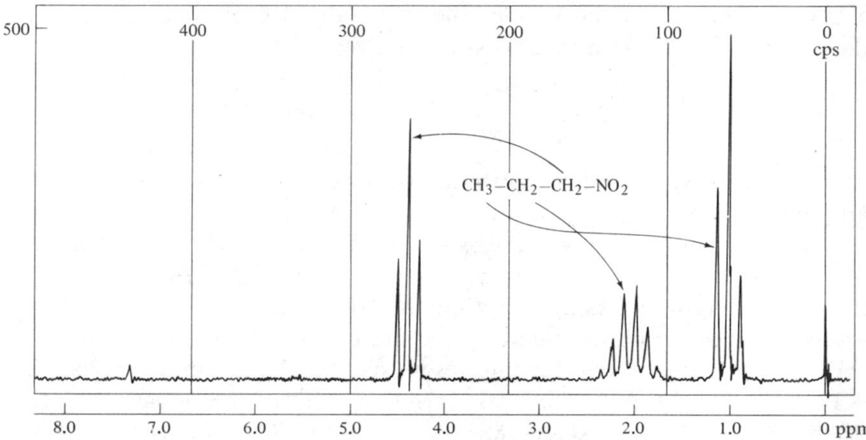

Figure 20.2 / *The proton resonance spectrum of* 1-*nitropropane,* 7 *percent solution in deuterated chloroform. Assignment of peaks to specific protons is indicated in the figure. The small peak at the far left arises from* $CHCl_3$ *impurity in the* $CDCl_3$, *while the peak at the right comes from* $(CH_3)_4Si$ *added as a reference.* SOURCE: Varian Associates, Palo Alto, Calif.

chemical shift was discovered in fluorine compounds in independent experiments by Dickinson. The following year Arnold, Dharmatti, and Packard were able to distinguish three closely spaced, but nonetheless distinct, peaks in the proton resonance of ethanol. The resolution of these peaks, separated by only a few milligauss in a field of several kilogauss, represents the advent of high-resolution nmr. It is, of course, the chemical shift that gives the separation of the multiplets in Fig. 20.2, since there are three chemically distinct types of protons in the compound. The fact that these appear as multiplets is discussed in the next section.

In retrospect, it comes as no surprise that the conditions for a nuclear resonance transition should include specification of the magnetic field *at the nucleus*, not simply the laboratory field strength. As we shall see, the electrons in the compound can interact with the applied field in ways that can either diminish or augment the field at particular nuclei. It is this *shielding* of nuclei by electrons that is the physical basis of the chemical shift. For samples in fluid media, the chemical shift can be described by a dimensionless numerical constant called the *shielding* or *screening constant*.[1]

[1] In the most general sense, σ is a tensor quantity; thus, in solids the shielding is anisotropic and the resonance condition depends on the orientation of the molecule with respect to H_0. In fluid media, rapid, random tumbling of the molecules averages out this anisotropy, giving an average $\bar{\sigma}$. Note also that in solids there are magnetic dipole–dipole interactions among the magnetic nuclei that cause the resonance lines to be, in general, very broad. Rapid, random molecular motion in fluids averages these dipolar interactions to zero. For these reasons, high-resolution nmr is normally possible only with fluid samples, not solid samples.

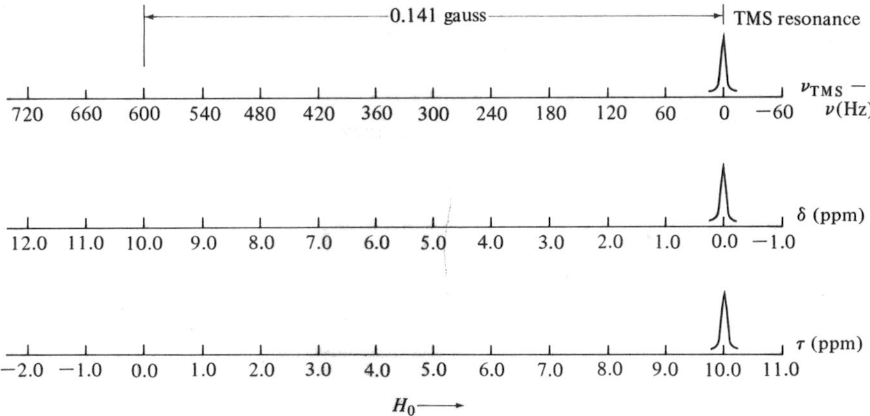

Figure 20.3 / *Comparison of the τ and δ scales of chemical shifts (in parts per million) with frequency displacement (in hertz) from* TMS *resonance for proton resonance at* 60 MHz. *Each unit of τ and δ, or* 60-Hz *displacement, corresponds to a magnetic field change of* 14 mG. *The magnetic field increases from left to right.*

The magnetic field at a nucleus is expressed by

$$H = (1 - \sigma)H_0 \tag{20.8}$$

where, following the usual symbology, σ represents the screening constant. In proton nmr, typical values of the screening constants are in the range of a few to about 16 ppm. Equation (20.8) shows that the absolute shift in magnetic field units depends on the external field strength. Since not all spectrometers operate at the same field strength, shifts are best expressed in parts per million so that the result is independent of the particular instrument.

In principle, it might be desirable to measure all chemical shifts with respect to a bare, unshielded proton; however, the impracticability of such a procedure has led to the use of standard compounds usually dissolved in the solvent along with the sample of interest. While many materials can serve as standards, the most generally accepted standard is tetramethylsilane, $(CH_3)_4Si$, abbreviated TMS. Two scales of chemical shifts based on TMS are in current use. These are the δ and τ scales; the parameters from which these scales derive their names are

$$\delta = \frac{10^6(H_{\text{TMS}} - H)}{H_{\text{TMS}}} \qquad \tau = 10.000 - \delta \tag{20.9}$$

A graphical example for the case of a spectrometer that operates at 60 MHz is given in Fig. 20.3. The resonant field strength at 60 MHz is 14,092 gauss, so each unit on these shift scales corresponds to about 14 mG.

Selection of TMS as the usual standard of choice came about because in addition to desirable physical and chemical properties (*vide infra*), its nmr absorption occurs at a field strength above that for most protons of other organic compounds. For this reason, the δ scale seems the preferable one of the two. A considerable number of compounds have resonance absorptions at $\delta > 10$, and these necessitate the use of negative τ values; use of the former scale requires negative numbers only in the rather limited number of cases for which resonance occurs at fields higher than that for TMS.

When a system of paired electrons is immersed in a magnetic field, orbital motions of the electrons are induced. These motions, identical to current loops, induce magnetic fields that oppose the external magnetic field. This interaction gives rise to the common phenomenon of *diamagnetism*, which causes most ordinary substances to be very slightly repelled by a magnetic field. Diamagnetism leads to the chemical shift because the induced fields oppose, and therefore reduce, the laboratory field at a resonant nucleus. To bring such a nucleus into resonance, the external field must be slightly stronger than it would be otherwise; such a nucleus is described as shielded (from the external field). Intuitively, therefore, we would expect nuclei in regions of high electron density to be more strongly shielded than those in regions of lower electron density. As a first approximation, then, the degree of shielding of protons attached to carbon atoms will depend on the inductive effect of the other substituents attached to the same carbon. Since silicon is less electronegative than carbon, the electron density at the protons in TMS is high, with the result that they are highly shielded and require an unusually high external field to bring them into resonance. The chemical shifts of protons bonded to heteroatoms will, of course, depend on the electronegativity of the particular heteroatom.

A further explanation is required, however, to account for all of the observed chemical shifts in organic compounds. For example, the protons on aromatic rings and aldehydic protons are found at *low* external fields (7–9 ppm below TMS). These protons are strongly deshielded, yet these molecules contain a high density of electrons in multiple bonds. A qualitative explanation of this phenomenon involves looking at the way in which the electrons in the π bonds move. In Fig. 20.4, we see that the circulation of electrons within the π system does, as expected, produce a magnetic field that opposes the external field; however, this opposition occurs in a region of space where the protons in question are not located. Since magnetic lines of force always form a closed loop, on the return side, the induced field *adds* to the external field, thus increasing the field at the protons and causing them to resonate at a smaller applied field than would have been anticipated.

20.3.2 / Spin-spin coupling

Magnetic interaction between groups of nuclei, called spin-spin coupling or spin-spin splitting, frequently results in resonance absorptions appearing

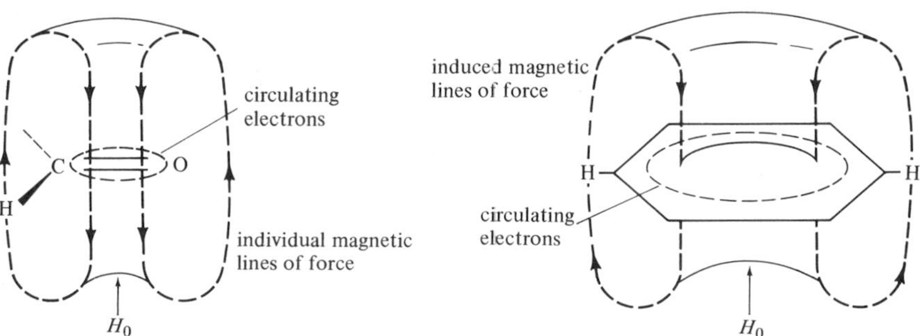

Figure 20.4 / *A qualitative picture of deshielding of aldehydic and aromatic protons by fields induced by electrons in π bonds.*

as multiplets rather than simply a single peak for each group of nuclei with a different chemical shift. The addition of spin-spin splitting complicates nmr spectra somewhat but provides additional information beyond that obtained from chemical shifts alone. The resulting multiplets exhibit characteristic patterns that depend on the number, distance, and symmetry of neighboring nuclei in the molecule and thereby aid in revealing the total molecular structure.

The mechanism of spin-spin coupling involves the electrons in the intervening bonds. The magnetic moments of nuclei in one group polarize the paired electron spins of bonding electrons to give a small net magnetic moment. This induced moment can then interact with a second group of nuclei. This "telegraphing" through bonds, as might be expected, falls off rapidly with an increase in the number of intervening bonds. Spin coupling usually becomes negligible for separations of four or more bonds unless multiple bonds with more delocalized electrons are involved. A quantum mechanical selection rule forbids observation of coupling between identical nuclei, and for these a single peak appears at the unperturbed chemical shift position. To a high order of approximation, spin-spin splittings are independent of the strength of the applied laboratory field, and a spin multiplet can be distinguished unambiguously from several lines with similar chemical shifts by observing the spectrum under two different resonant conditions. As noted previously, shifts are proportional to H_0 and will change, but spin couplings will not. Since spin couplings are independent of the spectrometer operating parameters, they are customarily reported directly in hertz and are denoted by the symbol J_{12}, where the subscripts identify the particular groups of nuclei that are coupled. Such splittings are reciprocal in the sense that if group 1 splits group 2 by 6 Hz, a splitting of 6 Hz will also be observed for the interaction of group 2 with group 1; the coupling is completely specified by a single J value. More concisely, $J_{12} = J_{21}$.

Further understanding of spin coupling can be obtained by inspection of the spectrum in Fig. 20.2. We note that the methyl proton absorption

($\delta = 1.03$) appears as a triplet as a result of spin coupling to the adjacent methylene group. The methylene group next to the nitro group (strongly electronegative) has its resonance shifted to low field ($\delta = 4.38$) since it is strongly deshielded. It, too, appears as a triplet due to coupling with the central methylene protons. The central methylene group resonance appears near the normal position for alkyl methylenes ($\delta = 2.07$) and is spin coupled to both the methyl protons and the protons of the first methylene group. We shall see that 12 peaks are predicted in this multiplet, but of these only six are resolved.

In a group of n equivalent nuclei of spin I, the total magnetic quantum number, m_I, may take on values that are the sums of the individual allowed values; there are $2nI + 1$ such values possible. Each of these possibilities has a different z component of magnetic moment that can be "telegraphed" to neighboring spins. A spin multiplet will therefore contain $2nI + 1$ peaks if they can be resolved. Two protons split an absorption by an adjacent group into a triplet $[(2 \times 2 \times \frac{1}{2}) + 1 = 3]$, while three protons split an adjacent absorption into four peaks. For $I = \frac{1}{2}$ nuclei, this generalization can be simplified to say merely that spin coupling results in one more peak than the number of nuclei that cause it. The central methylene group in 1-nitropropane is spin coupled to two groups, a methyl group that splits it into a quartet and a methylene group that splits each of these peaks into three, for a total of 12 peaks under ideal resolution conditions. Frequently, as shown in the figure, such multiple couplings result in peaks that fall on top of one another (they are said to *overlap*) so that a smaller number is observed than predicted.

As already indicated, the energy differences among states of various m_I values are so small that the states differ in population by trivially small amounts. Therefore, all the lines in a multiplet that result from coupling to a single nucleus are equally intense. If several magnetically equivalent spin $\frac{1}{2}$ nuclei couple their spins to a neighbor, the intensity of each line in the multiplet is proportional to the statistical probability of a particular value of the total magnetic quantum number in the group causing the splitting. As shown below, these statistical weights, or *degeneracies*, follow the coefficients of a common binomial expansion, the numbers that form Pascal's triangle. In Fig. 20.5(a), the multiplet structure of the spectrum of 1-nitropropane is analyzed. We represent the individual spins by arrows. For $I = \frac{1}{2}$ nuclei, there can be only two possible values of m_I, which are represented by the arrow pointing either up or down. For the case of two spins, both may be up or both may be down, but in either configuration there is one and only one combination of the two spins. By contrast, the situation that has one spin each way is *twice as likely* since the two nuclei are indistinguishable, and a up and b down is totally equivalent to a down and b up. As a result, the center peak of the triplet has twice the intensity of the outer peaks. Since the center peak arises from a combination of coupled protons that has zero magnetic moment, it is unshifted from the

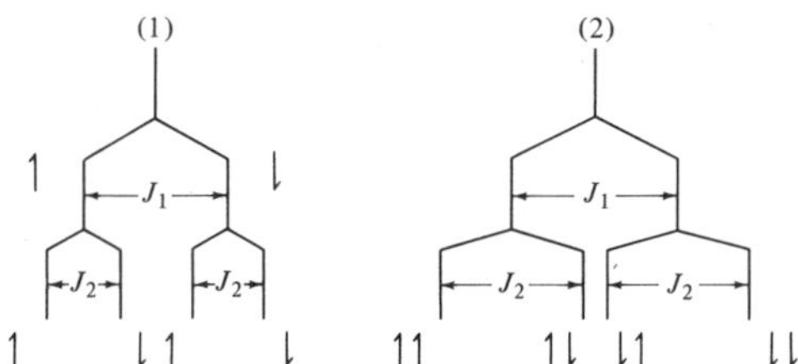

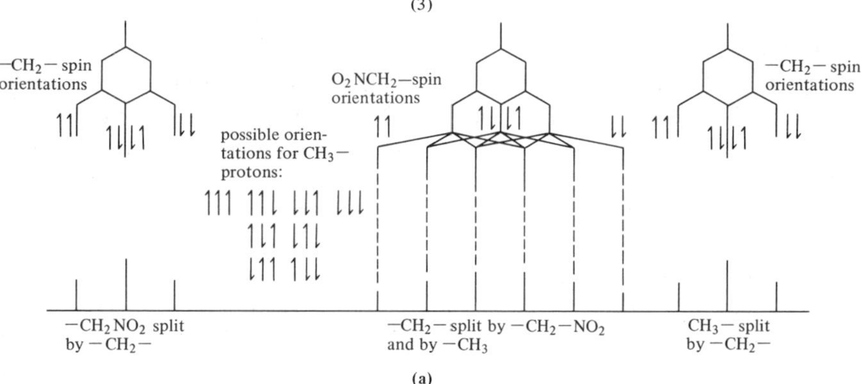

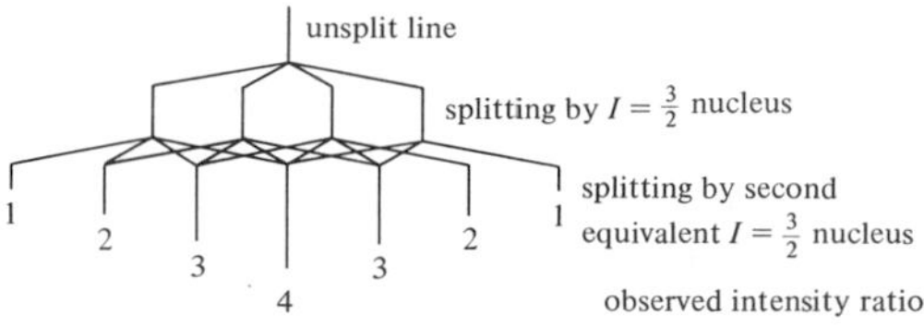

Figure 20.5 / (a) (1) *Pattern for spin coupling of a resonance line with two nonequivalent spin $\frac{1}{2}$ nuclei, $J_1 > J_2$. The two possible orientations of nucleus 1 give a doublet separated by J_1; the two orientations of nucleus 2 further split these lines into two, resulting in a four-line pattern with equal intensities.* (2) *When $J_1 = J_2$ (equivalent nuclei), the central two lines overlap exactly (shown separate for clarity), giving a triplet with* 1:2:1 *intensities.* (3) *Analysis of the spectrum of* 1-*nitropropane (Fig. 20.2). The methyl and low-field methylene resonances are split into triplets by the two central methylene protons. The resonance of the latter group is split into a triplet by the —CH_2NO_2 protons; each of these three lines is further split into a* 1:3:3:1 *quartet by the three methyl protons. Incomplete resolution of these 12 lines gives the observed six-line multiplet.*

(b) *Estimation of the splitting for a proton coupled to two equivalent nuclei of $I = \frac{3}{2}$.*

spectral position corresponding to the chemical shift for that group; chemical shifts are measured from the center of multiplets of this type. The three protons of a methyl group split a neighboring group's signal into four lines of intensity ratio 1:3:3:1. As seen in Fig. 20.5(a), there are three equivalent combinations in which one spin is up and two are down and three ways to obtain the one-down and two-up combination. Four protons produce a splitting with intensities 1:4:6:4:1, and so on, following the binomial coefficients.

Should one forget the binomial coefficients, the intensity ratio for any coupling pattern can easily be predicted by drawing the splitting pattern as indicated in Fig. 20.5. Although it is an uncommon happenstance in proton nmr, coupling to nuclei of $I > \frac{1}{2}$ may occur. When more than one such nucleus couples, the intensity ratio of the multiplet is determined by drawing the splitting pattern as shown in Fig. 20.5(b) for two nuclei of $I = \frac{3}{2}$. Only for spin $\frac{1}{2}$ nuclei do the intensity ratios follow the binomial expansion coefficients.

The remarks above concerning spin coupling apply strictly only to *first-order analysis* of an nmr spectrum. First-order analysis applies to spectra in which the resonance frequencies of the coupled nuclei differ by amounts that are large (resulting from large differences in chemical shift or coupling to a different nuclear species) compared to the spin-spin coupling constant. For the purposes of this comparison, the ratio $\Delta\nu/J$ is employed in which both quantities are expressed in hertz. When the value of the ratio is 15 or larger, strict first-order behavior is expected. In the region $5 < \Delta\nu/J < 15$, that is, where the spin coupling is small but not negligible in comparison to the difference in shifts, small variations in the first-order intensities may occur. In addition, some lines of the multiplets may be split further (called *second-order splitting*), although the additional splittings may not always be resolved. As a general rule, two groups of coupled nuclei give multiplets whose intensities are perturbed in such a way that the inner lines are augmented and the outer lines are weakened. This effect does not generally interfere with the straightforward interpretation of the spectra. Such an effect can, in fact, be noticed in Fig. 20.2 and accounts for the fact that the inner members of both triplets are somewhat more intense than the outer members; multiplets that arise from mutual spin-spin coupling "slant" toward each other, which is frequently an aid in interpretation. Under these conditions, corrections of various orders can be calculated for the line separations and intensities using perturbation theory; however, such calculations are not necessary to interpret spectra and are beyond the scope of this text.

When the ratio of chemical shift difference to spin coupling constant falls below 5, the resulting multiplet is significantly perturbed from its first-order appearance. The simplest illustration of this effect is shown in Fig. 20.6. Two doublets are anticipated for two coupled protons with rather different chemical shifts, as shown in the left-hand portion of the figure. As the

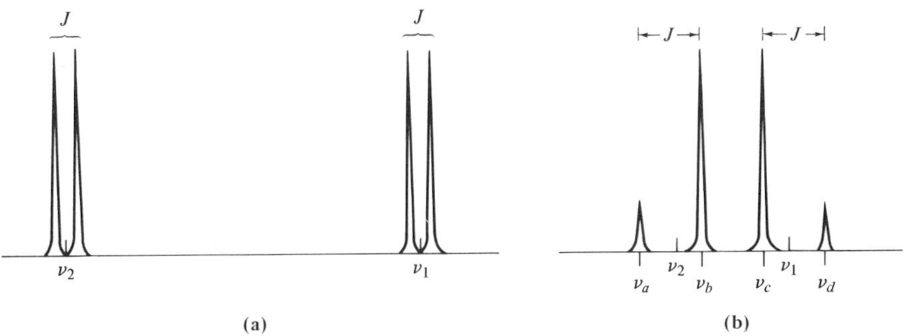

Figure 20.6 / *The change in the splitting pattern for two spin coupled protons as $\Delta\nu/J$ changes from* ~20 (a) *to* ~5 (b).

chemical shift decreases, the inner lines are enhanced and the other lines are diminished until the multiplet appears as shown in the right-hand portion of the figure. Note that while the coupling constant can be measured directly from the peak separation, the chemical shifts are no longer determinable from points midway between pairs of peaks. The correct points for measurement can be determined using Eq. (20.10):

$$\nu_a - \nu_c = \nu_b - \nu_d = [(\nu_1 - \nu_2)^2 + J^2]^{1/2} \tag{20.10}$$

From this relationship one determines that the lines of the multiplet will be equally spaced ($\nu_a - \nu_c = 2J$) when the chemical shift difference is $J/\sqrt{3}$. Near this ratio, it is quite easy to confuse such a multiplet with a quartet, which arises from coupling to three equivalent protons. As the chemical shift difference decreases further, the outer peaks disappear and the center peaks coalesce, eventually giving a singlet when the shift difference vanishes.

Except for relatively simple cases such as the example above, multiplets that involve small values of $\Delta\nu/J$ are difficult to unravel. One procedure involves solving the spin hamiltonian for the coupled system to obtain the transition frequencies and intensities, which may then be compared with the experimental spectrum. The details of this tedious procedure, which is frequently aided by computer techniques, are beyond the scope of this text. As we shall discuss further below, the use of the largest possible magnetic field, which leads to maximization of chemical shift differences, is the most desirable solution to this problem. As the field is increased, the fraction of spectra that yield to first-order analysis increases dramatically.

20.3.3 / Interpretation of spectra

In this section, we shall illustrate with a few selected examples methods of interpretation of nmr spectra. Of course, only rarely will the chemist be faced with a problem of identification in which the nmr spectrum is the only available datum. Nonetheless, unequivocal identification can sometimes be made on the basis of nmr data alone when simple molecules are involved.

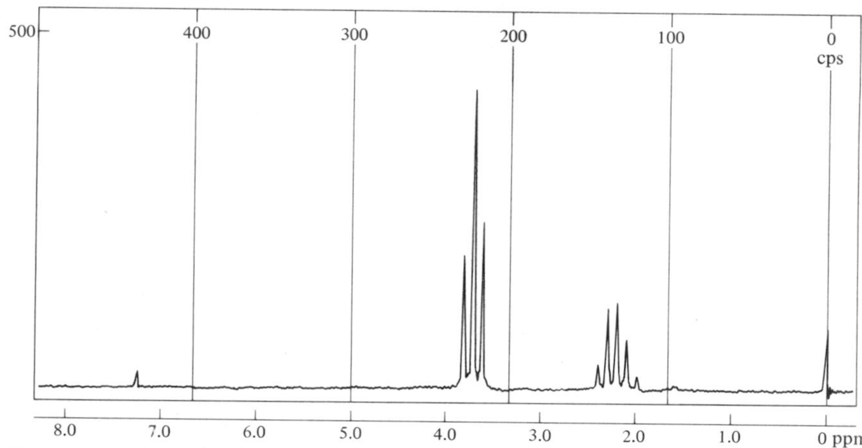

Figure 20.7 / nmr *Spectrum obtained at* 60 MHz *using a deuterated chloroform solution containing* TMS. *Sample is present at 7 percent by weight.*
SOURCE: Varian Associates, Palo Alto, Calif.

In addition to the chemical shift and spin coupling data, the area under a properly recorded nmr multiplet is directly proportional to the number of nuclei from which it arises. For this reason, most nmr spectrometers are equipped with electronic integrators that permit direct determination of the area under each peak or multiplet. Assuming that all the peaks are recorded under the same experimental conditions, the ratios of the integrated areas of the peaks give immediately the *ratios* of the numbers of different nuclei. Note that the absolute numbers of nuclei are not available and must be deduced from other data or, in some cases, from chemical intuition.

Consider the nmr spectrum shown in Fig. 20.7. We observe two coupled multiplets which are, respectively, a 1:2:1 triplet and a 1:4:6:4:1 quintet. Since the triplet arises from spin coupling to two protons, a methylene group is a reasonable postulate. The quintet requires coupling to four equivalent protons, which might easily be two identical methylene groups. If we are told that the sample is an alkane that contains chlorine, then it is easy to deduce the structure $Cl—CH_2CH_2CH_2—Cl$, 1,3-dichloropropane. The quintet is due to the central methylene group and is not shifted from the normal range of alkyl methylene resonance, whereas the triplet from the terminal groups is significantly deshielded by the electronegative chlorine substituents.

A second spectrum that is rather simple to interpret is shown in Fig. 20.8. We observe a singlet, septet, and doublet; the latter two are, of course, spin coupled. The chemical shifts of the coupled multiplets suggest the possibility of a methyl resonance ($\delta = 1.25$) and a methine resonance ($\delta = 2.90$). These ideas fit with the spin coupling pattern if we assume that an isopropyl group is present. The six equivalent methyl protons split the single methine proton into seven peaks, and it, in turn, splits the methyl resonance

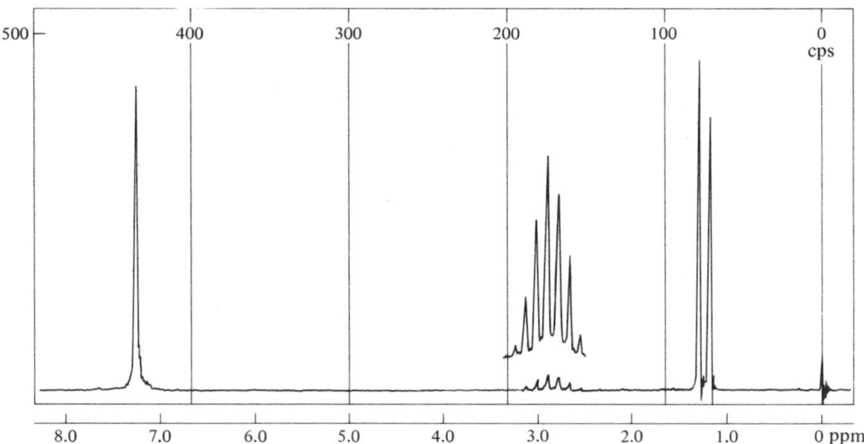

Figure 20.8 / nmr *Spectrum obtained at* 60 MHz *using a deuterated chloroform solution containing* TMS. *Sample is present at 7 percent by weight. The low-field singlet obscures the* $CHCl_3$ *peak. The second trace of the septet was run at higher gain and concentration.*
SOURCE: Varian Associates, Palo Alto, Calif.

into a doublet. The low-field peak is characteristic of phenyl protons. The fact that it appears as a singlet suggests a benzene ring with a substituent with relatively small inductive effect, since the difference in chemical shifts among the ring protons must be small in order to obtain a singlet. Thus, this spectrum is reasonably assigned to isopropyl benzene.

In Fig. 20.9 we note that three singlets appear. We must account, therefore, for three chemically distinct types of protons placed so that they are not coupled. The integrals show that, from low to high field, the relative peak areas are 4 : 1 : 9. The extreme low- and high-field peaks occur at chemical shift positions that suggest phenyl and methyl protons, respectively. To accommodate the third peak, a disubstituted benzene must be postulated that is consistent with four phenyl protons. Nine methyl protons are best explained by substitution of a tertiary butyl group. If we are also privy to the fact that the compound contains sulfur, the remaining peak can be explained by a —SH substituent. The singlet phenyl peak strongly suggests that the substitution is 1,4; the compound is actually 4-*t*-butylthiophenol.

At the outset in interpreting the spectrum given in Fig. 20.10, let us suppose that we know that two heteroatoms, chlorine and oxygen, are present. Integration of the three multiplets gives ratios of 4 : 2 : 3 protons, respectively, from left to right. The quartet and triplet should be immediately recognized as characteristic of an ethyl group. The methylene quartet is deshielded considerably, which indicates that it is bound to oxygen. The low-field multiplet that integrates for four protons is characteristic for a disubstituted benzene ring. This particular pattern is diagnostic for 1,4

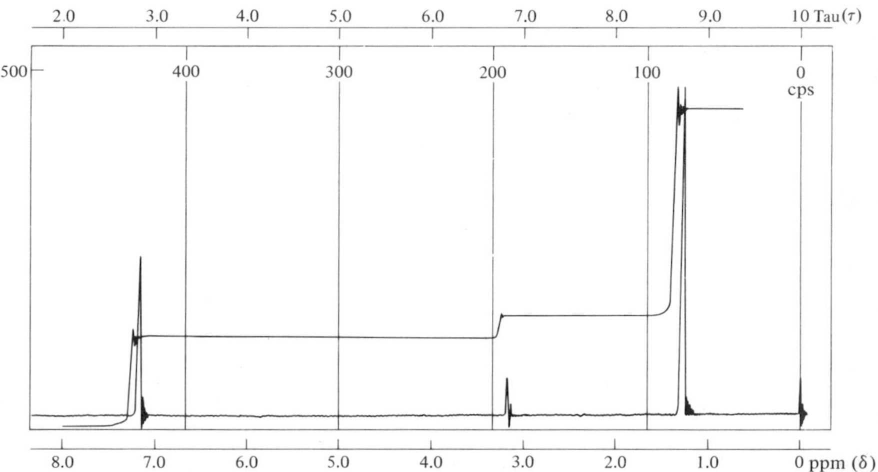

Figure 20.9 / nmr *Spectrum obtained at* 60 MHz *using a carbon tetrachloride solution containing* TMS. *The second trace is the output of the integrator, and the distance between successive flat portions of this trace is proportional to the peak area.*
SOURCE: © Sadtler Research Laboratories, Inc.

disubstitution with substituents that both perturb the ring significantly and are themselves inductively dissimilar. The sum of this information leads to assignment of this spectrum to 4-ethoxychlorobenzene.

To interpret the spectrum of Fig. 20.11, let us begin with the empirical formula, C_3H_8O. Inspection of the spectrum reveals that there are four

Figure 20.10 / nmr *Spectrum obtained at* 60 MHz *using a carbon tetrachloride solution containing* TMS. *As in the preceding figure, the second trace represents the integral of the peaks.*
SOURCE: © Sadtler Research Laboratories, Inc.

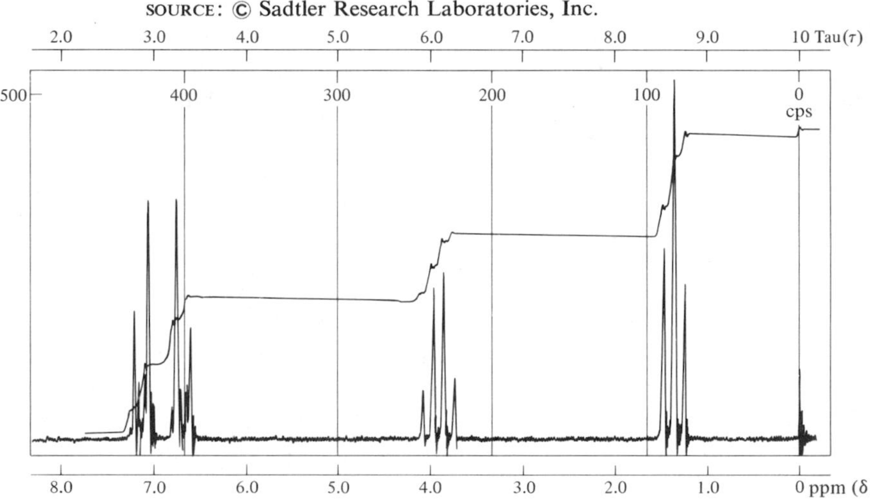

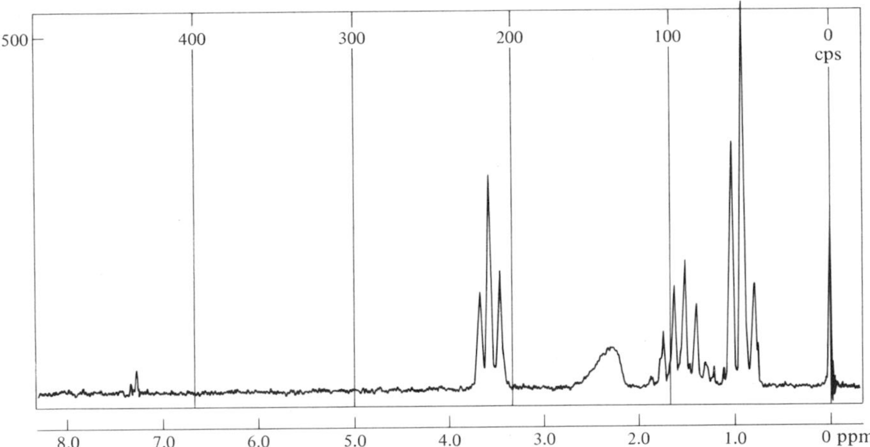

Figure 20.11 / nmr *Spectrum of* C_3H_8O *at* 60 MHz *using a* 7 *percent solution in deuterated chloroform containing* TMS.
SOURCE: Varian Associates, Palo Alto, Calif.

chemically distinct types of protons and three groups show spin coupling. The triplet at $\delta = 0.92$ suggests a methyl group spin coupled to a neighboring methylene, while the triplet at $\delta = 3.58$ could be a methylene group coupled to a second methylene and also bonded to oxygen to account for the observed deshielding. A logical assignment for the multiplet centered at $\delta = 1.57$ is a methylene group between a methyl and a methylene with approximately equivalent spin coupling to both. So far, we have postulated $CH_3CH_2CH_2O$—, and assignment of the broad singlet at $\delta = 2.3$ to an hydroxyl proton completes the assignment of this spectrum to *n*-propanol. It should be noted that exchangeable protons, such as the hydroxyl protons of the lower alcohols, give resonances for which both position and width are quite variable. These protons are subject to hydrogen bonding, which affects the resonance, as does the extent (solvent dependent) and the rate of exchange (*vide infra*).

Two further spectra are included to illustrate specific points. The spectrum of allyl alcohol in Fig. 20.12 demonstrates dramatically that it is not necessary to have large molecules to obtain complex nmr spectra. All that is necessary is a small number of inequivalent protons that are spin coupled. Second-order splittings are clearly observable in this spectrum, and we note the sharp singlet for the hydroxyl proton. In this case, rapid exchange decouples the hydroxyl proton from the methylene protons, and the exchange rate is sufficiently high that the peak is not broadened. The same comment may be made regarding the hydroxyl proton in the spectrum of tetrafluoropropanol shown in Fig. 20.13. This spectrum illustrates the effect of spin coupling of protons to a different species of magnetic nuclei. Coupling of protons to ^{19}F is particularly prominent in comparison to other common

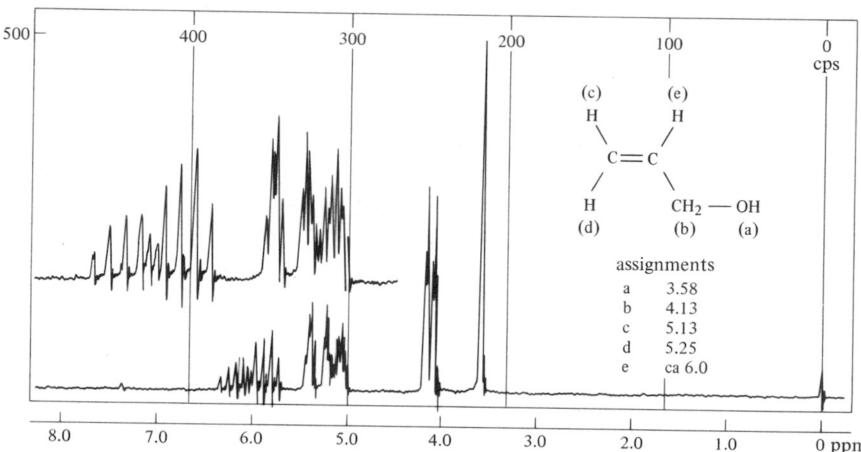

Figure 20.12 / nmr *Spectrum of allyl alcohol obtained at* 60 MHz *using a 7 percent solution in deuterated chloroform containing* TMS. *The low-field section is reproduced at higher gain with the x axis expanded in the upper trace.*
SOURCE: Varian Associates, Palo Alto, Calif.

heteroatoms. The low-field multiplet contains nine resolved peaks as a triplet of triplets. This results from coupling of the terminal hydrogen to both pairs of fluorine atoms with, of course, the larger coupling due to the fluorines attached to the same carbon atom. The methylene group is also coupled to both sets of fluorines, with the coupling from the farther pair just visible, but incompletely resolved. As in the case of coupled like nuclei,

Figure 20.13 / nmr *Spectrum of tetrafluoropropanol obtained at* 60 MHz *using a 7 percent solution in deuterated chloroform containing* TMS. *The small peak at* $\delta = 3.48$ *is due to an impurity.*
SOURCE: Varian Associates, Palo Alto, Calif.

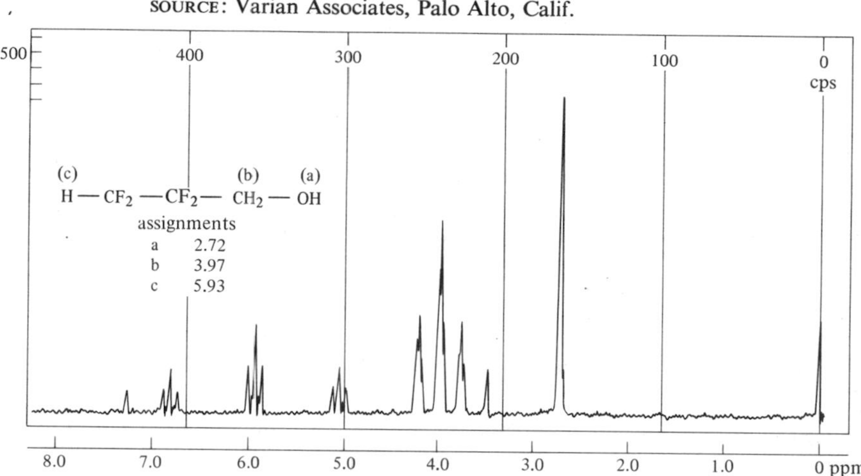

the protons also split the fluorines, but this is not observed unless the resonance conditions are modified to suit the magnetogyric properties of fluorine; that is, it is observed only via fluorine resonance experiment.

Chemical shifts anticipated for a variety of protons in organic molecules are summarized in Table 20.2.

20.3.4 / Preparation of samples

High-resolution nmr spectra are obtained in fluid media. The usual sample, a neat liquid or a solution in a suitable solvent, is contained in a glass tube, most often of 5 mm o.d. The tubes are constructed of thin glass to permit the maximum volume of sample to occupy the sensitive region of the spectrometer probe. Since the tube will be spun about its long axis, high-quality nmr tubes are ground to obtain axial symmetry, which prevents irregular spinning. Ordinarily, the sample consists of 0.5 ml of neat liquid or the same volume of solution in the concentration range 1–10 percent by weight. Sample weights of the order of 5–50 mg are required to meet these conditions. Specialized designs of tubes and spacers which are inserted in the tubes are available to increase sensitivity by restricting the sample to the sensitive area of the probe. With these devices, useful spectra can be obtained from samples of 1 mg in favorable cases. Instrumental techniques for sensitivity enhancement are also available and are discussed in Section 20.6.4.

To obtain optimum resolution, it is usual practice to degas the sample *in vacuo* and then seal the tube with a torch. This is done to remove dissolved atmospheric oxygen. Oxygen is a paramagnetic molecule (O_2 has a triplet ground state), and this relatively strong magnetism increases the relaxation rate (Section 20.5) of the protons, thereby broadening the resonance lines somewhat.

Because of the possibility of intermolecular interactions that may perturb the spectra, it is generally regarded as the best procedure to utilize ca. 10 percent solutions even if the sample is a liquid; solids must, of course, be dissolved. An ideal solvent would contain no protons and would be chemically unreactive toward the sample. Carbon tetrachloride and carbon disulfide fulfill these conditions, but unfortunately offer insufficient solubility for a majority of organic compounds. The most generally useful nmr solvent appears to be deuterated chloroform. It dissolves a wide range of compounds at acceptable concentrations and, since it contains relatively little deuterium, is not overly expensive. For specialized applications, almost all common solvents are now available commercially in the deuterated form at isotopic purity levels of 98+ percent. Salts and other species soluble in aqueous media can be run in D_2O, which is now in plentiful supply at moderate cost.

In addition to solvent and sample, it is necessary to add a reference standard with respect to which chemical shifts are measured. As already noted above, the usual reference is tetramethylsilane. In addition to its desirable resonance properties, TMS is soluble in most organic solvents, chemically

Table 20.2 *Characteristic nmr spectral shifts for protons in organic molecules*

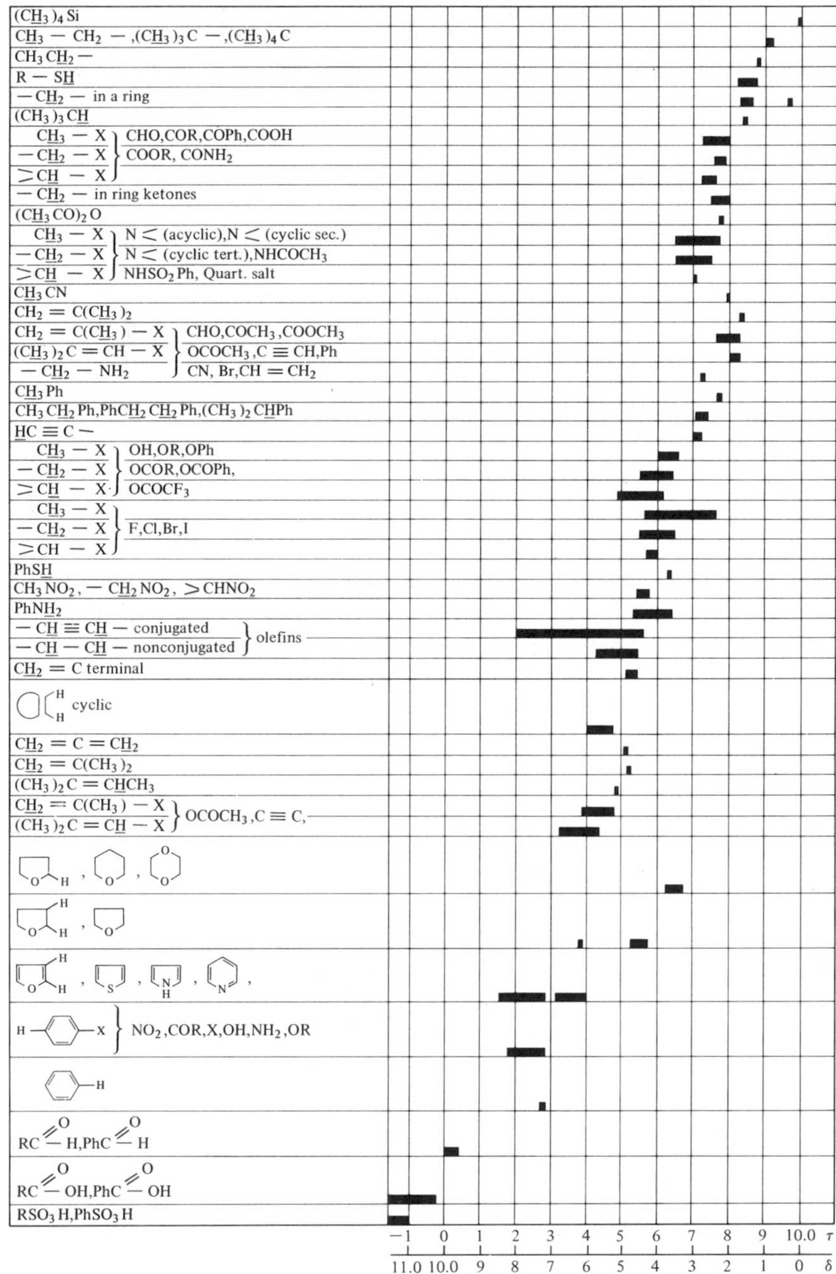

Source: E. Mohacsi, *J. Chem. Ed.* **41,** 38 (1964).

inert, and sufficiently volatile (b.p. 27°C) that it is easily removed. This latter point is frequently important when scarce or valuable samples are involved, since that ability to remove the solvent and TMS renders the nmr method wholly nondestructive. Although TMS is not soluble in water, the related compound, sodium 2,2-dimethyl-2-silapentane-5-sulfonate, $(CH_3)_3SiCH_2CH_2CH_2SO_3Na$ (DSS), is soluble in water and is frequently used as the standard in such experiments. The compound is used at low concentration so that the methyl proton singlet is observed, but spin coupled pattern from the methylene groups does not interfere with the spectrum of the sample. Unless specific chemical effects are present, shifts measured with respect to TMS in chloroform will be within a few hundredths of a part per million for the same peaks measured with respect to DSS in water. In some cases, of course, a suitable solvent peak that does not interfere with the spectrum of interest may be used as the internal standard; the benzene peak could serve as a reference for compounds that contain no phenyl protons.

20.4 / RESONANCE OF OTHER NUCLEI

As indicated in Table 20.1, many nuclei other than protons yield magnetic resonance spectra. Without exception, however, the intrinsic sensitivity of the experiments with other nuclei is inferior to that of proton resonance. The list of isotopes from which resonance has been observed is, of course, much longer than that in the table, but included there are the five nuclei of greatest interest to chemists: protons, fluorine, boron, phosphorus, and carbon. The relative amount of experimental work devoted to each of these nuclei has depended on two factors, the underlying chemical interest in the particular nucleus and the experimental difficulty of obtaining spectra.

Factors that affect the difficulty of the experiment include the magnitude of the nuclear magnetic moment; the spin number; and the abundance of the particular isotope. Considering these factors together, ^{19}F is the most suitable nucleus for high-resolution nmr following ^{1}H. Like ^{19}F, ^{31}P is a 100 percent-abundant spin $\frac{1}{2}$ nucleus, and is the third in the list of nuclei suited for high-resolution nmr. ^{11}B, with a large nuclear moment and 81 percent abundance, ranks next, its principal drawback being that $I = \frac{3}{2}$. Nuclei with spin numbers greater than $\frac{1}{2}$ possess electric quadrupole moments, which cause broadening of the resonance lines. As we shall see, ^{13}C resonance is of great interest, and the principal difficulty with this isotope is its low natural abundance. All of the nuclei heavier than protons exhibit much larger chemical shifts than the $\sim$10 ppm range observed for ^{1}H. This fortunate fact means that the stringent requirements of magnetic field homogeneity over the sample volume are relaxed somewhat. Larger samples can therefore be used with these nuclei, thereby partially compensating for the lower intrinsic sensitivity of the experiments. Some of the salient features of nmr of the heavier nuclei are now reviewed briefly.

20.4.1 / Fluorine resonance

Fluorine is monovalent and capable of forming a wide variety of inorganic salts and complexes, covalent compounds with numerous elements, and organic compounds in which it simply substitutes for hydrogen. Fluorine's nmr sensitivity is good, and since the ^{19}F resonance frequency is close to that for ^{1}H with the same applied magnetic field strength, the experiment is not difficult. In some instruments, the radio frequency oscillator may be tuned down from proton to fluorine frequency with no other adjustments required. Many common proton spectrometers that operate at 60 MHz also accommodate fluorine resonance at 56.4 MHz.

Fluorine resonance spectra appear similar in form to proton resonance spectra, since the approximately 10-fold increase in the range of chemical shifts is accompanied by an increase in spin coupling constants. These facts are associated, in part, with the availability of *p* orbitals on fluorine; by contrast, hydrogen bonds through the 1*s* orbital only. Fluorine nmr finds increasing application in structure determination and physicochemical studies of both inorganic and organic compounds. The use of fluorine-containing ligands in coordination chemistry has added to the utilization of ^{19}F nmr. Fluorine atoms may sometimes be intentionally substituted for hydrogen in organic molecules in order to take advantage of the larger chemical shifts, which simplify analysis of conformational and motional problems.

There is, unfortunately, no universally accepted standard compound to which ^{19}F chemical shifts are referred. F_2 has been used as the standard in inorganic applications (external!), while workers in organofluorine chemistry have used trifluoroacetic acid, hexafluorobenzene, 1,1-difluoroethylene, and fluorotrichloromethane. There seems to be some attempt to standardize on the last compound as reference for fluorine chemical shifts.

Some typical shift values for inorganic fluorine compounds are given in Table 20.3. F_2, the reference compound, was measured at 25 atm in the gas phase, while the other compounds were observed as pure liquids. The structure of nitryl fluoride, FNO_2, was confirmed by fluorine nmr. The fluorine nmr spectrum consists of a triplet of equal-intensity lines that arise from spin coupling to the ^{14}N nucleus ($I = 1$) and the coupling constant, 113 Hz, which is similar to that found in NF_3, leaves little doubt that the fluorine is bonded directly to nitrogen rather than to oxygen.

The chemistry of synthetic organofluorine compounds has become an active field of research (it is interesting to note that natural products containing fluorine are so exceedingly rare as to be nonexistent for practical

TABLE 20.3 / *Chemical shifts for inorganic fluorine compounds* (ppm)

F_2	0	CF_4	491.0
NF_3	285.0	BF_3	555.5
SF_6	375.6	HF	625.0

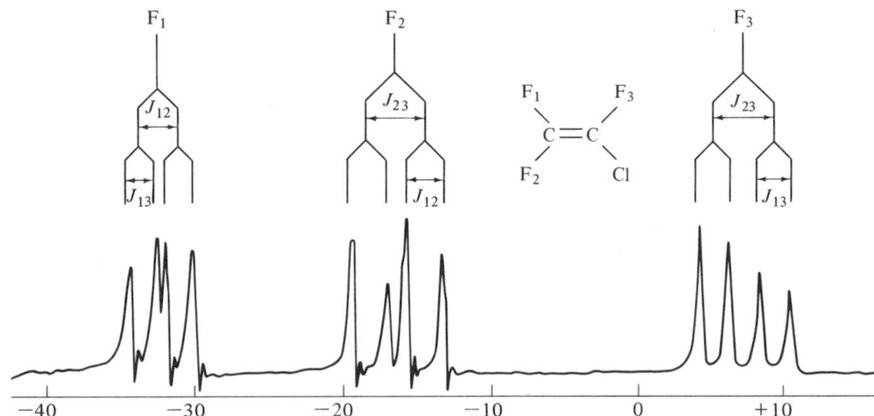

Figure 20.14 / *The* ^{19}F *spectrum of* $CF_2{=}CFCl$ *at* 30 MHz. *Chemical shifts are shown in parts per million relative to perfluorocyclobutane. Spin coupling constants are* $J_{12} = 78$, $J_{13} = 58$, *and* $J_{23} = 115$ Hz. *As with analogous hydrogen compounds, the* trans *spin coupling is larger than the* cis.
SOURCE: Varian Associates, Palo Alto, Calif.

purposes) and, just as with proton resonance, nmr is an important tool for structure determination and investigations of bonding. In addition, comparison of ^1H and ^{19}F spectral data aids the theoretical understanding of chemical shifts and spin coupling.

Although complex ^{19}F spectra do arise, the increase in chemical shifts over those for protons is greater than the increase in spin coupling, which means that first-order spectra are often observed in fluorine resonance when the analogous hydrogen compound would give rise to a complex, second-order spectrum. For example, the ^{19}F spectrum of trifluorochloroethylene shown in Fig. 20.14 is easily interpreted from first-order rules even at the relatively low frequency of 30 MHz. The analogous hydrogen compound, chloroethylene (vinylchloride) gives a complicated, second-order spectrum at 60 MHz. An example that illustrates the coupling of protons to fluorine is the nmr spectrum of 1,1-diphenyl-1,2-difluoroethane, $(C_6H_5)_2CFCH_2F$. The fluorine chemical shifts with respect to internal CCl_3F are for the 1 and 2 positions 156.6 and 226.3 ppm. The coupling of the two fluorines is $J_{FF} = 20$ Hz; the coupling of the 2-fluorine to the protons is $J_{HF} = 48$ Hz; and the coupling of the 1-fluorine to these same protons is $J_{HF} = 20$ Hz.

20.4.2 / Phosphorus and boron resonance

As seen in Table 20.1, phosphorus is a monoisotopic nucleus that offers nmr sensitivity of approximately 7 percent that of proton or fluorine resonance. With spin $\frac{1}{2}$, ^{31}P is nonetheless a good nucleus for high-resolution nmr since it gives narrow lines and a large range of chemical shifts. Phosphorus differs from protons and fluorine in that two stable valence states occur,

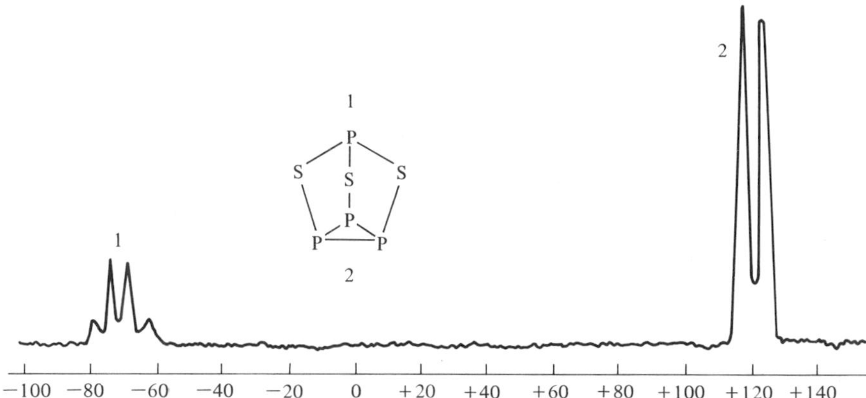

Figure 20.15 / *The* ^{31}P *resonance spectrum of phosphorus sesquisulfide at* 12.3 MHz. *Shifts are shown relative to* 85 *percent phosphoric acid. The splitting pattern establishes unequivocally that there are two kinds of phosphorus atoms in the molecule in the ratio of* 3:1.

SOURCE: C. F. Callis, J. R. van Wazer, J. N. Schoolery, and W. A. Anderson, *J. Amer. Chem. Soc.* **79,** 2719 (1957). Reprinted by permission of the American Chemical Society.

and these show distinct differences in chemical shifts. The range of shifts is approximately 100 ppm for phosphorus(V) and larger, spanning about 500 ppm, for phosphorus(III). Phosphorus shifts are routinely referred to the resonance of aqueous 85 percent phosphoric acid (H_3PO_4), although phosphorus oxide, P_4O_6, may, in fact, be a preferable standard. In contrast to the case of TMS in proton resonance, both positive and negative shifts with respect to H_3PO_4 occur with nearly equal frequency.

Spin coupling of phosphorus to other nuclei is frequently large and has proved extremely useful in elucidation of structures of numerous phosphorus compounds. For example, the fact that one unique proton is directly bound to phosphorus in phosphorus acid (H_3PO_3) is unequivocally established by the presence of a large doublet splitting (707 Hz) in the phosphorus resonance signal. Hence $H\overset{\overset{O}{\|}}{P}(OH)_2$ is the correct structure, not $P(OH)_3$. The structures of a large number of inorganic phosphates and polyphosphates have been studied in this way. The phosphorus resonance spectrum of P_4S_3 shown in Fig. 20.15 is typical of such results.

In addition, phosphorus forms a vast number of both natural and synthetic organophosphorus compounds. These materials are of great interest because they play central roles in the biochemistry and metabolism of both plants and animals and also serve as insecticides and nerve poisons. They find increasing application as ligands in coordination chemistry. Coupling of ^{31}P to protons occurs even through several intervening chemical bonds,

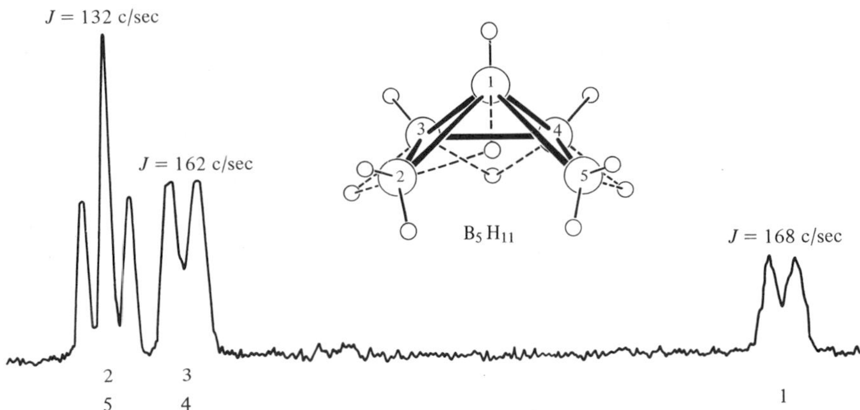

Figure 20.16 / *The* ^{11}B nmr *spectrum of pentaborane at* 64.2 MHz. *The spectrum shows clearly three different kinds of boron atoms in the molecule. Atoms labeled 1, 3, and 4 are bound to a single proton, while 2 and 5 are bound to two protons. Coupling to the bridging protons is too small to observe.*
SOURCE: R. E. Williams, F. J. Gerhart, and E. Pier, *Inorg. Chem.* **4**, 1239 (1965). Reprinted by permission of the American Chemical Society.

as illustrated by the nmr spectrum of trimethylphosphite, $(CH_3O)_3P$. The phosphorus resonance of this compound is a multiplet centered at -141 ppm relative to 85 percent H_3PO_4. It contains the anticipated 10 lines from spin coupling to 9 equivalent protons with the coupling constant, J_{PH}, about 20 Hz.

Of particular current research interest are applications of ^{31}P nmr to the identification and structure of metabolites and motional studies of organophosphorus ligands in structurally nonrigid organometallic compounds.

The nmr sensitivity of ^{11}B is somewhat better than that for phosphorus, but the lines are likely to be less well resolved due to quadrupole broadening if the nucleus is located in an unsymmetrical environment. Like phosphorus, much boron nmr has been done to reveal structures of inorganic compounds; applications to organoboron compounds have been limited.

The chemistry of boron is, of course, highlighted by the fascinating array of unusual geometric structures found in its compounds. Particular interest has been generated by the structures of the various boranes that contain three-center electron-deficient bonds in which a single hydrogen atom bridges two boron positions. In Fig. 20.16, a ^{11}B resonance spectrum is shown. It is typical in that the spin coupling and chemical shift patterns are used to elucidate a borane structure but atypical in that it was obtained with the use of a superconducting magnet (*vide infra*). There is as yet no reference standard in general use in boron nmr, and it appears that the number of peaks and spin coupling patterns are of more frequent use than the numerical values of the observed chemical shifts.

20.4.3 / Carbon resonance

After protons, carbon is the nucleus of greatest potential interest to chemists because of its presence in all organic compounds. ^{13}C nmr offers great promise as a tool for molecular structure analysis, perhaps even surpassing the utility of ^{1}H resonance: The latter gives information about the local environment of protons and only indirectly about the carbon skeleton, but carbon resonance reflects directly the structure of the carbon backbone. In spite of these reasons for wishing to study ^{13}C nmr, the field is much less well developed than proton or fluorine resonance because the low abundance of ^{13}C and its small magnetic moment combine to reduce the available signal-to-noise ratio by a factor of about 6000 in comparison to the former two nuclei. This unfavorable signal-to-noise ratio poses formidable experimental difficulties, but sufficient motivation exists that ways of circumventing these difficulties have been actively sought. Specific isotopic enrichment is a solution to the low abundance problem, but one seldom used because it involves not only high cost, but also laborious synthetic chemical effort prior to any nmr experiments. Electronic methods of signal-to-noise enhancement have been perfected to the point that carbon resonance at natural abundance levels is a completely routine matter in properly equipped laboratories. These techniques are discussed in the Section 20.6.4, but for the moment suffice it to say that they are expensive and it is likely that equipment for ^{13}C nmr will always cost two to three times more than equipment for routine proton resonance.

As with other heavy nuclei, ^{13}C chemical shifts are larger than those of protons by about an order of magnitude; carbon shifts span a range of about 350 ppm. Because of its low abundance, the probability of finding two neighboring ^{13}C nuclei in the same molecule is about one in 10^4, which means that spin coupling between ^{13}C nuclei is not observed. Coupling of carbon to adjacent protons does occur and may in some cases be useful as an aid to interpretation of the spectrum. In other cases, it is an unwanted complication and is routinely removed by a double-resonance technique (*vide infra*) known as *spin decoupling*. A spin decoupled ^{13}C nmr spectrum may be expected to show simply a single, well-resolved line for each magnetically distinct type of carbon atom in the molecule; in general, interpretation of such spectra is straightforward.

Carbon chemical shifts have been referred to the carbon resonances of benzene and carbon disulfide, but more recently the carbon resonance of TMS has come into use. TMS appears to be the best choice since, like its proton resonance, its carbon signal appears at higher field than the majority of other organic carbon resonances. It is therefore possible to use a shift scale in which resonances are characterized by a positive δ_C if they occur at fields lower than TMS resonance. In contrast, scales referred to benzene or CS_2 involve both positive and negative shifts, and these have been defined such that a positive shift means that the resonance is at higher field than the standard, and a negative shift implies lower field. The three scales

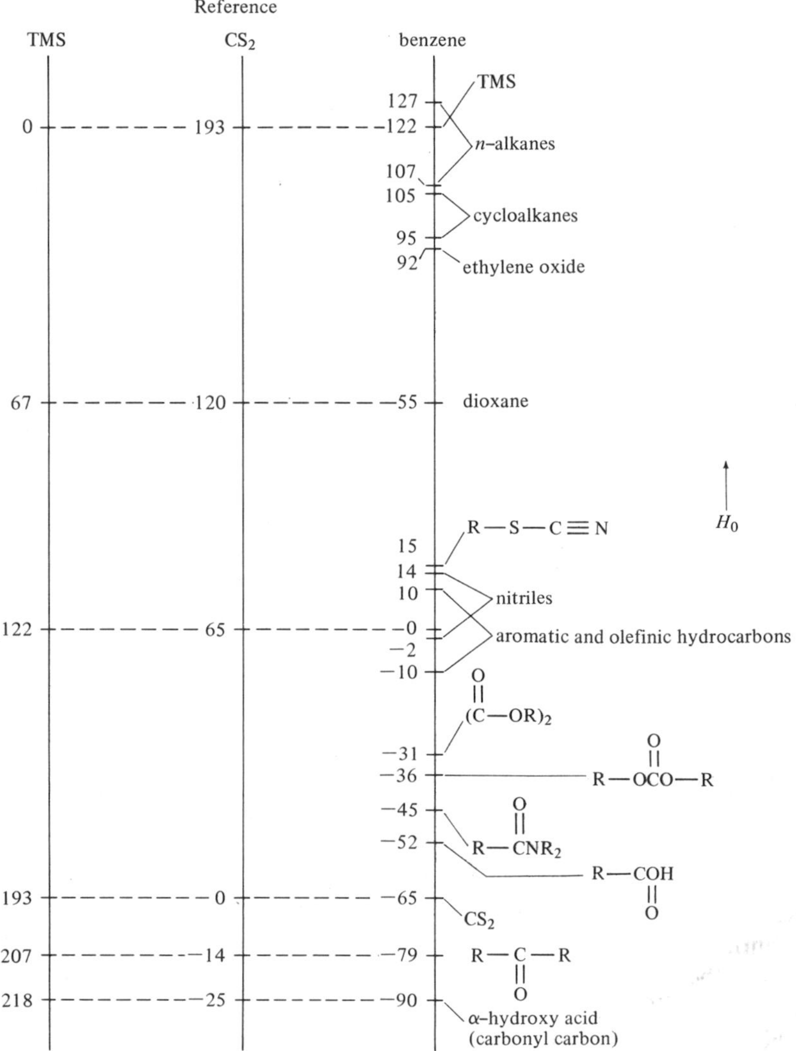

Figure 20.17 / *Approximate shifts in parts per million for various types of carbon atoms in organic molecules. Scales using benzene, carbon disulfide, and tetramethylsilane are compared. Note that on the scale referred to as* TMS, *shifts to lower field are defined as positive just as they are in proton resonance, while this is not true for the other two scales.*

are compared graphically in Fig. 20.17. Just as with proton resonance, carbon is shielded from the external field by circulating electrons, and carbon nuclei with relatively high local electron density have their resonance at the highest external field. Thus, only carbon atoms bonded to heavy, electron-rich atoms such as Br or I resonate appreciably above TMS, and the resonance of most simple hydrocarbons is shifted just below TMS. Carbon bonded to

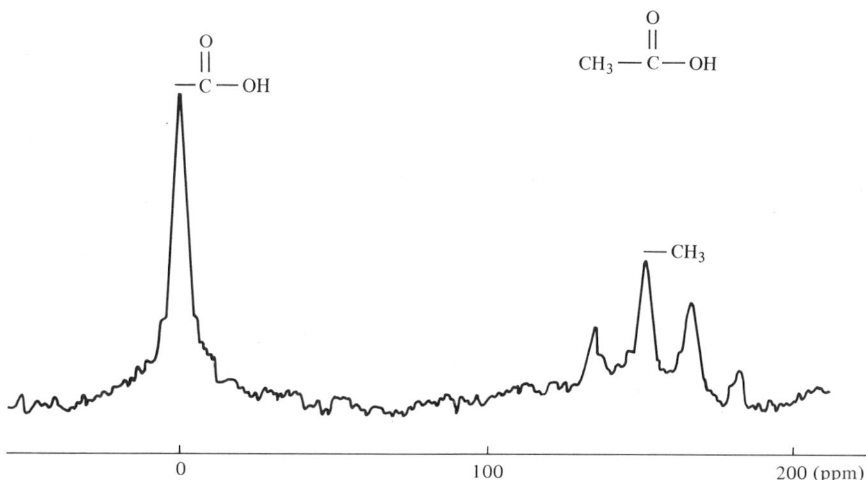

Figure 20.18 / *The* ^{13}C nmr *spectrum of acetic acid at* 8.5 MHz. *The sample is isotopically natural, so no carbon–carbon spin coupling is observed. Coupling of the three methyl protons to the methyl carbon splits the resonance as expected into a* 1:3:3:1 *quartet. Shifts are shown relative to the carboxyl carbon resonance.* SOURCE: P. Lauterbur, *Ann. N.Y. Acad. Sci.* **70**, 841 (1958).

electronegative atoms, particularly oxygen, is strongly deshielded, and these resonances appear more than 200 ppm below TMS.

The ^{13}C nmr spectrum of acetic acid is shown in Fig. 20.18. This spectrum was obtained without the use of proton decoupling, and the interaction of the methyl protons with the carbon atom is clearly visible.

The large range of ^{13}C shifts coupled with the sensitivity of the shift values to structural differences make these spectra extremely useful. Effects of neighboring groups on the shift of a particular carbon appear to be approximately additive, so that structure–shift correlations can be developed and used for "fingerprinting" molecules just as has been done with proton resonance. So sensitive is the carbon shift to environment that even in an unsubstituted hydrocarbon, shift differences of about 10 ppm occur for different positions in the chain. The sensitivity of ^{13}C resonance to extremely subtle effects is illustrated by the spectrum in Fig. 20.19. Not only is there a large shift difference for each of the different positions in the molecule, but three of the lines occur as doublets. This effect is due to the fact that the carbon atom at position 3 is asymmetric. The molecules can therefore occur as an unresolvable D,L pair plus the meso isomer. Apparently, the two diastereoisomers have slightly different shifts, and the effect extends to the carbons bonded to the asymmetric center. The carbons at position 1 are either accidentally equivalent in the two isomers or so little different that the effect is not resolved. The ability of ^{13}C nmr to reveal such subtle structural differences is now being exploited in structural studies of proteins, enzymes, and similar biologically important polymers. Such applications

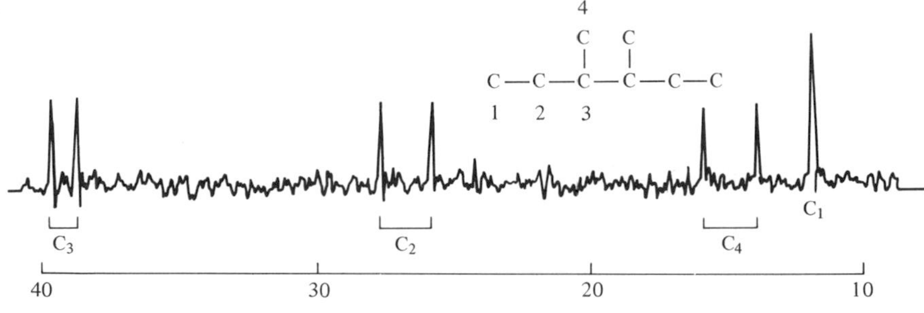

Figure 20.19 / *The proton-decoupled* ^{13}C nmr *spectrum of* 3,4-*dimethylhexane at* 25 MHz. *The doubling of the resonances for three of the carbon atoms is due to molecular asymmetry (see text).*
SOURCE: L. P. Lindeman and J. Q. Adams, *Anal. Chem.* **43**, 1245 (1971). Reprinted by permission of the American Chemical Society.

can only increase as the appropriate instrumentation becomes ubiquitously available.

Information not normally available from proton resonance can be obtained by observing coupling of naturally occurring ^{13}C to protons. These so-called ^{13}C satellites arise from molecules of the type shown in Fig. 20.20. In such a molecule, the presence of the ^{13}C renders the chemically equivalent protons magnetically inequivalent. Hence, the coupling constant J_{HH} can be deter-

Figure 20.20 / *The* 60-MHz ^{1}H nmr *spectrum of* 1,1,2,2-*tetrabromoethane at high gain. Satellite lines from molecules containing a* ^{13}C *atom in natural abundance can be seen. The satellites are doubled because the protons are inequivalent in the molecule, which contains different carbon isotopes and can therefore spin couple.*
SOURCE: F. A. Bovey, *Chemical and Engineering News*, Aug. 30, 1965, p. 110. Reprinted by permission of the American Chemical Society.

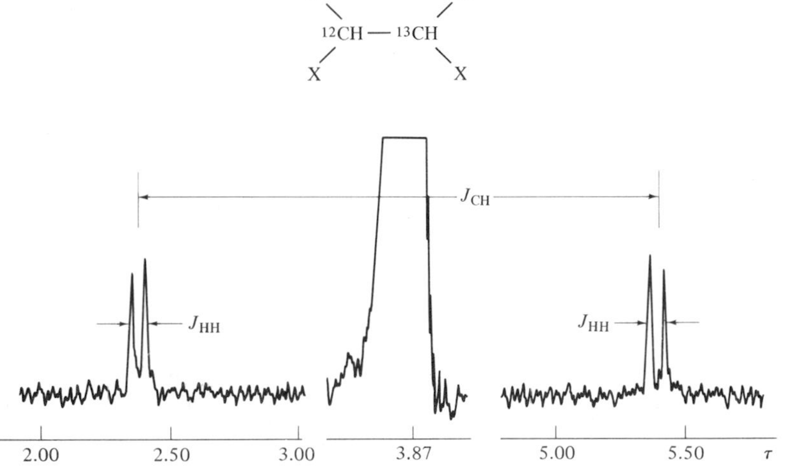

mined from the satellite lines as indicated in the figure. In the absence of the proton–proton coupling, only a single pair of satellites separated by J_{CH} would occur. One of the utilizations of such satellite lines is as an aid in structure determination, such as whether a particular molecule is cis or trans. As a general rule, the value of J_{HH} will be greater in the trans configuration. For example, in the case of 1,2-dichloroethylene, $J_{CH} = 198$ and $J_{HH} = 5.2$ Hz for the cis configuration, while for the trans configuration, $J_{CH} = 199$ and $J_{HH} = 11.1$ Hz.

20.5 / RELAXATION AND LINE SHAPES

In the discussion of magnetic resonance transitions in Section 20.2, we computed the frequencies at which these transitions would occur without regard for the finite width of the spectral lines. As the reader will have no doubt observed in the sections in which nmr spectra were presented, the resonance lines are indeed narrow, often less than 0.5 Hz in width. Nonetheless, the lines are sometimes too broad to afford sufficient resolution to obtain the maximum information from the spectra. In this section, we shall consider factors that affect the widths and shapes of magnetic resonance lines. Some understanding of these factors is essential if the chemist is to know how best to record magnetic resonance spectra. Line widths and spectral resolution are, of course, intimately related to both the underlying physics of the experiment and to the instrumentation employed. It is felt that by outlining some theoretical aspects now, the section concerned with instrumentation will be more meaningful.

20.5.1 / Saturation

It was indicated previously that the population difference between the energy states involved in magnetic resonance transitions is very small and that the spontaneous emission of radiation is too improbable a process at these frequencies to be of any help in preserving this population difference. Further, we know that the transition probabilities for absorption and induced emission are equal and that the *net* absorption of energy depends on the maintenance of a larger population in the lower state than in the upper state. Should the two populations become equal, the absorption will disappear; such a condition is termed *complete saturation.*

Let us now look quantitatively into the phenomenon of saturation. Consider a sample of spins immersed in a static magnetic field and irradiated with a radio frequency field, H_1, appropriate to excite transitions. Let P_+ be the probability of an absorption event, $n \rightarrow n^*$, and P_- be the probability for induced emission, $n^* \rightarrow n$. The time rate of change of the lower-state population is given by simple first-order kinetics:

$$\frac{dn}{dt} = n^*P_- - nP_+ = P(n^* - n) \qquad (20.11)$$

We next make use of the fact that the two probabilities are equal and use a single transition probability, P. If we let the population difference $(n^* - n) = \eta$ and the total population be N, Eq. (20.11) is easily transformed into

$$\frac{dn}{dt} = P\left[\frac{(N - \eta)}{2} - \frac{(N + \eta)}{2}\right] = -P\eta \tag{20.12}$$

In addition, we recognize that dn/dt can be expressed as $(d/dt)(N + \eta)/2 = d\eta/2\ dt$. Equating the two expressions for dn/dt on rearrangement yields

$$\frac{d\eta}{\eta} = -2P\ dt \tag{20.13}$$

which can be integrated directly. Let the limits of integration be 0 and t which represent, respectively, the time at which the rf field is turned on and an arbitrary time thereafter. The corresponding limits for η are $\eta(0)$ and $\eta(t)$, which represent the population difference at thermal equilibrium in the applied DC field, and the difference at time t after irradiation has begun. The result is simply $\eta(t) = \eta(0)\exp(-2Pt)$. The quantity measured experimentally is the power absorbed from the irradiating field, that is, the rate of energy absorption, $dE/dt = \eta P\,\Delta E = \eta Pg\mu_0 H_0$. Thus, we find that unless other factors intervene, our absorption will vanish exponentially after the beginning of the experiment. The two populations will move toward equality and complete saturation will result.

Before we discuss the factors that prevent every resonance experiment from ending with complete saturation, a few comments about a saturated system are in order. When saturated by the application of rf power, the system is *not* in thermal equilibrium with its surroundings at the laboratory temperature. This must be the case since the population ratio is not that computed from a Boltzmann distribution for the ambient temperature. One often sees such systems described using the Boltzmann expression and a fictious temperature, called the *spin temperature*, selected to fit the observed population ratio. A partially saturated system will have a spin temperature that is above ambient temperature; for complete equalization of the two populations, an infinite spin temperature is required. Should the population of the upper state exceed that of the lower state (a condition that *can* be produced experimentally), the spin temperature must be *negative*. These results are strictly artifacts of this method of describing the spin system and, since thermal equilibrium is not achieved under these circumstances, the third law of thermodynamics is not violated.

20.5.2 / Relaxation

The foregoing discussion ignores, of course, the obvious fact that the spin system can interact with its surroundings. Transfer of energy between a partially saturated (or "hot") spin system and the surroundings has the effect of "thermalizing" the spin system, that is, bringing it toward thermal equilibrium at the laboratory temperature. In the usual magnetic resonance terminology, the surroundings are called the *lattice* and the energy-transfer process is termed *spin-lattice relaxation.* The coupling of the spin system with the lattice occurs by means of fluctuating magnetic fields generated as the molecules execute random thermal motions. The coupling is very weak, since the lattice modes have energies of the order of thermal energy (kT) and the energy of a spin flip is very much smaller.

A quantitative approach to the relaxation process begins by defining w_+ and w_- as, respectively, the probability of an upward transition induced by lattice energy and a downward transition that transfers spin flip energy to the lattice. Using these transition probabilities, we can set up a first-order rate expression as we did previously for the case of rf-induced transitions. Suitable algebraic manipulation of that rate expression leads to

$$\frac{d\eta}{dt} = [-\eta + n(0)]\,(w_- + w_+) \tag{20.14}$$

Note that in contrast to P_+ and P_-, $w_+ \neq w_-$. This inequality is required since at thermal equilibrium in the absence of rf, $dn/dt = 0$, and the relation $n^*/n = w_+/w_- = \exp(-\Delta E/kT)$ must hold. The expression in Eq. (20.14) can be modified by defining the sum of the two transition probabilities (units of reciprocal seconds) as the inverse of a characteristic time, the *spin-lattice relaxation time*, $(w_- + w_+) = T_1^{-1}$. Hence, we can write Eq. (20.15) as the time rate of change of the population difference under the influence of the lattice:

$$\frac{d\eta}{dt} = -\frac{[\eta - \eta(0)]}{T_1} \tag{20.15}$$

This result is added to that obtained previously to obtain the rate of population change brought about by the combined influences of the rf radiation and the lattice. The result is

$$\frac{d\eta}{dt} = -2P\eta - \frac{[\eta - \eta(0)]}{T_1} \tag{20.16}$$

If we now consider the case of thermal equilibrium *in the presence of rf radiation* (experimentally, this means that the spectrum is swept very slowly; i.e., a *slow passage* experiment), the value of $d\eta/dt$ must be zero and Eq. (20.16) can be solved for η with the result shown in Eq. (20.17).

$$\eta = \frac{\eta(0)}{1 + 2PT_1} \tag{20.17}$$

Thus, the inclusion of spin-lattice relaxation causes the population difference to remain nonzero. It is necessary to ensure that the product PT_1 does not become excessively large. Since P is proportional to H_1^2, the rf power level must be kept quite small to avoid saturation. The power absorbed is given by

$$\frac{dE}{dt} = \eta(0)\,\Delta E\left(\frac{P}{1 + 2PT_1}\right) \tag{20.18}$$

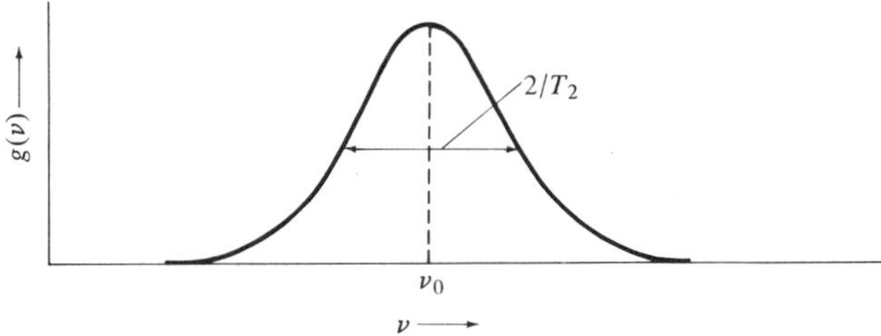

Figure 20.21 / *Amplitude vs. frequency plot for a magnetic resonance line of Lorentzian shape. The half-width at half intensity is* T_2^{-1}.

Spin-lattice relaxation times vary over a wide range of values in magnetic resonance. Typical values for esr are in the microsecond range, while for nuclei values of the order of a few seconds are common in solution. T_1 values as long as several hours have been observed for nuclear resonance in solids. Spin-lattice relaxation contributes to the width of the resonance lines by limiting the lifetimes of the states. The lifetime limitation introduces uncertainty broadening in accord with the Heisenberg principle and makes a contribution to the linewidth $\Delta\nu \simeq T_1^{-1}$. Of course, in the case where appreciable saturation occurs, rf-induced transitions are the limiting factor in the lifetimes of the spin states and the line is *saturation broadened.* In actual practice, the unsaturated width of a resonance line is almost always greater than T_1^{-1}, and other processes must be considered.

A number of processes occur that have the effect of varying the *relative energies* of the spin states rather than changing their lifetimes. Fluctuating magnetic fields due to the random motion of other nuclei in the solution, for example, modulate the energies of the spin states. Inhomogeneities in the applied DC field over the sample volume have a similar effect. As a result, transitions take place *from a range* of energies *to a range* of energies rather than between two precisely defined energy states, thereby broadening the line. These processes are taken together and characterized by a second relaxation time, T_2, called the *transverse* or *spin-spin* relaxation time. T_2 is also used in ad hoc manner as parameter to describe the shape of the resonance line.

The line shape will depend on the way the transition probability, P, varies with frequency. At this point, we assert that P contains a *shape function*, $g(\nu)$, which, for magnetic resonance in fluid media is usually a Lorentz shape. The Lorentz shape has the mathematical form

$$g(\nu) \propto \frac{T_2}{1 + T_2^{\,2}(\nu - \nu_0)^2} \tag{20.19}$$

where ν_0 is the frequency at the absorption maximum and ν represents frequencies elsewhere in the line. The line is, therefore, symmetric about its center and has the shape illustrated in Fig. 20.21. It is usual to define T_2 as the reciprocal of the half-width at half-height for a Lorentzian line. Thus, as shown in the figure, the full width at half-height is $2/T_2$.

20.5.3 / Line shapes: the Bloch equations

Thus far we have approached magnetic resonance from a microscopic viewpoint, examining the behavior of individual electrons or nuclei. It is a useful alternative view to examine these spin systems from a macroscopic perspective. By this we mean that we shall not be concerned with the individual electrons or nuclei, but only with the vector sum of their magnetic moments, the bulk, or *macroscopic magnetization.* The utility of this approach is that the macroscopic magnetization behaves exactly as one would anticipate from classical mechanics, whereas only quantum mechanics is useful in discussing the properties of the individual particles.

In classical electromagnetic theory, the magnetic induction, **B**, is given by $\mathbf{B} = \mathbf{H} + 4\pi\mathbf{M}$, where **H** is the applied magnetic field and **M** is the magnetization, or magnetic moment per unit volume. For nonferromagnetic materials, $\mathbf{M} = \chi_v\mathbf{H}$, where χ_v is known as the *volume magnetic susceptibility*. For an ensemble of magnetic dipoles, μ, the magnetization is obtained by summing all the individual moments with appropriate account taken of whether they are parallel or antiparallel to the imposed field. The expression for the net moment of an assembly of spin $\frac{1}{2}$ particles is given by

$$\bar{\mu} = \frac{(1 + \mu H_0/kT)\mu}{2} - \frac{(1 - \mu H_0/kT)\mu}{2} = \frac{\mu^2 H_0}{kT} \tag{20.20}$$

If there are N particles per unit volume, then $\mathbf{M} = \chi_v\mathbf{H}_0 = N\mu^2\mathbf{H}_0/kT$.

The behavior of a classical magnetic dipole, μ, with magnetogyric ratio γ, in a DC magnetic field in the absence of friction was examined by J. Larmor in 1900. Larmor showed that such a dipole could never come to rest along the field direction, but because of conservation of angular momentum would *precess* about the external field with angular frequency $\omega_0 = \gamma H_0$, as indicated in Fig. 20.22. The relationship between field and frequency derived by Larmor is identical to that obtained by quantum mechanical arguments in Eq. (20.6). The resonance frequency, ω_0, is therefore often termed the *Larmor frequency*, and nuclei are sometimes described as executing *Larmor precession.*

To complete this picture of the resonance experiment, we must specify the properties of the rf radiation more completely. This radiation will be plane polarized, and the plane of polarization must be perpendicular to $\mathbf{H}_0$. The reader will recall that plane-polarized radiation can be described equivalently as the sum of two oppositely rotating circularly polarized rays. One of these components is depicted in Fig. 20.22 as the vector $\mathbf{H}_1$, which rotates in the xy plane in the same sense as **M**; the second component, which rotates in the opposite direction, plays no part in the resonance phenomenon. The size of $\mathbf{H}_1$ is exaggerated for clarity in the figure and will be very much smaller than $\mathbf{H}_0$ in actuality.

The rf magnetic field has no component along the z direction; its other components have the forms $H_x = H_1 \cos \omega t$ and $H_y = H_1 \sin \omega t$, where ω is the angular frequency of the radiation. If $\mathbf{H}_1$ has a different frequency

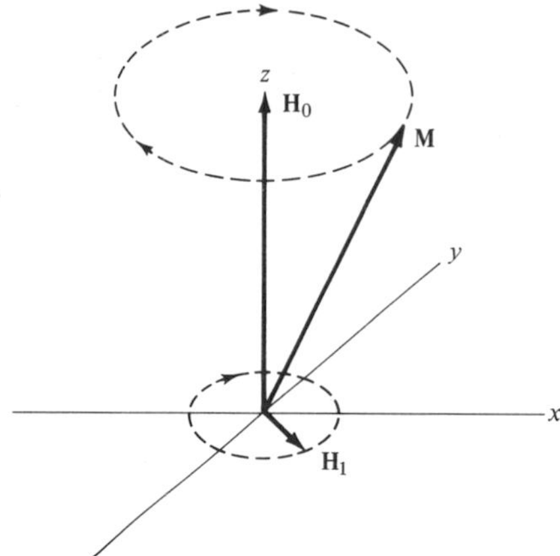

Figure 20.22 / *A pictorial representation of the macroscopic description of the magnetic resonance phenomenon. The DC field is shown along the z axis with the magnetization vector* **M** *precessing about it at angular frequency* $\omega_0 = \gamma \mathbf{H}_0$. *The magnetic component of the rf radiation,* $\mathbf{H}_1$, *is shown rotating about the z axis in the xy plane. Only when* **M** *and* $\mathbf{H}_1$ *rotate with the same frequency is a stable relationship between them maintained so that energy can be absorbed.*

from the precession frequency of **M**, there is no stable relation between the two vectors and only a small wobble of **M** can be superimposed on its precession about $\mathbf{H}_0$. By contrast, if $\mathbf{H}_1$ rotates at the Larmor frequency, it exerts a constant torque on **M** and can tip that vector away from $\mathbf{H}_0$ toward the *xy* plane. This process results in significant absorption of energy from the rf field and is responsible for the resonance signal in this model. The torques exerted on **M** by $\mathbf{H}_0$ and $\mathbf{H}_1$ are given by the vector products $\gamma(\mathbf{M} \times \mathbf{H}_0)$ and $\gamma(\mathbf{M} \times \mathbf{H}_1)$, where γ is the magnetogyric ratio of the particles whose magnetic moments constitute **M**. The torque on a dipole is defined as the time rate of change of its angular momentum, so these vector products yield a set of differential equations. Since **M** is a three-component vector, three such equations result. Of course, relaxation must be included in these equations. In Eqs. (20.21),

$$\begin{aligned}
\frac{dM_x}{dt} &= \gamma(M_y H_0 + M_z H_1 \sin \omega t) - \frac{M_x}{T_2} \\
\frac{dM_y}{dt} &= \gamma(M_z H_1 \cos \omega t - M_x H_0) - \frac{M_y}{T_2} \\
\frac{dM_z}{dt} &= \gamma(-M_x H_1 \sin \omega t - M_y H_1 \cos \omega t) - \frac{(M_z - M_0)}{T_1}
\end{aligned} \tag{20.21}$$

the terms in parentheses represent the interaction of **M** with the two magnetic fields and the additional terms represent relaxation. The *z* component of magnetization depends on the net excess of spins aligned with the DC field and therefore approaches its equilibrium value, M_0, in the same way in which the population difference does, with characteristic time T_1. In Eqs. (20.21), the assumption is introduced that the other two components of magnetization will also decay by a first-order process, but with a different characteristic time, T_2. Physically, decay of the x and y components amounts to having the precessing spins that constitute **M** get out of phase with one another so that their x and y components cancel. The treatment of resonance represented in Eqs. (20.21) was first introduced by Bloch, and the equations are generally known as *Bloch's phenomenological equations.* This terminology reflects the fact that these equations were postulated to explain resonance line shapes without regard for the microscopic processes that give rise to the lines.

In the nmr spectrometer, the signal observed is produced by either the x or y component of magnetization, depending on the adjustment of the instrument by its operator. Therefore, it will be of interest to examine the expressions for these components. Solution of the Bloch equations is most readily accomplished by first transforming them into a coordinate system that rotates with $\mathbf{H}_1$ about the z axis. The components of **M** in the rotating coordinate system are two new variables, u and v, which correspond, respectively, to the x and y components in Cartesian coordinates. The component u lies along $\mathbf{H}_1$ (rotates *in phase* with $\mathbf{H}_1$), while v rotates in the same plane, but at right angles to $\mathbf{H}_1$ (90° *out of phase* with $\mathbf{H}_1$). M_z is retained, since the z axis is unchanged by this transformation of coordinates. The equations transformed to the rotating frame and rearranged are

$$\frac{dM_z}{dt} + \frac{(M_z - M_0)}{T_1} - \gamma H_1 v = 0$$

$$\frac{du}{dt} + \frac{u}{T_2} + (\omega_0 - \omega)v = 0 \tag{20.22}$$

$$\frac{dv}{dt} + \frac{v}{T_2} - (\omega_0 - \omega)u + \gamma H_1 M_z = 0$$

The simplest solution of these equations is the so-called *steady state solution,* which corresponds to a slow-passage magnetic resonance experiment. The steady state condition is specified by setting all the time derivatives equal to zero, whereupon simple algebra yields

$$u = \frac{\gamma H_1 M_0 T_2^2(\omega_0 - \omega)}{1 + T_2^2(\omega_0 - \omega)^2 + \gamma^2 H_1^2 T_1 T_2}$$

$$v = \frac{-\gamma H_1 M_0 T_2}{1 + (\omega_0 - \omega)^2 T_2^2 + \gamma^2 H_1^2 T_1 T_2} \tag{20.23}$$

It is customary at this point to introduce the macroscopic radio frequency magnetic susceptibility, χ_{rf}. The rotating field, H_1, is one component of a linearly polarized field along the x axis, $H_{\text{rf}} = 2H_1 \cos \omega t$. The response of a sample of nuclear moments to excitation with this radiation can be de-

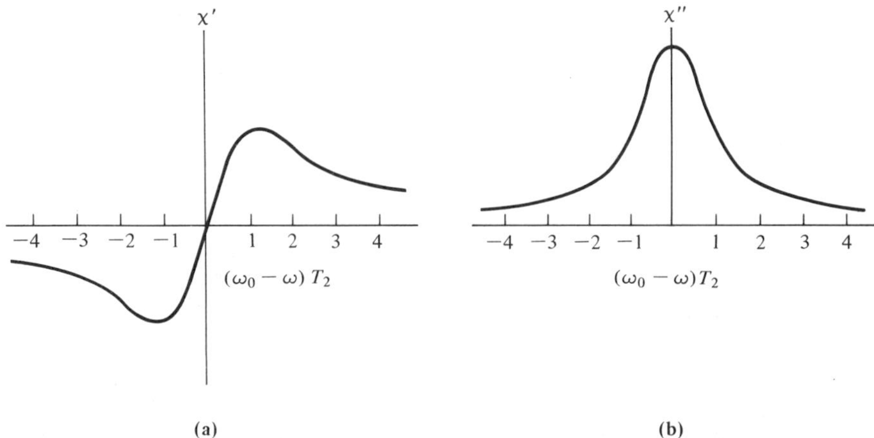

Figure 20.23 / *Line shapes that arise from* (a) *the real part and* (b) *the imaginary part of the* rf *susceptibility.*

scribed by a complex susceptibility, $\chi_{\text{rf}} = \chi'(\omega) - i\chi''(\omega)$. It is possible to associate the in-phase (u-mode) signal with $\chi'(\omega)$ and the out-of-phase (v-mode) signal with $\chi''(\omega)$. Specifically, $\chi'(\omega) = u/2H_1$ and $\chi''(\omega) = -v/2H_1$. Therefore, the expressions for these two susceptibilities with the substitutions $M_0 = \chi_0 H_0$ and $\gamma H_0 = \omega_0$ are

$$\begin{aligned} \chi'(\omega) &= \frac{\chi_0 \omega_0 T_2{}^2(\omega_0 - \omega)}{2[1 + T_2{}^2(\omega_0 - \omega)^2 + \gamma^2 H_1{}^2 T_1 T_2]} \\ \chi''(\omega) &= \frac{\chi_0 \omega_0 T_2}{2[1 + T_2{}^2(\omega_0 - \omega)^2 + \gamma^2 H_1{}^2 T_1 T_2]} \end{aligned} \tag{20.24}$$

Thus, the real part of the rf susceptibility gives rise to a line shape illustrated in Fig. 20.23(a), which is called the *dispersion mode* or u-mode signal, while the imaginary part of the susceptibility produces the usual *absorption mode* or v-mode signal shown in Fig. 20.23(b). Suitable adjustment of a magnetic resonance spectrometer permits observation of either absorption or dispersion signals.

Notice that when the third term of the denominator in the expression for $\chi''(\omega)$ is negligible, the absorption line has the form indicated in Eq. (20.19) and is of Lorentzian shape. The quantity $\gamma^2 H_1{}^2 T_1 T_2$ is known as the *saturation factor.* The line will differ from true Lorentzian shape when the saturation factor has values that are not small with respect to unity; that is, the condition for the absence of saturation is $\gamma^2 H_1{}^2 T_1 T_2 \ll 1$. Failure to fulfill this condition broadens the line and results in degradation of both resolution and the signal-to-noise ratio. We can make an order-of-magnitude estimate of the field, H_1, which can be employed with no appreciable saturation of a proton nmr line by solving for H_1 using $\gamma_P \simeq 2 \times 10^4$ gauss^{-1}-sec^{-1} and $T_1 = T_2 \simeq 1$ sec. The calculation yields 5×10^{-5} gauss for the amplitude of H_1, a very small value.

Some further aspects of the material in this section will be explored and utilized in the discussion of instrumentation in Section 20.6.1.

20.5.4 / Chemical exchange

As mentioned above in the discussion of the nmr spectra of certain alcohols, exchange of the hydroxyl proton affects the shape of the resonance line; the position of the line is also affected. Another type of chemical exchange that can also affect the position and shape of a resonance line is exchange of two sites in a molecule via internal motion. These effects are considered briefly in this section.

Consider the simple case of an nmr spectrum that exhibits no spin-spin coupling but is modified by an exchange process. Let us specify that there are two possible chemical sites, that the nuclei are spin $\frac{1}{2}$ (protons for convenience), and that the exchange of sites is a random statistical process. An example is a mixture of water and ethanol in which protons exchange between the hydroxyl position of the alcohol and water. If the exchange rate is very slow, the spectrum must consist of two lines, each exhibiting the chemical shift of its particular environment. Conversely, if the exchange is rapid, a single line will be observed at an appropriately averaged position. The parameter that defines "slow" and "rapid" in such a case is the *difference* between the chemical shifts of the two environments expressed in units of reciprocal frequency; that is, the lines are distorted for exchange rates of the order $(\nu_A - \nu_B)^{-1}$ as indicated in Fig. 20.24. It is a common misconception that the operating frequency of the spectrometer sets the time scale, but this is not so except as it influences $(\nu_A - \nu_B)$.

The Bloch equations may be employed to treat the case of chemical exchange. Two sets of equations are required, one for each of the separate resonances characterized by angular frequencies ω_A and ω_B. These equations are then modified to include the effects of exchange.

It is usual to assume that the nuclei remain in one position until a sudden jump is made to the new site and that the time involved in the exchange process is negligibly small. With this restriction, it is clear that jumps between *chemically equivalent* sites will have no effect on the nmr spectrum, and they are therefore neglected. Only jumps between sites A and B will be of concern. If τ_A is the average lifetime of the A sites and τ_B is the average lifetime of the B sites, then ${\tau_A}^{-1}$ is the probability of a jump from site A to site B; ${\tau_B}^{-1}$ is the probability of the reverse jump. These inverse lifetimes are the rate constant for the exchange process. The fractional populations of the A and B sites, p_A and p_B (clearly $p_B = 1 - p_A$) are related to the lifetimes by

$$p_A = \frac{\tau_A}{(\tau_A + \tau_B)} \qquad p_B = \frac{\tau_B}{(\tau_A + \tau_B)} \tag{20.25}$$

Evidently, in the case we are considering, the concentration of water in the

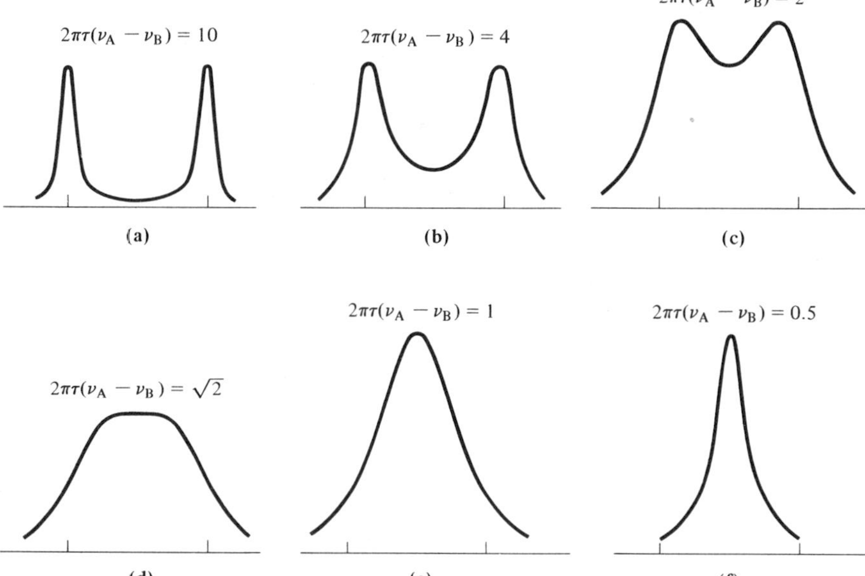

Figure 20.24 / *The change in the shape calculated for an* nmr *signal from a system undergoing chemical exchange between two positions with equal populations. The intensities of the various plots are not on the same scale. As defined here, τ is one-half the lifetime of either position.*
SOURCE: J. A. Pople, W. G. Schneider, and H. J. Bernstein, *High Resolution Nuclear Magnetic Resonance*, McGraw-Hill (New York, 1959).

alcohol is important; the larger this concentration, the more probable it is that the proton will be in a water site.

The solutions to the modified Bloch equations are simplest for the limiting cases of slow and fast exchange. In the slow-exchange limit, there are two lines and each is broadened by exchange. The solution for the v-mode signal, assuming negligible saturation, is given by

$$v = \frac{-\gamma H_1 M_0 p_A T'_{2A}}{1 + (T'_{2A})^2 (\omega_A - \omega)^2} \tag{20.26}$$

An identical expression is obtained for the second resonance by substitution of the subscript B in place of A. Thus, the A resonance is a broadened signal centered at the original frequency, ω_A, with width parameter $(T'_2)^{-1} = T_{2A}{}^{-1} + \tau_A{}^{-1}$. A corresponding statement can be made about the resonance due to the B sites. If T_2 is known for either resonance, measurement of the width of the broadened signal provides a means of estimating τ experimentally. In the case of rapid exchange, the solution for v, again in the absence of saturation, is given by

$$v = \frac{-\gamma H_1 M_0 T'_2}{1 + (T'_2)^2 (p_A\omega_A + p_B\omega_B - \omega)^2} \tag{20.27}$$

This result shows that a single resonance line centered at ω_{av} will be observed. Here, $\omega_{av} = p_A\omega_A + p_B\omega_B$ and the width parameter, $(T_2')^{-1} = p_A/T_{2A} + p_B/T_{2B}$. Thus, the observed line occurs with position and width given by averages of the two separate lines weighted by the probability of the individual sites. If the exchange process is insufficiently rapid, the exchange averaged line may be broader than the simple weighted sum of the two transverse relaxation times.

Solution of the modified Bloch equations for the general case of intermediate exchange rates results in a tediously complicated expression. A number of interesting cases can be treated, however, with the addition of two simplifying assumptions. These are (1) equal populations and lifetimes, that is, $p_A = p_B = \frac{1}{2}$ and $\tau_A = \tau_B = 2\tau$; and (2) large transverse relaxation times such that effectively $T_{2A}{}^{-1} = T_{2B}{}^{-1} \sim 0$. In other words, we are dealing with sites that are of equal probability and with signals whose widths in the absence of exchange are small compared to their separation. Under these conditions, the solution for v is

$$v = \frac{-\gamma H_1 M_0 \tau(\omega_A - \omega_B)^2}{4(\{[(\omega_A + \omega_B)/2] - \omega\}^2 + \tau^2(\omega_A - \omega)^2(\omega_B - \omega)^2)} \tag{20.28}$$

Since the signal shape depends only on the product $\tau|\omega_A - \omega_B|$, conversion to frequency units for comparison to experimental spectra and solution of Eq. (20.28) for the line shape can be carried out for various values of the parameter $2\pi\tau(\nu_A - \nu_B)$. Results of these calculations are shown in Fig. 20.24. When $\tau = \sqrt{2}/2\pi(\nu_A - \nu_B)$, the peaks have just coalesced into a single broad absorption [Fig. 20.24(d)].

Many molecules are known in which molecular motion causes site exchange at rates that can be studied by nmr. A classic example is that of N,N-dimethylformamide. In Fig. 20.25, spectra of this molecule at several temperatures are shown. The barrier to rotation about the C—N bond (variously estimated as 10–20 kcal) is such that at room temperature the two groups of methyl protons exchange environments slowly and give rise to two distinct resonance lines. As the temperature is increased, the rotation rate increases until eventually the two peaks coalesce. A case such as this can be treated with the simplifying assumptions made above. The two resonances are narrow and well separated at the lower temperatures, and the two sites are equally likely. Therefore, comparison of the observed and calculated line shapes permits estimation of the exchange rate as a function of temperature. Treatment of these data by standard methods of chemical kinetics allows estimation of the activation energy for the exchange process, which is equivalent to the barrier to internal rotation. Studies of this type have provided a fascinating and fruitful field of research.

In the foregoing discussion, the absence of spin coupling was assumed as a simplification. The effect of exchange on spin coupling is a dramatic one and is now considered. Above, exchange of equivalent sites was found to

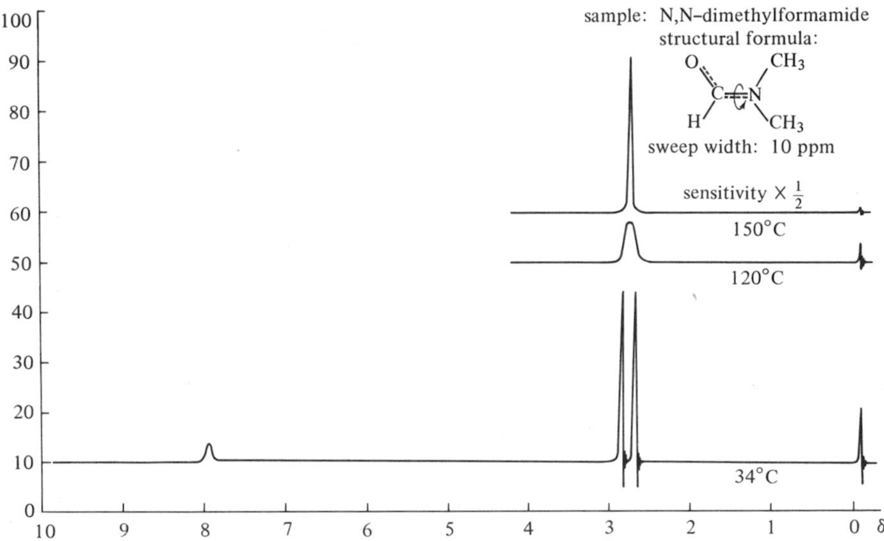

Figure 20.25 / *Proton magnetic resonance spectra of* N,N-*dimethylformamide at* 60 MHz *at various temperatures. The methyl resonance, a doublet at room temperature because of slow rotation about the* C—N *bond, collapses to a single line as the temperature is increased.*
SOURCE: The Perkin-Elmer Corporation.

have no effect, but in contrast, the collapse of spin multiplets can be brought about by exchange between chemically equivalent or nonequivalent sites.

Pure, dry ethanol exhibits a 1:2:1 triplet for the hydroxyl proton resonance due to spin coupling with the adjacent methylene group. In mixtures of water and ethanol, exchange of the hydroxyl proton leads to a single line with the average chemical shift weighted appropriately for the concentration of water. In dry ethanol, the exchange of hydroxyl protons among ethanol molecules can be catalyzed by the addition of minute amounts of acid. When sufficient catalyst is present, exchange among *like sites* will collapse the triplet to a single line while leaving the chemical shift unaltered.

The observation of the spin coupled multiplet depends on the fact that the methylene protons may exist in any one of four different spin states, two of which are degenerate. During exchange, it is equally likely that the hydroxyl proton that leaves a molecule with a given spin state will attach itself to another molecule with any one of the four possible spin states. With sufficiently rapid interchange, the effects of spin coupling are therefore averaged to zero and a singlet is observed. This result is predicted by solution of the appropriate form of the modified Bloch equations. Such calculations have been carried out, and in Fig. 20.26 the predicted line shape for the ethanol hydroxyl proton resonance is shown for several different values of τ, the mean time between exchange events. Comparison of calculated and experimental line shapes permits estimation of the rate of exchange.

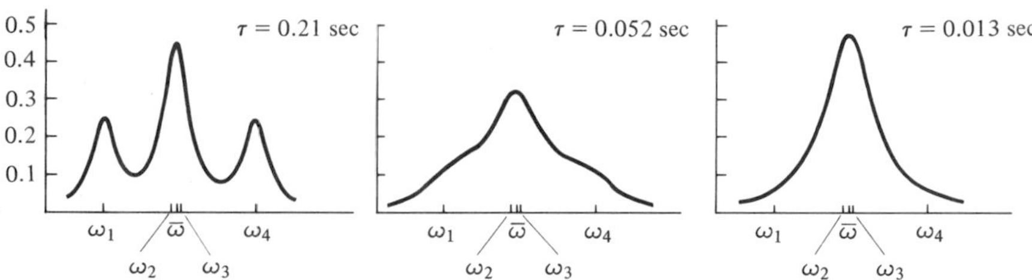

Figure 20.26 / *Predicted shapes of the hydroxyl proton resonance in ethanol for three values of the exchange time, τ. As the exchange rate increases, the spin coupling is averaged out.*
SOURCE: J. T. Arnold, *Phys. Rev.* **102**, 136 (1956).

20.5.5 / Spin decoupling

In the previous section, site exchange was found to collapse spin multiplets. An experiment known as spin decoupling produces the same end result using a double-irradiation technique. Consider a spectrum in which nuclear spins I are coupled to spins S with coupling constant J_{IS}. If the resonance of nucleus I is observed with a weak rf field at the appropriate resonance frequency, ω_I, while rapid transitions of the spins S are induced by a strong rf field at frequency ω_S, the multiplet structure of the I spins will collapse to a single line. Abragam has suggested the use of the term "spin stirring" to describe the induction of rapid transitions of the S spins and the use of "stirring" field to describe the strong rf field.

In spite of superficial similarity to exchange collapse of spin multiplets, rf decoupling is very different since the S spins are forced to undergo coherent motion under the influence of the stirring field in contrast to the random exchange process. Although in the limit of a strong stirring field the singlet produced appears the same as that in the fast-exchange case, if the stirring field is progressively increased in intensity from a small value at which the multiplet is unchanged to a large value at which collapse is complete, the intermediate patterns are very different from those observed when the rate of exchange is progressively increased by stepwise addition of a catalyst.

Analysis of the spin stirring problem requires solution of the spin hamiltonian in varying degrees of approximation depending on the sophistication of the treatment. We shall merely state the results for a simple example. If two spin $\frac{1}{2}$ nuclei are coupled, the nmr pattern will consist of two doublets. If we observe the doublet due to nuclei I as we progressively increase H_{1S}, the rf field stirring the nuclei S at ω_S, the following will occur. The doublet separated by J_{IS} will at first show diminished intensity and a new line will appear at the *uncoupled* chemical shift, that is, at the center of the doublet. As H_{1S} is increased still further, the intensity of the central line will increase at the expense of the outer lines and the latter will move increasingly farther apart. Eventually, a strong central line flanked by two barely detectable satellites will result.

A somewhat different sequence of events occurs if the stirring field is kept at high power but is initially not at ω_S. As the stirring field is brought progressively closer to the Larmor frequency of the S nuclei, the I doublet lines move together and coalesce to a singlet when the stirring field has come precisely into resonance.

The primary utilization of decoupling is to facilitate analysis of complicated spectra. Spin stirring allows simplification of overlapping multiplets by collapsing one of them to a single line, a process that may be repeated *seriatim* for each multiplet. At the same time, assignment of which multiplets are indeed coupled is verified, since the stirring field must be fixed precisely on the resonance frequency of one set of nuclei to collapse multiplets produced by it. Spin stirring experiments fall into two categories. If nuclei of the same kind are decoupled—for example the methyl protons in ethanol might be decoupled from the methylene protons—the term *homonuclear* decoupling is applied. Decoupling the proton from the fluorine in HF is termed *heteronuclear* decoupling. In principle, the two techniques are the same, however, in practice the latter is easier to accomplish since the stirring and observing frequencies are farther apart. Homonuclear decoupling requires very precisely fixed frequencies, the stirring frequency must be capable of very fine adjustment, and the observing detector must be tuned to a very narrow bandpass to avoid pickup from the much more powerful stirring field. In spite of these requirements, homonuclear spin decoupling is, with modern instruments, a relatively routine experiment widely exployed in the analysis of complex proton nmr spectra.

A variant of heteronuclear decoupling that is routinely used in ^{13}C nmr experiments is known as *broad-band decoupling* (sometimes called noise decoupling). Since the range of proton chemical shifts is small, it is possible to irradiate all the proton resonances of a sample simultaneously. It is necessary only to employ a stirring field that provides sufficient power over the required bandwidth. Thus, all the carbon–proton spin couplings are eliminated. This procedure not only simplifies the ^{13}C spectrum, but in many instances enhances the signal-to-noise ratio as well.

It is an interesting sidelight that the double-irradiation technique makes it possible to measure the resonance frequency of nuclei without having available any electronic detection device for that frequency. In the example given above, consider observation of the proton resonance of HF using a 60-MHz proton spectrometer. The signal is a doublet separated by J_{HF}. Spin stirring with a second frequency can now be employed, and the second frequency varied until the spin multiplet in the proton spectrum collapses to a singlet. Measurement of the stirring oscillator frequency yields the resonance frequency of the fluorine nucleus. The success of this experiment is, incidentally, not at all dependent on a population difference between the levels that are stirred.

Finally, it should be noted that the description often found, of a double-irradiation experiment as a saturation of the S spins, is incorrect. The quantitative condition for decoupling is $|\gamma_S H_{1S}| \gg J$, a condition that is much more stringent than the condition for saturation, $\gamma_S^2 H_1^2 T_{1S} T_{2S} > 1$. In other words, while indeed the S spins are saturated in the stirring experiment, the stirring field must be considerably greater than that required merely to saturate if decoupling is to occur.

20.6 / MAGNETIC RESONANCE INSTRUMENTATION

The general requirements for performing a magnetic resonance experiment are the same as in any other branch of spectroscopy except for the addition of the magnet. Required, therefore, are a source of monochromatic radiation, a detection system, and a method of producing the sweep through the spectral region of interest. In these regions of the spectrum, the source and monochromator are one, a crystal-controlled oscillator in the radio frequency region for nmr, and the klystron tube (Section 19.2.1) in the microwave region for esr.

Detectors for radio frequencies are simply coils of wire tuned to the appropriate frequency region, which function as radio antennas. Microwave radiation is detected by solid state devices, usually silicon–tungsten crystals that have the property of converting microwave radiation to direct current.

It is possible to employ frequency sweep to probe the spectral region of interest in nmr, although it is still more common to employ magnetic field sweep in routine instrumentation. This is so because it proved simpler in practice to produce very stable *fixed* frequencies than very stable variable frequencies. The magnetic field is brought into the general region of interest using a large permanent or electromagnet, with the choice between these two being largely the preference of the design engineer. The small sweep required for nmr is then produced by varying the current passed through a set of auxiliary coils, known as sweep coils or Helmholtz coils. Because large variations in magnetic field (sometimes thousands of gauss) are required in esr experiments, it is not practical to employ permanent magnets or auxiliary sweep coils. In these systems, the sweep is produced by varying the current from the main supply to the electromagnet.

As we shall see below, the sensitivity of an nmr experiment increases with the square of the applied magnetic field strength. Of course, in any magnetic resonance experiment, the difference between the upper and lower spin-state populations will increase with increasing applied field, thereby increasing the sensitivity. It is desirable, therefore, to operate one's spectrometer at the highest practical field strength. In nmr, the factor that limits the tendency toward higher magnetic fields is the technical ability to produce magnets with both very strong fields and the necessary homogeneity over the volume of the sample. As the strength of the field is increased, it becomes increasingly difficult to achieve the required homogeneity.

In electron resonance, the limiting factor in the drive toward higher fields and frequencies is the technology for handling very high microwave frequencies. As the wavelength is decreased, so must the size of the sample be decreased. Thus, the practical operating limits of frequency and field in esr are usually set by the configurations of sample that are acceptable to the investigator.

20.6.1 / Instrumentation for proton magnetic resonance

Because the chemical shifts associated with protons are smaller than those of other nuclei, the most stringent requirements for instrumentation will apply to spectrometers designed for high-resolution proton resonance. A design that suffices for protons will, in most respects, be suitable for any of the heavier nuclei.

It is presently assumed that resolution of the order of 0.3 Hz should be achievable with a modern proton resonance spectrometer. Assuming that the operating frequency is 60 MHz, this level of resolution requires stability of both field and frequency to 5 parts in 10^9 and, in addition, homogeneity over the sample volume at this same level. Achievement of these goals is a formidable task for the engineer.

Evidently, the electronics of the oscillator circuit and those of the magnet power supply must be designed and constructed with utmost care. The physical design of the magnet is also critical. Some idea of the care required in building such a magnet can be had by noting that the pressure of one finger on the magnet yoke may produce changes in the field at the part in 10^7 level! Magnet pole caps must be fabricated from metallurgically uniform iron, machined with great accuracy, and finally polished to optical flatness. Manufacturers find reject rates on finished pole caps as high as 90 percent. Even the best of magnets are sensitive to changes in temperature. Permanent magnets show a negative temperature coefficient of 1 part in 5000 per degree; electromagnets are somewhat less temperature sensitive. Thus, nmr spectrometer function best in carefully air-conditioned laboratories. Many spectrometers are equipped with elaborate insulation and/or thermostatting devices for the purpose of stabilizing the magnet temperature.

In spite of enormous effort, it is not possible to achieve the required homogeneity over the volume of the normal nmr sample by magnet design alone. It is necessary to add electrical coils to the pole caps (called *shim coils*) so that small currents can be used to balance out residual inhomogeneity. In practice, even so, the requisite homogeneity is achieved only along the long axis of the nmr sample tube (the y direction). The field will be constant to $\sim$0.1 milligauss over that length of the sample from which the nmr signal is derived. The homogeneity required normal to the sample tube is somewhat less stringent, 1–2 milligaus, because the spins can be fooled into believing that the field is more homogeneous than it really is.

This is accomplished by using an air-driven turbine to spin the sample tube rapidly about its long axis. In this way, the residual inhomogeneity in the z and x directions is averaged to an acceptable value on the time scale of the measurement.

Even with the best of modern design and construction, it is impossible to eliminate drift and instability completely. Under excellent operating conditions, the combined drift from all sources will usually not be less than 0.25 Hz/min. Temperature and line voltage fluctuations or component degradation will, of course, increase this value. Drift introduces errors in spectral data, particularly with the use of the slow sweeps frequently required to attain good resolution and signal to noise, and it renders difficult or impossible the application of signal accumulation (*vide infra*). Only after a solution was found to the problems of drift and instability did nmr spectrometers become available that could be routinely used by relatively unskilled operators.

The reader will observe that, in part, these problems lie in the independence of the radiation source and the magnetic field. Problems of this type were greatly reduced by the technique of *field–frequency locking*, which was introduced commercially on the Varian A-60 spectrometer in 1960. All commercial high-resolution nmr spectrometers now employ some variation of this field–frequency lock technique. The *external lock* is the simplest form of the technique and involves the use of a separate sample in addition to the analytical sample. The lock sample is usually a small amount of water that is made a permanent part of the spectrometer probe. Both the lock sample and the analytical sample are irradiated with rf from the same master oscillator. A separate detection circuit is used to observe the resonance of the lock sample, and in some designs the dispersion mode is used. As seen in Fig. 20.23(a), when the field and frequency are exactly at the resonance condition, the dispersion-mode signal is zero. Any drift away from resonance will result in a signal, the sign of which will depend on the direction of the drift. This error signal is amplified and fed back to correct either the field or frequency to maintain the resonance condition, $\omega_0 = \gamma H_0$ for the lock sample. So far as observation of the analytical sample is concerned, it is a matter of indifference to the operator whether either the field or the frequency is a precisely fixed quantity, so long as their ratio can be maintained.

The external lock method suffers from the disadvantage that the lock sample and the analytical sample are of necessity placed in slightly different regions of the magnetic field. This disadvantage is avoided by use of the *internal lock* technique, in which one peak of the sample spectrum is used to provide the lock signal. In practice, it is most common to lock onto a peak from an added standard compound, usually TMS, although any sharp singlet in the spectrum could be used. It is not possible to observe the peak used for an internal lock while the spectrometer is in the internal lock mode. The principal disadvantage of internal lock operation is that considerably

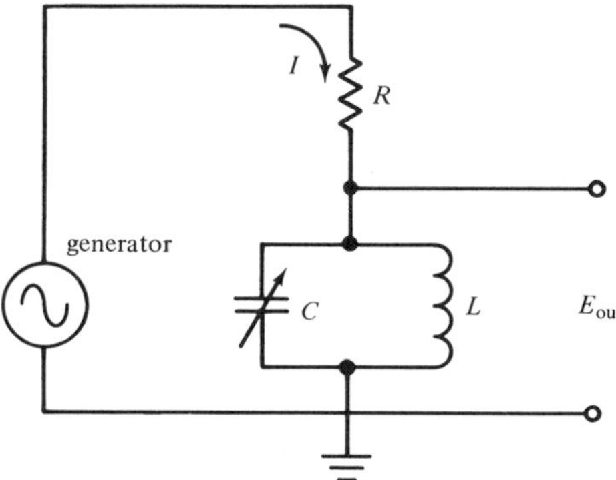

Figure 20.27 / *A simple single-coil* nmr *detector.*

more operator skill is required than for external lock operation. The former method, however, is preferred for the highest-quality spectra.

A / *Detection systems for nuclear magnetic resonance*

Detection circuits for observing the absorption of rf energy by a sample of nuclear spins fall into two classes: *single-coil*, or rf bridge detectors, and *cross-coil* devices, which operate on the principle known as *nuclear induction*. Both types are in common use in commercial instruments, with the choice dictated in part by the designer's preference and in part by a number of factors too specialized for the present discussion. The single-coil system will be described first.

Consider the circuit of Fig. 20.27, which consists of a rf generator, a resistor, a coil of inductance L within which the sample tube is placed, and a tuning capacitor C. The coil is tuned to be resonant at the rf frequency by adjustment of the capacitor and can therefore be represented as an impedance, $Z = \omega LQ$, where Q is the quality factor of the coil. The coil and contained sample are of course immersed in the homogeneous magnetic field H_0. The output of the rf generator is of sinusoidal form, $E = E_0 \sin \omega t$. The system functions simply as an rf voltage divider. If the resistor R is large with respect to Z, a constant AC current will flow in the circuit and the output voltage across the sample coil will be $E_{out} = iZ$. At resonance, the absorption of energy by the nuclear spins will lower the impedance across the coil slightly by decreasing Q, and the output voltage will be lowered proportionally. The disadvantage of this simple system is that it must detect small changes in a large rf voltage, with the result that minute fluctuations in the output of the rf source appear as noise.

A bridge circuit such as is shown in Fig. 20.28 may be used to eliminate this difficulty. In this circuit, the value of R_2 is adjusted so that $iZ = iR_3$; thus, the value of E_{out} is zero at balance. Now only when the coil impedance is altered by nmr absorption is the bridge unbalanced; the unbalance appears as the output voltage. Real circuits are, of course, more complex than the

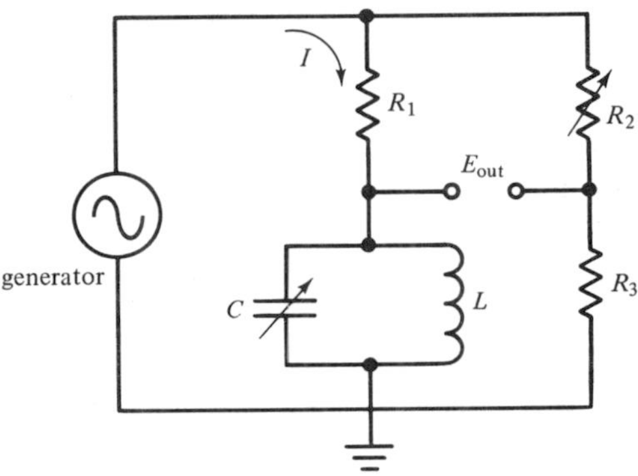

Figure 20.28 / *A bridge-type single-coil* nmr *detector.*

one shown. Because of the inevitable presence of stray couplings, at the minimum, two controls are required to balance a bridge circuit, one for amplitude and a second for phase.

In Fig. 20.29 the cross-coil circuit is shown. In this configuration, the sample is placed inside the *receiver coil* and that assembly is placed at right angles to the *transmitter coil*, which is connected to the rf oscillator. Because of the orthogonality of these two antennas, there is, ideally, no coupling between them. At resonance, however, the spins cause coupling between the coils because the magnetization of the sample acquires a nonzero component of sinusoidal form along the direction of the receiver coil. This time-varying magnetic field induces a voltage in the receiver coil—hence the name nuclear induction to describe this experiment. The cross-coil detector has the intrinsic advantage of having a separate output coil, which can be chosen for maximum signal pickup and impedance matched to the first stage of amplification that follows.

In real cross-coil systems, there is always some coupling between the two coils no matter how well they are designed. The complete system usually

Figure 20.29 / *A simplified schematic diagram of a cross-coil* nmr *detector.*

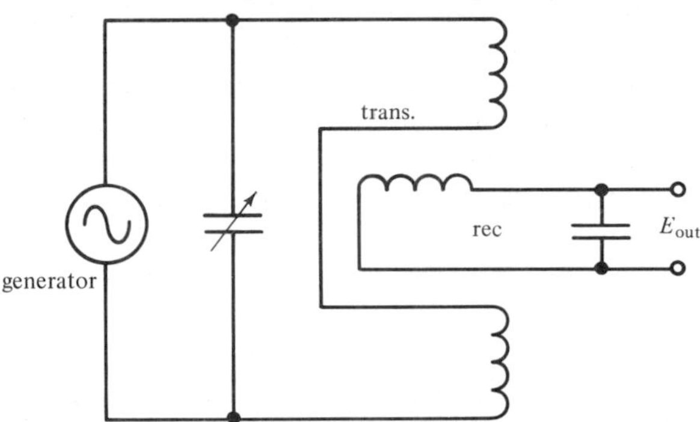

contains, therefore, two additional independent coils, the orientation of which can be adjusted. These coils, called *paddles*, may be used to complete the decoupling of the two coils, or to add coupling of any phase desired by the operator. The paddles may therefore be used to select either the absorption or dispersion mode for signal presentation.

B / *Theory of the nuclear induction detector*

The nuclear induction experiment is relatively simple to treat within the Bloch-equation formalism. Assume that the radiation provided by the transmitter coil is polarized along the x direction and that the signal is detected as a voltage induced in the receiver coil along the y direction, reserving, as is customary, the z direction for H_0. The induced voltage, V, appears because the coil of wire is subjected to an oscillating magnetic field, in particular the y component of nuclear magnetization, M_y. The value of V is obtained from Faraday's law of electromagnetic induction, $V = -k(dM_y/dt)$. Here k is a proportionality constant that will depend on the geometry of the sample and the size of the receiver coil.

We may obtain $M_y(t)$ in terms of the rf susceptibilities from the Bloch equations as indicated in Eq. (20.29),

$$M_y(t) = 2H_1(\chi'' \cos \omega t - \chi' \sin \omega t) \tag{20.29}$$

and differentiation with respect to time gives Eq. (20.30).

$$\frac{dM_y}{dt} = 2H_1\omega(-\chi'' \sin \omega t - \chi' \cos \omega t) \tag{20.30}$$

Therefore, as shown in Eq. (20.31), V contains a component that is in phase with the exciting rf radiation and also a component that is out of phase, that is, both u-mode and v-mode signals.

$$V = 2kH_1\omega(\chi'' \sin \omega t + \chi' \cos \omega t) \tag{20.31}$$

The analytical form of the dependence of the susceptibilities on frequency has been given in Eqs. (20.24). Since it is most common to obtain nmr spectra in the absorption mode, only the v-mode signal will be considered further.

Clearly, the maximum v-mode signal occurs at the resonance frequency, ω_0, so the maximum observed induced voltage has the functional form

$$V_{\text{max}} \propto \frac{H_1\chi_0\omega_0{}^2T_2}{1 + \gamma^2H_1{}^2T_1T_2} \tag{20.32}$$

Note that this voltage tends toward zero as the value of H_1 is increased without limit. This is, of course, what we expect as a manifestation of saturation of the resonance. Another important implication of Eq. (20.32) is the dependence of V_{max} on $\omega_0{}^2$. This shows that the sensitivity of an nmr spectrometer increases with the square of the operating frequency and therefore with the square of the applied magnetic field strength.

It is easy to show that the voltage at the absorption maximum is maximized when $(\gamma H_1)^2 T_1 T_2 = 1$ and that the value of this maximum voltage has the form indicated in Eq. (20.33).

$$V_{\mathrm{max(max)}} \propto \frac{\chi_0 \omega_0 H_0}{2} \left(\frac{T_2}{T_1}\right)^{1/2} \tag{20.33}$$

An important implication of Eq. (20.33) is that peak heights in an nmr spectrum depend on T_2. Since there is no a priori reason to suppose that the different peaks in a spectrum have the same T_2, peak heights are not used as measures of χ_0 to obtain the number of nuclei. The area under the peak is of the form

$$A \propto \frac{\chi_0 H_1}{(1 + \gamma^2 H_1{}^2 T_1 T_2)^{1/2}} \tag{20.34}$$

which is independent of relaxation times when the saturation factor is small. Integrals of nmr peaks are therefore suitable for estimation of the relative numbers of nuclei that give rise to particular absorptions.

20.6.2 / Instrumentation for electron spin resonance

As pointed out previously, the principles of esr are identical to those of nmr, but the large difference in frequency necessitates different engineering approaches. Microwave radiation from a klystron tube is moved about the esr spectrometer in wave guides in order to avoid the very large resistive losses introduced by skin effects, which appear with ordinary cables at these high frequencies (Section 19.2.2).

In place of the resonant coil used in nmr, the sample in the esr spectrometer is placed in a *resonant cavity*. The coil approach is not feasible at microwave frequencies because of the skin effects just mentioned and also because the dimensions of the circuit elements become comparable to the wavelength of the radiation, which introduces large losses by radiation. A microwave resonance cavity is simply a box made with high-conductivity inner walls and that has dimensions comparable to the wavelength in use. Near its resonance frequency, the cavity is able to sustain microwave oscillations (standing waves) from the superposition of microwaves multiply reflected from the cavity walls. Cavities of many types have been devised, but the ones in common use in esr spectrometers at present are nearly all of the reflection type and either rectangular or cylindrical in shape. Like the induction coil, a cavity can be described by an impedance Z and a quality factor Q. Just as in the nmr case, the absorption of energy by the sample in the cavity is observed as a decrease in Q with the consequent unbalancing of a bridge circuit.

The heart of the microwave bridge circuit is, in modern spectrometers, a ferrite circulator. This device employs ferromagnetic inserts to rotate the plane of polarization of the microwave radiation so that with proper

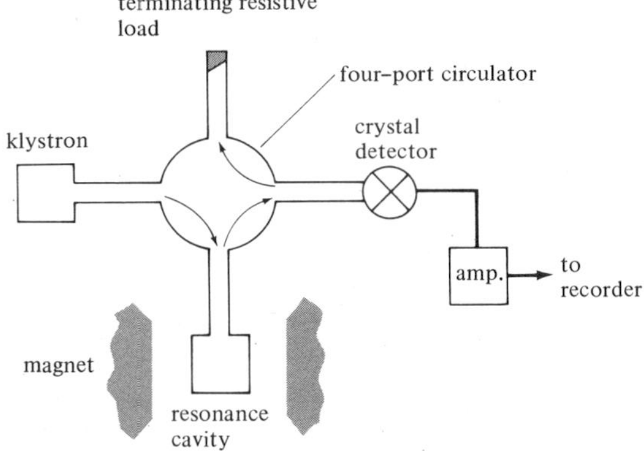

Figure 20.30 / *A simple esr spectrometer based on a four-port ferrite circulator. Arrows indicate the allowed direction of microwave travel. Unwanted reflections from the detector are allowed to fall on a nonreflecting termination.*

orientation of the wave guides the microwave power entering a given port is transmitted entirely to the next adjacent port and nowhere else. A simple circulator spectrometer is shown schematically in Fig. 20.30.

The cavity can be impedance matched to the rest of the circuit so that, in the absence of resonance absorption by the sample, no power is reflected and the crystal detector observes null. In practice, this is not done since it is necessary to bias the crystal to a region of its operation in which its output current is a linear function of the microwave power falling on it. The bridge is therefore adjusted so that a predetermined bias power (selected to suit the particular type of crystal) reaches the crystal when the sample is off resonance; at resonance, the cavity impedance and bridge balance change and these changes alter the current from the crystal. These changes in crystal current determine the esr spectrum.

The requirements for magnetic field homogeneity and stability are somewhat less stringent for esr than for high-resolution nmr, since the widths and separations of esr lines are much larger than the corresponding nmr parameters. Typically, line separations will be from a few tenths of a gauss to several hundred gauss, and observation of lines narrower than 20 milligauss (60 kHz at $g = 2$) is quite rare. Typical esr spectrometers operate at 9.5 GHz with fields of $\sim$3400 gauss for $g = 2$ samples.

20.6.3 / Magnetic field modulation in magnetic resonance

In the simplified nmr and esr systems described above, the signal would be detected at rf and thereafter handled as a change in DC level as the magnetic field is slowly swept through the spectrum. Difficulties are encountered with such a procedure. One serious difficulty would be the need to employ DC amplification of the signal. In Section 4.12.3, it was shown that AC amplifiers enjoy significant advantages in freedom from noise and drift.

It is therefore desirable to amplify magnetic resonance signals as AC voltages if possible. In both esr and nmr, the bridge circuit balance will change with both frequency and variation of component temperature, making base-line drift inevitable. Base-line drift is particularly annoying when the resonance signal is to be integrated. If the drift is integrated along with the desired signal, the result will contain large errors in peak areas.

The crystal detectors employed in esr generate thermal noise somewhat like that in resistors, but of larger magnitude. This noise amplitude is found to be inversely proportional to frequency, so significant improvement in the signal-to-noise ratio can be achieved by obtaining the spectral information at high frequency rather than at DC.

These *desiderata* are routinely attained by the use of magnetic field modulation. Therefore, in addition to the main magnet and slow-sweep coils, a separate set of coils is introduced to which a sinusoidal voltage is applied in order to produce a small-amplitude sinusoidal magnetic field at the sample. The net field at the sample is, then, excluding the linear sweep, $H_0 + H_m \cos \omega_m t$, where H_m is the peak amplitude and ω_m is the frequency of the modulation. Under certain conditions, this will result, as H_0 is swept slowly through a line, in turning the resonance on and off at ω_m in the manner of the chopper used in optical spectroscopy. The magnetic resonance information is then coded at frequency ω_m and, following detection at the rf frequency, it can be amplified by a narrow-band AC amplifier. Normally, this amplifier is of the phase-sensitive lock-in type (Section 6.4.3), so a portion of the modulating signal is fed to the amplifier as the phase reference. With such a system, only signals that have both the frequency and phase of the modulation are amplified, and all others are rejected.

As with other aspects of the two experiments, here again the differences between esr and nmr are more apparent than real, the principles of field modulation being identical in both techniques. The results appear quite different, however, because of the large differences in the linewidths observed in the two experiments. In any magnetic resonance experiment, it is necessary to recall that magnetic field and frequency are inextricably connected by the relation $\omega = \gamma H$. Hence, not only the *amplitude*, but also the *frequency* of the modulation employed will be important. The detailed effects of field modulation depend on answers to the following questions: Is the width of the resonance line large or small with respect to the amplitude of the modulation? And equally important, is the linewidth (in frequency units) large or small with respect to the modulation frequency?

Since the minimum esr linewidths will be several tens of kilohertz, both the amplitude and frequency of modulation can be made small on that scale. In nmr, by contrast, with lines of the order of 0.5 Hz in width, any useful modulation must be of much larger frequency and may also be of larger amplitude. For these reasons, the two experiments are discussed separately below. Note that the discussion of modulation in esr is equally applicable to broad-line nmr, a technique not discussed here, but of considerable

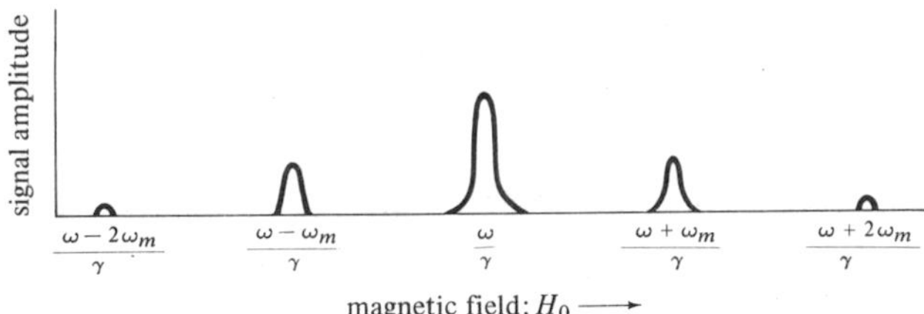

Figure 20.31 / *Single-line* nmr *spectrum showing sidebands resulting from modulation of the magnetic field at frequency* ω_m. *Harmonics with* $n > 1$ *are unimportant so long as* $\gamma H_m \ll \omega_m$. *In practice, the upper and lower sidebands may differ in phase by* 180°.

importance in physicochemical studies of solids. Linewidths encountered in such experiments are comparable to those observed in esr.

A / *Modulation in high-resolution nuclear magnetic resonance*

Nuclear magnetic resonance spectrometers employ audio frequency modulation with 5 kHz a typical value. The frequency is chosen for ease of amplification, with the restriction that it should be larger than the largest chemical shift one expects to observe. Use of modulation results in the appearance of sidebands separated from the resonance signal by the modulating frequency, and if this value is too small, sidebands from one resonance line may interfere with observation of another line. Quantitatively, modulation such that the field takes the form $H_0 + H_m \cos \omega_m t$ results in the appearance of signals not only at $H_0 = \omega_0/\gamma$, but also at $H_0 = (\omega \pm n\omega_m)/\gamma$, where n is any integer. Under the usual conditions, harmonics with $n > 1$ have very small amplitude and are neither useful nor bothersome. A possible sideband pattern is shown in Fig. 20.31.

The mathematical analysis of this modulation technique is unfortunately complex since, in spite of the fact that the amplitude of the magnetic field is varied, it becomes a *frequency modulation* technique. The results are, in fact, identical to those that could be obtained by frequency modulation of the irradiating rf. Field modulation, however, enjoys the significant advantage that the phase-sensitive detector observes null in the absence of a resonance line, whereas if the rf field were modulated, a signal would always be present and resonance would correspond to a difference in level. A complete analysis of the problem shows that the nmr signal can be detected satisfactorily in either the center band *or* in the sidebands, contrary to the case of amplitude modulation discussed in Chapter 6. Instruments in current use are divided among those that employ center band detection and those that employ first sideband detection. In either case, the absorption (or dispersion) signal is obtained directly.

Additional sidebands are sometimes introduced for the purpose of making accurate measurement of chemical shifts. Consider a spectral fragment consisting of two lines, A and B; it is desired to measure their separation precisely. Sidebands can be produced using an adjustable, low-frequency audio oscillator. As the oscillator frequency is increased from zero, eventually the frequency separation of A and B is matched, at which point one sideband from each line is exactly coincident with the other line itself. The chemical shift difference is obtained by counting the oscillator output with an electronic frequency counter.

B / *Modulation in electron spin resonance*

While it rectifies microwave radiation, the crystal detector will pass reasonably high frequencies. Since the noise generated by the crystal itself is lower at the higher frequencies (so-called 1/f noise), esr spectrometers commonly employ magnetic field modulation at 100 kHz. That frequency has a magnetic field equivalent of 35 milligauss at $g = 2$ and, as a result, sidebands appear at this distance above and below the center of the resonance line. Consequently, lines narrower than about 100 milligauss are distorted by the presence of the unresolved sidebands. Most spectrometers are equipped with a second, lower-frequency, modulation oscillator and phase detector in order to deal with very narrow lines.

Under the conditions of the esr experiment, it is also necessary to keep the amplitude of the modulation small. The peak-to-peak amplitude of the modulation should, if possible, be 0.1 or less of the linewidth in order to prevent distortion of the line shape. This rule is frequently violated, however, since the signal to the phase detector increases linearly with modulation amplitude until the latter is about equal to the linewidth. It is often necessary to sacrifice precise rendition of the line shape to obtain the additional intensity.

With the restriction of small amplitude, the effect of modulation is illustrated in Fig. 20.32(a). It is seen easily from the figure that the signal amplitude is proportional to both the modulation amplitude and the slope of the absorption curve. The crystal detector passes the spectral information at the modulation frequency to the amplifiers and phase-sensitive detector. The signal is rectified in the phase detector as described in Chapter 6, except that the mechanical synchroverter is replaced by a set of diode switches, which are switched between a conducting and nonconducting condition by the phase reference signal. The natural output of a phase detector under the conditions specified here is the first harmonic of the absorption, as illustrated in Fig. 20.32(b). Under conditions of small modulation amplitude (0.1 linewidth or less) this output is reasonably accurately the *first derivative* of the absorption; thus, double integration to obtain spin concentrations gives accurate results only when the modulation amplitude is suitably small.

20.6.4 / Techniques for signal-to-noise enhancement

Since nmr is a branch of spectroscopy that deals with extremely weak absorptions, one is more frequently faced with a need for signal-to-noise enhancement in this field than in most other types of spectroscopy. With

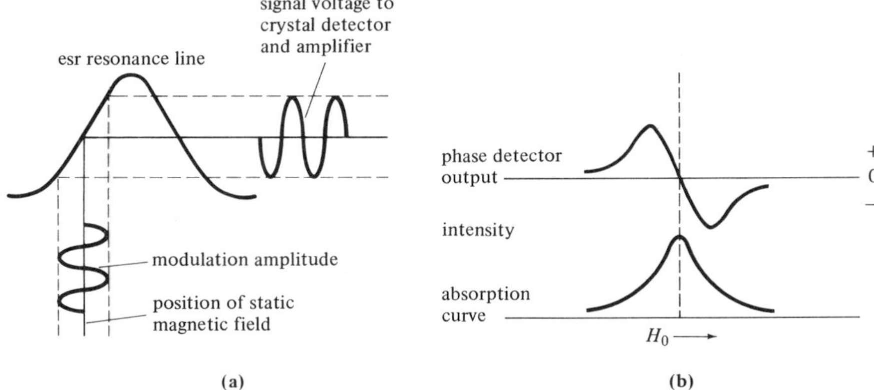

Figure 20.32 / (a) *The form of the signal that results from field modulation of an esr line when the amplitude of modulation is smaller than the linewidth. The signal to the amplifier is easily seen to be proportional to both the amplitude of modulation and the slope of the absorption line.*
(b) *The form of the output from the phase detector which, under these conditions, is approximately the first derivative of the absorption.*

proton spectra, S/N enhancement is required only when the amount of sample is extremely small. In nmr of other nuclei, enhancement techniques are common since the sensitivity of the experiments may be drastically reduced by low isotopic abundance, small nuclear moment, or a combination of the two. For example, carbon nmr, which is destined to play a large role in organic structure determination (*vide supra*), requires S/N enhancement on a routine basis.

A / *Signal averaging*

As described in Sections 6.4.2 and 9.4, repetitive scanning of a spectral region with accumulation and averaging of the signals results in enhancement of the signal-to-noise ratio because the signals are coherent and add together, but the noise is random and has a long-term average value of zero. It can be shown that a signal-averaging procedure leads to an increase in S/N that is proportional to the square root of the number of replicate scans.

This is a relatively simple procedure in both nmr and esr and requires a small digital computer with at least 1024 storage channels, an analog-to-digital converter (ADC) to accept the spectral information and convert it to digital form, and a digital-to-analog converter (DAC) to convert the stored digital information to analog form suitable to run a chart recorder. It is, of course, absolutely necessary to ensure that the repetitive scans fall exactly on top of one another. Failure to obtain exact coincidence broadens lines, degrades resolution, and leads to less than the theoretical enhancement. The success of the averaging procedure depends, therefore, on how well the spectral scan is locked to the address advance of the computer. In many modern instruments, the scan is generated by pulse methods. One such method uses a stepping motor drive in which the motor advances a certain fraction of a revolution each time an electrical pulse is applied to it. In this case, use of that same pulse to advance the address that is accepting the spectral data locks the sweep and computer together very effectively.

In Fig. 20.33, the progressive improvement of S/N is easily seen as the number of scans increases. This very large number of scans indicated in the figure, which leads to a 40-fold improvement in S/N, is practical only when the spectral region of interest is sufficiently narrow that the scans can be repeated quite quickly.

Unavoidable instrumental drift sets the upper limit to times over which signal averaging can be carried out without loss of effectiveness or degradation of resolution to 20–30 hr. Since a typical high-resolution scan may require about 10 min and 100 scans are necessary for only a 10-fold improvement in S/N, the instrumentally imposed limitations are serious. In addition, of course, limiting the use of an instrument costing $100,000 to a single sample per day is regarded as undesirable in most laboratories. It is for these reasons that an additional method of S/N enhancement has been sought for use in nmr.

B / *Fourier transform nuclear magnetic resonance*

As suggested in Section 6.4.4, Fourier transform spectroscopy can provide significant savings of instrument time since the entire spectrum can be observed simultaneously. Application of the Fourier method to nmr is discussed in this section.

As indicated in Section 6.4.4, Fourier transform spectroscopy requires simultaneous irradiation of the sample at all the frequencies of interest. Unlike the IR source, which provides these frequencies quite naturally, a rf source is inherently highly monochromatic. For this reason, the Fourier transform nmr experiment appears to be very different from the IR interferometry experiment described in Section 11.6.2; however, the apparent differences are artifacts of the nature of the radiation sources and the experiments are in principle identical.

As discussed in Section 1.3.3, any periodic function can be constructed using a Fourier series summation of sine and cosine functions. This is illustrated in Fig. 1.17, where a square wave of frequency ω is constructed using the odd harmonics of ω and sines alone. In the case of such a periodic function, nonzero amplitudes occur only at discrete frequencies and the graphical presentation, amplitude vs. frequency, is a line spectrum such as that shown in Fig. 1.18. By extension of this method, nonperiodic time functions can be represented by *Fourier integrals* in which the sinusoidal constituents have all possible frequencies. It is therefore possible to represent any real time response function as a continuous frequency spectrum.

For reasons of compactness, it is usual to express sine and cosine functions in complex exponential form and to extend the lower limit of integration

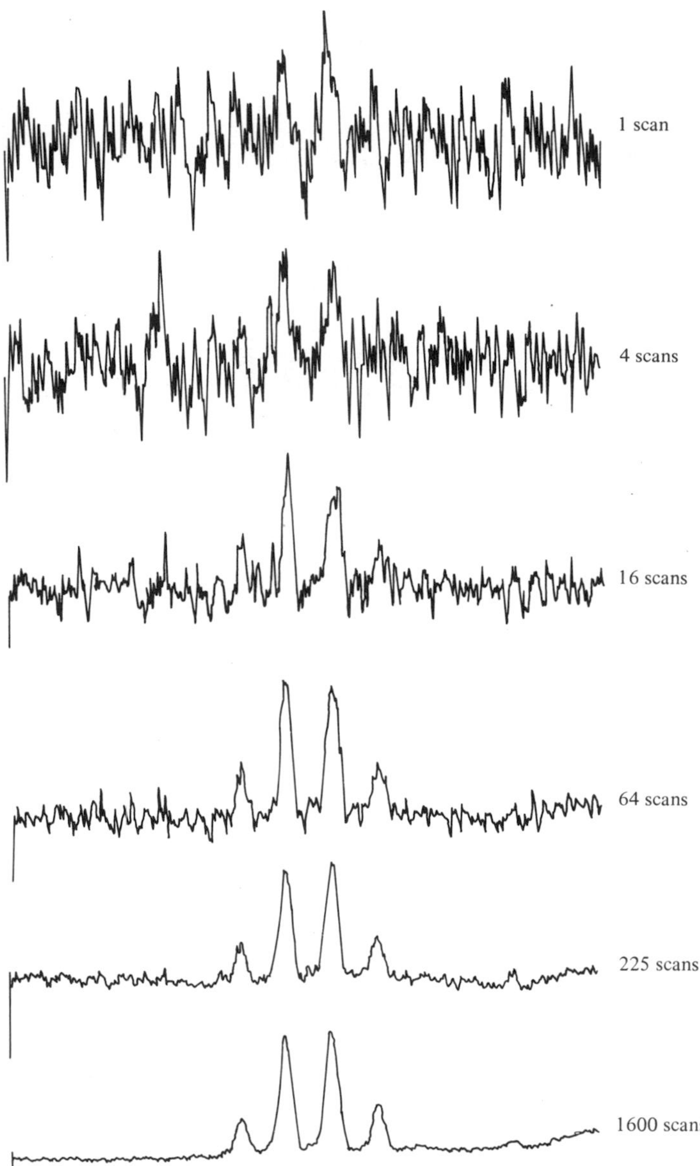

Figure 20.33 / *An example of signal averaging. These are spectra of the methylene quartet of ethylbenzene in very dilute solution* (0.1 *percent in* CCl_4) *shown after 1, 4, 16, 64, 225, and 1600 scans have been summed by a 1024-channel computer.*

SOURCE: F. A. Bovey, *Chemical and Engineering News*, August 30, 1965, p. 110. Reprinted by permission of the American Chemical Society.

from zero to negative infinity, which permits use of only one exponential rather than two. The resulting integrals,

$$f(t) = \int_{-\infty}^{\infty} F(\omega)e^{i\omega t}\, d\omega \tag{20.35a}$$

$$F(\omega) = \int_{-\infty}^{\infty} f(t)e^{-i\omega t}\, dt \tag{20.35b}$$

are known as *Fourier transforms* since they permit transformation of a time response function, $f(t)$, to a frequency spectrum, $F(\omega)$, and the reverse. These transforms have two important implications for nmr. First, the excitation in Fourier transform nmr is a square pulse of rf energy, that is, a truncated wave train of frequency ω_0. A Fourier transform of such a square pulse shows that it contains a spectrum of frequencies centered at ω_0 rather than a single frequency. Second, we shall see that the time response function that results when a sample of nuclear spins is subjected to such excitation can be converted, by a Fourier transform, to the usual amplitude–frequency presentation of an nmr spectrum.

Consider a sample of magnetically equivalent nuclei that would give a single absorption line under continuous wave (cw) nmr conditions. This sample may be subjected to a short pulse of high power rf at its resonance frequency by application of a suitably high-voltage rf pulse to the transmitter coil of a cross-coil probe. If the output of the receiver coil is examined, a signal will be observed that decays exponentially with time from the conclusion of the pulse. This signal is called the *free induction decay*. Its origin may be understood with the aid of Fig. 20.34, which depicts the situation in a coordinate system that rotates with H_1. In the figure, the magnetization vector **M** is shown initially aligned with the external field, $\mathbf{H}_0$, along the z axis. The strong rf pulse provides the field $\mathbf{H}_1$ along the x direction as shown in part (b) of the figure. As discussed above, this rf field exerts a torque on **M** that tips it toward the xy plane. If the rf power and pulse duration, τ, fulfill the condition $\gamma H_1 \tau = \pi/2$, **M** is brought exactly into the xy plane (tipped 90°) and the pulse is termed a *90° pulse.* At this point [part (c) of the figure], the z component of **M** is zero and the y component is a maximum. It is, of course, the y component that induces the voltage in the receiver coil. As the individual nuclear spin vectors that sum to make **M** begin to get out of phase with one another [part (d) of the figure], the y component of **M** decays. This dephasing is a T_2 process which, according to the Bloch formulation, is expected to be a first-order decay. Therefore, the signal induced in the receiver coil will decay as $\exp(-t/T_2)$, where t is measured from the end of the pulse. The recovery of the z component of **M** is a T_1 process, and the original value of this component will be restored in a time of the order of T_1. Note that the exciting rf pulse must be short with respect to T_2 in order that no dephasing of the spins occurs prior to cessation of the pulse.

If the sample contains two or more magnetically nonequivalent nuclei

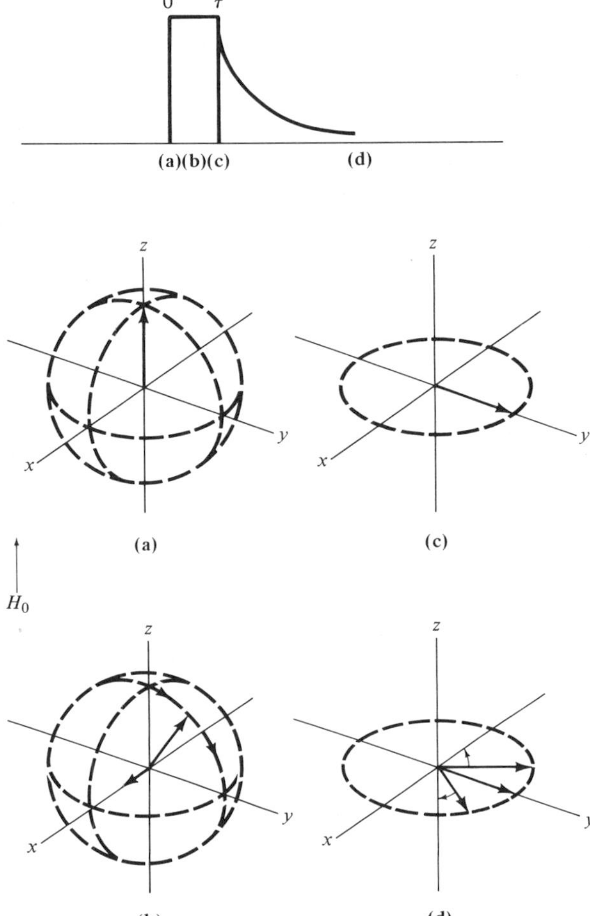

Figure 20.34 / *The origin of the free induction decay* nmr *signal that follows a 90° rf pulse. The magnetization vector is tipped from its initial orientation along H_0 (a) by the torque of the strong rf pulse applied along the x direction (b). At the end of the pulse, the maximum y component has developed (c), which induces a voltage in the receiver coil. As the spin vectors that compose M get out of phase, the y component decays (d). The figure is drawn in the rotating coordinate system.*

such that the cw nmr spectrum would contain several lines, and if all the resonances are simultaneously subjected to pulse excitation, a characteristic free induction decay curve will result for each type of nucleus. These several curves combine to form a beat pattern that is similar to the interferogram obtained from an optical interferometer. The frequency spectrum of this time response function, or interference pattern, is the ordinary nmr spectrum; it can be recovered by application of a Fourier transform of the kind shown in Eq. (20.35b).

It is not necessary to provide separate pulses of different frequencies to excite nuclei with different chemical shifts. Application of a transform of the type (20.35a) to a square pulse of rf centered at the nominal resonance frequency shows that this pulse contains a distribution of frequencies about ω_0. In Fourier transform nmr, the rf frequency is chosen so that all the spectral lines lie on one side of ω_0. The extent of this distribution in the frequency domain is inversely proportional to the duration of the pulse in the time domain (actually $\sim 1/4\tau$); hence, the broader the region of spectral interest, the shorter must be the pulse. Fortunately, nmr spectra are quite narrow, and the pulse times required for simultaneous excitation of all of the resonances in a spectrum are within a range that is easily accessible with modern electronics. However, the power available in a particular frequency interval falls off rapidly as that interval becomes further removed from ω_0. The magnitude of chemical shift that can be investigated is limited by the amount of rf power that can be applied to the transmitter coil without damage to the coil or probe. The large widths of common esr spectra render the Fourier transform method inapplicable.

In practice, a pulse of 50–100 μsec width is used for ^{1}H Fourier transform nmr and, since a larger spectral region must be observed for ^{13}C spectra, the pulse duration is shorter, about 10 μsec. In either case, the peak power in the pulse may provide an rf field as large as 200 gauss. While the free induction decay curve extends to infinity, it is necessary to observe it for at most a few seconds. The resolution achievable in this experiment is approximately the inverse of the observation time, so to obtain 1-Hz resolution, an observation time of 1 sec suffices. The usual procedure is to store the free induction decay signals in a computer and average over a large number of repetitive pulses, thus increasing the signal-to-noise ratio.

Notice that since these data are acquired much more rapidly than in cw spectroscopy, a faster ADC is required. A sampling rate of at least 6 kHz is required if the spectral region to be examined in 1 sec is 3 kHz wide. For maximum flexibility, a digitizing rate of 15–20 kHz is to be preferred.

Since both T_1 and T_2 are involved in the detailed examination of the response of a spin system to pulse excitation, in practice the pulse width, power, and repetition rate are adjusted to suit the particular sample in an empirical manner. Often it is found advantageous to use more complex, multiple-pulse sequences rather than the single-pulse method described here; this is particularly true when measurement of T_1 and T_2 is the objective.

Once the free induction decay data are stored in the computer, a computer program is employed to effect numerically the Fourier transform of Eq. (20.35b). This transformation has become practical for small laboratory computers only since the rediscovery by Cooley and Tukey in 1965 of a particularly efficient algorithm known now as the fast Fourier transform (FFT). Using this algorithm, a 4000-point free induction decay curve can be transformed in about 30 sec by a typical small computer.

Fellget's advantage (Section 11.6.4) is realized in Fourier transform nmr, with the result that a 100-fold improvement in S/N can be realized over cw spectroscopy in a constant-time experiment. Alternatively, for equal S/N, the Fourier transform method can acquire equivalent data in at least 100-fold shorter time. As a conservative example, a natural abundance ^{13}C spectrum may be obtained in less than 20 min when 20 hr would be required to obtain the equivalent result using cw methods.

20.6.5 / State-of-the-art instrumentation

The interest in exploiting the potential of ^{13}C nmr has provided the impetus to develop the Fourier transform technique and the required instrumentation. This method is inherently so desirable for any nmr experiment that one can predict with great confidence that the popularity of pulsed FFT nmr will grow rapidly and that it will become the standard method for routine nmr.

The desire for greater magnetic field strength has been present since production of the initial commercial instruments, which operated at 10 kG and 40 MHz for proton resonance. In a relatively short time, 14 kG and 60 MHz became the standard conditions for ^{1}H spectra, but these systems are now obsolescent. The new standard, 23.5 kG and 100 MHz, seems destined to remain unchallenged for a considerable period, since magnet designers believe that this represents the approximate upper limit to fields that can be produced by iron electromagnets with the requisite homogeneity. The limitation is a result of the limitations of strengths of materials, as one can understand by realizing that a force of 7 tons is exerted on the pole pieces of a 12-in. magnet operating at 14 kG.

Superconducting solenoids can produce very strong magnetic fields without the problems that beset iron pole pieces and with essentially no power consumption. Unfortunately, so far the highest temperature at which such a solenoid can operate is about 18°K, which means that it must be cooled with liquid helium. Several nmr instruments are available that employ superconducting solenoids. These operate between 220 and 300 MHz with 50–70 kG fields. The high initial cost and large operating cost of supplying liquid helium somewhat limit the market for such instruments. Slower growth in the numbers of these instruments than has been the case for iron magnet spectrometers can be anticipated unless a dramatic reduction in the price of liquid helium should occur. The discovery of a material that has its transition to the superconducting state above the boiling point of nitrogen (77°K) would, of course, alter this picture essentially immediately.

20.7 / CHEMICAL APPLICATIONS OF ELECTRON SPIN RESONANCE

Electron spin resonance spectroscopy can be used to study a variety of different chemical species that have unpaired electrons. These include organic free radicals that have a single unpaired electron; transition metal ions and complexes that may have one or more unpaired electrons; and triplet or other excited state species with two or more unpaired electrons. Of these, in general only the transition metal-containing species are stable. Some free radicals are sufficiently stable for examination in fluid media at normal temperatures, while others persist long enough for study only when isolated in an inert matrix or at low temperatures. Radiation damage and photolysis frequently lead to the production of free radicals that can be studied under appropriate experimental conditions.

20.7.1 / Spectral characteristics

The discussion of esr spectra in preceding sections of this chapter indicates only that an absorption line will be observed. In the usual fixed-frequency experiment, the magnetic field strength at which that line appears depends on the g value for the particular species. In addition, magnetic nuclei may interact with the unpaired electron and split this absorption into many lines. The rules for this splitting in esr are the same as those governing spin-spin splitting in nmr. There will be $(2I + 1)$ lines due to the interaction of the electron with a group of magnetically equivalent nuclei of total nuclear spin I. For example, the four equivalent protons in the *p*-benzosemiquinone radical (cf. Section 23.6.5) split the resonance into five equally spaced lines. As in nmr, the intensity ratios of splitting patterns from spin $\frac{1}{2}$ nuclei follow the binomial coefficients, and in this example the intensities are in the ratios 1:4:6:4:1. Nuclei with spin numbers larger than $\frac{1}{2}$ follow the same rule for the number of absorption lines—for example, two nitrogen nuclei ($I = 1$) give five lines; the intensity ratios, however, do not follow the binomial expansion and must be derived as shown in Fig. 20.5(b).

Restricting our attention to species in solution, the separation of lines split by interaction with nuclei is called the *isotropic hyperfine coupling constant*, usually reported in units of gauss. This quantity reflects the "amount" of odd electron in an s orbital in direct contact with the nucleus in question. The same "amount" of unpaired electron will give a different coupling with different nuclei, since it is also proportional to the nuclear moment.

20.7.2 / Electron spin resonance and analysis

While the discovery of a free radical participating in a particular reaction mechanism could be classed as analysis, that discussion is reserved for the following section. It is possible to use esr as both a qualitative and quantitative analytical tool, although the former application is both more common and usually more successful than the latter. An example is the analysis of oil refinery feed stock for V^{2+}. This ion, which has the undesirable property of poisoning certain catalysts, has a characteristic seven-line esr spectrum ($I = \frac{7}{2}$). Thus, esr examination of the entering crude oil can inform the operator of the presence and approximate quantity of V^{2+}. The schedule of catalyst replacement can be altered accordingly rather than simply waiting for a failure.

Various quantitative methods have been developed for analysis of paramagnetic transition metal ions. These depend usually on formation of a complex that is soluble in a nonaqueous solvent, such as ether, followed by esr examination of the solution. Difficulties in reproducing spectral intensities, which are very sensitive to the way the instrument is tuned, as well as the difficulty in obtaining accurate double integration of the spectra limit the accuracy of these methods. While they have the advantages of being simple and rapid and may be very satisfactory for some specific applications, there are almost always more accurate methods available for ordinary concentration levels. The esr methods are very sensitive and become competitive when the concentrations are in the 10^{-6} F range.

A very clever esr method has been developed for the analysis of morphine in urine. A nitroxide moiety is bound reversibly to a proprietary chemical species that, on reaction with morphine, liberates the nitroxide free radical ($R_2NO\cdot$). Estimation of the nitroxide concentration by esr thus indicates the morphine level immediately. The test requires only a few minutes and less than a milliliter of sample. It is so sensitive that morphine use more than 24 hr previously can be detected.

20.7.3 / Free radicals

Much research has been done investigating free radicals in solution. For reasons of radical stability, organic solvents have been more commonly employed than has water, although a number of radicals can be studied in an aqueous environment. In some instances, unambiguous demonstration of the participation of a free radical in a particular reaction sequence is in itself important. More commonly, however, the emphasis is on the evaluation of the isotropic coupling constants which, in turn, may permit determination of the spin-density distribution in the radical. Such estimates of spin densities have frequently served as the touchstone for examination and refinement of molecular orbital theory.

In addition, esr data can reveal unique conformational and structural information about free radicals in solution. Such information may be obtained from several aspects of the spectra, including the magnitudes of coupling constants, the number of coupling constants, and sometimes the observation of two overlaid spectra that arise from different isomeric forms of the same radical. Such investigations form the basis of a fascinating field of research, but are too specialized to discuss at length in this text. The reader is referred to the Suggested Readings at the end of this chapter.

The form in which esr data appear is indicated in Fig. 20.35. The spectrum shown is that obtained from the monofluorosemiquinone radical. Introduction of a fluorine in place of one hydrogen destroys the symmetry of the radical so there are four spin $\frac{1}{2}$ nuclei and each is magnetically distinct. The spectrum therefore contains lines that result from four successive doublet splittings (2^4 or 16 lines if all were resolved). Assignment of the coupling constants that make up the spectrum is indicated in the figure

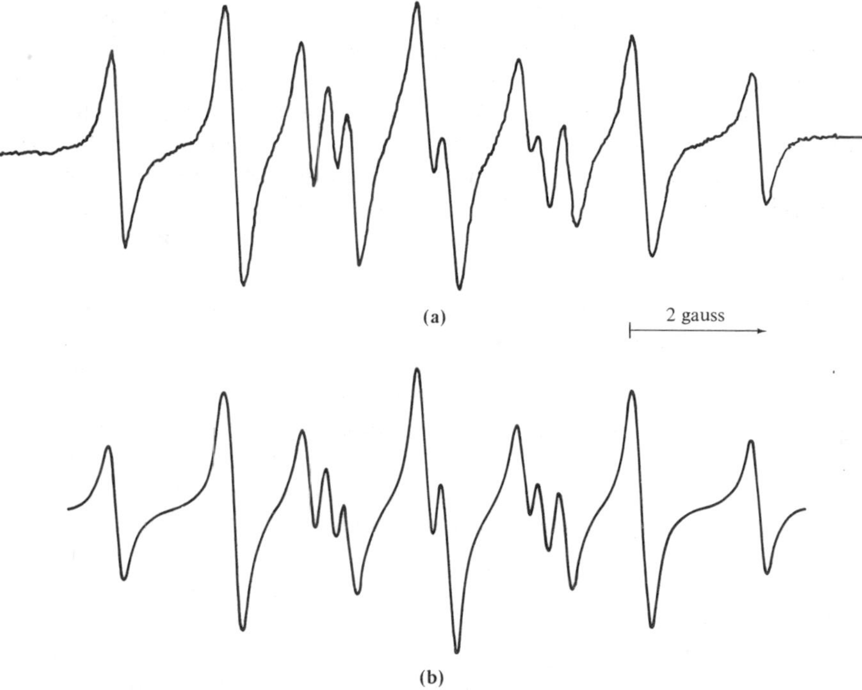

Figure 20.35 / (a) *The first-derivative esr spectrum of monofluorosemiquinone. Radicals were generated by electrolysis of a* 1 m*M solution of the parent quinone in acetonitrile at a platinum gauze cathode in a flow system.*
(b) *Spectrum computed with four spins of* $\frac{1}{2}$ *with coupling constants of* 1.85, 1.95, 3.15, *and* 3.50 G. *A Lorentzian line shape was employed and a line width of* 0.260 G.
SOURCE: W. E. Geiger and W. M. Gulick, *J. Amer. Chem. Soc.* **91**, 4657 (1969). Reprinted by permission of the American Chemical Society.

caption and has been confirmed by entering these parameters into a computer programmed to generate the lower spectrum for purposes of comparison. Such comparisons are used routinely when lines are incompletely resolved.

SUGGESTED READINGS

General

A. Carrington and A. D. McLachlan, *Introduction to Magnetic Resonance*, Harper & Row (New York, 1967).

C. P. Slichter, *Principles of Magnetic Resonance*, Harper & Row (New York, 1963).

NMR and EPR Spectroscopy, Varian Third Annual Workshop Proceedings, Pergamon (Elmsford, N.Y., 1960).

D. H. Whiffen, *Spectroscopy*, Wiley (New York, 1966), Chapters 3 and 5 (paperbound).

Nuclear magnetic resonance

R. M. Silverstein and G. C. Bassler, *Spectrometric Identification of Organic Compounds*, 2nd ed., Wiley (New York, 1967), Chapter 4.

P. Laszlo and P. Stang, *Organic Spectroscopy*, Harper & Row (New York, 1971), Chapters 3–5, 8, and 9.
E. R. Andrew, *Nuclear Magnetic Resonance*, Cambridge University Press (Cambridge, 1958).
J. A. Pople, W. G. Schneider, and H. J. Bernstein, *High Resolution Nuclear Magnetic Resonance*, McGraw-Hill (New York, 1959).
L. M. Jackman and S. Sternhell, *Applications of Nuclear Magnetic Resonance Spectroscopy in Organic Chemistry*, 2nd ed., Pergamon (Elmsford, N.Y., 1969).
A. Abragam, *The Principles of Nuclear Magnetism*, Oxford University Press (London, 1961).
T. C. Farrar and E. D. Becker, *Pulse and Fourier Transform NMR*, Academic (New York, 1971).

Electron spin resonance

D. J. E. Ingram, *Free Radicals*, Butterworths (London, 1958).
G. E. Pake, *Paramagnetic Resonance*, W. A. Benjamin (Menlo Park, Calif., 1962) (paperbound).
P. W. Atkins and M. C. R. Symons, *The Structure of Inorganic Radicals*, Elsevier (New York, 1967).
D. H. Geske, "Conformation and Structure as Studied by Electron Spin Resonance Spectroscopy," in *Progress in Physical Organic Chemistry*, Vol. 4, A. Streitwieser, Jr., and R. W. Taft, Eds., Wiley-Interscience (New York, 1967).
D. J. E. Ingram, *Biological and Biochemical Applications of Electron Spin Resonance*, Plenum (New York, 1969).
C. P. Poole, Jr., *Electron Spin Resonance*, Wiley-Interscience (New York, 1967).

PROBLEMS

1 / Calculate the resonance frequency, ν_0, required for ^{13}C, ^{11}B, and ^{15}N if the magnetic field is 20 kG.

2 / If the only oscillator and associated electronics available gives rf at $\nu_0 = 25$ MHz, what magnetic fields will be required for ^{1}H, ^{31}P, and ^{15}N resonance?

3 / The usual 100-MHz proton resonance spectrometer operates with the magnetic field at the maximum value that can be obtained from the electromagnet. It is common to equip such an instrument for ^{13}C by addition of a second oscillator and probe. At what frequency will ^{13}C be observed? Using the ^{13}C electronics, how would you go about observing ^{31}P resonance?

4 / Although 5 kHz is a common modulation frequency in proton resonance, it is unsuitable for ^{13}C experiments. Estimate the minimum modulation frequency required for a ^{13}C spectrometer that operates at 25 MHz.

5 / Compute the difference in population of the upper and lower spin states for protons in a 10-kG field at 300°K. By what percentage will this value change if the sample temperature is lowered to that of liquid nitrogen (77°K)?

6 / Two proton resonance singlets are observed to be separated by 10 Hz. If these lines arise from two equally probable sites that can undergo random exchange, what rate of exchange will just bring about collapse of the signal to a single line?

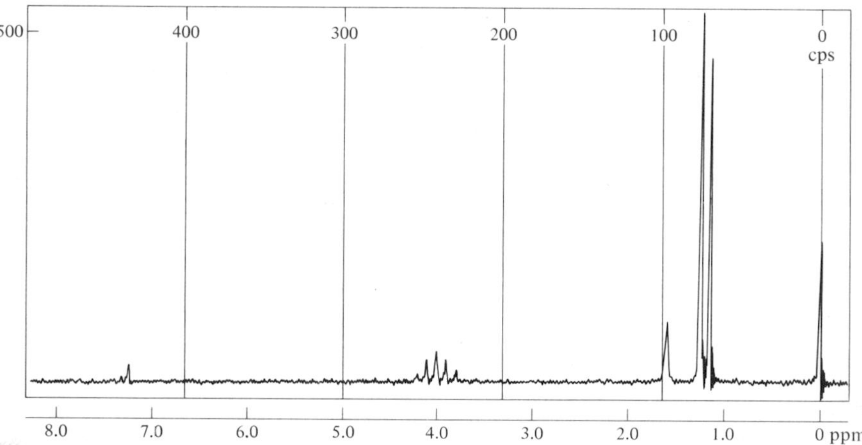

Figure 20.36 / *Spectrum for Problem 7.*
SOURCE: Varian Associates, Palo Alto, Calif.

7 / Assign a structure for the compound C_3H_8O, which has the 60-MHz proton nmr spectrum shown in Fig. 20.36. The small absorption near $\delta = 7.3$ is due to $CHCl_3$.

8 / A compound C_3H_7NO gives the 60-MHz proton nmr spectrum shown in Fig. 20.37. The peak areas, from left to right, are 1:1:1.5. Assign the structure consistent with these data.

9 / Assign a reasonable structure for a compound $C_8H_{11}NO$ that has the 60-MHz proton nmr spectrum shown in Fig. 20.38.

10 / A compound $C_{10}H_{12}O_2$ gives the 60-MHz proton nmr spectrum shown in Fig. 20.39. The peak areas, from left to right, are 5:2:2:3. Assign a structure consistent with these data.

Figure 20.37 / *Spectrum for Problem 8.*
SOURCE: Varian Associates, Palo Alto, Calif.

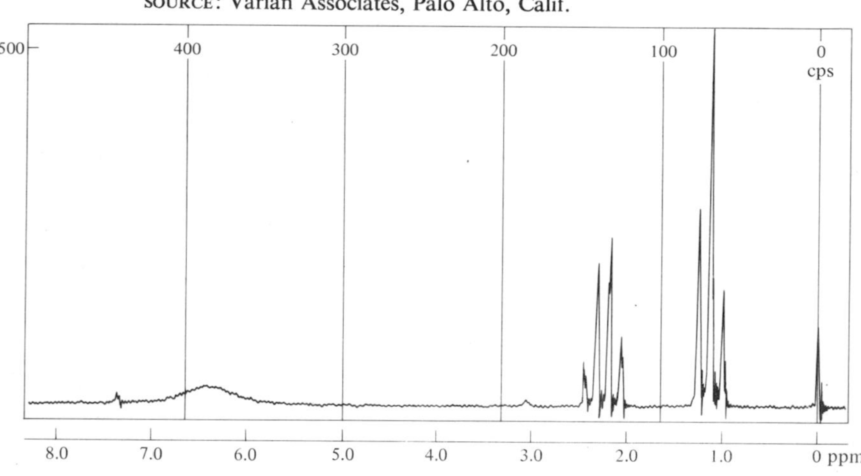

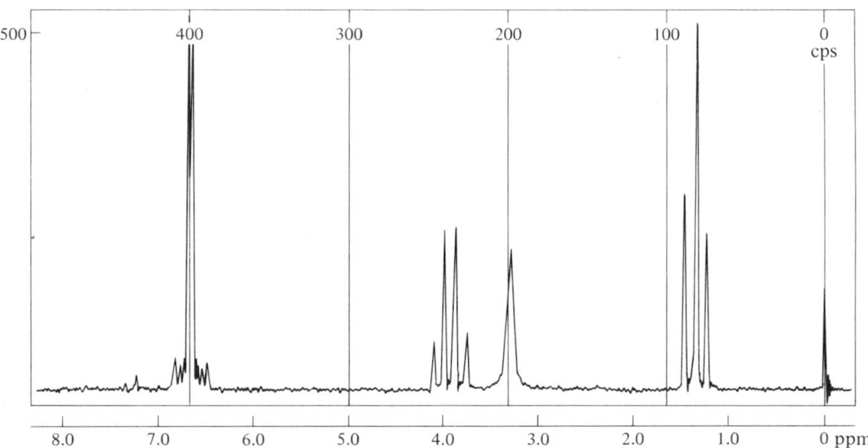

Figure 20.38 / *Spectrum for Problem 9.*
SOURCE: Varian Associates, Palo Alto, Calif.

Figure 20.39 / *Spectrum for Problem 10.*
SOURCE: Varian Associates, Palo Alto, Calif.

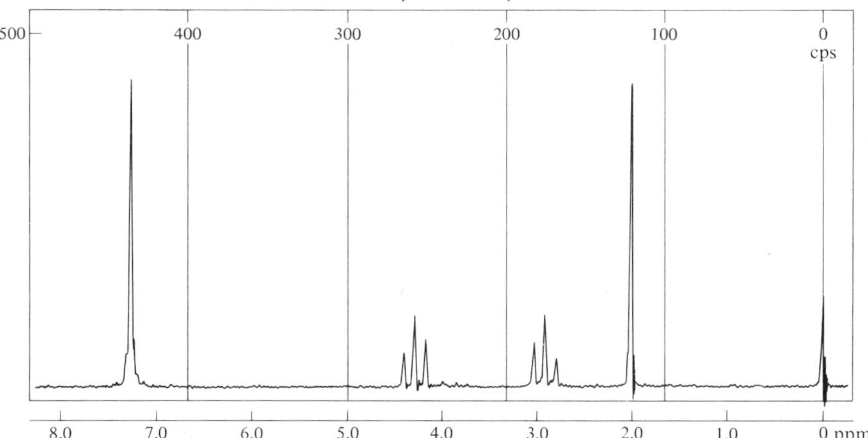

chapter 21 / Principles of mass spectrometry

chapter 21

21.1 / INTRODUCTION

Mass spectrometry is defined as the separation of charged molecular and submolecular species formed *in vacuo* according to their mass-to-charge ratios (m/e), and a mass spectrum is a record of the relative abundances of these species as a function of m/e. While it is entirely reasonable to term such a record a "spectrum," the reader will see at once that this technique is in principle quite different from the usual spectroscopic techniques, which depend on the absorption or emission of electromagnetic radiation. A highly simplified diagram of one type of mass spectrometer is given in Fig. 21.1.

In chemistry, mass spectrometry is applied to the elucidation of the structures and gas-phase reactions of organic molecules; to analyses of mixtures, including isotopic analyses; and to the determination of ionization and appearance potentials. Although the low volatility of most inorganic materials makes application of mass spectrometry to these substances a specialized problem, the technique finds wide application in the area of trace analysis.

This chapter deals with the principles that govern the design and use of mass spectrometers, interpretation of mass spectra, and some of the applications of this powerful tool in modern chemistry.

21.2 / INSTRUMENTATION

To obtain a mass spectrum, the following must be done:

1. Provision must be made for suitable pumps to keep the entire mass spectrometer under vacuum.
2. A satisfactory arrangement is required for introduction of the sample into the instrument; this is simple in principle, but often a complex array of "plumbing," heaters, and so on in practice. This part of the instrument is termed the *inlet system.*

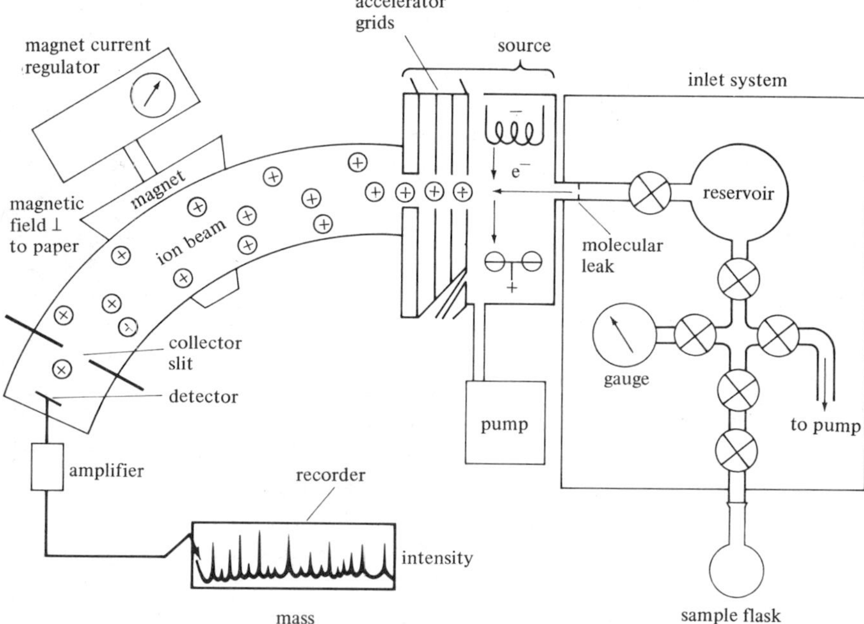

Figure 21.1 / *A simplified schematic diagram of a* 60° *magnetic deflection mass spectrometer. An inlet system suitable for volatile materials is shown on the right. The sample bleeds through the molecular leak into the source, where the molecules are bombarded with electrons. Electron bombardment causes both ionization and fragmentation of the sample molecules. Positive ions so produced are accelerated, subjected to magnetic mass analysis, and focused on the detector.*

3. The neutral molecules or atoms must be converted into charged species. As we shall see, it is desirable to produce species with only a single charge if possible. While mass spectra can be obtained from either positive or negative ions, most instruments and the bulk of the work done to date deal with positive ions. The section of the instrument in which the ions are produced is called the *ionization chamber*.
4. The ions must be removed from the ionization chamber. This is accomplished by use of an electrostatic potential, which accelerates the ions and produces a monodirectional ion beam. Together, the ionization chamber and accelerator constitute the *source*.
5. After the ions emerge from the source, they pass into the *analyzer section*, in which a variety of physical techniques may be employed to separate them according to m/e.
6. Finally, the ions impinge either on a *detector*, which produces an electrical signal proportional to the number of ions that strike it or a photographic plate that is darkened in proportion to the number of ions. Available detectors are capable of counting only; they cannot distinguish one ion from another, so it is necessary to provide the analyzer with a means of scanning various values of m/e, bringing one at a time into focus on the detector. However, a photographic plate may be used as the detector to collect simultaneously the entire spectrum. In this case, the position of the line on the plate must be related to m/e, just as plate position is related to wavelength in emission spectrography.

21.2.1 / The inlet system

The inlet system consists of a device for introducing the sample, a device for determining the amount of sample introduced (a known volume reservoir and a micromanometer), a device for metering the sample into the ionization chamber (literally a small hole customarily known as a *molecular leak*), and a vacuum pump with appropriate control valves and stopcocks to maintain the pressure in the range 10^{-3}–10^{-1} torr.

Introduction of gases is simply a matter of transferring the gas from a container (frequently a Pyrex bulb equipped with a ground-glass joint and vacuum stopcock) to the reservoir volume of the spectrometer. Liquids that have appreciable vapor pressure at room temperature can be handled in the same manner as gases. Usually the liquid is first frozen and the air is pumped out of the container; then the sample is remelted and introduced into the spectrometer. Alternatively, liquids may be introduced by means of various glass break-off devices, which are inserted through a suitable seal into the vacuum system and then crushed by a metal tool. Another liquid-sampling technique involves an entrance to the inlet system that is sealed with a sintered-glass disc covered with a layer of liquid mercury. The liquid sample is delivered from a micropipet by inserting the tip through the mercury layer until it touches the glass frit. Liquids can also be introduced by injection with a syringe through a silicone rubber septum as is done in gas chromatography.

Heated inlet systems are used for less volatile liquids and solids that have sufficient vapor pressure. Less volatile materials may be introduced directly into the ionization chamber by trapping a small amount of sample in the hollow end of a suitable probe, which is then inserted through an O-ring seal. The heat generated by the production of the electron beam used for ionization is often sufficient to provide volatilization. Many high molecular weight compounds, including polysaccharides and peptides, can be introduced in this way. The basic limitation is that the compound must be stable at a temperature such that its vapor pressure is of the order of 10^{-7}–10^{-6} torr. Sample sizes for liquids and solids may range from a few milligrams to a few nanograms, depending on the method of introduction and the sensitivity of the instrument. Techniques for dealing with nonvolatile inorganic materials will be discussed in the section dealing with spark sources (*vide infra*).

21.2.2 / Ion sources

By far the most common way of producing ions for organic mass spectrometry is *electron bombardment*. Positive ions are produced by passing a beam of electrons through the gaseous sample, usually at pressures of 10^{-6}–10^{-4} torr. The energy of the electron beam is controlled at a value (typically 70 V) that exceeds the ionization potential of the sample (ca. 10 eV for most organic molecules), so the electrons cause both ionization and fragmentation of the sample molecules. A simplified diagram of an electron bombard-

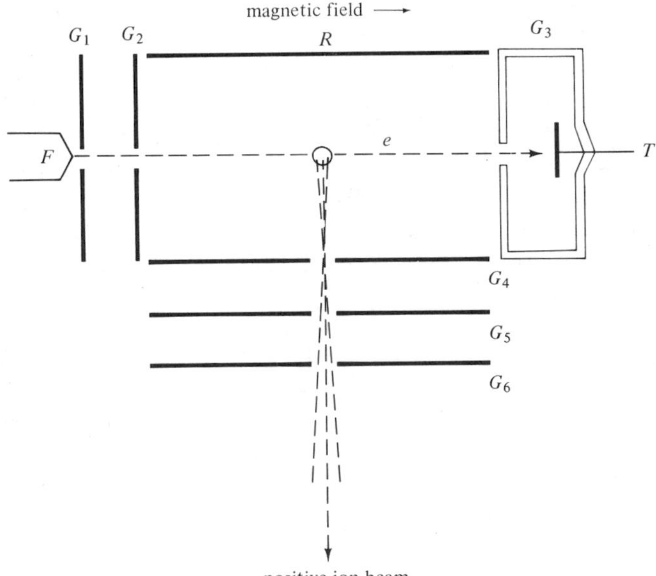

Figure 21.2 / *Schematic diagram of an electron bombardment ion source.*
SOURCE: R. W. Kiser, *Introduction to Mass Spectrometry and Its Applications*, Prentice-Hall (Englewood Cliffs, N.J., 1965), p. 34.

ment source is given in Fig. 21.2. Electrons are emitted from the heated filament F, and accelerated and collimated by the grids G_1 and G_2. The energy of the electron beam is fixed by the total potential drop between F and G_2. The electrons are collected at the anode, or target, T. To prevent secondary electrons that are emitted from the collector from reaching the ionization area, the grid G_3 is maintained at a potential slightly negative with respect to T. In addition, it is common to employ a magnetic field of about 100 gauss in order to align the electron beam along the desired path.

A weak electric field is applied between the rear wall of the ionization chamber (repeller, R) and the front wall, G_4. This field serves to draw out the positive ions. The positive ions that emerge from G_4 are then accelerated and collimated by a potential of several kilovolts applied between the slits G_5 and G_6.

The electron bombardment source appears to be the most reliable of all ion sources. The energy spread of the emergent beam is very small (0.1–5 eV), and a very steady ion beam is produced. Variations in the intensity of the ion beam with time are, of course, important if quantitative measurements of peak intensities are to be meaningful. For most purposes, the energy of the electron beam is variable up to about 100 eV, and 70 eV is rather commonly employed in routine measurements.

Other methods of ionization are employed for specialized applications.

For some sample molecules, fragmentation by electron bombardment proceeds to such an extent that the molecular ion is not observed. In such cases, a less energetic method of ionization may be desirable. One such technique is the *field emission* ion source, which uses a very intense electric field to ionize molecules and atoms. These very intense fields (ca. 10^8 V cm^{-1}) can be obtained by holding fine metal points or knife edges at high potentials. Acceleration of the ions produced is accomplished in the same way as with the electron bombardment source. Field emission sources usually give much simpler spectra than electron bombardment sources, and frequently only the molecular ion is observed.

Another less energetic method for ionization is the *chemical ionization* source. In this source, a reactant gas such as methane is ionized by electron bombardment at a pressure somewhat higher than that used in the usual source. The positive ions produced are then allowed to mix with and react chemically with the sample gas. Again, simplified spectra are obtained, and the molecular ion of the sample is usually the predominate ion observed. This technique finds application in the union of gas chromatographs with mass spectrometers. If methane is used as the carrier gas in the chromatographic experiment, it can also serve as the reagent gas in the source of the mass spectrometer.

When the sample to be examined is nonvolatile, as is the case with metals and inorganic solids, two problems must be solved. The material must be both vaporized and ionized; ideally, the vaporization procedure should be such that there is as little fractionation of various components as possible. Mass spectrometers designed for this purpose most often employ a radio frequency *spark source*. The source consists of two electrodes in place of the electron beam; a rf spark discharge is excited between these. The sample may be introduced in a cavity prepared in one of the electrodes, it may be ground up with graphite and compressed to form one or both of the electrodes, or if it is a metal it may be an electrode. The spark provides abundant energy to both vaporize and ionize the sample. Unlike sources described previously, the spark produces an abundance of multiply charged ions, and the ion beam produced has a large spread of energy (ca. 500 eV). Because of this large spread of energy in the ion beam, a more elaborate mass analyzer is necessary in order to obtain acceptable resolution when a spark source is employed.

The spark source provides excellent sensitivity for detection of most of the elements in the periodic table. The detection limit is seldom poorer than 0.1 ppm and many elements can be analyzed in the parts per billion range, although the precision obtainable is not high. Because the spark energy is high, the entire sample often volatilizes homogeneously, and this source is well rated for its general freedom from matrix effects. Nonetheless, for quantitative work, especially when relatively volatile elements (alkali metals, halogens, etc.) are present in a refractory matrix (e.g., oxides of tungsten and similar heavy metals), matrix effects may be important and require corrections through the use of appropriate standard samples.

Spark sources are often used in conjunction with photographic plate detection, since the spark source is less stable (there are greater time variations in the strength of the ion beam) than the electron bombardment source. The photographic plate detects the entire spectrum simultaneously and integrates over the exposure time, thus minimizing the effects of fluctuations in the ion beam intensity. By contrast, electronic detection of single m/e values at different times during a scan of m/e in which short-term fluctuations in the beam intensity occur could cause appreciable quantitative errors.

21.2.3 / The time-of-flight mass analyzer

Consider a positive particle of charge e and mass m formed in a uniform positive electric field of total potential drop V. The particle is accelerated according to Eq. (21.1), which equates the electrostatic energy and the kinetic energy of the particle:

$$eV = \frac{mv^2}{2} \tag{21.1}$$

The velocity of the ion is readily obtained by rearrangement of Eq. (21.1) to the form of Eq. (21.2),

$$v = \left(\frac{2eV}{m}\right)^{1/2} \tag{21.2}$$

If the ions emerge from the source into a field-free region maintained at high vacuum ($<10^{-6}$ torr), called a *drift tube*, their transit times to the detector will be inversely proportional to their velocities or, as one can deduce from Eq. (21.2), directly proportional to $(m/e)^{1/2}$. While mass discrimination via transit times may appear simple in principle, it poses some serious practical difficulties. Consider the following example. The transit time is given by Eq. (21.3), where d is the length of the drift tube.

$$t = d\left(\frac{m}{2eV}\right)^{1/2} \tag{21.3}$$

We shall assume d = 100 cm, m = 100 amu, and an accelerating potential of V = 3000 V. A singly charged ion will also be assumed.

Recalling that 1 electron charge = 4.802×10^{-10} esu, the reader should analyze the dimensions of Eq. (21.3) and satisfy himself that after m has been converted to grams, the appropriate unit for the product eV is ergs. The required conversion is brought about by noting that 300 practical volts = 1 erg-esu^{-1}, whence 3000 V = 10 erg-esu^{-1}.

Numerical evaluation of t proceeds as shown in Eq. (21.4).

$$t = 10^2 \text{ cm} \left[\frac{(100/6.02) \times 10^{-23} \text{ g}}{2 \times 10 \text{ erg-esu}^{-1} \times 4.802 \times 10^{-10} \text{ esu}} \right]^{1/2} \tag{21.4}$$

$$= 13.15 \times 10^{-6} \text{ sec}$$

Note that this result implies an ion velocity of 76.05×10^7 cm-sec^{-1}. Repeating this calculation for mass 101 gives a transit time of 13.22×10^{-6} sec, which shows that such an instrument must be able to resolve time differences of 0.07 μsec in order to obtain unit mass resolution at mass 100. Clearly, with a continuous ion beam, the detector will be subjected to a continual bombardment of ions and no spectrum can be obtained. Therefore, some form of synchronization between the source and detector must be introduced in order to obtain useful information from this instrument. In the commercial time-of-flight mass spectrometer, the ion beam is energized repetitively for 0.25 μsec during each 100 μsec. The detector output is observed during the 100-μsec interval using an oscilloscope that is triggered synchronously with the source. The rapidly repeating traces give the appearance of a stable display of the mass spectrum on the scope, and the time axis can be converted to m/e.

The principal advantage of the time-of-flight instrument is the rapidity with which an entire spectrum can be obtained. These devices are well suited to the study of gas-phase reactions and to the analysis of the fractions that emerge from a gas chromatograph. As indicated in the calculations above, these are not high-resolution instruments, and the complexity of the fast electronics necessary for their function has contributed to their general lack of acceptance for routine mass spectrometry.

21.2.4 / Magnetic deflection mass analysis

Magnetic discrimination of m/e values enjoys the widest application. If the ion beam that emerges from the source passes into a uniform magnetic field, the ions will be subjected to a force given by the vector product of their velocity and the magnetic field in accordance with the laws of electromagnetism. This relation is stated in Eq. (21.5),

$$f = e\mathbf{V} \times \left(\frac{1}{c}\right) \mathbf{H} \tag{21.5}$$

where the force is given in dynes and the factor $(1/c)$ (c is the speed of light *in vacuo*) is necessary to convert the common units of magnetic field strength (gauss or emu) to electrostatic units. The vector product is a vector perpendicular to the plane containing the two original vectors, so the magnetic force acts in a direction perpendicular to the direction in which the ions are originally traveling. If, as is normally the case in mass spectrometers, the magnetic field is perpendicular to the direction of the ion beam, this force will be given in the scalar notation of Eq. (21.6).

$$f = \frac{HeV}{c} \tag{21.6}$$

The action of a constant force perpendicular to the original velocity will result in deflection of the ion beam into a circular trajectory such that the magnetic force is just balanced by the centrifugal force of the ions. Therefore, we can write Eq. (21.7), which is readily rearranged to Eq. (21.8).

$$\frac{mv^2}{r} = \frac{HeV}{c} \tag{21.7}$$

$$v = \frac{Her}{mc} \tag{21.8}$$

Combination of Eq (21.8) with Eq. (21.1) gives Eq. (21.9) as the relation between the mass-to-charge ratio and the other parameters.

$$\frac{m}{e} = \left(\frac{H^2}{V}\right)\left(\frac{r^2}{2c^2}\right) \tag{21.9}$$

Since the source and detector will be located in fixed orientations on a single radius of curvature, the mass-to-charge ratio of ions that will follow this particular path and be detected is controlled by the ratio (H^2/V). Thus, increasing the magnetic field will bring progressively heavier ions into register, while increasing the accelerating potential will focus successively lighter ions on the detector.

As an example, let us inquire what magnetic field will be required to focus the accelerated ion of mass 100 from the previous example on a detector that lies on a 30-cm radius of curvature with respect to the source. Solution of Eq. (21.9) for the field gives Eq. (21.10), which may easily be evaluated.

$$H^2 = \frac{(m/e)2c^2V}{r^2} \tag{21.10}$$

$$= \frac{(100/6.02) \times 10^{23}\ \text{g} \times 2 \times (3 \times 10^{10}\ \text{esu-emu}^{-1})^2 \times 10\ \text{erg-esu}^{-1}}{4.082 \times 10^{-10}\ \text{esu} \times (30\ \text{cm})^2}$$

$$= 6.917 \times 10^6\ \text{g-cm}^{-2}\text{-erg-emu}^{-2}$$

$$H = 2630\ \text{g-sec}^{-1}\text{-emu}^{-1}\ \text{(gauss)}$$

Note that the final units, gram-second^{-1}-emu^{-1}, are equivalent to the usual units of magnetic induction, gauss. The unit emu as used here refers explicitly to the electromagnetic unit of charge, 1 emu = 10 coulombs.

The initial design of the magnetic deflection mass spectrometer was due

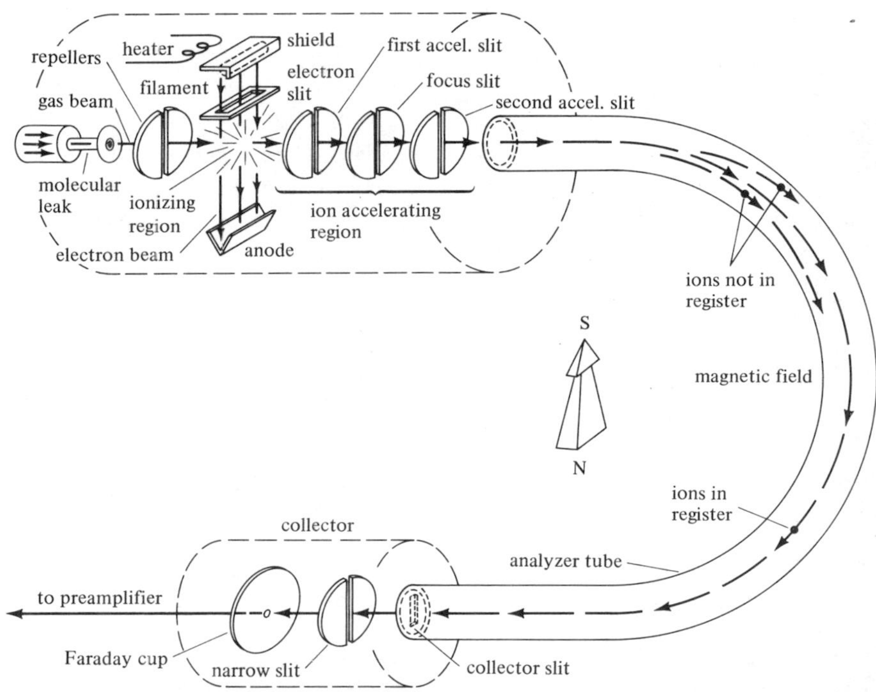

Figure 21.3 / *Schematic diagram of a mass spectrometer that employs* 180° *magnetic deflection of the ion beam.*

to Dempster (1918) and employed 180° magnetic deflection. A schematic diagram of a 180° deflection instrument is given in Fig. 21.3. Subsequent designs have employed 180°, 90°, and 60° magnetic sectors. The analyzer section of a 60° instrument is shown schematically in Fig. 21.4. Identical performance is, in principle, achieved with any one of the three designs. The advantage of the smaller angular deflection is that the magnet required is less cumbersome and less costly. The 180° design has the advantage of the shortest path from source to detector, which results in the smallest broadening of the ion beam due to collisions during its travel. Instruments that employ magnetic mass analysis can be made to resolve m/e values that differ by 1 amu over the range 1 to ca. 1000 amu.

There are a number of reasons why the ability of a mass spectrometer to discriminate between different m/e values is limited. It is not possible to give a detailed account of these here, but some of the principal limitations include mechanical imperfections in the dimensions and assembly of the spectrometer and edge effects of the magnetic field. An additional limitation is the fact that the ion beam is made up of ions with velocities distributed about v, rather than all having identical velocities. Refinements of manufacturing techniques and holding closer tolerances can minimize mechanical imperfections. It is not possible to produce magnets that have no lines of

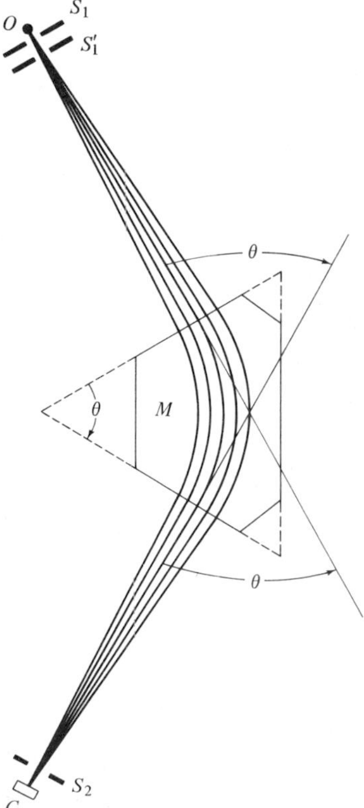

Figure 21.4 / *Focusing of a divergent ion beam by a* 60° *magnetic sector* (θ = 60°). *O = source, M = magnet, C = collector,* S_1 *= entrance slit, and* S_2 *= exit slit.* SOURCE: R. W. Kiser, *Introduction to Mass Spectrometry and Its Applications,* Prentice-Hall (Englewood Cliffs, N.J., 1965), p. 49.

force outside the gap between the pole pieces. Thus, the magnetic field is neither uniform nor constant near the edges of the gap, and this effect is difficult to overcome.

The ion beam may diverge slightly as it exits from the source, as indicated in Fig. 21.4. Proper design of the magnetic analyzer, however, can result in focusing the divergent beam on the detector. Nonetheless, the beam is of finite width, governed by the widths of the source and detector slit assemblies. In addition, not all of the ions have exactly the same kinetic energy, a result of the facts that the gas molecules are traveling in random directions with velocities distributed according to the Maxwell-Boltzmann equation at the time they are ionized and also that kinetic energy may be imparted to the ions during ionization. (As indicated earlier, the spread in energy of the ion beam depends strongly on the type of source used.) The effect of inhomogeneity in the beam energy may be largely corrected by use of a velocity analyzer in conjunction with the magnetic analyzer.

21.2.5 / Double-focusing instruments

If charge e passes through a uniform electric field of strength E perpendicular to the direction of its travel, it will experience a force, $\mathbf{f} = e\mathbf{E}$. Just as in the case of magnetic deflection, this constant force will deflect the particle into a circular orbit. The radius of the orbit can be obtained by equating the force applied with the centrifugal force of the particle as indicated in Eq. (21.11), where the radius of curvature for the electrostatic deflection is r_e.

$$eE = \frac{mv^2}{r_e} \qquad (21.11)$$

The kinetic energy of the particle is given as stated previously by eV, and substitution into Eq. (21.11) gives

$$eE = \frac{2eV}{r_e} \qquad r_e = \frac{2V}{E} \qquad (21.12)$$

To obtain r_e in centimeters, we require the accelerating potential to be in electrostatic units and the deflecting electric field in electrostatic units-volts-centimeter^{-1}. It is evident from Eq. (21.12) that the mass of the particle is not a parameter, so an electrostatic analyzer does *not* analyze mass. Since V is a parameter, the instrument actually analyzes energy. Therefore, use of a radial electrostatic analyzer enables one to select ions of a given energy. Under the proper conditions of geometry, the energy-selected beam may then be subjected to magnetic deflection analysis. The resulting instrument is called a *double-focusing* mass spectrometer. Such instruments are capable of resolving ions for which m/e differs by only 1 part or less in 30,000. These instruments can detect the difference between ions of the same *nominal* m/e value over the entire mass range of usual interest. For example, N_2^+ and CO^+, both of nominal $m/e = 28$, have actually values of 28.0062 and 27.9949 and are readily distinguished by a double-focusing spectrometer. The utility of such instruments in the analysis of organic compounds is very great, since precise determination of m/e values can lead directly to the elemental composition of the ions (*vide infra*). As pointed out above, double focusing is required to obtain useful resolution with a spark source because of the large energy spread of the ion beam.

Two popular geometries for double-focusing spectrometers are shown in Figs. 21.5 and 21.6. In Fig. 21.5, the design due to J. Mattauch and R. Herzog is shown. In this design, the dispersed ion beam is brought to focus on a plane surface, where a photographic plate may be used to detect the entire spectrum simultaneously. Alternatively, individual m/e values may be detected electronically at the position shown by varying the magnetic field and/or accelerating potential. The geometry depicted in Fig. 21.6 is that of E. G. Johnson and A. O. Nier, which brings the ion beam to focus at a single slit. This instrument uses electronic detection only and requires scanning either H or V to obtain a spectrum.

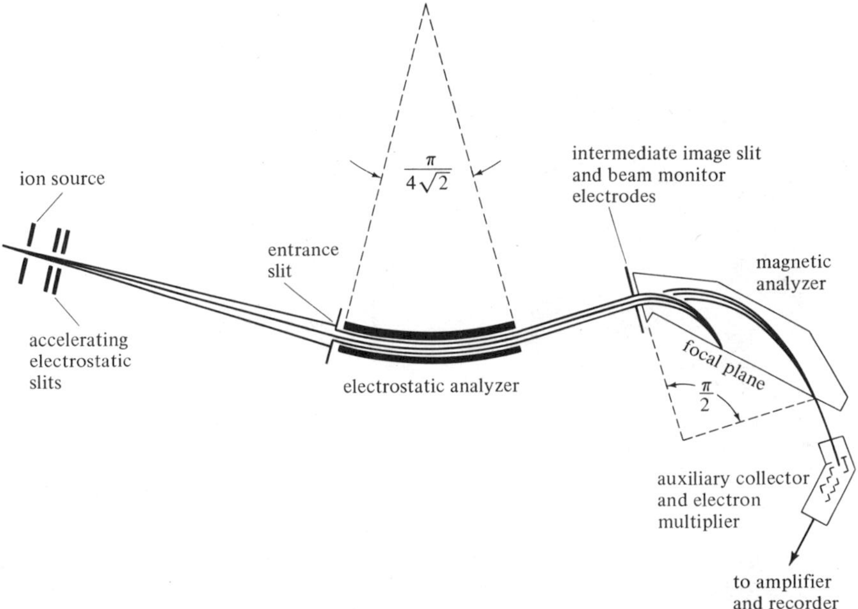

Figure 21.5 / *The Mattauch-Herzog double-focusing mass spectrometer shown with provision for either photographic or electronic detection of the ion beam.*

21.2.6 / Quadrupole mass analysis

This method of mass analysis employs a radio frequency electric quadrupole field to obtain m/e discrimination and, hence, no magnet is required. The analyzer consists of four long, precisely parallel electrodes of uniform cross section, which may be either hyperbolic as shown in Fig. 21.7 or cylindrical. As indicated in the figure, opposite pairs of electrodes are connected together electrically. Both a DC voltage, U, and a rf voltage,

Figure 21.6 / *The Nier-Johnson double-focusing spectrometer. Both the electrostatic and magnetic sectors are of the* 90° *design.*

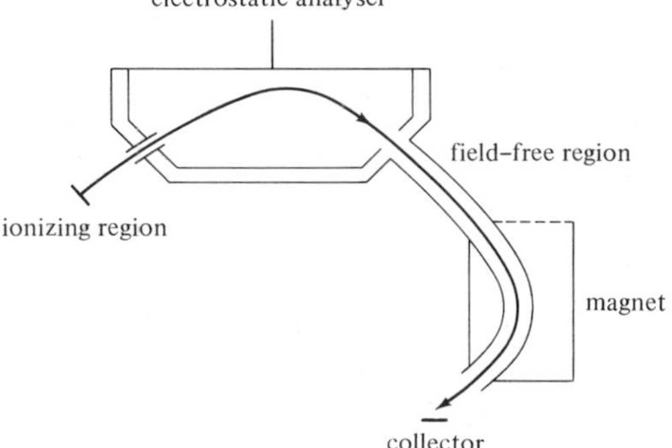

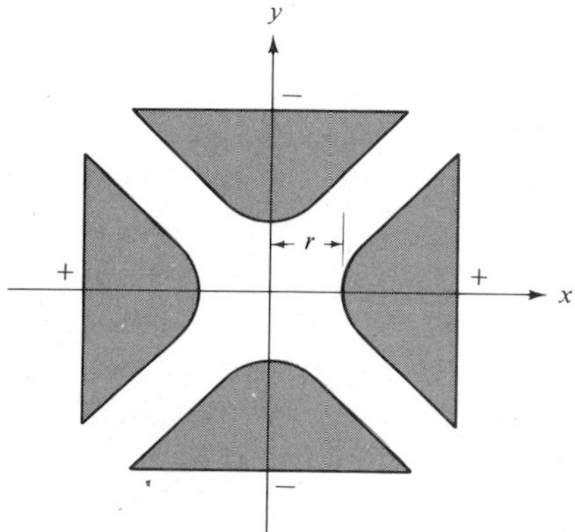

Figure 21.7 / *Cross-sectional view of a quadrupole mass analyzer. The ions travel with constant velocity in the direction perpendicular to the paper* (z).

$V^0 \cos \omega t$, are applied to the quadrupole array. Ions, which are injected from a conventional source along the z direction, continue to move in this direction with their initial velocity since the electric field has no component in the z direction. In the x and y directions, however, the ions are caused to undergo oscillatory motions. Solution of the rather complex equations of motion of these ions shows that for particular values of ω, U, and V^0, only ions with one particular m/e value will pass through the quadrupole analyzer without striking one of the electrodes and thereby reach the detector. For this one value of m/e, the induced oscillations remain quite small, while for all other values the amplitude of oscillation increases exponentially while traversing the quadrupole field. Mass scanning is accomplished either by varying U and V^0 while keeping their ratio constant, or by varying the rf frequency. The former method is the more common one. Typical values of the frequency are in the range of a few hundred kilohertz when the electrodes are separated by approximately 6 cm (i.e., $r \sim 3$ cm in the figure). Quadrupole spectrometers can be made to attain m/e resolution of about 1 part in 15,000. No detector slit is required in analyzers of this type and, as a result, a relatively large number of ions may pass through, a fact that has caused the instrument to receive serious consideration as an isotope separator. The absence of a magnet makes the quadrupole spectrometer a relatively inexpensive instrument. With its rather good resolution, it is therefore attractive for dedicated use with gas chromatographs.

21.2.7 / Other mass analyzers

Several other types of m/e analyzers have been designed and have found limited application. If the ions are injected from the source into a region of space in which a homogeneous magnetic field is perpendicular to a homogeneous electrostatic field, the ions will describe cycloidal paths. Proper design of such a device produces perfect focus as well as spatial mass dispersion. Instruments of this type constructed to date are of only moderate resolution and limited to values of $m/e \sim 200$. A similar design employs a constant magnetic field and a rf electric field perpendicular to it. This device, known as the *omegatron* mass spectrometer, functions on the same principles as a cyclotron. The ions must be formed at the center of the crossed fields, and those that are in resonance describe ever larger circular paths as they pick up energy from the rf field. Eventually, they spiral into the detector placed at one edge of the unit. Mass scanning is accomplished by varying either the frequency or the magnetic field. The omegatron has the advantages of being very compact and quite sensitive. It has found considerable use as a research tool in the study of gas-phase ion–molecule reactions, but not for general use in mass spectrometry. The only commercial instrument of this design encountered technical problems and found relatively poor acceptance in the market place.

Another specialized mass analyzer is a variation of the time-of-flight spectrometer known as the *Bennett tube.* In this analyzer, instead of a drift tube, a series of grids to which rf voltages are applied is located in the path of the ion beam. Appropriate adjustment of the rf frequency allows the passage of particular m/e ions to the detector; real m/e discrimination is obtained, and the detector need not be synchronized with the source. The principal advantage to this device is that it may be extremely compact, weighing less than 25 lb complete with the required electronics; however, its resolution is quite low. Applications in which small size and weight are primary considerations, such as space exploration, would seem appropriate for the Bennett tube.

21.2.8 / Ion detectors

When a photographic emulsion is struck by a beam of ions, it becomes sensitized just as it does when struck by photons. Thus, a photographic plate can be used as a detector in mass spectrometry in the same way as it is used in emission spectroscopy (see Section 11.2.5). After development of the plate, a series of parallel blackened lines is observed, each of which corresponds to a different m/e value. As noted in the discussion of spark sources, photographic detection has the advantages of obtaining the entire spectrum simultaneously and of integrating out fluctuations and instabilities in the source. Plates provide the most precise method of mass measurement and have been widely used in the determination of the isotopic masses of the elements. The technique has been refined to the point that isotopic masses can be measured to 1 part in 10^8, so that the remaining uncertainties in the atomic weights of the elements are now largely due to fluctuations in isotopic composition among different samples. The principal limitations of mass accuracy in the photographic plate itself are image widening due to scattering within the emulsion, the grain size of the emulsion, and inhomogeneities of sensitive elements in the emulsion. Of course, in addition, one must consider the precision of the comparators used to measure the displacements of the lines on the plate.

The disadvantages of plates include the necessity for development and reading prior to obtaining usable data and the low inherent accuracy of plates in ion abundance measurements. Photographic emulsions often have nonlinear sensitivity with the energy of the ion beam, the time of exposure, and the intensity of the ion beam. In addition, the image density may vary with the mass of the ion detected. The density of a line image can be measured with great precision with a microdensitometer, so if the difficulties inherent in the emuslion can be overcome by calibration, useful intensity measurements will result. Of course, only mass spectrometers that focus the dispersed ion beam on a focal plane are suitable for use with photographic plate detectors.

For scanning mass spectrometers, the simplest detector is a metal collector against which the ion beam impinges. This detector is known as a *Faraday cup* and frequently includes provision to prevent interference from secondary electrons emitted when ions strike it. The current generated by the ion beam usually is in the range 10^{-17}–10^{-9} A; it is passed to ground through a large resistance. The sensitivity of the instrument is governed by the ability of the detector amplifier to amplify accurately the voltage generated in the resistor to a useful level.

Both the Faraday cup and the electron multiplier detector described below function as current sources that exhibit very large internal impedances. It is therefore possible to measure their outputs accurately by observing the voltage dropped between these devices and ground through a very large load resistor using a voltage amplifier with a suitably high input impedance as described in Section 7.1.2.

In the past, this voltage was applied to the grid of an electrometer tube, which served as the first stage of a DC amplifier. Ion currents as small as 10^{-15} A could be measured this way, but the usual difficulties of instability associated with DC amplifiers were always present. A device known as a vibrating-reed electrometer proved much more stable and gave two orders of magnitude better sensitivity at the cost of increased electronic complexity. Both of these methods are now obsolescent, because FET-input solid state electrometers, which are both very stable and sensitive, are commercially available and can usually be substituted for the older devices with little difficulty.

The ultimate sensitivity for mass spectrometry is achieved using an *electron multiplier* detector, which permits detection of beams consisting of from a single ion up to ion currents ca. 10^{-9} A. In principle, the electron multiplier shown in Fig. 21.8 is identical to a photomultiplier tube (Section 11.2.5) without a photocathode. There are some differences in details of construction. The ion beam strikes the cathode and causes emission of secondary electrons which are accelerated toward the first dynode; cascade amplification occurs with more electrons emitted at each dynode. Typically, signal

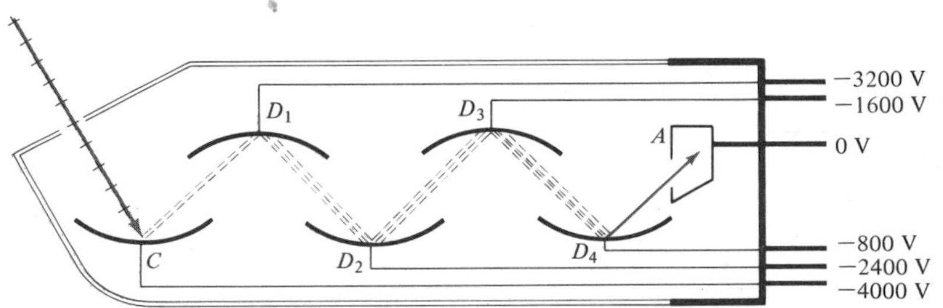

Figure 21.8 / *Schematic representation of an electron multiplier. Positive ions are shown falling on the cathode, C, which emits secondary electrons. Cascade amplification occurs as the electrons strike the dynodes, D, which are separated by* 800 V *per stage. The electrons are finally collected at the anode, A.*

multiplication by factors of 10^5–10^6 is achieved with a 10–12 dynode multiplier. All modern general-purpose high-resolution mass spectrometers that do not employ photographic detection use electron multipliers. The electron multiplier is not completely satisfactory for ion abundance measurements, however, since its characteristics may change with time and the type of sample.

The type of detector to be used must be determined by priorities established by the user. The electron multiplier is ideal in terms of rapid response and high sensitivity so long as the highest precision in abundance measurements is not required; it is well suited to most scanning spectrometers and can readily be used in conjunction with computer-assisted data-reduction systems. When the relative abundance of ions is of crucial importance, a well-designed modern electrometer detector is probably preferable.

Ion abundance is of particular importance in the examination of quantitative isotopic content, for example in kinetics or tracer studies. Detectors designed especially for the purpose of measuring the abundance *ratio* of isotopes are in use, and spectrometers so equipped are termed *isotope ratio mass spectrometers.* To minimize errors that might be introduced by the necessity of scanning two isotopically distinct ions across a single detector, an isotope ratio detector has two slits and two detectors located at an appropriate separation for the desired m/e difference. The two ions $C_7H_7^+$ and $C_7H_6D^+$ could, for example, be detected simultaneously by the two detectors. The two signals are then compared externally on a recorder that displays their difference. A precision resistor voltage divider is used to reduce the larger signal until the two are exactly equal and their difference is zero. The ratio of the two beam currents, which is the isotope ratio, is simply read from the voltage divider.

21.2.9 / Resolution

Resolution is defined as the ability of the mass spectrometer to differentiate m/e values. It is useful to have a quantitative definition of this term to describe the performance of an instrument.

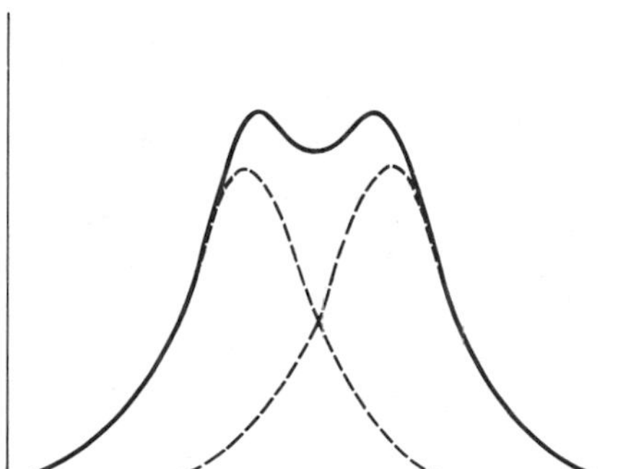

Figure 21.9 / *Two adjacent peaks "just resolved."*

If two adjacent peaks are "just resolved," we are only barely able to distinguish that two peaks are, in fact, present. Such a condition is depicted in Fig. 21.9. Two adjacent peaks are said to be *resolved* when the peaks due to the two ions are separated to a given degree, usually defined by 10 percent overlap. The *resolving power* of the instrument refers to the relative difficulty of the separation and is stated in terms of $M/\Delta M$, where M is the mass of a given ion and ΔM is the smallest difference in mass to an adjacent peak that can be considered resolved. The definition of ΔM is therefore subject to somewhat arbitrary definitions of what constitutes acceptable resolution between peaks and, unfortunately, several different definitions are in current use. It should be evident that the width of the peaks warrants consideration, since for two peaks that have a particular m/e difference, the degree to which they are separated will vary inversely as their widths.

Some definitions of ΔM are based on the ratio of the height of two equi-intensity peaks to the height of the valley between them, $\Delta I/I$ in Fig. 21.10. For many purposes, it is entirely satisfactory to consider two peaks separated when this ratio is 0.1, the so-called 10 percent valley. Thus, if two peaks of equal intensity are observed at 100 and 101 amu, respectively, and the valley between them is 10 percent of their height, the resolving power of the spectrometer is 100 at that m/e value.

The principal difficulty in using this approach to specify resolving power is the necessity of obtaining the two adjacent peaks of equal intensity—a *desideratum* that may not be realized in experiment. Furthermore, even if two such peaks are observed, what is to be done if the valley intensity exceeds 10 percent as it does in Fig. 21.10? These problems have led to definitions of resolving power based on the width of a single peak.

Two nearly equivalent definitions are currently in widespread use. The peak width as measured at 5 percent of its total intensity is defined equal to ΔM; or the width is measured at half-intensity and twice that value is equated to ΔM. If the peaks are of Gaussian shape, the two methods give nearly equivalent results (if 2.08 $W_{1/2}$ is used, the results are identical); both methods

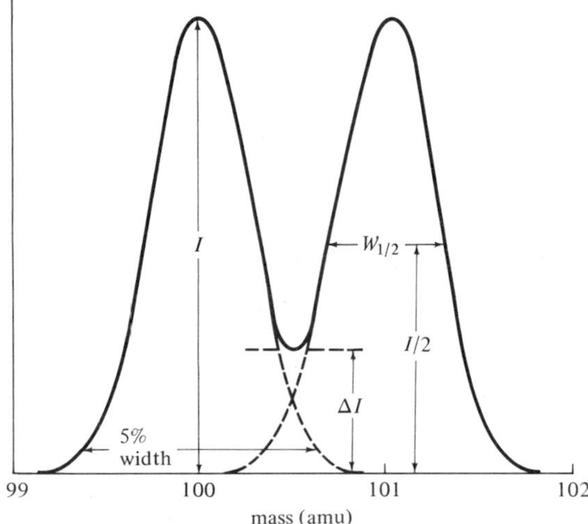

Figure 21.10 / *Two adjacent peaks of equal intensity with a valley between that corresponds to approximately* 30 *percent of the peak intensity. To what extent are they resolved?*

give results equivalent to the 10 percent valley method. For example, if a peak is observed at $m/e = 200$ and it has a width of 0.5 amu at half-intensity, the resolving power of the spectrometer is $200/(2 \times 0.5) = 200$.

21.3 / CHEMICAL APPLICATIONS OF MEDIUM-RESOLUTION MASS SPECTROMETRY

In this section, we shall consider the applications of mass spectrometry to chemistry that are feasible using single-focusing instruments. Single-focusing magnetic deflection devices are still by far the most commonly employed, primarily because of their lower cost compared to double-focusing spectrometers, and a longer history of commercial availability. By medium resolution we shall mean that the spectrometer is able to differentiate m/e values that differ by at least 1 amu throughout its useful mass range.

21.3.1 / The mass spectrum

Many mass spectrometers are equipped with galvanometer recorders that use light-beam recording and UV-sensitive paper that does not require wet development. One of the common recorders in use employs a five-element galvanometer system with different sensitivities in order to obtain a larger dynamic range in the course of a single scan. The sensitivity levels decrease from top to bottom in the ratio 1:3:10:30:100, and intensities are obtained by reading the peak height on the most sensitive scale that remains in range and multiplying by the appropriate sensitivity factor. The high-mass section of the spectrum of toluene (m.w. = 92) recorded on such a

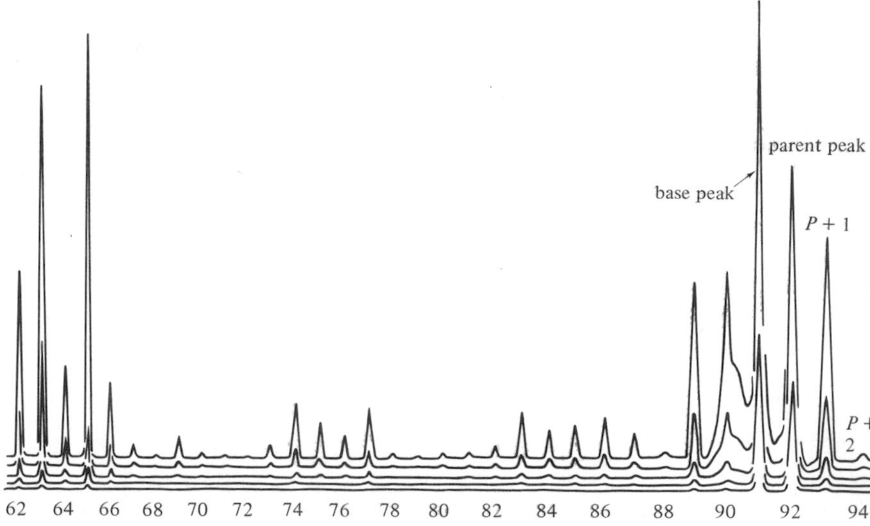

Figure 21.11 / *High-mass section of the mass spectrum of toluene (m.w. = 92) recorded by a five-element galvanometer.*

five-element recorder is shown in Fig. 21.11. Since the mass dispersion of magnetic deflection instruments is not constant but decreases approximately with the square root of mass, it is necessary to calibrate the trace to obtain m/e values. Generally, one can simply count unit m/e separations, since the instrument background will be sufficient to show a small peak on the most sensitive scale at each unit mass interval. When this technique is insufficient, a calibration compound may be added to the sample. Perfluorokerosene is a common calibration standard.

In the chemical literature, mass spectra are frequently presented in a tabular form that lists m/e values and the intensities of the observed peaks relative to the most intense peak (called the *base peak* or *analytical* peak) in the spectrum that is assigned a value of 100. Graphical presentations are also used and have the advantage of presenting patterns that can be quickly recognized by experienced workers. These are usually presented in the form of bar graphs in which peaks of less than 3 percent of the base peak are omitted unless they are isotope peaks (*vide infra*) or possess some other special significance. Such a graphical presentation of the spectrum of toluene is given in Fig. 21.12 along with its tabular equivalent.

In the simplest possible case, an electron bombardment source effects the removal of an electron from the sample molecule to give a cation radical as indicated in Eq. (21.13).

$$\text{R—R} + e^- = \text{R—R}^{+\cdot} + 2e^- \tag{21.13}$$

This process gives rise to the *parent* (this mass is designated P) or molecular ion. To observe a peak due to the molecular ion, it must have a lifetime of at

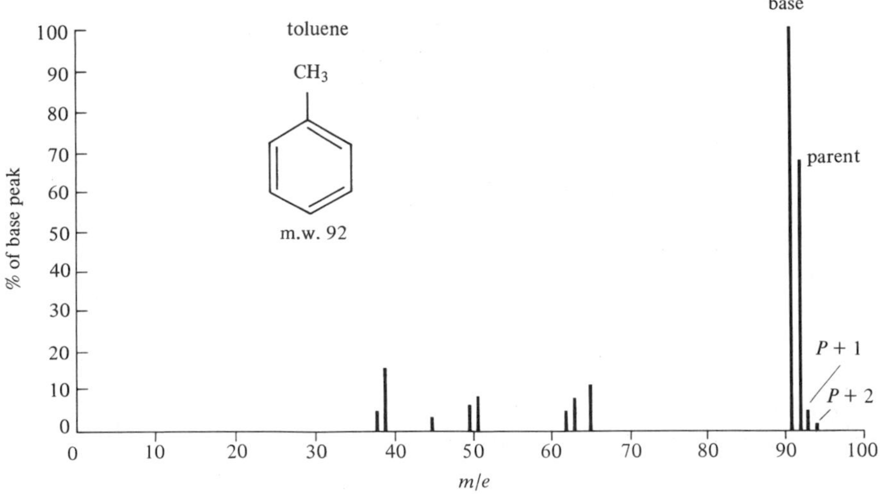

m/e	% of Base Peak		Isotope Abundances m/e	% of P
38	4.4		92 (P)	100
39	16		93 ($P+1$)	7.37
45	3.9		94 ($P+2$)	0.29
50	6.3			
51	9.1			
62	4.1			
63	8.6			
65	11			
91	100	(base)		
92	68	(parent)		
93	5.3	($P+1$)		
94	0.21	($P+2$)		

Figure 21.12 / *Bar graph and tabular presentations of the mass spectrum of toluene. The peak due to the parent or molecular ion is designated P.*

least a few microseconds, which is frequently the case. However, as we have seen above, the energy of the electron beam is usually well in excess of the ionization potential of the molecule, and as a result, many of these molecular ions undergo fragmentation in 10^{-10}–10^{-8} sec. The simplest fragmentation process is a simple bond rupture, as indicated in Eq. (21.14).

$$\text{R—R}^{+\cdot} = \text{R}^{\cdot} + \text{R}^{+} \tag{21.14}$$

Since most molecules can rupture at one of several places, numerous fragment ions may be produced. These can, in turn, cleave to give smaller fragments. These processes give rise to the peaks at m/e values smaller than the parent peak and produce the *fragmentation pattern.* It should be noted that identification of the parent peak is of considerable importance, since it yields the molecular weight of the compound. Medium-resolution mass spectrometry gives the molecular weight to the nearest whole atomic mass

TABLE 21.1 / *Isotope abundances for some common elements*

Elements		*Percent Abundance*	
carbon	^{12}C 100	^{13}C 1.08	
hydrogen	^{1}H 100	^{2}H 0.016	
nitrogen	^{14}N 100	^{15}N 0.38	
oxygen	^{16}O 100	^{17}O 0.037	^{18}O 0.20
silicon	^{28}Si 100	^{29}Si 5.10	^{30}Si 3.35
sulfur	^{32}S 100	^{33}S 0.78	^{34}S 4.40
chlorine	^{35}Cl 100	^{37}Cl 32.5	
bromine	^{79}Br 100	^{81}Br 98.0	

unit number, not simply an approximate value as is obtained from all wet-chemical molecular weight determinations.

21.3.2 / Determination of molecular formulas

The molecular formula of a compound can be determined from sufficiently accurate measurement of the mass of the parent ion; however, this is the province of high-resolution mass spectrometry, which will be discussed subsequently. In this section, determination of molecular formulas with unit mass resolution through the use of isotope peaks will be considered. Table 21.1 lists the common elements that have more than one stable isotope and gives the abundances of these isotopes relative to the lowest-mass isotope, which is assigned a value of 100 percent. The natural abundance of deuterium, ^{2}H, is sufficiently small that it is usually neglected. A compound that contains a single carbon atom will show a $P + 1$ peak of 1.08 percent of the intensity of the parent peak as a result of the fact that 1.08 percent of carbon atoms are present as ^{13}C. Similarly, a compound that contains one chlorine atom will have the intensity at $P + 2$ equal to 32.5 percent of the parent peak. A compound containing a single sulfur will have a small $P + 1$ peak, but a $P + 2$ peak that is 4.4 percent of the parent peak intensity.

The contribution to the isotope peaks increases with the number of equivalent atoms of a single element. In benzene, for example, there are six identical carbon atoms, any one of which may be ^{13}C. It is therefore six times as probable to find a molecule containing ^{13}C in a sample of benzene than in a sample of a compound containing only a single carbon. Therefore, we anticipate the $P + 1$ peak for benzene will have intensity 6×1.08 percent = 6.48 percent of the parent peak. Molecules that contain two ^{13}C atoms exist, but of course these are much more rare. For cases in which the uncommon isotope is present at the level of only a few percent, it is sufficient to say that molecules containing two atoms of the uncommon isotope will be present at a level approximately the square of the abundance; that is, molecules containing two ^{13}C atoms will be of about 0.01 percent abundance and not generally of interest. This approach is only approximate and may lead to serious misconceptions when the chemist encounters cases in which two natural isotopes are of nearly equal abundance, or cases in which specific isotopic enrichment has appreciably increased the isotope content. Accurate computation of the abundance of molecules of various isotopic mass must be made by considering the absolute probability of the occurrence of a particular combination.

Consider, for example, NO_2 enriched to 30 atom percent in ^{18}O. Ignoring the contribution of ^{15}N, which is quite small, there are three different molecules that will contribute molecular ion peaks to the mass spectrum: $^{14}N^{16}O^{16}O$, $^{14}N^{16}O^{18}O$, and $^{14}N^{18}O^{18}O$. The probability of finding the first combination is given by the product of the individual probabilities, $1.0 \times 0.7 \times 0.7 = 0.49$ and, similarly, the third combination is found with probability $1.0 \times 0.3 \times 0.3 = 0.09$. The molecule containing the mixed oxygen isotopes is found with probability $2 \times 1.0 \times 0.7 \times 0.3 = 0.42$. The factor 2 enters because the oxygen atoms are degenerate; that is, there are two equally probable ways to achieve the configuration. The reader will observe that the factors introduced for the oxygen isotopes correspond to the terms of a binomial expansion $(a + b)^n$ in which a is the probability (fractional abundance) of one isotope, b is the probability of the second, and n is the number of equivalent sites that the element may occupy. Using such a binomial expansion, the abundance of ions of various compositions may be predicted accurately if the overall isotope analysis is known and the isotopes are distributed statistically.

Selection of a likely molecular formula from the molecular ion peak and natural abundance isotope peaks has been of sufficient importance that tables have been compiled, notably by Beynon, giving the anticipated intensities for $P + 1$ and $P + 2$ for compounds containing C, H, O, and N. These tables include molecules with molecular weights up to 500, although the use of isotope peaks to determine molecular formulas begins to lose effectiveness above mass 250. Since these tables deal only with four elements, when other elements are present their kind and number must be determined and their mass subtracted from the molecular weight. The composition of the remainder of the molecule can then be determined with the aid of the tables. These methods all depend on correct identification of the parent peak. Examples will be withheld until identification of the parent peak has been discussed.

21.3.3 / Recognition of the parent ion

Usually the chemist will have some idea of what his sample is and will anticipate the parent peak in a certain m/e region. Two possible difficulties may arise: The parent peak may be weak or absent, or the parent peak may be present but as one of several adjacent or nearly adjacent peaks that are equally or perhaps more prominent. In the first case, the most obvious (although not always effective) solution is to increase the sensitivity to its maximum and increase the sample size. It may be necessary to accept a reduction in resolution in order to obtain the required intensity when the parent is extremely weak. In some cases, it may be possible to deduce the parent mass from the fragmentation pattern. This is particularly true of molecules that can lose small, stable fragments, such as CO and H_2O. Sometimes a suitable chemical derivative of the unknown will more easily yield a parent peak.

To distinguish among several possible candidates for the parent peak, reduction of the energy of the bombarding electron beam is frequently a good test. As the electron energy is reduced toward the value at which ionization can just be observed (called the *appearance potential*), a smaller total number of molecules is ionized and the overall spectral intensity decreases; however, the extent of fragmentation decreases more rapidly, so that the net effect is to increase the intensity of the parent peak relative to the remainder of the spectrum. In some cases, only the parent peak will remain detectable with sufficient reduction of the ionization energy. In other cases, the parent peak can be distinguished indirectly by determination of which is the $P + 1$ peak. This method depends on increasing the frequency of bimolecular collisions in the ionization chamber and is brought about either by reducing the repeller voltage and/or by increasing the sample size. The most common bimolecular reaction of molecular ions containing an electronegative heteroatom (O, N, S) is the transfer of a hydrogen atom from the neutral molecule to the molecular ion, thus adding intensity at $P + 1$. We recognize that the molecular ion must contain an odd number of electrons, so that all even-electron ions can be excluded from consideration for the parent. Many peaks can, of course, be ruled out as the parent ion from consideration of reasonable structures as discussed in the next section.

It should be noted that the presence in the sample of appreciable quantities of impurities may cause difficulty in location of the parent peak. Reduction of the ionization voltage will cause a relative increase in intensity of both the desired parent peak and also the parent peak of the impurities.

21.3.4 / The molecular formula and molecular structure

A generalization that is of assistance both in determining structures and also in ascertaining which peak is the parent peak in the mass spectrum is the *nitrogen rule.* This rule states that all organic compounds having an even molecular weight must contain either an even number of nitrogen atoms or no nitrogen; those having an odd molecular weight must contain an odd number of nitrogen atoms. As is done throughout mass spectrometry, the molecular weight is taken to be the weight of the combination containing only the most abundant isotopic species of each element. The nitrogen rule applies for all compounds containing C, H, N, O, S, and the halogens, as well as compounds that contain alkaline earths, silicon, phosphorus, and/or arsenic. The rule depends on the fact that the most abundant isotope of most elements having an odd-numbered valence also has an odd mass number, and the corresponding isotope of elements of even valency has an even mass number. Nitrogen is the common exception, since it has an odd valency but an even atomic weight of 14. If the ion observed at the largest m/e value is at an odd mass number, it must contain an odd number of nitrogen atoms in order to be the molecular ion. A further generalization of the nitrogen rule that applies to all the ions in the spectrum is that an odd-electron ion will occur at an even mass number unless it contains an odd number of nitrogen atoms.

Because of the rules of chemical valence, the total number of rings plus double bonds (unsaturated sites, R) can be determined once the molecular formula has been established. The number R is determined as indicated in Eq. (21.15) for molecules containing C, H, N, and O.

$$R = 1 + \text{carbons} - \frac{\text{hydrogens}}{2} + \frac{\text{nitrogens}}{2} \tag{21.15}$$

If other elements are present, they are counted as additional atoms of the type to which they correspond in valency. Halogens, for example, would be added to the number of hydrogens, and silicon would be added to the number of carbons.

With these generalizations in mind, we may now proceed to examine two examples of structure determination from mass spectral data. Consider the mass spectrum given in Fig. 21.13. We begin by assuming that $m/e = 73$ corresponds to the molecular ion and note that the intensity at $P + 1$ is 4.8 percent of the parent peak. Inspection of tables of the type compiled by Beynon for mass 73 suggests the following possible formulas.

Mass 73

Formula	$P + 1$
1. $C_2H_7N_3$	3.42
2. $C_3H_5O_2$	3.40
3. C_3H_7NO	3.77
4. $C_3H_9N_2$	4.15
5. C_4H_9O	4.51
6. $C_4H_{11}N$	4.88

Since the parent mass is odd, formulas 2, 4, and 5 may be excluded immediately on the basis of the nitrogen rule. Of those remaining, the best fit to the experimental data clearly is 6, so $C_4H_{11}N$ is taken as the molecular formula. Calculation of unsaturated sites proceeds as indicated in Eq. (21.16),

$$1 + 4 - \tfrac{11}{2} + \tfrac{1}{2} = 0 \tag{21.16}$$

and shows that the molecule is saturated. Therefore, an alkyl amine is suggested and the remaining structural problem is the size and number of the alkyl groups. The base peak at $m/e = 58$ corresponds to $P - 15$ or loss of a methyl group which is, evidently, a very probable process. Another prominent peak occurs at $P - 29$, which can arise by loss of C_2H_5 from the parent ion and also loss of C_2H_4 from the prominent $P - 1$

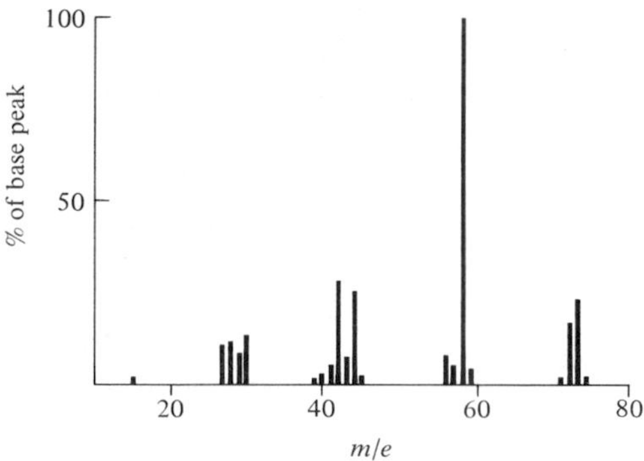

m/e	% of base peak
15	1.3
27	10.
28	11.
29	8.1
30	13.
31	0.27
32	—
33	—
39	1.2
40	2.1
41	4.5
42	28.
43	7.2
44	25.
45	1.8
56	7.3
57	5.0
58	100.
59	3.9
71	1.0
72	17.
73	23.
74	1.1

Figure 21.13 / *Mass spectrum of ethyldimethylamine.*

ion. Since extended alkyl groups usually show fragmentation patterns that contain ions due to progressive breakage of C—C bonds, this spectrum suggests that the largest single alkyl group in the compound is ethyl. Only two structures are possible with this formula that contain C_2H_5: These are $C_2H_5N(CH_3)_2$ and $(C_2H_5)_2NH$. In the second formula, the predominate fragmentation would be loss of ethyl and the base peak would be expected at $m/e = 44$. In the first formula, loss of methyl predominates, and it is therefore consistent with the observed spectrum.

As a second example, consider the spectrum shown in Fig. 21.14. The peak at $m/e = 86$ is too intense relative to the mass 85 peak for the latter to be the molecular ion (Table 21.1). If we pick the base peak at $m/e = 84$ as the molecular ion, the $P + 2$ intensity suggests the presence of a single sulfur atom. In addition, the prominent molecular ion suggests an aromatic

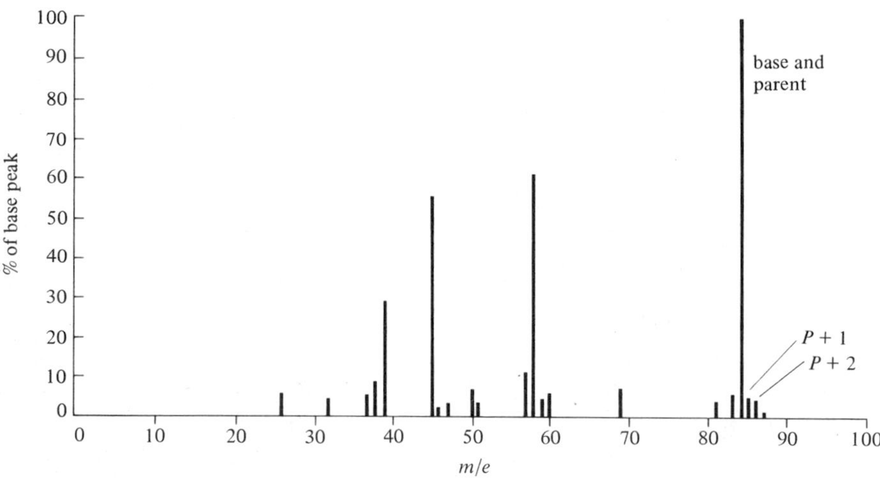

m/e	% of base peak	m/e	% of base peak
1	2.93	45	57.8
2	0.14	46	1.41
12	1.90	47	2.55
13	1.34	48	0.67
14	0.43	49	2.97
15	0.22	50	6.01
24	0.66	51	3.79
25	2.06	52	0.23
26	5.55	56	1.62
27	1.64	57	12.9
28	0.15	58	64.8
29	0.12	59	2.44
32	3.65	60	2.83
33	0.87	61	0.10
34	0.39	69	7.24
36	1.37	81	3.80
37	6.62	82	2.70
38	7.71	83	6.32
39	28.6	84	100
40	2.43	85	5.05
44	2.06	86	4.38
		87	0.17

Figure 21.14 / *Mass spectrum of thiophene,* S

structure. Proceeding on the assumption of one sulfur, we subtract the atomic weight of sulfur, 84 − 32 = 52, and consult tables for mass 52, where the following data are found:

Mass 52

Formula	$P + 1$	$P + 2$
1. C_2N_2	2.92	0.03
2. C_3H_2N	3.66	0.05
3. C_4H_4	4.39	0.07

We now subtract the sulfur contribution to $P + 1$, $5.05 - 0.78 = 4.47$; and $P + 2$, $4.38 - 4.40 \simeq 0$, indicating that formula 3 is a reasonable fit to the experimental data. Our tentative molecular formula, therefore, is C_4H_4S. Calculation of unsaturation is indicated in Eq. (21.17).

$$R = 1 + 4 - \tfrac{4}{2} = 3 \qquad (21.17)$$

Using this result, the most reasonable structure is thiophene, an aromatic in accord with the prominent molecular ion.

21.3.5 / Something about fragmentation

A detailed discussion of the fragmentation patterns of organic compounds is beyond the scope of this text; however, we shall briefly outline some of the behavior of common compounds. In general, the factors that influence the ways in which the molecular ion decomposes are bond strengths and the relative stabilities of both the charged and uncharged fragments that result. Cleavage is favored, for example, at branched carbon atoms rather than unsubstituted carbons because the carbonium ions formed have relative stabilities $R_3C^+ > HR_2C^+ > RCH_2{}^+$.

Similarly, the presence of double bonds, cyclic structures, or aromatic rings helps to stabilize the molecular ion since the positive charge may be delocalized. Resonance stabilization of an ion fragment may favor loss of that moiety. Thus, in general, the presence of a double bond favors allylic cleavage, as indicated in Eq. (21.18).

$$CH_2{=}\overset{+\,\cdot}{C}H{-}CH{-}R \xrightarrow{-R\cdot} \overset{+}{C}H_2{-}CH{=}CH_2 \leftrightarrow CH_2{=}CH{-}\overset{+}{C}H_2 \qquad (21.18)$$

Saturated rings can also stabilize positive charge and therefore tend to lose alkyl side chains at the α-position. Alkyl benzenes, in contrast, more often cleave at the β-position with rearrangement of the ring to the highly stable tropylium ion.

$$\left[\triangle_{R}\right]^{+\cdot} \longrightarrow \triangle^{+} + R\bullet \qquad C_6H_5^{+}{-}CH_2CH_3 \longrightarrow C_7H_7^{+} + CH_3\bullet \qquad (21.19)$$

Tropylium can then lose a neutral acetylene molecule, C_2H_2, to give the cyclopentyl cation, with the result that alkyl benzenes give prominent peaks at both $m/e = 91$ ($C_7H_7{}^+$) and $m/e = 65$ ($C_5H_5{}^+$).

The loss of small, stable neutral molecules is quite common. In particular, if a molecular ion can eliminate H_2O, CO, NH_3, HCN, or similar di- or triatomic species, it is likely that a prominent peak at $P - 18$, $P - 28$, and so on will be observed.

Positive charge can be stabilized by the presence of electronegative heteroatoms, which lead to cleavage of the C—C bond next to the heteroatom, as shown in Eq. (21.20).

$$CH_3 \text{—} CH_2 \text{—} \overset{+\cdot}{O} \text{—} R \xrightarrow{-CH_3\cdot} CH_2 \text{—} \overset{+}{O} \text{—} R \leftrightarrow \overset{+}{C}H_2 \text{—} \ddot{O} \text{—} R \qquad (21.20)$$

Aliphatic ethers cleave in this manner, although the ion R^+ is also usually observed. The parent peak is usually weak for ethers, and an increase in the size of the sample usually emphasizes the $P + 1$ peak by hydrogen atom transfer as mentioned previously, giving the ion $R\text{—}\overset{+}{\underset{H}{O}}\text{—}R'$.

Saturated straight-chain hydrocarbons have low-intensity molecular ions. The fragmentation patterns consist of clusters of peaks separated by 14 amu, which correspond to successive losses of —CH_2— groups. Branched-chain hydrocarbons fragment preferentially at the branch, and a prominent $P - 15$ peak is indicative of a methyl branch. Olefins give distinct parent peaks; however, location of the double bond is difficult because it can easily migrate via electron shifts in the fragment ions.

Aliphatic alcohols give spectra in which the parent peak is weak or absent; the molecular ion is not observed for tertiary alcohols. A prominent $P - 18$ peak may occur through loss of H_2O, particularly with primary alcohols. Also prominent in the spectra of primary alcohols is a peak at mass 31, which arises from the reaction indicated in Eq. (21.21).

$$R \text{—} CH_2 \text{—} \overset{+\cdot}{\underset{\cdot\cdot}{O}}H \rightarrow CH_2 \text{=} \overset{+}{\underset{\cdot\cdot}{O}}H + R\cdot \qquad (21.21)$$

A similar fragmentation process leads to a peak at mass 30 for aliphatic amines. Secondary and tertiary alcohols fragment similarly, with the largest R group being the most readily lost.

The nitro group is easily lost from aliphatic nitro compounds, so the parent peak is weak or absent. The major peaks found are hydrocarbon peaks up to $P - 46$, which corresponds to loss of NO_2. The presence of the nitro group is indicated by appreciable intensity at $m/e = 30$ (NO^+) and at $m/e = 46$ (NO_2^+). Aromatic nitro compounds, by contrast, give strong parent peaks. In addition, $ArNO_2$ is indicated by peaks at $P - 30$, $P - 46$ as for alkyl compounds and also by a prominent peak for Ar^+, which arises from loss of neutral NO_2.

In concluding the section on fragmentation, it should be noted that a small number of peaks may be observed at nonintegral mass values. Some of these peaks, usually of low intensity, arise from doubly charged ions. Others, which are also of low intensity and characteristically quite broad, are due to *metastable ions*. Such a peak appears between mass 90 and mass 91 in Fig. 21.11. Metastable ions are those that decompose during the period of a few microseconds between their formation and the time they reach the detector. With certain restrictions, the apparent mass of such an ion can be quite simply related to the mass of the original ion and the mass of the new ion following decomposition.

Suppose an ion of mass m_1 is formed by electron bombardment and accelerated through potential V_1 prior to its decomposition into an ion of mass m_2, which is accelerated through the remaining potential drop $(V - V_1)$. If the decomposition takes place with very little release of internal energy, the new ion will retain a fraction of the kinetic energy of the original ion in the ratio of their masses; to that energy is added the energy acquired by the ion of mass m_2 as it is accelerated through the remaining potential drop. The total kinetic energy of the new ion is given by Eq. (21.22), which is readily rearranged to the form of Eq. (21.23).

$$\frac{m_2 v^2}{2} = \frac{m_2}{m_1} eV_1 + e(V - V_1) \tag{21.22}$$

$$m_2 v = \left[2\left(\frac{m_2{}^2}{m_1}\right) eV_1 + 2m_2 e(V - V_1)\right]^{1/2} \tag{21.23}$$

Solution of Eq. (21.8) for the radius of curvature followed by an ion gives $r = (c/He)mv$ which, with suitable algebraic manipulation of Eq. (21.23), gives the value of r indicated in Eq. (21.24).

$$r = \left(\frac{2Vc^2}{H^2 e}\right)^{1/2} \left[\left(\frac{m_2{}^2}{m_1}\right)\left\{1 + \frac{(m_1 - m_2)(V - V_1)}{m_2 V}\right\}\right]^{1/2} \tag{21.24}$$

Rearrangement of Eq. (21.9) gives $r = (2Vc^2/H^2e)^{1/2} m^{*1/2}$ as the radius of curvature traversed by a normal ion of mass m^*. Thus, we can identify the apparent mass m^* with the quantity in square brackets in Eq. (21.24) by comparison of the two relations. Therefore, the metastable transition gives rise to a peak in the spectrum at position m^* on the mass scale, and m^* is given by

$$m^* = \left(\frac{m_2{}^2}{m_1}\right)\left[1 + \frac{(m_1 - m_2)(V - V_1)}{m_2 V}\right] \tag{21.25}$$

Only a very small release of internal energy will be required to discriminate strongly against ions formed at large values of $(V - V_1)$ and prevent their passing through the exit slit of the ionization chamber, so that in practice the ions most likely to be detected will appear at $m^* \simeq m_2{}^2/m_1$. Of course, dissociation can occur anywhere between the source and collector, so not all ions are properly treated by these equations and many such ions are not, in fact, recorded. It should be noted that the observation of metastable ions is strongly influenced by differences in instrumentation; the most obvious such difference is the travel time from source to collector. As a consequence, it is difficult to compare results from spectrometers of different designs. The systematic study of metastable ions can aid in elucidation of structures and also contribute to fundamental studies of bonding.

21.3.6 / Appearance and ionization potentials

As noted previously, the energy of the electron beam in an electron bombardment source can be varied by altering the potential through which the electrons are accelerated. The value of the smallest potential that produces a detectable peak is termed the *appearance potential* for that particular ion. To determine the appearance potential, a plot of intensity vs. accelerating potential is constructed, and the linear portion of the plot is extrapolated to zero intensity. Usually, the molecular ion will exhibit a smaller appearance potential than will its fragment ions. In some cases, the appearance potential for the molecular ion is nearly the same as the ionization potential of the molecule, while appearance potentials of fragment ions can be related to bond strengths. For such reasons, these quantities have been the subject of considerable study by mass spectrometry.

It must be pointed out that the appearance potential of the molecular ion is not usually identical to the ionization potential, and the magnitude of the difference may be difficult to ascertain. The ionization potential is the energy difference between the ground vibrational levels of the lowest electronic states of the molecule and the molecular ion. If the molecular ion is formed in an excited state, the appearance and ionization potential are not the same.

A 50-eV electron has a velocity of 4.2×10^{8} cm-sec^{-1} and will traverse the usual molecular dimension of about 10 Å in 2.4×10^{-16} sec, a time much shorter (by a factor of $\sim$40) than the most rapid molecular vibration (C—H stretch). As a result, the ionization process takes place before the nuclei can readjust to their equilibrium positions in the molecular ion which are, in general, different from those in the molecule. Thus, this "vertical" or Franck-Condon process usually leads to formation of the molecular ion in an excited state, and specification of the particular excited state is required to convert the appearance potential to the ionization potential.

21.3.7 / Mass spectrometric analysis of gas chromatography fractions

If a mass spectrum can be obtained for each fraction separated by a prior gas chromatography (GC) experiment, a potent tool for the analysis of complex mixtures results. As mentioned previously, the chemical ionization source can be combined with a gas chromatograph that employs methane carrier gas to achieve this result. In this section, however, combination of the conventional electron bombardment source mass spectrometer with a standard chromatograph (Section 22.5) will be discussed.

Since it is unlikely that compounds of molecular weight greater than 600 amu can be passed through the gas chromatograph, a spectrometer with a range of m/e 1–600 is usually judged adequate for this task. Since, however, the sample size is uniformly small and there may be considerable variation in the size of fractions within the sample, high sensitivity is required, as is a large dynamic range for the detection system. The dynamic range should be of the order of $10^4:1$, and will necessitate the use of a multichannel recording system. In addition, since the GC fractions may be eluted from the chromatograph in a few seconds, it is essential to be able to scan the desired mass range very rapidly and to record the mass spectrum obtained without appreciable distortion. Note that in addition to the small sample size, the fast scan rate that permits only a small number of ions of any one m/e value to be collected by the detector contributes to the requirements for high sensitivity. It has been shown that using a scan rate of 1 sec per mass decade with a spectrometer of resolving power 1000, the bandwidth of the recording system must exceed 5 kHz. The usual galvanometer UV recorder has a maximum bandwidth of about 4 kHz. The detector bandwidth may be increased, however, to 10 kHz by recording the data on magnetic tape instead. In typical magnetic sector instruments adapted for use with GC, the fastest magnetic scan is about 2 sec per mass decade and is limited by the inductance of the electromagnet.

The inlet system required to permit connection of a gas chromatograph to a mass spectrometer must overcome the inherent contradiction between the continuous flow of eluent gas and the requirement of high vacuum within the mass spectrometer. The basic function of any such system is to transfer the sample from the column to the ion source with the maximum efficiency consistent with the limitations on permissible gas flow into the ion source. This maximum flow is limited by considerations such as loss of resolving power due to gas scattering and space charge effects and, in the limit of large flows, breakdown of the ion accelerating voltage. Helium is usually the carrier gas of choice because of its high ionization potential. In a typical mass spectrometer, the upper limit of helium flow into the ion source is approximately 0.15 ml-min^{-1}.

The simplest adapter system consists of a leak that drops the pressure from atmospheric to the source pressure without the need for additional pumps. The leak consists of about 1 cm of 0.001-in. i.d. glass capillary tubing which is heated to prevent condensation of the sample. This simple device functions better with capillary GC columns than with packed columns. A typical carrier flow rate for a capillary column is of the order of 1 ml-min^{-1}. Since the inlet leak is arranged to pass 0.15 ml-min^{-1}, the efficiency of sample transfer is 15 percent. Use of such a leak with a packed column with a carrier flow of, say 30 ml-min^{-1}, would give a transfer efficiency of only 0.5 percent. Thus, a molecular separator will be required for use with packed columns.

There are a number of different designs of molecular separators, all of which work by preferentially pumping off the helium carrier gas while admitting as much sample as possible into the ion source of the mass spectrometer. One form of the separator designed by the Biemann group at M.I.T. is shown in Fig. 21.15(a). Restrictions to flow are required at both input and output ends of the fritted tube and are shown in the form of suitable glass capillary leaks. While the bulk of the helium diffuses through the porous walls of the fritted tube and is pumped away, a significant fraction of sample is admitted to the ion source. Sample transfer efficiencies of 10–50 percent have been reported with this device. The Ryhage separator, which uses stainless steel jets rather than a permeable barrier, is shown in Fig. 21.15(b). Transfer efficiencies up to 90 percent have been reported for this separator. It has the disadvantage of requiring two vacuum pumps in addition to those required for the spectrometer, whereas the Biemann separator requires only one additional pump.

The conventional GC detector is inadequate for the combined MS-GC system due to the delay between the response of the column detector and the arrival of the sample in the ion source. Such a delay would make it

Figure 21.15 / (a) *A molecular separator of the Biemann design.* (b) *The Ryhage separator.*

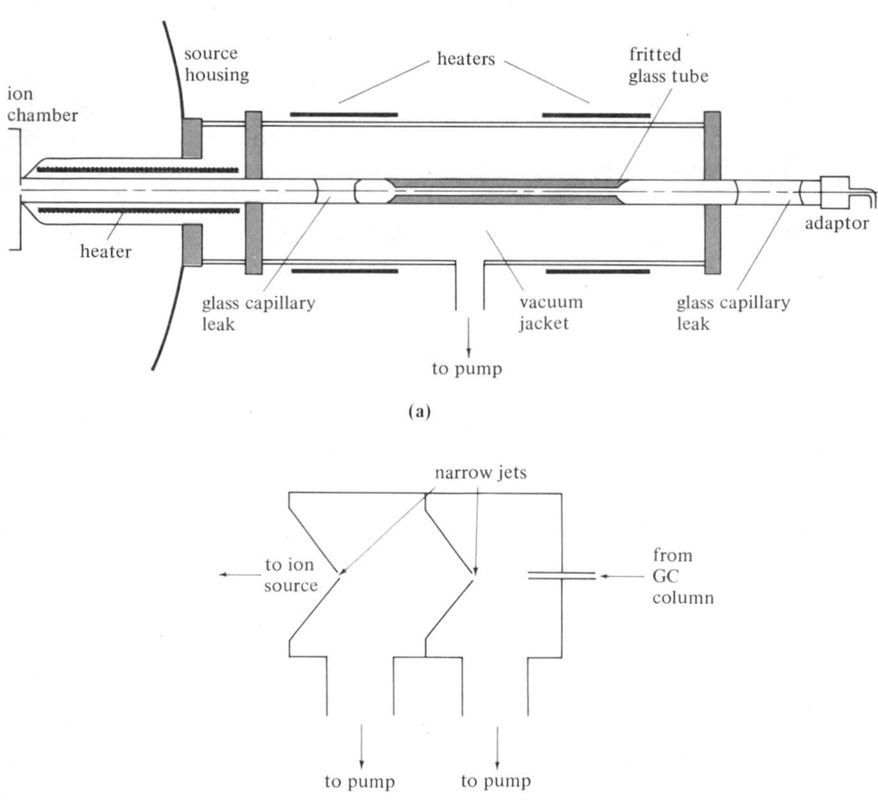

impossible for the operator to know when to scan the spectrum. It is therefore necessary to employ a peak detector within the ion source, the simplest of which is a total ion current monitor. Since the sample may be at low concentration in the entering helium gas and the carrier flow rate is subject to variations in time, suppression of the ion current due to helium is necessary. This can be accomplished by operating the electron beam at 20 eV, a value below the ionization potential for helium but above that for most organic samples. The ion current can be read out on a pen recorder, which yields a plot very similar to the ordinary gas chromatogram. In some instruments, the electron beam energy is automatically switched to 70 eV when the mass spectrum is scanned. This provides both an increase in sensitivity and spectra obtained under the usual conditions to facilitate recognition of the fragmentation pattern obtained.

21.3.8 / Computer-assisted data handling

The combination of gas chromatography with mass spectrometry can lead to production of prodigious amounts of data. Such an instrument is capable of producing 400 complete mass spectra during a half-hour chromatograph run. While these data can be recorded automatically and presented in digital or other form as desired, it is still necessary for the chemist to identify the sample materials by a detailed interpretation of the spectra. Biemann has pointed out that in a complex gas chromatogram, an appreciable fraction of the components will be previously encountered substances, and for many of these mass spectra have been obtained and interpreted previously. Thus, since any interpretation is always confirmed by comparison of the spectrum with a spectrum of an authentic sample of the compound, the time-consuming interpretive step is necessary only as a means of selecting the file spectrum for comparison in the case of known materials.

Approximately 10,000 reference spectra are available. A number of laboratories are now perfecting computer search routines that will permit the computer to compare the mass/intensity data from a new spectrum with the spectra on file, pick the nearest matches, and calculate an error of fit for each. It should be noted that the relative intensities of peaks in a mass spectrum will vary with the design of the instrument on which the spectrum is obtained. Since the available file spectra come from many sources, any useful search routine cannot place too stringent error limits on the comparisons of intensity data.

To economize on computer time, mass spectra are usually abbreviated prior to the comparison-search step. One such method of abbreviation designed by Biemann involves retaining the two most intense peaks in each 14-amu interval beginning at $m/e = 6$. Because 14 amu represents the smallest group usually lost (CH_2), this method is thought to avoid loss of structurally significant peaks. Computerized interpretation is, of course, not limited to GC-MS, and is widely applied to medium-resolution mass spectrometry.

21.4 / HIGH-RESOLUTION MASS SPECTROMETRY

No new principles are involved in the interpretation of high-resolution mass spectra. Once the data have been reduced to a suitable presentation, the molecular ion is identified and the structure elucidated in the same way as has been described above for spectra at unit mass resolution. High-resolution data, of course, permit more precise determination of the mass of each ion and can, in many cases, lead to direct and unequivocal determination of the elemental composition of these ions. In this way the interpretation of the spectra is simplified, since there need be no guesswork regarding composition. However, the number of peaks in the mass spectrum can be very much larger than in a unit resolution spectrum, a result that may complicate interpretation by inundating the chemist with more data than can be sorted and interpreted by hand. In addition, the separation of signal into a larger number of peaks results in a decrease in sensitivity compared with that of a unit resolution spectrum. Figure 21.16 is a partial listing of the possible ions of nominal mass 310 that might be resolved under high-resolution conditions. It is natural in such a situation that the chemist should employ computer techniques to assist him in coping with the data reduction and interpretation problems that arise. The ways in which computers are used will be governed by the method of data collection chosen, as discussed in the following section.

21.4.1 / Data acquisition

The method of automatic data collection and processing used in high-resolution mass spectrometry will depend on whether the instrument in use is of the photographic or electron-multiplier recording type. In the latter method, the lines of the mass spectrum are recorded sequentially by varying the strength of the deflecting magnetic field. Ion intensity is represented by a voltage and line position as a time value measured relative to the beginning of the scan. In photographic recording, as indicated above, the entire spectrum is recorded simultaneously while the analyzing fields are held constant. Ion intensity is manifested after development of the plates as an areal blackening of a line on the photoplate, and this blackening is converted to a voltage using a photometer or densitometer. Line position is measured as a lateral distance along the plate, using a comparator. In both methods, the first requirement to make the data acceptable to a computer is conversion of both the x and y analog signals to digital form.

The major difference between the two methods is the rate of data transmission. In the photographic method, the conversion of the image data on the plate to digital form can proceed at any convenient rate selected on the basis of compatibility with the computing equipment at hand. With electrical recording, however, the computing equipment should be able to accept data input at the scan rate chosen by the mass spectroscopist. To take full advantage of the high sensitivity of electron-multiplier detection and conserve small samples, as well as simply for convenience, the spectroscopist

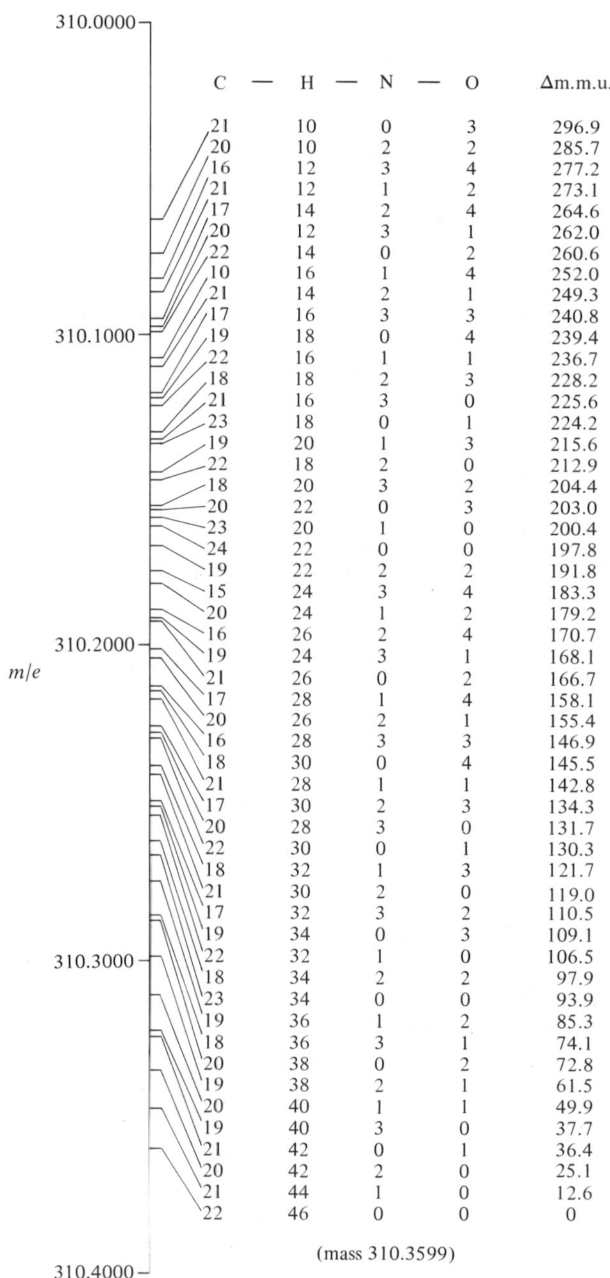

C	H	N	O	Δm.m.u.
21	10	0	3	296.9
20	10	2	2	285.7
16	12	3	4	277.2
21	12	1	2	273.1
17	14	2	4	264.6
20	12	3	1	262.0
22	14	0	2	260.6
10	16	1	4	252.0
21	14	2	1	249.3
17	16	3	3	240.8
19	18	0	4	239.4
22	16	1	1	236.7
18	18	2	3	228.2
21	16	3	0	225.6
23	18	0	1	224.2
19	20	1	3	215.6
22	18	2	0	212.9
18	20	3	2	204.4
20	22	0	3	203.0
23	20	1	0	200.4
24	22	0	0	197.8
19	22	2	2	191.8
15	24	3	4	183.3
20	24	1	2	179.2
16	26	2	4	170.7
19	24	3	1	168.1
21	26	0	2	166.7
17	28	1	4	158.1
20	26	2	1	155.4
16	28	3	3	146.9
18	30	0	4	145.5
21	28	1	1	142.8
17	30	2	3	134.3
20	28	3	0	131.7
22	30	0	1	130.3
18	32	1	3	121.7
21	30	2	0	119.0
17	32	3	2	110.5
19	34	0	3	109.1
22	32	1	0	106.5
18	34	2	2	97.9
23	34	0	0	93.9
19	36	1	2	85.3
18	36	3	1	74.1
20	38	0	2	72.8
19	38	2	1	61.5
20	40	1	1	49.9
19	40	3	0	37.7
21	42	0	1	36.4
20	42	2	0	25.1
21	44	1	0	12.6
22	46	0	0	0

(mass 310.3599)

Figure 21.16 / *The exact masses of a variety of possible ions of nominal mass* 310 *containing* C, H, *and not more than three N atoms or four O atoms.*

SOURCE: F. W. McLafferty, *Interpretation of Mass Spectra*, copyright © 1966, W. A. Benjamin, Inc., Menlo Park, Calif., p. 16.

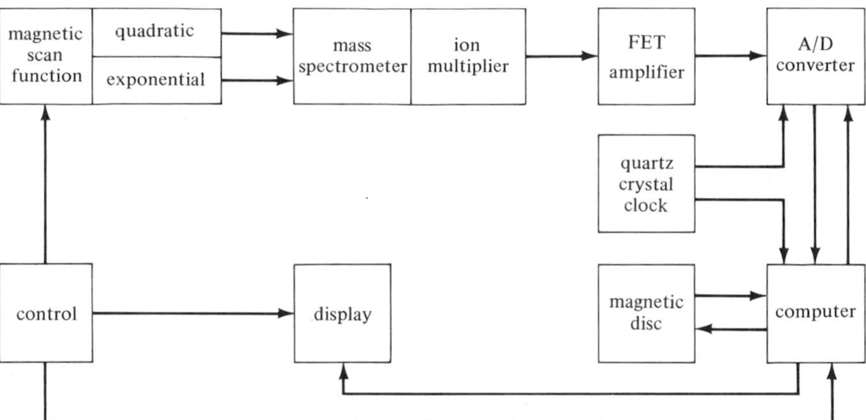

Figure 21.17 / *Block diagram of data-acquisition hardware.*

will generally wish to use the most rapid scan possible that does not degrade the resolving power of the instrument. Scan rates of 10–30 sec per mass decade are common. It has been shown that a spectrometer that operates at a resolving power of 10,000 requires a digital sampling rate of 50 kHz to avoid loss of information when a 10–sec per decade scan is employed. In the usual case of recording the shape of each peak, each sample must contain at least 12 bits of information. This necessitates data transmission at 600,000 bits/sec, a rate that can be handled only by computers with a memory cycle time of ≤ 1 μsec. Obviously, in addition, a large amount of storage is required. In practice, sampling rates of about half this value are in use.

A block diagram of a data-acquisition system is shown in Fig. 21.17. The quartz crystal clock provides timing pulses which define the time scale for calibration of the spectrum. The output from the electron-multiplier amplifier is fed to an analog-to-digital converter, which is operated on a sample-and-hold basis as indicated in Fig. 21.18. During each hold mode, the signal from the previous sample period is digitized for storage or processing by the computer. If the computer is programmed to reject data below a preset threshold intensity, that is, to find and record peaks but ignore the base line, it is usually possible to decrease the amount of data to be stored or processed by at least a factor of 10. Such peak-finding routines are normally employed. Counting the timing pulses provides a direct digital calibration link between the recorded peaks and the mass scan of the spectrometer. With this data-acquisition method, the computer can begin processing the data during the scan; it is for this reason described as an on-line method. Alternatively, the spectral data may be recorded on magnetic tape in either analog or digital form for later, off-line processing.

With photographic recording, the photoplate itself stores the spectral data in analog form. For digitization, the plate is processed in a precision

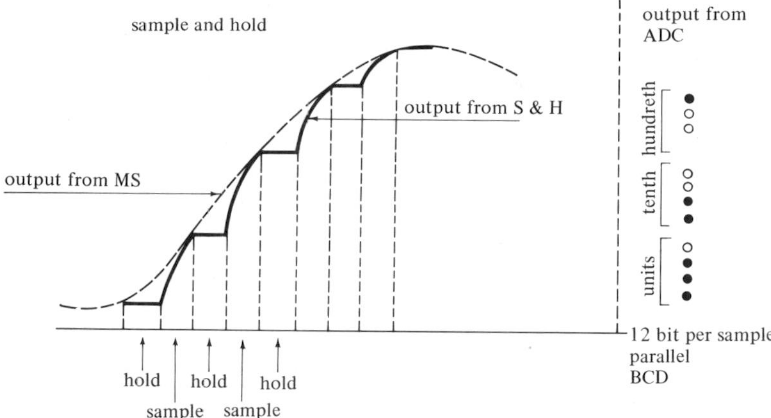

Figure 21.18 / *Diagrammatic representation of the sample-and-hold method of evaluating a mass spectral peak. The analog voltage observed during the sample cycle is converted to digital form for use by a computer in binary coded decimal (BCD) format.*

comparator that employs two A/D converters. The first of these is a digital photometer, which measures the areal blackening of the plate using the sample-and-hold method. This A/D converter is actuated by a second A/D converter, which is part of the distance-measuring system. The distance drive delivers a pulse when the photoplate has traversed a fixed small distance. In typical systems, pulses are supplied at 0.2-μm intervals. The comparator may be interfaced to a computer and the spectrum evaluated in about 5 min of processing time.

21.4.2 / Data reduction

The digitized mass spectrum may consist of as many as 5 million bits of information. The purpose of the data-reduction step is to have the computer reduce the data to a pair of digital values for each peak, one representing mass and the other intensity. While there are differences in detail, these steps are carried out in similar fashion for both methods of data acquisition.

The digital representation of the peaks is smoothed by fitting a curve or polynomial to the data points using the method of least squares. The peak position is then computed, usually by finding the center of its area, or centroid, rather than the top of the peak. The intensity of the peak is also computed on the basis of its area. In both electrical and photographic recording, a calibration must be employed to correct for variations of the detector response with mass. Such a calibration curve is readily incorporated into the computer analysis program. At this stage, the peaks are located only on the arbitrary time or distance axis, which must finally be converted into m/e. This is accomplished by addition of a standard compound, usually perfluorokerosene (PFK), for which the exact masses and relative locations

CALCULATED MASS	ERR	C12/13	H	N	O	P	MEASURED MASS	NO. PTS	INTENSITY
258.1256	1.15	16/0	18	0	3	0	258.1267	11	+++++++
257.1284	-2.01	19/1	16	0	0	0	257.1264	20	++++++++++
257.1261	.36	12/1	21	0	3	1			
256.1206	1.89	19/1	15	0	0	0	256.1225	30	+++++++++++++
256.1228	- .24	13/0	21	0	3	1			
255.1128	1.03	19/1	14	0	0	0	255.1138	18	++++++++
255.1149	-1.09	13/0	20	0	3	1			
NO COMP CALC							247.9867	9	++++++
242.1050	.42	18/1	13	0	0	0	242.1054	17	+++++++
242.1071	-1.67	12/0	19	0	3	1			
241.0971	1.06	18/1	12	0	0	0	241.0982	24	+++++++++
241.0993	-1.06	12/0	18	0	3	1			

Figure 21.19 / *A portion of the printed mass spectral data for diethyl-2,6-dimethylbenzylphosphonate,* $C_{13}H_{21}O_3P$. *The molecular ion is* $m/e = 256$. *Errors are given in millimass units, and each + indicates a power of 2 in intensity. The number of points is the number of samples per peak and is proportional to the width of the peak.*

of peaks are stored in the computer. These peaks are located in the experimental spectrum, and the computer then calibrates the intervening regions by interpolation and assigns a mass value to each peak in the spectrum on this basis. The PFK peaks are then deleted from the spectrum. The computer then examines the mass value for the remaining peaks and computes possible elemental compositions that fall within a preselected error limit of each observed mass. Finally, the data are printed as a list of exact calculated masses and possible elemental compositions, the measured mass for each peak, and the difference between the observed and calculated mass in millimass units. In addition, the intensity and width of each experimental peak are indicated. A section of a mass spectrum presented in this tabular form is shown in Fig. 21.19.

The compound from which this spectrum was obtained is diethyl-2,6-dimethylbenzylphosphonate, $C_{13}H_{21}O_3P$; the molecular weight is 256. That excellent fits are obtained for the parent peak and the peak at $P + 1$ is seen from the errors, which are only a few tenths of a millimass unit. Peaks at 255 and 242 correspond, respectively, to loss of H and loss of CH_3. For the peak observed at $m/e = 248$, the computer has been unable to find a suitable elemental composition and has indicated this by printing "NO COMP CALC." The widths of the peaks are proportional to the number of above-threshold data points listed in the output. Each + indicates a power of 2 in intensity.

When electrical recording is used, in addition to this tabular output, the spectrum is also available on the usual multichannel galvanometer recorder. In many systems, the computer program contains an additional routine which sums the intensities at each unit mass and produces, with the aid of a plotter, a bar graph unit resolution spectrum from the high-resolution data. Such a graphical display is useful, since from it the chemist may recognize familiar fragmentation patterns. Other methods of data display in which fragment ions are arranged according to composition have been devised to assist in interpretation. This technique is known as "element mapping."

For any laboratory contemplating the purchase of high-resolution equipment, the question arises as to whether it should be of the photographic or electrical recording type. Advocates for each type abound. Mass measurement precision has long been the forte of photographic recording, and precisions of 0.1 millimass unit are obtained with well-designed computer-comparator systems. The comparators are, however, an expensive additional piece of hardware. At sampling rates of the order of 25 kHz, precisions of mass measurement using electrical recording approach the 0.1 mmu level in the hands of skilled operators, leaving little to choose between the two methods. Intensity data from electrical recording, however, are generally better than those from photographic recording. Despite the fact that quantitative intensity data are little used, when the spectra are compared to unit resolution spectra via a bar graph plot, the results from electrical recording are more similar to the unit resolution spectra, thus facilitating recognition of familiar patterns. In addition, the convenience of the on-line technique appears to be an overwhelming advantage, which will make it ultimately the more widely employed.

21.4.3 / Inorganic trace analysis

Inorganic trace analysis is properly discussed in the section on high-resolution mass spectrometry because it involves the application of double-focusing instruments. Nearly all the inorganic mass spectrometry presently carried out employs the rf spark source which, as discussed previously, requires the use of an electrostatic energy selector in addition to a magnetic mass analyzer in order to obtain useful resolution. Because of the large energy spread of the ion beam, even with double focusing, the obtainable resolving power is inferior to that achieved routinely in organic high-resolution work.

Spark source mass spectrometry enjoys the advantages of high sensitivity and great generality. As mentioned previously, all the elements of the periodic table may be determined, usually with detection limits of the order of 10 ppb. As a result, this technique has made notable contributions to the advancement of semiconductor technology by affording a method for the determination of trace impurities in transistor materials such as silicon and germanium. The electrical properties of these elements are very significantly altered by the presence of small traces of other elements. More recently, it has proved useful in the broad-spectrum analyses of samples of various extraterrestial materials. On the negative side, the precision of spark source analyses is relatively poor, with typical standard deviations in the 20–50 percent range; major components can be difficult to determine and may interfere with detection of less abundant materials. More precise methods are almost always available for the major components themselves.

As pointed out previously, instabilities in the spark source led, in the past, to nearly universal use of photographic detection in this branch of mass spectrometry; the difficulty in obtaining accurate quantitative intensity data from photoplates has been alluded to already. The sensitivity of the photographic emulsion is a function of both the energies of the impinging ions and their masses, so careful calibration is necessary when quantitative intensity data are required. The use of accurately known natural abundance ratios for polyisotopic elements has formed the basis of many such calibration techniques. The most important limitation of sensitivity generally involves an intense halo formed on the high-mass side of lines due to the major component matrix. This interference is due to secondary emission processes, including low-energy ions and electrons ejected from the emulsion under bombardment by primary ions of approximately 20 keV energy. Considerable research is under way in the attempt to reduce this problem by modification of the photo emulsion and/or development processes.

A number of problems relate to proper exposure of the photographic plate. The exposure time is usually estimated by the use of a total ion beam monitor. This device consists of a faraday collector placed at the entrance slit of the magnetic sector, where it collects a known fraction of the entering ion beam. The total charge collected may be related to the required exposure time. Multiple exposures may be made by altering the position of the photoplate, and a series of graded exposures is generally necessary to obtain useful data when the sample contains several components at diverse concentrations. Overexposure of the plate leads to general fogging, with reduction in sensitivity and precision for weak lines. As a result, the amount of sample consumed by the spark during a typical exposure may be only a few micrograms. Many workers have traced a substantial part of the large observed standard deviations in analysis results to sample inhomogeneity at this small sample size. Analytical precision is improved by the use of an electrostatic ion beam chopper, which permits an adjustable fraction of the ion beam to reach the detector. Thus, longer spark times and greater sample consumption are possible without overexposure. This technique is capable of reducing standard deviations by at least a factor of 2 when sample inhomogeneity is the limiting factor. It should be noted that the general convenience of electrical detection and the persistent difficulties with photographic emulsions are leading to development of electrical detection systems for spark source work. These employ complex and sophisticated electronics that attempt to overcome the inherent difficulties of the spark source.

Finally, it should be noted that in addition to the usual interferences of overlapping mass values, in spark source work explicit consideration of multiply charged ions is required. Consider, for example, analysis of traces of sulfur in zinc. Since zinc is the predominate matrix material, strong lines will exist for the doubly charged ions $^{64}Zn^{2+}$, $^{66}Zn^{2+}$, and $^{68}Zn^{2+}$ and these, unfortunately, correspond to the same m/e values as $^{32}S^{+}$, $^{33}S^{+}$, and $^{34}S^{+}$. The real mass differences are all less than 9 mmu and cannot be adequately resolved. In some cases, such a difficulty can be circumvented by using multiply charged trace ions for analysis, but with, of course, a degradation in the limit of detection.

SUGGESTED READINGS

R. M. Silverstein and G. C. Bassler, *Spectrometric Identification of Organic Compounds*, 2nd ed., Wiley (New York, 1967), Chapter 2.

R. W. Kiser, *Introduction to Mass Spectrometry and Its Applications*, Prentice-Hall (Englewood Cliffs, N.J., 1965).

H. Budikiewicz, C. Djerassi, and D. H. Williams, *Interpretation of Mass Spectra of Organic Compounds*, Holden-Day (San Francisco, 1967).

F. W. McLafferty, *Interpretation of Mass Spectra*, W. A. Benjamin (Menlo Park, Calif., 1966) (paperbound).

J. H. Beynon, *Mass Spectrometry and Its Applications to Organic Chemistry*, Elsevier (Amsterdam, 1960).

A. J. Ahearn, Ed., *Mass Spectrometric Analysis of Solids*, Elsevier (New York, 1966).

G. W. A. Milne, Ed., *Mass Spectrometry: Techniques and Applications*, Wiley (New York, 1972).

PROBLEMS

1 / Calculate the velocity of (*a*) an electron accelerated by a 70-V potential drop and (*b*) a monopositive ion of mass 150 accelerated through a 5-kV potential drop. Compare these velocities with the velocity of light. For which species will relativistic effects become important at low energy?

2 / A magnetic focusing mass spectrometer uses an analyzer with a radius of curvature of 15 cm and a permanent magnet of 7000 gauss. What acceleration voltage will be required to bring $m/e = 100$ amu into focus at the detector?

3 / Predict the relative intensities of peaks at $m/e = 46, 47, 48, 49, 50$ in the mass spectrum of a sample of NO_2 that has been enriched to 10 atom-percent in ^{17}O and 20 atom-percent in ^{18}O. Consider the natural abundance of $^{15}N = 0.4$ percent.

4 / What resolving power is required of a mass spectrometer to separate the following pairs of ions:
(*a*) $C_{12}H_{10}O^{+}$ and $C_{12}H_{11}N^{+}$
(*b*) $^{64}Zn^{2+}$ and $^{32}S^{+}$ (exact isotopic masses required)
(*c*) N_2^{+} and CO^{+}

5 / On what radius of curvature will an ion that has been accelerated through a 3-kV potential drop travel when deflected by a radial electrostatic field of 600 V-cm^{-1}?

6 / Assign reasonable structures to the organic compounds for which mass spectra are given in Fig. 21.20. The parent peak is identified in each spectrum; only C, H, O, N, and S are present in these compounds.

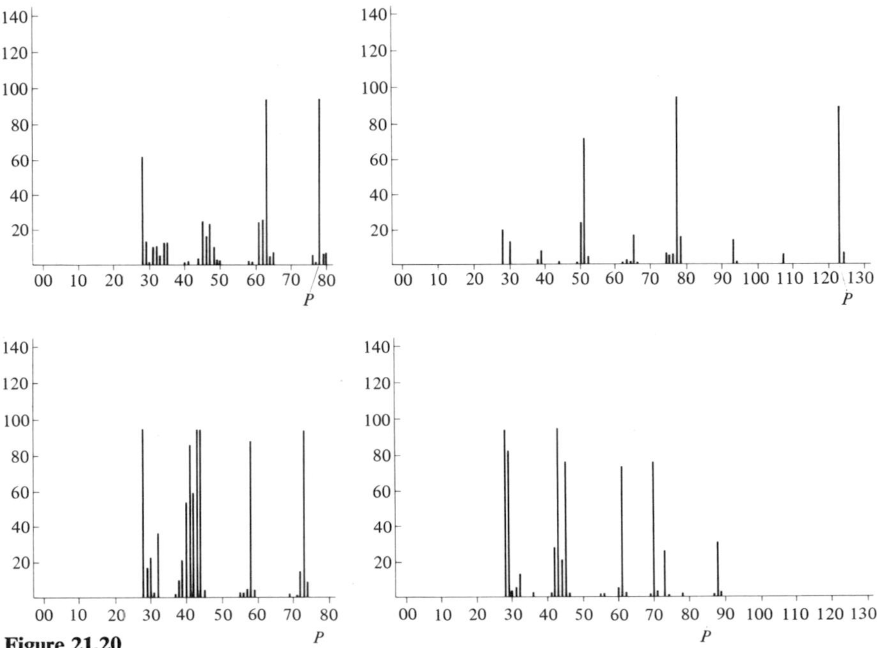

Figure 21.20

chapter 22 / Chromatography

chapter 22

As a generalization, chromatography may be described as a process by which two phases, in physical contact, are caused to move relative to each other. If certain components of these phases are capable of selective partition between them, then the essential conditions for separation by chromatography are established.

The name chromatography is derived from two Greek words, *chroma*, color, and *graphein*, to write. The first such experiments were the separations of colored constituents of an extract of plant leaves reported in 1906 by the Russian chemist, Tswett. In these experiments, the crude extract was allowed to pass through a column of calcium carbonate and differentiation of the components was accomplished visually by observation of the separations of colors; hence the name, which means literally "color writing."

Chromatography typically involves movement of a gas or liquid past a stationary liquid or solid. Selective partition of components of a sample between the phases may be based on solubility differences, on differences in equilibrium constants for chemical reactions of sample components with one of the phases, or on differences in extent of physical adsorption of sample components on the stationary phase.

22.1 / INFLUENCE OF THE SORPTION PROCESS

In the discussion immediately following, the term *sorption* is used to designate the process by which sample components interact with the stationary phase. The term carries no implication about the actual mechanism of interaction. Sample components that are subject to sorption are termed *eluates* when in the moving phase and *sorbates* when in contact with the stationary phase. The material of the moving phase is referred to as the *eluent* and that of the stationary phase as the *sorbent*. Some of these terms are given more precise definitions as the discussion is developed below.

22.1.1 / Linear isotherms

Consider a chromatography experiment in which a finely divided solid sorbent is packed in a tube through which a liquid eluent passes. At the inlet, a small slug of a solute, A, is introduced into the eluent stream, which carries it onto the top of the column. Suppose that A is capable of interacting with the stationary phase as indicated in Eq. (22.1) and that equilibrium is maintained continuously throughout the column. In Eq. (22.1), C_e is the concentration of A in the moving phase and C_s is the concentration of A in the stationary phase.

$$C_e \rightleftharpoons C_s \tag{22.1}$$

If reaction (22.1) is very complete, the effect of this process will be simply to deposit the sample of A at the top of the bed. If, however, the reaction reaches equilibrium without going to completion, then continued movement of the liquid phase will carry eluate away from the top of the bed, thereby reducing C_s in that part of the column; to reestablish the equilibrium, sorbate will be transferred back to the moving phase, increasing C_e. This process will be repeated continuously, with the effect that component A will move in a band along the column, the rate of band movement controlled by the flow rate of the eluant and by the equilibrium constant for Eq. (22.1). The more nearly complete the equilibrium, the slower will the band move along the column for a given eluent flow rate.

When equilibrium is being established, selection of the molecules of sorbate that are desorbed in a given moment takes place according to the energies possessed by individual molecules, a process that must be dealt with statistically. It may be expected, therefore, that as the band moves along the column, its profile in terms of C_e as a function of location will assume a Gaussian distribution, if the equilibrium constant of Eq. (22.1) is indeed a constant and not a function of solute concentration.

When this is true, it can be said that the partition process follows a linear isotherm, illustrated in Fig. 22.1(a). The slope of the curve in Fig. 22.1(a) is

Figure 22.1 / *Partition between phases described by isotherms.*
(a) *Linear isotherm.*
(b) *Langmuirian isotherm.*
(c) *Anti-Langmuir isotherm.*

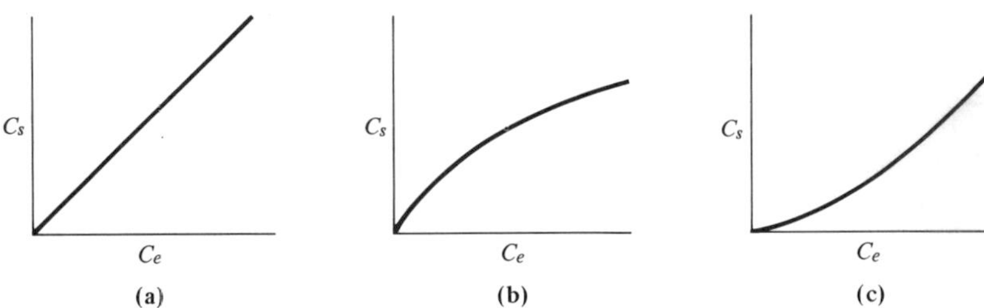

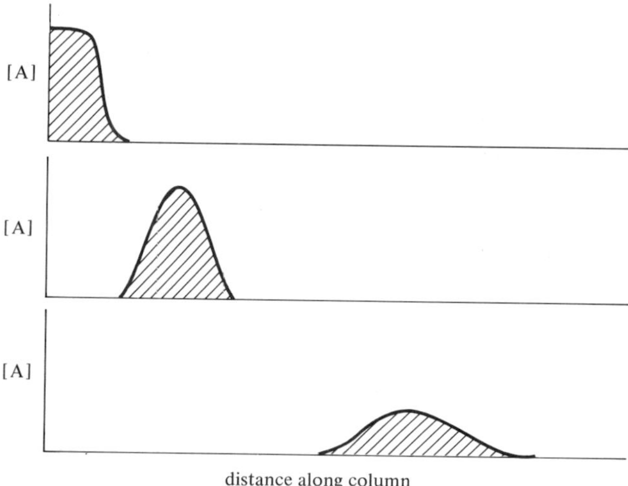

Figure 22.2 / *Elution chromatography with a linear isotherm.*

proportional to the distribution coefficient for the species, which is defined in Eq. (22.2).

$$K_d = \frac{C_s}{C_e} \tag{22.2}$$

The progress of a band along a column can be described by showing the variation of concentration of A as a function of the distance along the column. Such plots for the case of sorption involving a linear isotherm are shown in Fig. 22.2. The statistical nature of the sorption–desorption process produces a Gaussian band shape which continually spreads as the band moves. Under ideal conditions, with equilibrium established at every point in the column, the band width increases with the square root of distance of travel along the stationary phase.

22.1.2 / Nonlinear isotherms

It is generally true that linear isotherms are realized over only limited ranges of sorbate concentrations. At sufficiently high concentrations, the stationary phase behaves as though it were saturated with sorbate, producing a nonlinear isotherm as shown in Fig. 22.1(b). When this happens, the band shape is not Gaussian, but has an extended trailing edge and a sharp front, as shown in Fig. 22.3(a). This occurs because, at high values of C_s, K_d is reduced and the rate of band movement is accordingly increased compared with what it would be with lower values of C_s. That portion of a band with highest concentration tends to move along more rapidly than the portions with lower concentrations. With an isotherm curving in the reverse direction, as in Fig. 22.1(c), a band with a sloping front and sharp trailing edge, as in Fig. 22.3(b), will be observed.

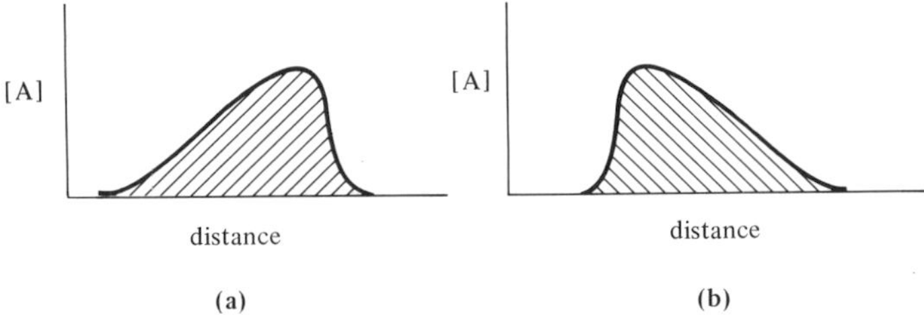

Figure 22.3 / *Elution with nonlinear isotherms.*
(a) *Langmuirian isotherm.*
(b) *Anti-Langmuir isotherm.*

22.2 / ELUTION, DISPLACEMENT, AND FRONTAL DEVELOPMENT

Chromatographic operations are categorized according to the physical states of the components and according to the sorption mechanism. As an introduction to that discussion, we shall consider a broader categorization according to the reactivities and quantities of the components, without regard to the mechanism of reaction or to the physical state of the reactants.

22.2.1 / Elution

The process described by Figs. 22.1–22.3 is one in which the affinity of the stationary-phase material to the solute carried in by the moving phase is greater than to other components of the system, that is, greater than to the moving phase itself or to any other component of it. Chromatography involving movement of a defined band of a strongly sorbed material along a stationary phase is termed *elution*—hence the use of the term *eluent* for the moving phase and *eluate* for the material being eluted.

From the example above, we may deduce that if the slug of solute introduced were composed of more than one substance and if the K_ds for each of these differed, then each should move along the column at a different rate. That being true, it is therefore possible in principle to effect isolation of the components of such a sample from each other by using a sufficiently long column. The smaller the difference in K_d values, the longer the column required for separation. However, even in the ideal case, bands spread as they move, causing reduction of eluate concentration when long columns are used to separate very similar substances. This dilution may constitute the limiting factor in certain applications.

22.2.2 / Displacement

If the same type of experiment is performed, involving introduction of a limited amount of sample into the moving stream but under conditions such that the moving phase, or a major component of it, is more strongly sorbed than the sample, the result is very different from elution.

Suppose, using the liquid–solid illustration as above, that a sample containing components A and B were introduced in a stream of water, with the

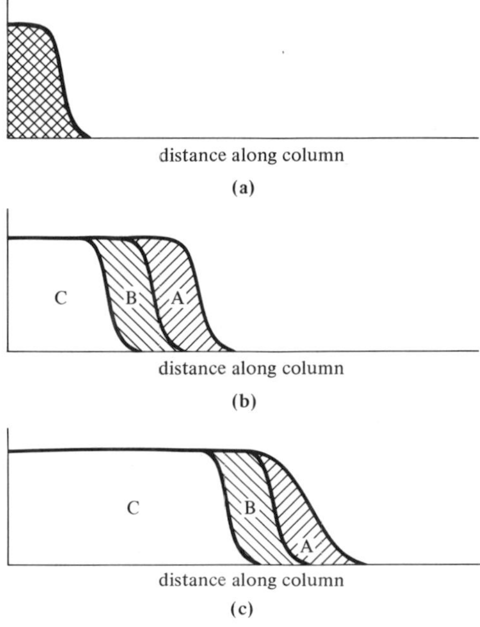

Figure 22.4 / *Fractionation of a two-component mixture by displacement.*

result that both sample components were quantitatively sorbed at the top of the column. This situation is illustrated in Fig. 22.4(a). If the composition of the influent stream is then changed to include a fixed concentration of a third substance, C, which is more strongly sorbed than either A or B, C will compete with A and B for sorption sites. Since C is more reactive than A or B, exchange of C for A or B will be very efficient and there will be little overlapping of the zone containing C with the bands containing A and B. If there is also a difference in the K_d values for sorption of A and B, then the band containing these components will split into two bands, as shown in Figs. 22.4(b) and 22.4(c). That component having the smallest K_d will be concentrated in a band at the front of the advancing zone. The other components will be arranged according to the magnitudes of their distribution coefficients. In Fig. 22.4, the order of increasing K_d values is

$$K_{d_A} < K_{d_B} < K_{d_C}$$

This process is termed *displacement.* The more reactive substance displaces the less reactive. From this illustration, it may be noted that displacement chromatography differs from elution chromatography in various ways. Displacement permits fractionation and purification of sample components but does not afford a means of isolation of one from another.

Unlike the situation realized on performing elution, displacement does not necessarily involve a continual reduction in the concentration of material in the moving band. Zone boundaries formed as a result of displacement of a component by a more reactive one, as between A and B or between B and C in Fig. 22.4, are self-sharpening and, in properly designed columns, can be maintained indefinitely. Boundaries that do not involve displacement are not self-sharpening and, even under ideal conditions, will become diffuse. This is always true of elution and occurs in displacement experiments when the front of a solute band moves along an unreacted stationary phase, as indicated in Fig. 22.4.

Displacement chromatography is used to produce pure samples from mixtures when it is not necessary to effect a quantitative isolation. A monitoring procedure must be provided to detect zone boundaries. By this procedure it is possible to utilize quite small differences in K_d values to effect fractionation, since with proper design, very long columns can be used. The fractionation of rare earth mixtures on a large scale is a notable example of the use of displacement chromatography.

22.2.3 / Frontal development

A third possible type of operation, called *frontal development*, involves the removal of a component from a solution. Typically, the solution comprises the moving phase. A stationary phase is used that reacts effectively with the component to be removed. Addition of material is continued until that component saturates the stationary phase and passes through the column. This can be used to remove an unwanted component from a system, for example, removal of trace metals from distilled water. If the sorption process can be reversed, it can be used to concentrate, for example, recovery of valuable trace metals from solution. Frontal development can also be used to isolate a less strongly sorbed component from more strongly sorbed components.

Of these types of chromatographic operations, elution is most often used for analytical applications because it can provide both qualitative and quantitative information. The volume of carrier required to produce breakthrough of a particular component is a measurable quantity that may be used as an aid in identification. The material in individual bands may readily be measured as a means of quantitative analysis. Bands may be collected in order to apply other techniques of identification. The discussion below is restricted to various types of elution chromatography.

22.3 / THE PLATE MODEL IN CHROMATOGRAPHY

Using mathematical models, it is possible to simulate the behavior of chromatographic systems. For example, the length of column required for complete isolation of the components of a mixture by elution can be evaluated if certain aspects of behavior of the system are known. In practice, however, this approach is usually limited to engineering studies for large-scale applications involving economically important quantities of materials. Conditions for analytical and laboratory-scale preparative operations are almost always established on the basis of trial and error because of the considerable effort required for a design based on theoretical considerations. For this reason, only a very brief consideration of one of the existing models is undertaken here. This is the theoretical plate model, originally developed to describe fractional distillation.

For the plate model, the system is imagined to be subdivided into a number of sections or *theoretical plates*. Each phase within a plate is assumed to be homogeneous, but not necessarily of the same composition as that of neighboring plates. In the operation of the model, the phases comprising each plate are brought into contact and equilibrium is established between them. Then, without any further alteration of their compositions, the two phases are separated and the moving phase of each plate is placed in contact with the stationary phase of the next, whereupon the process is repeated. Knowledge of the distribution coefficient for each component of the sample permits calculation of the composition of each plate at each stage of the process.

This model can be expected to be a valid representation only for a system in which several restrictive conditions obtain. The eluate and sorbate must not move relative to the moving and stationary phases, respectively. There must be complete establishment of equilibrium in the system, and transfer of material between the phases must occur instantaneously. The path length experienced by any of the material of the moving phase in a given column cross section must be the same as that experienced by any other sample taken from the same cross section. In applying this model, a theoretical plate is identified with the length of column required to effect the same change in concentration that would occur in the hypothetical plate. *The height equivalent to a theoretical plate*, or *HETP*, is used as a measure of column efficiency. The smaller the HETP, or conversely, the larger the number of plates for a column, the higher the efficiency.

No real system shows rigorous compliance with the limitations inherent in this model, although the deviations are often not overwhelming. It is, however, generally not practical to start with fundamental information and calculate the HETP. Instead, one generally measures the plate height by observing the performance of a column. Various methods have been used; these are based on some measure of band width in relation to the amount of carrier required to move the band out of the column and ordinarily take the form of a calculation of N, the number of theoretical plates for a column, where $N = \text{HETP}/l$, when l is the column length.

An expression for estimating N from experimental observations, which can be derived from plate theory, is stated in Eq. (22.3).

$$N = 8\left(\frac{V_r - V_0}{\beta}\right)^2 \tag{22.3}$$

Here V_r is the retention volume, the volume of eluent required for appearance of the peak concentration of the eluted band at the column outlet and

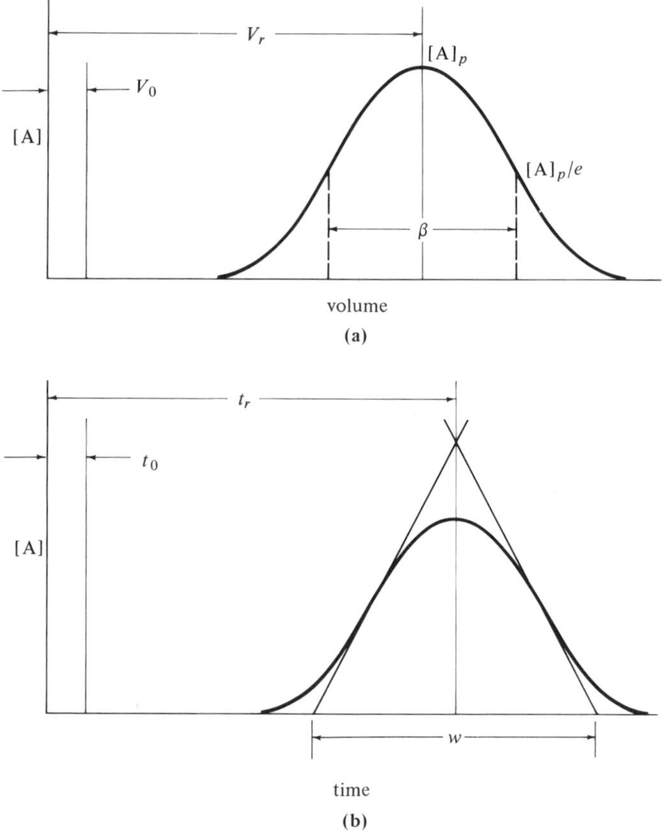

Figure 22.5 / *Measurement of quantities needed for HETP estimation.*
(a) *Eluent volume is measured.*
(b) *Elution time is measured.*
Symbols are defined in the text.

V_0 is the void volume[1] of the stationary bed. β is the band width in volume units measured between the two points at which eluate concentration equals the peak concentration divided by the natural logarithm base, e. This measurement procedure is illustrated in Fig. 22.5(a). An alternative method involves measurement of the band width in terms of time required for elution. The expression is given in Eq. (22.4).

$$N = 16\left(\frac{t_r - t_0}{w}\right)^2 \tag{22.4}$$

[1] The void, or dead, volume is the interstitial volume in the column and is equal to the total internal column volume less the total volume of packing. It may be determined by measuring the volume of moving phase required to move an unsorbed material through the column.

Here t_r, the *retention time*, is the time required for appearance of the peak; t_0 is the time required to move a slug of eluent equal to the void volume through the column, and w is the peak width as defined in Fig. 22.5(b). This approach is used with gas and high-pressure liquid chromatography when elution data are automatically read out by a strip-chart recorder.

Such measurements have little fundamental use in comparing different types of separations, because a theoretical plate of one system does not necessarily represent the same separation capability as a plate of a different system. However, HETP measurements do afford a convenient means of comparing different units of the same type of system. Especially, variation of HETP gives a valuable indication of the types of system alterations that may be useful in improving column efficiency.

22.4 / CHROMATOGRAPHIC OPERATIONS

22.4.1 / Chromatographic systems

Chromatographic systems may involve liquid (LC) or gaseous (GC) moving phases. Either of these may be used with liquid (LLC or GLC) or with solid (LSC or GSC) stationary phases. The stationary phase may be retained within a tube and the moving phase passed through it. Alternatively, the bed may be in a flat layer, the edge of which is dipped in a pool of the liquid that is to serve as the moving phase. In this type of experiment, thin-layer chromatography (TLC) or paper chromatography, the liquid moves by capillary attraction. In any case, provision must be made to observe the results of the elution; the type of detection system used depends on the details of the chromatographic system employed.

Of the possible types of chromatographic experiments, certain ones such as GC, LC, and TLC receive much broader analytical application than the others. Major factors leading to wide acceptance are rapid operation, high efficiency without the requirement of great skill on the part of the operator, and ready availability of reliable equipment.

22.4.2 / Operational parameters

It was pointed out in Section 22.3 that measurements of variables may be made in terms of eluent volume or in terms of elution time, depending on which is more convenient. The HETP can be estimated on either basis. Similarly, chromatographic parameters used to describe the rate of band movement may be expressed either way.

The distribution coefficient, defined by Eq. (22.2) as the ratio of concentrations of sorbate to eluate, may be measured by means of Eq. (22.5).

$$K_d = \frac{V_r - V_0}{V_s} \tag{22.5}$$

Here V_s is the volume of the stationary phase, total volume less void volume.

An additional parameter, the *capacity factor*, k', is useful in describing the performance of LC operation. It is the ratio of the total amounts of the sample in the two phases and is defined by Eq. (22.6). Combining Eq. (22.5) with Eq. (22.6) gives Eq. (22.7).

$$k' = K_d \frac{V_s}{V_0} \tag{22.6}$$

$$V_r = V_0 (1 + k') \tag{22.7}$$

When k' is zero, the sample moves through the bed at the rate of the moving phase and $V_r = V_0$. When the retention time, rather than retention volume, is measured, k' can be estimated with the aid of Eq. (22.8).

$$k' = \frac{t_r - t_0}{t_0} \tag{22.8}$$

If a sample is to be fractionated, its components must show different k' values. For a pair of compounds this may be expressed in terms of the ratio of the distribution coefficients, α. The larger the value of α for two compounds, the easier the separation will be.

$$\alpha = \frac{K_{d_2}}{K_{d_1}} = \frac{k_2'}{k_1'} = \frac{t_{r_2} - t_0}{t_{r_1} - t_0} \tag{22.9}$$

This is indicated quantitatively in Table 22.1.

22.5 / GAS CHROMATOGRAPHY

Gas chromatography (GC) is one of the most widely applied analytical separation techniques. It provides a method whereby complex mixtures can, in one operation, be fractionated and qualitatively and quantitatively analyzed. The process is rapid, typically requiring 5–30 min, and is well suited for use with milligram-scale samples. It is ordinarily applied to substances that exhibit appreciable vapor pressures at temperatures below 300°C. Gas chromatography can readily be combined with a trapping procedure to isolate small quantities of pure components for analysis or other use. It can also be combined with mass spectrometry to afford rapid analysis of individual components (see Section 21.3.7).

TABLE 22.1 / *Number of theoretical plates required for separation[a] of a two-component mixture*

α	N
1.01	160,000
1.05	6800
1.10	1940
1.15	940
1.2	575

[a] Separation is defined as 2 percent overlap of Gaussian peaks.

SOURCE: J. J. Kirklands, Ed., *Modern Practice of Liquid Chromatography*, Wiley-Interscience (New York, 1971), p. 19.

22.5.1 / Gas chromatography stationary phases

Gas chromatography involves interaction of a sample in the vapor phase swept through the column by an inert gaseous carrier, with either a liquid or solid stationary phase. A liquid phase may be simply applied to the walls of the column, or it may be dispersed on the surface of a finely divided solid. In either case, the separation depends on differences in solubility of sample components in the liquid phase.

Selection of a liquid phase for a particular use is ordinarily made on an empirical basis; there is a voluminous literature of GLC applications. It is evident from the foregoing discussion that a suitable liquid will be one capable of dissolving at least some components of the sample to be fractionated; otherwise the sample will simply move through the column with the carrier. The stationary phase chosen must show a negligibly small vapor pressure and must be a liquid at operating temperatures. While GLC can be performed with capillary columns having a layer of liquid on the wall, packed columns are more often used. The liquid phase is dispersed on a finely divided solid supporting material, which ideally should be entirely unreactive with the sample being chromatographed.

A variety of solid stationary phases have been used in GSC. Some of these, such as activated charcoal and molecular sieves, depend on physical adsorption as a fractionating process. These are mainly restricted to use in separations of permanent gases because the adsorption–desorption rates for larger molecules tend to be slow, resulting in low column efficiency.

Much more generally applicable is a class of synthetic stationary phases in which compounds having properties suitable for use as liquid phases in GLC are chemically bonded to inert cross-linked polymers. The result is a product that has the desirable properties of liquid phases, especially high resolution and availability of a varity of materials with differing properties to deal with different types of separation problems. These materials have very low vapor pressures and therefore do not present the difficulties caused by column bleeding (loss by volatilization in the carrier stream) that very often occur with liquid phases and that cause trouble especially when a sample is to be collected after being chromatographed.

22.5.2 / Gas chromatography moving phases

A variety of gases are used as carriers in GC experiments. Helium, hydrogen, argon, and nitrogen are examples of commonly used gases. Selection of the carrier gas depends on several factors, particularly the type of detector to be used and the type of sample to be fractionated. Additional comments are included in the discussion of GC detectors below.

22.5.3 / Factors affecting gas chromatography efficiency

In this context, it is instructive to recall the assumptions that are implicit in the plate model. One of these is that the eluate should not move relative to the carrier stream. Since both the eluate and carrier are gases in the GC experiment, it is evident that this restriction can be approximated only for rather short periods of time, since the band of eluate will constantly spread by diffusion in the carrier stream. We therefore expect and find that operation of GC columns is not satisfactory when the retention times become excessively long.

Another restriction is that the sorbate must not move relative to the stationary phase. If a column is operated at sufficiently fast flow rates to avoid band spreading by gaseous molecular diffusion, there is ordinarily no difficulty from diffusion of solute in the liquid phase. Where solid stationary phases are used, this is, of course, not a problem.

For the plate model to be valid, there must be complete and instantaneous establishment of equilibrium between the phases. Instantaneous response and complete establishment of equilibrium are, of course, idealizations in a moving system. However, the rapid transport possible in the gas phase makes it much easier to achieve a reasonable approximation than is generally true for other types of chromatography. It is accordingly possible to operate GC columns on a short time scale without the rigorous precautions against stationary bed imperfections required in liquid chromatography.

Some band spreading in addition to that predicted by theory would be expected to be caused by eddy currents in the gas stream. This effect will be magnified when a column packed with solid stationary phase material is used. The moving phase must flow around the particles of packing, and the path length experienced by different segments of the moving phase will evidently be different. The use of finely divided packing of uniform particle size, uniformly packed in the column, will minimize difficulties from eddy diffusion. This problem is less pronounced in GC than in LC because of the lower viscosity of the moving phase.

The interaction of these factors has been expressed by J. J. Van Deemter in an empirical equation, Eq. (22.10), which bears his name.

$$\text{HETP} = A + \frac{B}{u} + Cu \qquad (22.10)$$

Here u is volume flow rate of the eluent gas. A is a term independent of flow rate, which primarily reflects eddy diffusion and is therefore affected by uniformity of column packing and particle size. B is ascribed to molecular diffusion. This process will contribute to the broadening of peaks only in proportion to the time the sample spends in the column, since the rate of diffusion is a function only of temperature for a given system—hence the inverse relation of column efficiency and flow rate. The factor C reflects mass-transport phenomena; that is, at high flow rates, efficiency is lost due to a failure to attain equilibrium between the phases. C is proportional to the square of the thickness of the liquid film that coats the solid support. The contribution of each of these factors, eddy diffusion, molecular diffusion, and mass transport, to the observed variation of HETP with flow rate is illustrated in Fig. 22.6.

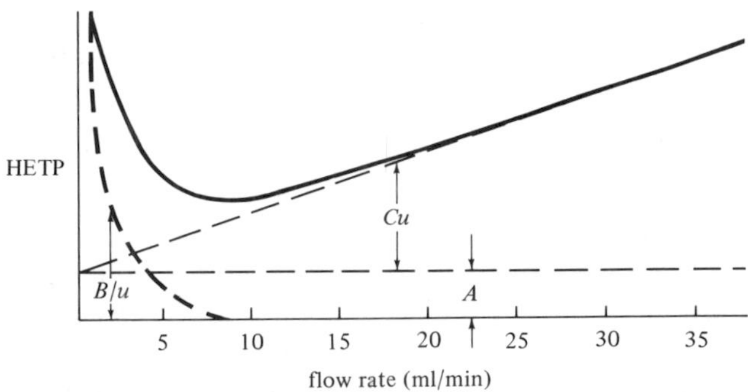

Figure 22.6 / *Effect of flow rate on column efficiency. A, B, and C are parameters from the Van Deemter equation (22.10).*

22.5.4 / Gas chromatography apparatus

The operation of a gas chromatograph is carried out by introducing the sample into a stream of carrier gas. This is usually done by syringe injection of the sample, or a solution of it, through a rubber septum onto the column or into a specially designed vaporization chamber. The sample size is typically 0.001–0.1 ml. The sample is carried through the column, typically 2–15 ft long, which is usually housed in an oven to permit operation above room temperature. At the outlet of the column, a detector is provided to monitor the effluent. The detector signal is amplified if necessary and transmitted to a recorder. A simplified diagram of a gas chromatograph is given in Fig. 22.7.

22.5.5 / Effect of varying gas chromatography conditions

The operator of a GC has several variables under his control. We shall consider their effects on the performance of GC separations.

A / *Temperature*

The possible operating temperature range is delimited at the lower end by the vapor pressure of the sample and the melting point of the liquid phase, if one is used. The sample must exhibit an appreciable vapor pressure or it

Figure 22.7 / *Principal elements of a gas chromatograph.*

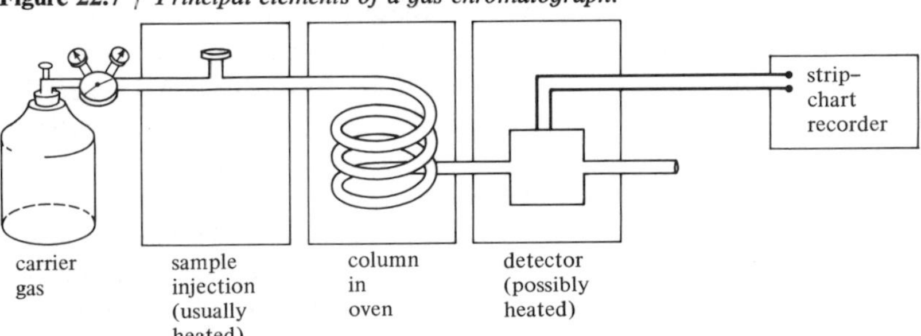

will not move through the column. The liquid phase must in fact be liquid or there will be no separation. The upper temperature limit is controlled by decomposition of the sample, decomposition of the stationary phase, and loss of liquid from the column owing to evaporation of the stationary phase in the carrier stream—"bleeding." Operating temperatures may range from below room temperature to several hundred degrees. Most often it is in the range 50–300°C.

Within the safe operating temperature range, one finds that an increase in temperature will usually reduce the retention time, carrier flow rate being held constant. This reflects an increase in volatility, hence of lowered solubility of the sample in the liquid phase in GLC and of a shift in the adsorption equilibrium in GSC.

It can be shown from thermodynamic considerations that, other factors being equal, the best separations will be obtained at the lowest column temperature consistent with reasonable values of t_r. This is a result of the fact that the volatilities of materials tend to converge; that is, the relative volatility tends toward unity as the temperature is increased. This can be seen from the Clausius-Clapeyron equation given as Eq. (22.11),

$$\frac{d(\ln P)}{d(1/T)} = \frac{-\Delta H}{R} \tag{22.11}$$

where P is the vapor pressure, T is the absolute temperature, R is the gas constant, and ΔH is the enthalpy of vaporization for the particular substance. This equation suggests that a plot of ln P vs. $1/T$ will be a straight line of negative slope; moreover, since different materials have differing values of enthalpy of vaporization, such a plot for several substances will be a series of straight lines with different slopes. Since these lines are not parallel, they must converge. For example, the relative volatility for two hydrocarbons (C_{17} and C_{18}) varies from 1.2 at 273°C to 2.1 at 85°C. The effectiveness of separation, as measured by the ratio of K_d values, would therefore be expected to be improved by nearly a factor of 2 by lowering the temperature by 150°. This will, of course, increase the length of time required for separation.

It is frequently true that variation of temperature during a run may allow a great decrease in time without sacrificing resolution. In isothermal operation, the temperature required for resolution of the most volatile components must be used. If the sample contains components with a wide range of volatilities, it may be appropriate to raise the column temperature after the most volatile components are eluted. This may be done by adjusting the temperature to a new value or by a gradual continual increase, referred to as *temperature programming*.

B / *Flow rate*

It was pointed out in Section 22.5.3 that molecular diffusion will cause excessive band spreading at low flow rates and that failure to establish equilibrium occurs with high flow rates. Typical results in terms of HETP vs. flow rate are presented in Fig. 22.6. Contributions attributed to each factor in the Van Deemter equation (22.10) are shown. There is usually a fairly wide range over which fluctuations in flow rate have little effect; hence, flow rate adjustment is not critical. It is of course true that retention time is a direct function of flow rate; therefore, when t_r is used in identification of a component, care must be taken to avoid significant fluctuation.

C / *Column length*

It has been pointed out in Section 22.2 that the extent of separation of components in a mixture is ideally directly proportional to column length. That is, the centers of peaks representing various components spread apart in proportion to the length of the column. However, the widths of the peaks, even under ideal conditions, spread as the square root of column length. In practice, conditions are not ideal, so that band spreading with increasing column length is an important factor. Also, increased column length implies increased time for operation of the column. In many cases if the use of a 6-ft packed column is unsuccessful, it will be more fruitful to change other conditions, such as using a different stationary phase or different sample workup, than to increase the column length.

D / *Column packing*

Gas chromatography is operated with empty tube columns and with packed columns. Empty tube columns or *capillary* columns, have a liquid phase dispersed in a layer on the wall of a tube of 0.25–1 mm diameter. Transfer between gas and liquid is facilitated, since sorbate must diffuse only through the thin, uniform, liquid layer. Eddy diffusion effects are minimal. The column diameter must, however, be quite small to afford intimate contact between the phases. The volume of liquid phase in a given length of column is necessarily limited, and very long columns (100–300 ft) are used. Accordingly, sample sizes must be restricted to 1 μl or less in order to avoid saturating the stationary phase. Sample sizes in this range are difficult to measure. Moreover, measurement of eluent concentrations becomes a problem owing to inadequate sensitivity of widely used detectors. Accordingly, packed columns are more commonly used than capillary columns, although better separations are sometimes obtained with the latter.

A packed column for GLC consists of a tube uniformly packed with particles of uniform size, typically in the range of 100 mesh (150 μm). The particles consist of a solid support on which the liquid phase is uniformly dispersed. The choice of solid support and the identity and amount of liquid phase are under the control of the chromatographer.

The solid support should be entirely inert, since the process of adsorption of vapor on solid surfaces is often undesirably slow and in many cases not completely reversible. Eluent-support interaction ordinarily causes severe band spreading. In addition, the solid support can catalyze decomposition of samples. For many purposes, firebrick or diatomaceous earth performs satisfactorily.

Difficulties owing to support reactivity are usually encountered with polar, rather than nonpolar compounds. Various treatments to minimize adsorption are used. These include washing with acids and bases to minimize subsequent reactivity with each of these classes of compounds. Silanizing, another such treatment, involves reaction of a silane with active sites on the support. Difficulties with basic compounds may be minimized by adding an inorganic base to the liquid phase. For certain separations, the use of glass or Teflon beads is effective.

Solid supports are ordinarily coated with 3–40 percent by weight of liquid. Several factors must be considered in determining the loading of liquid phase. Light loading ordinarily improves efficiency and permits the use of higher flow rates without sacrifice of efficiency. More rapid separations, accordingly, are possible.

On the other hand, sample size must be reduced in proportion to the column loading to avoid saturating the liquid phase. This causes difficulties in detection, especially of minor constituents of a sample. A small amount of bleeding by a lightly loaded column can cause a significant percentage change in extent of loading and thus an alteration of column behavior. If the support adsorbs the sample, this behavior will be more pronounced with light loading. The upper limit of usable column loading is determined by the time required for operation and by the necessity of using a free-flowing mixture in packing a column. Usually a compromise between these factors is sought; 10 percent loading is a typical value.

In selecting a liquid phase, the chromatographer has available a wide range of choices. The selection is ordinarily made on the basis of experience; however, some generalizations can be made. The liquid phase must be capable of dissolving the sample to be fractionated; hence, nonpolar liquids, such as silicone grease, are likely choices if nonpolar compounds are to be handled. Relative retention times quite often follow molecular weights for these systems, the smallest molecule exhibiting the shortest t_r. These liquid phases may show inadequate selectivity for polar solutes. In this case, use of a polar solvent, such as a polyglycol or polyester, may be more effective. The Carbowaxes are commonly used examples. Retention times for polar compounds frequently do not follow the order of their molecular weights, but may depend on various factors, just as is the case with their boiling points. The Suggested Readings at the end of the chapter provide detailed discussions of selection of liquid phases and preparation of columns.

22.5.6 / Gas chromatography detectors

Gas chromatography detectors produce either a current or a voltage signal that responds to changes in the composition of the effluent stream. Detectors may be conveniently classified in two categories: those sensitive to concentration of the sample and those that respond to the rate at which the sample is introduced into the detector. Both types are widely used. Perhaps the most significant difference is that concentration flow detectors are subject to error if flow rate changes during a peak, while mass flow detectors are unaffected by flow rate fluctuations.

With either type, the quantity of sample vapor is generally measured by integrating detector response with respect to time. If detector response is plotted against time, integration amounts to measurement of the area under the peak. Chromatographic peak heights can be used as a measure of concentration; however, this requires rigid control of conditions, since peaks must be Gaussian or the relationship between height and amount of solute must be established empirically. In addition, chromatograms must be produced under conditions of constant HETP. If peak height, rather than area, is used, mass flow detectors will be affected by flow rate changes.

In selecting a GC detector, several factors are of concern to the chromatographer. These include types of compounds to which the detector responds, sensitivity of response, dynamic range, and the mechanism of operation, destructive or nondestructive. While a variety of detectors have been used, those based on measurement of thermal or electrical conductivity of the effluent gas stream are most often encountered.

A / *Thermal conductivity detectors*

The thermal conductivity of a gas can be discussed using the coefficient of thermal conductance, κ, derived from kinetic theory. The coefficient is related to fundamental properties of the gas as indicated in Eq. (22.12),

$$\kappa = \frac{\rho \bar{c} \lambda C_v}{2} \tag{22.12}$$

where ρ is the gas density in gram-centimeters^{-3}, $\bar{c}$ is the average molecular velocity in centimeters-second^{-1}, λ is the mean free path in centimeters, and C_v is the gram specific heat at constant volume with units calories-gram^{-1}-°C^{-1}. The coefficient of thermal conductance, therefore, has dimensions of calories-centimeter^{-1}-second^{-1}-°C^{-1}. We note that the density and mean free path vary in the opposite directions with change in pressure, so κ is approximately independent of pressure as long as it is not too low. Some values of κ are given in Table 22.2.

TABLE 22.2 / *Thermal conductance coefficients of gases at* 0°C

Compound	κ (cal-cm^{-1}-sec^{-1}-°C^{-1})
H_2	416
He	348
N_2	58
CH_4	72
C_2H_6	44
$(CH_3)_2CO$	24

The thermal conductivity (TC) detector consists of a filament in contact with the GC gas stream. The element is electrically heated to maintain it at a temperature higher than that of its surroundings so that heat loss, primarily by thermal conductance by the gas, is balanced by electrical heat input. Any change in gas-stream conductivity, owing to change in composition or to other factors, will cause a change in element temperature and therefore a change in element resistance. Thermal conductivity detector elements desirably provide large temperature coefficients of resistance and are chemically inert. Tungsten alloy wires are very often used.

This type of detector will respond to any stream composition change that alters the thermal conductivity. Optimum sensitivity is obtained when the carrier and sample conductance differ greatly. Hydrogen and helium are ordinarily the gases of choice, since they offer the largest thermal conductivities, as indicated in Table 22.2. Thermal conductivities vary so that for mixtures of similar compounds, detector response for a particular component relative to total sample response is proportional to the weight percent of that component in the sample. In most applications, the TC detector is nondestructive.

Attempts to provide quantitative estimates of detector sensitivities are not very satisfactory because the various relevant factors can rarely be delineated. Some notion of the range in which the detector operates can be conveyed by pointing out that under normal chromatographic conditions, with no special precautions it is reasonable to expect a readily detectable response from 10^{-8} mole of an organic compound. Quantitative use of the detector will probably be more satisfactory with larger samples.

Thermal conductivity detectors generally consist of four elements connected in a bridge arrangement. Two elements are in contact with the stream of carrier gas and the other two with the effluent stream. Thus, one pair of elements functions as a reference, the other as a sample detector with the bridge output proportional to differences in the resistance of the two pairs of elements so long as the differences are small. This mode of operation optimizes performance, since the effects of variables such as fluctuation of flow rate, ambient temperature, and atmospheric pressure are largely canceled out. Any change in the balance of the bridge is attributed to alteration of the composition of the sample stream. The reader is referred to the discussion of difference measurements in Section 7.2.4 and to the discussion of bridge measurements in Section 7.2.6.

In GC, provision is generally made for readout through a suitable attenuator of the bridge unbalance voltage to a recorder. With large sample, bridge voltages can be as large as 1 V.

B / *Ionization detectors*

While gas streams are normally not electrical conductors, they can be ionized and rendered conductive in a variety of ways. In general, the resistance of an ionized gas sample is affected by its composition and thus can be made to serve as the basis for detection. Several types of ionization processes have been used. This discussion is limited to two of these: the flame ionization detector (FID) and the electron capture (EC) detector. All of the ionization detectors are similar in that the ionized gas stream passes between pairs of electrodes that have a voltage imposed. Detector response takes the form of variation in current flow. The reader is referred to Sections 7.1.4 and 7.3.3 for discussions of this type of measurement.

The flame ionization detector. The flame ionization detector consists of a pair of electrodes with a hydrogen–air flame between them. Provision is made to introduce the column effluent into the flame. In the absence of organic components in the column effluent, the flame produces very few ions, hence little current flows. Carbon-containing compounds produce ions in quantities that are approximately proportional to the concentration of carbon in the effluent. The principal advantage of the FID is that its sensitivity toward many organic compounds is very great. It is possible to detect 10^{-12} g of benzene; all of the components in a sample of petroleum ether can be detected in a 10^{-11}-g sample.

The FID is an example of a mass flow detector. The sample is consumed on introduction into the sensitive volume of the detector, hence it cannot respond to steady state concentrations. The detector is, of course, destructive of the sample and is not used if it is desired to recover material from the effluent. Its dynamic range is very large; it is capable of linear response to changes of concentration over seven orders of magnitude. The FID differs from the TC detector in that it is insensitive to a number of common substances. These include most inorganic compounds such as oxygen, nitrogen, water, carbon dioxide, and carbon disulfide. It shows reduced sensitivity for organics with large percentages of heteroatoms, such as formaldehyde.

The electron capture detector. The electron capture detector differs from the FID in that ionization is caused by the presence of a short-range, β-emitting source such as ^{63}Ni. In operation, the β source produces a steady stream of electrons that is collected by the electrodes, providing the steady state current output of the detector. Introduction of a compound having a significant cross section for capture of thermal electrons results in the formation of ions having very much smaller mobilities than that of electrons. As a result, the current output of an EC detector decreases in the presence of a compound to which it is sensitive.

Accordingly, the response of an EC detector cannot be linear. It can be described by an equation in the same form as that of Beer's law:

$$I_b = I_a e^{-KC} \tag{22.13}$$

In Eq. (22.13), I_a is the steady state current observed with carrier gas only. I_b is current in the presence of a sample component, K is a constant describing the detector performance, and C is the concentration of the component responsible for electron capture.

The value of this detector lies in its ability to show very high sensitivities for a relatively small number of compounds having large electron-capture cross sections, while being quite insensitive to a great many other compounds. It is therefore suitable for detecting very small concentrations of those compounds to which it is sensitive, especially halogenated organics and polyaromatics, in the presence of large amounts of other components, such as saturated hydrocarbons. The EC detector is particularly useful for determining traces of pesticide residues.

22.6 / LIQUID CHROMATOGRAPHY

Liquid chromatography may involve either solid (LSC) or liquid (LLC) phases; however, when both phases are liquids, the term *partition* chromatography may be used. The stationary phase is retained in a tubular column on a finely divided support as is done in GLC. Solid stationary phases take the form of finely divided particles in a column or a flat open bed. When a flat bed is used (TLC), the liquid phase moves parallel to the surface of the bed by capillary attraction.

22.6.1 / High-pressure operation

The earliest chromatographic experiments involved LSC carried out in a column with the eluent moving under the force of gravity. By comparison with GC, the separations obtained were very slow, owing to the necessity for limited flow rates and long columns. Improvement in efficiency requires a reduction in stationary phase particle size. However, with particle sizes below about 150-μm diameter, flow rates commensurate with reasonable separation times cause substantial pressure drops. Accordingly, rapid efficient LC operation has become possible only with the availability of pumping systems capable of applying 1000–3000 psi to the inlet of a column. This effect is illustrated in the data of Table 22.3. Column efficiency is expressed in terms of effective plates per second, N/t, to include the time factor specifically.

22.6.2 / Liquid–solid chromatography

Liquid–solid chromatography may involve fractionations based on selective adsorption, selective exclusion according to molecular size (gel permeation chromatography, GPC), or ion exchange. Each of these is discussed briefly. Detailed treatments are given in the Suggested Readings at the end of the chapter.

A / *Adsorption chromatography*

Polar solids such as silica gel or alumina provide surfaces on which physical adsorption of organic molecules can take place. Selectivity of adsorption

TABLE 22.3 / *Column efficiencies in liquid chromatography*

Column Type	*Average particle size* (μm)	N/t (sec^{-1})
classical packed	150	0.02
silica gel	20	2
Corasil[a]	31	2
Zipax[b]	27	8

[a] Silica bead produced by Waters, Associates.
[b] Silica bead produced by Du Pont.

SOURCE: J. J. Kirkland, *Anal. Chem.* **43**(12), 37A (1971),

is based on differences in polarity of adsorbates, the more polar being more strongly retained. Liquid chromatography offers an additional variable, as compared with GC, in that the moving phase also can participate by competing with sample molecules for adsorption sites.

Therefore, the observed capacity factor depends not only on the properties of the sample and stationary phase, but also on the polarity of the solvent. In general, increasing moving phase polarity leads to decreased k' values (more rapid band movement). Conditions are selected to ensure that all components of the sample are adsorbed; otherwise, selectivity is lost, but not so completely as to prolong unduly the time required for separation. Values of k' in the range 1–10 with columns for which $N/t > 1$ lead to satisfactory results, provided that adequate separation factors are obtained (see Table 22.1).

Solvent polarity can be controlled in a systematic way by using mixtures of varying composition. For example, the solvent polarity can be changed gradually from nonpolar (weak) to polar (strong) by using a series of mixtures of pentane and methylene chloride. Pure pentane would be the weakest solvent in the series. The polarity range available can be extended beyond what can be obtained with CH_2Cl_2–pentane by changing to acetonitrile–benzene mixtures. Other series that have been used include isopropyl chloride–pentane, ether-pentane, and methanol–ether as well as benzene–pentane and ethyl acetate–pentane.

Satisfactory performance of adsorption chromatography usually involves partial deactivation of the adsorbent. Silica gel that has been activated by heating exhibits such very strong adsorptive properties that its linear capacity is very limited. That is, all components of samples tend to show very large capacity factors (k') and small ratios of distribution coefficients. Partial deactivation by adding a controlled amount of water can cause significant selective decreases in k' values, which result in increased α values and decreased separation times.

Adsorption chromatography yields separations based on compound functionality rather than molecular weight. A mixture may be split into fractions, each of which consists of homologous compounds having the same functional groups, rather than into fractions containing individual compounds. Adsorption chromatography is ordinarily applied to systems of organic compounds with nonaqueous solvents. Solvent systems that contain significant amounts of water deactivate the adsorbent.

B / *Gel permeation chromatography*

Gel permeation chromatography depends on selective retention of the components of a mixture on the basis of their ability to penetrate into pores in the column packing. Assuming a porous packing material in equilibrium with a solvent, introduction of a solute will result in its diffusion into the absorbed solvent provided that the physical size of the molecules permits. Under chromatographic conditions, replacement of the liquid phase by fresh solvent will cause the solute to diffuse back into the unabsorbed liquid. Hence, the band of solute would move along the bed under conditions of a linear isotherm as long as no sorption mechanism other than pore diffusion obtains.

In fact, pore sizes are not uniform, but exhibit a range of values for any sample of packing material. Accordingly, solutes can be fractionated into three groups by this mechanism. Solute molecules too large to penetrate any pores will be eluted with the first volume of liquid corresponding to the interstitial void space. For a given range of pore diameters, there will exist a range of solvated molecular sizes that will penetrate some but not all of the pores. The elution volume for these solutes will exceed the interstitial void volume and will be a function of molecular diameter. These solutes experience selective permeation and will be fractionated according to size. All solutes having smaller diameters than these will experience total permeation. For these, elution volume is simply a function of total solvent volume, and there will be no fractionation according to molecular size. It is evident that GPC effectiveness is largely determined by control of the distribution of pore sizes in a given sample of packing material.

Packing materials for GPC are available for use with aqueous and nonaqueous systems. A gel will be suitable for use only with solvents that wet it and hence can be absorbed by it. With organic solvents, gels made from polystyrene, polyvinyl acetate, and polymethyl methacrylate are suitable. Starch, polyacrylamide, polydextran, and sulfonated polystyrene gels can be used with water.

Gels are categorized as soft, semirigid, and rigid. For substances of the same chemical composition, this distinction reflects the degree of cross-linking that exists. Soft gels provide a very large capacity. However, they are very subject to deformation under the pressures required for high flow rates. Rigid gels offer fixed pore sizes, a desirable attribute for accurate physical measurement. They do, however, tend to adsorb solutes that may interfere with measurements of molecular size. Semirigid gels offer intermediate properties. They can be obtained with a wide range of pore sizes, from those that can accept only small molecules to those with exclusion limits of molecular weights of a million.

Gel permeation chromatography, or as it is termed when aqueous systems are used, gel filtration chromatography, can be used to fractionate a sample according to molecular weights. It is primarily of interest in the study of high molecular weight material of either synthetic or biological origin.

C / *Ion exchange*

An ion exchanger consists of an ionic substance having one macromolecular ion paired with a micromolecular ion. If such a substance is brought into contact with a solution that is capable of penetrating its structure, ions in solution of the same charge type as the micromolecular ion may exchange with it.

Ion exchangers used for chromatography are usually synthetic cross-linked polymers with ionic functional groups chemically bound to the polymer. Very often these are polystyrene divinyl–benzene copolymers to which either positively or negatively charged functional groups are attached. If the group is negative, the counter ion will be a cation and the substance functions as a cation exchanger. With positively charged functional groups, an anion exchanger is produced.

Cation exchangers in the hydrogen form are acids. With sulfonate groups they can function as strong acids; carboxylate groups form weak acids. Anion exchangers in the hydroxide form are bases. They may be made with substituted ammonium groups. A quaternary ammonium ion will act as a strong base; a tertiary ammonium ion forms a weak base.

Ion exchangers can be used to remove one or both types of ion from solution, as in deionizers and water softeners. They also exhibit varying affinities for various ions, which can be made the basis for chromatographic separations. The most important uses of ion exchangers have been in large-scale purification or synthetic operations. Analytical applications are less important, although their use in amino acid analyzers can be cited as an important exception.

Ion exchange resins that consist of homogeneous beads of material capable of swelling in a solvent are disadvantageous in that mass transport associated with exchange in the interior of the bead limits the N/t that can be obtained. This problem can be circumvented by the use of pellicular resin. This is a finely divided material having solid cores with a film of ion exchange material on the surface. These materials naturally offer only limited capacities.

22.6.3 / Liquid–liquid chromatography

Development of techniques for high-speed efficient liquid chromatography have been most notable in application to liquid–liquid partition chromatography. Liquid–liquid chromatography offers great versatility, rather in analogy to GC, but without the restrictions regarding vapor pressure and thermal stability that must be observed in using GC. The stationary phase is dispersed on a solid support. Chromatography involves partition of the sample between the moving and stationary phases based on solubility differences.

A / *Solid supports*

The solid supports are finely divided inert materials. Silica gel, diatomaceous earth, and silica beads are used. To achieve high efficiency in high-speed operation, particle sizes must be restricted as is indicated in Section 22.6.1. In addition, supporting materials that exhibit only superficial porosity, such as the silica beads commercially available under the trade names Corasil and Zipax (see Table 22.3) are used to minimize the time required for equilibration between the phases.

B / *Choice of liquid phase*

The choice of liquid phase for LLC involves similar questions to those discussed in Section 22.5.1 dealing with GC stationary phases and in Section 22.6.2 dealing with moving phases for adsorption chromatography. If selective partition is to take place, the samples must be soluble in both phases. As with LSC, k' values must be in the range 1–10 to permit operation on a reasonable time scale. As before, choices are empirical, and a large number of examples have been described in the literature. If, on trying a system, the k' value is too large, a moving phase of increased polarity is indicated. Too small a k' indicates the need to reduce solvent polarity.

It is, of course, necessary that the moving phases selected show at most limited solubility for the stationary phase. To avoid the stripping that would otherwise occur, it is necessary as a routine to saturate the carrier with stationary-phase material before use. This is especially disadvantageous when isolated components of samples are to be collected, since the stationary phase constitutes an unavoidable contaminant.

C / *Bonded stationary phases*

Difficulties inherent in the use of liquid stationary phases can be circumvented by using packing materials having the partitioning substance bonded chemically to a support. These have been produced in forms suitable for efficient LC by esterification of the surfaces of silica beads with alcohols. These materials, unfortunately, are very sensitive to hydrolytic cleavage and can be used only under nonaqueous conditions. Polymeric silicones, bonded to the surface of silica beads, provide stationary phases that are more resistant to hydrolytic and thermal degradation. In preparation, silica surfaces are reacted with silanes which are then polymerized to produce silicones. Both types of column packings are commercially available.

22.6.4 / Apparatus for high-speed liquid chromatography

The apparatus used for high-speed LC bears some resemblance to that used for GC. The techniques of operation and readout of data are therefore similar.

The reservoir provided for the carrier is connected to the inlet of a high-pressure pump which forces the liquid through the column. Sample is injected at the inlet of the column either via a syringe specially constructed to operate at high pressure or through a valving arrangement. Column effluent passes through a detector and to an outlet. There may be provision for automatic recycling of the column effluent.

It is pointed out in Section 22.6.1 that efficient operation of LC requires the use of column packing with particle size less than 100 μm. Under these conditions, it is possible to use high flow rates, which make it feasible to operate on the same time scale as GC, but only at the cost of a large pressure drop in the column. Accordingly, the pump is a critical component in LC apparatus. It should be capable of operating at least at 1000 psi and preferably to 3000–5000 psi. Provision must be made to obtain a smooth fluid flow, either through the pump design or through incorporation of adequate damping in the solvent train. Several effective designs are offered commercially.

Operation of LC columns at $N/t > 1\ \text{sec}^{-1}$ requires that very great care be taken with a number of operational variables. These factors are important to the efficient operation of any form of chromatography, but for a given level of performance, they are more critical in LC than in GC. Among the most important factors are the design and the packing of columns. Great care must be taken to ensure uniform packing and to minimize dead space in the solvent train. Much of the difficulty in achieving high efficiency in LC is caused by nonuniform flow along the column wall. The difficulty can be minimized by reducing the packing particle size and avoiding too narrow columns. However, the high cost of much of the available packing material, and also the need to avoid large system volumes when small samples are to be handled, makes it desirable in many cases to avoid large columns. Efficient columns can be prepared with inner diameters as small as 2 mm.

The pressures involved make it necessary to use steel, rather than copper, in fabricating columns. Stainless steel is ordinarily used to minimize corrosion. Fittings and connecting tubing must be selected to avoid dead space; accordingly, connecting tubing with i.d. $\simeq$ 0.2 mm is used. This makes it necessary to take rigorous precautions against introduction of particulate matter into this part of the system. Solvent must be filtered and kept scrupulously clean. Care must be taken to avoid loss of packing material from columns and to avoid any possibility of forming precipitates within the system.

Another factor that may have a more pronounced effect on LC column efficiency than on GC efficiency is the mode of introduction of the sample.

Ideally, it should be placed on an infinitely thin layer of the packing at the top of the column. As an approximation, on-column injection may be used. This involves insertion of the sample syringe needle into the center of the top of the bed and depositing the sample directly. The other alternative for syringe injection is to deposit the sample in the carrier stream to be carried on to the column. This procedure inevitably involves some dilution of the sample and causes a reduction of column efficiency.

A variety of detectors has been used with LC. These may be categorized as bulk property and specific property detectors. Those most commonly encountered are the refractive index and the ultraviolet detectors.

Refractive index is a bulk property, the value of which will be affected by all components of a solution. Accordingly, it is almost universally applicable, but the measurement must be made at very high sensitivities ($\pm 10^{-5}$ RI units) in order to detect concentration changes of interest in solutions. This is troublesome, because obtaining a stable reading of $\pm 10^{-5}$ RI units requires temperature control to $\pm 0.001°C$. This difficulty is minimized by using a difference measurement so that, rather than control the temperature of the effluent so closely, the temperature differential between a sample and a reference is controlled. This can be done satisfactorily by controlling the ambient temperature to about $\pm 1°$ and arranging to bring both liquid streams into intimate thermal contact.

The ultraviolet detector, a specific property detector, is not so widely applicable as the RI detector. However, very many organic compounds show very strong absorbances in the 250–280 nm region, which is conveniently obtained from a mercury vapor lamp. For many applications, therefore, the UV detector shows excellent sensitivity. It tolerates larger temperature fluctuations than does the RI detector. Use of an UV detector imposes a restriction on the choice of solvent, since it cannot absorb at the wavelength of interest.

22.7 / THIN-LAYER CHROMATOGRAPHY

Where sample volatility permits, the technique of thin-layer chromatography (TLC) offers a very efficient method of separation. It is used primarily for qualitative purposes. A bed is prepared, usually by depositing a controlled amount of a stationary-phase material on a glass plate in a uniform layer which can be used in a small-scale variant of adsorption chromatography.

The stationary phase most often used is silica gel. Alumina, diatomaceous earth, and cellulose are also used. These are commercially available compounded with a binder such as Plaster of Paris, which strengthens the dried layer. Plates are prepared by evaporating the solvent from a slurry of the stationary phase. One convenient method is to use microscope slides. Two slides held together can be dipped into the slurry, separated, and dried to provide two plates. The sample is placed in a small spot near one edge of the bed. For application, the sample is usually dissolved in a volatile solvent

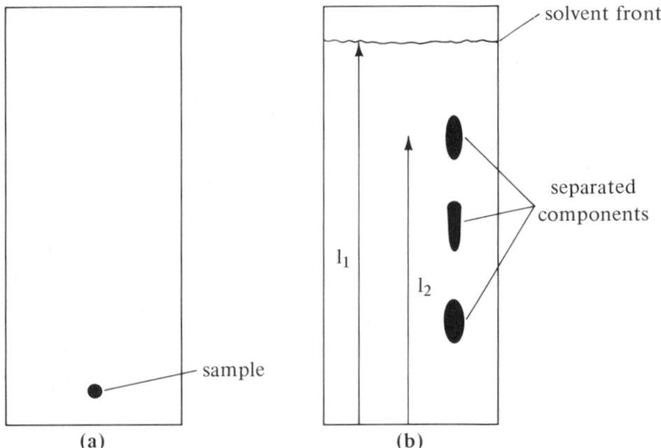

Figure 22.8 / *Thin layer chromatography.*
(a) *TLC plate before development*
(b) *Developed TLC plate.* $l_2/l_1 = R_F$ *for uppermost spot.*

so that the initial area can be effectively minimized by introduction of small increments with evaporation of the solvent after each addition. This is the analog of on-column loading in column chromatography. The plate is eluted or *developed* by immersing it in the carrier with the sample just above the surface. This is done in a closed container, the atmosphere of which is saturated with vapors of the carrier to avoid changes in composition caused by evaporation. When the carrier has moved up the plate, the chromatogram is dried and given some treatment to render the fractionated sample visible. Development of a TLC plate is illustrated in Fig. 22.8.

Selection of conditions involves the same types of variables that are discussed in Section 22.6.1. In fact, TLC can be conveniently used to evaluate conditions that may be applied to column LSC, because the apparatus is very much simpler and the effort required to perform an experiment very much less. The detection step, of course, depends on the characteristics of the system. Colored compounds require no treatment. Some compounds fluoresce under UV. Others absorb UV and can be visualized, if a fluorescent dye is incorporated in the stationary phase, as spots that do not fluoresce. Iodine vapor reacts with many adsorbed compounds to produce visible spots. With inorganic stationary phases, organic compounds can be visualized by spraying the developed plate with sulfuric acid, which chars them.

Thin-layer chromatography is widely used because it provides a simple and very versatile method for fractionating complex mixtures of non-volatile compounds. It can be performed rapidly, ordinarily less than half an hour for development of a plate is required, and only very small amounts of sample are needed.

Sample movement on TLC can be quantified by comparing the distance moved by the compound with that moved by the solvent front. This is expressed by the parameter R_F, which is the ratio of the distance of sample to solvent movement. R_F values are usually not very reproducible, so identification is most effectively made by chromatographing a valid sample on the same plate with the unknown.

A variation of this technique can be used to fractionate mixtures in sufficient amounts to permit recovery of the fractions. This is called preparative TLC or sometimes thick-layer chromatography. To accommodate larger sample sizes, the thickness of the bed must be increased to 1 mm or larger. Devices can be purchased to use in casting carefully controlled layers of slurry on plates. A simple alternative consists in placing several layers of masking tape along the edges of a plate. The slurry is poured on the plate and spread to the thickness defined by the tape with the aid of a rod. Sample is introduced in a thin, straight line along one edge of the plate, and it is developed in the same fashion as for small plates. The developed plate can be sectioned and the fractions extracted from the adsorbant material. As is generally true of chromatographic procedures, scaling small experiments upward to the preparative scale requires care to avoid alteration of conditions that may reduce efficiency.

SUGGESTED READINGS

S. Dal Nogare and R. S. Juvet, *Gas-Liquid Chromatography*, Wiley-Interscience (New York, 1962).

A. B. Littlewood, *Gas Chromatography*, Academic (New York, 1970).

J. M. Bobbitt, A. E. Schwarting, and J. R. Gritter, *Introduction to Chromatography*, Van Nostrand Reinhold (New York, 1968).

J. J. Kirkland, Ed., *Modern Practice of Liquid Chromatography*, Wiley-Interscience (New York, 1971).

J. J. Kirkland, *Anal. Chem.* **43** (12), 37A (1971).

K. H. Altgelt and L. Segal, Eds., *Gel Permeation Chromatography*, Marcel Dekker (New York, 1971).

J. A. Dean, *Chemical Separation Methods*, Van Nostrand Reinhold (New York, 1969).

chapter 23 / Electrochemical methods

chapter 23

23.1 / INTRODUCTION

There are several possible ways to categorize electrochemical procedures; one of the more rational depends on the amount of current employed in the particular method. In this chapter, we shall begin with a brief review of some fundamental concepts and then discuss selected electrochemical techniques in order of increasing current flow. Thus, potentiometry (ideally a zero current method) will be discussed first, followed by voltammetry (small current) and finally coulometry and controlled potential electrolysis (larger currents). The chapter concludes with a brief discussion of the kinetics of electrode reactions.

23.2 / REVIEW OF FUNDAMENTALS

23.2.1 / Electromotive force of galvanic cells

Electrochemical procedures generally depend on the occurrence of oxidation and reduction reactions at the surface of an electrode in solution. We distinguish between *galvanic* cells, which operate spontaneously and convert chemical potential energy into electrical energy that is then available for use (e.g., battery action), and *electrolysis* cells, which involve just the reverse. In an electrolysis cell, electrical energy is supplied from an external source and used to force a reaction to occur. The galvanic cell will be considered first. Consider, for example, the chemical reaction

$$Zn(s) + Cu^{2+}(aq) + SO_4{}^{2-}(aq) \rightarrow Zn^{2+}(aq) + Cu(s) + SO_4{}^{2-}(aq) \tag{23.1}$$

Here metallic zinc reacts with aqueous copper sulfate solution. Zinc is oxidized, cupric ion is reduced and, if the reactants are simply mixed in a beaker, the chemical energy is released as heat. If, instead, a galvanic cell

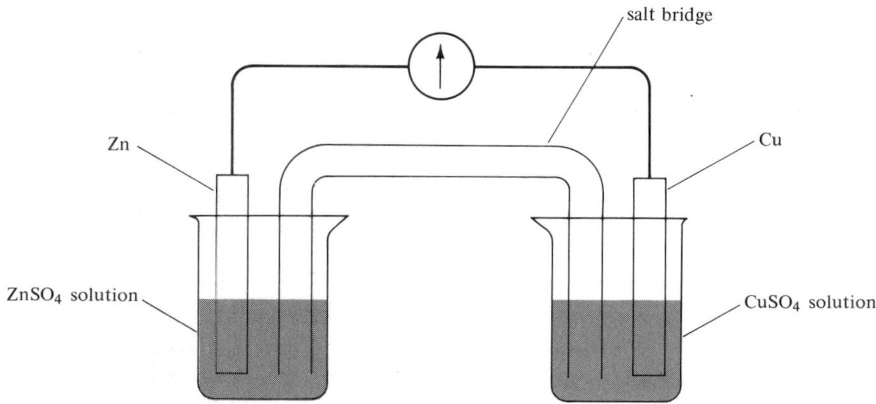

Figure 23.1 / *A simple galvanic cell.*

is constructed such as that shown in Fig. 23.1, a substantial fraction of the chemical energy can be made available as electrical energy to do useful work—in this example, the work appears as movement of the voltmeter needle. Instead of depicting galvanic cells in this cumbersome way, a schematic representation is used for convenience. The cell in Fig. 23.1 is represented by Eq. (23.2).

$$\mathrm{Zn|ZnSO_4}(C_1)\|\ |\mathrm{CuSO_4}(C_2)|\mathrm{Cu} \tag{23.2}$$

A single vertical line indicates a phase boundary across which an electrical potential difference is developed, and a double vertical line represents a salt bridge. The chemical reaction implied by Eq. (23.2) is Eq. (23.1), proceeding from left to right.

The electromotive force (emf) or voltage developed by the cell is the sum of the potential differences.

$$E_{\text{cell}} = E_{\text{Zn,Zn}^{2+}} + E_{\text{lj}} + E_{\text{Cu}^{2+},\text{Cu}} \tag{23.3}$$

In general, we do not know the value of the term E_{lj}, the liquid junction potential. This potential is developed because of differences in mobilities of the various ions in solution and will be discussed in somewhat more detail in Section 23.3.2. For the present, suffice it to say that the use of a proper salt bridge *minimizes*, but can never eliminate, the contribution of the liquid junction potential. The electrode potentials themselves are functions of the concentrations of reacting species, the temperature, and the pressure. Under ordinary laboratory conditions, the latter two quantities are kept constant and only variations of concentration are of concern.

The quantity E_{cell} has both *magnitude* and *sign*, both of which are related to the change in the Gibbs free energy, ΔG, which accompanies the reaction. Reactions that proceed spontaneously at constant temperature and pressure do so with a *decrease* of free energy; that is, ΔG is negative. The relation between ΔG and E_{cell} is stated as Eq. (23.4), where $\mathscr{F}$ is the faraday and n represents the number of electrons transferred in the reaction.

$$\Delta G = -n\mathscr{F}E_{cell} \tag{23.4}$$

From Eq. (23.4), it is seen that the sign conventions of thermodynamics are such that a *spontaneous* reaction is associated with a *positive* cell emf. Thus, the sign of the cell emf will depend on the direction in which the cell is written. The cell (23.5) is simply the reverse of (23.2), since it implies a chemical reaction such as (23.1) proceeding from right to left.

$$\mathrm{Cu|CuSO_4}(C_2)|\,|\mathrm{ZnSO_4}(C_1)|\mathrm{Zn} \tag{23.5}$$

The chemical reaction does not proceed spontaneously in the direction written, and the cell emf is therefore *negative* but of the same numerical value as that for the cell written in the opposite sense.

23.2.2 / Sign conventions

It is of great importance to distinguish between the sign of the cell emf, which is a thermodynamic quantity, and the *relative polarity* of the two electrodes. Quite clearly, the relative polarity of the electrodes is a physical observable that does not depend on the relative position of the two beakers on the laboratory bench, and neither does it depend on the direction in which the cell is written. Rather, it depends only on the chemical process that actually occurs in the cell. In the present example, regardless of the way we write the cell, the spontaneous chemical processes that occur when the circuit is completed by an external load are the oxidation of zinc and the reduction of cupric ion. Therefore, at the zinc electrode, $Zn \rightarrow Zn^{2+} + 2e^-$, while at the copper electrode, $Cu^{2+} + 2e^- \rightarrow Cu$. We define the *anode* as the electrode at which oxidation occurs (in this case the zinc electrode), while the *cathode* is the electrode at which reduction occurs (copper in this example). No electrons flow in solution; all charge transport in solution occurs via movement of ions. Electrons, however, do flow in the external circuit. Thus, when a zinc atom leaves the electrode to enter solution as an ion, it leaves behind two electrons, while reduction of a cupric ion at the cathode consumes two electrons. In other words, electrons are evolved at the zinc electrode and consumed at the copper electrode so the former is negative and the latter is positive.

If the convention that a positive cell emf implies spontaneous reaction as written from left to right (and the cell is, in fact, written in this way) is adhered to, the following rules will obtain:

1. The left-hand electrode is the anode, and the right-hand electrode is the cathode.
2. The left electrode is (−) and the right electrode is (+).
3. Electrons flow from left to right in the external circuit.
4. Positive ions flow from left to right, negative ions flow from right to left in the cell.

All of these relations are reversed if the direction of spontaneous discharge is right to left.

23.2.3 / Reversibility

The meaning of "reversible process" in electrochemistry is frequently confusing, in part because of a tendency to use the same word to mean two different things. A cell reaction is *chemically reversible* if reversing the direction of current flow simply reverses the half-reactions. The copper–zinc example is chemically reversible.

$$Cu^{2+} + Zn \rightleftharpoons Zn^{2+} + Cu \tag{23.6}$$

The net reaction (23.6) and both half-reactions are reversed if the current flow is reversed. The net reaction is left-to-right on discharge (galvanic cell) and right-to-left when the cell is charged (electrolysis cell).

Consider, in contrast, the cell (23.7), consisting of a manganese electrode and a platinum electrode.

$$Mn|MnSO_4(aq)|\,|H_2SO_4(aq)|Pt \tag{23.7}$$

The spontaneous reactions that occur are

$$\begin{array}{rcll} Mn & = & Mn^{2+} + 2e^- & \text{(at the manganese electrode)} \\ 2H^+ + 2e^- & = & H_2 & \text{(at the platinum electrode)} \\ \hline 2H^+ + Mn & = & Mn^{2+} + H_2 & \text{(net spontaneous reaction)} \end{array}$$

When the cell is operated as an electrolysis cell, however, the following reactions occur:

$$\begin{array}{rcll} 2H^+ + 2e^- & = & H_2 & \text{(at the manganese electrode)} \\ 2H_2O & = & O_2 + 4H^+ + 4e^- & \text{(at the platinum electrode)} \\ \hline 2H_2O & = & O_2 + 2H_2 & \text{(net electrolysis reaction)} \end{array}$$

This system is chemically irreversible since, when the direction of current flow is reversed, totally different reactions take place. A chemically irreversible system can never be made to operate reversibly in any case.

The second kind of reversibility is *thermodynamic reversibility.* Chemically reversible cells can approach, but in practice never achieve, complete reversibility in the strict thermodynamic sense. A truly reversible process is one that proceeds via a series of equilibrium states and never deviates from chemical equilibrium. In practice, the only way to achieve any change is allow the system to depart from equilibrium; reversibility may be approached if this is done in a series of steps, each of which involves only a *differential* departure from equilibrium. However, this implies that the process must be carried out infinitely slowly. True thermodynamic cell potentials are therefore obtained only under conditions that permit *zero* current to flow through the cell. This condition can be approached quite well through the use of a very high input impedance measuring device. Any time that current is

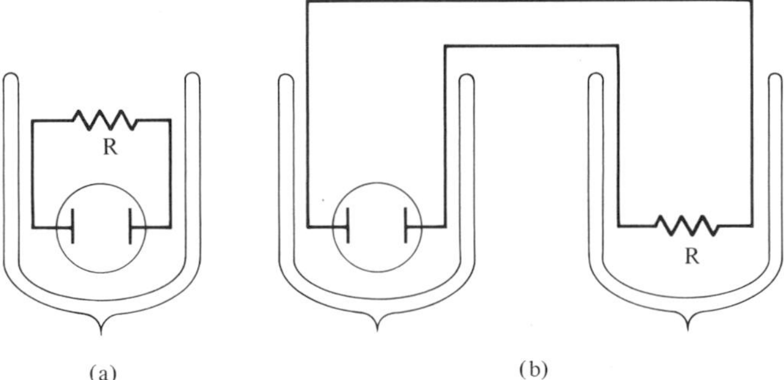

Figure 23.2 / *Calorimetric measurement of the energy output of a galvanic cell.*
SOURCE: J. J. Lingane, *Electroanalytical Chemistry*, 2nd ed., Wiley-Interscience Publishers (New York, 1958).

drawn from the cell, reversible behavior is no longer possible, and the larger the current, the larger the departure from strict reversibility.

23.2.4 / Cell emf and thermodynamics

If we specify that measurements are to be made under reversible conditions and that all products and reactants shall be present in their thermodynamic standard states, then the potentials that result are termed standard potentials and are given the symbol E^0. It is possible to use a variety of definitions of standard states, but the usual ones are as follows: gases, 1 atm partial pressure or, more strictly, unit fugacity; solids and liquids, the pure substance (when *any* of the pure substance is in contact with the chemical system, that material is taken to be present at unit activity); solutes in solution, unit activity. The standard potential is related to the *standard* free-energy change by Eq. (23.4), except that both are now specified as the "standard" quantities by affixing zeros:

$$\Delta G^0 = -n\mathcal{F}E^0 \tag{23.8}$$

When the reaction of cupric ion with zinc is carried out under standard conditions in a calorimeter, it is found to be exothermic to the extent of 51.8 kcal/mole. This is the enthalpy of reaction and, because of the specification of standard conditions, is the standard enthalpy change, ΔH^0, in this case. Suppose the reaction is instead carried out in the cell indicated in Eq. (23.9),

$$\mathrm{Zn}|\mathrm{Zn}^{2+}(a = 1)|\,|\mathrm{Cu}^{2+}(a = 1)|\mathrm{Cu} \tag{23.9}$$

and the cell and load resistor are both immersed in the same calorimeter as in Fig. 23.2(a). If the cell is allowed to discharge at some convenient rate, again we would find 51.8 kcal/mole liberated. If, instead, the setup of Fig. 23.2(b) is used, with the cell and load resistor in separate calorimeters and the two connected together by ideal wires (zero electrical resistance and zero thermal conductivity), then the sum of the heat evolved in the two calorimeters is always equal to ΔH^0; however, the disposition of the heat between the two vessels will depend on the *rate* at which the cell is discharged. As this rate approaches zero, the energy liberated in the cell itself decreases and that available for useful work (in this case simply dissipated in the load resistor) increases until in the limit of zero rate the latter becomes 50.7 kcal/mole and the former becomes 1.1 kcal/mole. In any real experiment, these limits will not actually be attained. The absolute maximum useful work derivable from any chemical reaction is simply, under conditions of constant temperature and pressure, the Gibbs free-energy change, and is in this case $\Delta G^0 = -50.7$ kcal/mole. Recalling the Gibbs-Helmholtz equation, Eq. (23.10),

$$\Delta G^0 = \Delta H^0 - T\,\Delta S^0 \tag{23.10}$$

we recognize that the 1.1 kcal/mole liberated in the cell itself is a result of the entropy change associated with the chemical reaction; at 298°K, this corresponds to -3.67 cal/mole-deg. This energy, which can never be made available for useful work, is the "overhead" demanded by the second law of thermodynamics.

It is shown from straightforward thermodynamic manipulations that, at constant pressure, the temperature dependence of ΔG is given by

$$\left(\frac{\partial\,\Delta G}{\partial T}\right)_P = -\Delta S \tag{23.11}$$

Differentiating the relation of cell potential and the Gibbs free energy, one obtains

$$n\mathscr{F}\left(\frac{dE}{dT}\right)_P = \left(\frac{-\partial\,\Delta G}{\partial T}\right)_P \tag{23.12}$$

whence

$$n\mathscr{F}\left(\frac{dE}{dT}\right)_P = \Delta S \tag{23.13}$$

Therefore, we must expect cell potentials to be temperature dependent, and we see that this temperature dependence can be used to determine ΔS of the reaction.

23.2.5 / Potentials and concentrations

When the activities of reacting species are not fixed at unity, the standard free-energy change is related to the free-energy change actually observed under the nonstandard conditions by

$$\Delta G^0 = \Delta G - RT\ln\left(\frac{[\text{products}]}{[\text{reactants}]}\right) \tag{23.14}$$

where R is the gas constant and T is the absolute temperature. Since the observed free-energy change must equal zero when equilibrium has been attained, we can write

$$\Delta G^0 = -RT\ln\left(\frac{[\text{products}]}{[\text{reactants}]}\right)_{\text{eq}} = -RT\ln K_{\text{eq}} \tag{23.15}$$

Substitution in these two equations, using Eqs. (23.4) and (23.8), gives

$$-n\mathscr{F}E^0 = -n\mathscr{F}E - RT\ln\left(\frac{[\text{products}]}{[\text{reactants}]}\right) \tag{23.16}$$

$$E = E^0 - \frac{RT}{n\mathscr{F}}\ln\left(\frac{[\text{products}]}{[\text{reactants}]}\right) \tag{23.17}$$

and

$$E^0 = \frac{RT}{n\mathscr{F}}\ln K_{\text{eq}} \tag{23.18}$$

Equation (23.17) for the emf of a cell as a function of the concentrations of reactants and products is, of course, the familiar Nernst equation. Equation (23.18) represents a special form of the Nernst equation that obtains at equilibrium and shows how equilibrium constants are calculated from standard potential data or vice versa. The common form of the Nernst equation results upon conversion to common logarithms and evaluation of the constant $RT/\mathscr{F}$ at 298°K.

$$E = E^0 - \frac{0.059}{n}\log\left(\frac{[\text{products}]}{[\text{reactants}]}\right) \tag{23.19}$$

While a thorough discussion of ionic activities is beyond the scope of this text, it is necessary to recall that the argument of the logarithm in the Nernst equation is in terms of activities rather than concentrations. The activity of a species in solution is related to the molar concentration by $a = \gamma C$, where γ is the activity coefficient and C is the concentration. In very dilute solutions activities and concentrations become approximately equal, as expressed by Eq. (23.20).

$$\lim \gamma_{C\to 0} = 1 \quad \text{and} \quad \lim a_{C\to 0} = C \tag{23.20}$$

For electrolytes in solution, activity coefficients usually decrease rapidly with increasing concentration and pass through a rather flat minimum in the region of a few molar, that is, in the normal working range, and then increase again. In the range of concentrations of normal interest, $\gamma < 1$ is to be anticipated. When solutions are of sufficiently low ionic strength, $<0.1\ M$, activity coefficients can be calculated approximately using the relations derived by Debye and Hückel, and the accuracy of the calculations increases as the concentration decreases.

23.2.6 / Half-cell potentials

In experimental reality, we always measure the emf of a complete electrochemical cell. Since it is not possible to obtain an oxidation without at the same time causing the reduction of some other substance, there is no way to measure the emf of a single oxidation or reduction process. Nonetheless, we speak of "half-cell potentials," and the tables of oxidation potentials list these for half-reactions. Of course, these numbers establish only a relative scale, or pecking order, of what oxidizes what, not a scale of absolute free energies. For convenience, an arbitrary zero point is established by defining the potential of the standard hydrogen electrode as zero for all temperatures.

$$H_2(1 \text{ atm}) = 2H^+(a = 1) + 2e^- \qquad E^0 \equiv 0 \tag{23.21}$$

Therefore, the statement of a half-cell potential or half-reaction emf implies that the half-reaction in question would have the stated potential when measured in a complete cell, the second half of which is a standard hydrogen electrode. Thus, the implication of Eq. (23.22),

$$Ag + Cl^-(a = 1) = AgCl + e^- \qquad E^0 = -0.222 \text{ V} \tag{23.22}$$

is that the cell (23.23),

$$Ag, AgCl(s)|Cl^-(a = 1), AgCl(sat)|\,|H^+(a = 1)|H_2(1 \text{ atm}), Pt \tag{23.23}$$

has been constructed for which $E^0 = -0.222$ V. (Note that the chemical reaction in this cell is spontaneous from right to left, hence the negative E^0.)

Obtaining half-reaction data from tables permits one to compute, after proper manipulation, equilibrium constants for chemical reactions, predict the course of some reactions, and make measurements of concentrations. As an example, let us calculate the potential for the zinc–copper cell (23.2) and then the equilibrium constant for the reaction (23.1). Consulting a table of standard potentials, we find that

$$Cu^{2+} + 2e^- = Cu \qquad E^0 = +0.337 \text{ V}$$

$$Zn^{2+} + 2e^- = Zn \qquad E^0 = -0.763 \text{ V}$$

The cell (23.2) and reaction (23.1) as written involve the reduction of cupric ion, but the oxidation of zinc. Therefore, it is necessary to reverse the direction of the second half-reaction, and this necessitates the reversal of the sign of E^0. Addition yields the cell emf, which is 1.100 V and, in accord with the previous discussion of the polarity of the electrodes, we would find the zinc electrode 1.1 V *negative* with respect to the copper electrode when measured with a voltmeter of suitably high input impedance. Here the liquid junction potential has been ignored.

Calculation of the equilibrium constant is carried out by inserting the relevant numbers into Eq. (23.18),

$$1.100 = \frac{0.059}{2} \log K_{eq} \tag{23.24}$$

whence K_{eq} is readily found to be $10^{37.3}$. Thus, as asserted previously, a positive E^0 is associated with a reaction that proceeds spontaneously (has a large equilibrium constant) from left to right. Another possible use of this type of cell is the measurement of the activity of one component when the other is known. Suppose we assemble a cell such as (23.2) in which we know that the activity of zinc ion is unity but that of cupric ion is unknown. The measured emf of the cell is 1.070 V (zinc is still the negative electrode). The calculation of the cupric ion activity requires the use of the Nernst equation in the form of Eq. (23.19).

$$1.070 = 1.100 - \frac{0.059}{2} \log \frac{1}{[Cu^{2+}]} \tag{23.25}$$

Solving, we find that $\log 1/[Cu^{2+}] = 1.0$, or $1/[Cu^{2+}] = 10$, so the activity of cupric ion is 0.1. This is an example of the technique of direct potentiometry, which is discussed in Section 23.4.

As a second example of this type of calculation, consider the estimation of the solubility product of silver chloride from electrochemical data. From the table of standard potentials, one finds these data:

$$Ag + Cl^- = AgCl + e^- \quad E^0 = -0.222 \text{ V}$$

$$Ag = Ag^+ + e^- \quad E^0 = -0.799 \text{ V}$$

Reversing the first reaction and the sign of its standard potential followed by algebraic addition of the two half-reactions yields

$$AgCl = Ag^+ + Cl^- \qquad E^0 = -0.577 \text{ V} \tag{23.26}$$

and insertion of these quantities into Eq. (23.18) gives

$$-0.577 = \frac{0.059}{n} \log \frac{[Ag^+][Cl^-]}{[AgCl]} \tag{23.27}$$

In problems of this type, where no electron transfer appears in the net reaction [Eq. (23.26)], the value of n used in the Nernst expression is that used in the half-reactions employed to obtain the net reaction, in this case $n = 1$. We recall that pure solids are taken to be at unit activity and recognize the activity product $[Ag^+][Cl^-]$ as the quantity called the solubility product, K_{sp}. Therefore,

$$\frac{-0.577}{0.059} = \log K_{sp}$$

and the desired constant is approximately 10^{-10}.

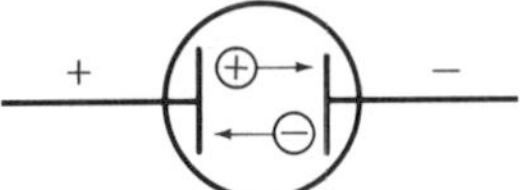

Figure 23.3 / *Schematic representation of a conductivity cell.*

It is frequently necessary to add two or more half-reactions together to obtain another half-reaction that is not listed in the particular table at hand. Under these circumstances, one cannot simply add the standard potentials. The fundamental quantity that is added algebraically is the free-energy change, $n\mathscr{F}E^0$, not the standard potential itself. This is intuitively understandable when one recalls that a potential is an intensive property that does not depend on the quantity of reaction, whereas ΔG values represent work and *do* depend on the quantity of reaction. Normally, ΔGs are given in calories per mole of reaction, and 1 mole of reaction will require the transfer of n electrons. Now, when half-reactions are added together to form a net reaction, the number of electrons liberated in the oxidation is necessarily the same as the number consumed in the reduction and, through this requirement, the E^0s *are* proportional to ΔG^0. This is no longer true when the quantity sought is the standard potential for another half-reaction. Since the faraday is a constant, nE^0 *is* proportional to ΔG^0, and these quantities *can* be added to obtain the desired result. This should be clear in the following examples:

	E^0	nE^0
$Zn = Zn^{2+} + 2e^-$	0.762	1.524
$2[AgCl + e^- = Ag + Cl^-]$	0.222	0.444
$2AgCl + Zn = Zn^{2+} + 2Cl^- + 2Ag$	$0.984 = E^0$	$1.968\ (= 2E^0)$

Here the sum is a net reaction and the standard potential can be obtained either by adding E^0s directly or by adding nE^0 values and dividing by the total number of electrons transferred, $1.968/2 = 0.984$. In the following case, only addition of nE^0 values gives the correct result:

	E^0	nE^0
$2I^- = I_2 + 2e^-$	-0.534	-1.068
$I_2 + 6H_2O = 2IO_3{}^- + 12H^+ + 10e^-$	-1.195	-11.95
$2I^- + 6H_2O = 2IO_3{}^- + 12H^+ + 12e^-$	$-1.729 \neq E^0$	$-13.018\,(= 12E^0)$

and therefore $-13.018/12 = -1.08$ V is the standard potential for the oxidation of iodide ion to iodate ion.

23.3 / SOLUTION CONDUCTIVITY

23.3.1 / Charge transport

As pointed out above, the electrical conductivity of a solution is due entirely to the movement of ions. Thus, in a cell such as that in Fig. 23.3, the transport of positive charge from left to right could be accomplished either by the

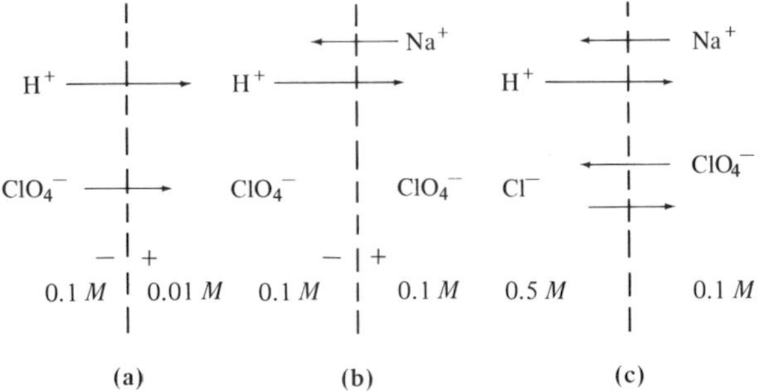

Figure 23.4 / *Liquid junctions showing the movement of ions that gives rise to the liquid junction potential.*

movement of positive ions from left to right or the movement of negative ions in the reverse direction. In actual fact, such charge transport involves both these processes; the movement of the ions occurs in such a way that the solution remains electrically neutral throughout, and the relative importance of the two current-carrying paths depends on the properties of the ions in the particular solution. The fraction of charge transported by the positive ions in, say, a calcium chloride solution, is given the symbol N_+ and called the transference number of Ca^{2+}. Similarly, N_- is the transference number of Cl^- and, by definition, $N_+ + N_- = 1$. The ability of a particular ion to transport charge can be discussed in terms of a number of different parameters, but the one most useful for the present discussion and probably the most fundamental to the physical properties of the electrolyte in solution is the *mobility* of the ion. The mobility of an ion is essentially its rate of travel through the solution under the influence of an imposed force, which could be an electric field or a concentration gradient. The mobility of an ion is a characteristic property that is affected by such factors as the charge, size, mass, and extent of solvation. Lighter ions have larger mobilities than heavier ones; in general, more highly solvated ions have smaller mobilities than similar ions less highly solvated, since the inner solvation sphere moves with the ion.

The particular concern at this juncture is the difference in rates of diffusion of ions under the influence of concentration gradients at the various interfaces of an electrochemical cell; these differences give rise to the liquid junction potentials referred to above.

23.3.2 / Liquid junction potentials

A quantitative treatment of liquid junction potentials is outside the scope of this text, but a qualitative picture will be given. Three distinct types of liquid junction are illustrated in Fig. 23.4. These are (a) two solutions containing the same electrolyte, but at different concentrations; (b) solutions containing electrolytes at the same concentration and having one ion in common; and (c) solutions that have no ion in common and have different concentrations. In case (a), both ions will diffuse toward the more dilute solution, but the protons will do so more rapidly since their mobility is much greater than that of perchlorate ions. A slight charge separation occurs as indicated. An equilibrium liquid junction potential will exist when the charge separation is of sufficient magnitude to retard the diffusion of positive ions so that they no longer cross the boundary in excess. Since perchlorate is present at equal concentration on both sides of the junction in (b), virtually no net transport occurs; however, here the larger mobility of the proton in comparison with sodium ion will result in the charges as shown in the figure. In case (c), since all the concentrations and mobilities are different, the situation is very complicated. It is possible to calculate the magnitude and sign of the junction potentials in case (a) exactly and in case (b) approximately, but such calculations are not feasible for case (c).

It is important to realize that almost all electrochemical cells contain at least a small liquid junction potential, which is generally of unknown magnitude. Only the voltages of amalgam concentration cells such as (23.28) do not contain such a contribution.

$$\underset{c_1}{\mathrm{Zn(Hg)}}|\underset{c_2}{\mathrm{Zn^{2+}}}|\underset{c_3}{\mathrm{Zn(Hg)}} \tag{23.28}$$

Such cells (and also those for which E_{lj} is very small) are termed "cells without transference." Even the cell (23.29), which is commonly spoken of as being without transference, actually has a small (although probably negligible) liquid junction potential since the two solutions are not strictly identical; that on the left is saturated with silver chloride, while that on the right is saturated with hydrogen gas.

$$\mathrm{Ag|AgCl(s)HCl}(C_1)|\mathrm{HCl}(C_1)|\mathrm{H_2(1\ atm)\ Pt} \tag{23.29}$$

The two most important facts to remember about liquid junction potentials are that they can almost never be completely eliminated, but they can be minimized. It is an experimental fact that the use of a concentrated salt bridge between two nonidentical solutions greatly reduces the liquid junction potential. Because the mobilities of potassium and chloride ions are nearly equal, the usual choice for a salt bridge (if not contravened by other chemical aspects of the cell) is saturated KCl. Experimental measurements have been carried out on the cell (23.30), which, since the right- and left-hand electrodes are identical, should have an observed emf = 0 (assuming the activity of Cl^- is the same in 0.1 *F* HCl as it is in 0.1 *F* KCl) in the absence of liquid junction phenomena. The observed emf was measured as a function of the concentration, *C*, of KCl in the salt bridge with the results indicated.

$$\mathrm{Hg|Hg_2Cl_2,\ HCl(0.1)|KCl}(C)|\mathrm{KCl(0.1),\ Hg_2Cl_2|Hg} \tag{23.30}$$

C (molar)	0.1	0.2	0.5	1.0	2.5	3.5
emf (mV) = E_{lj}	26.8	20.0	12.6	8.4	3.4	1.1

23.3.3 / Conductance measurements

Since ionic mobilities are directly related to the ability of electrolyte solutions to conduct electric current, a brief discussion of the measurement of conductivity and its use in analytical chemistry is appropriate at this point. The conductance of a solution can be expressed by

$$\frac{1}{R} = k(c_1\lambda_1 + c_2\lambda_2 + c_3\lambda_3 + \cdots + c_i\lambda_i)\ \Omega^{-1} \tag{23.31}$$

Here the c_i are the concentrations of various ions in solution, λ_i are numerical constants characteristic of the ions, and k is a proportionality factor that takes account of the geometry of the cell in which the measurement is made. Because of interionic interactions and solvent–ion interactions, the values of λ_i depend slightly on concentration, and extrapolation to infinite dilution is required to obtain values that are physical constants of the individual ions. To obtain analytical data from a single measurement of the conductance of a solution, it is first necessary to obtain the concentration dependence of λ_i; it is also necessary to know a definite relation between concentrations of various ionic species in the solution. Thus, with proper calibration, a single measurement of conductance can give the concentration of a sodium chloride solution (since $c_{Na^+} = c_{Cl^-}$); this is not possible when there is no such convenient relationship among concentrations. Thus, the analytical utility of direct conductance measurements is limited. One use commonly made of conductance measurements is the monitoring of the quality of distilled water or other purified solvents. When the conductance of the distillate increases beyond a specified level, the still is known to be operating improperly.

23.3.4 / Conductometric titrations

If the conductivity of a solution is monitored during a titration, the changes noted can be used to establish the equivalence point; the method is referred to as a conductometric titration. Conductometric titrations depend on either a change in the net number of ions in solution during the course of the procedure, or the replacement of very mobile ions by less mobile ions or vice versa. The titration of hydrochloric acid with sodium hydroxide is an example of the type of titration that can be followed by conductance measurements. During the course of the neutralization reaction,

$$HCl + NaOH = H_2O + NaCl \tag{23.32}$$

very mobile protons are removed from solution and replaced by less mobile sodium ions. Consequently, prior to the equivalence point, the conductance of the solution decreases as the titration progresses, while after the equivalence point, continued addition of sodium hydroxide increases the conductivity of the solution by addition of more ions. Therefore, a plot of conductance vs. titrant volume will appear as indicated in Fig. 23.5(a). A

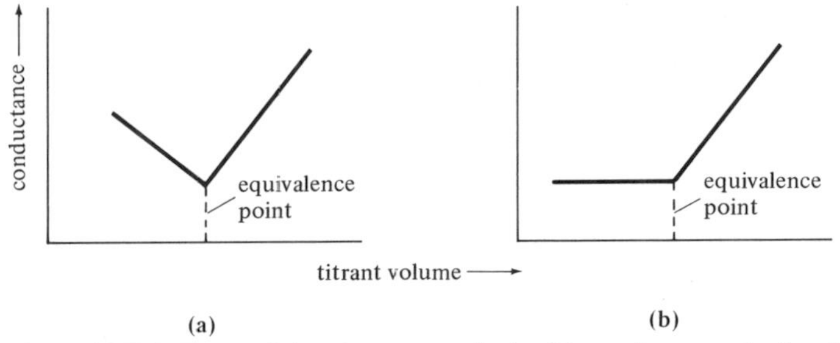

Figure 23.5 / *Types of titration curves obtained in conductometric titrations.*

precipitation titration, such as that indicated in Eq. (23.33),

$$NaCl + AgNO_3 = NaNO_3 + AgCl(s) \qquad (23.33)$$

results in replacement of silver ions by sodium ions prior to the equivalence point. This will have only a small effect on the conductance, since the two ions differ only slightly in mobility. After equivalence, the addition of more ions will increase the conductivity and the titration curve will appear as in Fig. 23.5(b). Other shapes of titration curves are possible depending on the specific situation. In general, the V-shaped curves provide the most unequivocal determination of the equivalence point and, of course, the steepness of the inflection in any of the curves governs the precision with which this determination can be made.

Solution conductivities are usually measured with alternating current using the cell as one arm of a Wheatstone bridge (see Sections 7.2.6 and 7.3.2). The use of AC is necessary to ensure that the current-carrying ability of the solution actually limits the flow of current, not the rate of an electrolytic reaction at one of the electrodes. Use of 1-kHz AC generally ensures that no significant electrode reaction can occur before the direction of the current flow is reversed.

While a number of both experimental and theoretical complications arise when very high frequency AC currents are used, it has proved feasible to carry out conductometric titrations with megahertz frequencies. The unusual aspect of this method is that it is not necessary to place the electrodes in the solution, but only on the outside of the titration vessel. One can see that such a method could have specific utility in the case of corrosive, toxic, or radioactive solutions.

23.4 / DIRECT POTENTIOMETRY

23.4.1 / Measurement of pH

Probably the ideal device for the analysis of solutes in solution would consist of a series of probes that could simply be placed in the solution and that would permit the concentration of the substance to be read from an appropriate meter. Direct potentiometry provides the means of doing just this for a limited number of ionic species. By far the most common usage of the direct potentiometric method is the measurement of hydrogen ion concentration with the ordinary glass electrode pH meter. In recent years, the development of a number of ion-specific electrodes has increased the scope of the method and resulted in renewed vigor in research in this area. In this section, we shall first discuss the measurement of pH and then extend the discussion to the determination of other species in solution.

A thermodynamically rigorous definition of pH is embodied in Eq. (23.34), but since, in practice, we cannot measure single-ion activities, we do not achieve the ideal in experiment.

$$\text{pH} = -\log a_{\text{H}^+} \tag{23.34}$$

Careful experimental work over a number of years has established an empirical or operational pH scale that approaches the ideal relation quite closely. The most obvious way of measuring the activity of hydrogen ion with an electrochemical cell would be to use a hydrogen electrode in an arrangement such as (23.35).

$$\text{Pt, } \text{H}_2(P)|\text{H}^+(\text{unknown})|\,|\text{reference electrode} \tag{23.35}$$

The emf of such a cell is given by Eq. (23.36).

$$E_{\text{cell}} = E_{\text{ref}} - 0.059 \log a_{\text{H}^+} + \frac{0.059}{2} \log P_{\text{H}_2} + E_{\text{lj}} \tag{23.36}$$

Thus, the expression for the emf, when we have fixed the partial pressure of hydrogen gas at some convenient value, still contains two unknowns, a_{H^+} and E_{lj}. Since rigorous elimination of the liquid junction potential has not proved possible, we obtain a good approximation, not a rigorous measure of a_{H^+}.

With the use of a good salt bridge, for a series of similar solutions, E_{lj} will be minimized and nearly constant. If the same reference electrode is employed, then $\varepsilon = E_{\text{lj}} + E_{\text{ref}}$ will be nearly constant. Therefore, if we arrange to make the partial pressure of hydrogen 1 atm, we obtain

$$E_{\text{cell}} = \varepsilon - 0.059 \log a_{\text{H}^+} = \varepsilon + 0.059\,\text{pH} \tag{23.37}$$

Differentiation of the relationship between E_{cell} and pH in Eq. (23.37) yields $dE/d(\text{pH}) = 0.059$ V at 25°C. An electrode that exhibits a potential change of 0.059 V for a 10-fold change in the concentration of the species to which it is sensitive is said to give *Nernstian response*.

23.4.2 / Accuracy of potentiometric measurements

It is of interest at this point to examine the inherent uncertainty associated with the direct potentiometric determination of ionic activities. Differentiation of the relation between activity and emf gives

$$E = \varepsilon - \frac{RT}{\mathscr{F}} \ln a \tag{23.38}$$

$$dE = \frac{-RT}{\mathscr{F}}\left(\frac{da}{a}\right) \tag{23.39}$$

which is easily rearranged, without regard to sign and assuming $T = 300°K$, to give

$$\frac{da}{a} = \frac{\mathscr{F}}{RT}\,dE = 39\,dE \tag{23.40}$$

With scrupulous care, it is possible to measure cell potentials with an uncertainty of 0.1 mV, that is $dE = \pm 0.1$ mV. From Eq. (23.40), we see that the relative uncertainty in the activity determination is 39 times this large; that is, $da/a = \pm 3.9 \times 10^{-3}$ or approximately ± 0.4 percent. A more typical uncertainty in the emf measurement might be 1 mV, and ± 4 percent is a more realistic assessment of the accuracy of the method. Further uncertainty may be introduced in converting activity to concentration. Because the liquid junction potentials are the same, repeated measurements on the same solution, or a series of very similar solutions, may well give reproducibility much better than this estimate, but accuracy and reproducibility should not be confused.

While the inherent accuracy of the direct potentiometric method is not high, there are many circumstances when it might be the analytical method of choice. While there are many methods capable of much higher accuracy in the determination of, say 0.1 *M* silver ion, a determination good to 4 percent using direct potentiometry at the subpart-per-million level would be more than acceptable competition for titration methods.

Measurement of cell emf to ± 0.1 mV suggests that [Eq. (23.37)] the uncertainty in pH measurements could be as low as ± 0.002 pH. However, an additional uncertainty is introduced by the necessity to estimate single ion activity coefficients in order to obtain the single ion activity of H^+, and this uncertainty limits the accuracy of pH measurements to ± 0.01 pH. In spite of the claims of some instrument manufacturers, significance may be attached to a third decimal place in pH measurements *only* when one is measuring pH *changes* in the same test solution under conditions that permit activity coefficients to remain essentially constant.

23.4.3 / The glass electrode

The disadvantage of the use of the hydrogen electrode for pH measurement is evident: It is inconvenient to require a source of gaseous hydrogen. A number of other chemical cells have been devised that are responsive to changes in a_{H^+}, and the primary advantage offered by any of these is that

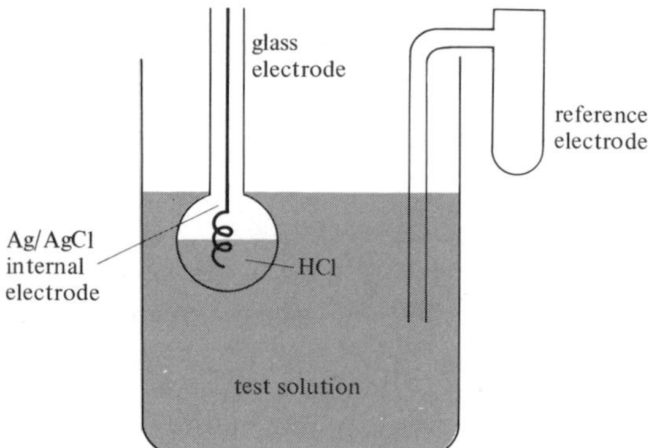

Figure 23.6 / *Diagrammatic representation of a glass electrode* pH-*measuring cell.*

the measurement can be made with a lower input impedance device than that required for a glass electrode. Nonetheless, the glass electrode has almost entirely supplanted other pH electrodes, and the reader is referred elsewhere for discussions of the chemical electrodes.

The glass electrode cell for pH determination can be represented schematically by cell (23.41) and pictorially in Fig. 23.6.

$$\text{Ag}|\text{AgCl(s)}, \text{HCl}(C_1)\overset{\text{glass}}{|}\text{ test solution}|\,|\text{reference electrode} \qquad (23.41)$$

The glass electrode consists of a thin-walled glass bulb containing an internal electrode (in this case a silver–silver chloride electrode, although several others are also in common use). This bulb is immersed in the test solution, and the potential of the electrode is measured with respect to a reference electrode connected to the test solution with a good salt bridge. Since a glass membrane now forms part of the current-carrying circuit, the result is an extremely high impedance source. Typically, the impedance of such cells is 1–100 MΩ. It has been found experimentally that a soft glass membrane gives Nernstian response to a_{H^+} over the range of pH = 0–10.

The mechanism of the glass electrode response to changes in pH has been investigated over a major part of the twentieth century. Early work appeared to indicate that the thin glass wall formed a semipermeable barrier through which only protons could pass. It is well known that water is essential to the electrode's function and that this function is impaired or lost when the electrode is dehydrated by long storage or soaking in a dehydrating solvent such as absolute ethanol. This process is reversible, and several hours soaking in water will restore the electrode to useful pH response.

These observations have led to the belief that an outer layer of the glass is hydrated forming what is called the "swollen layer" and that this layer is, in fact, responsible for the electrode's function. Ions move about in this layer and from the layer to the solution surrounding and reverse. Quantitative studies have shown that protons penetrate into the swollen layer at a rate of about 3 Å per hour and eventually reach a depth of 50–500 Å. This process cannot be the source of the electrical response, however, since the potential of the electrode is established quite rapidly, usually in only a few seconds.

It is now believed that ion exchange equilibria between the swollen layer and the solution are the source of the electrode potential, for example, reaction (23.42), in which A^+ is an arbitrary monovalent cation.

$$A^+(\text{glass}) + H^+(\text{sol}) \rightleftharpoons A^+(\text{sol}) + H^+(\text{glass}) \qquad (23.42)$$

The existence of such equilibria has been confirmed experimentally by measurement of the pickup of cations from solution by the glass electrode through the use of radioactive tracers.

For a long time it had been realized that the Nernstian behavior of a glass electrode broke down in strongly basic media. In solutions of pH > 10, a so-called alkaline error occurred; the observed emf corresponded to a negative error of nearly one pH unit at pH = 12 when the electrode membrane was composed of Corning 015 glass (72 percent SiO_2, 22 percent Na_2O, 6 percent CaO) and the solution contained 1 *M* Na^+. The magnitude of the error was reduced to -0.3 pH if the concentration of Na^+ was diminished to 0.1 *M*, and smaller errors were obtained when other cations were substituted for sodium. In addition, it was found that the error could be reduced by changing the composition of the glass. For example, substitution of Li_2O for Na_2O makes possible an electrode that shows only about 0.1 pH error at pH = 12.8 even in the presence of 2 *M* Na^+ in solution. These observations showed that the glass electrode was responding to cations other than H^+ in the solution and that the response could be altered by changing the composition of the glass, thus suggesting the possibility of producing an electrode that would give a useful analytical response to components of the test solution other than protons.

23.4.4 / Ion-selective glass electrodes

Since all types of glass electrodes show *some* response to metal ions, the simple Nernst equation involving only a_{H^+} is, at best, only an approximation to reality. Thus, for a solution containing cations A^+ and B^+, the logarithm in the Nernst equation should be written as $\log(a_{A^+} + Ka_{B^+})$. Here A^+ and B^+ represent two arbitrary monovalent cations to which the electrode responds and K is the *selectivity ratio*, which gives the quantitative relation between the response to A^+ and that to B^+. The selectivity ratio can be evaluated for a particular electrode by measuring the emf developed in solutions of known activity containing only A^+ or only B^+. Complications arise, however, since the majority of glass electrodes show response to a large number of metallic cations as well as protons.

Considerable variety in response has been achieved using glass compositions involving only variation of three components, Al_2O_3, Na_2O, and SiO_2. If less than 1 percent alumina is used, the electrode has little metal ion response and is a good pH electrode. The composition 5 percent alumina and 27 percent sodium oxide produces an electrode with general cation response, whereas 18 percent alumina and 11 percent sodium oxide gives a highly sodium-selective electrode. Roughly speaking, glass electrodes can be divided into three major categories:

	Response
1. pH type	$H^+ \ggg Na^+ > K^+ > Rb^+, Cs^+ \gg Ca^{2+}$
2. general cation	$H^+ > K^+ > Na^+ > NH_4{}^+ > Li^+ \gg Ca^{2+}$
3. sodium selective	$Ag^+ > H^+ > Na^+ \gg K^+, Li^+ \gg Ca^{2+}$

It has proved possible to obtain selectivity ratios for the alkali metal of interest of about 10^3 except for the case of potassium. The best potassium-sensitive glass electrodes thus far developed have selectivity ratios $K_{K^+/Na^+} \sim 20$. Anions do not seem to be involved in the ion exchange processes and, in general, anions do not give response on glass electrodes.

The available experimental data suggest that the response of a glass electrode depends largely on the fact that the hydrated glass lattice contains anionic sites that are attractive to cations of the appropriate charge-to-size ratio. It seems reasonable to assume that adsorption phenomena are involved. If, therefore, it were possible to tailor-make electrode membranes with a variety of specific adsorption sites, the production of ion-selective electrodes would be facilitated and, if cationic sites can be prepared, then anion sensitive electrodes would be possible. To a considerable extent these possibilities have been realized; however, the development has involved membranes other than glass.

23.4.5 / Other membrane electrodes

Typically, insoluble inorganic salts adsorb their own ions, so these materials are likely candidates for electrode materials—for example, AgCl for a chloride-sensitive electrode or $BaSO_4$ for a sulfate electrode. The major problem is one of mechanical fabrication, since the salts do not have very suitable mechanical properties. One solution to this problem has been devised by Pungor and coworkers in Hungary. A finely divided silver halide is dispersed in a silicone rubber monomer prior to polymerization and the resulting rubber is used as the active membrane. Since the membrane must have some electrical conductivity, the halide particles must be in physical contact, a requirement that places limits on the relative amounts of the halide and rubber that may be used. The electrode is formed by sealing the rubber membrane to an inert electrode body that contains an internal reference electrode as shown in Fig. 23.7. Electrodes of this type are commercially available for Cl^-, Br^-, I^-, and S^{2-}. An alternative construction by Orion, Inc., utilizes cast pellets of silver halide in place of the impregnated membrane. Both types give essentially identical response. It should be noted that the ion exchange type of electrode avoids one problem

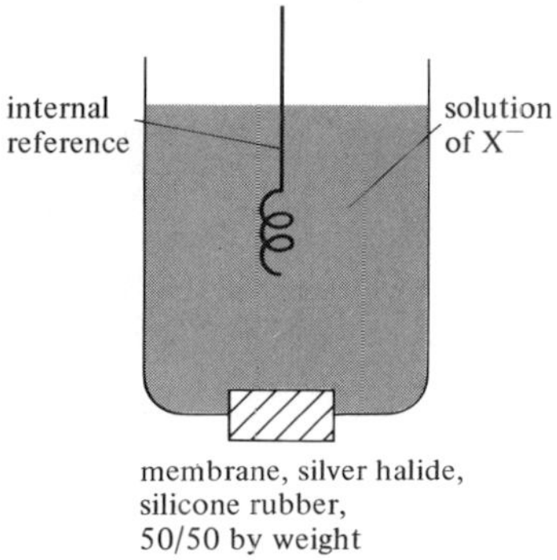

Figure 23.7 / *Schematic representation of a silicone rubber membrane electrode.*

inherent in determination of, say Cl^-, with a silver–silver chloride electrode. The latter, a redox electrode, will respond to changes in oxidation potential as well as chloride activity, whereas the ion exchange system will not. Some values of the selectivity ratios are given in Table 23.1; these are the values of K for the ion that is a constituent of the membrane in the presence of an equal concentration of the second ion. The preferential adsorption by the membrane of its own anions leads to a larger emf.

A number of organic liquids exist that are insoluble in aqueous media but have the property of being able to exchange ions with the aqueous solution. Such materials have been applied to a growing number of commercially available electrodes. The major difficulties are again of a mechanical nature. The ion exchange liquid must make electrical contact with the sample, yet physical mixing of the two fluids must be avoided or at least minimized. One design of electrode employing a liquid ion exchanger is illustrated in Fig. 23.8. As an example, some data will be given for a calcium electrode that uses a proprietary calcium organophosphorus ion exchange liquid. This electrode gives Nernstian response down to 5×10^{-5} F calcium and a nonlinear, but useful (with calibration) response an order of magnitude lower. The lower limit of useful response of any such electrode is limited by the solubility in water of the ion exchange material. Such solubility, of course, always results in some contamination of the sample with ions from

TABLE 23.1 / *Selectivity ratios*

	Membrane material		
Pair of ions	AgI	AgBr	AgCl
Br^-/I^-	2.1×10^2	1.3×10^2	—
Cl^-/I^-	1.7×10^5	—	3.6×10^2
Cl^-/Br^-	—	1.0×10^2	15

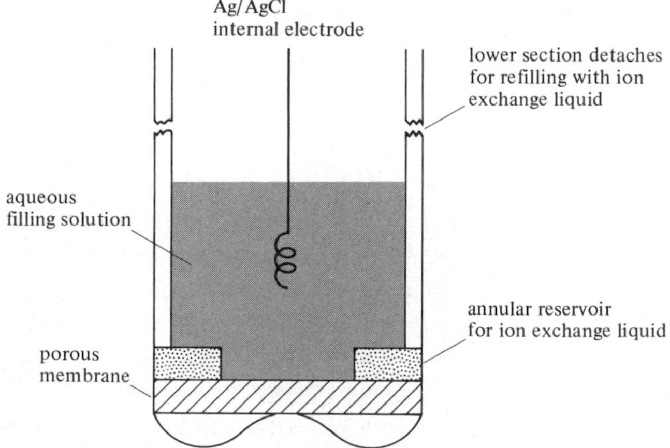

Figure 23.8 / *Schematic representation of a liquid ion exchange electrode.*

the electrode, and when the contamination and sample concentration are of the same order of magnitude, the electrode is no longer useful. The calcium electrode shows the selectivity ratios indicated in Table 23.2. An electrode of similar design, which gives nearly equal response to both Ca^{2+} and Mg^{2+}, is called a divalent ion electrode and forms a simple direct method of determining water hardness. Liquid ion exchange electrodes are commercially available for NO_3^-, ClO_4^-, Cl^-, and Cu^{2+}.

Probably the most nearly ideal specific ion electrode yet devised is the Orion fluoride electrode, which employs a section of a single crystal of lanthanum fluoride as the active membrane. The fluoride ions in this crystal are quite mobile, but the lattice requirements are such that essentially only fluoride can be exchanged. Thus, the emf of this electrode is linear with fluoride activity over the range 1 F to 10^{-6} F, and selectivity ratios of 1000 or greater are observed for all ions except hydroxide. Since the determination of fluoride is one of the more difficult problems in analytical chemistry, the availability of this electrode represents a significant advance.

Despite some of the obvious limitations of electrodes that respond to Na^+, K^+, and Ca^{2+}, it should be recognized that these species, too, are relatively difficult to determine by ordinary chemical means, so the ion-selective electrodes will find considerable application, for example, in clinical

TABLE 23.2 / *Selectivity ratios for the calcium electrode*

Cation	*Selectivity for* Ca^{2+}
Sr^{2+}	70
Mg^{2+}	200
Ba^{2+}	630
Na^+, K^+	3200

chemistry. A number of manufacturers are now producing miniaturized electrodes for use *in vivo*.

23.5 / POTENTIOMETRIC TITRATIONS

23.5.1 / Acid–base and EDTA titrations

In a potentiometric titration, the change in potential of a suitable indicator electrode is observed as a function of the volume of added titrant. Since this procedure requires accurate measurement of a *change* in potential rather than the absolute value of the potential, it frequently affords greater accuracy than the direct potentiometric method. There are, of course, the usual drawbacks of a titration procedure: The titrant concentration must be accurately known, usually involving a standardization; and the volume of titrant delivered must be precisely measured. Glass electrodes and ion-selective electrodes find numerous applications in following potentiometric titrations. Thus, for example, any feasible acid–base titration can be carried out with a pH meter and glass electrode. Potentiometric endpoint determinations can similarly be used employing specific ion electrodes for Ca^{2+} or Cu^{2+} when these species are titrated with ethylenediaminetetraacetic acid (EDTA). Since the electrode responds to the logarithm of the concentrations of species in solution, the common S-shaped titration curve results naturally.

23.5.2 / Precipitation titrations

Precipitation titrations can be followed in similar fashion using specific ion electrodes. For example, the titration of fluoride ion with La^{3+} is conveniently followed with the fluoride ion electrode. Conventional redox electrode systems may also be used, of course, when appropriate to the particular system under investigation. The silver–silver chloride electrode is useful in potentiometric chloride titrations since the potential of the half-reaction [Eq. (23.22)] depends on the concentration of chloride. Thus, the potential of a cell such as (23.43) is given by (23.44).

$$\text{Ag, AgCl}|\text{test solution}|\,|\text{H}^+|\text{Pt, H}_2 \tag{23.43}$$

$$E_{\text{cell}} = E^0_{\text{Ag,AgCl}} - 0.059\left\{\log[\text{Cl}^-]^{-1} - \log\frac{(P_{\text{H}_2})^{1/2}}{[H^+]}\right\} \tag{23.44}$$

If the right-hand half-cell is the standard hydrogen electrode, the second log term is zero, and if another reference electrode is used, its constant potential is added to E_{cell}. Since we are now interested in ΔE, we have written concentrations rather than activities in the expression for E_{cell}. Since $E^0_{\text{Ag,AgCl}} = -0.222$ V, the direction of spontaneous change is from right to left and the silver electrode is positive. If the indicator electrode reaction is always written as a reduction, however, then its E^0 will have the

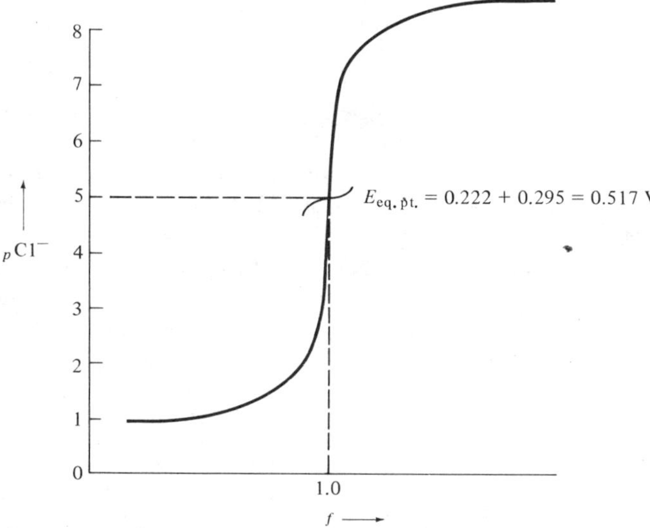

Figure 23.9 / *Titration curve for* 0.1 *F chloride titrated with* 0.1 *F silver.*

same *sign* as the *polarity* of the electrode with respect to the standard hydrogen electrode. Therefore, the measured potentials will be given by

$$E_{\text{cell}} = E^0_{\text{AgCl/Ag}} - 0.059 \log[\text{Cl}^-] = (0.222 + 0.059\, p\text{Cl}^-)\ \text{V} \qquad (23.45)$$

The usual precipitating reagent for chloride is Ag^+ (it is only a coincidence that silver is involved both in the indicator electrode and the titration reaction), and the titration reaction is

$$\text{Ag}^+ + \text{Cl}^- = \text{AgCl} \qquad K = K_{\text{sp}}^{-1} = 1 \times 10^{10} \qquad (23.46)$$

It is easy to see that the concentration of chloride at the stoichiometric equivalence point is simply that in a saturated solution of AgCl, 10^{-5} F. From elementary stoichiometry and equilibrium considerations, it is easily shown that the concentration of chloride at any point during such a titration is given by

$$[\text{Cl}^-] = C^0\gamma(1 - f) + \frac{K_{\text{sp}}}{[\text{Cl}^-]} \qquad (23.47)$$

where C^0 is the initial chloride concentration, f is the "fraction titrated," defined as the ratio of titrant volume added at any point to the volume required for stoichiometric equivalence, and γ is a numerical dilution factor defined as the ratio of the initial volume to the total volume at any point. Solution of this equation permits the computation of pCl as a function of f, and gives the analytical shape of the titration curve that is observed. The ordinate can be expressed in either pCl or volts vs. NHE computed using Eq. (23.45). In the case that $C^0 = 0.1$ F and the titrant concentration is also 0.1 F, the curve would appear as shown in Fig. 23.9.

23.5.3 / Oxidation–reduction titrations

This terminology is generally reserved for titrations in which there is a change of oxidation state for both titrant and titrate. As in other potentiometric procedures, the course of the titration is followed by an electrode that exhibits a potential that is a function of the concentrations of species in solution. This indicator electrode need not, and generally does not, involve reaction with any of the species of interest. Most often, the indicator electrode of choice is an inert metallic electrode, which serves only to transfer electrons to and from the test solution. Because of its chemical inertness, platinum is frequently the metal of choice. Platinum electrodes frequently give Nernstian response to the oxidizing power of a solution; however, the mechanism by which the potential of the electrode is established is not, as yet, fully understood.

In systems that involve strong reducing agents, a mercury indicator electrode is preferred over platinum. Since the equilibrium constant for reaction (23.48) is approximately 10^{13}, the fact that chromous ion can be used as a titrant in aqueous media is due entirely to the small rate of this reaction.

$$2Cr^{2+} + 2H^{+} = 2Cr^{3+} + H_2 \tag{23.48}$$

A platinum surface catalyzes the evolution of hydrogen, making the measurement of accurate potentials impossible; however, mercury does not effect this catalysis.

We shall now review briefly the concepts required for the calculations of potentiometric titration curves using the titration of iron(II) with manganese(VII) as the example. The first question that must be asked about any titration procedure is whether it is "quantitative," that is, whether the reaction proceeds sufficiently toward completion to give analytically useful accuracy. This question is answered in part by evaluating the equilibrium constant, which, in the present example, is conveniently done using the half-cell potentials.

$$Fe^{3+} + e^{-} = Fe^{2+} \qquad E^0 = +0.76\ V \tag{23.49}$$

$$8H^{+} + MnO_4^{-} + 5e^{-} = Mn^{2+} + 4H_2O \qquad E^0 = +1.52\ V \tag{23.50}$$

These half-reactions combine to yield the overall titration reaction (23.51).

$$8H^{+} + MnO_4^{-} + 5Fe^{2+} = Mn^{2+} + 4H_2O + 5Fe^{3+} \qquad E^0 = +0.76\ V \tag{23.51}$$

Using the method outlined previously, the standard potential for the overall reaction permits estimation of the equilibrium constant as $\sim 3 \times 10^{64}$, a value large enough to ensure essentially complete reaction. This is not the only concern, since a titration is not feasible if the rate at which the reaction proceeds is too small. In the present example, however, it is an

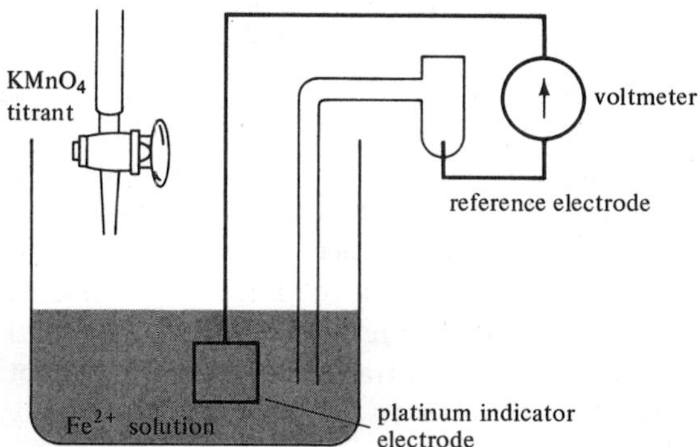

Figure 23.10 / *Typical equipment for a potentiometric redox titration.*

experimental fact that the reaction proceeds rapidly. The experimental arrangement for a typical potentiometric redox titration is shown in Fig. 23.10. An indicator electrode and reference electrode are placed in the sample solution, and titrant solution of accurately known concentration is delivered from the buret. The potential of the indicator electrode is monitored during the course of the titration and, from the resulting plot of potential vs. volume of titrant, the stoichiometric equivalence point is determined.

While the overall reaction (23.51) is the *chemical* reaction that takes place in the vessel, it is important to realize that this is *not* the electrochemical reaction that produces voltage readout on the meter. There are two equivalent ways to represent the cell that does give rise to the observed voltage. These are given as (23.52) and (23.53), where, for convenience, it is assumed that the standard hydrogen electrode is used as the reference.

$$\text{Pt}|\text{Fe}^{3+}(C_1),\ \text{Fe}^{2+}(C_2)|\ |\text{H}^+(a = 1)|\text{Pt},\ \text{H}_2(1\ \text{atm}) \qquad (23.52)$$

$$\text{Pt}|\text{MnO}_4^-(C_1),\ \text{Mn}^{2+}(C_2),\ \text{H}^+(C_3)|\ |\text{H}^+(a = 1)|\text{Pt},\ \text{H}_2(1\ \text{atm}) \qquad (23.53)$$

In practice, this is unlikely to be the case, but use of the NHE reference simplifies the calculations. Since the species on the left of the salt bridge in both (23.52) and (23.53) are in chemical equilibrium, the redox potential calculated from either cell must be the same. Stated another way, any given solution can have only one value for its redox potential. This potential can be calculated by application of the Nernst equation to either cell as shown in Eqs. (23.54) and (23.55).

$$E = E^0_{\text{Fe}^{3+}/\text{Fe}^{2+}} - 0.059 \log \frac{[\text{Fe}^{2+}]}{[\text{Fe}^{3+}]} \qquad (23.54)$$

$$E = E^0_{\text{MnO}_4^-/\text{Mn}^{2+}} - \frac{0.059}{5} \log \frac{[\text{Mn}^{2+}]}{[\text{MnO}_4^-][\text{H}^+]^8} \qquad (23.55)$$

In both equations, the reactions must be written as reductions in order that the calculated potentials have the same sign as that physically measured for the platinum indicator electrode. Note that as the concentration of the oxidized species becomes larger, the potential of the indicator electrode becomes more positive, or oxidizing. Either of these equations can, in principle, be used to calculate the entire titration curve, save for the initial point which cannot be calculated, and the equivalence point for which one must use both equations or one equation and the equilibrium constant; the first of these approaches is generally the more convenient. The initial point cannot be calculated because the concentration of iron(III) is not known. It cannot, of course, be zero, since that would imply an infinitely reducing potential. Even if it were possible to prepare a perfectly pure iron(II) solution, water or hydrogen ion will oxidize ferrous ion to regenerate a small concentration of ferric ion in any such solution.

23.5.4 / Calculation of potentials

While either equation *can* be used, it is much more convenient to calculate the cell potential using the ferric–ferrous half-cell prior to the equivalence point and the manganese half-cell thereafter. This is so because, in view of the near completeness of the reaction, the concentrations of the iron species are easily obtained to high accuracy prior to equivalence. Thus, at any point on the titration curve for which $0 < f < 1$, these concentrations are given by Eq. (23.56).

$$[Fe^{2+}] \simeq \frac{(\text{moles } Fe^{2+} \text{ initially present } - 5 \text{ moles } MnO_4^- \text{ added})}{\text{total volume}}$$

$$[Fe^{3+}] \simeq \frac{(5 \text{ moles } MnO_4^- \text{ added})}{\text{total volume}} \tag{23.56}$$

It is evident that the assumption of complete reaction also implies that the value of $[Mn^{2+}]$ can be obtained the same way; however, the concentration of remaining permanganate ion can be obtained only by a somewhat tedious equilibrium calculation. In a similar fashion, the concentrations of Mn^{2+} and excess MnO_4^- are readily obtained for points on the titration curve corresponding to $f > 1$; however, in this region an equilibrium calculation is required to obtain $[Fe^{2+}]$. Therefore, it is usual to calculate the points on the curve beyond the equivalence point from the titrant couple.

Computation of the equivalence-point potential is accomplished by making use of the fact, mentioned above, that the solution can have only a single redox potential which must be given by both Eqs. (23.54) and (23.55). Therefore, after multiplying both sides of (23.55) by 5, the two may be added together to obtain

$$6E = E^0_{Fe^{3+}/Fe^{2+}} + 5E^0_{MnO_4^-/Mn^{2+}} - 0.059 \log \frac{[Fe^{2+}][Mn^{2+}]}{[Fe^{3+}][MnO_4^-][H^+]^8} \tag{23.57}$$

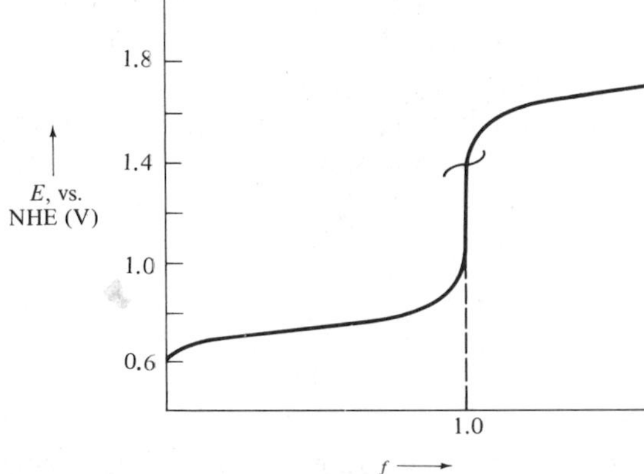

Figure 23.11 / *Potentiometric titration curve for the titration of* Fe^{2+} *with* MnO_4^-.

At the equivalence point, by definition, exactly the stoichiometric amount of titrant has been added, and from Eq. (23.51) we may write

$$[Fe^{3+}] = 5[Mn^{2+}] \quad \text{and} \quad [Fe^{2+}] = 5[MnO_4^-] \tag{23.58}$$

Substitution of Eq. (23.58) into Eq. (23.57) permits elimination of most of the factors in the argument of the logarithm, and by rearranging we obtain

$$E = \frac{E^0_{Fe^{3+}/Fe^{2+}} + 5E^0_{MnO_4^-/Mn^{2+}}}{6} + \frac{0.059}{6}\log[H^+]^8 \tag{23.59}$$

The method outlined is completely general for computation of the equivalence-point potential in any redox titration. Only in cases such as this example, in which the reaction is symmetric with respect to products and reactants, will all such concentrations cancel from the logarithm. Thus, for example, when $Cr_2O_7^{2-}$ or I_3^- are used as oxidants (with products, respectively, $2Cr^{3+}$ and $3I^-$), a concentration factor will always remain in the expression for the equivalence-point potential, and an exact calculation of the potential will require previous computation of the remaining concentration from equilibrium considerations. Frequently, as in the present example, the concentration of a species such as H^+ must be included. Since these concentrations appear as a small multiple of a logarithmic term, they are sometimes small. It is therefore fair to say that the equivalence-point potential is *dominated*, but not given exactly except in special cases, by the *average* of the standard potentials for the titrant and titrate couples *weighted* by the number of electrons that appear in each half-reaction.

The titration curve for the titration of ferrous ion with permanganate in 1 M H^+ is shown in Fig. 23.11. It is worth noting that the portion of the curve prior to equivalence depends only on the titrate couple and will be the same no matter what titrant is employed; the portion of the curve beyond equivalence, of course, depends on the titrant as does the location of equivalence-point potential. Only in the case of titrant and titrate that react in 1:1 molar ratio do the equivalence-point potential and the point of maximum slope of the titration curve coincide. In most cases, however, the accuracy of the method is not limited by the error introduced by taking the point of steepest slope as the equivalence point. As in any titration procedure, the precision and accuracy of potentiometric titrations are usually limited by the measurement of the volume of titrant delivered and the care exercised in its standardization.

23.6 / VOLTAMMETRY

23.6.1 / Mass-transport phenomena

In the methods discussed this far, we have intentionally tried to prevent flow of current in order that potential measurements could be made under conditions approaching equilibrium as nearly as possible. In the present section, we shall consider several experiments classed as voltammetry. As the name implies, both a current and a potential are measured, and as a consequence, an electrochemical reaction is allowed, or caused, to proceed.

Briefly, the voltammetric experiment involves an indicator electrode and a reference electrode. A potential difference is imposed between these electrodes, and the current that flows as a result of electrochemical reactions is measured; the output consists of a record of current as a function of indicator electrode potential. In all such experiments, one deals with heterogeneous reactions that occur at the surface of the working or indicator electrode. Since material must get from the bulk of solution to the electrode in order to react, a primary consideration in voltammetry is mass transport in solution. It is necessary, therefore, to examine mass-transport phenomena in some detail if voltammetry is to be properly understood.

There are three mechanisms for the transport of charge particles in solution: migration, convection, and diffusion. Migration, in the simplest terms, refers to the movement of ions under the influence of an electric field gradient and results in the attraction of positive ions to a negatively charged electrode and negative ions to a positively charged electrode. Convection is a mixing process, which arises from mechanical agitation or vibration of the solution, thermally induced differences in density of various portions of the solution, or density differences resulting from the electrochemical reaction. If, for example, the product of the reaction should be less (more) dense than the reactant, convective stirring at the electrode surface could result. Diffusion is the movement of dissolved species from regions of high concentration to regions of lower concentration, which results from the fact that the free energy (more strictly, the chemical potential) is proportional to concentration. The fact that diffusion and concentration are intimately connected is the basis of the analytical utility of voltammetry. The experiments are carried out in such a way that only diffusion makes an important contribution to the movement of material to the electrode.

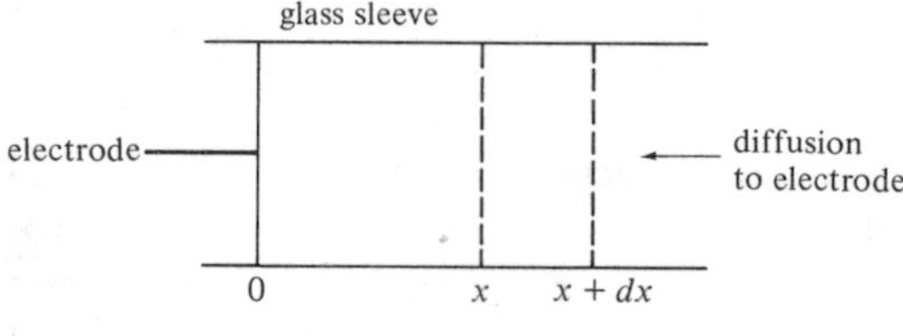

Figure 23.12 / *Diffusion to a shielded, planar electrode from a single direction.*

In experimental practice, the contribution of migration to the total current is minimized by adding to the solution a background, or supporting, electrolyte at a concentration 50–100 times that of the material of interest. The supporting electrolyte makes the solution conductive (lowers its resistance) and thereby minimizes electric field gradients. A supporting electrolyte that does not undergo reaction at the electrode is chosen, and therefore the current that flows at the electrode is caused solely by the electroactive material. In the bulk of the solution, on the contrary, current is carried approximately in proportion to concentration, that is, predominantly by the supporting electrolyte. At the same time, convection is minimized by taking care that the electrolysis cell is not vibrated, and immersion of the cell in a thermostat bath is frequently employed to ensure uniform temperature.

23.6.2 / Diffusion to a planar electrode

Only the process of diffusion remains to transport material to the electrode. The relations between diffusion, concentration, and current will now be examined. The case that is simplest to treat mathematically, and is frequently at least an approximation to experiment, is that of diffusion to a planar electrode surface from a single direction, that is, linear diffusion. This is illustrated approximately in Fig. 23.12, where diffusion is taken to occur along the x direction. The problem is formulated by asking how many moles of particles of electroactive material diffuse across a boundary parallel to the electrode at distance x in unit time. The rate at which such particles reach the electrode obviously governs the flow of current, the quantity we are ultimately seeking.

We define the *flux* of particles, a function of two variables, distance from the electrode, and time in Eq. (23.60),

$$f(x, t) = \frac{1}{A}\frac{dN(x, t)}{dt} \tag{23.60}$$

where A is the electrode area and $N(x, t)$ is the number of moles of particles that cross the boundary at distance x. It is assumed that the particle flux is simply directly proportional to the concentration gradient at x; this is stated as Eq. (23.61), which is known as *Fick's first law of diffusion.*

$$\frac{1}{A}\frac{dN(x, t)}{dt} = D\frac{\partial C(x, t)}{\partial x} = f(x, t) \tag{23.61}$$

The proportionality constant, D, is called the *diffusion coefficient* and is primarily a function of the nature of the diffusing species and the medium through which it moves. As with all molecular motion, the rate of diffusion increases with increasing temperature and in general is inversely proportional to the molecular weight. The latter relation may break down if the electroactive material is tightly solvated, since an appreciable number of solvent molecules must then move with it. In different solvents, the value of D varies approximately inversely as the viscosity. In a completely rigorous treatment, D is also a function of concentration; however, for the purposes of this discussion, D will be taken to be independent of C, x, and t. When it is noted that the usual units of concentration in problems of this sort are moles-centimeter^{-3}, one readily deduces from Eq. (23.61) that the proper units for D are centimeters^2-second^{-1}.

We wish now to obtain an expression for the time rate of change of concentration within a layer of solution of thickness dx, that is, between x and $x + dx$. This is done in the following manner. Over the short distance dx, we assume that the concentration gradient is linear in x; we may therefore write the concentration at $x + dx$ in terms of $C(x, t)$, the gradient, and dx, as indicated in Eq. (23.62).

$$C(x + dx, t) = C(x, t) + \frac{\partial C}{\partial x}\, dx \tag{23.62}$$

Since we have seen that the flux is given by the product of the diffusion coefficient and the partial derivative of C with respect to x, the flux at the second boundary is given by

$$f(x + dx, t) = D\left(\frac{\partial}{\partial x}\right) C(x + dx, t) = D\,\frac{\partial C(x, t)}{\partial x} + D\left(\frac{\partial}{\partial x}\right)\left[\frac{\partial C(x, t)}{\partial x}\right] dx \tag{23.63}$$

Evidently, the flux into and out of the volume element of thickness dx governs the rate at which the concentration changes. Equation (23.63) simplifies to

$$f(x + dx, t) = D\,\frac{\partial C(x, t)}{\partial x} + D\left[\frac{\partial^2 C(x, t)}{\partial x^2}\right] dx \tag{23.64}$$

and we note that the first term on the right-hand side is simply $f(x, t)$. Thus, Eq. (23.64) can be rearranged to obtain

$$\frac{f(x + dx, t) - f(x, t)}{dx} = D\,\frac{\partial^2 C(x, t)}{\partial x^2} \tag{23.65}$$

Recalling that flux has units of moles-centimeter^{-2}-second^{-1}, then the left-hand side of Eq. (23.65) must have units moles-centimeter^{-3}-second^{-1}, or concentration-second^{-1}. Thus, the left-hand side of Eq. (23.65) is the time rate of change of concentration, and we may write

$$\frac{\partial C(x, t)}{\partial t} = D \frac{\partial^2 C(x, t)}{\partial x^2} \tag{23.66}$$

which is known as *Fick's second law of diffusion.* Solution of Eq. (23.66) will provide an analytical description of the spatial and time dependencies of the concentration gradient, and therefore also of the particle flux and current under diffusion-limited conditions.

23.6.3 / Solution of the Fick's second-law equation[1]

Closed-form solutions of *Fick's second-law equation* cannot be found in general. We must specify *initial* and *boundary* conditions that are imposed by experimental reality in order that the equation may be solved. In discussing the initial conditions, we define $t = 0$ as the time at which electrolysis begins; therefore, nothing has changed in the solution until $t > 0$. We shall define $C(0, 0) = C^0$ as the concentration of electroactive material at the electrode surface prior to initiation of the electrolysis. The initial condition is the quite reasonable one that the solution is homogeneous; that is, $C(x, 0) = C^0$ for all values of x. The boundary conditions pertain to all $t > 0$, that is, at any time after initiation of the electrolysis. We shall assume that the electrode is maintained at a constant potential such that the electrode reaction is diffusion limited. Diffusion-limited processes are those for which the electrode reaction is much more rapid than diffusive transport of material to the electrode surface. Since the electroactive material can react more rapidly than it is supplied, for all $t > 0$, the electrode surface concentration is maintained at zero, or $C(0, t) = 0$. The second boundary condition is that at sufficiently large distances from the electrode, there is no change in the concentration during the electrolysis. In other words, $C(\infty, t) = C^0$ for all values of t.

We shall now proceed to solve Eq. (23.66) by a convenient, if somewhat indirect, method. We rewrite the equation in terms of a dummy variable, $U(x, t) = C^0 - C(x, t)$ and obtain

$$D \frac{\partial^2 U(x, t)}{\partial x^2} = \frac{\partial U(x, t)}{\partial t} \tag{23.67}$$

The fact that derivatives with respect to both x and t appear in this equation makes it necessary to eliminate one such derivative. This is conveniently done using the method of *Laplace transforms.* The Laplace transform is represented by $f(P) = L[F(t)]$, where $F(t)$ is the original function, $f(P)$ is the transformed function (a new function of a new variable), and L is a linear operator that represents the Laplace transform. For our purposes, these transforms can be used in the same way as logarithms; they are avail-

[1] Readers who wish to do so may skip to Section 23.6.4 without loss of continuity; however, even those who do not relish the mathematics in this section should understand the formulation of the initial and boundary conditions.

able in tables. A Laplace transform is analogous to obtaining the logarithm of a number, and obtaining the inverse transform is analogous to taking an antilogarithm. In the present case, we wish to know $f(P)$, the transformed function, when the original function is d/dt or $F'(t)$. Consulting tables, one finds that $L[F'(t)] = Pf(P) - F(0)$, where $F(0)$ is the original function evaluated at $t = 0$. Note that the transformation is with respect to one variable only, in this case time. Using $\overline{U}$ for the transformed function, Eq. (23.67) becomes significantly simplified by removal of the time derivative, as seen in Eq. (23.68).

$$\frac{\partial^2 \overline{U}(x, P)}{\partial x^2} - \frac{P}{D}\overline{U}(x, P) = -\frac{1}{D}U(x, 0) \tag{23.68}$$

Recalling the definition $U(x, t) = C^0 - C(x, t)$, we find that $U(x, 0) = 0$ for all x, further simplifying the equation.

$$\frac{\partial^2 \overline{U}(x, P)}{\partial x^2} - \frac{P}{D}\overline{U}(x, P) = 0 \tag{23.69}$$

Since Eq. (23.69) states that the function $\overline{U}$ is equal to its second derivative except for a constant multiplier, this suggests that $\overline{U}$ is an exponential. Let $\overline{U} = e^{ax}$; then $\partial^2\overline{U}/\partial x^2 = a^2e^{ax}$, and Eq. (23.69) can be rewritten as

$$a^2e^{ax} - \frac{P}{D}e^{ax} = 0 = e^{ax}\left(a^2 - \frac{P}{D}\right) \tag{23.70}$$

Since the exponential does not vanish, Eq. (23.70) can be satisfied only if $(a^2 - P/D) = 0$, which leads to the conclusion $a = \pm(P/D)^{1/2}$. Thus, there are two solutions and the most general solution will be a linear combination of the two possibilities as given in Eq. (23.71).

$$\overline{U}(x, P) = \alpha \exp\left[\left(\frac{P}{D}\right)^{1/2}x\right] + \beta \exp\left[-\left(\frac{P}{D}\right)^{1/2}x\right] \tag{23.71}$$

The coefficients α and β must be determined from the initial and boundary conditions. To return to the original variable, we require the transform of the definition of U, that is, $\overline{U}(x, P) = L[C^0 - C(x, t)] = C^0/P - \overline{C}(x, P)$. Inserting this relation and absorbing the negative signs into the undetermined coefficients, we obtain

$$\overline{C}(x, P) = \alpha \exp\left[\left(\frac{P}{D}\right)^{1/2}x\right] + \beta \exp\left[-\left(\frac{P}{D}\right)^{1/2}x\right] + \frac{C^0}{P} \tag{23.72}$$

which is a nearly general solution, since only the condition of homogeneous solution has been imposed at this point.

We now make use of the boundary conditions to evaluate the coefficients. The condition $C(\infty, t) = C^0$ has the Laplace transform $\overline{C}(\infty, P) = C^0/P$. Evaluating Eq. (23.72) with $x = \infty$ shows that this condition can be satisfied only if $\alpha = 0$. The remaining constant is evaluated from the other boundary condition, that is, $C(0, t) = 0$, for which the transform is simply $\bar{C}(0, P) = 0$. Consequently, $0 = \beta + C^0/P$, and the complete solution of the transformed equation is given as

$$\bar{C}(x, P) = \frac{-C^0}{P} \exp\left[-\left(\frac{P}{D}\right)^{1/2} x\right] + \frac{C^0}{P} \tag{23.73}$$

We seek now from tables the inverse transform of this expression. One finds that $(1/P) \exp(-kP^{1/2})$ is the transform of $\text{erfc}(k/2t^{1/2})$, so letting $k = x/D^{1/2}$ and multiplying by C^0 will effect the desired inverse transformation. The inverse transform of C^0/P is simply C^0. Thus, the solution is expressed as

$$C(x, t) = C^0 - C^0 \,\text{erfc}\left(\frac{x}{2D^{1/2}t^{1/2}}\right) \tag{23.74}$$

and since $\text{erfc}(\lambda) = 1 - \text{erf}(\lambda)$, simple substitution after factoring C^0 gives

$$C(x, t) = C^0 \,\text{erf}\left(\frac{x}{2D^{1/2}t^{1/2}}\right) \tag{23.75}$$

Here, erf is the error function, and erfc is the error function complement. The error function has the form $\text{erf}(\lambda) = (2/\pi^{1/2}) \int_0^{\lambda} \exp(-z^2)\, dz$. Since the current is proportional to the concentration gradient, we must differentiate Eq. (23.75) with respect to x. The derivative of the error function is available in tables and may be combined with the result of differentiation of Eq. (23.75). The result is given in Eq. (23.76).

$$\frac{\partial C(x, t)}{\partial x} = \exp\left(\frac{-x^2}{4Dt}\right) \frac{C^0}{(\pi D t)^{1/2}} \tag{23.76}$$

23.6.4 / The diffusion current

Equation (23.76), the solution to the Fick's second-law expression, gives the x and t dependence of the concentration gradient from which the time dependence of the diffusion current is readily deduced. Current flow is governed by the particle flux *at the electrode*, that is, at $x = 0$. Evaluating Eq. (23.76) at $x = 0$ and inserting this result into Eq. (23.61), we obtain

$$\left(\frac{1}{A}\right) \frac{dN(0, t)}{dt} = \frac{C^0 D^{1/2}}{(\pi t)^{1/2}} \tag{23.77}$$

as the flux at the electrode surface. The current that flows is, by Faraday's law, $n\mathscr{F}\, dN(0, t)/dt$. Using Eq. (23.77), we obtain Eq. (23.78) for the diffusion-limited current, i_d.

$$i_d = \frac{n\mathscr{F} A C^0 D^{1/2}}{(\pi t)^{1/2}} \tag{23.78}$$

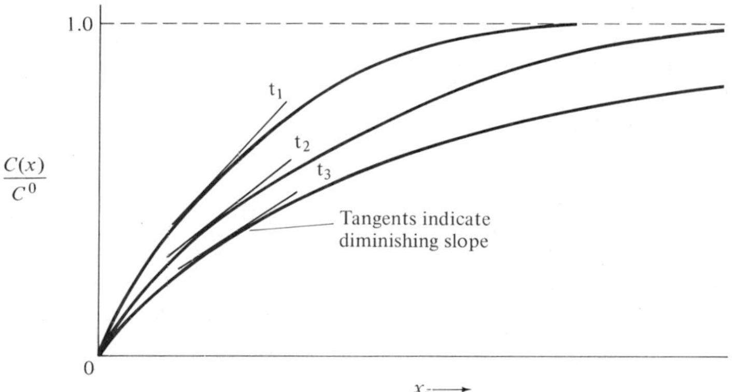

Figure 23.13 / *Concentration profiles at a planar electrode under diffusion-limited conditions.*

The physical meaning of Eq. (23.76) is illustrated in Fig. 23.13. The family of curves plotted indicates the normalized concentration as a function of distance from the electrode at progressively larger values of t. For any particular value of t, the slope of the curve decreases exponentially with x^2, so that at sufficiently large distance from the electrode, the slope approaches zero and C approaches C^0. For a particular value of x, the slope decreases with increasing time; and for the point $x = 0$, the slope diminishes with the square root of time. Therefore, for diffusion to a planar electrode under the conditions assumed, the diffusion current, i_d, diminishes toward zero with the square root of time, as stated in Eq. (23.78). The analytical utility of voltammetry lies in the direct proportion between i_d and C^0; however, the difficulty of accurately measuring a quantity that is decaying with time suggests that the experiment should be modified to avoid this condition.

23.6.5 / Polarography

The first specific case of voltammetry that will be considered is polarography. The term polarography is applied to a voltammetric experiment carried out using a dropping mercury electrode (dme) as the working electrode. This technique was invented at the Charles University in Prague by Jaroslav Heyrovsky, who published the first paper on the subject in 1922. Heyrovsky's achievement was recognized by the Nobel Prize in chemistry in 1959.

The dme consists of a 10–15 cm section of capillary tubing of inside diameter 0.02–0.04 mm. One end is connected to a mercury reservoir capable of applying about 50 cm head and the other end is immersed in the test solution. Mercury issues from the capillary as a series of small, nearly spherical drops, which grow to a critical weight governed by surface tension and then finally fall. A typical experimental arrangement for polarography is shown in Fig. 23.14.

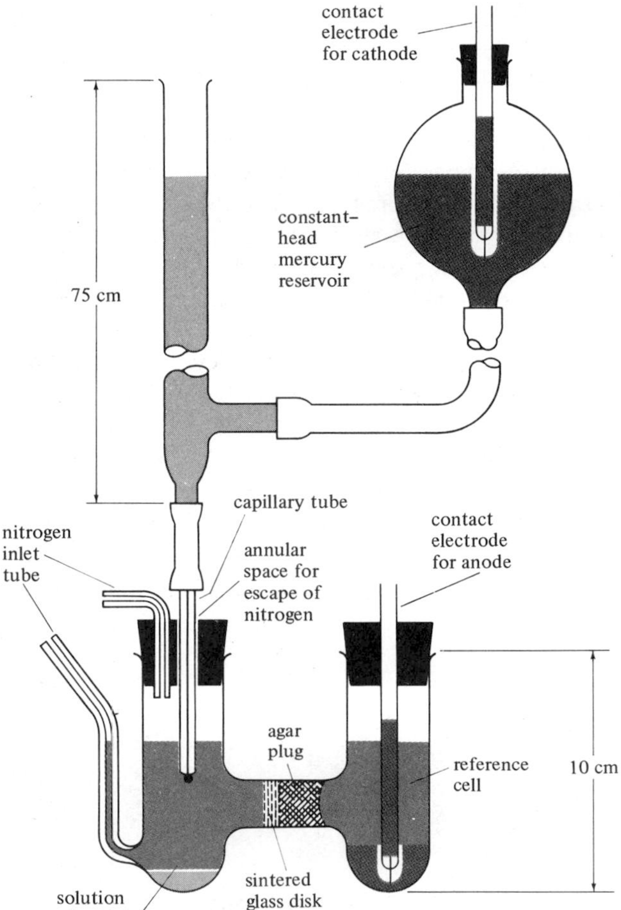

Figure 23.14 / *Experimental apparatus for polarography. Shown here is a cell employing a permanent external anode and arrangement of the dropping electrode according to Lingane and Laitinen.*

Use of a dme may, at first, seem an additional complication in an already complicated experiment, but in fact it offers some unique advantages. In contrast to a stationary electrode, use of a dme means that the electrode surface is renewed with each new drop, so that the previous history of the electrode does not affect the results. In addition, most of the solution depleted by electrolysis is carried away with the falling drop, thereby permitting each drop to be in contact with fresh solution. The result of these factors is that diffusion currents observed with a dme are more reproducible and more easily given quantitative interpretation than those obtained with solid electrodes. Mercury is the electrode of choice for cathodic processes since its large overpotential (*vide infra*) for the reduction of H^+ makes accessible an especially large range of cathode potentials. As we shall see, the growth of the drop more than compensates for the time decay of the diffusion current, thereby eliminating the difficulty of measuring a slowly decaying quantity.

To elucidate the current–time characteristics of the dme, we shall apply the theory of linear diffusion we have developed with the inclusion of the time variation of the electrode area. This procedure is, of course, an approximation, since we are dealing now with an expanding spherical electrode. A rigorous treatment of the expanding-sphere electrode is mathematically difficult, and the improvement obtained over the approximate treatment is only at the 10 percent level.

Consider a dme held at constant potential such that the current is diffusion limited. Let m be the mass of mercury that exits from the capillary in 1 sec in milligrams per second and ρ be the density of mercury in milligrams per centimeter^3. At time t after a drop has begun to form, its mass is mt and its volume at that time, $V(t) = mt/\rho$. Since the electrode is spherical, $mt/\rho = \frac{4}{3}\pi r^3$, where r is the radius of the sphere. Since the area of a sphere is $4\pi r^2$, the area can be expressed as a function of time by Eq. (23.79), which contains the capillary characteristic, m, time, and numerical constants which are collected as k.

$$A(t) = 4\pi \left(\frac{3mt}{4\pi\rho}\right)^{2/3} = km^{2/3}t^{2/3} \tag{23.79}$$

Combination of Eqs. (23.78) and (23.79) gives

$$i_d(t) = \frac{n\mathscr{F}C^0D^{1/2}km^{2/3}t^{2/3}}{\pi^{1/2}t^{1/2}} = k'nD^{1/2}C^0m^{2/3}t^{1/6} \tag{23.80}$$

where k' again represents a collection of numerical constants. Since m is constant for a constant head of mercury, the diffusion current at a dme *increases* with the sixth root of time until the drop eventually falls. (To avoid later confusion of units, note that $t^{-2/3}$ lies concealed in $m^{2/3}$.) The time variation of i_d is illustrated in Fig. 23.15. The time at which the drop falls is termed the *drop time* and is symbolized τ. As seen in Fig. 23.15, the maximum current provides the most suitable point for accurate measurement, as opposed to points at which the current is changing. In the older literature of polarography, average currents were frequently reported, while with modern instruments equipped with strip-chart recorders the maximum current is always chosen. The change is due simply to changes in instrumentation. Older polarographs employed light-beam galvanometer recording which, with proper damping, produced small oscillations. The average deflection was an accurate measure of the average current during the life of a drop. Most strip-chart recorders are unable to do this, so average currents cannot be determined accurately with these devices. The average current is obtained by dividing the total current by the drop time and is

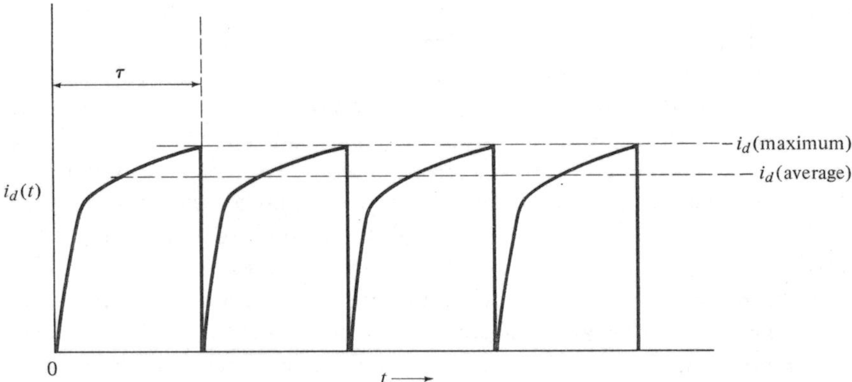

Figure 23.15 / *Diffusion current as a function of time for a dropping mercury electrode.*

given by $(1/\tau) \int_0^\tau i(t)\, dt = (1/\tau)k'' \int_0^\tau t^{1/6}\, dt$. Evaluation of the indicated integral gives $i_{av} = (6k''/7)\tau^{7/6}/\tau = (6/7)k''\tau^{1/6}$ and we recognize $k''\tau^{1/6} = i_{max}$. Therefore, the average current is simply six-sevenths of the maximum current.

Not all of the units used in the derivation of Eq. (23.80) are convenient in practical polarography; currents are commonly measured in microamperes (μA) and concentrations are expressed in millimoles per liter (mM) with retention of the units previously given for m, t, and D. The constant k' in Eq. (23.80) will be evaluated in these units. Equation (23.80) was first derived by D. Ilkovic in 1934, and it is usually referred to as the *Ilkovic equation.* While Ilkovic did not treat the dme as an expanding sphere, he did include the expansion of the electrode, approximated as an expanding plane. Such a treatment results in multiplying k' by $(\frac{7}{3})^{1/2}$. Physically, this factor takes into account the movement of the electrode surface toward the solution as the electrode expands. Including the $(\frac{7}{3})^{1/2}$ factor, $k' = 708$ for maximum currents and 607 for average currents. Thus, to about 10 percent accuracy, the maximum diffusion current is given by

$$i_{d(\max)} = 708nC^0D^{1/2}m^{2/3}\tau^{1/6} \tag{23.81}$$

As mentioned previously, m depends on the capillary used and the drop time is also dependent on the capillary. In addition, τ depends on the interfacial tension between the mercury and the test solution and this, in turn depends on both the solution composition and the electrode potential. We have previously noted the temperature dependence of D. Therefore, assuming constant temperature, for a particular substance, the quantity i_d/C will be constant so long as the same capillary is used with the same head of mercury. The form of the dependence of i_d on mercury pressure is easily derived. The flow rate, m, is directly proportional to the applied pressure or hydrostatic head, H. For a particular applied voltage and solvtion composition, the interfacial tension is constant and, therefore, so is the critical weight of the mercury drop. Since the weight of the drop is constant, the drop time must be inversely proportional to H. Inserting $m \propto H$ and $\tau \propto H^{-1}$ into $i_d \propto m^{2/3}\tau^{1/6}$, we find that $i_d \propto H^{1/2}$. The accurate variation of i_d with the square root of mercury head is sometimes used as a criterion of a diffusion-controlled electrode process. The fact that i_d/C depends on the capillary and cannot be reproduced from one laboratory to another caused Lingane to suggest use of the diffusion-current constant, $I = i_d/Cm^{2/3}t^{1/6}$, which is independent of the capillary used to the extent that the relations of i_d to concentration and capillary characteristics predicted by the Ilkovic equation are valid, that is, at the 10 percent level. Note that the common units for I are microamperes (millimoles/liter)$^{-1}$ milligram$^{-2/3}$-second$^{-1/2}$.

There have been, of course, numerous refinements of Ilkovic's treatment that incorporate varying degrees of rigor in solution of the spherical diffusion problem. These studies are of great interest in testing the validity of the concept of diffusion-controlled processes. In a practical analytical problem, however, one rarely attempts to calculate diffusion currents, but rather uses the usual analytical techniques of standard solutions, working curves, or standard additions. About 1 percent accuracy is expected.

23.6.6 / Current–potential curves

Thus far, the discussion has been limited to the case of a constant electrode potential. The familiar technique in polarography is, however, measurement of current as a function of electrode potential as the latter is slowly varied. The rate at which the potential is varied must be sufficiently small that the electrode potential is approximately constant over the life of a single drop, for example, 0.1 V/min. A typical chart-recorder display of a polarogram is given in Fig. 23.16. The curve shown is for the one-electron reduction of *para*-benzoquinone to the semiquinone radical and was obtained in acetonitrile solution. The electrode reaction is that of Eq. (23.82).

$$\text{O=C}_6\text{H}_4\text{=O} + e^- \rightleftharpoons \left[\text{O=C}_6\text{H}_4\text{=O}\right]^{\dot{-}} \tag{23.82}$$

As shown in Fig. 23.16, it is conventional to take cathodic currents as positive and to display increasing cathodic potentials to the right. Since the large majority of polarograms involve reduction of material at potentials negative with respect to the reference electrode, this convention represents a convenience, however irrational it first may seem. The diffusion current that results from electrolysis of the species of interest is the difference of the

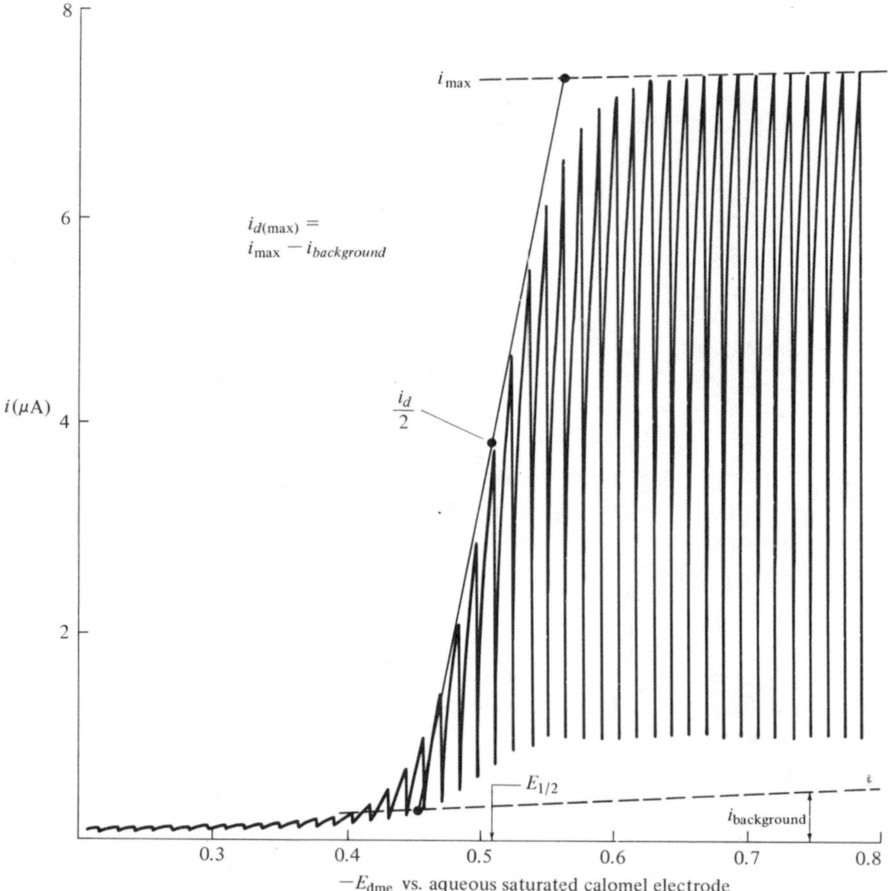

Figure 23.16 / *Polarogram of p-benzoquinone in acetonitrile solution. Tetraethylammonium perchlorate at* 0.1 *F concentration was employed as supporting electrolyte, and the quinone concentration was* 1.4 m*F. The drop time was approximately* 6 *sec.*

maximum current and the background, or residual current as indicated in Fig. 23.16. The residual current results in part from capacitive charging of the growing drop surface and in part from traces of electroactive impurities in the solution.

Current limited by diffusion is oberved only on the top, or *diffusion plateau* of the polarographic wave. Elsewhere along the wave, the electrode potential is insufficiently cathodic to reduce the equilibrium concentration of reactant essentially to zero, and the condition $C(0, t) = 0$ is not fulfilled. The shape of the current–potential curve can be calculated from the Nernst equation if the assumption is made that at each value of electrode potential the concentrations of the oxidized and reduced species *at the electrode surface* are those required by thermodynamic equilibrium.

Consider a generalized electrochemical reduction such as

$$\text{Ox} + ne^- = \text{Red} \tag{23.83}$$

Let the bulk solution concentration of Ox = C^0_{Ox} and the initial concentration of Red = 0. As indicated in Eq. (23.83), Red is the only reaction product. At any point on the wave, the observed current is proportional to the concentration gradient at the electrode surface, $\partial C(0, t)/\partial x$, and the square root of the diffusion coefficient, $D^{1/2}_{\text{Ox}}$. The overall concentration gradient is $C^0_{\text{Ox}} - C_{\text{Ox}}(0)$, that is, the difference between the bulk and surface concentrations. Note that as $C_{\text{Ox}}(0)$ decreases with increasingly cathodic potential, the gradient *increases*; this accounts for the rise in current along the wave. The current at any point can be expressed by

$$i = \kappa D^{1/2}_{\text{Ox}}[C^0_{\text{Ox}} - C_{\text{Ox}}(0)] \tag{23.84}$$

Here the proportionality constant, κ, includes $m^{2/3}\tau^{1/6}$ so that when $C_{\text{Ox}}(0) \rightarrow 0$, Eq. (23.84) becomes identical to Eq. (23.81). Therefore, when the indicated multiplication is carried out, the first term on the right is identically i_d, so we may write

$$i = i_d - \kappa D^{1/2}_{\text{Ox}} C_{\text{Ox}}(0) \tag{23.85}$$

which is easily solved to obtain $C_{\text{Ox}}(0) = (i_d - i)/\kappa D^{1/2}_{\text{Ox}}$.

The amount of the species Red produced is proportional to the current that flows and from the symmetry of the process we may write an expression analogous to that for $C_{\text{Ox}}(0)$. Of course, the diffusion coefficient for the reduced species must be used, and the result is $C_{\text{Red}}(0) = i/\kappa D^{1/2}_{\text{Red}}$. Using these concentrations in a Nernst expression gives

$$E_{\text{dme}} = E^0_{\text{Ox,Red}} - \frac{RT}{n\mathscr{F}} \ln\left\{\frac{(i/\kappa D^{1/2}_{\text{Red}})}{(i_d - i)/\kappa D^{1/2}_{\text{Ox}}}\right\} \tag{23.86}$$

which is readily rearranged to Eq. (23.87), the equation of the current–potential curve.

$$E_{\text{dme}} = E^0_{\text{Ox,Red}} - \frac{RT}{n\mathscr{F}} \ln\left[\frac{i}{i_d - i}\right] - \frac{RT}{n\mathscr{F}} \ln\left(\frac{D_{\text{Ox}}}{D_{\text{Red}}}\right)^{1/2} \tag{23.87}$$

If, at some point on the polarographic wave, the logarithmic term in Eq. (23.87) becomes zero, the potential of the dme will be equal to the standard potential and independent of current. Approximately this occurs at the point $i = i_d/2$, or halfway up the wave, where the first log term in Eq. (23.87) vanishes. The potential at that unique point is termed the *half-wave potential*, symbolized $E_{1/2}$. Since the half-wave potential differs from E^0 by the term that contains the logarithm of the square root of the ratio of the diffusion coefficients, the values of $E_{1/2}$ and E^0 usually are not grossly different. In some cases, the diffusion coefficients of the oxidized and reduced forms are virtually equal, and $E_{1/2}$ then becomes of thermodynamic significance; that is, $E_{1/2} = E^0$. Since $E_{1/2}$ is defined by

$$E_{1/2} = E^0 - \frac{RT}{n\mathcal{F}} \ln\left(\frac{D_{\mathrm{Ox}}}{D_{\mathrm{Red}}}\right)^{1/2} \tag{23.88}$$

we can substitute Eq. (23.88) into Eq. (23.87) to obtain Eq. (23.89). The evaluation of the constant factors at 298°K yields Eq. (23.90).

$$E_{\mathrm{dme}} = E_{1/2} - \frac{RT}{n\mathcal{F}} \ln\left(\frac{i}{i_d - i}\right) \tag{23.89}$$

$$= E_{1/2} - \frac{0.059}{n} \log\left(\frac{i}{i_d - i}\right) \tag{23.90}$$

Thus, a plot of E_{dme} vs. $\log[i/(i_d - i)]$ is expected to give a straight line with a slope of $-59/n$ mV and an intercept of $E_{1/2}$.

23.6.7 / Criteria for reversibility

Obtaining such a straight line is a widely used criterion for the reversibility of the electron transfer process. In order to use this criterion, the value of n must be estimated independently from comparison of i_d with known systems or, preferably, from coulometry. This caution is needed since it is possible to obtain a slope quite accurately -59 mV for, say, an irreversible two-electron process. A convenient modification of this test of reversibility is embodied in evaluation of Eq. (23.90) at two points on the wave, those for which $i = 3i_d/4$ and $i = i_d/4$ followed by comparison of the actual difference in electrode potential and that predicted by Eq. (23.91).

$$\begin{aligned} E_{3/4} - E_{1/4} &= \frac{-0.059}{n}\left\{\log\left[\frac{3/4}{1/4}\right] + \log\left[\frac{1/4}{3/4}\right]\right\} \\ &= \frac{-0.059}{n} \log 9 = \frac{-56}{n} \text{ mV} \end{aligned} \tag{23.91}$$

Both of these tests provide necessary, but not sufficient, conditions for a reversible process. A sufficient condition is that the polarogram of an equimolar mixture of Ox and Red gives a composite wave with anodic and cathodic branches. The potential at which the current is midway between the anodic and cathodic diffusion plateaus must be identical to $E_{1/2}$ for either the cathodic or anodic process studied individually. In addition, there must be no inflection in the curve as it passes through the zero of current.

Assessment of the reversibility of a particular electrode reaction is of great value when the object of the study is elucidation of the mechanism of the electrode process; however, the analytical utility of polarography does not depend on reversibility. The ramifications of irreversible behavior will be discussed later; for the moment we shall point out that the usual result is merely to draw the wave out as indicated in Fig. 23.17. In the

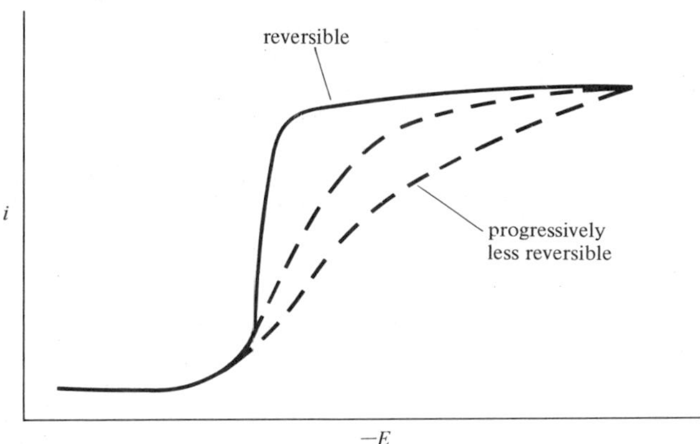

Figure 23.17 / *The effect of irreversibility on the shape of a polarographic wave.*

irreversible case, the limiting current is still proportional to concentration. For an irreversible wave $E_{1/2}$ may sometimes vary with concentration, while for reversible processes $E_{1/2}$ is independent of concentration.

23.6.8 / Polarographic analysis

The primary analytical utility of polarography is therefore the quantitative analysis of electroactive materials, with applications largely to metal ions. Qualitative information can also be obtained, since $E_{1/2}$ is characteristic of a particular ion in a specified medium (solvent, supporting electrolyte). More than one material may be reduced in the same potential region, however, and the "resolution" of different substances by polarography is considerably inferior to that obtained by spectroscopic methods. The unique advantages of voltammetric analysis are the ability to handle easily small volumes of dilute solutions and to do so in very nearly a nondestructive way. Currents of several microamperes are normally observed for one-electron reductions of species present at the 1-mM level. It is not particularly difficult to obtain polarographic data from 10^{-5} M solutions, and the volume of solution required is only enough to immerse the dme and reference electrode probe, frequently less than 1 ml. The amount of material consumed is usually trivial. Since typical conditions might involve a current of the order of 1μA for 100 sec, 10^{-4} C of electricity are consumed, or 10^{-9} mole of material for a one-electron process.

Polarographic half-wave potentials for metal ions are particularly sensitive to two effects: formation of amalgams with the mercury electrode and the formation of complex ions with ligands in solution. The reduction of metals that form amalgams is facilitated by the lower chemical potential of the metal in the amalgam compared to the pure metal. For example, both sodium and potassium decompose aqueous solutions by reduction of water, yet both can be reduced on a dme within the potential range accessible in water. The potential required to reduce the ions Na^+ and K^+ to the pure metals are -2.71 and -2.92 V, respectively, whereas both show $E_{1/2}$ values about -1.9 V at the dme (all potentials vs. NHE). By contrast, complex ion formation shifts the polarographic wave to potentials more cathodic than those for the free metal ion. Thus, $MX_n{}^{+m}$ is more difficult to reduce than $M(H_2O)_n^{+m}$ because of enhanced stability of the complex ion. Ligands likely to cause cathodic shifts include halide, cyanide, ammonia, and the multidentate ligands such as EDTA. If the cathodic shift of the half-wave potential is observed as a function of ligand concentration, both the number of ligands and the stability constant of the complex ion can be determined.

Polarographic analysis can be used to determine organic compounds which, like the example of Fig. 23.16, contain reducible functional groups such as carbonyl, nitro, cyano, and numerous others. In addition, polarographic studies have been widely applied to examine the mechanisms of the electroreduction of organic compounds. Organic polarography is frequently carried out in other than aqueous media for reasons of solubility and also the greater accessible potential range of certain organic solvents. Generally, solvents of high dielectric constant are required to obtain reasonably conductive solutions and among these, acetonitrile, dimethylformamide, and ethanol are widely used.

Invariably, it is necessary to remove dissolved oxygen from solutions prior to a polarographic experiment, since oxygen is reduced at about -0.3 V vs. NHE in a neutral aqueous solution and the oxygen wave will obscure reduction of most materials of interest. With aqueous solutions, 5 min of nitrogen bubbling is usually sufficient to remove the dissolved oxygen; organic solvents hold oxygen more tenaciously, and nitrogen purging for 30–40 min is sometimes required for complete removal of the oxygen wave.

23.6.9 / Surface phenomena at the dropping mercury electrode

As noted previously, the drop time depends on the potential of the dme. This is so because the attractive van der Waals forces, which tend to minimize the drop area (surface tension), are opposed by the repulsion between like charges on the drop surface. The charge on the drop is a function only of the electrode potential so long as no constituents of the solution are specifically adsorbed on the mercury surface. A plot of surface tension vs. the potential of the dme is called the *electrocapillary curve* and is characterized by a maximum value of surface tension at approximately -0.5 V vs. a calomel reference electrode. At this potential, the surface charge on the drop is zero, while a positive charge is present at more anodic potentials and a negative charge is found at more cathodic potentials. Since the zero charge condition occurs at only a single potential and the surface charge must be neutralized, an electrical *double layer* exists at the drop surface under most experimental conditions. Ions of charge opposite to that on the dme surround it in almost sufficient numbers to neutralize this surface charge; these form what is known as the *Helmholtz layer.* The remainder of ions required to neutralize the charge are much more loosely held and form what is known as the *diffuse double layer.*

Two important effects of the formation of the double layer occur. First, a certain amount of current flows as the drop grows in order to provide its surface charge. This can be viewed as charging the capacitor formed by the drop–solution interface and is called the *charging current.* The existence of the charging current is the principal limitation on the sensitivity of polarographic analysis, since it obscures very small diffusion currents. A second effect that is of importance only in the study of electrode processes is to introduce some uncertainty regarding the potential at which the process occurs. Electrode reactions occur either within the double layer or at the Helmholtz plane, and whatever fraction of the electrode potential is consumed in orienting the double layer cannot be effective in the charge-transfer process. Both of these effects vanish at the point of zero charge; however, since it is impossible to work at a single potential, this observation serves only to explain the commonly observed phenomenon that the charging current passes through zero and changes sign at this potential.

The structure of the double layer has been the subject of extensive investigation since the invention of polarography, and the reader is referred to the original literature for a more complete discussion. At this point, we note only that a double layer exists at all electrodes, not just the dme, and that the phenomenon becomes much more complex when a constituent of the solution is specifically adsorbed on the electrode.

23.6.10 / Polarographic maxima

Polarography is frequently complicated by the appearance of a reproducible current spike on the wave. Such spikes are called *maxima of the first kind.* The current that flows at the top of the spike usually appears to be limited only by the cell resistance, indicating that some process, in addition to diffusion, is bringing electroactive material to the dme. Careful visual observation has shown that these maxima are associated with vigorous streaming of solution past the electrode surface. This is believed to result from motion of the electrode–solution interface due to asymmetry in the electric field surrounding the drop. In accord with this rationale, maxima do not occur at the zero charge potential.

Maxima of the second kind appear as anomalous polarographic waves, usually at potentials more cathodic than maxima of the first kind. The feature that distinguishes these waves from real polarographic waves is that with increasingly cathodic potentials, the limiting current decreases faster than would be expected from the decrease in drop time. It is known that these maxima are also a result of stirring of the drop–solution interface, but the cause is believed to be turbulence within the mercury drop.

Turbulence resulting from the flow of mercury into the drop is transmitted to the surface, thereby stirring the interface.

Maxima make accurate measurement of the wave parameters difficult and are avoided by the addition of maximum suppressors to the solution. Compounds that function as maximum suppressors are large organic molecules such as gelatin or the detergent Triton X-100 (the polyethyleneglycol ether of monoisooctylphenol), which are known to become adsorbed on the mercury surface. It is possible that the adsorbed material immobilizes the double layer and thereby prevents stirring. Only of the order of 0.002 percent by weight maximum suppressor is needed, and use of excessive amounts will produce erroneously small values of the diffusion current by insulating part of the electrode surface.

23.6.11 / Anodic voltammetry

Anodic voltammetry has proven extremely valuable, particularly for organic compounds, since many of these contain functional groups that can be oxidized but not reduced. The information obtainable from oxidative voltammetry is the same as that obtained from polarography; however, the ease of oxidation of mercury itself ($\sim$0.6 V vs. NHE) requires the use of a different electrode system. Platinum, gold, and carbon anodes have been employed. To obtain reproducible mass-transport conditions, the electrode is usually rotated. The rotating platinum microelectrode (rpe) has found considerable acceptance. This electrode consists of a glass tube closed at one end with a small piece of platinum wire sealed through the wall at right angles to the axis near the closed end. The tube is filled with mercury to make electrical contact with the platinum wire and permit contact at the open end while the electrode is rotated by a hollow spindle motor. Although the details of the mass transport to such an electrode are complex, it is found in practice that a well-defined voltammetric wave can be obtained with the limiting current (so called because it is not truly a diffusion current, but limited by a combination of diffusion and hydrodynamics) proportional to concentration and independent of time. Since the limiting current depends on the rate of rotation, a constant speed motor is necessary. Unlike the dme, the rpe is subject to fouling by deposited material and requires careful cleaning prior to each use.

23.6.12 / Instrumentation

The instrumentation required for voltammetry can be very simple. The circuit shown in Fig. 23.18 suffices for a manually operated instrument. The essential features are (1) a means of applying a known, variable DC voltage to the cell and (2) a method of measurement of the current that flows, usually in the microampere range. In the figure, a battery and variable resistor, R_1, provide the applied voltage, and the current is measured by observing the voltage drop across a precision resistor, R_2. Use of a double-

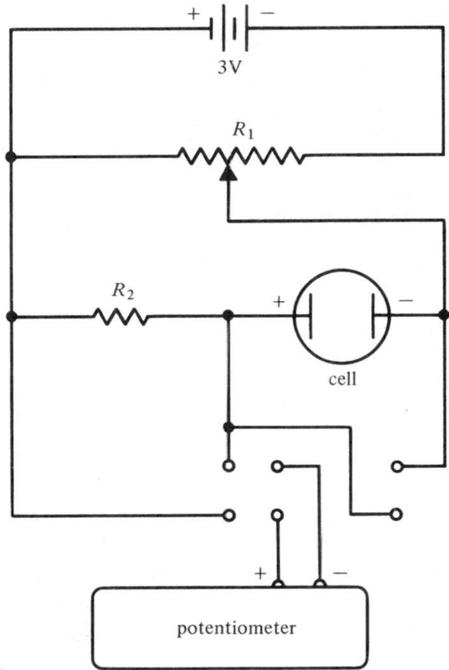

Figure 23.18 / *Simple manual circuit for polarographic analysis.*

pole, double-throw switch permits the voltages to be measured alternately with the same potentiometer. In this circuit, and in any voltammetry experiment carried out using only two electrodes, the cell voltage consists of two terms, $E_{\text{dme}} + iR_{\text{cell}}$. With aqueous solutions, the cell resistance is frequently so small that the *iR* correction is negligible; however, this is only infrequently the case for nonaqueous experiments. Since the current increases up the voltammetric wave, the *iR* correction is greater near the top of the wave than near the foot, and its effect is to tip reduction waves toward more cathodic potentials and oxidation waves toward more anodic potentials. The more sophisticated circuit described in Section 8.4.2 avoids the necessity of correcting for cell resistance by using three electrodes. In such a circuit, essentially all the current passes between the working electrode and a third, or auxiliary, electrode. The reference electrode is part of a very high impedance measuring circuit which draws negligible current. Hence, the measured potential, working vs. reference, contains, in principle, no *iR* drop. A dme, however, shows about 100 Ω resistence. One realizes, in addition, that there *is* an *iR* drop between the working and auxiliary electrodes, and as a result there is a potential gradient in the intervening solution. The existence of such a potential gradient dictates that the reference electrode probe should be placed physically as close as feasible to the working electrode. Nonetheless, some uncompensated *iR* is unavoidable and must be considered when the most accurate measurements are required. Electronic iR compensation techniques are known and are incorporated in modern commercial instruments.

23.6.13 / Rapid-scan methods

As described above, in a polarographic experiment the voltage sweep is sufficiently slow (say 0.1 V/min) that we treat each point on the polarogram as if the electrode potential were constant. The experiments we now wish to describe briefly are *fast-scan voltammetry* and *cyclic voltammetry*. These experiments are usually carried out using a solid electrode, or sometimes a hanging mercury drop; both are stationary electrodes. In both experiments, the rate of potential sweep may be as high as hundreds of volts per second; the cyclic experiment involves use of a triangular sweep, first in one direction and then symmetrically in the reverse direction. The sweep is begun at potentials corresponding to the foot of the polarographic wave, and the width of the sweep is sufficient to reach potentials that correspond to the diffusion plateau. As the potential is swept through the wave, the current first increases as the surface concentration of electroactive material is driven toward zero and the concentration gradient increases. Eventually, the current decreases as the gradient decreases in a manner similar to that predicted by Eq. (23.76). The result is that the current passes through a maximum value, or peak. The peak current is denoted i_P, and the potential at that point is termed the peak potential, E_P. Fast-scan voltammetry is somewhat more sensitive than DC voltammetry; the sensitivity increases with increasing sweep rate within the limits imposed by the capacitance of the electrode, since i_P is proportional to the square root of the scan rate. If, at some point beyond E_P, the potential scan is reversed, a cyclic voltammogram is the result. In Fig. 23.19, a cyclic voltammogram for the *para*-benzoquinone reduction is shown as presented on an xy plotter with the x axis driven by the voltage sweep. The anodic peak on the reverse sweep is due to oxidation of material that was reduced on the forward sweep.

Whether or not the reverse sweep is employed, for a reversible process, E_P (forward) is related to the polarographic half-wave potential by

$$E_P - E_{1/2} = \frac{-28.5}{n}\ \text{mV} \tag{23.92}$$

It can also be shown that the half-peak potential is $28.0/n$ mV more anodic than $E_{1/2}$. The potential is exactly $E_{1/2}$ when $i = 0.852 i_P$. For an oxidation, the relative positions are the same, and only interchange of the words anodic and cathodic is required.

The position of the reverse peak is slightly affected by the potential at which the sweep is reversed. For switching potentials more than 100 mV cathodic of $E_{1/2}$, the reverse peak position is about $30/n$ mV anodic of $E_{1/2}$. How far the sweep is continued beyond the forward peak governs the extent of electrolysis in the solution surrounding the electrode and therefore the

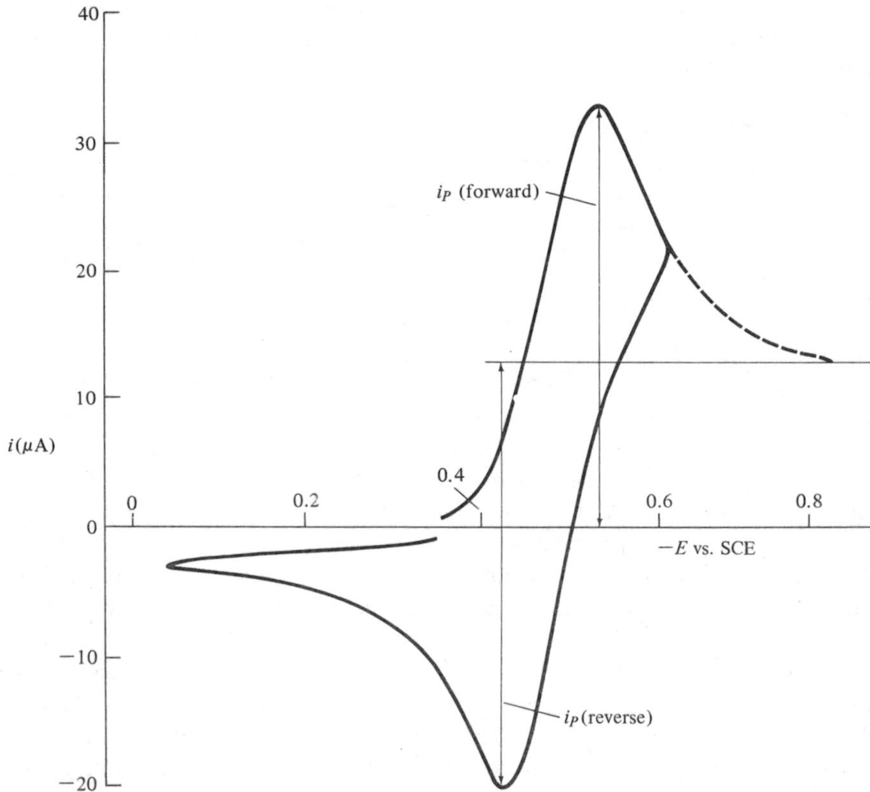

Figure 23.19 / *Cyclic voltammogram for the reversible reduction of p-benzoquinone in acetonitrile. The solution is the same as described for Figure 23.16.*

amount of material available for the back reaction. If the display is measured as shown in Fig. 23.19, where the reverse current base line is taken as the extrapolation of the forward curve, then for a reversible process, i_p (reverse)/i_p (forward) $= 1$. If the ratio is not unity, some type of irreversible behavior is indicated.

A large amount of information about the detailed nature of the electrode process may be obtained from cyclic voltammetry. The solution of the various boundary-value problems is very difficult, and generally one must resort to numerical solutions obtained from a computer. One example will be given to show the type of information that can be obtained. Consider an electrode process represented by Eq. (23.93),

$$\begin{aligned} &\text{Ox} + ne^- \rightleftharpoons \text{Red} \\ &\qquad \text{Red} \xrightarrow{k} \text{Z} \end{aligned} \tag{23.93}$$

in which the electron transfer step is reversible but followed by decomposition of the reduced form with rate constant k to give a product that is electrochemically inert. The reverse current will now depend on the relative values of the sweep rate and k, which together determine how much of the reduced form has decomposed prior to reaching the reverse peak potential. The reverse current is equal to the forward current at rapid-scan rates and/or if k is small. As k becomes large in comparison to the sweep rate, the reverse current decreases until no anodic peak is detectable. Removal of Red facilitates the forward reaction, as one would expect, and the resulting shift of E_P (forward) is related to k. Thus, instead of lagging 28.5 mV behind $E_{1/2}$, it can be shown that E_P and $E_{1/2}$ are related by

$$E_P = E_{1/2} - \frac{RT}{n\mathscr{F}}\left[0.780 - \ln\left(\frac{k}{a}\right)^{1/2}\right] \tag{23.94}$$

where $a = (n\mathscr{F}/RT)v$ and v is the sweep rate.

A rapid-scan technique that is especially designed for analysis of very dilute solutions is known as *anodic stripping voltammetry*. This procedure usually involves a hanging mercury drop electrode and requires that the substance to be analyzed can be deposited on, or amalgamated with, the electrode. A solution containing, for example, 10^{-8} M Cd^{2+} might be analyzed in this way. Since the solution is too dilute for ordinary voltammetry or rapid-scan methods, the cadmium is first deposited in the hanging drop by electrolysis. This step amounts to a preconcentration. After cessation of the electrolysis, an anodic voltage sweep is applied to the drop, whereupon the deposited metal is reoxidized. The peak current in the anodic voltammogram is limited by the concentration of cadmium in the mercury amalgam. About two orders of magnitude improvement in sensitivity can normally be obtained by the use of this technique.

23.6.14 / Conclusion

In concluding the section on voltammetry, let us consider the occasional case of a reduction wave that occurs at *positive* potentials (or an oxidation wave that occurs at *negative* potentials). These conditions are not terribly common, but when encountered generally provoke confusion. "I thought the cathode was *always* negative in an electrolysis cell!" is a typical reaction. Of course, a reduction that is observed at a potential positive with respect to the reference electrode in a two-electrode cell is occurring spontaneously, and the cathode is always *positive* with respect to the anode in a galvanic cell. Thus, the problem is solved by remembering that it is occasionally possible to observe spontaneous reactions in voltammetric experiments.

23.7 / CHRONOPOTENTIOMETRY

23.7.1 / Experimental procedure

As the name implies, chronopotentiometry is a technique that involves the observation of electrode potential as a function of time. In addition, the

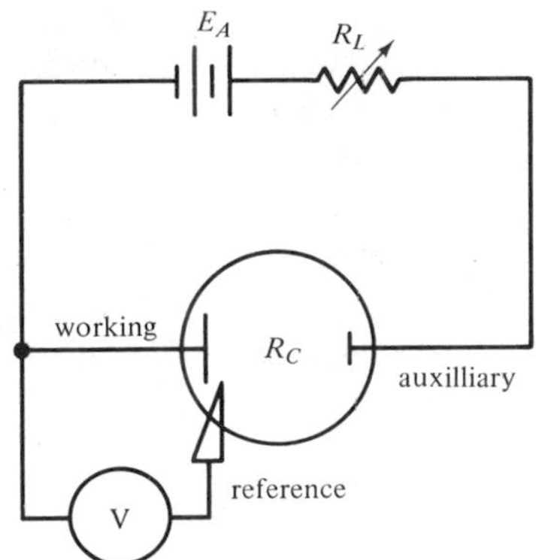

Figure 23.20 / *Essential instrumentation for chronopotentiometry. In addition to this equipment, an electrical stopclock or stopwatch is required.*

technique requires the passage of constant current and the assumptions introduced previously that the electrode process involves mass transport only by diffusion. Although chronopotentiometry can be used to determine concentrations of electroactive materials in solution, it has found little utilization for that purpose and has been restricted primarily to the study of electrode processes. An understanding of the function of this technique will prove useful toward understanding of the succeeding sections, which deal with controlled-potential and constant-current electrolyses.

Chronopotentiometry is usually carried out with a three-electrode system, but the circuitry need not be complicated. The simple circuit illustrated in Fig. 23.20 will suffice nicely. Here E_A represents the applied voltage and R_L is a current-limiting resistor in series with the cell. If R_C is the cell resistance and if the back emf of the cell is neglected, the current is simply $E_A/(R_L + R_C)$. When R_L is chosen to be $\gg R_C$, then small changes in R_C will have little effect and $i \simeq E_A/R_L$ = constant. In the figure, V represents a high impedance meter which monitors the potential of the working electrode during the course of the experiment.

The solution for chronopotentiometry will contain the electroactive material, usually at the millimolar level, an excess of supporting electrolyte, and be unstirred, just as described for polarography. When the constant current is applied, the electrode will rapidly assume some constant potential, which will be maintained for a time that depends on the value of i and the concentration of electroactive species. After time τ, here called the *transition time*, a potential step will be observed as illustrated in Fig. 23.21. This step is toward more cathodic potentials if the electrode reaction is a reduction and toward more anodic potentials for an oxidation.

23.7.2 / The potential–time curve

We shall now describe this observation both mathematically and physically with the assistance of the theory of linear diffusion already developed.

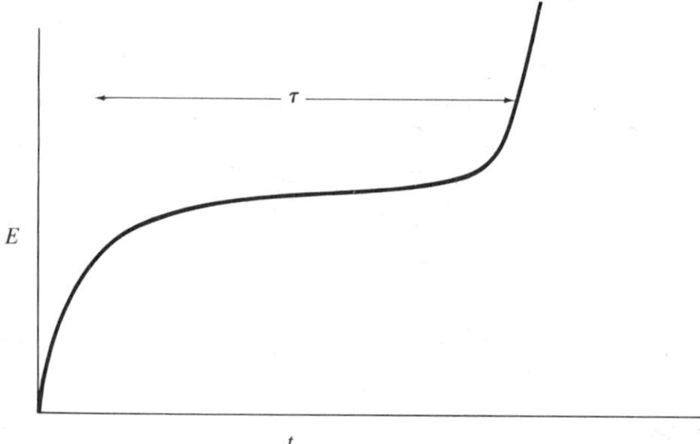

Figure 23.21 / *A typical chronopotentiogram. The transition time, τ, is measured as shown.*

The general solution in the Laplace plane given in Eq. (23.72) serves as the starting point; we need only supply the appropriate boundary conditions for the new experimental situation. One of the boundary conditions is the same as for voltammetry, namely, that the solution is not depleted of electroactive material at sufficiently large distances from the electrode, or $C(\infty, t) = C^0$. The second boundary condition is that of constant current. We have already seen that the current is proportional to the concentration gradient at the electrode surface, and $i = n\mathscr{F}AD\ \partial C(0, t)/\partial x$. Since the current is constant, the concentration gradient must also be constant, $\partial C(0, t)/\partial x = i/n\mathscr{F}AD$.

The first boundary condition requires that the positive exponential term in Eq. (23.72) must vanish. The remaining equation is differentiated with respect to x, and the Laplace transform of the second boundary condition is employed to evaluate β. The inverse transform is then obtained from tables, and the complete solution is given in Eq. (23.95).

$$C(x, t) = C^0 - \frac{2it^{1/2}}{n\mathscr{F}AD^{1/2}\pi^{1/2}} \exp\left(\frac{-x^2}{4Dt}\right) + \frac{ix}{n\mathscr{F}AD} \operatorname{erfc}\left(\frac{x}{2D^{1/2}t^{1/2}}\right) \tag{23.95}$$

When this result is evaluated at $x = 0$, the third term vanishes and the exponential becomes unity, leaving only Eq. (23.96).

$$C(0, t) = C^0 - \frac{2it^{1/2}}{n\mathscr{F}AD^{1/2}\pi^{1/2}} \tag{23.96}$$

At the time defined as the transition time, the second term on the right of Eq. (23.96) becomes equal to C^0, and the surface concentration of electroactive material becomes zero. At this time, since constant current still passes through the cell, the electrode potential must move to a new potential that will allow a new electrode reaction to carry the current. Solving Eq.

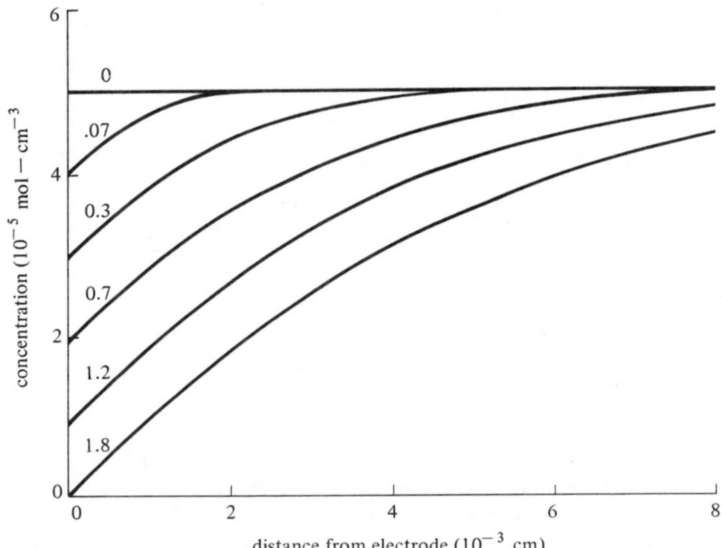

Figure 23.22 / *Concentration profiles at an electrode surface under constant current conditions. Here* $i = 10$ mA-cm^{-2}, $n = 1$, $C^0 = 5 \times 10^{-2}$ *M, and* $D = 10^{-5}$ cm^2-sec^{-1}. *Each curve is labeled with the time in seconds after initiation of the electrolysis.*
SOURCE: P. Delahay, *New Instrumental Methods in Electrochemistry*, Wiley-Interscience (New York, 1954).

(23.96) for the transition time, we obtain

$$\tau^{1/2} = \frac{n\mathscr{F}AD^{1/2}\pi^{1/2}}{2i} C^0 \tag{23.97}$$

which can be stated more simply as $i\tau^{1/2}/C^0$ = constant. It is this relation that permits analytical use of chronopotentiometry. If the electrode process is diffiusion controlled, Eq. (23.97) will hold for a large range of different constant currents. Failure of $i\tau^{1/2}/C^0$ to remain constant with variation of i is indicative of adsorption or kinetic complications in the electrode reaction.

In Fig. 23.22, the concentration profiles near the electrode at various times after initiation of the electrolysis are shown. Each member of this family of curves must, by the second boundary condition, intersect $x = 0$ with exactly the same slope. Notice that if we define the diffusion layer as that portion of the solution for which $C(x, t) < C^0$, this is really quite thin, and its thickness is given approximately by $(Dt)^{1/2}$.

23.7.3 / Relations to voltammetry

An analytical description of the potential–time curve is obtained by use of the Nernst equation with the assumption of reversible behavior as in the derivation of the current–potential curve for voltammetry. We have already obtained an expression for the concentration of the reactant, Ox, in terms of current in Eq. (23.96). Note that the second term on the right of Eq. (23.96), the amount of Ox consumed, must be equal to the amount of product, Red. These values of concentration give the Nernst equation

$$E = E^0 - \frac{RT}{n\mathscr{F}} \ln\left(\frac{D_{\text{Ox}}}{D_{\text{Red}}}\right)^{1/2} - \frac{RT}{n\mathscr{F}} \ln\left(\frac{kt^{1/2}}{C^0 - kt^{1/2}}\right) \tag{23.98}$$

in which $k = 2i/n\mathscr{F}AD_{\text{Ox}}^{1/2}\pi^{1/2}$. Note that the first two terms on the right of Eq. (23.98) are identically the polarographic half-wave potential as defined in Eq. (23.88). Since Eq. (23.97) shows that we can write $C^0 = k\tau^{1/2}$, Eq. (23.98) simplifies to

$$E = E_{1/2} - \frac{RT}{n\mathscr{F}} \ln\left(\frac{t^{1/2}}{\tau^{1/2} - t^{1/2}}\right) \tag{23.99}$$

The argument of the logarithm vanishes when $t = \tau/4$, so the potential of a chronopotentiometric curve is exactly the polarographic half-wave potential after a time one-quarter of the transition time. The equation of the current–time curve predicts that a plot of E vs. $\log[t^{1/2}/(\tau^{1/2} - t^{1/2})]$ will be a straight line with slope $-59/n$ mV and an intercept $= E_{1/2}$ for a reduction. This has been amply verified experimentally.

If the direction of current flow is reversed at the transition time, then reduced material in the vicinity of the electrode will be reoxidized and a second transition is expected, but with the potential step in the opposite, or anodic, sense. It has been shown that if the forward and reverse currents are equal and the electrode process is uncomplicated, then the second transition time will be one-third the first, that is, $\tau_2 = \frac{1}{3}\tau_1$. There is a slight asymmetry in the potentials, and for the reverse transition $E = E_{1/2}$ at $0.222\tau_2$. Comparison of the potential at this point with the quarter-transition potential for the forward transition provides another test of the reversibility of the electrode process.

23.8 / MACROSCOPIC ELECTROLYSES

23.8.1 / Experimental methods

In contrast to the techniques described previously in which we assumed no change in the bulk concentration of electroactive material, it is the intention of a macroscopic electrolysis to electrolyze completely a particular chemical species. There are three possible ways to carry out such an electrolysis: (1) with passage of a constant current through the cell; (2) with constant voltage applied to the cell; (3) with constant working electrode potential. From the discussion of chronopotentiometry, it is clear that if we demand constant current, the electrode potential will vary with time. Consideration of voltammetry suggests that with constant electrode potential, the current varies with time. We shall see that because of variations in the product of current and cell resistance, the constant applied voltage experiment will

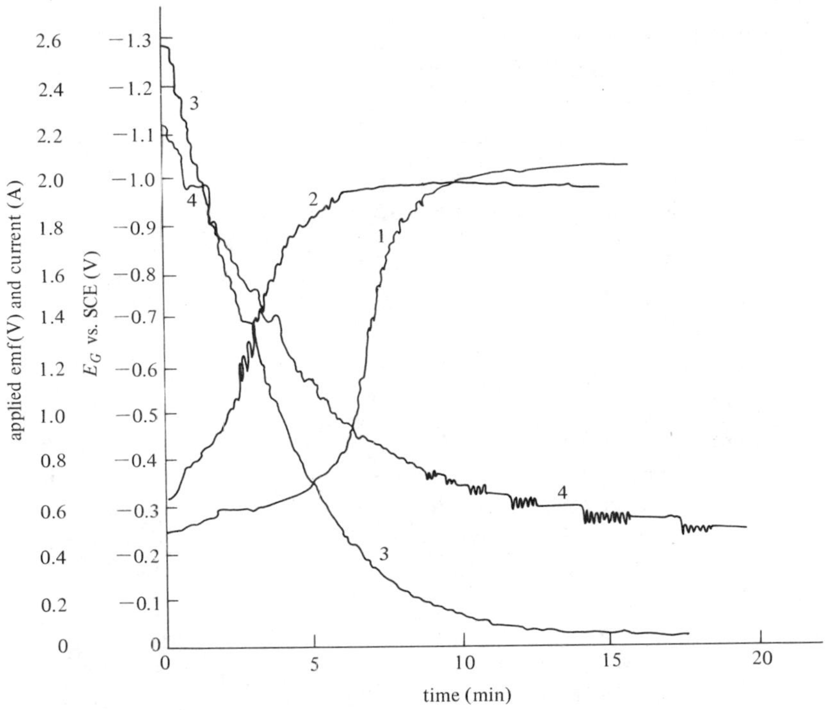

Figure 23.23 / *Electrodeposition of* 0.2 g *of copper onto a platinum cathode* (*area* 160 cm^2) *from* 200 cc *of* 0.5 *M tartrate solution of pH* = 4.5 *in the presence of hydrazine.* (1) *Change of cathode potential with time with current constant at* 1.00 A. (2) *Change of cathode potential when total voltage applied to the cell was held constant at* 2.00 V. (3) *Current decay when cathode potential was kept constant at* −0.36 V *vs. saturated calomel reference electrode.* (4) *Change of total applied voltage required to maintain cathode potential constant at* −0.36 V. *The curves were recorded automatically.*
SOURCE: J. J. Lingane, *Electroanalytical Chemistry*, 2nd ed., Wiley-Interscience (New York, 1958).

result in changes of both the electrode potential and the current during the experiment.

At the outset, it is necessary to realize that only one of the three variables can be controlled. The current that flows in a cell is the result, not the cause, of the electrode reactions, and the magnitude of the current is a manifestation of the rate of the electrode reactions. The electrode potential governs which electrode reactions will occur and, frequently, the rates at which they occur. Note that the iR drop in the cell makes it impossible for the applied voltage to equal the electrode potential.

Figure 23.23 compares data for the three possible experimental situations for the same electrode process, the reduction of cupric ion from 0.5 *M* tartrate at pH = 4.5. The constant-current case, curve 1, is similar to the potential–time relation obtained in chronopotentiometry. When the supply of Cu^{2+} at the electrode surface becomes insufficient to carry the current, the electrode potential drifts cathodically until another reaction begins and again stabilizes the potential. In this instance, the second reduction is that of H^+. This observation is simply understood by writing a Nernst equation for the cathode potential such as Eq. (23.100).

$$E_C = E^0 - \frac{0.059}{n} \log\left(\frac{C_{\mathrm{Red}}}{C_{\mathrm{Ox}}}\right) \tag{23.100}$$

As the concentration of oxidized form decreases at the electrode surface, the cathode potential must become more reducing and will do so without limit until another electrochemical reaction commences to limit the drift. Curve 2 shows the potential–time behavior for the constant applied voltage experiment, and a similar shift of potential is observed. The effect of reactant concentration operates here, but so does an effect of the iR drop of the cell. The applied voltage divides among three terms, $E_A = E_{\mathrm{anode}} - E_{\mathrm{cathode}} + iR$. Since the potential of the auxiliary anode is frequently nearly constant, any decrease in current, and consequent decrease in the voltage dissipated as iR, will also cause cathodic drift of the working electrode potential. Even for low-resistance cells, this shift can be appreciable. If an initial current of 1 A were passed through a cell of 4 Ω resistance, the cathode potential could drift 2 V in the reducing direction if the current decreased to 0.5 A during the experiment and the anode potential remained constant.

Curves 3 and 4 pertain to the case in which the cathode potential is controlled. Curve 3 shows the decrease in current and suggests an exponential time decay. Curve 4 shows the adjustment required in the applied voltage to maintain the constant working cathode potential, compensating for the decrease in iR.

23.8.2 / Comparison of the methods

The primary advantage of the controlled-potential method is *selectivity*. With several reducible species present, it is frequently possible to analyze these *seriatim* by sufficiently precise potential control. This selectivity is lost in the constant-current method, although this may not be a serious loss when a single species is to be determined. The constant applied voltage method is little used, since in order to obtain reasonable selectivity, the initial current and applied voltage must be rather small, with the result that the electrolysis time is prolonged unduly.

The constant-current method provides simplicity in calculation of the results. Except in the case of analysis by electrodeposition, in which the electrode is weighed both before and after electrolysis, the analytical datum from an electrolysis experiment is the number of coulombs that have passed. This is simply the constant current multiplied by the time of the electrolysis, $Q = it$. By contrast, when the potential is controlled and the current is not constant, the number of coulombs must be calculated from the relation $Q = \int_0^t i(t)\,dt$ either by graphical or instrumental integration. The simplicity of the constant-current method has provided the incentive to devise procedures that circumvent its inherent difficulties, and a large number of constant-current coulometric methods are known. The problem is essentially to discover a chemical means of preventing the drift of the electrode potential from causing unwanted reactions. If the number of coulombs is to be meaningful, the reaction of analytical interest must be the only reaction that consumes electricity; that is, it must proceed with 100 percent current efficiency. This condition can be brought about by the use of an added chemical substance, which can be viewed as a "redox buffer" or electrogenerated titrant. Consider, for example, the oxidation of iron(II) to iron(III). Under constant-current conditions, the anodic drift of the working electrode will eventually cause the oxidation of water, $2H_2O = O_2 + 4H^+ + 4e^-$, to compete with the analytical reaction and the current efficiency will drop below 100 percent. If, however, a relatively large excess of Ce(III) is added to the solution (the concentration need not be known), the anodic drift of potential will be limited by the oxidation $\text{Ce(III)} = \text{Ce(IV)} + e^-$, rather than by the oxidation of water. Ce(IV) is well known to react with Fe(II) rapidly and quantitatively, $\text{Ce(IV)} + \text{Fe(II)} = \text{Ce(III)} + \text{Fe(III)}$. Thus, despite the fact that two electrode reactions occur, the net result is the same as if the oxidation of iron had proceeded with 100 percent efficiency. The rather obvious relation between this method and the titration of iron(II) with Ce(IV) has caused such procedures to be called coulometric titrations. As in any titration, a method of endpoint detection is required, and in the present example either a separate circuit that measures the electrode potential, or a visual indicator, could be used exactly as they are employed in a buret titration. This method enjoys several distinct advantages over buret titrations. No standard solutions are required; unstable reagents may be generated *in situ*; it is much better suited to very small samples. The last advantage is a reflection of the fact that small currents and times are easier to measure accurately than are small weights and/or volumes. Thus, it is entirely feasible to employ a current of 10^{-7} A for 10 sec with the precision limited by that of an inexpensive stopclock, ± 0.1 sec, that is, 1 percent. Such an analysis involves 10^{-6} C or about 10^{-11} mole of material for a one-electron process. If the molecular weight of the material is 100, the analysis corresponds to 1 ng with 1 percent precision. Equivalent results with a buret would require dispensing 10^{-4} ml of 10^{-4} M reagent with a precision of $\pm 10^{-6}$ ml, a procedure that, while possible, is far from easy or routine.

23.8.3 / Instrumentation

For both constant-current and controlled-potential electrolytic experiments, it is normally necessary to separate physically the working electrode from the auxiliary electrode to prevent the product of the electrode reaction from

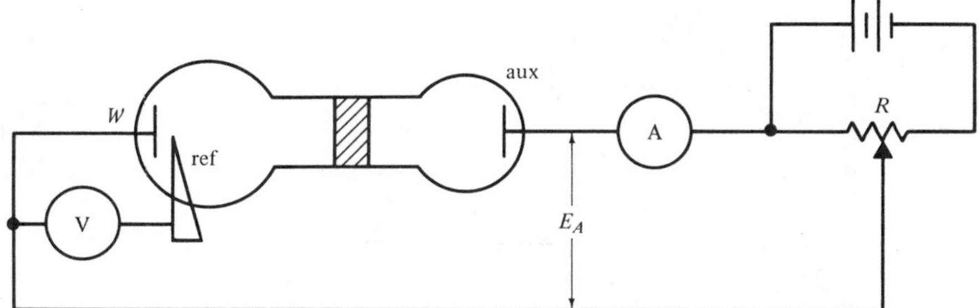

Figure 23.24 / *Essential equipment for macroscopic controlled-potential electrolysis experiments.*

undergoing the reverse reaction at the other electrode. This is conveniently accomplished through the use of sintered glass discs that permit electrical conduction but prevent gross mixing of anolyte and catholyte. Such a cell is shown in Fig. 23.24, along with a circuit for manual control of the working electrode potential. In use, the operator varies the applied voltage, E_A, by periodically adjusting the resistor, R, in order to maintain the desired electrode potential observed with the high-impedance meter V. The current, as measured by A, can be noted at various times. This design can be automated in a number of ways, all of which operate on similar principles. The potential difference between the working and reference electrodes is compared electronically with a preset voltage, and the error signal is amplified and used to adjust E_A. Such a device is termed a *potentiostat* and frequently includes an automatic integrator to convert the current–time curve directly to coulombs. The latter is accomplished by passing the electrolysis current through a precision resistor. The resulting voltage is presented to an operational amplifier wired as an integrator (Section 8.3.4); the output voltage of the amplifier is then proportional to the integral of the current–time curve.

23.8.4 / The current–time curve for controlled-potential experiments

We shall now describe the current–time curve for controlled-potential electrolysis with the assumption that the electrode potential is set at a value that corresponds to the diffusion plateau of the voltammetric wave, that is, that the current is limited by diffusion and convection. A significant difference between this experiment and voltammetry is that the solution will be efficiently stirred. Under these conditions, the bulk of the solution will be of uniform concentration except for a thin layer near the electrode which, because of viscous drag, remains essentially undisturbed. As indicated in Fig. 23.25, the assumption of diffusion-limited conditions permits setting the electrode surface concentration of reactant equal to zero, while at the distance δ from the electrode which corresponds to the thickness of the diffusion layer, efficient stirring requires that the concentration gradient

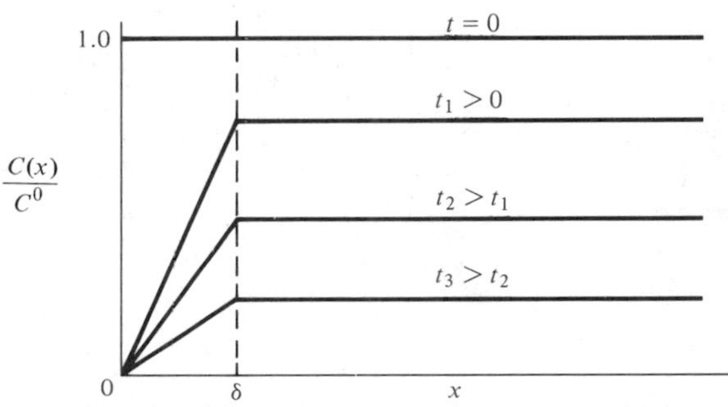

Figure 23.25 / *Concentration profiles at the electrode surface in a controlled-potential electrolysis at increasing times after initiation of the electrolysis.*

must vanish. Therefore, the entire gradient can be assumed to be within the diffusion layer and approximated as a straight line as shown in the figure. Note that it is not necessary to specify detailed properties of the diffusion layer; in fact, considerable disagreement about the nature of this layer exists in the literature. The ensuing treatment depends only on the fact that the concentration gradient at the electrode surface is linearly proportional to the bulk solution concentration, and the results based on this assumption have been amply verified by experiment.

We begin with Fick's first law, Eq. (23.61). Since the current is proportional to the flux at the electrode surface, that is, $i = n\mathscr{F}\, dN(0, t)/dt$, the current can be expressed by

$$i(t) = n\mathscr{F}AD \frac{\partial C(0, t)}{\partial x} \tag{23.101}$$

We shall assume, as indicated in Fig. 23.25, that the concentration gradient is constant within the diffusion layer and can be written as the difference between the concentration at δ and that at the electrode surface divided by the distance, that is, $\partial C(0, t)/\partial x = [C(\delta, t) - 0]/\delta$. Since we assume that the concentration is uniform at distances $\geq \delta$, this will be designated simply $C(t)$ and $\partial C(0, t)/\partial x = C(t)/\delta$. The current is therefore given by

$$i(t) = \frac{n\mathscr{F}ADC(t)}{\delta} \tag{23.102}$$

In addition, we know that the current is proportional to the particle flux. Since concentration times volume is the number of moles of electroactive material present, $dN(0, t)/dt = d[C(t)V]/dt$, where V is the total volume of solution. Thus, the current is given by

$$i(t) = -n\mathscr{F}V \frac{dC(t)}{dt} \tag{23.103}$$

The negative sign must be introduced because $dC(t)/dt$ is inherently negative, since the concentration of the reactant diminishes as the electrolysis proceeds. The right-hand members of Eqs. (23.102) and (23.103) are equated and after rearrangement and integration between the limits C^0 and $C(t)$ and 0 and t, respectively, we obtain Eq. (23.104a). Since it is proportional to $C(t)$ the current follows the same functional dependence as indicated in Eq. (23.104b).

$$C(t) = C^0 \exp\left(\frac{-DAt}{V\delta}\right) \tag{23.104a}$$

$$i(t) = i(0) \exp\left(\frac{-DAt}{V\delta}\right) \tag{23.104b}$$

We therefore expect that a plot of log i vs. t will be a straight line with a negative slope, a prediction verified by experiment. The slope, and thus the rate of electrolysis, will depend on the diffusion coefficient and the ratio $A/V\delta$. The greatest efficiency of electrolysis will be obtained when the ratio of electrode area to solution volume is maximized and the diffusion-layer thickness is minimized. The latter is accomplished by maximizing the efficiency of stirring. When the working electrode is a mercury pool, the goal of maximizing A/V suggests that the simple conical flask is a reasonably good cell design.

Since both current and concentration decay exponentially, neither can ever be made exactly zero. Prolonging the electrolysis until $i(t)$ reaches $10^{-3}i(0)$ is, however, indicative of "complete" electrolysis at the part-per-thousand level. Note that for a given electrode potential, the theoretical extent of electrolysis, that is, the equilibrium ratio of product and reactant concentrations, is easily calculated from Eq. (23.100). Use of such a Nernst equation for each of several electroactive species in the same solution permits estimation of both the feasible degree of separation and also selection of optimum control potentials for analyses of the several substances.

As with other electrochemical techniques, in addition to the straightforward analytical applications, controlled-potential electrolysis finds increasing use in the study of electrode reactions. Appreciable deviations from the straight-line behavior of the log i vs. time plot are indicative of chemical complications to the electrode process. The essentially infinite variety of oxidizing and reducing power available with electrolytic techniques makes possible a degree of selectivity in redox reactions that is not possible with chemical oxidizing or reducing agents. In the past, this advantage of controlled-potential electrolysis has not been exploited, and only recently has it begun to cause acceptance of controlled-potential methods for synthesis on both the laboratory and industrial scales.

23.9 / ELECTRODE KINETICS

23.9.1 / Types of irreversibility

In Section 23.2.3, we distinguished between chemical reversibility and thermodynamic reversibility. In the discussions of electrochemical techniques, we have frequently referred to reversible and irreversible processes and tests to distinguish one from the other. We must understand these terms as used by electrochemists before discussing the special type of irreversibility that is the subject of this section.

So long as the concentrations of reactant and product at the electrode surface maintain rather closely the values required for equilibrium with the electrode potential as computed using a Nernst equation, the electrode reaction will appear to be reversible using any of the tests previously described. Any process that causes these concentrations to differ significantly from their equilibrium values introduces irreversibility. Such processes include the chemical decomposition of the product of the electrode reaction and a chemical reaction between the product and the reactant. As indicated in the discussion of cyclic voltammetry, the effects of such chemical reactions depend on their rates. When these complicating chemical processes are slow, the electrode reaction will appear reversible even when actually it is chemically irreversible. The time scale, or definition of "fast" and "slow" in an electrochemical experiment, is relative and set ultimately by comparison with the rate of mass transport by diffusion in solution.

More subtle, perhaps, is irreversibility that occurs when the electron-transfer reaction, rather than mass transport, is the rate-limiting step in an electrode reaction. In this case, the electrode surface concentrations do not reach their equilibrium values. It is the kinetically limited electrode reaction and its result, *overpotential*, that will be discussed here. At the outset, we must distinguish overpotential from a related phenomenon, *concentration polarization*. The latter does not depend on electrode kinetics and can occur with electrode reactions that are reversible in every sense. When current flows at an electrode, the consumption of reactant and liberation of product may occur more rapidly than diffusion can equilibrate these concentrations with the bulk of the solution. When this happens (which is more likely the larger the current), the concentrations at the electrode surface are not the same as those in the bulk solution. Although the electrode potential is in equilibrium with the surface concentrations, its value is different from that anticipated from the bulk concentrations. This difference is the manifestation of concentration polarization. The reader will see that a number of the techniques we have discussed depend on this phenomenon for their function; it is, however, anathema for potentiometry and must be avoided if accurate results are to be obtained by that method.

23.9.2 / The electrochemical rate equation

We may write the generalized electrode reaction in the form of Eq. (23.105), where we include explicitly the idea that equilibria are dynamic.

$$\mathrm{Ox} + ne \underset{k_b}{\overset{k_f}{\rightleftharpoons}} \mathrm{Red} \qquad (23.105)$$

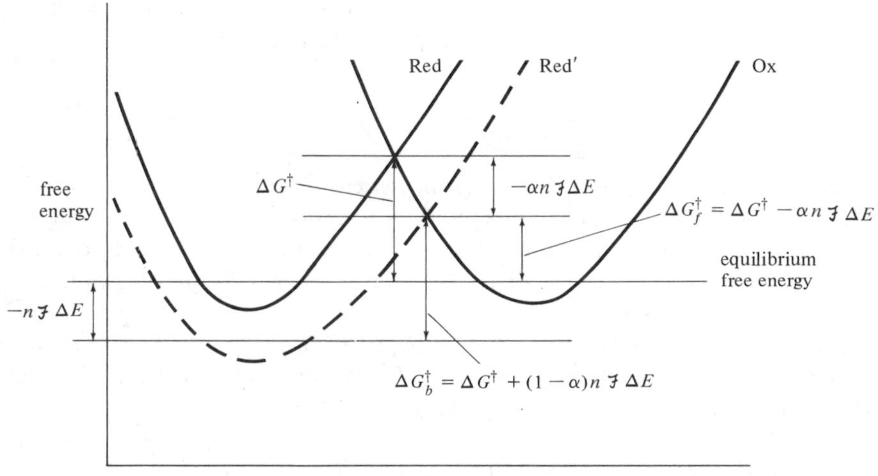

Figure 23.26 / *Potential energy diagram showing the energy of activation of an electrode reaction.*

Thus, although no *net* change occurs after equilibrium is attained, both forward and back reactions continue to occur with equal rates. The forward rate corresponds to the rate of disappearance of Ox, which must equal the rate of appearance of Red. Since Eq. (23.105) is a first-order reaction, we write the formal rate law, Eq. (23.106).

$$\frac{-dN_{\text{Ox}}}{dt} = \frac{dN_{\text{Red}}}{dt} = k_f C_{\text{Ox}} - k_b C_{\text{Red}} \tag{23.106}$$

Note that since the reaction occurs at an electrode surface, that is, is heterogeneous, the proper units for the rate constants are centimeters-second^{-1}, *not* simply second^{-1}. Just like any chemical reaction, an electrode reaction involves an energy of activation. The electrode reaction differs in that the activation energy can be modified by changing the electrode potential. Consider the reaction (23.105) at equilibrium with an electrode of potential E^0. Since the forward and reverse rates are equal, so must the activation energies be equal. This condition is depicted by the potential energy diagram in Fig. 23.26. Only those molecules that have free energy equal to or greater than $\Delta G^{\dagger}$ can cross the activation barrier from one potential energy surface to the other as shown by the solid curves. Now let the electrode potential be shifted in the reducing sense by an amount $\Delta E = (E - E^0)$. This will lower the potential energy of the reduced form, as indicated by the dashed curve denoted Red′. The curve for the oxidized form is unchanged, but lowering the curve for the reduced form shifts the point at which the two curves cross, thereby altering the activation energies. The forward reaction will be facilitated and the back reaction hindered as shown in the figure. The new activation energies at the new potential, E, are given by

$$\begin{aligned} \Delta G_f^{\dagger} &= \Delta G^{\dagger} - \alpha n\mathscr{F}(E - E^0) \\ \Delta G_b^{\dagger} &= \Delta G^{\dagger} + (1 - \alpha)n\mathscr{F}(E - E^0) \end{aligned} \tag{23.107}$$

where the quantity α is called the *transfer coefficient* and depends on the symmetry of the potential energy curves. The transfer coefficient determines what fraction of the change in electrode potential facilitates the forward reaction and what fraction hinders the back reaction. Ordinarily, values of α near 0.5 are observed.

Since the rate of a chemical reaction depends on the *fraction* of molecules that have at least the activation energy, a Boltzmann factor is expected in the rate expressions, which may be formulated as

$$\begin{aligned} k_f &= k_f^{\,0} \exp\left[\frac{-\alpha n\mathscr{F}(E - E^0)}{RT}\right] \\ k_b &= k_b^{\,0} \exp\left[\frac{(1 - \alpha)n\mathscr{F}(E - E^0)}{RT}\right] \end{aligned} \tag{23.108}$$

In these equations, k^0 is the rate constant at the equilibrium potential and includes the equilibrium activation energy, $\Delta G^{\dagger}$. The other symbols have their usual meanings. Combination of Eq. (23.106) with Eq. (23.108) and the required constants permits writing Eqs. (23.109) for the forward and reverse currents.

$$\begin{aligned} i_{\text{cathodic}} &= n\mathscr{F}AC_{\text{Ox}}k_f^{\,0} \exp\left[\frac{-\alpha n\mathscr{F}(E - E^0)}{RT}\right] \\ i_{\text{anodic}} &= -n\mathscr{F}AC_{\text{Red}}k_b^{\,0} \exp\left[\frac{(1 - \alpha)n\mathscr{F}(E - E^0)}{RT}\right] \end{aligned} \tag{23.109}$$

As usual, the anodic current is defined to be negative.

23.9.3 / Overpotential

Equations (23.109) can be combined to obtain an expression for the net current as a function of the displacement of the electrode potential from its equilibrium value, $E - E^0$. The net current is given by

$$\begin{aligned} i = i_{\text{cathodic}} + i_{\text{anodic}} &= n\mathscr{F}A\left\{C_{\text{Ox}}k_f^{\,0} \exp\left[\frac{-\alpha n\mathscr{F}(E - E^0)}{RT}\right]\right. \\ &\quad \left. - C_{\text{Red}}k_b^{\,0} \exp\left[\frac{(1 - \alpha)n\mathscr{F}(E - E^0)}{RT}\right]\right\} \end{aligned} \tag{23.110}$$

Consider a special case of Eq. (23.110), that of equilibrium at the standard potential. The net current is now zero and $E = E^0$. Under those conditions, the exponential factors become unity and $C_{\text{Ox}}k_f^{\,0} = C_{\text{Red}}k_b^{\,0}$. Since the concentrations must be equal at the standard potential, the two rate constants must also be the same. Hence, the process can be described by a single rate constant, usually symbolized k_s, and Eq. (23.110) can be rewritten as

$$i = n\mathscr{F}Ak_s\left\{C_{\text{Ox}} \exp\left[\frac{-\alpha n\mathscr{F}(E - E^0)}{RT}\right] - C_{\text{Red}} \exp\left[\frac{(1 - \alpha)n\mathscr{F}(E - E^0)}{RT}\right]\right\} \quad (23.111)$$

Equation (23.111) is a perfectly general equation for the current for any values of concentrations and potential. Now suppose that concentrations C_{Ox} and C_{Red} other than the standard state concentrations are present. At equilibrium ($i = 0$), the potential must be the equilibrium potential, E_{eq}, as given by the Nernst equation. Substitution of $i = 0$ and $E = E_{\text{eq}}$ in Eq. (23.111) produces, on rearrangement,

$$E_{\text{eq}} = E^0 - \frac{RT}{n\mathscr{F}} \ln\left(\frac{C_{\text{Red}}}{C_{\text{Ox}}}\right) \quad (23.112)$$

which is the Nernst equation as anticipated.

As stated above, at the equilibrium potential the current given by Eq. (23.111) is zero by definition, but as soon as the potential differs from the equilibrium potential, a current will be observed. The magnitude of the current that corresponds to a departure η from the equilibrium potential depends on the value of the rate constant k_s. The quantity η is termed the *overpotential*, since it is the amount by which the electrode potential must exceed its equilibrium value to cause a given current to flow. To obtain a given current density, i/A, a greater displacement from the equilibrium potential will be required the smaller the value of k_s. Since k_s is characteristic of the electrode process and independent of potential, we can categorize electrode reactions with k_s values large with respect to diffusion rates as "fast" and those with small k_s values as "slow." For a particular electrode reaction (k_s fixed), larger displacements from the equilibrium potential are required to obtain larger currents.

In Table 23.3, some values of overpotential are given for selected values of k_s under the conditions $C_{\text{Ox}} = C_{\text{Red}} = 10^{-3}$ F, $n = 1$, $\alpha = 0.5$, $T =$

TABLE 23.3 / *Overpotential values*

k_s (cm-sec^{-1})	η(V)
10^{-3}	± 0.0002
10^{-5}	0.024
10^{-8}	0.36
10^{-10}	0.59
10^{-14}	1.06

298°K, and $i/A = 1 \times 10^{-6}$/A-cm^{-2}. Since η always displaces the potential from zero, the + sign applies to anodic processes and the − sign to cathodic processes. Note that while the first entry in the table corresponds to an essentially reversible reaction at the current density 10^{-6} A-cm^{-2}, keeping the other conditions constant, η increases to 0.12 V when the current density is increased to 10^{-3} A-cm^{-2}. As pointed out previously, it is the large overpotential for the evolution of hydrogen on mercury (1.2 V at 10^{-2} A-cm^{-2}) that makes mercury so useful as a cathode.

The distinction between reversible and irreversible processes on the basis of k_s values is, of course, artificial. No electrode reaction is infinitely rapid and, as we have seen, a process that appears reversible under some conditions of electrolysis may become irreversible when the flow of current is sufficiently increased. As an operational definition, a reaction that proceeds without a measurable overpotential is reversible; conversely, an electrode reaction involving a measurable overpotential is irreversible.

It is important to realize that the difference between reversible and irreversible electrochemical reactions is a quantitative one dependent on relative rates, rather than a qualitative distinction, regardless of whether chemical or kinetic irreversibility is considered. Despite these limitations, the terms "reversible" and "irreversible" are useful in describing electrode reactions and are in wide use among electrochemists. As Delahay has pointed out, the use of these terms "is harmless provided that the significance attached to them is well understood."

SUGGESTED READINGS

General reading on electrochemical methods in order of increasing difficulty

L. Meites and H. C. Thomas, *Advanced Analytical Chemistry*, McGraw-Hill (New York, 1958), Chapters 2–7.

P. Delahay, *Instrumental Analysis*, Macmillan (New York, 1957), Chapters 2–7.

J. J. Lingane, *Electroanalytical Chemistry*, 2nd ed., Wiley-Interscience (New York, 1958). (Detailed discussions of instrumentation and applications; some of the instrumentation now verges on obsolescence.)

P. Delahay, *New Instrumental Methods in Electrochemistry*, Wiley-Interscience (New York, 1954). (Heavily theoretical.)

Thermodynamics and electrochemistry

G. N. Lewis and M. Randall, *Thermodynamics*, 2nd ed., revised by K. S. Pitzer and L. Brewer, McGraw-Hill (New York, 1961), Chapters 22–24.

pH measurement

Malcolm Dole, *The Glass Electrode*, Wiley (New York, 1941). (Historical interest.)

R. G. Bates, *Determination of pH*, Wiley (New York, 1964). (Definitive work on the subject.)

Ion-selective electrodes

G. A. Rechnitz, "Chemical Studies at Ion-Selective Membrane Electrodes," *Accts. Chem. Res.* **3**, 69 (1970).

R. A. Durst, Ed., *Ion-Selective Electrodes*, National Bureau of Standards Special Publication 314 (Washington, D.C., 1969).

Polarography

I. M. Kolthoff and J. J. Lingane, *Polarography*, 2nd ed., Wiley-Interscience (New York, 1952). (Volume I treats the theory; volume II provides a wealth of experimental data.)
Charles Perrin, "Mechanisms of Organic Polarography," *Progr. Phys. Org. Chem.* **3,** 165 (1965).

Solid electrode voltammetry

R. N. Adams, *Electrochemistry at Solid Electrodes*, Marcel Dekker (New York, 1969). (A very competent overview of the subject.)
R. S. Nicholson and I. Shain, "The Theory of Stationary Electrode Voltammetry," *Anal. Chem.* **36,** 706 (1964). (The definitive work on the theory; the authors include interpretation and experimental tests for interpreting results.)

Reference electrodes

D. J. G. Ives and G. I. Janz, Eds., *Reference Electrodes*, Academic (New York, 1961). (Useful information on theory and practice of preparation of nonpolarizable reference electrodes.)

PROBLEMS

1 / Compute the standard potential for the half-reaction

$$ClO_4^- + 8H^+ + 7e^- = \tfrac{1}{2}Cl_2 + 4H_2O$$

given the following data:

$$ClO_4^- + 2H^+ + 2e^- = ClO_3^- + H_2O \qquad E^0 = +1.19\text{ V}$$
$$ClO_3^- + 6H^+ + 5e^- = \tfrac{1}{2}Cl_2 + 3H_2O \qquad E^0 = +1.49\text{ V}$$

Ans. +1.40 V

2 / Compute E^0 for the half-reaction

$$IO_3^- + 6H^+ + 2Cl^- + 4e^- = ICl_2^- + 3H_2O$$

given

$$IO_3^- + 6H^+ + 5e^- = \tfrac{1}{2}I_2 + 3H_2O \qquad E^0 = +1.195\text{ V}$$
$$ICl_2^- + e^- = \tfrac{1}{2}I_2 + 2Cl^- \qquad E^0 = +1.06\text{ V}$$

Ans. +1.23 V

3 / Compute the dissociation constant for the $SnCl_4^{2-}$ ion, that is, for the reaction

$$SnCl_4^{2-} = Sn^{2+} + 4Cl^-$$

given

$$Sn^{2+} + 2e^- = Sn \qquad E^0 = -0.14\text{ V}$$
$$SnCl_4^{2-} + 2e^- = Sn + 4Cl^- \qquad E^0 = -0.19\text{ V}$$

4 / What is the solubility product of cuprous iodide, given

$$Cu^{2+} + I^- + e^- = CuI \qquad E^0 = +0.86\text{ V}$$
$$Cu^{2+} + e^- = Cu^+ \qquad E^0 = +0.153\text{ V}$$

Ans. 1.0×10^{-12}

5 / What is the partial pressure of chlorine in equilibrium with a solution in which the concentration of free chlorine is 0.5 *F* and the concentration of chloride ion is 1 *F*, given

$$Cl_2(g) + 2e^- = 2Cl^- \qquad E^0 = +1.3595 \text{ V}$$

$$Cl_2(aq) + 2e^- = 2Cl^- \qquad E^0 = +1.3905 \text{ V}$$

Ans. 5.6 atm assuming ideal gas behavior

6 / Compute the standard potential at 25°C for the half-reaction

$$Sb + 3H_2O + 3e^- = SbH_3 + 3OH^-$$

given that

$$SbH_3 = Sb + 3H^+ + 3e^- \qquad E^0 = +0.570 \text{ V}$$

7 / The cell

$$Pb|Pb(NO_3)_2(0.1\ F)|\ |K_2SO_4(0.2\ F),\ PbSO_4(s)|Pb$$

has an emf of -0.179 V at 25°C.

(*a*) Which electrode is negative?

(*b*) What is the value of the solubility product of lead sulfate estimated from this cell?

8 / The following cell is assembled:

$$Pb,\ PbSO_4(s)|H_2SO_4(1\ F)|PbO_2(s),\ Pb$$

What is the net reaction that occurs on discharge? What is the emf of the cell? Which electrode is the anode and which electrode is positively charged when the cell discharges?

$$PbSO_4(s) + 2e^- = Pb + SO_4^{2-} \qquad E^0 = -0.356 \text{ V}$$

$$PbO_2 + 4H^+ + 2e^- = Pb^{2+} + 2H_2O \qquad E^0 = +1.455 \text{ V}$$

9 / The emf of the following galvanic cell at 25°C was 0.100 V.

$$-\ Hg|Hg_2Cl_2,\ Cl^-|\ |M^{n+}\ |M\ +$$

When the M^{n+} solution was diluted 50-fold, the emf of the cell was 0.050 V. What is the value of n?

10 / Calculate the emf of the following galvanic cell. What is the polarity of the silver electrode?

$$Ag|AgCl(s),\ NaCl(10^{-2}F)|\ |Cu(ClO_4)_2 = 10^{-1}\ F|Cu$$

The entire cell is maintained at 50°C. Standard electrode potentials at 50°C:

	E^0 vs. NHE
$Cu^{2+} + 2e^- = Cu$	+0.335 V
$Ag^+ + e^- = Ag$	+0.799 V

Assume that the solubility product of AgCl is 10^{-10}.

$$\frac{2.30\ R}{\mathcal{F}} = 1.98 \times 10^{-4} \text{ V/°C}$$

11 / A standard method of determining iron is addition of excess KI to a solution of the unknown in the ferric state. I^- is oxidized to I_2 and Fe^{3+} is reduced to Fe^{2+}; the iodine liberated is then titrated with standard thiosulfate.

(*a*) What is the potential (vs. NHE) of a platinum indicator electrode immersed in a solution composited from 0.1 mole KI, 0.05 mole $Fe_2(SO_4)_3$ and distilled water to make 1.00 liter. Assume that all the I_2 produced remains in solution as $I_2(aq)$.

(*b*) What is the standard potential for the reaction

$$I_2 + 2FeY^{2-} = 2I^- + 2FeY^-?$$

(Y^{-4} is the fully ionized species of EDTA.)

$$Fe^{3+} + e^- = Fe^{2+} \qquad E^0 = +0.72\text{ V}$$

$$I_2(aq) + 2e^- = 2I^- \qquad E^0 = +0.62\text{ V}$$

$$Fe^{3+} + Y^{-4} = FeY^- \qquad K = 1.0 \times 10^{25}$$

$$Fe^{2+} + Y^{-4} = FeY^{2-} \qquad K = 2.0 \times 10^{14}$$

Ans. (*a*) +0.71 V

12 / A titration is carried out in which 50 ml of 0.1 *N* Sn^{2+} is oxidized with 0.1 *N* I_3^-. Calculate the potential (vs. NHE) of a platinum indicator electrode observed after the following volumes of titrant have been added: 49, 50, 51 ml. The following data are given:

$$Sn^{4+} + 2e^- = Sn^{2+} \qquad E^0 = +0.15\text{ V}$$

$$I_3^- + 2e^- = 3I^- \qquad E^0 = +0.536\text{ V}$$

Ans. +0.20 V, +0.369 V, +0.538 V

13 / Derive an equation of the simplest possible form for the potential of a platinum indicator electrode (vs. NHE) at the equivalence point in a titration of iron(II) with $Cr_2O_7^{2-}$ in acid solution. Recall that the reduction of dichromate produces Cr^{3+}.

14 / What will be the diffusion current (average current, in microamperes) when a 4-m*F* solution of a substance that has a diffusion coefficient of 8×10^{-6} $cm^2\text{-}sec^{-1}$ undergoes a two-electron reduction at a dme with the following characteristics: drop time = 4.0 sec, mercury flow rate = 1.5 $mg\text{-}sec^{-1}$?

Ans. 22.8 μA

15 / A substance undergoes a reversible two-electron reduction at a dme and gives a diffusion current = 6.00 μA. When the potential of the dme was −0.612 V, the current was 1.50 μA. What is the half-wave potential?

Ans. −0.626 V

16 / The following measurements were made on a *reversible* polarographic reduction wave at 25°C:

E (V *vs.* SCE)	*Maximum current* (μA)
−0.395	0.56
−0.406	1.13
−0.415	1.70
−0.422	2.26
−0.431	2.83
−0.445	3.40

The maximum diffusion current was 3.78 μA. Calculate the number of electrons involved in the reduction and the standard reduction potential (vs. NHE). Assume that the activity coefficients and diffusion coefficients of the reducible and reduced species are the same.

17 / An *average* diffusion current of 6.72 μA was obtained in the electrolysis of a given substance under the following conditions: $n = 1$, $m = 1.37$ mg-sec^{-1}, $\tau = 3.93$ sec, $C^0 = 1.20$ mF. Calculate the polarographic diffusion coefficient of this substance.

18 / According to Lingane and Loveridge (*J. Amer. Chem. Soc.* **72** 438, 1950), the equation

$$i_d = 706nm^{2/3}\tau^{1/6}D^{1/2}C\left(\frac{1 + 39D^{1/2}\tau^{1/6}}{m^{1/3}}\right)$$

is less approximate than the usual Ilkovic equation. Calculate the ratio of "Ilkovic current" to "Lingane current" for the reduction of a 1 mF solution of a cadmium ion (which has a diffusion coefficient of 1×10^{-5} cm^2-sec^{-1}) using a capillary that discharged 600 mg of mercury in 100 sec (20 drops).

19 / Offer an explanation for the observation that the half-wave potential for reduction of cadmium ion is -0.65 V vs. SCE when a platinum electrode is used, but when a dropping mercury electrode is used $E_{1/2} = -0.58$ V vs. SCE.

20 / A common error in polarographic analysis is the inadvertent omission of the supporting electrolyte. Compare the effects of this mistake for the following two cases:

(*a*) The solution to be analyzed is approximately 0.05 F aqueous cadmium nitrate.

(*b*) The solution is approximately 0.05 F nitrobenzene in acetonitrile.

21 / Consider the following polarographic data for reduction of a 0.88 mF solution of species A.

i(μA)	E, V *vs.* SCE
0.31	-0.419
0.62	-0.451
0.77	-0.462
1.24	-0.491
1.86	-0.515
2.48	-0.561
2.79	-0.593

The diffusion current is 3.10 μA. The mercury flow rate is 1.26 mg-sec^{-1} and the drop time is 3.53 sec.

(*a*) Estimate the quantity $708nD^{1/2}$.

(*b*) Evaluate $E_{1/2}$.

(*c*) Is this a "polarographically reversible" electrode process?

22 / The electrooxidation of M,

$$M \rightarrow M^+ + H^+ + 2e^-$$

is examined at 25°C in an aqueous solution buffered at a pH of 2.0. The half-wave potential is +0.100 V vs. SCE. Assuming that the electrochemical reaction is reversible, what is the half-wave potential when the pH is increased to 3.0?

23 / The zinc content of a certain solution was determined by controlled-potential electrolytic deposition. The cathode potential was controlled to such a value that you may assume that the only electrode reaction was reduction of zinc. The electrolysis current was passed through a 20.00-Ω precision resistor. The voltage developed across the resistor was initially 0.500 V. This voltage was observed to decrease *linearly with time*, reaching 0.00 V after 5.00 min. What weight of zinc did the solution contain?

Ans. 1.27 mg

24 / What cathode potential should be chosen if one desired to separate silver from a 0.005 F solution of cupric ion by controlled-potential electrolysis? If the initial silver concentration is 0.05 F, how long will the electrolysis take assuming $\delta = 2 \times 10^{-3}$ cm; $D = 7 \times 10^{-5}$ cm^2-sec^{-1}; $V = 200$ ml; and $A = 150$ cm^2?

Ans. 0.26 V vs. NHE; approximately 4 min to remove 99.9 percent of the silver.

25 / The half-wave potential for the polarographically reversible reduction of vanadic ion (V^{3+}) to vanadous ion (V^{2+}) in 1 F perchloric acid is −0.508 V vs. SCE. When 50.0 ml of a solution of vanadic ion in 1 F perchloric acid was electrolyzed with a mercury cathode maintained at a potential of −0.600 V vs. SCE, the current fell to zero after 250.0 C had been passed. Calculate the concentration of vanadic ion at the *beginning* and at the *end* of the experiment.

Ans. 5.33×10^{-2} F and 1.6×10^{-3} F

26 / (*a*) A mixture of Pb^{2+} and Ni^{2+} is to be analyzed by controlled-potential coulometry. What cathode potential (vs. NHE) should be selected to obtain the optimum separation (i.e., the maximum deposition of Pb that can be obtained without significant electrolysis of Ni^{2+}) if the solution contains initially $[Pb^{2+}] = 0.1$ F and $[Ni^{2+}] = 0.1$ F? Assume that the activity of deposited metal is unity and that there are no overpotentials.
(*b*) Calculate the percentage Pb^{2+} that will remain unelectrolyzed at the cathode potential selected in (*a*).
(*c*) If the sample consists of 30 ml of solution, how many coulombs will have passed at the completion of the electrolysis of Pb^{2+} at the potential selected in (*a*)?

$$Pb^{2+} + 2e^- = Pb \qquad E^0 = -0.126 \text{ V}$$

$$Ni^{2+} + 2e^- = Ni \qquad E^0 = -0.240 \text{ V}$$

27 / Show that for a controlled-potential electrolysis in which the current accurately obeys the relation $i(t) = i(0)e^{-kt}$ the number of coulombs obtained in the limit of 100 percent electrolysis is simply $Q = i(0)/k$.

28 / A controlled-potential reduction of cadmium ion is carried out under conditions such that the current obeys the relation $i(t) = i(0)e^{-kt}$. The cell constant, k, has been determined to be 3.63×10^{-3} sec^{-1}. If the initial current is 1.75 A, the volume of solution 50.0 ml, and the initial cadmium concentration is 0.050 F, how long will be required to remove the cadmium quantitatively at the part-per-thousand level?

Ans. 1880 sec

29 / Indicate the polarity of each of the liquid junctions given below:

$H^+ClO_4^-(0.001\ F)|H^+ClO_4^-(0.010\ F)$

$Na^+ClO_4^-(0.001\ F)|H^+ClO_4^-(0.001\ F)$

30 / A 0.5 m*F* solution of cadmium nitrate shows a chronopotentiometric transition time (τ) of 25 sec when the electrolysis current is 20 mA. What is the concentration of a silver nitrate solution for which a transition time of 49 sec is observed when the electrolysis current is 30 mA? The same electrode is used in both experiments, and you may assume equal diffusion coefficients for cadmium and silver ions.

Ans. 2.1 m*F*

31 / A cyclic voltammetry experiment is performed at a planar electrode. The process investigated is

$$\text{Ox} + 2e^- = \text{Red}$$

which is completely reversible. If $E_{1/2}$ was found to be -0.53 V vs. SCE, what will be the values of E_p and $E_{p/2}$? Sketch the shape of the wave for a full triangular cycle of ± 0.5 V about $E_{1/2}$. What is the ratio of i_p(anodic) to i_p(cathodic)? How would the voltammogram differ in shape if the reaction were instead

$$\text{Ox} + 2e^- = \text{Red} \xrightarrow{k_1} \text{electroinactive product}$$

32 / Consider diffusion to a planar electrode for which the solution is bounded at $x = 0.1$ cm. It is convenient to write the diffusion equation in terms of the function

$$U(x, t) = C^0 - C(x, t)$$

The electrode is maintained at a potential such that $C(0, t) = 0$. Obtain a solution for $\overline{U}(x, p)$. It will be necessary to recall that $\partial C(0.1, t)/\partial x \equiv 0$.

appendix A / Units, symbols, and physical constants

Units and symbols employed in the text have been selected on the basis of present usage in the field and are generally consistent with those recommended for use in American Chemical Society journals. An International System of Units (SI) was defined and given official status by the General Conference on Weights and Measures in 1960. The following tables summarize features of the SI units and report numerical values of selected physical constants in SI and other commonly used units.

SI BASE UNITS

Physical Quantity	Name	Symbol
length	meter or metre	m
mass	kilogram	kg
time	second	s
electric current	ampere	A
thermodynamic temperature	kelvin	K
luminous intensity	candela	cd
amount of substance	mole	mol

SPECIAL NAMES AND SYMBOLS FOR CERTAIN SI-DERIVED UNITS

Physical Quantity	Name	Symbol	Definition
force	newton	N	$kg\text{-}m\text{-}s^{-2}$
pressure	pascal	Pa	$kg\text{-}m^{-1}\text{-}s^{-2}$ (= $N\text{-}m^{-2}$)
energy	joule	J	$kg\text{-}m^{2}\text{-}s^{-2}$ (1 volt-coulomb = 1J)

Physical Quantity	Name	Symbol	Definition
power	watt	W	$kg\text{-}m^2\text{-}s^{-3}$ (= $J\text{-}s^{-1}$)
electric charge	coulomb	C	A-s
electrical potential difference	volt	V	$kg\text{-}m^2\text{-}s^{-3}\text{-}A^{-1}$
electrical resistance	ohm	Ω	$kg\text{-}m^2\text{-}s^{-3}\text{-}A^{-2}$
electrical conductance	siemens	S	$kg^{-1}\text{-}m^{-2}\text{-}s^3\text{-}A^2$ (= Ω^{-1})
electrical capacitance	farad	F	$A^2\text{-}s^4\text{-}kg^{-1}\text{-}m^{-2}$
magnetic flux	weber	Wb	$kg\text{-}m^2\text{-}s^{-2}\text{-}A^{-1}$
inductance	henry	H	$kg\text{-}m^2\text{-}s^{-2}\text{-}A^{-2}$
magnetic flux density	tesla	T	$kg\text{-}s^{-2}\text{-}A^{-1}$
luminous flux	lumen	lm	cd-sr
illumination	lux	lx	$cd\text{-}sr\text{-}m^{-2}$
frequency	hertz	Hz	s^{-1}

SUPPLEMENTARY UNITS

The SI units for plane and solid angle are the radian (rad) and steradian (sr), respectively.

PREFIXES FOR FRACTIONS AND MULTIPLES OF SI UNITS

Fraction	Prefix	Symbol	Multiple	Prefix	Symbol
10^{-1}	deci	d	10	deka	da
10^{-2}	centi	c	10^2	hecto	h
10^{-3}	milli	m	10^3	kilo	k
10^{-6}	micro	μ	10^6	mega	M
10^{-9}	nano	n	10^9	giga	G
10^{-12}	pico	p	10^{12}	tera	T
10^{-15}	femto	f			
10^{-18}	atto	a			

DECIMAL FRACTIONS AND MULTIPLES OF SI UNITS HAVING SPECIAL NAMES

Physical Quantity	Name	Symbol	Definition
length	angstrom	Å	10^{-10} m
force	dyne	dyn	10^{-5} N
pressure	bar	bar	10^5 $N\text{-}m^{-2}$
energy	erg	erg	10^{-7} J
kinematic viscosity	stokes	St	10^{-4} $m^2\text{-}s^{-1}$
viscosity	poise	P	10^{-1} $kg\text{-}m^{-1}\text{-}s^{-1}$
magnetic flux	maxwell	Mx	10^{-8} Wb
magnetic flux density	gauss	G	10^{-4} T

UNITS EVENTUALLY TO BE ABANDONED

Physical Quantity	Name	Symbol	Definition
pressure	atmosphere	atm	101,325 N-m^{-2}
	torr	torr	133.322 N-m^{-2}
	millimeter of mercury	mm Hg	133.322 N-m^{-2}
energy	British thermal unit	BTU	1055.056 J
	kilowatt-hour	kWh	3.6×10^6 J
	thermochemical calorie	cal_{th}	4.184 J

SELECTED PHYSICAL CONSTANTS

Quantity	Symbol	Value
electronic charge	e	1.60210×10^{-19} C, 4.803×10^{-10} cm$^{3/2}$-g$^{1/2}$-s^{-1}
electron mass	m_e	9.1091×10^{-28} g
proton mass	m_p	1.6725×10^{-24} g
speed of light in vacuum	c	2.9979×10^8 m-s^{-1}
Avogadro constant	N_A	6.02252×10^{23} mol^{-1}
Faraday constant	$\mathscr{F}$	9.64870×10^4 C-mol^{-1}, 2.8926×10^{14} cm$^{3/2}$-g$^{1/2}$-s^{-1}-mol^{-1}
Planck constant	h	6.6256×10^{-34} J-s, 6.6256×10^{-27} erg-s
	$\hbar$	1.05450×10^{-34} J-s, 1.05450×10^{-27} erg-s
gyromagnetic ratio of proton	γ	2.67519×10^8 rad-s^{-1}-T^{-1}, 2.67519×10^4 rad-s^{-1}-G^{-1}
	$\gamma/2\pi$	4.25770×10^7 Hz-T^{-1}, 4.25770×10^3 s^{-1}-G^{-1}
Bohr magneton	β, μ_B	9.2732×10^{-24} J-T^{-1}, 9.2732×10^{-21} erg-G^{-1}
nuclear magneton	μ_0, μ_N, β_N	5.0505×10^{-27} J-T^{-1}, 5.0505×10^{-24} erg-G^{-1}
gas constant	R	8.3143 J-K^{-1}-mol^{-1}, 8.3143×10^7 erg-K^{-1}-mol^{-1}
Boltzmann constant	k	1.3805×10^{-23} J-K^{-1}, 1.3805×10^{-16} erg-K^{-1}, 8.617×10^{-5} eV-K^{-1}
Wien displacement constant	b	2.8978×10^{-1} cm-K
Stefan-Boltzmann constant	σ	5.6697×10^{-8} W-m^{-2}-K^{-4}, 5.6697×10^{-5} erg-cm^{-2}-s^{-1}-K^{-4}
electron volt	eV	1.60210×10^{-19} J-eV^{-1}, 1.60210×10^{-12} erg-eV^{-1}

appendix B / Standard potentials of important half-reactions

Half-reactions	E^0 (V vs. NHE)
Aluminum	
$Al^{3+} + 3e^- = Al$	−1.66
$H_2AlO_3^- + H_2O + 3e^- = Al + 4OH^-$	−2.35
$AlF_6^{3-} + 3e^- = Al + 6F^-$	−2.07
Antimony	
$SbO^+ + 2H^+ + 3e^- = Sb + H_2O$	+0.212
$SbO_2^- + 2H_2O + 3e^- = Sb + 4OH^-$	−0.66
$Sb_2O_5 + 6H^+ + 4e^- = 2SbO^+ + 3H_2O$	+0.581
$SbO_3^- + H_2O + 2e^- = SbO_2^- + 2OH^-$(10 *M* NaOH)	−0.589
$Sb + 3H^+ + 3e^- = SbH_3$	−0.51
Arsenic	
$HAsO_2 + 3H^+ + 3e^- = As + 2H_2O$	+0.2475
$AsO_2^- + 2H_2O + 3e^- = As + 4OH^-$	+0.68
$H_3AsO_4 + 2H^+ + 2e^- = HAsO_2 + 2H_2O$	+0.559
$AsO_4^{3-} + 2H_2O + 2e^- = AsO_2^- + 4OH^-$	−0.67
$As + 3H^+ + 3e^- = AsH_3$	−0.60
$As + 3H_2O + 3e^- = AsH_3 + 3OH^-$	−1.43
Barium	
$Ba^{2+} + 2e^- = Ba$	−2.90
Beryllium	
$Be^{2+} + 2e^- = Be$	−1.85
$Be_2O_3^{2-} + 3H_2O + 4e^- = 2Be + 6OH^-$	−2.62

Bismuth

$BiO^+ + 2H^+ + 3e^- = Bi + H_2O$	+0.32
$BiCl_4^- + 3e^- = Bi + 4Cl^-$	+0.16
Bi_2O_4 ("sodium bismuthate") $+ 4H^+ + 2e^- = 2BiO^+ + 2H_2O$	+1.59
$BiO^+ + e^- = BiO$	+0.39
$BiO + 2H^+ + 2e^- = Bi + H_2O$	+0.285
$Bi + 3H^+ + 3e^- = BiH_3$	ca. −0.8

Boron

$H_3BO_3 + 3H^+ + 3e^- = B + 3H_2O$	−0.87
$H_2BO_3^- + H_2O + 3e^- = B + 4OH^-$	−1.79

Bromine

$Br_2(l) + 2e^- = 2Br^-$	+1.065
$Br_2(aq) + 2e^- = 2Br^-$	+1.087
$Br_3^- + 2e^- = 3Br^-$	+1.05
$BrO_3^- + 6H^+ + 5e^- = \frac{1}{2}Br_2 + 3H_2O$	+1.52
$HBrO + H^+ + e^- = \frac{1}{2}Br_2 + H_2O$	+1.59
$BrO^- + H_2O + 2e^- = Br^- + 2OH^-$	+0.76

Cadmium

$Cd^{2+} + 2e^- = Cd$	−0.403
$Cd(CN)_4^{2-} + 2e^- = Cd + 4CN^-$	−1.09
$Cd(NH_3)_4^{2+} + 2e^- = Cd + 4NH_3$	−0.61
$2Cd^{2+} + 2e^- = Cd_2^{2+}$	ca. −0.6
$Cd_2^{2+} + 2e^- = 2Cd$	ca. −0.2

Calcium

$Ca^{2+} + 2e^- = Ca$	−2.87

Carbon

$2CO_2 + 2H^+ + 2e^- = H_2C_2O_4$	−0.49
$\frac{1}{2}C_2N_2 + H^+ + e^- = HCN$	+0.37
$CNO^- + H_2O + 2e^- = CN^- + 2OH^-$	−0.97
$HCNO + H^+ + e^- = \frac{1}{2}C_2N_2 + H_2O$	+0.33

Cerium

$Ce^{4+} + e^- = Ce^{3+}$ (1 *M* $HClO_4$)	+1.70
(1 *M* HNO_3)	+1.61
(1 *M* H_2SO_4)	+1.44

Cesium

$Cs^+ + e^- = Cs$	−2.923

Chlorine

$Cl_2 + 2e^- = 2Cl^-$	+1.3595
$ClO_4^- + 2H^+ + 2e^- = ClO_3^- + H_2O$	+1.19
$ClO_3^- + 6H^+ + 5e^- = \frac{1}{2}Cl_2 + 3H_2O$	+1.47
$ClO_3^- + 2H^+ + e^- = ClO_2 + H_2O$	+1.15
$HClO_2 + 2H^+ + 2e^- = HClO + H_2O$	+1.64
$HClO + H^+ + e^- = \frac{1}{2}Cl_2 + H_2O$	+1.63
$HClO + H^+ + 2e^- = Cl^- + H_2O$	+1.49
$ClO^- + H_2O + 2e^- = Cl^- + 2OH^-$	+0.89

Chromium

$Cr_2O_7^{2-} + 14H^+ + 6e^- = 2Cr^{3+} + 7H_2O$	+1.33
$CrO_4^{2-} + 4H_2O + 3e^- = Cr(OH)_3 + 5OH^-$	−0.13
$Cr^{3+} + e^- = Cr^{2+}$	−0.41
$Cr^{3+} + 3e^- = Cr$	−0.74
$Cr^{2+} + 2e^- = Cr$	−0.91

Cobalt

$Co^{3+} + e^- = Co^{2+}$	+1.842
$Co(OH)_3 + e^- = Co(OH)_2 + OH^-$	+0.17
$Co(NH_3)_6^{3+} + e^- = Co(NH_3)_6^{2+}$	+0.1
$Co(CN)_6^{3-} + e^- = Co(CN)_6^{4-}$	−0.8
$Co^{2+} + 2e^- = Co$	−0.277

Copper

$Cu^{2+} + 2e^- = Cu$	+0.337
$Cu^{2+} + e^- = Cu^+$	+0.153
$Cu^{2+} + I^- + e^- = CuI$	+0.86
$Cu^{2+} + 2CN^- + e^- = Cu(CN)_2^-$	+1.12
$Cu(CN)_2^- + e^- = Cu + 2CN^-$	−0.43
$Cu(NH_3)_4^{2+} + e^- = Cu(NH_3)_2^+ + 2NH_3$	−0.01
$Cu(NH_3)_2^+ + e^- = Cu + 2NH_3$	−0.12
$CuCl_3^{2-} + e^- = Cu + 3Cl^-$ (1 *M* HCl)	+0.178

Fluorine

$F_2 + 2e^- = 2F^-$	+2.87

Germanium

$H_2GeO_3 + 4H^+ + 4e^- = Ge + 3H_2O$	−0.131
$HGeO_3^- + 2H_2O + 4e^- = Ge + 5OH^-$	−1.0
$H_2GeO_3 + 4H^+ + 2e^- = Ge^{2+} + 3H_2O$	ca. −0.3
$Ge^{2+} + 2e^- = Ge$	ca. 0.0
$Ge + 4H^+ + 4e^- = GeH_4$	ca. −0.3

Gold

$Au^{3+} + 2e^- = Au^+$	ca. +1.41
$Au^{3+} + 3e^- = Au$	+1.50
$Au(OH)_3 + 3H^+ + 3e^- = Au + 3H_2O$	+1.45
$AuCl_4^- + 3e^- = Au + 4Cl^-$	+1.00
$AuBr_4^- + 3e^- = Au + 4Br^-$	+0.87
$Au(CN)_2^- + e^- = Au + 2CN^-$	−0.60

Hafnium

$HfO_2 + 4H^+ + 4e^- = Hf + 2H_2O$	−1.57

Hydrogen

$2H^+ + 2e^- = H_2$	0
$2H_2O + 2e^- = H_2 + 2OH^-$	−0.828
$\frac{1}{2}H_2 + e^- = H^-$	−2.25
$2D^+ + 2e^- = D_2$	−0.0034

Indium

$In^{3+} + 3e^- = In$	-0.342
$In(OH)_3(s) + 3e^- = In + 3OH^-$	-1.0
$In^{3+} + e^- = In^{2+}$	ca. -0.45
$In^{2+} + e^- = In^+$	-0.35

Iodine

$I_2(s) + 2e^- = 2I^-$	$+0.5345$
$I_2(aq) + 2e^- = 2I^-$	$+0.6197$
$I_3^- + 2e^- = 3I^-$	$+0.5355$
$HIO_4 + H^+ + 2e^- = IO_3^- + H_2O$	ca. $+1.6$
$IO_3^- + 5H^+ + 4e^- = HIO + 2H_2O$	$+1.14$
$IO_3^- + 6H^+ + 5e^- = \frac{1}{2}I_2 + 3H_2O$	$+1.20$
$ICl_2^- + e^- = \frac{1}{2}I_2 + 2Cl^-$	$+1.06$
$ICl(aq) + e^- = \frac{1}{2}I_2 + Cl^-$	$+1.19$
$IBr_2^- + e^- = \frac{1}{2}I_2 + 2Br^-$	$+0.87$
$IBr + e^- = \frac{1}{2}I_2 + Br^-$	$+1.02$

Iridium

$IrCl_6^{2-} + 4e^- = Ir + 6Cl^-$	$+0.835$
$IrCl_6^{2-} + e^- = IrCl_6^{3-}$	$+1.017$
$IrBr_6^{3-} + e^- = IrBr_6^{4-}$	$+0.99$
$IrO_2 + 4H^+ + 4e^- = Ir + 2H_2O$	ca. $+0.93$
$Ir^{3+} + 3e^- = Ir$	ca. $+1.15$

Iron

$Fe^{3+} + e^- = Fe^{2+}$	$+0.771$
$Fe^{2+} + 2e^- = Fe$	-0.440
$Fe(CN)_6^{3-} + e^- = Fe(CN)_6^{4-}$	$+0.36$
$Fe(CN)_6^{3-} + e^- = Fe(CN)_6^{4-}$ (1 *M* HCl or $HClO_4$)	$+0.71$
$FeO_2^- + e^- = FeO_2^{2-}$ (10 *M* NaOH)	-0.68
$FeO_4^{2-} + 2H_2O + 3e^- = FeO_2^- + 4OH^-$ (10 *M* NaOH)	$+0.55$
$FeCl_4^- + e^- = Fe^{2+} + 4Cl^-$ (1 *M* HCl)	$+0.70$
$Fe^{3+} + e^- = Fe^{2+}$ (0.5 *M* H_3PO_4–1 *M* H_2SO_4)	$+0.61$

Lanthanum

$La^{3+} + 3e^- = La$	-2.52

Lead

$Pb^{2+} + 2e^- = Pb$	-0.126
$HPbO_2^- + H_2O + 2e^- = Pb + 3OH^-$	-0.54
$PbSO_4 + 2e^- = Pb + SO_4^{2-}$	-0.3563
$PbO_2 + 4H^+ + 2e^- = Pb^{2+} + 2H_2O$	$+1.455$
$PbO_3^{2-} + H_2O + 2e^- = PbO_2^{2-} + 2OH^-$ (8 *M* NaOH)	$+0.208$
$PbO_2 + SO_4^{2-} + 4H^+ + 2e^- = PbSO_4 + 2H_2O$	$+1.685$

Lithium

$Li^+ + e^- = Li$	-3.045

Magnesium

$Mg^{2+} + 2e^- = Mg$	-2.37
$Mg(OH)_2 + 2e^- = Mg + 2OH^-$	-2.69

Manganese

$Mn^{2+} + 2e^- = Mn$	-1.18
$MnO_4^- + e^- = MnO_4^{2-}$	$+0.564$
$MnO_4^- + 4H^+ + 3e^- = MnO_2 + 2H_2O$	$+1.695$
$MnO_4^- + 8H^+ + 5e^- = Mn^{2+} + 4H_2O$	$+1.51$
$MnO_2 + 4H^+ + 2e^- = Mn^{2+} + 2H_2O$	$+1.23$
$Mn^{3+} + e^- = Mn^{2+}$ (aq. H_2SO_4)	ca. $+1.5$

Mercury

$2Hg^{2+} + 2e^- = Hg^{2+}$	$+0.920$
$Hg^{2+} + 2e^- = Hg$	$+0.854$
$Hg_2^{2+} + 2e^- = 2Hg$	$+0.789$
$HgO + H_2O + 2e^- = Hg + 2OH^-$	$+0.098$
$Hg(CN)_4^{2-} + 2e^- = Hg + 4CN^-$	-0.37
$HgCl_4^{2-} + 2e^- = Hg + 4Cl^-$	$+0.48$
$Hg_2Cl_2(s) + 2e^- = 2Hg + 2Cl^-$	$+0.2676$
$Hg_2Cl_2(s) + 2K^+ + 2e^- = 2Hg + 2KCl(s)$	$+0.2415$
$Hg_2SO_4 + 2e^- = 2Hg + SO_4^{2-}$	$+0.6151$

Molybdenum

$H_2MoO_4(aq) + 2H^+ + e^- = MoO_2^+ + 2H_2O$	$+0.4$
$MoO_2^+ + 4H^+ + 2e^- = Mo^{3+} + 2H_2O$	0.0
$Mo^{6+} + e^- = Mo^{5+}$ (2 M HCl)	$+0.53$
$Mo^{5+} + 2e^- = Mo^{3+}$ (g) (2 M HCl)	-0.25
$Mo^{5+} + 2e^- = Mo^{3+}$ (r) (2 M HCl)	$+0.11$
$Mo(CN)_6^{3-} + e^- = Mo(CN)_6^{4-}$	$+0.73$
$Mo^{3+} + 3e^- = Mo$	ca. -0.2

Nickel

$Ni^{2+} + 2e^- = Ni$	-0.24
$NiO_2 + 4H^+ + 2e^- = Ni^{2+} + 2H_2O$	$+1.68$
$NiO_2 + 2H_2O + 2e^- = Ni(OH)_2 + 2OH^-$	$+0.49$
$Ni(CN)_4^{2-} + e^- = Ni(CN)_4^{3-}$	-0.82

Nitrogen

$NO_3^- + 3H^+ + 2e^- = HNO_2 + H_2O$	$+0.94$
$NO_3^- + 4H^+ + 3e^- = NO + 2H_2O$	$+0.96$
$NO_3^- + 2H^+ + e^- = NO_2 + H_2O$	$+0.80$
$NO_2 + H^+ + e^- = HNO_2$	$+1.07$
$NO_2 + 2H^+ + 2e^- = NO + H_2O$	$+1.03$
$HNO_2 + H^+ + e^- = NO + H_2O$	$+1.00$
$NH_3OH^+ + 2H^+ + 2e^- = NH_4^+ + H_2O$	$+1.35$
$N_2 + 5H^+ + 4e^- = N_2H_5^+$	-0.23

Osmium

$OsO_4 + 8H^+ + 8e^- = Os + 4H_2O$	$+0.85$
$HOsO_5^- + 4H_2O + 8e^- = Os + 9OH^-$	$+0.02$
$OsO_4 + 8H^+ + 6Cl^- + 4e^- = OsCl_6^{2-} + 4H_2O$	$+1.0$
$OsCl_6^{2-} + e^- = OsCl_6^{3-}$	$+0.85$
$OsCl_6^{3-} + e^- = Os^{2+} + 6Cl^-$	$+0.4$
$Os^{2+} + 2e^- = Os$	$+0.85$

Oxygen

$O_3 + 2H^+ + 2e^- = O_2 + H_2O$	+2.07
$O_2 + 4H^+ + 4e^- = 2H_2O$	+1.229
$O_2 + 2H^+ + 2e^- = H_2O_2$	+0.682
$O_2 + H_2O + 2e^- = HO_2^- + OH^-$	−0.076
$H_2O_2 + 2H^+ + 2e^- = 2H_2O$	+1.77
$H_2O_2 + 2e^- = 2OH^-$	+0.88

Palladium

$Pd^{2+} + 2e^- = Pd$	+0.987
$Pd(OH)_2 + 2e^- = Pd + 2OH^{2-}$	+0.07
$PdCl_4^{2-} + 2e^- = Pd + 4Cl^-$	+0.62
$PdBr_4^{2-} + 2e^- = Pd + 4Br^-$	+0.6
$PdCl_6^{2-} + 2e^- = PdCl_4^{2-} + 2Cl^-$	+1.288
$Pd(OH)_4 + 2e^- = Pd(OH)_2 + 2OH^-$	ca. +0.73
$PdO_3 + H_2O + 2e^- = PdO_2 + 2OH^-$	+1.2

Phosphorus

$H_3PO_4 + 2H^+ + 2e^- = H_3PO_3 + H_2O$	−0.276
$H_3PO_3 + 2H^+ + 2e^- = H_3PO_2 + H_2O$	−0.50

Platinum

$Pt^{2+} + 2e^- = Pt$	ca. +1.2
$Pt(OH)_2 + 2H^+ + 2e^- = Pt + 2H_2O$	+0.98
$PtCl_4^{2-} + 2e^- = Pt + 4Cl^-$	+0.73
$PtBr_4^{2-} + 2e^- = Pt + 4Br^-$	+0.58
$PtO_2 + 2H^+ + 2e^- = Pt(OH)_2$	ca. +1.0
$PtCl_6^{2-} + 2e^- = PtCl_4^{2-} + 2Cl^-$	+0.68
$PtBr_6^{2-} + 2e^- = PtBr_4^{2-} + 2Br^-$	+0.59
$PtO_4^{2-} + 4H_2O + 2e^- = Pt(OH)_6^{2-} + 2OH^-$	ca. +0.4

Plutonium

$PuO_2^{2+} + e^- = PuO_2^+$	+0.93
$PuO_2^{2+} + 4H^+ + 2e^- = Pu^{4+} + 2H_2O$	+1.067
$PuO_2^+ + 4H^+ + e^- = Pu^{4+} + 2H_2O$	+1.15
$Pu^{4+} + e^- = Pu^{3+}$	+0.97
$Pu^{3+} + 3e^- = Pu$	+2.03

Potassium

$K^+ + e^- = K$	−2.925

Radium

$Ra^{2+} + 2e^- = Ra$	−2.92

Rhenium

$ReO_4^- + 8H^+ + 7e^- = Re + 4H_2O$	+0.363
$ReO_4^- + 4H^+ + 3e^- = ReO_2 + 2H_2O$	+0.51
$ReO_2 + 2H^+ + 4e^- = Re + 2H_2O$	+0.252
$ReCl_6^{2-} + e^- = ReCl_4^- + 2Cl^-$ (1 *M* HCl)	ca. −0.25
$Re + e^- = Re^-$	−0.4

Rhodium

$RhO_4^{2-} + 6H^+ + 2e^- = RhO^{2+} + 3H_2O$	+1.46
$RhO^{2+} + 2H^+ + e^- = Rh^{3+} + H_2O$	+1.40
$Rh^{3+} + 3e^- = Rh$	ca. +0.8
$RhCl_6^{2-} + e^- = RhCl_6^{3-}$	ca. +1.2
$RhCl_6^{3-} + 3e^- = Rh + 6Cl^-$	+0.44
$Rh^{2+} + e^- = Rh^+$	ca. +0.6
$Rh^+ + e^- = Rh$	ca. +0.6

Rubidium

$Rb^+ + e^- = Rb$	−2.925

Ruthenium

$RuO_4 + e^- = RuO_4^-$	+0.9
$RuO_4^- + e^- = RuO_4^{2-}$	+0.6
$RuCl_5OH^{2-} + H^+ + 4e^- = Ru + H_2O + 5Cl^-$	+0.6
$RuCl_5^{2-} + 3e^- = Ru + 5Cl^-$	+0.4

Scandium

$Sc^{3+} + 3e^- = Sc$	−2.08

Selenium

$SeO_4^{2-} + 4H^+ + 2e^- = H_2SeO_3 + H_2O$	+1.15
$SeO_4^{2-} + H_2O + 2e^- = SeO_3^{2-} + 2OH^-$	+0.05
$H_2SeO_3 + 4H^+ + 4e^- = Se + 3H_2O$	+0.740
$SeO_3^{2-} + 3H_2O + 4e^- = Se + 6OH^-$	−0.366
$Se + 2H^+ + 2e^- = H_2Se$	−0.40
$Se + 2e^- = Se^{2-}$	−0.92

Silver

$AgO^+ + 2H^+ + e^- = Ag^{2+} + H_2O$ (4 *M* HNO_3)	ca. +2.1
$Ag^{2+} + e^- = Ag^+$ (4 *M* HNO_3)	+1.927
$Ag^{2+} + e^- = Ag^+$ (4 *M* $HClO_4$)	+1.970
$Ag_2O_3 + H_2O + 2e^- = 2AgO + 2OH^-$	+0.74
$2AgO + H_2O + 2e^- = Ag_2O + 2OH^-$	+0.57
$Ag^+ + e^- = Ag$	+0.7995

Sodium

$Na^+ + e^- = Na$	−2.714

Strontium

$Sr^{2+} + 2e^- = Sr$	−2.89

Sulfur

$S_2O_8^{2-} + 2e^- = 2SO_4^{2-}$	+2.01
$SO_4^{2-} + 4H^+ + 2e^- = H_2SO_3 + H_2O$	+0.17
$SO_4^{2-} + H_2O + 2e^- = SO_3^{2-} + 2OH^-$	−0.93
$4H_2SO_3 + 4H^+ + 6e^- = S_4O_6^{2-} + 6H_2O$	+0.51
$2H_2SO_3 + 2H^+ + 4e^- = S_2O_3^{2-} + 3H_2O$	+0.40
$2SO_3^{2-} + 3H_2O + 4e^- = S_2O_3^{2-} + 6OH^-$	−0.58
$S_4O_6^{2-} + 2e^- = 2S_2O_3^{2-}$	+0.08

Sulfur (continued)

Reaction	E (V)
$SO_3^{2-} + 3H_2O + 6e^- = S + 6OH^-$	−0.66
$S_2^{2-} + 2e^- = 2S^{2-}$	−0.48
$S + 2H^+ + 2e^- = H_2S$	+0.141
$S + 2e^- = S^{2-}$	−0.48

Tantalum

Reaction	E (V)
$Ta_2O_5 + 10H^+ + 10e^- = 2Ta + 5H_2O$	−0.81

Tellurium

Reaction	E (V)
$H_6TeO_6 + 2H^+ + 2e^- = TeO_2 + 4H_2O$	+1.02
$TeO_4^{2-} + H_2O + 2e^- = TeO_3^{2-} + 2OH^-$	ca. +0.4
$TeOOH^+ + 3H^+ + 4e^- = Te + 2H_2O$	+0.559
$TeO_2(s) + 4H^+ + 4e^- = Te + 2H_2O$	+0.529
$TeO_3^{2-} + 3H_2O + 4e^- = Te + 6OH^-$	−0.57
$Te + 2H^+ + 2e^- = H_2Te$	+0.72
$Te + 2e^- = Te^{2-}$	−1.14
$2Te + 2e^- = Te_2^{2-}$	−0.84

Thallium

Reaction	E (V)
$Tl^{3+} + 2e^- = Tl^+$	+1.25
$Tl(OH)_3(s) + 2e^- = Tl^+ + 3OH^-$	−0.05
$TlCl_4^- + 2e^- = Tl^+ + 4Cl^-$ (1 *M* HCl)	+0.77
$Tl^+ + e^- = Tl$	−0.3363

Thorium

Reaction	E (V)
$Th^{4+} + 4e^- = Th$	−1.90

Tin

Reaction	E (V)
$Sn^{4+} + 2e^- = Sn^{2+}$	+0.154
$SnCl_6^{2-} + 2e^- = SnCl_4^{2-} + 2Cl^-$ (1 *M* HCl)	+0.14
$Sn(OH)_6^{2-} + 2e^- = HSnO_2^- + H_2O + 3OH^-$	−0.93
$HSnO_2^- + H_2O + 2e^- = Sn + 3OH^-$	−0.91
$Sn^{2+} + 2e^- = Sn$	−0.136
$SnCl_4^{2-} + 2e^- = Sn + 4Cl^-$ (1 *M* HCl)	−0.19

Titanium

Reaction	E (V)
$TiO^{2+} + 2H^+ + 2e^- = Ti^{2+} + H_2O$	+0.1
$TiOCl^+ + 2H^+ + 3Cl^- + e^- = TiCl_4^- + H_2O$ (1 *M* HCl)	−0.09
(6 *M* HCl)	+0.24
$Ti^{4+} + e^- = Ti^{3+}$ (5 *M* H_3PO_4)	−0.15
$Ti^{3+} + e^- = Ti^{2+}$	−0.37
$Ti^{2+} + 2e^- = Ti$	−1.63

Tungsten

Reaction	E (V)
$WO_3(s) + 6H^+ + 6e^- = W + 3H_2O$	−0.09
$WO_4^{2-} + 4H_2O + 6e^- = W + 8OH^-$	−1.05
$2WO_3(s) + 2H^+ + 2e^- = W_2O_5(s) + H_2O$	−0.03
$W_2O_5(s) + 2H^+ + 2e^- = 2WO_2 + H_2O$	−0.043
$WO_2(s) + 4H^+ + 4e^- = W + 2H_2O$	−0.12
$W^{6+} + e^- = W^{5+}$ (12 *M* HCl)	+0.26

$W^{5+} + 2e^- = W^{3+}$(r) (12 *M* HCl)	−0.2
$W^{5+} + 2e^- = W^{3+}$(g) (12 *M* HCl)	+0.1
$W^{5+} + e^- = W^{4+}$ (12 *M* HCl)	−0.3
$W(CN)_8{}^{3-} + e^- = W(CN)_8{}^{4-}$	+0.57

Uranium

$UO_2{}^{2+} + e^- = UO_2{}^+$	+0.05
$UO_2{}^{2+} + 4H^+ + 2e^- = U^{4+} + 2H_2O$	+0.334
$UO_2{}^+ + 4H^+ + e^- = U^{4+} + 2H_2O$	+0.62
$U^{4+} + e^- = U^{3+}$	−0.61
$U^{3+} + 3e^- = U$	−1.80

Vanadium

$VO_2{}^+ + 2H^+ + e^- = VO^{2+} + H_2O$	+1.000
$VO^{2+} + 2H^+ + e^- = V^{3+} + H_2O$	+0.361
$V^{3+} + e^- = V^{2+}$	−0.255
$V^{2+} + 2e^- = V$	−1.18

Zinc

$Zn^{2+} + 2e^- = Zn$	−0.763
$ZnO_2{}^{2-} + 2H_2O + 2e^- = Zn + 4OH^-$	−1.216
$Zn(NH_3)_4{}^{2+} + 2e^- = Zn + 4NH_3$	−1.04
$Zn(CN)_4{}^{2-} + 2e^- = Zn + 4CN^-$	−1.26

Zirconium

$Zr^{4+} + 4e^- = Zr$	−1.53

appendix C / Carbon-13 nmr chemical shifts

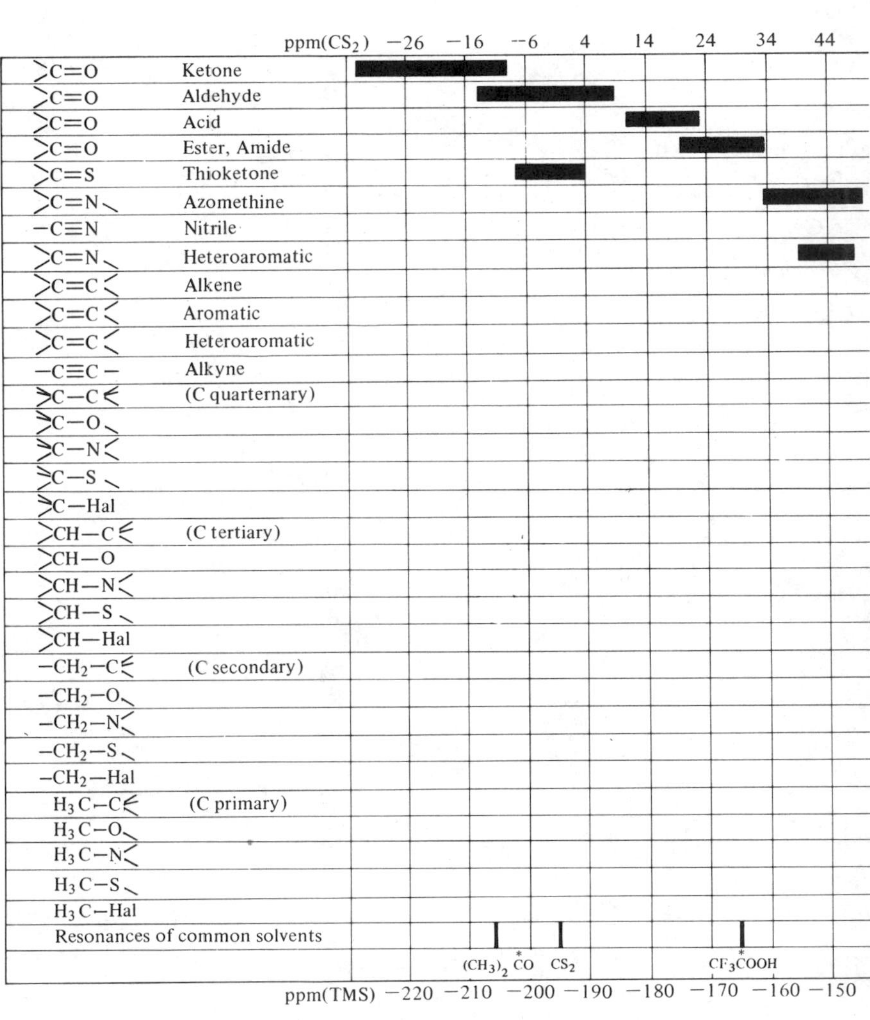

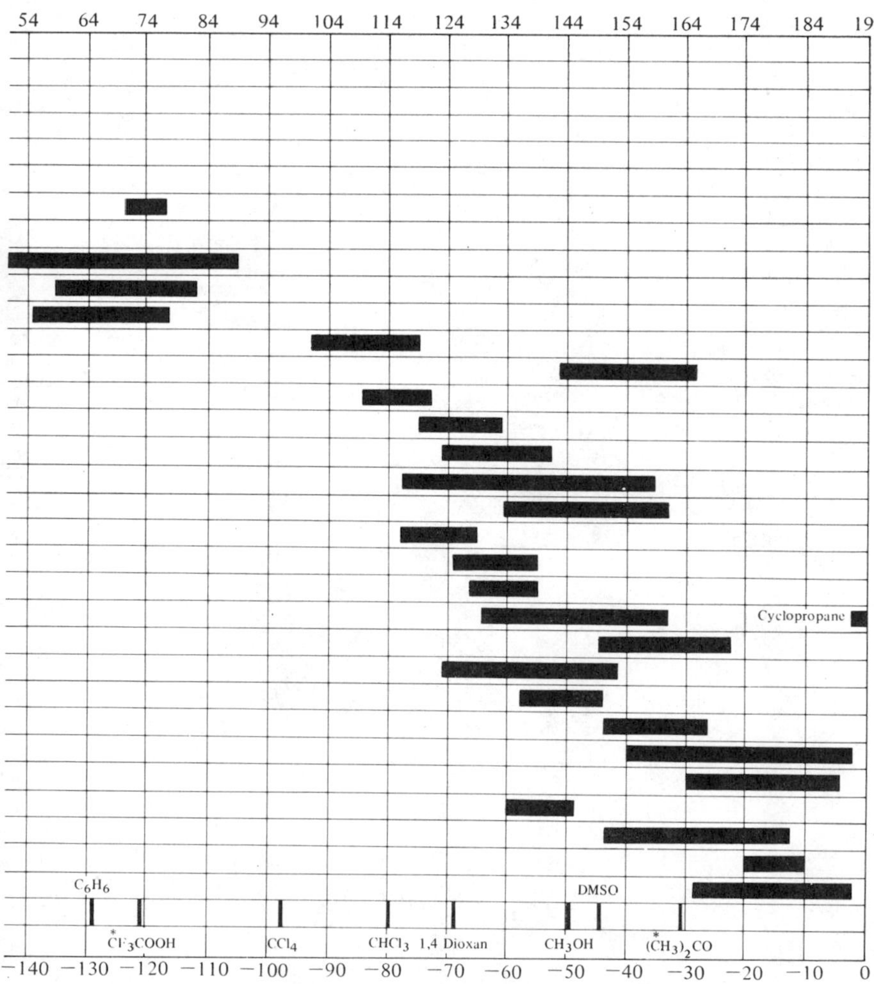

SOURCE: Courtesy of Bruker Magnetics, Burlington, Mass.

Index